ZOOLOGY
THE ANIMAL KINGDOM

Stephen A. Miller
The College of the Ozarks

John P. Harley
Eastern Kentucky University

Wm. C. Brown Publishers
Dubuque, IA Bogota Boston Buenos Aires Caracas Chicago
Guilford, CT London Madrid Mexico City Sydney Toronto

Book Team

Editor *Margaret J. Kemp*
Developmental Editor *Kathleen R. Loewenberg*
Production Editor *Jane E. Matthews*
Designer *Kristyn A. Kalnes*
Art Editor *Kathleen M. Timp*
Photo Editor *Janice Hancock*
Permissions Coordinator *Gail I. Wheatley*

Wm. C. Brown Publishers

President and Chief Executive Officer *Beverly Kolz*
Vice President, Publisher *Kevin Kane*
Vice President, Director of Sales and Marketing *Virginia S. Moffat*
Vice President, Director of Production *Colleen A. Yonda*
National Sales Manager *Douglas J. DiNardo*
Marketing Manager *Thomas C. Lyon*
Advertising Manager *Janelle Keeffer*
Production Editorial Manager *Renée Menne*
Publishing Services Manager *Karen J. Slaght*
Royalty/Permissions Manager *Connie Allendorf*

A Times Mirror Company

Copyedited by Linda Gomall

Cover photo © Tony Stone Images

The credits section for this book begins on page 365 and
is considered an extension of the copyright page.

Library of Congress Catalog Card Number: 94–76845

ISBN 0–697–29036–0

Printed in the United States of America by Times Mirror Higher Education Group, Inc.,
2460 Kerper Boulevard, Dubuque, IA 52001

10 9 8 7 6 5 4 3 2 1

brief CONTENTS

expanded

CONTENTS

chapter four

Animal Classification, Phylogeny, and Organization 49

chapter five

Animallike Protists: The Protozoa 63

chapter six

Multicellular and Tissue Levels of Organization 81

endpaper **one**

chapter seven

The Triploblastic, Acoelomate Body Plan 107

chapter eight

The Pseudocoelomate Body Plan: Aschelminths 123

chapter nine

Molluscan Success 141

chapter ten

Annelida: The Metameric Body Form 161

chapter eighteen

Birds: Feathers, Flight, and Endothermy 303

chapter nineteen

Mammals: Specialized Teeth, Endothermy, Hair, and Viviparity 321

preface

The traditional, and very popular, approach to a general zoology course is to survey the animallike protists and the animal kingdom. Other approaches to this course often involve covering biological principles, especially when there is no prerequisite general biology course, and/or aspects of animal structure and function in an organ system-by-organ system approach. Most general zoology textbooks are written to be adapted to any of these course formats. The goal of writing *Zoology: The Animal Kingdom* was to provide a thorough, yet concise, textbook for use in courses that are devoted solely to surveying the animallike protists and the animal kingdom.

We believe that this goal has been achieved. Reviewers have praised this textbook for its clear, concise, and friendly writing style. Many reviewers have commented that it is a book written with the student in mind. The same reviewers complement the concise, yet thorough, coverage of the protists and animal phyla. Its manageable size (approximately 400 pages) and full-color format will help make reading this book enjoyable for students.

Organization and Content

Zoology: The Animal Kingdom is written with an ecological and evolutionary focus. Chapter 1 is an introduction to zoology and a review of ecological principles. Students need to be reminded that animals function in the context of communities and ecosystems. This chapter is not intended as an exhaustive discourse on ecology, but as a review of important ecological principles that students should recall when studying the animal phyla.

Chapters 2 and 3 provide an introduction to, or review of, evolutionary principles that unite all organisms. All chapters of this textbook are loaded with evolutionary concepts. Chapters 2 and 3 will help students understand evolutionary mechanisms and appreciate the fact that evolution is the basis for understanding animal diversity.

Chapter 4 is an introduction to animal classification and body organization. It includes a discussion of numerical taxonomy, evolutionary systematics, and phylogenetic systematics (cladistics). The principles of cladistics are discussed in detail to help students interpret the cladograms that are presented in chapters 5–19.

The heart of this textbook, chapters 5–19, is a survey of the animallike protists and the animal phyla. This survey includes coverage of the classification, structure, function, and ecology of each group being discussed. The evolutionary focus of these chapters is unmistakable.

Pedagogy

Zoology: The Animal Kingdom is filled with useful pedagogy that makes the use of this book easier and more enjoyable for students and professors. Each chapter begins with a chapter outline, chapter concepts, and "Would You Like To Know" questions. The latter are designed to spark students' interest in the content of a chapter and are referenced at the appropriate place within the chapter. "Stop and Ask Yourself" questions are placed at strategic places within each chapter to help students review concepts they have just read. Evolutionary concepts are highlighted within each chapter by a distinctive font. Boldface type is used to emphasize new terms and concepts, and boxed readings provide interesting, informative, and current side-lights to the text material. In the survey chapters (5–19), lists of characteristics, a taxonomic summary, and (for most chapter) a cladogram are provided for each phylum. Each chapter ends with a summary, a list of selected key terms, and critical thinking questions. Suggested readings and a page-referenced glossary, including pronunciation guides, are also provided. A poster depicting the evolutionary relationships within the animal kingdom is available to instructors using this textbook. It includes some of the drawings, many of the cladograms, and the phylum characteristics lists from the survey chapters. It is attractively designed in full color and is a convenient study aid for the office or classroom wall. All of these options will provide students with an attractive, easy-to-use learning package.

Supplementary Materials

The following materials have been prepared for *Zoology* and can be adapted for use with *Zoology: The Animal Kingdom* as well:

1. An **Instructor's Manual/Test Item File** prepared by Eric Nelson, Ohio Northern University, provides examples of lecture/reading schedules for courses with various emphases. In addition, each chapter contains a detailed outline, purpose, objectives, key terms, summary, sources for audiovisual materials and computer software, and approximately 50 multiple-choice test questions.

2. A **Student Study Guide,** prepared by Jay M. Templin, contains chapter summaries, outlines, key terms with phonetics, pretest assessment questions, various learning activities, and mastery tests.
3. A set of 100 full-color acetate **transparencies** and 150 **transparency masters** is available and may be used to supplement classroom lectures.
4. **General Zoology Laboratory Manual,** third edition, by Stephen A. Miller, is an excellent corollary to the text and incorporates many of the same learning aids. This edition offers improved illustrations and laboratory exercises that emphasize animal adaptations and ecology. A **Laboratory Resource Guide** is also available and contains additional information about materials and procedures, and the answers to lab exercises.
5. The **Customized Laboratory Manual.** Each lab manual exercise is also available individually as offprints, so students need buy only those exercises used in the laboratory. Contact your local sales representative for more details.
6. **Life Science Animations** is a series of five videotapes featuring more than 50 animations of key physiological processes spanning the breadth of concepts covered in a typical life science course. Of particular interest to professors teaching zoology are tape three, *Animal Biology: Part I*, and tape four, *Animal Biology, Part II*. These full-color animations enable the student to more fully and easily grasp concepts such as the formation of the myelin sheath, saltatory nerve conduction, signal integration, reflex arcs, and the organ of static equilibrium. (ISBNs 25050 and 25071)
7. **How to Study Science** by Fred Drewes, offers students valuable tips on note-taking, how to interpret text figures, how to manage time, how to prepare for tests, and how to overcome "science anxiety." (ISBN 14474)
8. **The Life Science Living Lexicon CD-ROM,** by William Marchuk, provides comprehensive coverage of all the life science disciplines—biology, anatomy, physiology, botany, zoology, environmental science, and microbiology—by combining complete lexicon components:

    - Overview of word construction and how to use the textual material,
    - Glossary of common biological root words, prefixes, and suffixes,
    - Glossary of descriptive terms,
    - Glossary of common biological terms, characterized by discipline,
    - Section describing the classification system,

    in a powerful interactive CD-ROM with more than 1,000 vivid illustrations and animations of key processes and systems—including histology micrographs, and an interactive quizzing program. (ISBN 12133)

9. **Critical Thinking: A Collection of Readings,** by David J. Stroup and Robert D. Allen, is available to instructors who are working to integrate critical thinking into their curricula. This inexpensive text will help in the planning and implementation of programs intended to develop students' abilities to think logically and analytically. The reader is a collection of articles that provide instruction and examples of current programs. The authors have included descriptions and evaluations of their personal experiences with incorporating critical thinking study in their coursework. (ISBN 14556)

Acknowledgments

We wish to express our thanks to the reviewers who provided detailed criticism and analysis of the textbook during development. In the midst of their busy teaching and research schedules, they took time to read our manuscript and offer constructive advice that greatly improved the final text.

Reviewers

Thomas P. Buckelew
California University of Pennsylvania
DuWayne C. Englert
Southern Illinois University at Carbondale
Mary Sue Gamroth
Joliet Junior College
Michael C. Hartman, Ph.D.
Dan F. Ippolito
Anderson University
Eddie Lunsford
Tri County Community College
Murphy, North Carolina
Eric V. Nelson
Ohio Northern University
Richard E. Trout
Oklahoma City Community College

The production of a textbook requires the efforts of many people. We are grateful for the work of our colleagues at Wm. C. Brown Publishers, who have shown extraordinary patience, skill, and commitment to this textbook. Marge Kemp, our editor, and Kevin Kane, publisher, have helped shape Zoology from its earliest planning stages. Our developmental editor, Kathy Loewenberg, helped make the production of the third edition remarkably smooth. Jane Matthews, our production editor, kept us on schedule and the production moving in the plethora of directions that are nearly unimaginable to us.

Finally, but most importantly, we wish to extend appreciation to our families for their patience and encouragement. Our wives, Carol A. Miller and Jane R. Harley, have been supportive from the beginning of this project. We appreciate the sacrifices that our families have made during the writing of this textbook. We dedicate this book to them.

Stephen A. Miller
John P. Harley

special FEATURES

Second edition adopters and reviewers made it clear that they found *Zoology* to be clear, concise, and intriguing—and we think you will find the third edition even more so. *Zoology* is unique in its emphasis on learning aids for the student. Consistent features include:

Chapter-Opening Outlines

Chapter-opening **Outlines** present a brief structure of the chapter as a preview to the student.

Chapter-Opening Concepts

The chapter-opening **Concepts** help prepare the student for major concepts covered in the chapter and also serve as a study tool for review.

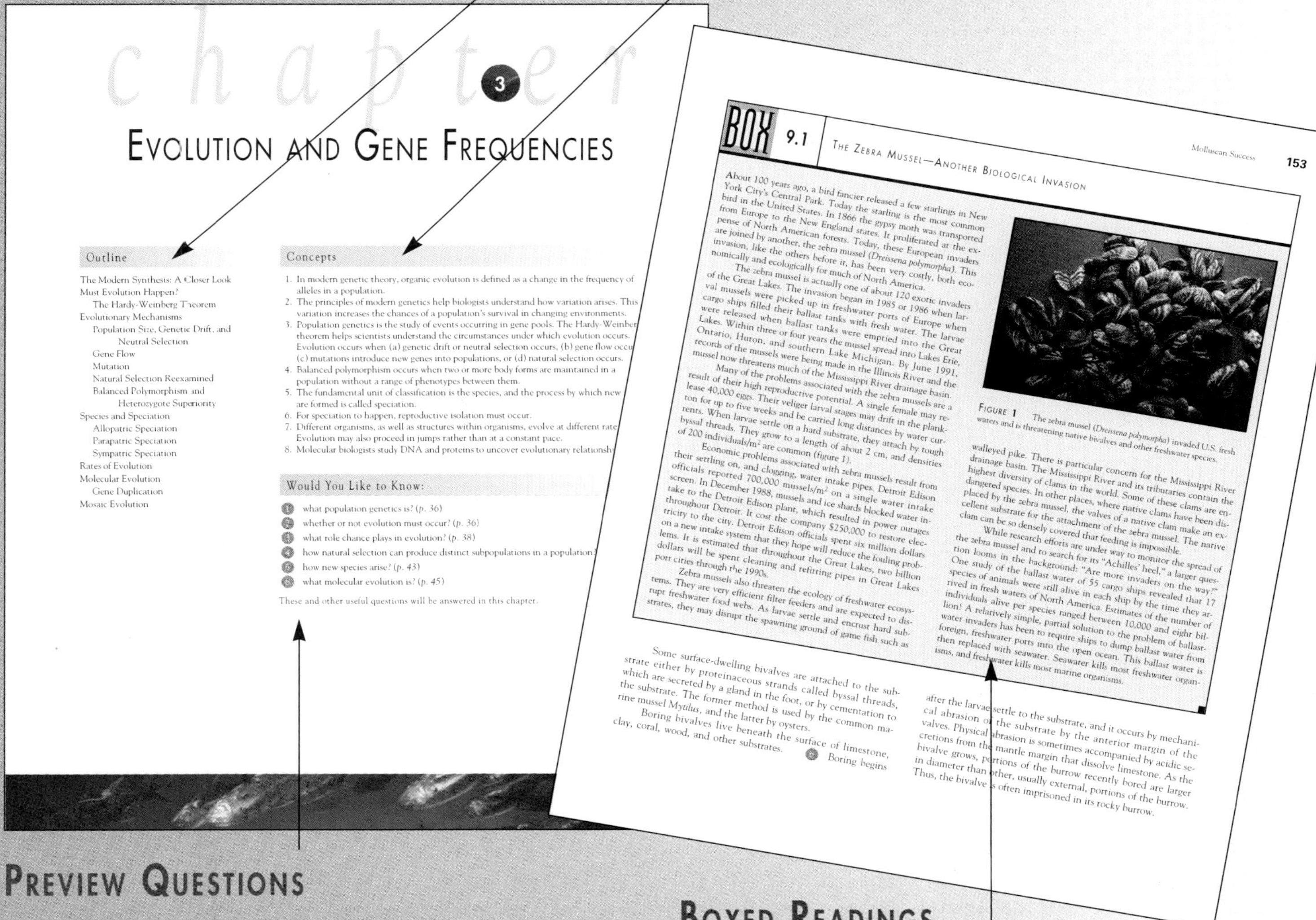

chapter 3

EVOLUTION AND GENE FREQUENCIES

Outline

The Modern Synthesis: A Closer Look
Must Evolution Happen?
The Hardy-Weinberg Theorem
Evolutionary Mechanisms
Population Size, Genetic Drift, and Neutral Selection
Gene Flow
Mutation
Natural Selection Reexamined
Balanced Polymorphism and Heterozygote Superiority
Species and Speciation
Allopatric Speciation
Parapatric Speciation
Sympatric Speciation
Rates of Evolution
Molecular Evolution
Gene Duplication
Mosaic Evolution

Concepts

1. In modern genetic theory, organic evolution is defined as a change in the frequency of alleles in a population.
2. The principles of modern genetics help biologists understand how variation arises. This variation increases the chances of a population's survival in changing environments.
3. Population genetics is the study of events occurring in gene pools. The Hardy-Weinber[g] theorem helps scientists understand the circumstances under which evolution occurs. Evolution occurs when (a) genetic drift or neutral selection occurs, (b) gene flow occu[rs,] (c) mutations introduce new genes into populations, or (d) natural selection occurs.
4. Balanced polymorphism occurs when two or more body forms are maintained in a population without a range of phenotypes between them.
5. The fundamental unit of classification is the species, and the process by which new [...] are formed is called speciation.
6. For speciation to happen, reproductive isolation must occur.
7. Different organisms, as well as structures within organisms, evolve at different rate[s.] Evolution may also proceed in jumps rather than at a constant pace.
8. Molecular biologists study DNA and proteins to uncover evolutionary relationsh[ips.]

Would You Like to Know:

1. what population genetics is? (p. 36)
2. whether or not evolution must occur? (p. 36)
3. what role chance plays in evolution? (p. 38)
4. how natural selection can produce distinct subpopulations in a population?
5. how new species arise? (p. 43)
6. what molecular evolution is? (p. 45)

These and other useful questions will be answered in this chapter.

Molluscan Success 153

BOX 9.1 THE ZEBRA MUSSEL—ANOTHER BIOLOGICAL INVASION

About 100 years ago, a bird fancier released a few starlings in New York City's Central Park. Today the starling is the most common bird in the United States. In 1866 the gypsy moth was transported from Europe to the New England states. It proliferated at the expense of North American forests. Today, these European invaders are joined by another, the zebra mussel (*Dreissena polymorpha*). This invasion, like the others before it, has been very costly, both economically and ecologically for much of North America.

The zebra mussel is actually one of about 120 exotic invaders of the Great Lakes. The invasion began in 1985 or 1986 when larval mussels were picked up in freshwater ports of Europe when cargo ships filled their ballast tanks with fresh water. The larvae were released when ballast tanks were emptied into the Great Lakes. Within three or four years the mussel spread into Lakes Erie, Ontario, Huron, and southern Lake Michigan. By June 1991, records of the mussels were being made in the Illinois River and the mussel now threatens much of the Mississippi River drainage basin.

Many of the problems associated with the zebra mussels are a result of their high reproductive potential. A single female may release 40,000 eggs. Their veliger larval stages may drift in the plankton for up to five weeks and be carried long distances by water currents. When larvae settle on a hard substrate, they attach by tough byssal threads. They grow to a length of about 2 cm, and densities of 200 individuals/m² are common (figure 1).

Economic problems associated with zebra mussels result from their settling on, and clogging, water intake pipes. Detroit Edison officials reported 700,000 mussels/m² on a single water intake screen. In December 1988, mussels and ice shards blocked water intake to the Detroit Edison plant, which resulted in power outages throughout Detroit. It cost the company $250,000 to restore electricity to the city. Detroit Edison officials spent six million dollars on a new intake system that they hope will reduce the fouling problems. It is estimated that throughout the Great Lakes, two billion dollars will be spent cleaning and refitting pipes in Great Lakes port cities through the 1990s.

Zebra mussels also threaten the ecology of freshwater ecosystems. They are very efficient filter feeders and are expected to disrupt freshwater food webs. As larvae settle and encrust hard substrates, they may disrupt the spawning ground of game fish such as walleyed pike. There is particular concern for the Mississippi River drainage basin. The Mississippi River and its tributaries contain the highest diversity of clams in the world. Some of these clams are endangered species. In other places, where native clams have been displaced by the zebra mussel, the valves of a native clam make an excellent substrate for the attachment of the zebra mussel. The native clam can be so densely covered that feeding is impossible.

While research efforts are under way to monitor the spread of the zebra mussel and to search for its "Achilles' heel," a larger question looms in the background: "Are more invaders on the way?" One study of the ballast water of 55 cargo ships revealed that 17 species of animals were still alive in each ship by the time they arrived in fresh waters of North America. Estimates of the number of individuals alive per species ranged between 10,000 and eight billion! A relatively simple, partial solution to the problem of ballast-water invaders has been to require ships to dump ballast water from foreign, freshwater ports into the open ocean. This ballast water is then replaced with seawater. Seawater kills most freshwater organisms, and freshwater kills most marine organisms.

FIGURE 1 The zebra mussel (*Dreissena polymorpha*) invaded U.S. fresh waters and is threatening native bivalves and other freshwater species.

Some surface-dwelling bivalves are attached to the substrate either by proteinaceous strands called byssal threads, which are secreted by a gland in the foot, or by cementation to the substrate. The former method is used by the common marine mussel *Mytilus*, and the latter by oysters.

Boring bivalves live beneath the surface of limestone, clay, coral, wood, and other substrates. 4 Boring begins after the larvae settle to the substrate, and it occurs by mechanical abrasion of the substrate by the anterior margin of the valves. Physical abrasion is sometimes accompanied by acidic secretions from the mantle margin that dissolve limestone. As the bivalve grows, portions of the burrow recently bored are larger in diameter than other, usually external, portions of the burrow. Thus, the bivalve is often imprisoned in its rocky burrow.

Preview Questions

Approximately 5–8 **"Would You Like to Know"** questions are presented at the beginning of each chapter and are intended to pique the student's interest in the topic. Answers are found within the chapter and are indicated by a number corresponding to the question.

Boxed Readings

Nearly every chapter contains at least one boxed reading covering an interesting subject relevant to the surrounding text. For example, topics include **"The Zebra Mussel—Another Biological Invasion"**; **"Malaria Control—A Glimmer of Hope"**; **"Mammalian Echolocation"**; and **"Planktonic Tunicates."**

Evolutionary Concepts

Incorporated into the design of the third edition are visual cues to let the reader know when concepts of evolution are being discussed.

Boldfaced Terms

Important **Key Terms** are emphasized in bold type and clearly defined when they are first presented.

222 Chapter 13

Evolutionary Perspective

If one could visit 400-million-year-old Paleozoic seas, one would see representatives of nearly every phylum studied in the previous eight chapters of this textbook. In addition, one would observe many representatives of the phylum Echinodermata (i-ki'na-dur"ma-tah) (Gr. *echinos*, spiny + *derma*, skin + *ata*, to bear). Many ancient echinoderms were attached to their substrate and probably lived as filter feeders—a feature found in only one class of modern echinoderms (figure 13.1). Today, we know this phylum by the relatively common sea stars, sea urchins, sand dollars, and sea cucumbers. In terms of numbers of species, echinoderms may seem to be a declining phylum. Studies of fossil records indicate that about 12 of 18 classes of echinoderms have become extinct. That does not mean, however, that living echinoderms are of minor importance. Members of three classes of echinoderms have flourished and often make up a major component of the biota of marine ecosystems (table 13.1).

Characteristics of the phylum Echinodermata include the following:

1. Calcareous endoskeleton in the form of ossicles that arise from mesodermal tissue
2. Adults with pentaradial symmetry and larvae with bilateral symmetry
3. Water-vascular system composed of water-filled canals used in locomotion, attachment, and/or feeding
4. Complete digestive tract that may be secondarily reduced
5. Hemal system derived from coelomic cavities
6. Nervous system consisting of a nerve net, nerve ring, and radial nerves

Relationships to Other Animals

Most zoologists believe that echinoderms share a common ancestry with hemichordates and chordates. Evidence of these evolutionary ties is seen in the deuterostome characteristics that they share (*see figure 9.3*): an anus that develops in the region of the blastopore, a coelom that forms from outpockets of the embryonic gut tract (vertebrate chordates are an exception), and radial, indeterminate cleavage. Unfortunately, no fossils have been discovered that document a common ancestor for these phyla or that demonstrate how the deuterostome lineage was derived from ancestral diploblastic or triploblastic stocks (figure 13.2).

Although adults are radially symmetrical, it is generally accepted that echinoderms evolved from bilaterally symmetrical ancestors. Evidence for this relationship includes bilaterally symmetrical echinoderm larval stages and extinct forms which were not radially symmetrical.

Echinoderm Characteristics

There are approximately 7,000 species of living echinoderms. They are exclusively marine and occur at all depths in all oceans. Modern echinoderms have a form of radial symmetry, called **pentaradial symmetry,** in which body parts are arranged

FIGURE 13.1
Phylum Echinodermata. This feather star (*Comanthina*) uses its highly branched arms in filter feeding. Although this probably reflects the original use of echinoderm appendages, most modern echinoderms use arms for locomotion, capturing prey, and scavenging the substrate for food.

TABLE 13.1 Classification of the Phylum Echinodermata

Phylum Echinodermata (i-ki'na-dur"ma-tah).
The phylum of triploblastic, coelomate animals whose members are pentaradially symmetrical as adults, possess an endoskeleton covered by epithelium, and possess a water-vascular system. Pedicellaria often present.
Class Crinoidea (kri-noi'de-ah)
Free-living or attached by an aboral stalk of ossicles; flourished in the Paleozoic era; approximately 230 living species. Sea lilies, feather stars.
Class Asteroidea (as'te-roi"de-ah)
Rays not sharply set off from central disk; ambulacral grooves with tube feet; suction disks on tube feet; pedicellariae present. Sea stars.
Class Ophiuroidea (o-fe-u-roi"de-ah)
Arms sharply marked off from the central disk; tube feet without suction disks. Brittle stars.
Class Concentricycloidea (kon-sen'tri-si-kloi"de-ah)
Two concentric water-vascular rings encircle a disklike body; no digestive system; digest and absorb nutrients across their lower surface; internal brood pouches; no free-swimming larval stage. Sea daisies.
Class Echinoidea (ek'i-noi"de-ah)
Globular or disk shaped; no rays; movable spines; skeleton (test) of closely fitting plates. Sea urchins, sand dollars.
Class Holothuroidea (hol'o-thu-roi"de-ah)
No rays; elongate along the oral-aboral axis; microscopic ossicles embedded in a muscular body wall; circumoral tentacles. Sea cucumbers.

This listing reflects a phylogenetic sequence; however, the discussion that follows begins with the echinoderms that are familiar to most students.

Taxonomic Summaries

Tables within the text include **Taxonomic Summaries** where appropriate. The summaries provide pronunciation guides and descriptive outlines of taxonomic sequences.

Interior Chapter Review Questions

Interspersed throughout the text are small sets of **"Stop and Ask Yourself"** questions directly correlated to material the student has just read. They allow students to test their knowledge and understanding of the section's factual material before continuing their reading. Visually highlighted, the questions are easy to find and make excellent study tools after the complete chapter has been read.

Cladograms

New to this edition are **cladograms,** which introduce the student to cladistic analysis and help them to visualize evolutionary pathways.

Chapter Summaries

These **End-of-Chapter Summaries** reiterate important chapter concepts and are intended to serve as a guide for study.

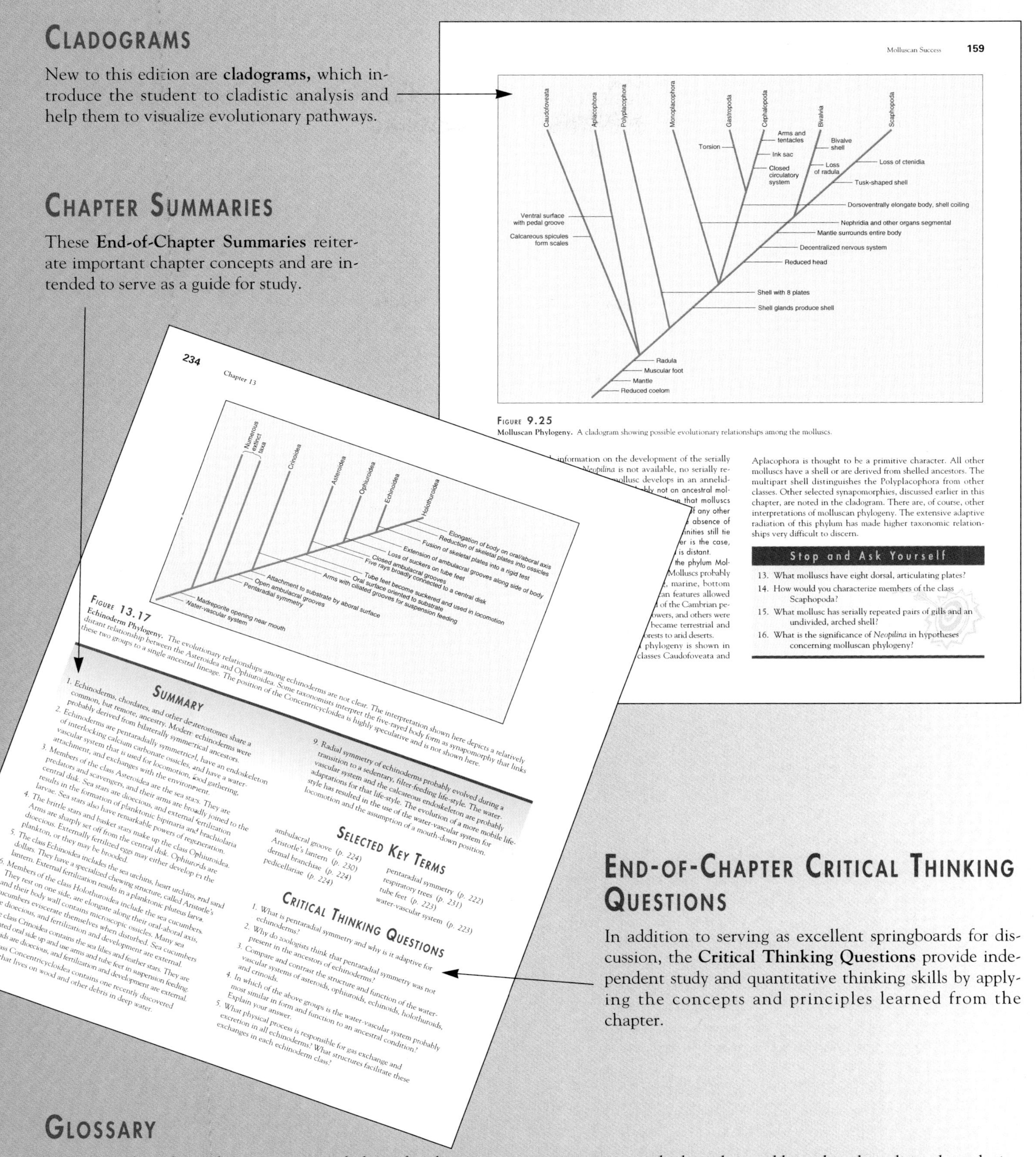

Figure 9.25
Molluscan Phylogeny. A cladogram showing possible evolutionary relationships among the molluscs.

Aplacophora is thought to be a primitive character. All other molluscs have a shell or are derived from shelled ancestors. The multipart shell distinguishes the Polyplacophora from other classes. Other selected synapomorphies, discussed earlier in this chapter, are noted in the cladogram. There are, of course, other interpretations of molluscan phylogeny. The extensive adaptive radiation of this phylum has made higher taxonomic relationships very difficult to discern.

Stop and Ask Yourself

13. What molluscs have eight dorsal, articulating plates?
14. How would you characterize members of the class Scaphopoda?
15. What mollusc has serially repeated pairs of gills and an undivided, arched shell?
16. What is the significance of *Neopilina* in hypotheses concerning molluscan phylogeny?

234 Chapter 13

Figure 13.17
Echinoderm Phylogeny. The evolutionary relationships among echinoderms are not clear. The interpretation shown here depicts a relatively distant relationship between the Asteroidea and Ophiuroidea. Some taxonomists interpret the five-rayed body form as synapomorphy that links these two groups to a single ancestral lineage. The position of the Concentricycloidea is highly speculative and is not shown here.

Summary

1. Echinoderms, chordates, and other deuterostomes share a common, but remote, ancestry. Modern echinoderms were probably derived from bilaterally symmetrical ancestors.
2. Echinoderms are pentaradially symmetrical, have an endoskeleton of interlocking calcium carbonate ossicles, and have a water-vascular system that is used for locomotion, food gathering, attachment, and exchanges with the environment.
3. Members of the class Asteroidea are the sea stars. They are predators and scavengers, and their arms are broadly joined to the central disk. Sea stars are dioecious, and external fertilization results in the formation of planktonic bipinnaria and brachiolaria larvae. Sea stars also have remarkable powers of regeneration.
4. The brittle stars and basket stars make up the class Ophiuroidea. Arms are sharply set off from the central disk. Ophiuroids are dioecious. Externally fertilized eggs may either develop in the plankton, or they may be brooded.
5. The class Echinoidea includes the sea urchins, heart urchins, and sand dollars. They have a specialized chewing structure, called Aristotle's lantern. External fertilization results in a planktonic pluteus larva.
6. Members of the class Holothuroidea include the sea cucumbers. They rest on one side, are elongate along their oral-aboral axis, and their body wall contains microscopic ossicles. Many sea cucumbers eviscerate themselves when disturbed. Sea cucumbers are dioecious, and fertilization and development are external.
7. The class Crinoidea contains the sea lilies and feather stars. They are oriented oral side up and use arms and tube feet in suspension feeding. Crinoids are dioecious, and fertilization and development are external.
8. The class Concentricycloidea contains one recently discovered species that lives on wood and other debris in deep water.
9. Radial symmetry of echinoderms probably evolved during a transition to a sedentary, filter-feeding life-style. The water-vascular system and the calcareous endoskeleton are probably adaptations for that life-style. The evolution of a more mobile life-style has resulted in the use of the water-vascular system for locomotion and the assumption of a mouth-down position.

Selected Key Terms

ambulacral groove (p. 224)
Aristotle's lantern (p. 230)
dermal branchiae (p. 224)
pedicellariae (p. 224)
pentaradial symmetry (p. 222)
respiratory trees (p. 231)
tube feet (p. 223)
water-vascular system (p. 223)

Critical Thinking Questions

1. What is pentaradial symmetry and why is it adaptive for echinoderms?
2. Why do zoologists think that pentaradial symmetry was not present in the ancestors of echinoderms?
3. Compare and contrast the structure and function of the water-vascular systems of asteroids, ophiuroids, echinoids, holothuroids, and crinoids.
4. In which of the above groups is the water-vascular system probably most similar in form and function to an ancestral condition? Explain your answer.
5. What physical process is responsible for gas exchange and excretion in all echinoderms? What structures facilitate these exchanges in each echinoderm class?

End-of-Chapter Critical Thinking Questions

In addition to serving as excellent springboards for discussion, the **Critical Thinking Questions** provide independent study and quantitative thinking skills by applying the concepts and principles learned from the chapter.

Glossary

Helping students learn the extensive vocabulary of zoology is a continuing concern, which we have addressed in this edition by reducing the number of unnecessary technical terms. We also define new terms when they are first introduced, with word derivations and a pronunciation guide. A Glossary, which has undergone careful revision and expansion for the third edition, is included at the end of the text.

chapter 1

Zoology: An Evolutionary and Ecological Perspective

Outline

Concepts

1. Evolutionary processes are the source of diversity within the animal kingdom. Zoology, the study of animals, is a very broad field with many subdisciplines.
2. Evidence documenting the evolution of the animal kingdom comes from biogeography, paleontology, comparative anatomy, and molecular biology.
3. Ecology is the study of the relationships of organisms to their environment, and to other organisms. In part, it involves the study of how abiotic factors such as energy, temperature, moisture, light, geology, and soils influence individuals.
4. Ecology also involves the study of populations. Populations grow, and growth is regulated by population density, the carrying capacity of the environment, and interactions between members of the same population.
5. Ecology also involves the study of individuals interacting with members of their own and other species. Herbivory, predator-prey interactions, competition for resources, and other kinds of interactions influence the makeup of animal populations.
6. All populations living in an area make up a community. Communities have unique attributes that can be characterized by ecologists.
7. Communities and their physical surroundings are called ecosystems. Energy flow through an ecosystem is one way. Energy that comes into an ecosystem must support all organisms living there before it is lost as heat. Nutrients, on the other hand, cycle through an ecosystem and are reused by organisms.
8. Concepts of community and ecosystem ecology provide a basis for understanding many of our ecological problems including human population growth, pollution, and resource depletion.

Would You Like to Know:

1. why boa constrictors have remnants of pelvic appendages? *(p. 8)*
2. whether or not bears really hibernate? *(p. 9)*
3. why a parasite usually does not kill its host? *(p. 12)*
4. why we have an energy crisis? *(p. 14)*
5. whether or not the earth's current human population can be supported for many years to come? *(p. 17)*
6. why predators are often the first to feel the effects of poisons released into the environment? *(p. 19)*

These and other useful questions will be answered in this chapter.

This chapter contains evolutionary concepts, which are set off in this font.

Humans have studied animals throughout prehistoric and historic times. This study of animals is now called **zoology** (Gr. *zoon*, animal + *logos*, to study), and the application of knowledge derived from this study has helped to shape human cultures throughout the world. Human history has been a story of humans interacting with other animals, and any human culture has its own unique zoological saga.

No cultures have captivated more interest than those found on the Polynesian islands. These islands, like the Galápagos Islands visited by Charles Darwin, have a unique and rich evolutionary history. The chance invasion of these islands by plants and animals and millions of years of evolution have resulted in populations found nowhere else on the earth (figure 1.1). Humans, relative newcomers to these islands, began colonization from southeast Asia about 4,000 years ago. The islands—New Caledonia, Fiji, Tonga, Samoa, and Hawaii—have human cultures that have shaped, and have been shaped by, their animal inhabitants.

Polynesian colonists, known as the Lapita, navigated hundreds of kilometers of open ocean in outrigger or double canoes and brought with them horticultural, hunting, and fishing technologies (figure 1.2). Their settlements were densely populated, politically complex, and depended on animal resources to sustain the colonists. Recent studies show that Polynesians relied heavily on birds, turtles, molluscs (clams, oysters, snails), fish, and domesticated mammals for food, but animals were much more than a source of food. The social organization of the new inhabitants of the Polynesian islands was a chiefdom. The chief's authority was derived from his ability to control access to animals. This control was accomplished through the use of taboos that limited the consumption of pigs and dogs to the highest ranked individuals in the society. The chief also maintained the control of hunting and fishing tool production, of fishing privileges, of fish pond construction, and of the fleets of canoes used in fishing.

As was the case in other colonization events by humans, the effects of human colonization on the native animal populations of the Polynesian islands were severe. The most dramatic changes in the islands' animal populations occurred in terrestrial animal species. Numerous species of lizards and birds declined to extinction. In the first 1,000 years after colonization, birds were a major source of food on some of the islands. These islands were free of mammalian predators for millions of years, and native species of birds had little fear of, and no defense against, human hunters. In fact, hunting birds must have been no more difficult than picking fruit from a tree. A decline in the importance of birds in the diet of Polynesians about 3,000 years ago is probably a result of the extinction of important food species. Human dependence on native animals changed these islands in ways that can never be recovered.

The zoological saga of the Polynesian islands is not exceptional—even in our modern world. Modern humans are equally dependent on animals, and that dependence too often leads to exploitation. We are dependent on animals for food, medicines, and clothing. We also depend on animals in very subtle ways. This dependence may not be noticed until human activities upset the delicate ecological balances that have evolved over millions of years. Dependence becomes obvious when pollution kills a predator or when a barrier to a predator is removed. Our dependence becomes obvious when exotic species are introduced into new habitats and natural ecological controls are thwarted. If human culture is to continue to reap the benefits of living with other animals, and if other animals are simply given the right to live, it is so very important that we appreciate animal diversity and the effects of our actions on that diversity.

FIGURE 1.1
The Polynesian Islands of Tahiti. Mt. Otomanu on the island of Bora Bora is shown here. The Polynesian islands were colonized about 4,000 years ago by people from southeast Asia. Polynesian cultures have shaped, and have been shaped by, the animal inhabitants of the islands.

The purpose of this textbook is to help you begin a journey of discovery into the animal kingdom. The ecological and evolutionary focus at the beginning of this textbook is intended to help you understand the origin of, and the importance of preserving animal diversity. Both of these themes will be reinforced throughout the book and are the most important reasons for studying the animal kingdom.

What Is Zoology?

Zoology is one of the broadest fields in all of science. The diversity of the subdisciplines within zoology reflects the breadth of

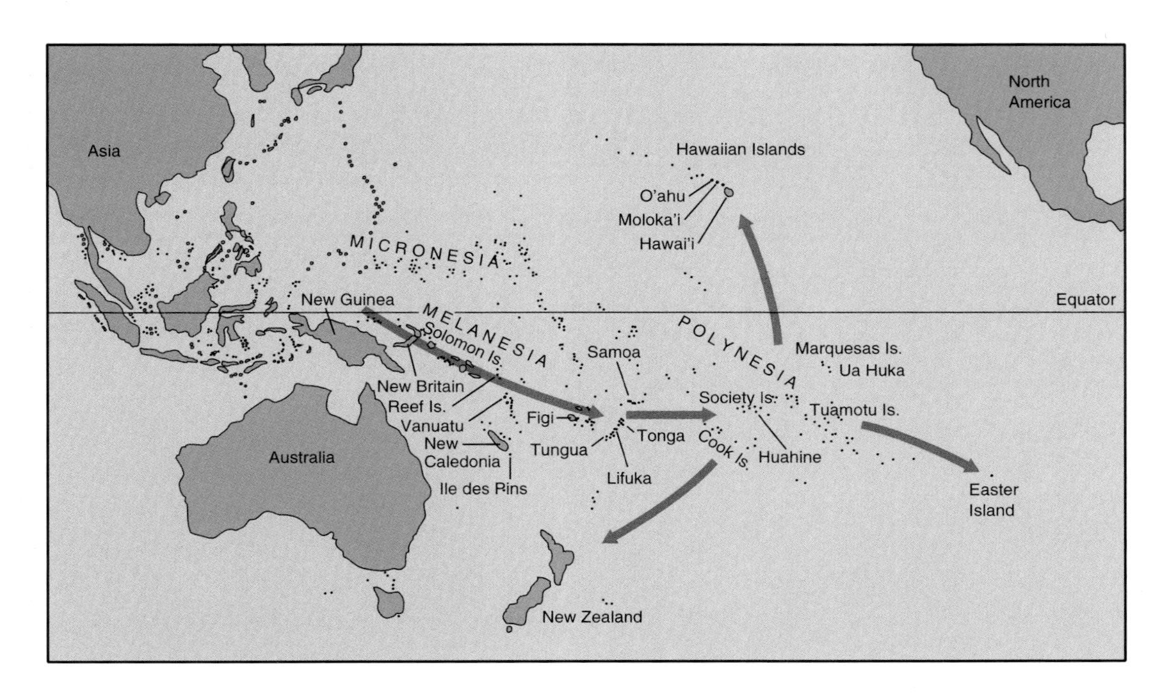

FIGURE 1.2

Polynesian Colonization. The migration route of the Lapita from southeast Asia to the Polynesian islands. The migration began about 4,000 years ago and ended with colonization of the Hawaiian islands about 3,000 years ago. *Source: Tom Dye and David W. Steadman, "Polynesian Ancestors and Their Animal World,"* American Scientist *78(3), May-June 1990, p. 280.*

the field. These subdisciplines are based on particular functional, structural, or ecological interests that span many animal groups (table 1.1). It is also possible to specialize in the biology of a particular group of animals (table 1.2). When one considers the size of some animal groups, it is no wonder that further specialization is also common. For example, there are approximately 300,000 described (and many undescribed) species of beetles! One person cannot possibly be an expert in all areas of beetle biology. Therefore, one can specialize even further and become a beetle taxonomist, physiologist, or ecologist. Obviously, the subdisciplines within zoology are not separated by sharply defined boundaries, and much information is shared among them.

TABLE 1.1 EXAMPLES OF SPECIALIZATIONS IN ZOOLOGY

SUBDISCIPLINE	DESCRIPTION
Anatomy	The study of the structure of entire organisms and their parts
Cytology	The study of the structure and function of cells
Ecology	The study of the interaction of organisms with their environment
Embryology	The study of the development of an animal from the fertilized egg to birth or hatching
Genetics	The study of the mechanisms of transmission of traits from parents to offspring
Histology	The study of tissues
Molecular biology	The study of subcellular details of animal structure and function
Parasitology	The study of animals that live in or on other organisms at the expense of the host
Physiology	The study of the function of organisms and their parts
Systematics	The study of the classification of, and the evolutionary interrelationships between, animal groups

EVOLUTION—THE ORIGIN OF THE ANIMAL KINGDOM

The theory of organic evolution is the concept that organisms change over time. Evolutionary processes are the source of diversity in all of life—including the animal kingdom. Evolutionary processes are remarkable for their relative simplicity, yet they are awesome because of the effects they have had on life-forms. Evolutionary processes have resulted in an estimated 4 to 30 million species of organisms living today. (Only 1.4 million species have been described.) Many more animals existed in the past and have become extinct. Modern evolutionary theories are described in chapters 2 and 3. The following discussion summarizes some of the sources of evidence for evolution.

TABLE 1.2	EXAMPLES OF SPECIALIZATIONS IN ZOOLOGY BY TAXONOMIC CATEGORIES
Entomology	The study of insects
Herpetology	The study of amphibians and reptiles
Ichthyology	The study of fishes
Mammology	The study of mammals
Ornithology	The study of birds
Protozoology	The study of protozoa

BIOGEOGRAPHY

Biogeography is the study of the geographic distribution of plants and animals. Biogeographers attempt to explain why organisms are distributed as they are. Biogeographic studies have shown that life-forms in different parts of the world have had distinctive evolutionary histories. For example, large areas of the world often have similar climatic and geographic factors. Each of these geographic regions usually has distinctive plants and animals that have particular roles in the environment. Compare the large meat-eating animals of North America with those of Africa (figure 1.3). Their similar form suggests a distant common ancestry, and they have similar life-styles. Obvious differences, however, result from millions of years of independent evolution.

(a)

(b)

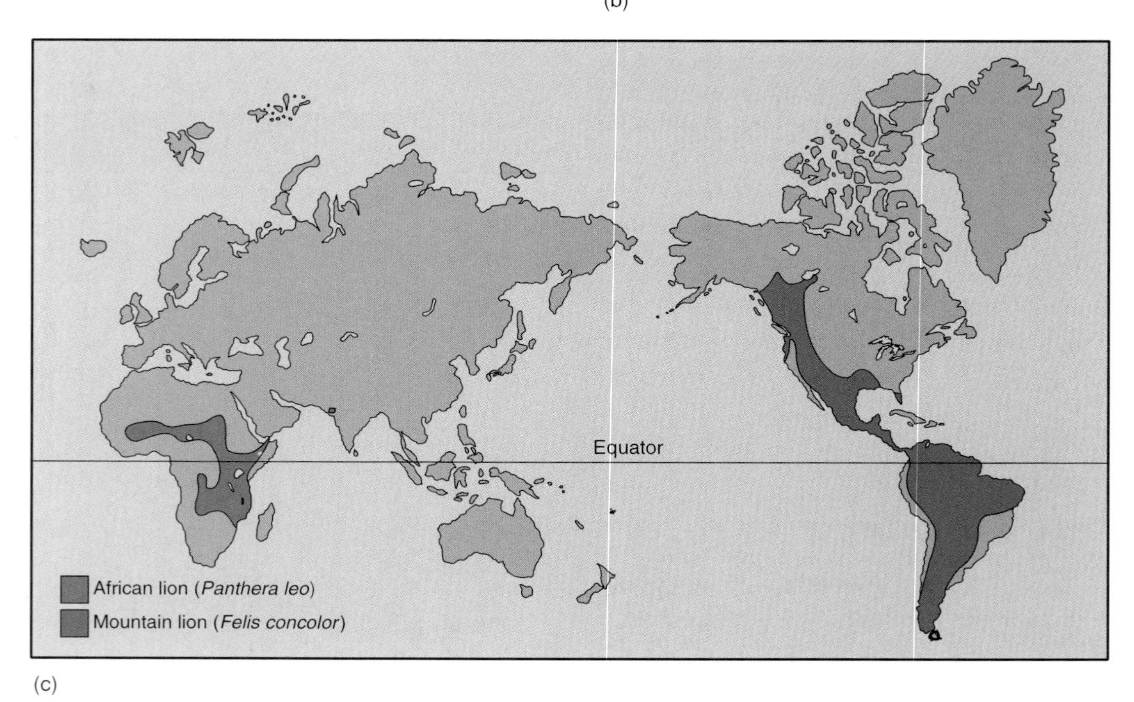

(c)

FIGURE 1.3

Biogeography as Evidence of Evolutionary Change. (*a*) The mountain lion (or cougar, *Felis concolor*) of North and South America has a similar ecological role as the (*b*) lion (*Panthera leo*) of Africa. Their similar form suggests a distant common ancestry and similar life-styles. Obvious differences, however, result from millions of years of independent evolution.

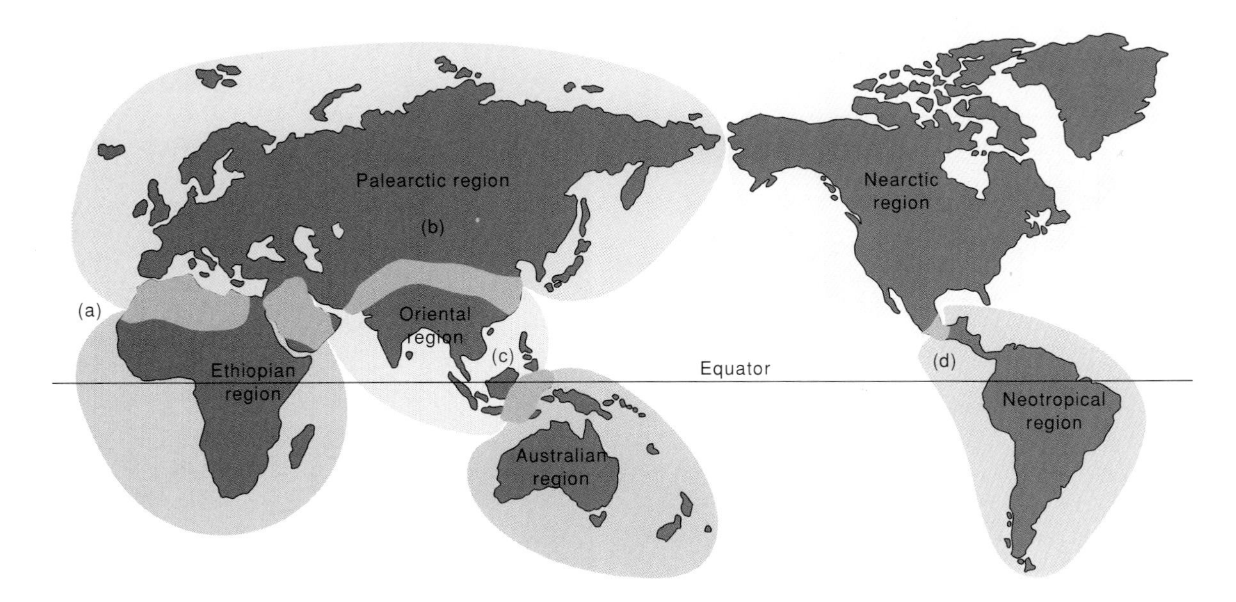

FIGURE 1.4

Biogeographic Regions of the World. Biogeographic regions of the world are separated from one another by barriers, such as oceans, mountain ranges, and deserts. The Ethiopian region is separated from the Palearctic by the Sahara and Arabian Deserts (*a*), the Palearctic region is separated from the Oriental region by the Himalayan Mountains (*b*), the Oriental and Australian regions are separated by deep ocean channels (*c*), and the Nearctic and Neotropical regions are separated by the mountains of southern Mexico and the tropical lowlands of Mexico (*d*).

Recognizing that plants and animals in different parts of the world have distinctive evolutionary histories, biogeographers have divided the world into six major biogeographic regions (figure 1.4). Each region has a characteristic group of plants and animals, and even though one may move from one region to another and experience similar climates, the different plants and animals encountered may make it seem as if one has entered another world.

PALEONTOLOGY

Paleontology (Gr. *palaios*, old + *on*, existing + *logos*, to study), which is based on the study of the fossil record, has provided some of the most direct evidence for evolution. **Fossils** (L. *fossilis*, to dig) are evidence of plants and animals that existed in the past and have become incorporated into the earth's crust (e.g., as rock or mineral) (figure 1.5). For fossilization to occur, an organism must be quickly covered by sediments to prevent scavenging, and in a way that seals out oxygen and slows decomposition. Fossilization is most likely to occur in aquatic or semiaquatic environments. The fossil record is, therefore, more complete for those groups of organisms living in or around water and for organisms with hard parts. This documentation provides some of the most convincing evidence for evolution. In spite of gaps in the fossil record, paleontology has resulted in nearly complete understanding of many evolutionary lineages (figure 1.6). Paleontologists have estimated the age of earth (about 4.6 billion years old), as well as the ages of many rocks and fossils (table 1.3).

FIGURE 1.5

Paleontological Evidence of Evolutionary Change. Fossils, such as this trilobite (*Phacops rana*), are direct evidence of evolutionary change. Trilobites were in existence about 500 million years ago and became extinct about 250 million years ago. Fossils form when an animal dies and is covered with sediments. Water dissolves calcium from hard body parts and replaces calcium with another mineral, forming a hard replica of the original animal. This process is called mineralization.

COMPARATIVE ANATOMY

A structure in one animal may resemble a structure in another animal because of a common evolutionary origin. **Comparative anatomy** is the subdiscipline of zoology that is fundamentally

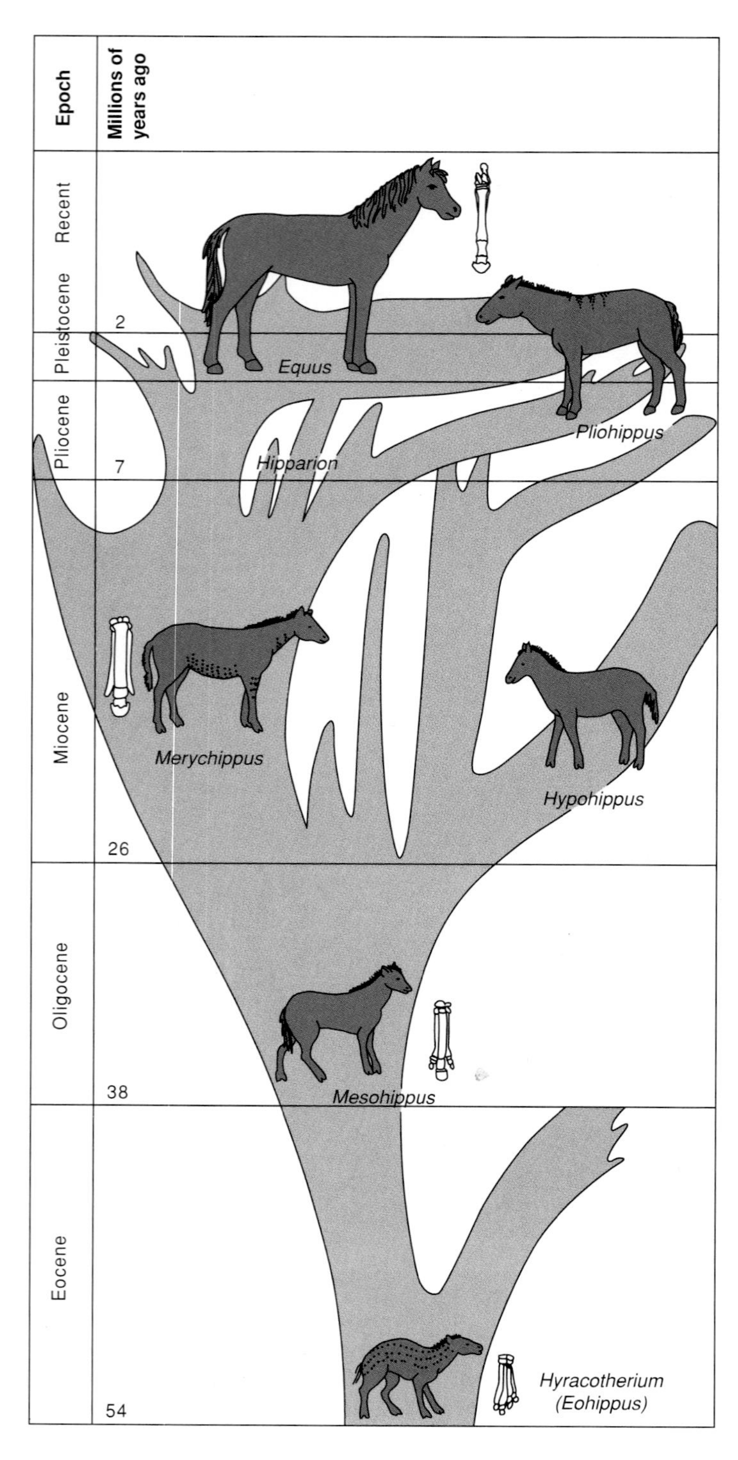

FIGURE 1.6

Reconstruction of an Evolutionary Lineage Based on Evidence in the Fossil Record. Evolution of the horse has been traced back about 60 million years using the fossil record. *Hyracotherium* was a dog-sized animal with four prominent toes on each foot. A single middle digit of the toe and vestigial digits on either side of that remain in modern horses. About 17.7 million years ago there was a very rapid evolutionary diversification that resulted in the grazing life-style of modern horses. About 15 million years ago 10–12 contemporaneous species of fossil horses occurred in North America. Note that evolutionary lineages are seldom simple ladders of change. Instead, numerous evolutionary side branches often meet with extinction.

TABLE 1.3 THE HISTORY OF THE EARTH: GEOLOGICAL ERAS, PERIODS, AND MAJOR BIOLOGICAL EVENTS*

Era	Period	Age (Millions of Years)	Major Biological Events
Cenozoic	Quaternary	0.01	Subtropical forests gave way to cooler forests and grassland areas.
	Tertiary	65	Modern orders of mammals evolved. Evolution of humans in the last 5 million years.
Mesozoic	Cretaceous	135	Continental seas and swamps spread. Extinction of ancient birds and reptiles.
	Jurassic	195	Climate warm and stable. High reptilian diversity. Birds first appeared.
	Triassic	240	Climate warm. Extensive deserts. Mammal-like reptiles replaced by dinosaurs. First true mammals.
Paleozoic	Permian	285	Climate cold early, but then warmed. Mammal-like reptiles common. Widespread extinction of amphibians.
	Carboniferous	375	Warm and humid with extensive coal-producing swamps. Arthropods and amphibians were very common. First reptiles appeared.
	Devonian	420	Land high and climate cool. Freshwater basins developed. Fish diversified. Early amphibians appeared.
	Silurian	450	Extensive shallow seas. Warm climate. First terrestrial arthropods. First jawed fish.
	Ordovician	520	Shallow extensive seas. Climate warmed. Many marine invertebrates. Jawless fish widespread.
	Cambrian	570	Extensive shallow seas and warm climate. Trilobites and brachiopods were common. Earliest vertebrates were found late in the Cambrian.
Proterozoic		2,000	Multicellular organisms appeared and flourished. Many invertebrates. Eukaryotic organisms appeared (1,500 million years ago). Oxygen accumulated in the atmosphere.
Archean		4,600	Prokaryotic life appeared (3,500 million years ago). Origin of the earth (4,600 million years ago).

*Note that the time scale in the Proterozoic and Archean eras are greatly compressed.

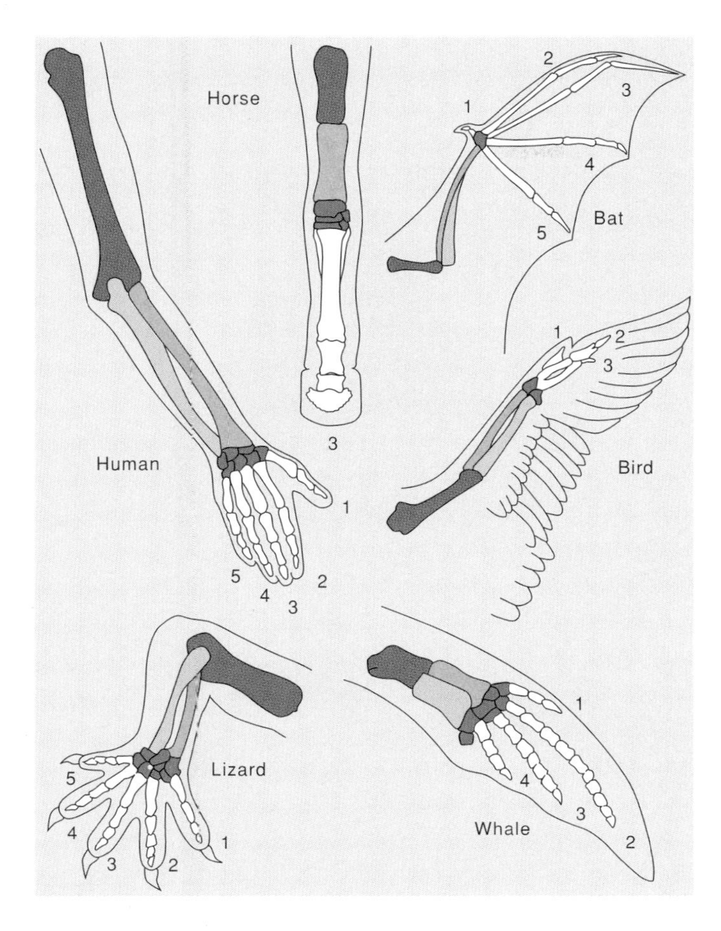

FIGURE 1.7

The Concept of Homology. The forelimbs of vertebrates evolved from an ancestral pattern. Even in vertebrates as dissimilar as whales and bats, the same basic arrangement of bones can be observed. The digits (fingers) are numbered 1 (thumb) to 5 (little finger). Homologous bones are indicated by color coding.

(a)

(b)

FIGURE 1.8

The Concept of Analogy. The wings of birds (*a*) and insects (*b*) are analogous. They are specialized for a similar function, and that similar function has coincidentally led to flat, planing surfaces required for flight. Both kinds of wings arose independently and, therefore, are not homologous.

based on this relationship. Comparative anatomists study the structure of fossilized and living animals, looking for similarities that could be indications of close evolutionary relationships. Structures derived from common ancestry are said to be **homologous** (Gr. *homolog* + *os*, agreeing) (i.e., having the same or a similar relation). Some examples of homology are obvious. For example, the appendages of vertebrates have a common arrangement of similar bones, even though the function of the appendages may vary (figure 1.7). Along with other evidence, this similarity in appendage structure indicates that the vertebrates evolved from a common ancestor.

Not all such similarities indicate homology. **Convergent evolution** occurs when two unrelated organisms adapt to similar conditions, resulting in superficial similarities in structure. For example, the wing of a bird and the wing of an insect are both adaptations for flight, but are not homologous (figure 1.8). Any similarities are simply reflections of the fact that, to fly, an animal must have a broad, flat gliding surface. Instead of being homologous, these structures are said to be **analogous** (i.e., having a similar function but dissimilar origin).

Structures are often retained in an organism, even though the structures may have lost their usefulness. They are often poorly developed, and are called **vestigial structures.**

1 For example, boa constrictors have minute remnants of hind (pelvic) limb bones that have no function in these snakes. They are left over from appendages of their reptilian ancestors. Such remnants of once useful structures are clear indications of change, hence evolution.

Molecular Biology

Recently, **molecular biology** has yielded a wealth of information on evolutionary relationships. ust as animals can have homologous structures, animals may also have homologous processes. Ultimately, structure and function are based on the genetic blueprint found in all living animals, the DNA molecule. Related animals have DNA derived from their common ancestor. Because DNA carries the codes for proteins that make up each animal, related animals are expected to have similar proteins. With the modern laboratory technologies now available, zoologists can extract and analyze the structure of proteins from animal tissue, and compare the DNA of different animals. By looking for dissimilarities in the structure of related proteins and DNA, and by assuming relatively constant mutation rates, molecular biologists can estimate the elapsed time since divergence from a common ancestral molecule.

The above fields of study have generated impressive documentation of evolution since the initial studies of Darwin. There is no doubt in the minds of the vast majority of scientists as to the reality of evolution. Evolutionary theory has impacted biology like no other single theory. It has impressed scientists with the fundamental unity of all of biology. As you progress through this text, you will continually be reminded of the unity that exists within life because of its common origin.

Stop and Ask Yourself

1. What is zoology? Why do most zoologists choose to specialize in some field within zoology?
2. What is the study of amphibians and reptiles called? Of insects? Of fishes?
3. What is evolution?
4. What is biogeography? How does it provide evidence of evolution?
5. What is homology? How does it provide evidence of evolution?

Ecological Principles—Preserving the Animal Kingdom

All animals have certain requirements for life. In searching out these requirements, animals come into contact with other organisms and their physical environment. These encounters result in a multitude of interactions that influence all organisms. Even the physical environment is altered by the organisms that live in it. **Ecology** is the study of the relationships of organisms to their environment and to other organisms. Understanding basic ecological principles helps us to understand why animals live in certain places, why animals eat certain foods, and why animals interact with other animals in specific ways. It is also the key to understanding how human activities can harm animal populations and what we must do to preserve animal resources. The following discussion is a brief review of the ecological principles that are central to understanding how animals live in their environment.

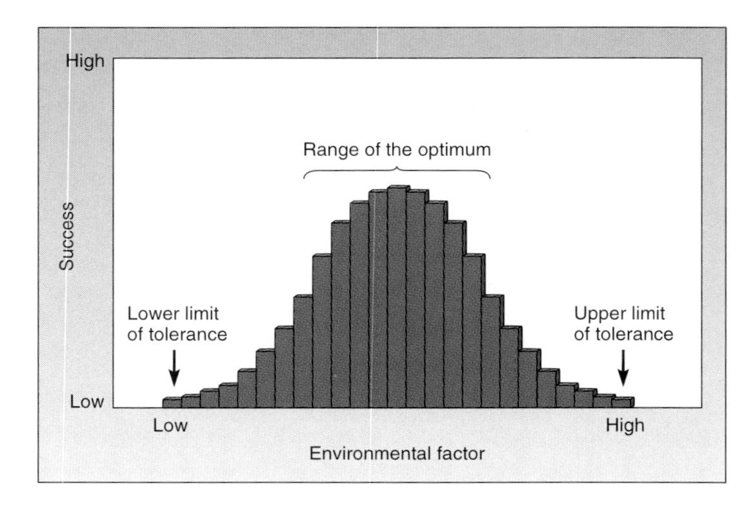

Figure 1.9
Tolerance Range of an Animal. The tolerance range of an animal can be depicted by plotting changes in an environmental factor versus some index of success (perhaps egg production, longevity, or growth). The graphs that result are often, though not always, bell-shaped. The range of the optimum is the range of values of the factor within which success is greatest. The range of tolerance and range of the optimum may vary depending on an animal's stage of life, health, and activity. *Source: Richard R. Brewer, The Science of Ecology, copyright 1988 Saunders College Publishing.*

Animals and Their Abiotic Environment

An animal's **habitat** (environment) includes all living (biotic) and nonliving (abiotic) characteristics of the area in which the animal lives. Abiotic characteristics of a habitat include factors such as availability of oxygen and inorganic ions, light, temperature, and current or wind velocity. For any of these abiotic characteristics, animals live within a certain range, called a **tolerance range.** At either limit of the tolerance range, one or more essential functions cease. A certain range of values within the tolerance range, called the **range of the optimum,** defines the conditions under which an animal is most successful (figure 1.9). Combinations of abiotic factors are necessary for an animal to survive and reproduce. When one of these factors is out of the range of tolerance for an animal, that factor becomes a limiting factor.

There are numerous abiotic factors that are important for animals and have the potential to become limiting factors. Some of the most important abiotic factors include: energy, temperature, moisture, light, geology, and soils.

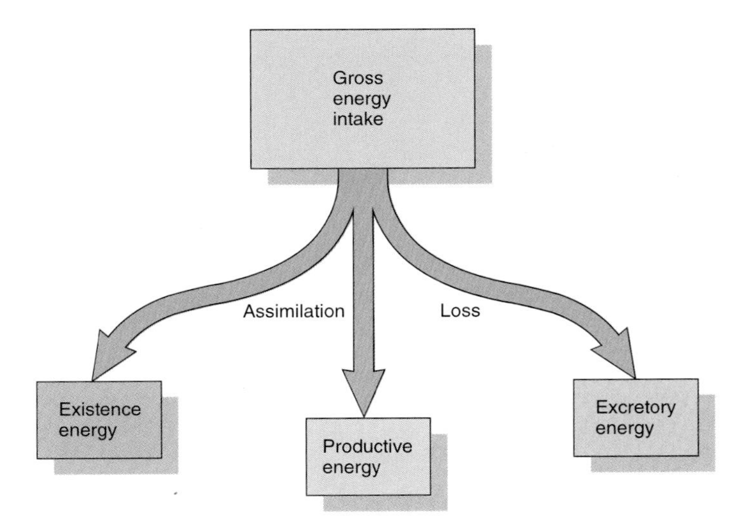

FIGURE 1.10

Energy Budgets of Animals. The gross energy intake of an animal is the sum of energy lost in excretory pathways, plus energy assimilated for existence and productive functions. The relative sizes of the boxes in this diagram are not necessarily proportional to the amount of energy devoted to each function. The total energy in an energy budget of an animal, and the amount of energy devoted to productive functions, depends upon various internal and external factors (e.g., time of year, reproductive status).

Energy

Energy is the ability to do work, which includes everything from foraging for food to moving molecules within cells. To supply their energy needs, animals ingest other organisms—that is, animals are **heterotrophic** (Gr. *hetero,* other + *tropho,* feeder). **Autotrophic** (Gr. *autos,* self) organisms (e.g., plants) carry on photosynthesis or other carbon-fixing activities that supply their food. An accounting of the total energy intake of an animal, and a description of how that energy is used and lost, is referred to as an **energy budget.** Notice in figure 1.10 that the total energy contained in the food eaten by an animal is called gross energy intake. Some of this energy is lost in feces and through excretion (excretory energy); some energy is devoted to minimal maintenance activities such as pumping blood and exchanging gases (existence energy); and some energy may be left for growth, mating, and caring for young (productive energy). Survival requires that individuals acquire enough energy to supply these productive functions. Favorable energy budgets are sometimes difficult to attain, especially in temperate regions where winter often makes food supplies scarce.

Temperature

Part of an animal's existence energy is expended in regulating body temperature. Temperature influences rates of chemical reactions in animal cells (metabolic rate) and affects the overall activity of the animal. The body temperature of an animal seldom remains constant because of an inequality between heat loss and heat gain. Heat energy can be lost to objects in an animal's surrounding as infrared and heat radiation, to the air around an animal through convection, and as evaporative heat loss. On the other hand, heat is gained from solar radiation, infrared and heat radiation from objects in the environment, and relatively inefficient metabolic activities that generate heat as a by-product of cellular functions. Thermoregulatory needs may influence many habitat requirements such as the availability of food, water, and shelter.

When food becomes scarce, or when animals are not feeding for other reasons, they are subject to starvation. Under these circumstances, metabolic activities may decrease dramatically. **Torpor** is a time of decreased metabolism and lowered body temperature that occurs in bats, hummingbirds, and other animals who must feed almost constantly when they are active. Torpor allows these animals to conserve energy when they are not feeding. **Hibernation** is a time of decreased metabolism and lowered body temperature that may last for weeks or months. It occurs in small animals such as rodents, shrews, and bats. During these times, body temperatures drop to about 20° C, but thermoregulation is not suspended. **Winter sleep** occurs in larger animals that are sustained through periods of winter inactivity by large energy reserves. (2) The body temperature of a sleeping bear does not drop substantially, and the bear can wake and become active very quickly. **Aestivation** is a period of inactivity in some animals that must withstand extended periods of drying. The animal usually enters a burrow as its environment begins to dry and does not eat or drink until moisture returns to the environment. Some aquatic invertebrates, amphibians, and reptiles withstand seasonal droughts by aestivating.

Other Abiotic Factors

Other important abiotic factors for animals include moisture, light, geology, and soils. All life's processes occur in the watery environment of the cell. Water that is lost must be replaced. The amount of light and the length of the light period in a 24-hour day is an accurate index of seasonal change. Animals use light for timing many activities such as reproduction and migration. Geology and soils often directly or indirectly affect organisms living in an area. Characteristics such as texture, amount of organic matter, fertility, and water-holding ability directly influence the number and kinds of animals living in or on the soil. These characteristics also influence the plants upon which animals feed.

POPULATIONS

Populations are groups of individuals of the same species that occupy a given area at the same time and have unique attributes. Two of the most important attributes involve the potential for population growth.

One attribute of populations concerns the rate at which individuals die. This attribute can be summarized with survivorship curves (figure 1.11). Juvenile mortality in most natural animal populations is very high. For individuals that reach adulthood, however, mortality rates decline. This decline is seen in the type III survivorship curve in figure 1.11. Most invertebrates and fishes

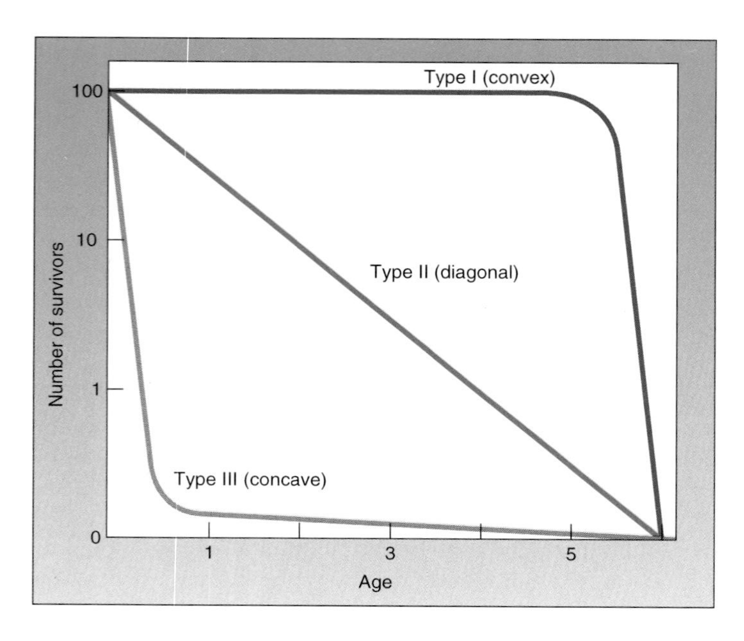

Figure 1.11

Survivorship. Survivorship curves are a plot of number of survivors (a logarithmic plot) versus age. Type I curves apply to populations in which individuals are likely to live out their potential life span. Type II curves apply to populations in which mortality rates are constant throughout age classes. Type III curves apply to populations in which mortality rates are highest for the youngest cohorts. *Source: Richard R. Brewer,* The Science of Ecology, *copyright 1988 Saunders College Publishing.*

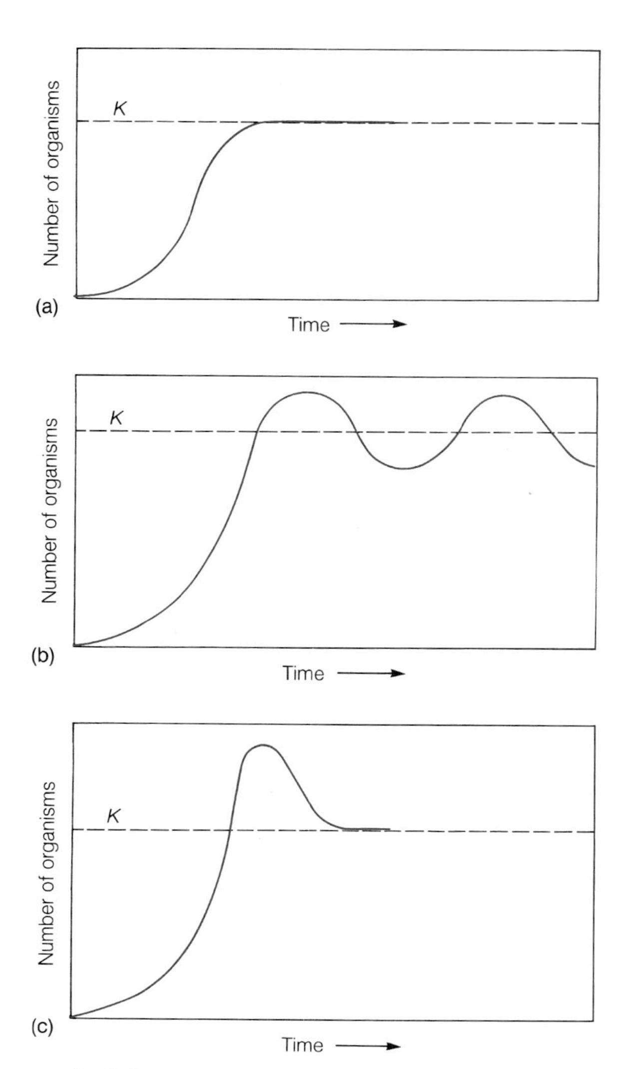

Figure 1.12

Logistic Population Growth. Logistic growth curves take into consideration the fact that limited resources place an upper limit on population size. (*a*) When carrying capacity (*K*) is reached, population growth levels off, creating an S-shaped curve. (*b,c*) During its exponential growth phase, a population may overshoot carrying capacity because demand on resources may lag behind population growth. When that happens, numbers may cycle on either side of *K*, or form a J-shaped curve.

are characterized by type III survivorship curves. For some other animals (e.g., some birds and rodents), the probability of dying is relatively constant because the environment is no harsher on juveniles than on adults. They are characterized by type II survivorship curves. For a few animals, environmental factors are relatively unimportant in influencing mortality, and most individuals live their entire life span. Human populations in most developed countries of the world are characterized by type I survivorship curves.

A second attribute of populations concerns population growth. The potential for a population to increase in numbers of individuals is remarkable. Populations tend to grow by a constant ratio per unit time rather than by adding a constant number of individuals to the population in every generation. In other words, populations experience **exponential growth.** The rate of growth is unique for every population because it is influenced by factors such as the number of offspring produced, the duration of the reproductive period, and the likelihood of reaching reproductive age. Exponential growth cannot occur indefinitely. Constraints placed on populations by climate, food, space, and other environmental factors (such as chemicals) limit the population size that a particular environment can support. This limit is called the environment's **carrying capacity.** In these situations the growth curves assume a sigmoid, or flattened S shape, and the population growth is referred to as **logistic population growth** (figure 1.12).

Interspecific Interactions

Members of other species can affect all characteristics of a population. Interspecific interactions include: herbivory, predation, competition, coevolution, and symbiosis. Animals, however, are rarely limited by artificial categories that zoologists create to help organize life's complexity. As you study the following material, realize that animals often do not interact with other animals in only one way. The nature of interspecific interactions may change as an animal matures, as seasons change, or as the environment changes.

Herbivory and Predation

Animals that feed on plants by cropping portions of the plant, but usually not killing the plant, are called herbivores. Herbivores convert a plant's production into animal flesh. This conversion

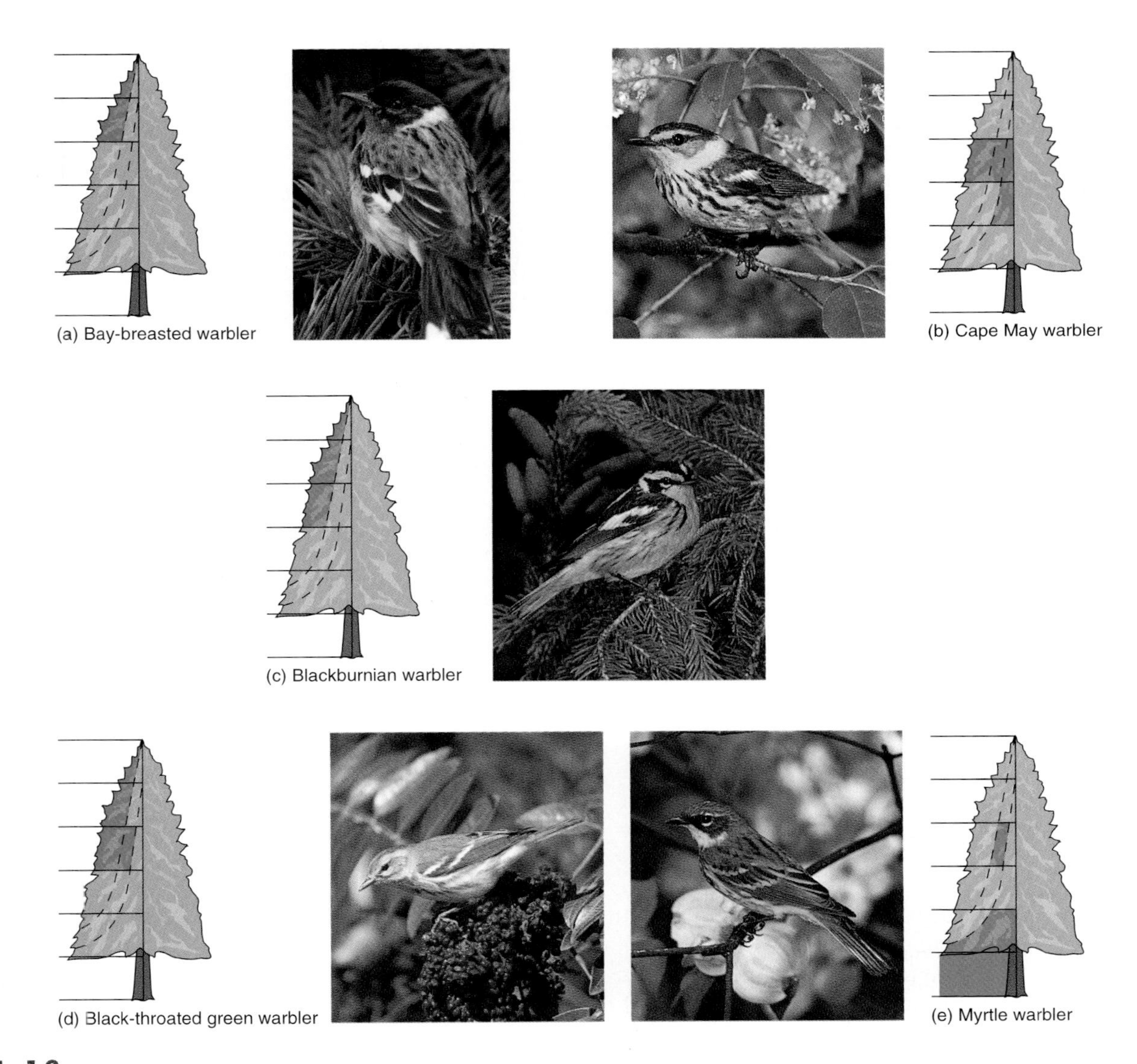

FIGURE 1.13
Coexistence of Competing Species. Robert MacArthur found that five species of warblers (*a–e*) coexisted by partitioning resources. Partitioning occurred by dividing up spruce trees into preferred foraging regions. Preferred foraging regions are shown in dark green.

provides food for predators that feed on other animals. Interactions between plants and herbivores, and predators and prey, are complex and affected by many characteristics of the environment. Many of these interactions will be described in the chapters that follow.

Interspecific Competition

When competition for resources occurs between members of different species, one species may be forced to move or become extinct, or the two species may share the resources and coexist. While the first two options (moving or extinction) have been documented in a few instances, most studies have shown that competing species usually can coexist. Coexistence can occur when species utilize resources in slightly different ways and when the effects of interspecific competition are less severe than the effects of animals of the same species competing for a resourse (intraspecific competition). These ideas can be illustrated by the studies of Robert MacArthur on five species of warblers that all used the same caterpillar prey. Warblers partitioned their spruce tree habitats by dividing a tree into preferred regions for foraging. Although there was some overlap of foraging regions, competition was limited, and the five species coexisted (figure 1.13).

Coevolution

The evolution of ecologically related species is sometimes coordinated so that each species exerts a strong selective influence on the other. This occurrence is called **coevolution.** Coevolution may occur when species are competing for the same resource, during predatory-prey interactions, or in the relationship between some flowering plants and their animal pollinators.

In predator-prey relationships, natural selection may favor the development of protective characteristics in the prey species. Similarly, selection favors characteristics in predators that allow them to become better at catching prey. Coevolution occurs when a change toward increased elusiveness of prey is countered by greater predator efficiency.

Coevolution is most obvious in plant/animal interactions. Flowers attract pollinators with a variety of elaborate olfactory and visual adaptations. Insect-pollinated flowers are usually yellow or blue because insects best see these wavelengths of light. In addition, petal arrangement often provides perches for pollinating insects. Flowers pollinated by hummingbirds, on the other hand, are often tubular and red. Hummingbirds have a poor sense of smell but see red very well. The long beak of hummingbirds is an adaptation that allows them to reach far into tubular flowers. Their hovering ability means they have no need of a perch.

Symbiosis

Some of the best examples of adaptations arising through coevolution are seen when two different species live in continuing, intimate associations. These relationships are called **symbiosis** (Gr. *sym,* together + *bios,* life). Such interspecific interactions influence the species involved in dramatically different ways. Many examples of symbiotic interactions are described in the chapters that follow.

Parasitism is a very common form of symbiosis in which one organism lives at the expense of a second organism, called a host. The host usually survives at least long enough for the parasite to complete one or more life cycles. **Commensalism** is a symbiotic relationship in which one member of the relationship benefits, and the second is neither helped nor harmed. The distinction between parasitism and commensalism is somewhat difficult to apply in natural situations. Whether or not the host is harmed often depends on factors such as the nutritional state of the host. Thus, symbiotic relationships can be parasitic in some situations and commensalistic in others. **Mutualism** is a symbiotic relationship in which both members of the relationship benefit. Examples of mutualism abound in the animal kingdom, and many examples will be described in chapters that follow.

Other Interspecific Interactions

Many other characteristics of animals have been shaped by interspecific interactions. **Camouflage** occurs when an animal's color patterns help hide the animal or the developmental stage from another animal (figure 1.14). **Cryptic coloration** (L. *crypticus,* hidden) is a type of camouflage that occurs when an animal takes on color patterns in its environment to prevent the animal from being seen by other animals. **Countershading** is a type of camouflage common in eggs of frogs and toads. These eggs are darkly pigmented on top and lightly pigmented on the bottom. Whey viewed by a bird from above, the dark top side hides the eggs from detection against the darkness below. On the other hand, when viewed from below by a fish, the light undersurface blends with the bright air-water interface.

Some animals that protect themselves by being dangerous or distasteful to predators advertise their condition by conspicuous coloration. The sharply contrasting colors of a skunk and bright colors of poisonous snakes give a clear message to other animals—stay away! These color patterns are examples of warning or **aposematic coloration** (Gr. *apo,* away from + *sematic,* sign).

Figure 1.14
Camouflage. The color pattern of this tiger (*Panthera tigris*) provides effective camouflage that helps when stalking prey.

Resembling conspicuous animals may also be advantageous. **Mimicry** (L. *mimus,* to imitate) occurs when a species resembles one, or sometimes more than one, other species and gains protection by the resemblance (figure 1.15).

Animal Communities

All populations living in an area make up a **community.** Communities are characterized by the numbers and kinds of organisms that comprise them. Most communities have certain members that are of overriding importance in determining

Stop and Ask Yourself

6. What is an energy budget? Why is productive energy necessary if a population is maintained through multiple generations?
7. What is a population? What would the survivorship curve look like for a population of mosquitoes?
8. What kinds of factors limit exponential growth of a population? What term is used to designate these kinds of environmental constraints?
9. What is coevolution? How could predatory-prey interactions be produced through coevolution?

FIGURE 1.15
Müllerian Mimicry. These six species of butterflies (*Heliconius* spp.) are all distasteful to bird predators. A bird that consumes any member of the six species will be likely to avoid all six species in the future.

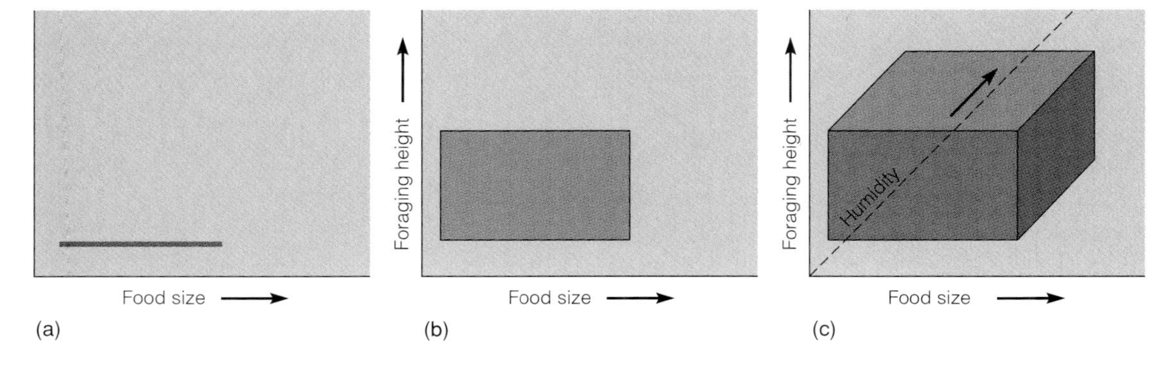

FIGURE 1.16
The Niche as a Multidimensional Space. Each characteristic of an animal's niche can be plotted on a separate axis of a graph. A visualization of a three-dimensional niche involving (*a*) food size, (*b*) foraging height, and (*c*) humidity is shown here. Additional dimensions would be added to portray the complete niche.

community characteristics. For example, a stream may have a large population of rainbow trout that helps determine the makeup of invertebrate populations of the stream. Species that are responsible for establishing community characteristics are called **dominant species.** Communities are also characterized by the variety of organisms present in the community. This variety is called **community (species) diversity,** or richness. Forces of nature and human activities influence community diversity. Factors that promote high diversity include a wide variety of resources, high productivity, climatic stability, moderate levels of predation, and moderate levels of disturbance from outside the community. Pollution often reduces community diversity.

The Ecological Niche

The **ecological niche** is an important concept of community structure. The niche of any species includes all the attributes of an animal's life-style: where it looks for food, what it eats, where it nests, and what conditions of temperature and moisture it requires (figure 1.16). Theoretically, competition occurs when the niches of

FIGURE 1.17
Succession. Primary succession on a sand dune. Beach grass is the first to become established on a sand dune. It stabilizes the dune so that shrubs, and eventually trees, can become established.

two species overlap. Although the niche concept is difficult to quantify, it is valuable in helping us perceive community structure. It illustrates that members of a community tend to partition resources rather than compete for them. It also helps us visualize the role of an animal in the environment.

Community Stability

As with individuals, communities are born and they die. Between those events is a time of continual change. Some changes are brought about by climatic or geological events. Others are brought about by the members of the community. The latter often bring about changes in predictable ways in a process called **succession** (L. *successio,* to follow). Communities begin in areas nearly devoid of life. The first community to become established is called the **pioneer community** (figure 1.17). The death and decay of its members add nutrients to the community. These nutrients support more life. Over thousands of years nutrients accumulate, and the characteristics of the ecosystem change. Each successional stage is called a **seral stage,** and the entire successional sequence is a **sere** (ME. *seer,* to wither). Succession occurs because the dominant life-forms of a sere gradually make the area less favorable for themselves but more favorable for the next successional stage. The final community is called the **climax community.** It is different from the seral stages that preceded it because it can tolerate its own reactions. The accumulation of products of life and death no longer make the area unfit for the individuals living there. Climax communities usually have high species diversity.

TROPHIC STRUCTURE OF ECOSYSTEMS

Communities and their physical environment are called **ecosystems.** 4 One important fact of ecosystems is that energy is constantly being used, and once it leaves the ecosystem this energy is never reused. Energy supports the activities of all organisms in the ecosystem. It usually enters the ecosystem in the form of sunlight, is incorporated into the chemical bonds of molecules within living and decaying tissues, and is eventually lost as heat (box 1.1).

The sequence of organisms through which energy moves is called a **food chain.** One relatively simple food chain might look like the following:

grass → grazing insects → shrews → owls

It is more realistic to envision complexly interconnected food chains, called **food webs,** that involve many kinds of organisms (figure 1.18). Because food webs can become very complex, it is convenient to group organisms based on the form of energy used. These groupings are called **trophic levels.**

Producers (autotrophs) obtain food from inorganic materials and an energy source. Producers form the first trophic level of an ecosystem. The producers that are most familiar to us are green plants. Other trophic levels are made up of consumers (heterotrophs). Consumers obtain their energy by eating other organisms. Herbivores (primary consumers) eat producers. Some carnivores (secondary consumers) eat herbivores, and other carnivores (tertiary consumers) eat the carnivores that ate the herbivores. Consumers also include scavengers that feed on large chunks of dead and decaying organic matter. Feeding at any consumer level is never 100% efficient. Decomposers break down organic matter left over from other trophic levels by digesting it extracellularly and absorbing the products of digestion.

The efficiency with which the animals of a trophic level convert food into new biomass (the sum of all living and decaying tissues in an ecosystem) depends on the nature of the food and the trophic level (figure 1.19). An average efficiency for the conversion of biomass at one trophic level to biomass at the next trophic level is approximately 10 percent.

CYCLING WITHIN ECOSYSTEMS

A second important lesson learned from the study of ecosystems is that matter is constantly recycled within an ecosystem. Matter moves through ecosystems in **biogeochemical cycles.** A nutrient is any element essential for life, and the cycling of nutrients within ecosystems is called **nutrient cycling.** Approximately 97% of living matter is made of oxygen, carbon, nitrogen, and hydrogen. Gaseous cycles involving these elements utilize the atmosphere or

BOX 1.1 Hydrothermal Vent Communities

What some have called the oceanographic discovery of the century was made in 1977 by a Woods Hole Oceanographic Institute expedition to the Galápagos Rift, in the Pacific Ocean (figure 1.1*a*). The rift (an opening made by splitting) is over 2,700 m (over 1.5 mi) below the surface and part of an extensive mid-oceanic ridge system that has developed where the tectonic plates of the earth's crust are moving apart (*see box 2.2*). In such places, a flow of lava (magma) occasionally emerges and **hydrothermal vents** spew out hot water rich in hydrogen sulfide and other minerals.

One unusual finding of the expedition was that the life of a vent community is based not on the "rain" of material generated by the producers in the surface zones, but on a rich community of chemolithotrophic bacteria that derive all of the energy they need from the oxidation of inorganic compounds, such as hydrogen sulfide. They live in total darkness here, because the vents are far below the level of light penetration.

The expedition also noted that substrate around each vent was covered with many clams, crabs, polychaete annelids, and one species (*Riftia pachyptila*) of pogonophoran (figure 1.1*b*). Like other pogonophorans (tube worms), *Riftia* is nourished in part by the endosymbiotic bacteria found in its trophosome (*see endpaper 2*). These bacteria can oxidize hydrogen sulfide to sulfate and reduce carbon dioxide to organic compounds, which nourish both the symbiont and host. The worms' hemoglobin carries oxygen and hydrogen sulfide, tightly bound to another protein. This chemical bonding keeps these two molecules from reacting in an unproductive fashion before they are delivered to the bacteria. They also protect the host's tissue from the toxic hydrogen sulfide. These vent communities are among the few on earth that do not depend on solar energy for life.

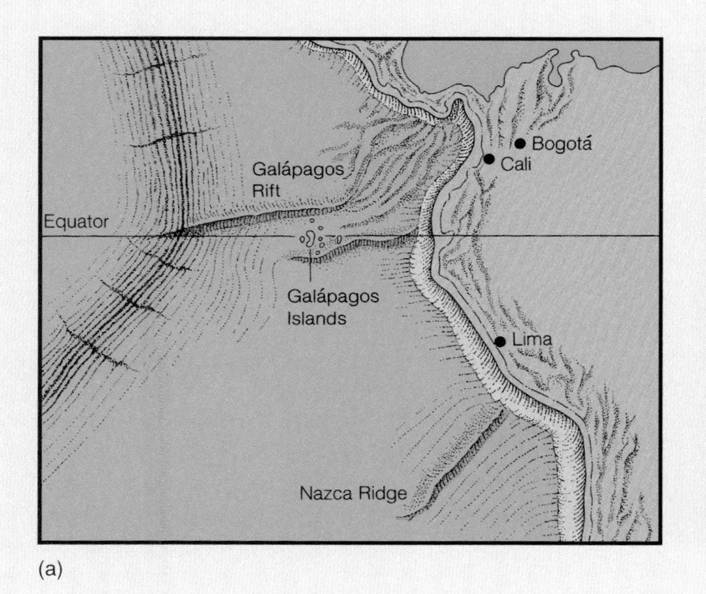

FIGURE 1 **Hydrothermal Vent Communities.** (*a*) The Galápagos Rift is the site of an extensive ocean-bottom community living on inorganic nutrients issuing from hydrothermal vents. (*b*) Life around these vents in the Galápagos Rift. Tube worms and crabs can be seen close to the vent.

oceans as a reservoir. Elements such as sulfur, phosphorus, and calcium are less abundant in living tissues than those with gaseous cycles, but they are no less important in sustaining life. The nonliving reservoir for these nutrients is the earth, and the cycles involving these elements are called sedimentary cycles.

To help you understand the concept of a biogeochemical cycle, study the carbon cycle in figure 1.20. Carbon is very plentiful on our planet and is rarely a limiting factor. The reservoir for carbon is carbon dioxide (CO_2) in the atmosphere or water. Carbon enters the reservoir when organic matter is oxidized to CO_2. Carbon dioxide is released to the atmosphere or water, where autotrophs incorporate it into organic compounds. In aquatic systems, some of the CO_2 combines with water to form carbonic acid. Because this reaction is reversible, carbonic

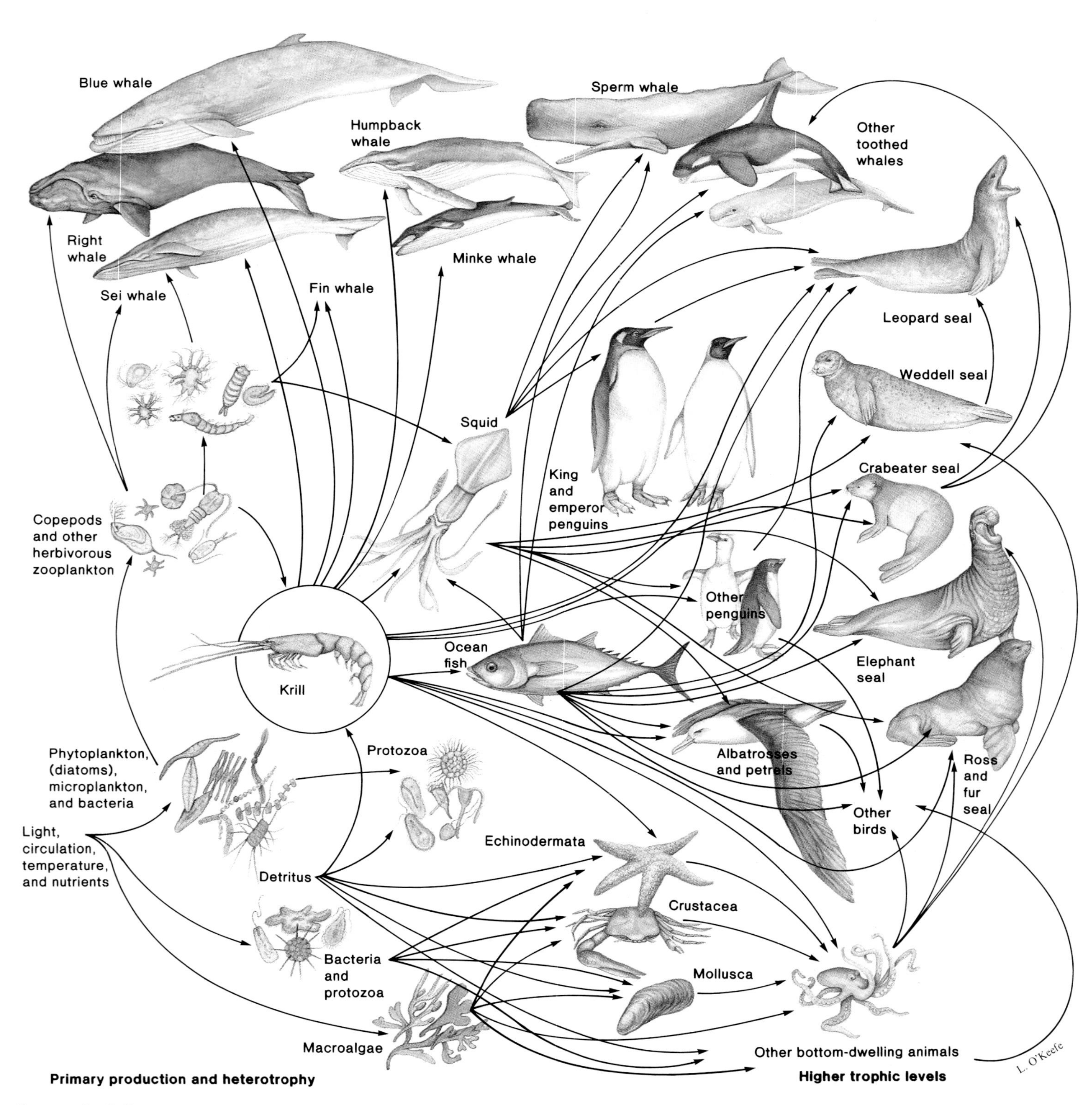

Figure 1.18

Food Webs. An Antarctic food web. Small crustaceans called krill support nearly all life in Antarctica. Krill are eaten by 6 species of baleen whales, 20 species of squid, over 100 species of fish, 35 species of birds, and 7 species of seals. Krill feed on algae, protozoa, other small crustaceans, and various larvae. To appreciate the interconnectedness of food webs, trace the multiple paths of energy from light (lower left) through krill, to the leopard seal.

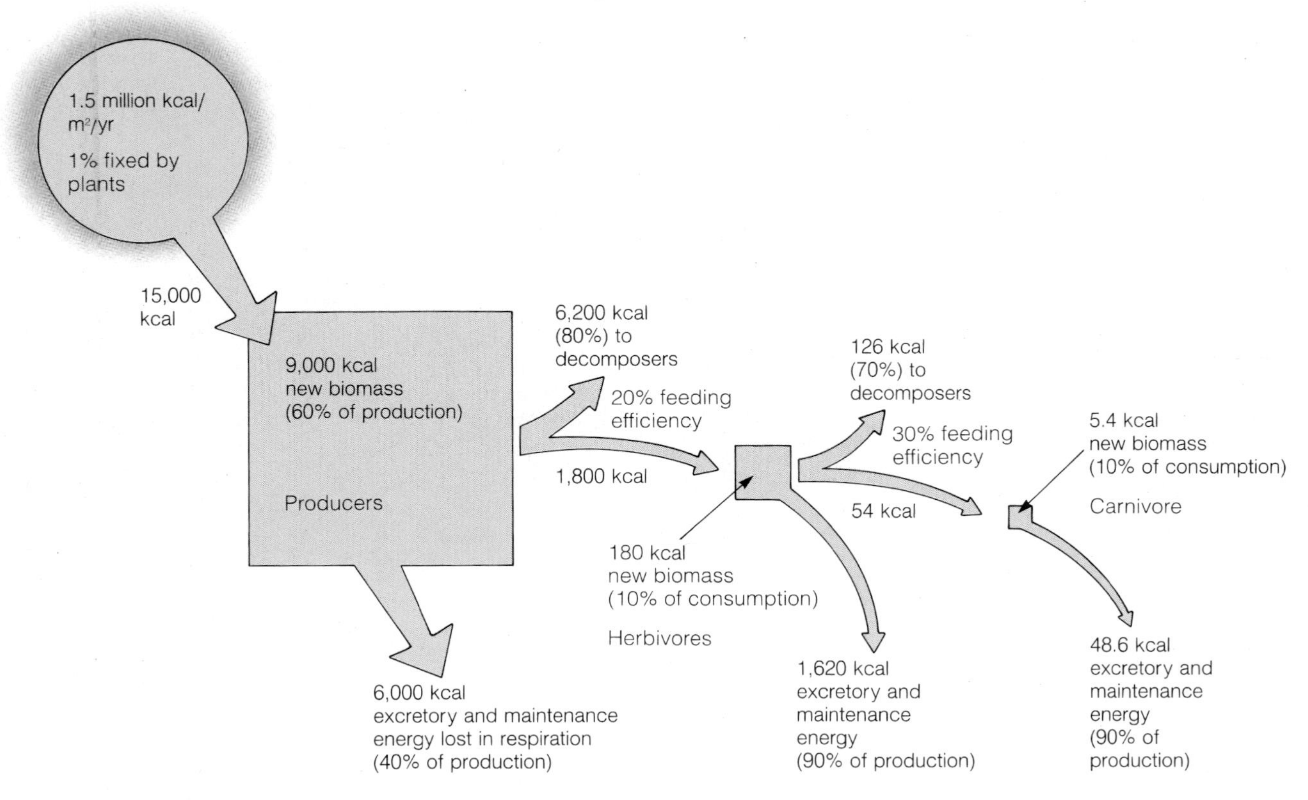

Figure 1.19

Energy Flow through Ecosystems. Approximately 1.5 million kilocalories of radiant energy strike a square meter of the earth's surface each year. Less than 1% (15,000 kcal/m^2/yr) is converted into chemical energy by plants. Of this, approximately 60% is converted into new biomass, and 40% is lost in respiration. The herbivore trophic level harvests approximately 20% of net primary production, and decomposers get the rest. Of the 1,800 kcal moving into the herbivore trophic level, 10% (180 kcal) is converted to new biomass, and 90% (1,620 kcal) is lost in respiration. Carnivores harvest about 30% of the herbivore biomass, and 10% of that is converted to carnivore biomass. At subsequent trophic levels, harvesting efficiencies of about 30% and new biomass production of about 10% can be assumed. All of these percentages are approximations. Absolute values depend on the nature of the primary production (e.g., forest vs. grassland) and characteristics of the herbivores and carnivores (e.g., ectothermic vs. endothermic).

acid (H_2CO_3) can supply CO_2 to aquatic plants for photosynthesis or to the atmosphere. Some carbon is tied up as calcium carbonate ($CaCO_3$) in the shells of molluscs and the skeletons of echinoderms. Accumulations of these shells and skeletons have resulted in limestone formations. Geological uplift, volcanic activity, and weathering returns much of this carbon to the earth's surface and the atmosphere. Other carbon is tied up in fossil fuels. The burning of fossil fuels returns large quantities of this carbon to the atmosphere as CO_2.

Ecological Problems and Biodiversity

In the last few hundred years of our history, humans have attempted to provide for the needs and wants of a growing human population. In our search for longer and better lives, however, humans have lost a sense of being a part of our world's ecosystems. Now that you have studied some general ecological principles, it should be easier to understand many of our ecological problems.

Human Population Growth

An expanding human population is the root of most of our other environmental problems. Humans, like other animals, have a tendency to undergo exponential population growth. The earth, like any ecosystem on it, has a carrying capacity and a limited supply of resources. When human populations achieve that carrying capacity, populations should stabilize. If they do not stabilize in a fashion that limits human misery, then war, famine, and/or disease is sure to be the vehicle that accomplishes our ecological destiny.

5 What is our planet's carrying capacity? The answer to that question is not simple. In part, it depends on the standard of living that we desire for ourselves and whether or not we expect resources to be distributed equally among all populations. Currently, the earth's population stands at 5.6 billion people. Virtually all environmentalists agree that number is too high if all people are to achieve the affluence of developed countries.

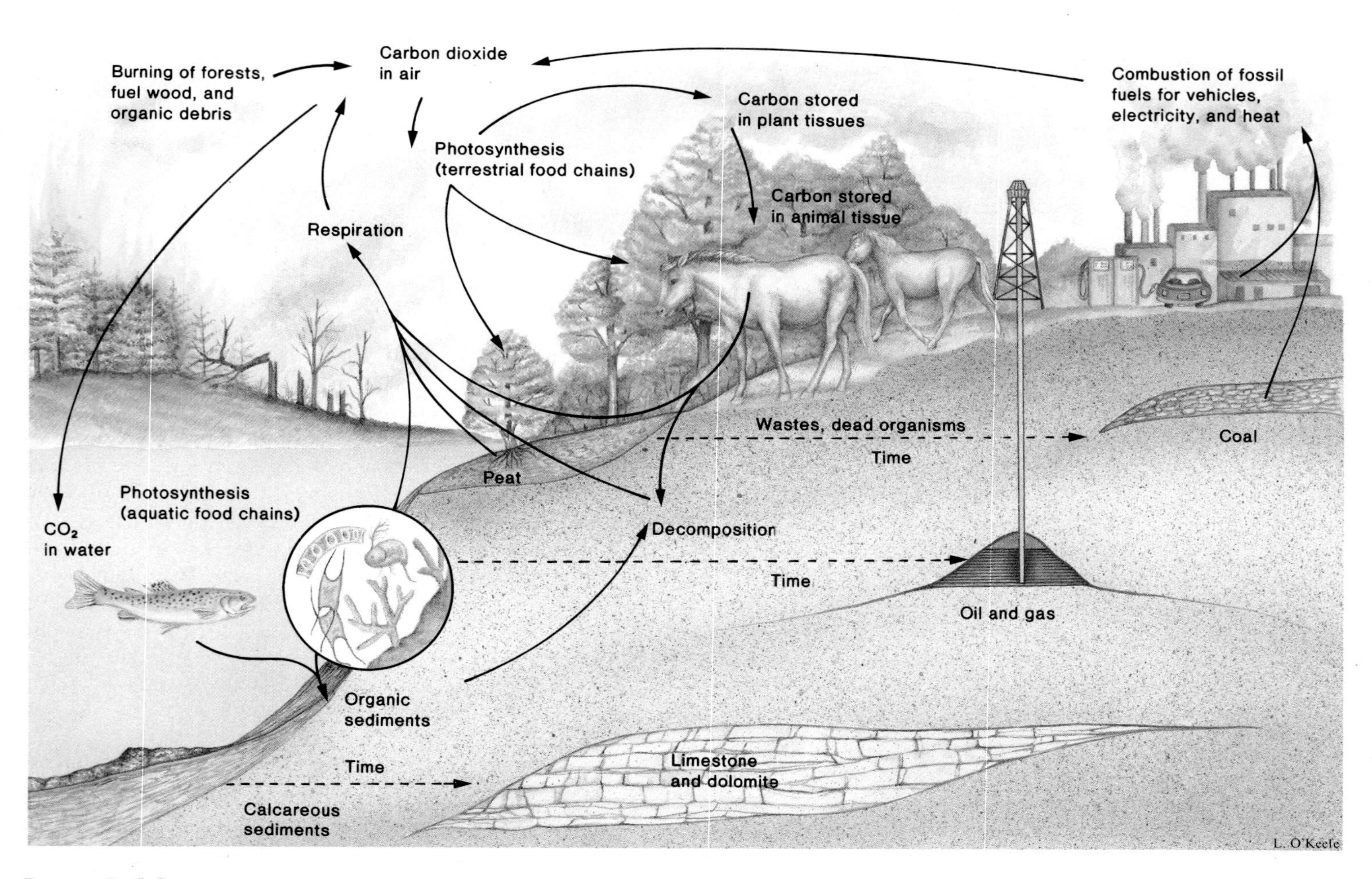

Figure 1.20
The Carbon Cycle. Carbon cycles between its reservoir in the atmosphere, living organisms, fossil fuels, and limestone bedrock.

Unless intense efforts are made to curb population growth, world populations could double in the next 50 years. Looking at the age characteristics of world populations helps to explain why human populations will grow rapidly. The **age structure** of a population is the proportion of a population that is in prereproductive, reproductive, and postreproductive classes. Age structure is often represented by an age pyramid. Figure 1.21 shows an age pyramid for a developed country and for a lesser developed country. In less developed countries, the age pyramid has a very broad base, indicating high birth rates. As in many natural populations, these high birth rates are offset by high infant mortality. However, what happens when less developed countries begin accumulating technologies that reduce prereproductive mortality and prolong the life of the elderly? Unless reproductive practices change, a population explosion occurs. Unfortunately, cultural practices change very slowly, and there has been a reluctance to use modern birth control practices.

In developed countries, population growth tends to be slower, and the proportion of the population in each reproductive class is balanced. Birth rates in the United States have decreased in recent years because of the use of modern birth control practices. In spite of decreased birth rate the U.S. population, now at about 258 million, is still growing. Immigration is currently the biggest factor influencing population growth in the United States. If one assumes 2 million legal and illegal immigrants per year, then the U.S. population may increase to 500 million by the year 2050. Given our current standard of living, even 200 million people living in the United States is too many. Problems of homelessness, hunger, resource depletion, and pollution all stem from trying to support too many people at our current standard of living.

Pollution

Pollution is any detrimental change in an ecosystem. Most kinds of pollution are the results of human activities. When human populations are large, and affluence demands more and more goods and services, pollution problems are compounded. After reading the sections of this chapter on ecosystem productivity and nutrient cycling, you should be able to understand why these problems exist.

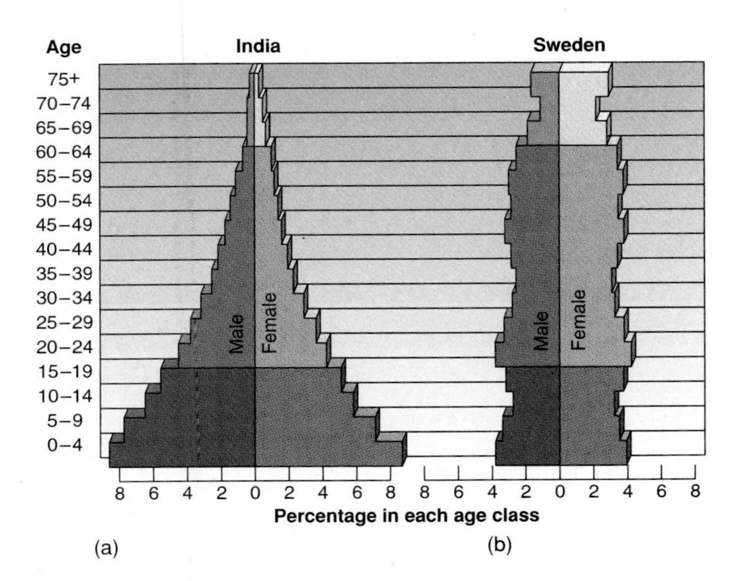

FIGURE 1.21

Human Age Pyramids. (*a*) In less developed countries, a greater proportion of the population is in the prereproductive age classes. High mortality in that age class compensates for high birth rates. As technologies reduce infant mortality and prolong the life span of the elderly, populations increase rapidly. (*b*) In developed countries, the age structure is more rectangular because of reduced mortality in all age classes. *Source: Robert Leo Smith,* The Ecology of Man, *2d ed., copyright 1976 Harper & Row, 1976.*

Pollution of our waters takes many forms. Industries generate toxic wastes, heat, and plastics, some of which will persist for centuries. Every household in the world generates human wastes that must be disposed of. All too often, industrial and human wastes find their ways into groundwater or into streams, lakes, and oceans. When they do, water becomes unfit for human consumption and unfit for wildlife.

Air pollution also presents serious problems. Burning fossil fuels releases sulfur dioxide and nitrogen oxides into the atmosphere. Sulfur dioxide and water combine to produce sulfuric acid, which falls as **acid rain.** Acid rain lowers the pH of lakes, often many miles from the site of sulfur dioxide production. Carbon dioxide released in burning fuels has accumulated in the atmosphere, causing the **greenhouse effect.** Carbon dioxide reflects solar radiation back to the earth, causing an increase in world temperature, polar ice caps to melt, and ocean levels to rise. The release of chlorinated fluorocarbons from aerosol cans, air conditioners, and refrigerators contributes to the depletion of the earth's ultraviolet filter—the atmospheric ozone layer. As a result, the incidence of skin cancer is likely to increase.

When wastes and poisons enter food webs, organisms at the highest trophic levels usually suffer the most. 6 Very tiny amounts of a toxin incorporated into primary production can quickly build up as carnivores feed on herbivores that have concentrated toxins in their tissues. This problem is especially severe when the material is not biodegradable (not broken down by biological processes). The accumulation of matter in food webs is called **biological magnification.**

Resource Depletion and Biodiversity

Other environmental problems arise because humans have been too slow to realize that an ecosystem's energy is used only once. When a quantity of energy is lost to outer space as heat, it is gone from the earth forever.

As with energy, other resources are also being squandered by human populations. Overgrazing and deforestation have led to the spread of our world's deserts. Exploitation of tropical rain forests has contributed to the extinction of many plant and animal species.

The variety of living organisms in an ecosystem is called **biodiversity.** No one knows for sure how many species there are in the world. About 1.4 million species have been described. Taxonomists estimate that there are 4 to 30 million more. Much of this unseen, or unnoticed, biodiversity is unappreciated for the free services it performs. Forests hold back flood waters and recycle CO_2 and nutrients. Insects pollinate crops and control insect pests, and subterranean organisms promote soil fertility through decomposition. Many of these undescribed species would, when studied, provide new food crops, petroleum substitutes, new fibers, and pharmaceuticals. All of these functions require not just remnant groups but large healthy populations. Large populations promote the genetic diversity required for surviving environmental changes. When genetic diversity is lost, it is lost forever. Our heroic attempts to save endangered species come far too late. Even when they succeed, they salvage only a tiny portion of an original gene pool.

The biodiversity of all natural areas of the world is threatened. Acid rain, pollution, urban development, and agriculture know no geographic or national boundaries. The main threats to biological diversity arise from habitat destruction by expanding human populations. Humans are either directly or indirectly exploiting about 40% of the earth's net primary production. Often this involves converting natural areas to agricultural uses, frequently substituting less efficient crop plants for native species. Habitat loss displaces thousands of native plants and animals.

Some of the most important threatened natural areas include tropical rain forests, coastal wetlands, and coral reefs. Of these, tropical rain forests have been an important focus of attention. Tropical rain forests cover only 7% of the earth's land surface, but they contain more than 50% of the world's species. Tropical rain forests are being destroyed rapidly, mostly for agricultural production. About 76,000 km^2 (an area greater than the area of the country of Costa Rica) is being cleared each year. At current rates of destruction, most tropical rain forests will be gone in the twenty-first century. According to some estimates, we are losing 17,500 rain forest species each year. Clearing of tropical rain forests achieves little, because the thin, nutrient-poor soils of tropical rain forests are exhausted within 2 years (figure 1.22). Sadly, rain forests are a nonrenewable resource. Seeds of rain forest plants germinate rapidly, but seedlings are unprotected on sterile, open soils. Even if a forest were able to become reestablished, it would take many centuries to return to a climax rain forest.

(a)

(b)

Figure 1.22

Tropical Rain Forests: A Threatened World Resource. A tropical rain forest before (*a*) and after (*b*) clear-cutting and burning to make way for agriculture. These soils will quickly become depleted, and will be abandoned for richer soils of adjacent forests. Loss of tropical forests results in extinction of many valuable forest species.

There are solutions to the problems of threatened biodiversity. None of the solutions, however, is quick and easy. First, more money needs to be appropriated for training taxonomists and ecologists, and for supporting their work. Second, all countries of the world need to realize that biodiversity, when preserved and managed properly, is a source of economic wealth. Third, we need a system of international ethics that values natural diversity for the beauty it brings to our lives. Anything short of these steps will surely lead to severe climatic changes and mass starvation.

Stop and Ask Yourself

10. What is community diversity?
11. How is a climax community different from the seral stages that preceded it?
12. If there were 1,000 units of energy in the producer trophic level of an ecosystem, approximately how many units of energy would be available to animals in the third trophic level?
13. What is an age pyramid? How do age pyramids help predict future population trends?
14. Why are higher trophic levels often most severely affected by poisons in the food web?

Summary

1. Zoology is the study of animals. It is a very broad field that requires zoologists to specialize.
2. Evidence for evolution comes from fields of biogeography, paleontology, comparative anatomy, and molecular biology.
3. Many abiotic factors influence where an animal may live. Animals have a tolerance range and a range of the optimum for environmental factors.
4. Energy for animal life comes from consuming autotrophs or other heterotrophs. Energy is expended in excretory, existence, and productive functions.
5. Temperature, water, light, geology, and soils are important environmental factors that influence animal life-styles.
6. Animal populations change in size over time. Changes can be characterized using survivorship curves.
7. Animal populations grow exponentially until the carrying capacity of the environment is achieved, at which point constraints such as food, chemicals, climate, and space restrict population growth.
8. Interspecific interactions influence animal populations. These interspecific interactions include: herbivory, predatory-prey interactions, interspecific competition, coevolution, mimicry, and symbiosis.
9. All populations living in an area make up a community.
10. Organisms have roles in their community. The ecological niche concept helps ecologists visualize those roles.
11. Communities often change in predictable ways. Successional changes often lead to a stable climax community.
12. Energy in an ecosystem is not recyclable. Energy that is fixed by producers is eventually lost as heat.
13. Nutrients are cycled through ecosystems. Cycles involve movements of material from nonliving reservoirs in the atmosphere or earth to biological systems and back to the reservoirs again.
14. Human population growth is the root of most of our environmental problems. Trying to support too many people at the standard of living found in developed countries has resulted in air and water pollution and resource depletion.
15. Pollution and resource depletion are important environmental problems that threaten life as we know it. Biodiversity is an important resource that is threatened by human activities.

Selected Key Terms

biodiversity (*p. 19*)
biogeochemical cycles (*p. 14*)
carrying capacity (*p. 10*)
coevolution (*p. 11*)
community (*p. 12*)
ecosystems (*p. 14*)
exponential growth (*p. 10*)
homologous (*p. 7*)
symbiosis (*p. 12*)
zoology (*p. 2*)

Critical Thinking Questions

1. What are some current issues that involve both zoology and questions of ethics or public policy? How can a background in zoology help resolve these issues?
2. Assuming a starting population of 10 individuals, a doubling time of 1 month, and no mortality, how long would it take a hypothetical population to achieve 10,000 individuals?
3. Why do you think that winter inactivity of many small mammals takes the form of hibernation, whereas winter inactivity in larger mammals is in the form of winter sleep?
4. Which of the following would be a more energetically efficient strategy for supplying animal protein for human diets? Explain your answers.
   a. Feeding people grain-fed beef in feed lots, or feeding people beef that has been raised in pastures.
   b. Feeding people sardines and herrings, or processing sardines and herrings into fish meal that is subsequently used to raise poultry, which is used to feed people.

chapter 2

Evolution: A Historical Perspective

Outline

Concepts

1. Organic evolution is the change of a species over time.
2. Although the concept of evolution is very old, the modern explanation of how change occurs was formulated by Charles Darwin. Darwin began gathering his evidence of evolution during a worldwide mapping expedition on the HMS *Beagle* and spent the rest of his life formulating and defending his ideas.
3. Darwin's theory of evolution by natural selection, although modified from its original form, is still a highly regarded account of how evolution occurs.
4. Modern evolutionary theorists apply principles of genetics, ecology, and geographic and morphological studies when investigating evolutionary mechanisms.
5. "Adaptation" may refer to either a process of evolutionary change or to the result of a change. In the latter sense, an adaptation is a structure or a process that increases an animal's potential to survive and reproduce in specific environmental conditions.

Would You Like to Know:

1. whether the idea of evolutionary change originated with Charles Darwin? *(p. 24)*
2. what circumstances led Charles Darwin to become a naturalist on the HMS *Beagle?* *(p. 26)*
3. how Charles Darwin's experiences in South America convinced him that evolution occurs? *(p. 27)*
4. what prompted Charles Darwin to publish his work in 1859, 23 years after returning from his voyage on the HMS *Beagle?* *(p. 30)*
5. why geography is important for biologists? *(p. 31)*

These and other useful questions will be answered in this chapter.

Questions of earth's origin and life's origin have been on the minds of humans since prehistoric times, when accounts of creation were passed orally from generation to generation. For many people, these questions centered around concepts of purpose. Religious and philosophical writings help provide answers to questions such as: Why are we here? What is human nature really like? How do we deal with our mortality?

Many of us are also concerned with other, very different, questions of origin. How old is the planet earth? How long has life been on earth? How did life arise on earth? How did a certain animal species come into existence? Answers for these questions come from a different authority—that of scientific inquiry.

The purpose of this chapter is to present the history of the study of organic evolution, and to introduce the **theory of evolution by natural selection. Organic evolution** (L. *evolutus*, unroll), according to Charles Darwin, is "descent with modification." This statement simply means that species change over time. Evolution by itself does not imply any particular lineage or any particular mechanism, and virtually all scientists agree that the evidence for change in organisms over long time periods is overwhelming (*see chapter 1*). Further, most scientists agree that natural selection, the mechanism for evolution outlined by Charles Darwin, is one explanation of how evolution occurs. In spite of the scientific certainty of evolution and an acceptance of a general mechanism, much is still to be learned about the details of evolutionary processes. Scientists will be debating these details for years to come (box 2.1).

Pre-Darwinian Theories of Change

1 The idea of evolution did not originate with Charles Darwin. Some of the earliest references to evolutionary change are from the ancient Greeks. The philosophers Empedocles (495–435 B.C.) and Aristotle (384–322 B.C.) described concepts of change in living organisms over time. Georges-Louis Buffon (1707–1788) spent many years studying comparative anatomy. His observations of structural variations in particular organs of related animals, and especially his observations of vestigial structures (*see chapter 1*), convinced him that change must have occurred during the history of life on earth. Erasmus Darwin (1731–1802), a physician and the grandfather of Charles Darwin, was intensely interested in questions of origin and change. He believed in the common ancestry of all organisms.

Lamarck: An Early Proponent of Evolution

Jean Baptiste Lamarck (1744–1829) was a distinguished French zoologist. His contributions to zoology include important studies of the classification of animals. Lamarck, however, is remembered more for a theory of how change occurs. He believed that species are not constant, and that existing species were derived from preexisting species.

Lamarck's rather elaborate explanation of how evolutionary change occurs involved a theory of inheritance that was widely accepted in the early 1800s called the **theory of inheritance of acquired characteristics.** Lamarck believed that organisms develop new organs or modify existing organs as environmental problems present themselves. In other words, organs change as the need arises. Lamarck illustrated this point with the often-quoted example of the giraffe. He contended that ancestral giraffes had short necks, much like those of any other mammal. Straining to reach higher branches during browsing resulted in their acquiring higher shoulders and longer necks. These modifications, produced in one generation, were passed on to the next generation. Lamarck went on to state that the use of any organ resulted in that organ becoming highly developed, and that disuse resulted in degeneration. Thus, the evolution of highly specialized structures, such as vertebrate eyes could be explained.

Lamarck published his theory in 1802 and defended it in the face of social and scientific criticism. Society in general was unaccepting of the ideas of evolutionary change, and evidence for evolution had not been developed thoroughly enough to convince most scientists that evolutionary change occurs. Thus, Lamarck was criticized in his day more for advocating ideas of evolutionary change than for the mechanism he proposed for that change. Today he is criticized for defending a mechanism of inheritance and evolutionary change that we now know lacks reasonable supporting evidence. For a change to be passed on to the next generation, it must be based on genetic changes in the germ cells. Changes in the giraffes' necks, as envisioned by Lamarck, could not be passed on because they did not originate as changes in the genetic material. Even though Lamarck's mechanism of change was incorrect, we should remember him for his steadfastness in promoting the idea of evolutionary change and his other accomplishments in zoology.

Stop and Ask Yourself

1. What is organic evolution?
2. What contributions to our concept of evolution were made by Buffon?
3. What is the theory of inheritance of acquired characteristics, and how did Lamarck use it to explain how evolution occurred?

Darwin's Early Years and His Journey

Charles Robert Darwin (1809–1882) was born on February 12, 1809. His father, like his grandfather, was a physician. During Charles Darwin's youth in Shrewsbury, England, his interests centered around dogs, collecting, and hunting birds—all popular pastimes in wealthy families of nineteenth century England.

BOX 2.1 The Origin of Life on Earth—Life from Nonlife

Geologists estimate from radioisotope dating that the earth is approximately 4.6 billion years old. The oldest fossils are of cyanobacteria that come from 3.5-billion-year-old rocks (stromatolites) from Australia and South Africa. Thus, it took no more than 1 to 1.5 billion years for life to originate.

In trying to explain how life may have arisen, scientists first needed to know the conditions that existed on the earth after its formation. In 1929, J. B. S. Haldane described the atmosphere of primordial earth as a reducing atmosphere (with little free oxygen present) containing primarily hydrogen, water, ammonia, and methane. In 1953, Stanley Miller and Harold Urey constructed a reaction vessel in which they duplicated the atmosphere described by Haldane. They heated the mixture to 80° C and provided the atmosphere with an electrical spark to simulate lightning. Over the course of a week, they removed samples from their system and found a variety of common amino acids and other organic acids.

The above scenario for the origin of the first organic compounds, although once widely accepted, is under increasing scrutiny. Recent evidence suggests that carbon dioxide and nitrogen gas, not methane and ammonia, were the major components of the earth's primitive atmosphere. These conditions are much less favorable for the formation of organic compounds using the Miller/Urey apparatus.

Scientists have begun to look for new explanations for the origin of the first organic chemicals and at older explanations, which are being revived. One of these is that life's beginnings may have occurred deep in the oceans, in underwater hot springs called hydrothermal vents. These vents could have supplied the energy and raw materials for the origin and survival of early life-forms. This vent hypothesis is supported by the presence of a group of bacteria, called archaebacteria, that tolerate temperatures up to 120° C and seem to have undergone less evolutionary change than any other living species.

Another explanation for the origin of the earth's first organic molecules is that they came from outer space. Astronomers are detecting an increasing diversity of organic compounds (such as amino acids and other hydrocarbons) in meteorites that have collided with the earth. Investigations of the most recent pass-by of Halley's comet revealed that comets may be relatively rich in organic compounds. Even though many scientists think it is possible that the first organic compounds could have come from space, no microbial life-forms have been detected in space, and conditions in outer space are incompatible with life as we know it.

A second step in the origin of life must have been the hooking together of early organic molecules into the polymers of living organisms: polypeptides, polynucleotides, and carbohydrates. Organic molecules may have become isolated in tidepools or freshwater ponds, and as water was lost through evaporation, condensation reactions could have occurred. Alternatively, reacting molecules may have been concentrated by adsorption on the surfaces of clay or iron pyrite particles, where polymerization could occur.

The final steps in the origin of life are the subject of endless speculations. In some way, organic molecules were surrounded by a membranelike structure, self-replication occurred, and DNA became established as the genetic material. A "chicken-or-the-egg" paradox emerges if one thinks of DNA as the first genetic material. DNA codes for proteins, yet proteins are required for DNA replication, transcription, and translation. A possible way around this paradox was suggested by Thomas R. Cech and Sidney Altman in the early 1980s. They discovered a certain type of RNA, which acts like an enzyme that cuts and splices itself into a functional molecule. The first organisms could have been vesicles of self-replicating RNA molecules. Other scientists think that proteins may have served as the first genetic material, and that DNA was established as the code-carrying molecule secondarily. No self-replicating proteins, however, have been found.

Early life would have been limited by the nutrients produced in the primordial environment. If life were to continue, another source of nutrients was needed. Photosynthesis, which is the production of organic molecules using solar energy and inorganic compounds, probably freed living organisms from a dwindling supply of nutrients. The first photosynthetic organisms probably used hydrogen sulfide as a source of hydrogen for reducing carbon dioxide to sugars. Later, water served this same purpose and oxygen liberated by photosynthetic reactions began to accumulate in the atmosphere. Earth and its atmosphere slowly began to change. Ozone in the upper atmosphere began to filter ultraviolet radiation from the sun, the reducing atmosphere slowly became an oxidizing atmosphere, and at least some living organisms began to utilize oxygen. About 420 million years ago enough protective ozone had built up to make life on land possible. Ironically, the change from a reducing atmosphere to an oxidizing atmosphere also meant that life could no longer arise abiotically.

These activities captivated him far more than the traditional education he received at boarding school. In 1825, he entered medical school in Edinburgh, Scotland. For two years, he enjoyed the company of the school's well-established scientists. Darwin, however, was not interested in a career in medicine because he could not bear the sight of pain. This problem prompted his father to suggest that he train for the clergy in the Church of England. With this in mind, Charles enrolled at Cambridge University in 1828 and graduated with honors in 1831. This training, like the medical training he received, was disappointing for Darwin. Again, his most memorable experiences were those with Cambridge scientists. During his stay at Cambridge, Darwin developed a keen interest in collecting beetles and made valuable contributions to beetle taxonomy.

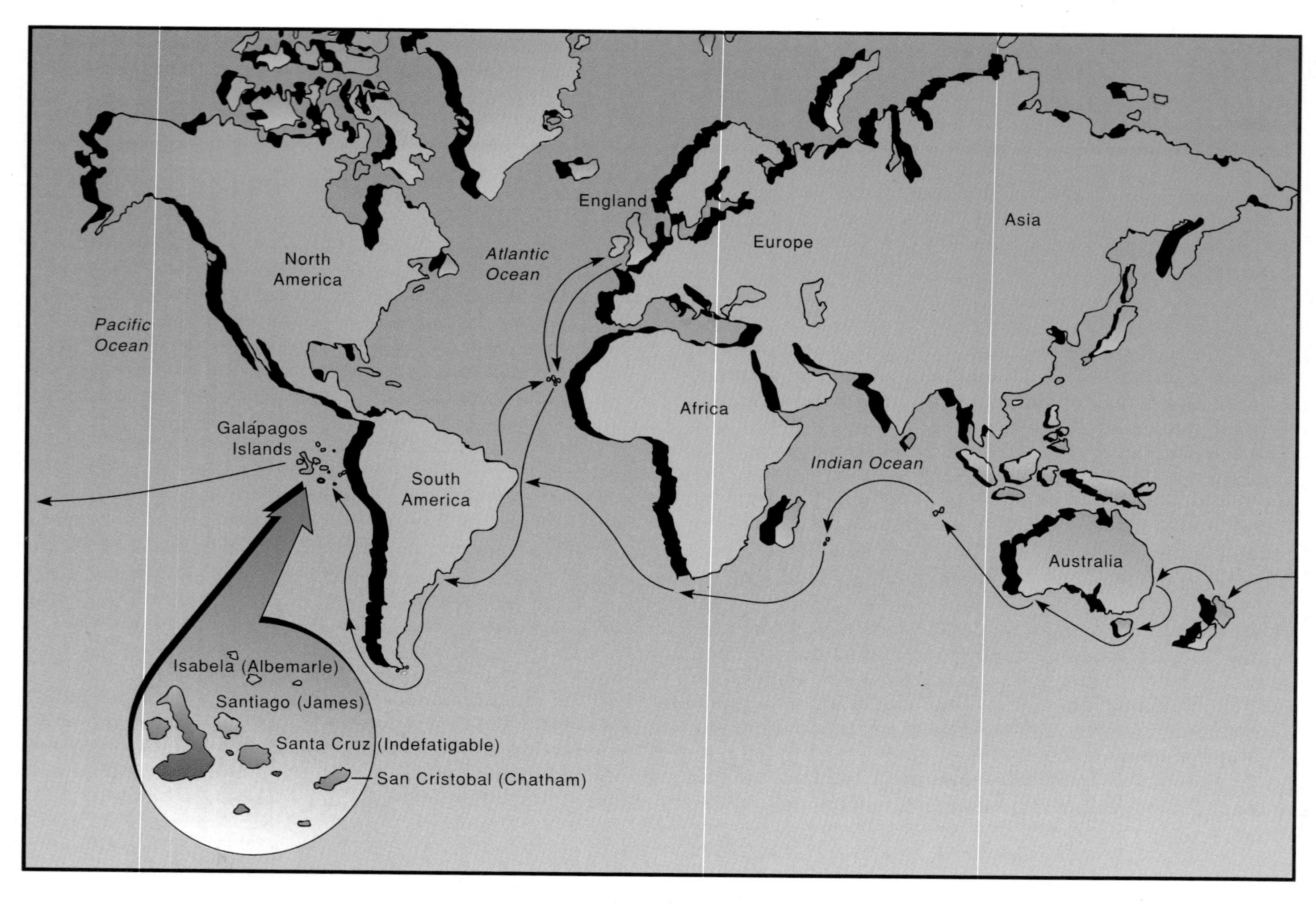

FIGURE 2.1

The Voyage of the *Beagle*. Charles Darwin served as a naturalist on a mapping expedition that lasted 5 years. Darwin's observations throughout those 5 years, especially those on the Galápagos Islands, served as the basis for the theory of evolution by natural selection.

VOYAGE OF THE HMS *BEAGLE*

2 One of his Cambridge mentors, a botanist by the name of John S. Henslow, nominated Darwin to serve as a naturalist on a mapping expedition that was to travel around the world. He was commissioned as a naturalist on the HMS *Beagle*, which set sail on December 27, 1831 on a voyage that lasted 5 years (figure 2.1). Darwin helped with routine seafaring tasks and made numerous collections, which he sent to Cambridge. The voyage gave him ample opportunity to explore tropical rain forests, fossil beds, the volcanic peaks of South America, and the coral atolls of the South Pacific. Most importantly, Darwin spent 5 weeks on the **Galápagos Islands,** a group of volcanic islands that lie 600 miles off the coast of Ecuador. Some of his most revolutionary ideas came from his observations of plant and animal life on these islands. At the end of the voyage, Darwin was just 27 years old. He spent the rest of his life examining specimens, rereading notes, making new observations, and preparing numerous publications. His most important publication, *On the Origin of Species by Means of Natural Selection*, revolutionized biology.

EARLY DEVELOPMENT OF DARWIN'S IDEAS OF EVOLUTION

The development of Darwin's theory of evolution by natural selection was a long, painstaking process. Darwin had to become convinced that change occurs over time. Before leaving on his voyage, Darwin accepted the prevailing opinion that the earth and its inhabitants had been created 6,000 years ago and had not changed since. Because his observations during his voyage suggested that change does occur, he realized that 6,000 years could not account for the diversity of modern species if they arose through gradual change. Once ideas of change were established in Darwin's thinking, it took about 20 years of study to conceive, and thoroughly document, the mechanism by which change occurs. Darwin died without knowing the genetic principles that support his theory.

Geology

Darwin began his voyage by reading Charles Lyell's (1779–1875) *Principles of Geology*. In this book, Lyell developed the ideas of another geologist, James Hutton, into the theory of **uniformitarianism.** His theory was based on the idea that the earth is shaped today by the forces of wind, rain, rivers, volcanoes, and geological uplift—just as it has been in the past. Lyell and Hutton contended that it was these forces, not catastrophic events, that shaped the face of the earth over hundreds of millions of years. This book planted two important ideas in Darwin's mind: (1) the earth could be much older than 6,000 years and (2) if the face of the earth changed gradually over long periods of time, could not living forms also change during that time?

Fossil Evidence

Once the HMS *Beagle* reached South America, Darwin spent time digging in the dry riverbeds of the pampas (grassy plains) of Argentina. He found the fossil remains of an extinct hippopotamuslike animal, now called *Toxodon*, and fossils of a horselike animal, *Thoantherium*. Both of these fossils were from animals that were clearly different from any other animal living in the region. Modern horses were in South America, of course, but they had been brought to the Americas by Spanish explorers in the 1500s. The fossils suggested that horses had been present and had become extinct long before the 1500s. Darwin also found fossils of giant armadillos and giant sloths. Except for their large size, these fossils were very similar to forms Darwin found living in the region.

Fossils were not new to Darwin. They were popularly believed to be the remains of animals that perished in catastrophic events, such as Noah's flood. (3) In South America, however, Darwin understood them to be evidence that the species composition of the earth had changed. Some species became extinct without leaving any descendants. Others became extinct, but not before giving rise to new species.

Galápagos Islands

On its trip up the western shore of South America, the HMS *Beagle* stopped at the Galápagos Islands, which are named after the very large tortoises that inhabit them (Sp. *galápago*, tortoise). The tortoises weigh up to 250 kg, have shells up to 1.8 m in diameter, and live for 200 to 250 years. It was pointed out to Darwin by the islands' governor that the shape of the tortoise shells from different parts of Albemarle Island differed. Darwin noticed other differences as well. Tortoises from the drier regions had longer necks than tortoises from wetter habitats (figure 2.2). In spite of their differences, the tortoises were quite similar to each other and to the tortoises on the mainland of South America.

(a)

(b)

Figure 2.2

Galápagos Tortoises. (*a*) Shorter-necked subspecies of *Geochelone elephantopus* live in moister regions and feed on low-growing vegetation. (*b*) Longer-necked subspecies live in drier regions and feed on high-growing vegetation.

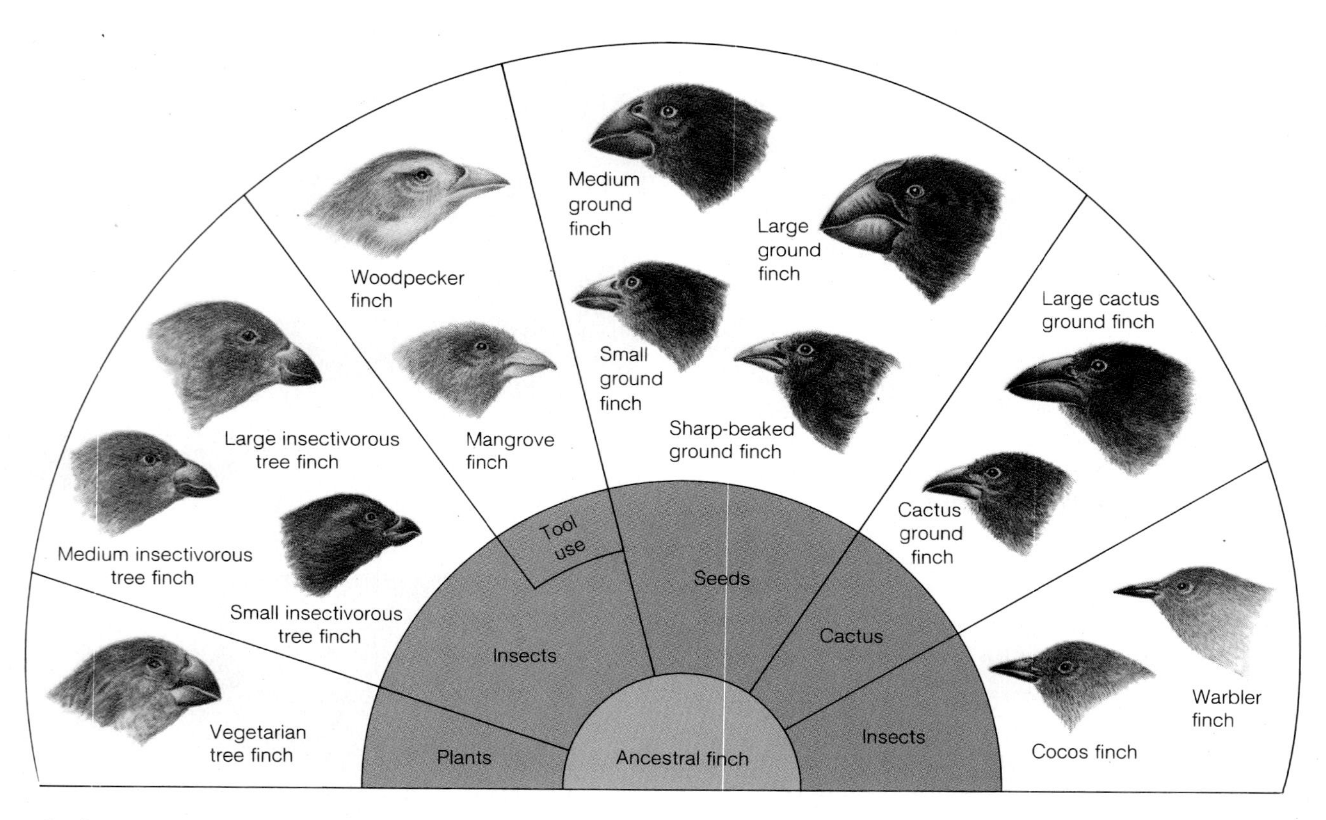

FIGURE 2.3
Adaptive Radiation of the Galápagos Finches. Ancestral finches from the South American mainland colonized the Galápagos Islands. Open habitats and few predators promoted the radiation of finches into most of the roles normally filled by birds.

How could these overall similarities be explained? Darwin reasoned that the island forms were derived from a few ancestral animals that managed to travel from the mainland, across 600 miles of ocean. (Because the Galápagos Islands are volcanic islands and arose out of the seabed, there was never any land connection with the mainland. One modern hypothesis is that tortoises floated from the mainland on mats of vegetation that regularly break free from coastal riverbanks during storms.) Without predators on the islands, tortoises gradually increased in number.

Darwin also explained some of the differences that he saw. In dryer regions, where vegetation was sparse, tortoises with longer necks would be favored because they could reach higher to get food. In moister regions, tortoises with longer necks would not necessarily be favored and the shorter-necked tortoises could survive.

Darwin made similar observations of a group of dark, sparrowlike birds. Although he never studied them in detail, Darwin noticed that the Galápagos finches bore similarities suggestive of common ancestry. Scientists now think that Galápagos finches also descended from an ancestral species that originally inhabited the mainland of South America. The chance arrival of a few finches, in either single or multiple colonization events, probably set up the first bird populations on the islands. Early finches encountered many different habitats, all empty of other birds and predators. Ancestral finches, probably seed eaters, multiplied rapidly and filled the seed-bearing habitats most attractive to them. Fourteen species of finches arose from this ancestral group, including one species found on small Cocos Island northeast of the Galápagos Islands. Each species is adapted to a specific habitat on the islands. The most obvious difference between these finches relates to dietary adaptations and is reflected in the size and shape of their bills (figure 2.3).

Darwin's experiences in South America and the Galápagos Islands convinced him that animals change over time and that, in the course of the history of life on earth, many kinds of animals have become extinct. Although he would not document his hypothesis for many years, these experiences helped him formulate ideas of how evolutionary change occurs.

The Theory of Evolution by Natural Selection

On his return to England in 1836 and for the next 17 years, Darwin worked diligently on the notes and specimens he had collected and made new observations. He was particularly interested in the obvious success of breeders in developing desired variations in plant and animal stocks (figure 2.4). He wondered if this artificial selection of traits could have a parallel in the natural world.

Initially, Darwin was unable to find a natural process similar to artificial selection. However, in 1838, he read an essay by Thomas Malthus (1766–1834) entitled *Essay on the Principle of Population*. Malthus believed that the human population has the potential to increase geometrically. However, because resources cannot keep pace with the increased demands of a burgeoning population, the influences of population-restraining factors, such as poverty, wars, plagues, and famine, begin to be felt. It occurred to Darwin that a similar struggle to survive occurs in nature. This struggle, when viewed over generations, could serve as a means of **natural selection.** Traits that were detrimental for an animal would be eliminated by the failure of the animal containing them to reproduce.

(a)

(b)

Figure 2.4

Artificial Selection. The artificial selection of domestic fowl has resulted in the diverse varieties that we see today. Breeders perpetuated the variations they desired by allowing only certain offspring to breed. (*a*) Bearded white Polish fowl. (*b*) Jungle fowl.

Natural Selection

By 1844 Darwin had formulated, but not yet published, his ideas on natural selection. The essence of his theory is as follows:

1. All organisms have a far greater reproductive potential than is ever realized. For example, a female oyster releases about 100,000 eggs with each spawning, a female sea star releases about 1 million eggs each season, and a female robin may lay four fertile eggs each season. What if all of these eggs were fertilized and developed to reproductive adults by the following year? A half million female sea stars (½ of the million eggs would produce females and ½ would produce males), each producing another million eggs, repeated over just a few generations would soon fill up the oceans! Even the adult female robins, each producing four more robins, would result in unimaginable resource problems in just a few years.
2. Inherited variations arise by mutation. Seldom are any two individuals exactly alike. Some of these genetic variations may confer an advantage to the individual possessing them. In other instances, variations may be harmful for an individual. In still other instances, particular variations may be neither helpful nor harmful. (These are said to be neutral.) These variations can be passed on to offspring.
3. Because resources are limited, there is a constant struggle for existence. Many more offspring are produced than resources can support; therefore, many individuals will die. Darwin reasoned that the individuals that die will be those with the traits (variations) that make survival and successful reproduction less likely. Other traits that promote successful reproduction are said to be adaptive.
4. Adaptive traits will be perpetuated in subsequent generations. Because organisms with maladaptive traits are less likely to reproduce, the maladaptive traits will become less frequent in a population, and eventually will tend to be eliminated.

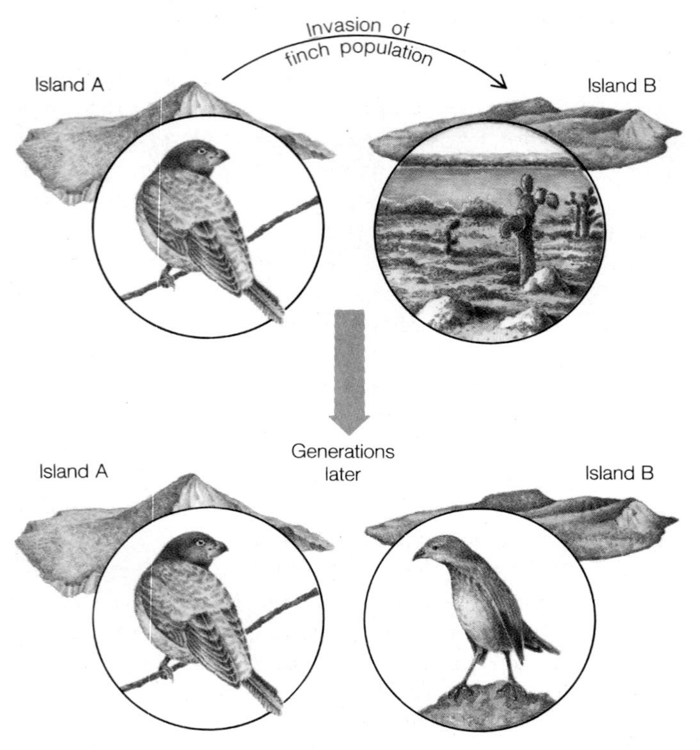

FIGURE 2.5

Natural Selection in Galápagos Finches. Natural selection of finches on the Galápagos Islands resulted in changes in bill shape. In this illustration, tree-feeding birds on island A invaded island B. The relatively treeless habitats of island B selected against birds most adapted to living in trees, and individuals that could exploit seeds on the ground in low-growing vegetation were more likely to survive and reproduce. Subsequent generations of finches on island B should show ground-feeding bill characteristics.

With these ideas, Darwin formulated a theory that explained how the tortoises and finches of the Galápagos Islands changed over time (figure 2.5). In addition, Darwin's theory explained how some animals, such as the ancient South American horses, could become extinct. What if a new environment is presented to a group of animals that are ill-adapted to that environment? Climatic changes, food shortages, and other environmental stressors could lead to extinction.

Adaptation

Adaptation occurs when a change in a phenotype increases an animal's chance of successful reproduction. It is likely to be expressed when an organism encounters a new environment, and may result in the evolution of multiple new groups if an environment can be exploited in different ways. No terms in evolution have been laden with more confusion than "adaptation" and "fitness or adaptedness." Adaptation is sometimes used to refer to a process of change in evolution. That use of the term is probably less confusing than when "adaptation" is used to describe the result of the process of change. For our purposes, adaptations are defined as characteristics that increase an organism's, or a species', potential to successfully reproduce in a specified environment. In a similar fashion, adaptedness or fitness is a measure of the capacity for successful reproduction in a given environment.

There has been a tendency to view every characteristic as an adaptation to some kind of environmental situation. The view has been that if a structure is now performing a specific function, it must have arisen for that purpose, and is, therefore, an adaptation. An extreme extension of this incorrect view is that evolutionary adaptations lead to perfection.

ALFRED RUSSEL WALLACE

Alfred Russel Wallace (1823–1913) was an explorer of the Amazon Valley, and led a zoological expedition to the Malay Archipelago, which is an area of great biogeographical importance. Wallace, like Darwin, was impressed with evolutionary change and had read the writings of Thomas Malthus on human populations. In the midst of a bout with malarial fever (*see box 5.2*), he synthesized a theory of evolution similar to Darwin's theory of evolution by natural selection. After writing the details of his theory, Wallace sent his paper to Darwin for criticism. (4) Darwin recognized the similarity of Wallace's ideas, and prepared a short summary of his own theory. Both Wallace's and Darwin's papers were published in the *Journal of the Proceedings of the Linnean Society* in 1859. Darwin's insistence on having Wallace's ideas presented along with his own shows Darwin's integrity. Darwin then shortened a manuscript he had been working on since 1856 and published it as *On the Origin of Species by Means of Natural Selection* in November, 1859. The 1,250 copies prepared in the first printing sold out the day the book was released.

In spite of the similarities in the theories of Wallace and Darwin, there were also important differences. Wallace, for example, believed that every evolutionary modification was a product of selection and, therefore, had to be adaptive for the organism. Darwin, on the other hand, admitted that natural selection may not explain all evolutionary changes. He did not insist on finding adaptive significance for every modification. Further, unlike Darwin, Wallace stopped short of attributing human intellectual functions and the ability to make moral judgments to evolution. On both of these matters, Darwin's ideas are closer to the views of most modern scientists.

Wallace's work provided an important spark that motivated Darwin to publish his own ideas. The theory of natural selection, however, is usually credited to Charles Darwin. The years of work given to the theory by Darwin, and the accumulation of massive evidence for the theory, led even Wallace to attribute the theory to Darwin. Wallace wrote to Darwin in 1864:

> I shall always maintain [the theory of evolution by natural selection] to be actually yours and yours only. You had worked it out in details I had never thought of years before I had a ray of light on the subject.

Stop and Ask Yourself

4. How did the following contribute to Charles Darwin's formulation of the theory of natural selection: Uniformitarianism? South American fossils? Galápagos tortoises?
5. What are four elements of the theory of evolution by natural selection?

Evolutionary Thought after Darwin

The most significant changes that occurred in evolutionary thought began in the 1930s and have continued to the present. The combination of population genetics with evolutionary theory is called the **modern synthesis** or **neo-Darwinism.**

Biogeography

5 In the tradition of Darwin and Wallace, biologists of the period of modern synthesis recognized the importance of geography as an explanation of the evolution and the distribution of plants and animals (*see chapter 1; figures 1.3 and 1.4*). One of the distribution patterns that biogeographers try to explain is how similar groups of organisms can live in places separated by seemingly impenetrable barriers. Recall that Darwin was struck by the presence of fossil horses in South America. This distribution was puzzling because modern horses were introduced to America from Europe. The fossil horses must have arisen in America or arrived by some unknown means, then became extinct and left no descendants. Biogeographers also try to explain why plants and animals, separated by geographical barriers, are often very different in spite of similar environments. For example, evolution may take different directions in different parts of the world; therefore, major predators in Africa and America might be expected to differ if they had no common ancestry. Finally, biogeographers try to explain why oceanic islands, such as the Galápagos, often have relatively few, but unique, resident species. They try to document island colonization and subsequent evolutionary events, which may be very different from evolutionary events in ancestral, mainland groups.

Modern evolutionary biologists recognize the importance of geological events, such as volcanic activity, the movement of great land masses, climatic changes, and geological uplift (mountain building), in creating or removing barriers to the movements of plants and animals. As scientists learned more about the geologic history of the earth, they understood more about animal distribution patterns and factors that played important roles in their evolution. Only in understanding how the surface of the earth came to its present form can we understand its inhabitants (box 2.2).

Information from the study of biogeography can be applied in the study of contemporary environmental problems. Prehistoric removal of geographic barriers allowed species from one region of the world to invade other regions, sometimes resulting in dramatic changes in species composition. Invading species tend to be highly competitive, with high reproductive potential, and have been important agents of extinction in regions that they invade. For example, during the Pliocene and Miocene epochs (about 10 million years ago) the Central American land bridge formed between North and South America (*see box 2.2*). Many species of plants and animals used this land bridge to move between continents. Information from the study of past events is now being applied to predict the effect of the removal of geographical barriers on native populations. For example, the proposed sea-level canal across Panama could bring an interchange of species between the Atlantic and Pacific Oceans, causing extinction of many fishes and other animals. In addition the voracious coral predator, *Acanthaster planci* (the crown-of-thorns sea star), could be introduced into the Caribbean region.

Evolution is one of the major unifying themes in biology because it helps explain both the similarities and diversity of life. In chapter 1 you learned of the sources of evidence for evolution—biogeography gives evidence of prehistoric climates, habitats, and animal distribution patterns; paleontology provides evidence of animals that existed in the past; comparative anatomy leads to the description of homologous structures; and molecular biology provides evidence of relatedness of animals based on their biochemical similarities. This evidence leaves little doubt that evolution has occurred in the past, and this chapter described the historical development of a theory that accounts for how evolution occurs. In chapter 3 you will learn how the incorporation of the principles of population genetics has affected scientific concepts of the mechanism of evolution.

Stop and Ask Yourself

6. What is modern synthesis or neo-Darwinism?
7. What are some of the distribution patterns that biogeographers attempt to explain?
8. How has the discovery of continental drift influenced our understanding of evolution?

BOX 2.2 CONTINENTAL DRIFT

It seems remarkable, but the earth's largest land masses are moving—about 1 cm a year! The 1960s revolutionized geology as **continental drift** became an accepted theory.

During the Permian period (about 250 million years ago), all of earth's land masses were united into a single continent called Pangaea (figure 1*a*). Soon after, Pangaea began to break up, as the huge crustal plates began to move apart. Approximately 200 million years ago, there were two great continents. The northern continent was Laurasia, and the southern continent was Gondwana (figure 1*b*). Seventy million years ago, Gondwana broke apart, followed later by the breakup of Laurasia (figure 1*c*).

Movement of these crustal plates continues today. Their study is known as **plate tectonics.** During these movements, new crustal material is thrust up from the ocean floor along the mid-Atlantic ridge, and flows in both directions from that ridge. The mid-Atlantic ridge runs from the Falkland Islands, at its southern end, to Iceland, at its northern end. As the huge crusts move away from each other in the Atlantic, old crustal material sinks back into the earth in deep oceanic trenches in the Pacific Ocean. Evidence of these movements is seen in the earthquakes along the western coast of North America.

The drifting of the continents has important implications for biogeographers and paleontologists. Continental drift explains why some fossils have a worldwide distribution. Land organisms in existence during the time that all continents were united as Pangaea had access to most of the land masses of the world. It should not be surprising, therefore, to find some fossils that are similar in all parts of the world, such as the fossils that Charles Darwin found in South America that were very similar to animals living in Africa. Were it not for continental drift, this pattern would be very difficult to explain.

Continental drift also explains how differences in organisms may develop. Because oceanic barriers were created as a result of continental drift, plants and animals were separated from one another when continents separated. Evolution then proceeded independently in each biogeographical region and resulted in species characteristic of that region.

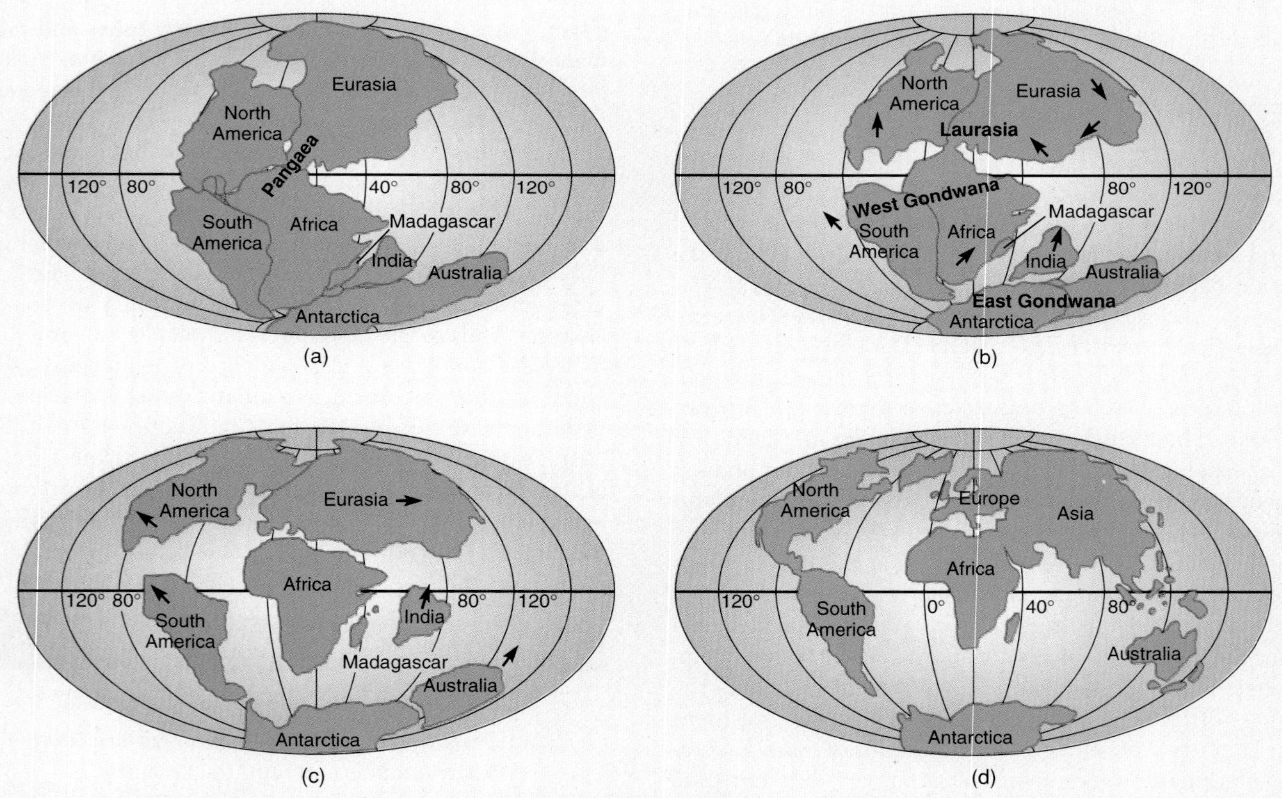

FIGURE 1 **Continental Drift.** (*a*) About 250 million years ago, the continents of the earth were joined into a single land mass that geologists call Pangaea. (*b*) About 150 million years ago, Pangaea broke up into the northern (Laurasia) and southern (Gondwana) continents. (*c*) This breakup was followed, about 70 million years ago, by the separation of continents in the Southern Hemisphere, and later by the separation of continents in the Northern Hemisphere. (*d*) The present position of continents. Note the complementary outlines of the eastern coast of South America and the western coast of Africa. The collision of India with Eurasia (*c*) resulted in the formation of the Himalaya Mountains.

Summary

1. Organic evolution is the change of a species over time.
2. Ideas of evolutionary change can be traced back to the ancient Greeks.
3. Jean Baptiste Lamarck was an eighteenth-century proponent of evolution, and proposed a mechanism—inheritance of acquired characteristics—to explain it.
4. Charles Darwin saw impressive evidence for evolutionary change while on a mapping expedition on the HMS *Beagle*. The theory of uniformitarianism, South American fossils, and observations of tortoises and finches on the Galápagos Islands convinced Darwin that evolution occurs.
5. After returning from his voyage, Darwin began formulating his theory of evolution by natural selection. In addition to his experiences on his voyage, later observations of artificial selection and Malthus' theory of human population growth helped shape his theory.
6. Darwin's theory of natural selection includes the following elements: (a) all organisms have a greater reproductive potential than is ever attained; (b) inherited variations arise by mutation; (c) there is a constant struggle for existence in which those organisms that are least suited to their environment will die; (d) the adaptive traits present in the survivors will tend to be passed on to subsequent generations, and the nonadaptive traits will tend to be lost.
7. "Adaptation" may refer to a process of change or a result of change. An adaptation is a characteristic that increases an organism's potential to survive and reproduce in a given environment.
8. All evolutionary changes are not adaptive, nor do all evolutionary changes lead to perfect solutions to environmental problems.
9. Alfred Russel Wallace outlined a theory similar to Darwin's, but never accumulated significant evidence documenting his theory.
10. Modern evolutionary theorists apply principles of genetics, ecological theory, and geographic and morphological studies to solving evolutionary problems.
11. The study of biogeography provides information on how similar groups of organisms can live in widely separated areas, accounts for the importance of geological events in evolution of plants and animals, and helps scientists predict the effects of removal of geographical barriers on extant populations.

Selected Key Terms

adaptation (*p. 30*)
continental drift (*p. 32*)
Galápagos Islands (*p. 26*)
modern synthesis (*p. 31*)
natural selection (*p. 29*)
neo-Darwinism (*p. 31*)
organic evolution (*p. 24*)
theory of evolution by natural selection (*p. 24*)
theory of inheritance of acquired characteristics (*p. 24*)
uniformitarianism (*p. 27*)

Critical Thinking Questions

1. Review the definition of "adaptation" in the sense of a result of evolutionary change. Imagine that two deer, A and B, are identical twins. Deer A is shot by a hunter before it has a chance to reproduce. Deer B is not shot and goes on to reproduce. According to our definition of adaptation, is deer B more fit for its environment than deer A? Why or why not?
2. Why is the stipulation of "a specific environment" included in the definition of "adaptation"?

chapter 3

Evolution and Gene Frequencies

Outline

Concepts

1. In modern genetic theory, organic evolution is defined as a change in the frequency of alleles in a population.
2. The principles of modern genetics help biologists understand how variation arises. This variation increases the chances of a population's survival in changing environments.
3. Population genetics is the study of events occurring in gene pools. The Hardy-Weinberg theorem helps scientists understand the circumstances under which evolution occurs. Evolution occurs when (a) genetic drift or neutral selection occurs, (b) gene flow occurs, (c) mutations introduce new genes into populations, or (d) natural selection occurs.
4. Balanced polymorphism occurs when two or more body forms are maintained in a population without a range of phenotypes between them.
5. The fundamental unit of classification is the species, and the process by which new species are formed is called speciation.
6. For speciation to happen, reproductive isolation must occur.
7. Different organisms, as well as structures within organisms, evolve at different rates. Evolution may also proceed in jumps rather than at a constant pace.
8. Molecular biologists study DNA and proteins to uncover evolutionary relationships.

Would You Like to Know:

1. what population genetics is? (*p. 36*)
2. whether or not evolution must occur? (*p. 36*)
3. what role chance plays in evolution? (*p. 38*)
4. how natural selection can produce distinct subpopulations in a population? (*p. 41*)
5. how new species arise? (*p. 43*)
6. what molecular evolution is? (*p. 45*)

These and other useful questions will be answered in this chapter.

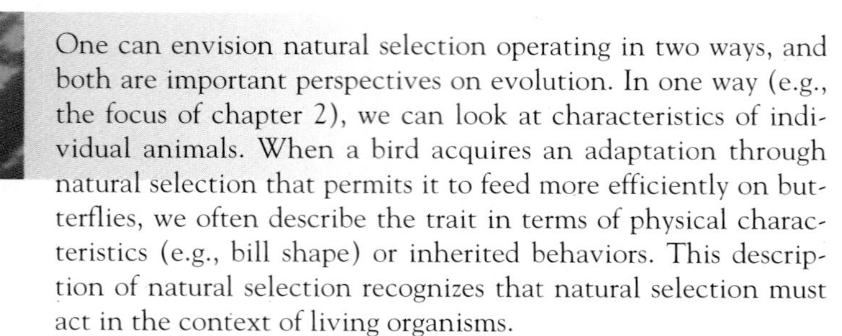

One can envision natural selection operating in two ways, and both are important perspectives on evolution. In one way (e.g., the focus of chapter 2), we can look at characteristics of individual animals. When a bird acquires an adaptation through natural selection that permits it to feed more efficiently on butterflies, we often describe the trait in terms of physical characteristics (e.g., bill shape) or inherited behaviors. This description of natural selection recognizes that natural selection must act in the context of living organisms.

The organism, however, must be viewed as a vehicle that permits the phenotypic expression of genes. In chapter 3, you will learn the second way that natural selection operates is upon the gene. A bird or a butterfly is not permanent—they die. The genes that they carry, however, are potentially immortal. The result of natural selection (and evolution in general) is reflected in how common, or how rare, specific genes are in a group of animals that are interbreeding—and, therefore, sharing genes. A group of individuals of the same species that occupy a given area at the same time and share a unique set of genes is called a **population.**

A more precise definition of organic evolution is a change in the frequency of alleles in a population. The frequency of an allele in a population is the abundance of that particular allele in relation to the sum of all alleles at that locus. Another way to express the same idea is that organic evolution is a change in the total genetic makeup of a population (the **gene pool**). This chapter examines evolution from the perspective of changes in gene pools and covers some of the mechanisms by which evolution occurs, as well as processes by which new species arise.

The Modern Synthesis: A Closer Look

Much of your background for understanding the modern synthesis comes from studying genetics because it explains why variations between individuals exist, and how they are passed to future generations. Genetic variation is important in evolution because some variations may confer an advantage to individuals, leading to natural selection. The potential for genetic variation in individuals of a population is unlimited. Even the relatively simple principles of inheritance described by Gregor Mendel provide for remarkable variation. In addition, crossing-over, multiple alleles, and mutations add to this variation. The result is that no two individuals, except identical twins, are genetically identical. Chance combinations of genes are likely to result in some individuals being better able to survive and reproduce in a given environment than other individuals.

Must Evolution Happen?

Evolution is central to all of biology, but is evolution always occurring in a particular population? As we will see, there are certainly times when the rate of evolution is very slow, and there are times when the rate of evolution is very rapid. But, are there times when evolution does not occur at all? (1) The answer to this question lies in the theories of **population genetics,** the study of the genetic events that occur in gene pools.

The Hardy-Weinberg Theorem

In 1908, an English mathematician, Godfrey H. Hardy, and a German physician, Wilhelm Weinberg, independently derived a mathematical model describing what happens to the frequency of alleles in a population over time. Their combined ideas became known as the Hardy-Weinberg theorem. It states that the mixing of alleles at meiosis and their subsequent recombination will not alter the frequencies of the alleles in future generations, as long as certain assumptions are met. If they are met, then evolution will not occur, because the allelic frequencies will not change from generation to generation, even though the specific mixes of alleles in individuals may vary.

The assumptions of the Hardy-Weinberg theorem are as follows:

1. The size of population must be very large. Large size ensures that the frequency of a gene will not change by chance alone.
2. Mating within the population must be random. Every individual must have an equal chance of mating with any other individual in the population. Expressed in a slightly different way, the choice of a mate must not be based on similarity or dissimilarity in a given trait. If this condition is not fulfilled, then some individuals are more likely to reproduce than others, and natural selection may occur.
3. There must not be any migration of individuals into, or out of, the population. Migration may introduce new genes into the gene pool, or add or delete copies of existing genes.
4. Mutations must not occur. If they do, mutational equilibrium must exist. Mutational equilibrium occurs when mutation from the wild-type allele to a mutant form is balanced by mutation from the mutant form back to the wild type. In either case, no new genes are introduced into the population from this source.

These assumptions must be met if allelic frequencies are not changing; that is, if evolution is not occurring. These assumptions are clearly very restrictive, and few, if any, real populations meet them. (2) This means that most populations are evolving. The theorem, however, does provide a

3.1 Genetic Equilibrium

To illustrate the Hardy-Weinberg theorem, consider a hypothetical population that meets all of the assumptions listed on p. 36. In this population, a particular autosomal locus has two alleles; allele A has a frequency of 0.8 [$f(A) = 0.8$], and allele a has a frequency of 0.2 [$f(a) = 0.2$]. Another way to say the same thing is that, in this population, 80% of the genes for this locus are allele A, and 20% are allele a. For the purpose of our example, assume dominance; however, it is irrelevant to the Hardy-Weinberg theorem.

Because random mating is assumed, one can easily determine the genotypic and phenotypic frequencies of the next generation using a Punnett square type of analysis (figure 1). Because we are assuming dominance, the phenotypic frequencies of the F_1 generation are:

$$f(\text{phenotype A}) = 0.64 + 0.32 = 0.96$$
$$f(\text{phenotype } a) = 0.04$$

Gene frequencies of the F_1 generation can be determined by adding the frequency of each homozygote to ½ the frequency of the heterozygote. (The frequency of an allele in the heterozygote is reduced by ½ because heterozygotes contain only a single copy of each allele.)

	f(allele A)	f(allele a)
AA individuals	0.64	0
Aa individuals	½(0.32) = 0.16	½(0.32) = 0.16
aa individuals	0	0.04
Overall frequency	0.80	0.20

Note that the overall frequency of each allele in the F_1 generation has not changed from that assumed for the parental generation. This example illustrates that if the assumptions of the Hardy-Weinberg theorem are met, the frequency of genes does not change from generation to generation. In other words,

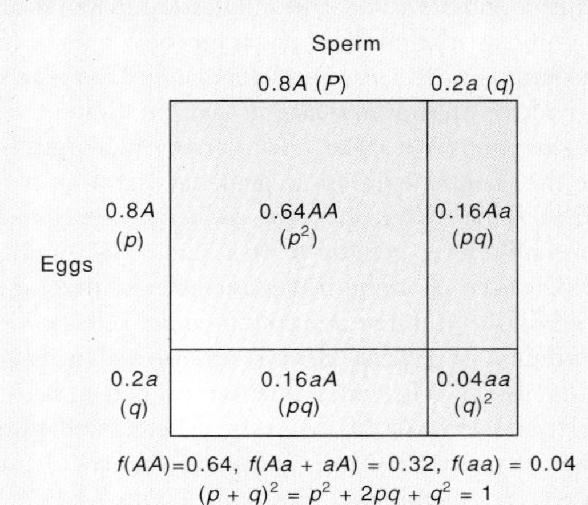

Figure 1 **Genotypic Frequencies.** Determining the genotypic frequencies of offspring from allelic frequencies in a parental generation. This analysis assumes random mating, and that each allele is equally likely to be incorporated into a viable gamete. The frequency of each allele in a sperm or egg is simply the frequency of that allele in the parental population. Allelic frequencies are often represented by p and q, and genotypic frequencies can be found by expanding the binomial $(p + q)^2$.

Hardy-Weinberg equilibrium is achieved, and evolution does NOT occur!

The Hardy-Weinberg theorem can be stated in general mathematical terms. Considering a locus with two interacting alleles, if p = the frequency of one allele, and q = the frequency of a second allele, then $p + q = 1$. The frequency of genotypes in the next generation can be found by algebraically expanding the binomial equation $(p + q)^2 = p^2 + 2pq + q^2 = 1$.

useful theoretical framework for examining changes in gene frequencies in populations (box 3.1)

In the next section, we will see how, when the assumptions are not met, evolutionary change occurs.

Stop and Ask Yourself

1. What is a gene pool?
2. What are the four assumptions of the Hardy-Weinberg theorem?
3. Are the assumptions listed in your answer to question 2 realistic for animal populations?
4. What conclusion can be drawn from the Hardy-Weinberg theorem?

Evolutionary Mechanisms

Evolution is neither a creative force working for progress, nor a dark force working to sacrifice individuals for the sake of the group. It is neither moral nor immoral. It has neither a goal, nor a mind to conceive a goal. Such goal-oriented thinking is said to be teleological. Evolution is simply a result of some individuals in a population surviving and being more effective at reproducing than others in the population, leading to changes in gene frequencies. In this section, we will examine some of the situations when the Hardy-Weinberg assumptions are not met—situations in which gene frequencies do change from one generation to the next and evolution occurs.

Population Size, Genetic Drift, and Neutral Selection

Chance often plays an important role in the perpetuation of genes in a population, and the smaller the population size, the more significant the effect of chance events may be. Fortuitous circumstances, such as a chance encounter between reproductive individuals, may promote reproduction. Some traits of a population survive, not because they convey increased fitness, but because they happened to be in gametes that were involved in fertilization. When chance events influence the frequencies of genes in populations, **genetic drift** is said to occur. Because gene frequencies are changing independently of natural selection, genetic drift is often referred to as **neutral selection.**

The process of genetic drift is analogous to flipping a coin. The likelihood of getting a head or a tail is equal. One is most likely to achieve the 50:50 ratio of heads and tails in a large number of tosses. In only ten tosses, one should not be surprised to get, for example, a disproportionate 7 heads and 3 tails. Similarly, the chance of one or the other of two equally adaptive alleles being incorporated into a gamete, and eventually into an individual in a second generation, is equal. In sampling gametes in a small population, unusual proportions of alleles may occur in any one generation of gametes because meiotic events, like tossing a coin, are random. Assuming that both alleles have equal fitness, these unusual proportions will be reflected in the phenotypes of the next generation. These chance events may result in a particular allele increasing in frequency or decreasing in frequency (figure 3.1*a*). In small populations, inbreeding is also common. Genetic drift and inbreeding are likely to reduce genetic variation within a population.

If a mutation introduces a new allele into a population and that allele is no more or less adaptive than existing alleles, genetic drift may permit the new allele to become established in the population (figure 3.1*b*), or the new allele may be lost because of genetic drift. The likelihood of genetic drift occurring in small populations suggests that a Hardy-Weinberg equilibrium will not occur in such a population.

Two special cases of genetic drift have influenced the genetic makeup of some populations. When a few individuals from a parental population colonize new habitats, they seldom carry a representative sample of the gene pool from which they came. The new colony that emerges from the founding individuals is likely to have a distinctive genetic makeup with far less variation than the larger population. This form of genetic drift is called the **founder effect.**

An often-cited example of the founder effect concerns the genetic makeup of the Dunkers of eastern Pennsylvania. They emigrated from Germany to the United States early in the eighteenth century, and for religious reasons, have not married outside their sect. Examination of certain traits (e.g., ABO blood type) in their population reveals very different gene frequencies from the Dunker populations of Germany. These differences are attributed to the chance absence of certain genes in the individuals that founded the original Pennsylvania Dunker population.

A similar effect can occur when the number of individuals in a population is drastically reduced. For example, cheetah populations in South and East Africa are endangered. The severe depletion in numbers experienced in these populations has reduced genetic diversity to the point that even if populations are restored, the recovered populations will have only a remnant of the original gene pool (figure 3.2). This form of ge-

	(a) Loss of genetic diversity through genetic drift	(b) Substitution of a new allele for an existing allele through genetic drift
Parental generation	AA Aa Aa aa	AA Aa Aa aa
		Neutral mutation introduces a new allele
Parental gametes	4/8 A A A A 4/8 a a a a	3/8 A A A 1/8 A′ 4/8 a a a a
	Chance selection of 8 gametes ↓	Chance selection of 8 gametes ↓
F₁ generation	Aa Aa aa aa	Aa A′a aa A′a
F₁ gametes	2/8 A A 6/8 a a a a a a	1/8 A 2/8 A′ A′ 5/8 a a a a a
	Chance selection of 8 gametes ↓	Chance selection of 8 gametes ↓
F₂ generation	Aa aa aa aa	A′a A′A′ aa aa
F₂ gametes	1/8 A 7/8 a a a a a a a	3/8 A′ A′ A′ 5/8 a a a a a

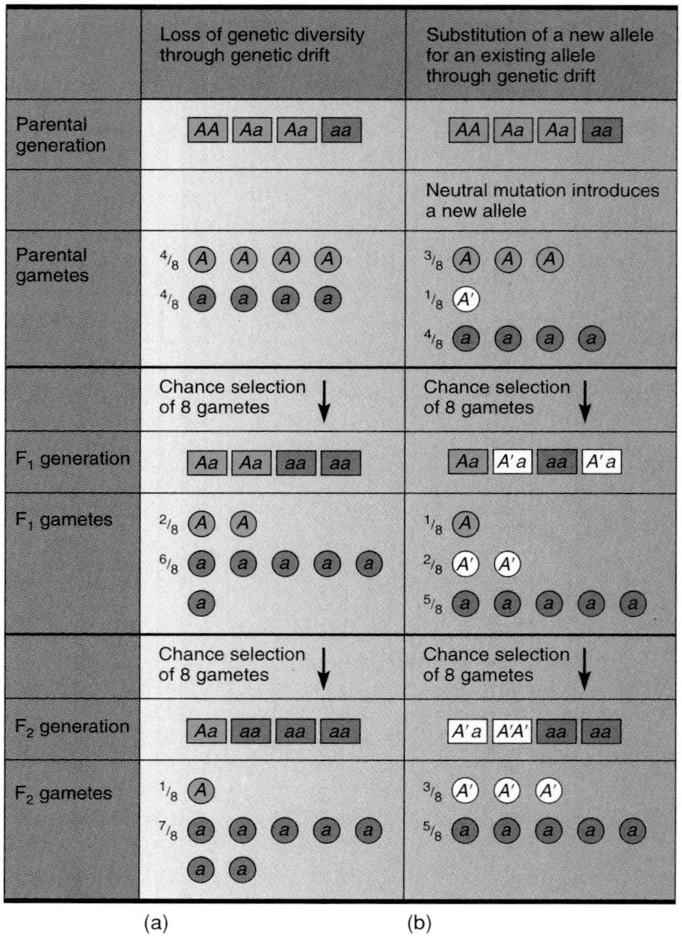

Figure 3.1

Genetic Drift. (*a*) Genetic diversity may be lost as a result of genetic drift. Assume alleles *a* and A are equally adaptive. Allele *a* might be incorporated into gametes more often than A, or it could be involved in more fertilizations. In either case, the frequency of *a* increases and the frequency of A decreases because of random events operating at the level of gametes. (*b*) A new, equally adaptive, allele may become established in a population as a result of genetic drift. In this example, the allele (A′) substitutes for an existing allele (A). The same mechanisms could also account for the loss of the newly introduced allele or the establishment of A′ as a third allele.

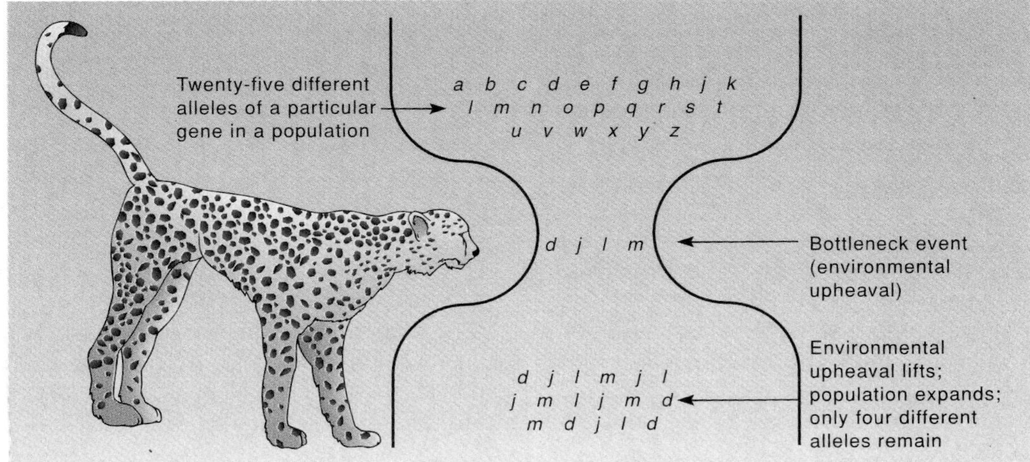

Figure 3.2
The Bottleneck Effect. (*a*) Cheetahs (*Acinonyx jubatus*) of South and East Africa have been endangered by human activities. (*b*) Severe reduction in the original population has caused a bottleneck effect that, even if population size recovers, has resulted in a loss of genetic diversity.

Figure 3.3

The Bottleneck Effect. The Northern elephant seal (*Mirounga angustirostris*) was severely overhunted in the late 1800s. Even though its population numbers are now increasing, its genetic diversity is very low.

netic drift is called the **bottleneck effect.** A similar example concerns the Northern elephant seal (figure 3.3). It was hunted to near extinction in the late 1800s. Legislation to protect the seal was enacted in 1922, and now the population is greater than 100,000 individuals. In spite of this relatively large number, the genetic variability in the population is very low.

The effects of bottlenecks are somewhat controversial. The traditional interpretation is that decreases in genetic diversity make populations less likely to withstand environmental stress and more susceptible to extinction. That is, in a population with high genetic diversity, it is more likely that some individuals will have a combination of genes that allows them to withstand environmental changes. In the case of the cheetah, recent evidence indicates that this cat's current problems may be more a result of predation by lions and spotted hyenas on cheetah cubs than a result of low genetic diversity. Most evolutionary biologists agree, however, that over evolutionary time frames, high genetic diversity makes extinction less likely.

Gene Flow

The Hardy-Weinberg theorem assumes that no individuals enter a population from the outside (immigrate), and that no individuals leave a population (emigrate). If immigration or emigration occurs, the Hardy-Weinberg equilibrium is upset, and changes in gene frequency (evolution) occur. These changes in gene frequency from migration of individuals are called **gene flow.** Although there are certainly some natural populations for which gene flow may not be significant, most populations experience changes in allelic frequency from this source.

Mutation

Changes in the structure of genes and chromosomes are called mutations. The Hardy-Weinberg theorem assumes that no mutations occur, or that mutational equilibrium exists. Mutations, however, are a fact of life. Most importantly, mutations are the origin of all new genes and a source of variation that may prove adaptive for an animal. Mutation counters the loss of genetic material that results from natural selection and genetic drift, and increases the likelihood that variations will be present that allow a group to survive future environmental shocks.

Mutations are random events, and the likelihood of a mutation is not affected by the usefulness of the mutation. Organisms have no special device to filter harmful genetic changes from advantageous changes before they occur. Mutations occur, and animals must take the bad along with the good. The effects of mutations vary enormously. Most are deleterious. Some mutations that are neutral or harmful in one environment may help an organism survive in another environment.

Mutational equilibrium occurs when a mutation from the wild-type allele to a mutant form is balanced by mutation from the mutant back to the wild type. This has the same effect on allelic frequency as if no mutation occurred. Mutational equilibrium rarely exists, however. **Mutation pressure** is a measure of the tendency for gene frequencies to change through mutation.

Natural Selection Reexamined

The theory of natural selection remains preeminent in modern biology. Natural selection occurs whenever some phenotypes are more successful at leaving offspring than other phenotypes, and the tendency for it to occur—and upset Hardy-Weinberg equilibrium—is called **selection pressure.** Although natural selection is simple in principle, it is quite diverse in actual operation.

Modes of Selection

Many populations have a range of phenotypes for certain traits. This range may be characterized using a bell-shaped curve, where phenotypic extremes are less common than the intermediate phenotypes. Natural selection may affect a range of phenotypes in three different ways.

Directional selection occurs when individuals at one phenotypic extreme are at a disadvantage compared to all other individuals in the population (figure 3.4*a*). In response to this selection, the deleterious gene(s) decreases in frequency and all other genes increase in frequency. Directional selection may occur when a mutation gives rise to a new gene, or when the environment changes to select against an existing phenotype.

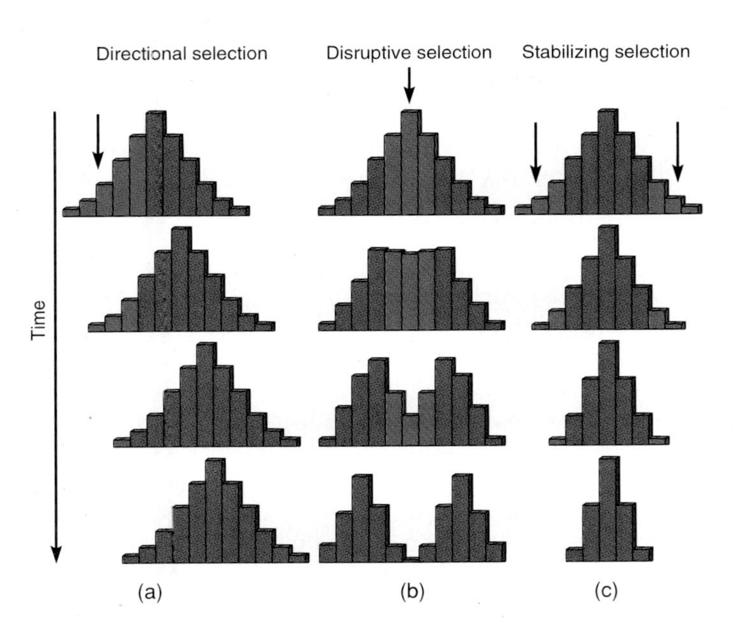

FIGURE 3.4

Modes of Selections. (*a*) Directional selection occurs when individuals at one phenotypic extreme are selected against. It results in a shift in phenotypic distribution toward the advantageous phenotype. (*b*) Disruptive selection occurs when an intermediate phenotype is selected against and results in the production of distinct subpopulations. (*c*) Stabilizing selection occurs when individuals at both phenotypic extremes are selected against, and results in a narrowing at both ends of the range. Arrows indicate selection against one or more phenotypes. The X-axis of each graph indicates the range of phenotypes for the trait in question.

Industrial melanism, a classic example of directional selection, occurred in England during the industrial revolution. Museum records and experiments document how environmental changes affected selection against one phenotype of the peppered moth, *Biston betularia*.

In the early 1800s, a gray form made up about 99% of peppered moth population. That form still predominates in nonindustrial northern England and Scotland. In industrial areas of England, the gray form was replaced by a black form over a period of about 50 years. In these areas, the gray form made up only about 5% of the population, and 95% of the population was black. The gray phenotype, previously advantageous, had become deleterious.

The nature of the selection pressure was understood when it was discovered that birds prey more effectively on moths resting on a contrasting background. Prior to the industrial revolution, the gray form was favored because gray moths blended with the bark of trees on which the moths rested. The black form contrasted with the lighter lichen-covered bark, and was easily spotted by birds (figure 3.5*a*). Early in the industrial revolution, however, factories used soft coal, and spewed soot and other pollutants into the air. Soot covered the tree trunks where the moths rested, and bird predators could easily pick out

(a)

(b)

FIGURE 3.5

Directional Selection. Two body forms of the peppered moth, *Biston betularia*: There is a gray and black form in each picture. (*a*) Prior to the industrial revolution, the black form was easily spotted by bird predators, and the gray form was camouflaged. (*b*) In industrial regions, after the industrial revolution, selection was reversed because of soot covering the bark of trees where moths rested. Note how clearly the gray form can be seen, whereas the black form is almost invisible.

gray moths against the black background. The black form was now effectively camouflaged (figure 3.5*b*).

In the 1950s, the British Parliament enacted air pollution standards that have helped reduce soot in the atmosphere. As would be expected, the gray form of the moth has had a small but significant increase in frequency.

Another form of natural selection may occur when circumstances select against individuals of an intermediate phenotype (figure 3.4*b*). 4 **Disruptive selection** produces distinct subpopulations. Consider, for example, what could happen in a population of snails having a range of shell colors between white and dark brown. Their marine tidepool habitat provides two background colors. The sand, made up of pulverized mollusc shells, is white, and rock outcroppings are brown. In the face of predation by shorebirds, what phenotypes are going to be most common? Although white snails may not actively select a white background, those present on the sand are less likely to be preyed on than intermediate phenotypes on either sand or rocks. Similarly, brown snails are less likely to be preyed on than intermediate phenotypes on either substrate. Thus, two distinct subpopulations, one white and one brown, could be produced through disruptive selection.

A third form of natural selection occurs when both phenotypic extremes are deleterious. This process is called **stabilizing selection,** and results in a narrowing of the phenotypic range (figure 3.4c). During long periods of environmental constancy, new variations that arise, or new combinations of genes that occur, are unlikely to result in more fit phenotypes than the genes that have allowed a population to survive for thousands of years, especially when the new variations are at the extremes of the phenotypic range.

A good example of stabilizing selection is the horseshoe crab (*Limulus*), which occurs along the Atlantic coast of the United States (*see figure 11.8*). Comparison of the fossil record with living forms indicates that this body form has changed very little over 200 million years. Apparently, the combination of characteristics present in this group of animals represents a very successful combination of characteristics for the horseshoe crab's environment.

Neutralist/Selectionist Controversy

Most biologists recognize that both natural selection and neutral selection occur, but may not be equally important in all circumstances. For example, during long periods when environments are relatively constant, and stabilizing selection is acting on phenotypes, neutral selection may operate at the molecular level. Certain genes could be randomly established in a population. Occasionally, however, the environment shifts, and directional or disruptive selection begins to operate, resulting in gene frequency changes (often fairly rapid).

The relative importance of neutral selection and natural selection in natural populations is debated and is an example of the kind of debates occurring among evolutionists. These debates concern the mechanics of evolution and are the foundations of science. They lead to experiments that will ultimately present a clearer understanding of evolution.

Balanced Polymorphism and Heterozygote Superiority

Polymorphism occurs in a population when two or more distinct forms exist without a range of phenotypes between them. **Balanced polymorphism** (Gr. *poly,* many + *morphe,* form) occurs when different phenotypes are maintained at relatively stable frequencies in the population and may resemble a population in which disruptive selection operates.

Sickle cell anemia results from a change in the structure of the hemoglobin molecule. Some of the red blood cells of persons with the disease are misshapen, and their ability to carry oxygen is reduced. In the heterozygous state, there are roughly equal quantities of normal and sickled cells. Sickle cell heterozygotes occur in some African populations with a frequency as high as 0.4. The maintenance of the sickle cell heterozygotes and both homozygous genotypes at relatively unchanging frequencies makes this trait an example of a balanced polymorphism.

Why hasn't such a seemingly deleterious gene been eliminated by natural selection? The sickle cell gene is most common in regions of Africa that are heavily infested with the malarial parasite, *Plasmodium falciparum*. Heterozygotes are less susceptible to malarial infections; if infected, they experience less severe symptoms than do homozygotes without sickled cells. Individuals homozygous for the normal allele are at a disadvantage because they experience more severe malarial infections, and individuals homozygous for the sickle cell allele are at a disadvantage because they suffer from severe anemia caused by the sickled cells. The heterozygotes, who usually experience no symptoms of anemia, are more likely to survive than either homozygote. This system is also an example of heterozygote superiority, which occurs when the heterozygote is more fit than either homozygote. Heterozygote superiority can lead to balanced polymorphism, because perpetuation of the alleles in the heterozygous condition maintains both alleles at a higher frequency than would be expected if natural selection acted only on the homozygous phenotypes.

Stop and Ask Yourself

5. What is neutral selection? Why are the effects of neutral selection more likely to show up in small populations, or in fringe groups of larger populations?
6. Why are mutations important evolutionary occurrences?
7. How would you define the terms directional selection, disruptive selection, and stabilizing selection? Which is more likely to occur in a time of climatic change?

Species and Speciation

Taxonomists classify organisms into groups based on their similarities and differences. This classification system is discussed in chapter 4. The fundamental unit of classification is the species. Unfortunately, it is difficult to formulate a universally applicable definition of species. According to a biological definition, a **species** is a group of populations in which genes are actually, or potentially, exchanged through interbreeding.

Although this definition is concise, it has problems associated with it. Taxonomists often work with morphological characteristics, and the reproductive criterion must be assumed based on morphological and ecological information. Also, some organisms do not reproduce sexually. Obviously, other criteria need to be applied in these cases. Another problem concerns fossil material. Paleontologists describe species of extinct organisms, but how are they to test the reproductive criterion? Finally, populations of similar organisms may be so isolated from each other that the exchange of genes is geographically impossible. To test a reproductive criterion, biologists can transplant individuals to see if mating can occur. Under such circumstances, however, one can never be certain that mating of transplanted individuals would really occur if animals were together in a natural setting.

Rather than trying to establish a definition of a species that solves all these problems, it is probably better to simply understand the problems associated with the biological definition. In describing species, taxonomists use morphological, physiological, embryological, behavioral, molecular, and ecological criteria, realizing that all of these have a genetic basis.

5 **Speciation** is the formation of new species. A requirement of speciation is that subpopulations are prevented from interbreeding. This is called **reproductive isolation.** When subpopulations are reproductively isolated, natural selection and genetic drift can result in evolution taking a different course in each subpopulation. Reproductive isolation can occur in different ways.

Premating isolation prevents mating from taking place. For example, subpopulations may be separated by impenetrable barriers such as rivers or mountain ranges. Other forms of premating isolation are more subtle. If courtship behavior patterns of two animals are not mutually appropriate, mating will not occur. Similarly, individuals that have different breeding periods, or that occupy different habitats, will be unable to breed with each other.

Postmating isolation prevents successful fertilization and development, even though mating may have occurred. For example, conditions in the reproductive tract of a female may not support the sperm of another individual, which prevents successful fertilization. Postmating isolation also occurs because hybrids are usually sterile (e.g., the mule produced from a mating of a male donkey and a mare is a sterile hybrid). Mismatched chromosomes cannot synapse properly during meiosis, and any gametes produced are not viable. Other kinds of postmating isolation include developmental failures of the fertilized egg or embryo.

Allopatric Speciation

Allopatric (Gr. *allos,* other + *patria,* fatherland) **speciation** occurs when subpopulations become geographically isolated from one another. For example, a mountain range or river may permanently separate members of a population. Adaptations to different environments or neutral selection in these separate populations may result in members not being able to mate successfully with each other, even if experimentally reunited. Allopatric speciation is believed by many biologists to be the most common kind of speciation (figure 3.6).

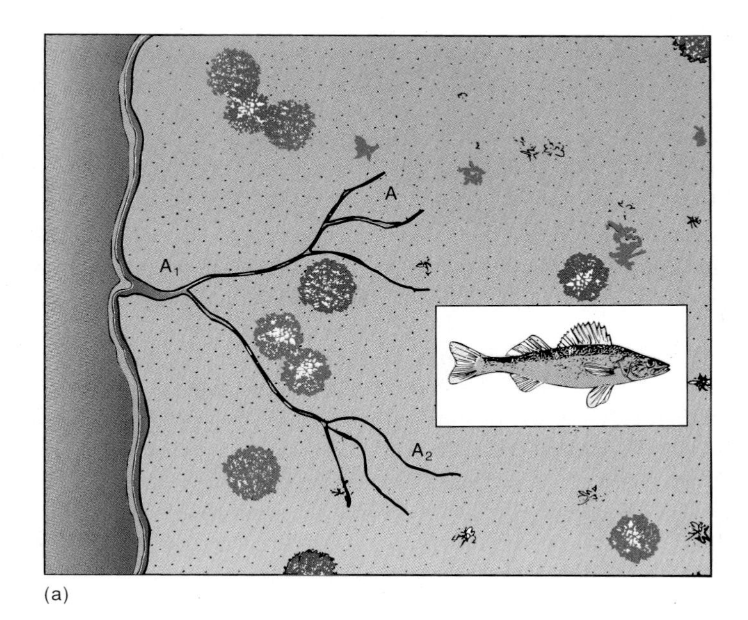

(a)

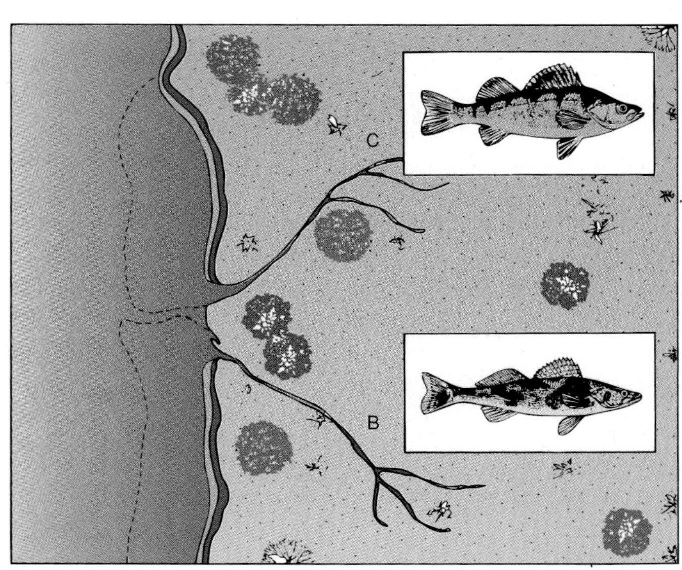

(b)

Figure 3.6

Allopatric Speciation. Allopatric speciation can occur when a geographic barrier divides a population. (*a*) In this hypothetical example, a population of freshwater fish in a river drainage system is divided into three subpopulations: A, A_1 and A_2. Genetic exchanges may occur between A and A_1, and between A_1 and A_2. Exchanges are less common between A and A_2. (*b*) A rise in the level of the ocean forces the breakup of A_1, and separates A and A_2 into separate populations. Genetic drift and different selection pressures in the two populations may eventually result in the formation of species B and C.

The finches that Darwin saw on the Galápagos Islands are a classic example of allopatric speciation, as well as an important evolutionary process called adaptive radiation. **Adaptive radiation** occurs when a number of new forms diverge from an ancestral form, usually in response to the opening of major new habitats.

Fourteen species of finches evolved from the original finches that colonized the Galápagos Islands. It is likely that ancestral finches, having emigrated from the mainland, were distributed among a few of the islands of the Galápagos. Populations became isolated on various islands over a period of time. Even though the original population probably displayed some genetic variation, even greater variation arose over time. The original finches were seed eaters, and after their arrival, they probably filled their preferred habitats very rapidly. Variations within the original finch population may have allowed some birds to exploit new islands and habitats where no finches had been. Mutations changed the genetic composition of the isolated finch populations, introducing further variations. Natural selection favored the retention of the variations that happened to promote successful reproduction.

The combined forces of isolation, mutation, and natural selection allowed the finches to diverge into a number of species with specialized feeding habits (*see figure 2.3*). Of the 14 species of finches, six have beaks specialized for crushing seeds of different sizes. Others feed on flowers of the prickly pear cactus or in the forests on insects and fruit.

Parapatric Speciation

Another form of speciation is called **parapatric** (Gr. *para,* beside) **speciation,** which occurs in small, local populations, called **demes.** For example, all of the frogs in a particular pond, or all of the sea urchins in a particular tidepool, make up a deme. Individuals of a deme are more likely to breed with one another than with other individuals in the larger population, and because they experience the same environment, they are subject to similar selection pressures. Demes are not completely isolated from each other because individuals, developmental stages, or gametes can move between demes of a population. On the other hand, the relative isolation of a deme may mean that its members experience different selection pressures than other members of the population. If so, speciation can occur. Although most evolutionists theoretically agree that parapatric speciation can occur, no certain cases are known. Parapatric speciation is therefore considered of less importance in the evolution of animal groups than allopatric speciation.

Sympatric Speciation

A third kind of speciation is called **sympatric** (Gr. *sym,* together) **speciation.** It occurs within a single population. Even though organisms are sympatric, they still may be reproductively isolated from one another. Many plant species are capable of producing viable forms with multiple sets of chromosomes. Such events could lead to sympatric speciation among groups in the same habitat. Sympatric speciation in animals, however, rarely, if ever, occurs.

Stop and Ask Yourself

8. Why is reproductive isolation necessary for speciation to occur?
9. What form of premating isolation is likely to promote allopatric speciation?
10. What is postmating isolation? What are three forms it may take?
11. What are sympatric, parapatric, and allopatric speciation?
12. What is adaptive radiation?

Rates of Evolution

Charles Darwin perceived evolutionary change as occurring gradually over millions of years. This concept, called **phyletic gradualism,** has been the traditional interpretation of the tempo, or rate, of evolution.

Some evolutionary changes, however, occur very rapidly. Studies of the fossil record show that many species do not change significantly over millions of years. These periods of stasis (Gr. *stasis,* standing still), or equilibrium, are interrupted when a group encounters an ecological crisis, such as a change in climate or a major geological event. Over the next 10,000 to 100,000 years, a variation might be advantageous that previously would have been selectively neutral or disadvantageous. Alternatively, geological events might result in new habitats becoming available. (Events that occur in 10,000 to 100,000 years are almost instantaneous in an evolutionary time frame.) This geologically brief period of change "punctuates" the previous million or so years of equilibrium, and eventually settles into the next period of stasis. In this model, the periods of stasis are characterized by stabilizing selection, and the periods of change are characterized by directional or disruptive selection (figure 3.7). This model of evolution in which long periods of stasis are interrupted by brief periods of change is called the **punctuated equilibrium model.**

Such rapid evolutionary changes in small populations have been observed in the field. The acquisition of pesticide and antibiotic resistance by insect pests and bacteria are examples of rapid natural selection that have been observed by humans. In a series of studies over a 20-year period, Peter R. Grant has shown that natural selection results in rapid morphological changes in the bills of Galápagos finches. A long, dry period from the middle of 1976 to early January 1978 resulted in birds with larger, deeper bills. Early in this dry period, smaller, easily cracked seeds were quickly consumed by birds. As birds were forced to turn to large seeds, birds with weaker bills were selected against, resulting in a measurable change in the makeup of the finch population of the island, Daphne Major.

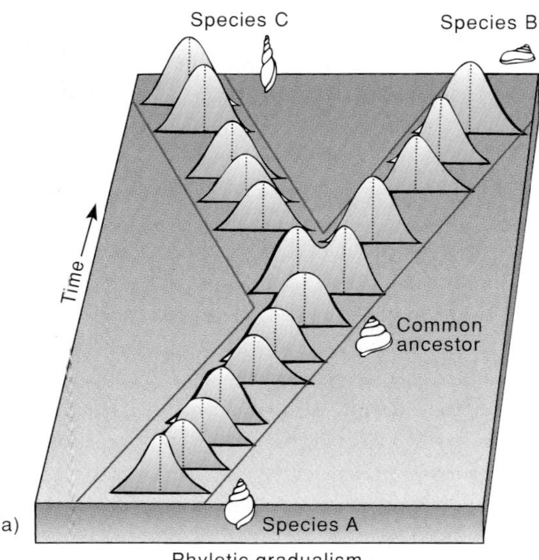

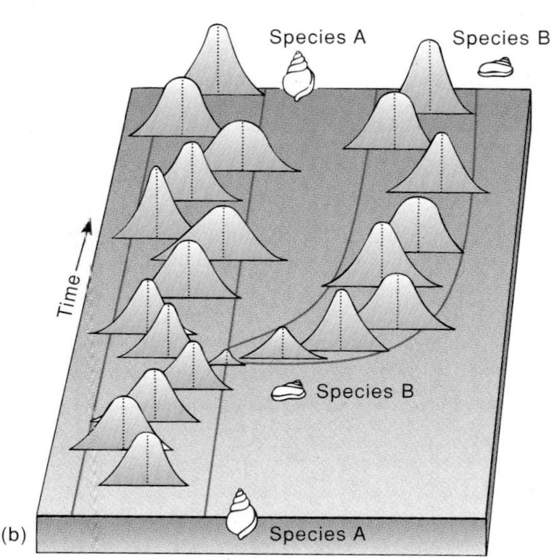

FIGURE 3.7

Rates of Evolution. A comparison of phyletic gradualism and punctuated equilibrium. (*a*) Phyletic gradualism is a model of evolution in which gradual changes occur over very long time periods. Notice that a continuous series of intermediate phenotypes are present as species B and species C gradually diverge from species A. (*b*) Punctuated equilibrium is a model of evolution in which long periods of stasis are interrupted by rapid periods of change. Notice the very rapid divergence of species B from species A, after which species B enters a period of relative stasis.

TABLE 3.1 AMINO ACID DIFFERENCES IN CYTOCHROME C FROM DIFFERENT ORGANISMS

ORGANISMS	NUMBER OF VARIANT AMINO ACID RESIDUES
Cow and sheep	0
Cow and whale	2
Horse and cow	3
Rabbit and pig	4
Horse and rabbit	5
Whale and kangaroo	6
Rabbit and pigeon	7
Shark and tuna	19
Tuna and fruit fly	21
Tuna and moth	28
Yeast and mold	38
Wheat and yeast	40
Moth and yeast	44

One advantage of the punctuated equilibrium model is its explanation for why the fossil record does not always show transitional stages between related organisms. Gradualists attribute the absence of transitional forms to the fact that fossilization is an unlikely event, and, therefore, many transitional forms disappeared without leaving a fossil record. Since punctuated equilibrium involves rapid changes in small, isolated populations, preservation of intermediate forms in the fossil record would be even less likely. The rapid pace (geologically speaking) of evolution resulted in apparent "jumps" from one form to another.

MOLECULAR EVOLUTION

Many evolutionists study changes in animal structure and function that are observable on a large scale—for example, changes in the shape of a bird's bill or in the length of an animal's neck. All evolutionary change, however, results from changes in the sequence of bases in DNA and amino acids in proteins. (6) Molecular evolutionists investigate evolutionary relationships between organisms by studying DNA and proteins. For example, cytochrome *c* is a protein present in the pathways of cellular respiration in all eukaryotic organisms (table 3.1). Organisms shown to be closely related from using other investigative tools, have similar cytochrome *c* molecules. The fact that cytochrome *c* has changed so little during hundreds of millions of years suggests that mutations of the cytochrome *c* gene are nearly always detrimental, and will be selected against. Because it has changed so little, cytochrome *c* is said to have been conserved evolutionarily.

Not all proteins are conserved as rigorously as cytochrome *c*. Although variations in highly conserved proteins can be used to help establish evolutionary relationships between distantly related organisms, less conserved proteins are useful for looking at relationships between more closely related animals. Because some proteins are conserved and others are not, the best information regarding evolutionary relationships is obtained by comparing as many proteins as possible in any two species.

Gene Duplication

Recall that most mutations are selected against. However, if an extra copy of a gene is present, one copy may be modified and, as long as the second copy is furnishing the essential protein, the organism is likely to survive. Gene duplication, the accidental duplication of a gene on a chromosome, is one way that extra genetic material can arise.

Vertebrate hemoglobin and myoglobin are believed to have arisen from a common ancestral molecule. Hemoglobin carries oxygen in red blood cells, and myoglobin is an oxygen storage molecule in muscle. The ancestral molecule probably carried out both functions. However, about 1 billion years ago, gene duplication followed by mutation of one gene resulted in the formation of two polypeptides: myoglobin and hemoglobin. Further gene duplications over the last 500 million years probably explain the fact that most vertebrates, other than primitive fishes, have hemoglobin molecules consisting of four polypeptides.

Mosaic Evolution

A previous section described how rates of evolution can vary both in populations and in molecules and structures. A species might be thought of as a mosaic of different molecules and structures that have evolved at different rates. Some molecules or structures are conserved in evolution, others change more rapidly. The basic design of a bird provides a simple example. All birds are easily recognizable as birds because of highly conserved structures, such as feathers, bills, and a certain body form. Particular parts of birds, however, are less conservative and have a higher rate of change. Wings have been modified for hovering, soaring, and swimming. Similarly, legs have been modified for wading, swimming, and perching. These are examples of **mosaic evolution.**

Stop and Ask Yourself

13. How do phyletic gradualism and punctuated equilibrium models differ?
14. How can amino acid sequencing and DNA base sequencing provide evidence of evolutionary relationships?
15. What does it mean to say that cytochrome *c* is an evolutionarily conserved protein?

Summary

1. Organic evolution is a change in the frequency of alleles in a population.
2. Unlimited genetic variation, in the form of new alleles and new combinations of alleles, increases the chances that a population will survive future environmental changes.
3. Population genetics is the study of events occurring in gene pools. The Hardy-Weinberg theorem describes the fact that, if certain assumptions are met, gene frequencies of a population remain constant from generation to generation.
4. The assumptions of the Hardy-Weinberg theorem, when not met, define circumstances under which evolution will occur. (a) Fortuitous circumstances may allow only certain alleles to be carried into the next generation. Such chance variations in allelic frequencies are called genetic drift or neutral selection. (b) Allelic frequencies may change as a result of individuals immigrating into, or emigrating from, a population. (c) Mutations are the source of new genetic material for populations. Mutational equilibrium rarely exists, and thus mutations usually result in changing allelic frequencies. (d) The tendency for allelic frequencies to change, due to differing fitness, is called selection pressure.

5. Selection may be directional, disruptive, or stabilizing.
6. Balanced polymorphism occurs when two or more body forms are maintained in a population. Heterozygote superiority can lead to balanced polymorphism.
7. According to a biological definition, a species is a group of populations within which there is potential for exchange of genes. Significant problems are associated with the application of this definition.
8. In order for speciation to happen, reproductive isolation must occur. Speciation may occur sympatrically, parapatrically, or allopatrically, although most speciation events are believed to be allopatric.
9. Premating isolation may prevent mating from occurring and/or postmating isolation may prevent the development of fertile offspring, if mating has occurred.
10. Phyletic gradualism is a model of evolution that depicts change as occurring gradually, over millions of years. Punctuated equilibrium is a model of evolution that depicts long periods of stasis interrupted by brief periods of relatively rapid change.
11. The study of rates of molecular evolution helps establish evolutionary interrelationships between organisms.
12. A duplicated gene may be modified by mutation, and by chance, come to serve a function other than its original role.
13. Any species can be thought of as a mosaic of different molecules and structures that have evolved at differing rates.

Selected Key Terms

adaptive radiation (*p.* 43)
Hardy-Weinberg equilibrium (*p.* 37)
neutral selection (*p.* 38)
phyletic gradualism (*p.* 44)
population genetics (*p.* 36)
punctuated equilibrium model (*p.* 44)
reproductive isolation (*p.* 43)
speciation (*p.* 43)
species (*p.* 42)

Critical Thinking Questions

1. Can natural selection act on variations that are not inherited? (Consider, for example, deformities that arise from contracting a disease.) If so, what is the effect of that selection on subsequent generations?
2. In what way does overuse of antibiotics and pesticides increase the likelihood that these chemicals will eventually become ineffective? This is an example of which one of the three modes of natural selection?
3. What are the implications of the "bottleneck effect" for wildlife managers who try to help endangered species, such as the whooping crane, recover from near extinction?
4. What does it mean to think of evolutionary change as being goal-oriented? Explain why this way of thinking is wrong.
5. Imagine that two species of butterflies resemble one another closely. One of the species (the model) is distasteful to bird predators, and the other species (the mimic) is not. How could directional selection have resulted in the mimic species evolving a resemblance to the model species?

chapter 4

Animal Classification, Phylogeny, and Organization

Outline

Concepts

1. Order in nature allows systematists to name animals and discern evolutionary relationships among them.
2. All organisms can be placed in one of five kingdoms based on whether their cells are prokaryotic or eukaryotic; whether they are truly multicellular or not; and whether they get their food through absorption, ingestion, or autotrophy.
3. Animal systematists use a variety of methods to discern evolutionary relationships. Evolutionary systematics and phylogenetic systematics (cladistics) are two widely used approaches in the study of evolutionary relationships.
4. Animal relationships are represented by branching evolutionary tree diagrams.
5. Animal body plans can be categorized based upon how cells are organized into tissues and how body parts are distributed within and around an animal.

Would You Like to Know:

1. why zoologists use scientific names, rather than common names, for the animals they study? (*p. 51*)
2. why zoology courses often cover organisms such as *Amoeba* and *Paramecium* when these organisms are really not animals? (*p. 51*)
3. why evolutionary tree diagrams can be misleading? (*p. 56*)
4. why sedentary animals, such as sea anemones, do not have heads and tails? (*p. 57*)
5. why body cavities are advantageous to animals that possess them? (*p. 59*)

These and other useful questions will be answered in this chapter.

This chapter contains evolutionary concepts, which are set off in this font.

One of the cornerstones of science that is virtually unchallenged is that there is order in nature. The order found in living systems is a natural consequence of the shared evolutionary processes that influence life. This chapter describes how the classification of animals and the basic organization of their bodies reflect that order.

Classification of Organisms

One of the characteristics of modern humans is our ability to communicate with a spoken language. Language not only allows us to communicate, but it also helps us encode and classify concepts, objects, and organisms that we encounter. To make sense out of life's diversity, we need more than just names for organisms. A potpourri of over a million animal names is of little use to anyone. To be useful, a naming system must reflect the order and relationships that arise from evolutionary processes. The study of the kinds and diversity of organisms and the evolutionary relationships among them is referred to as **systematics** (Gr. *systema,* system + *ikos,* body of facts) or **taxonomy** (Gr. *taxis,* arrangement + L. *nominalis,* belonging to a name). These studies result in the description of new species and the organization of animals into groups (taxa) based on degree of evolutionary relatedness. (Some biologists distinguish between systematics and taxonomy. These biologists prefer to think of taxonomy as the work involved with the original description of species, and systematics as the assignment of species into evolutionary groups. In this textbook, we will not make this distinction because of the extensive overlap between the two tasks.) **Nomenclature** (L. *nominalis,* belonging to a name + *calator,* to call) is the assignment of a distinctive name to each species.

A Taxonomic Hierarchy

Our modern classification system is rooted in the work of Karl von Linné (1707–1778). His binomial system is still used today. Von Linné also recognized that different species could be grouped into broader categories based on shared characteristics. A group of animals that shares a particular set of characteristics forms an assemblage called a **taxon.** For example, a housefly (*Musca domestica*), although obviously unique, shares certain characteristics with other flies (the most important of these being a single pair of wings). Based on these similarities, all true flies form a logical, more inclusive group. Further, all true flies share certain characteristics with bees, butterflies, and beetles. Thus, these animals form an even more inclusive taxonomic group. They are all insects.

Von Linné recognized five taxa. Modern taxonomists use those five, and have added two other major taxa. They are arranged hierarchically (from broader to more specific): **kingdom, phylum, class, order, family, genus,** and **species**

Table 4.1 Taxonomic Categories of a Human and a Dog

Taxon	Human	Domestic Dog
Kingdom	Animalia	Animalia
Phylum	Chordata	Chordata
Class	Mammalia	Mammalia
Order	Primates	Carnivora
Family	Hominidae	Canidae
Genus	*Homo*	*Canis*
Species	*sapiens*	*familiaris*

Figure 4.1
Classification of Organisms. Animals belong to one of five large groups of organisms called kingdoms. The grouping of organisms according to evolutionary relationships helps scientists make sense of life's diversity. The Sumatran tiger cubs shown here belong to the species *Panthera tigris*.

(table 4.1). Even though von Linné did not accept evolution, many of his groupings reflect evolutionary relationships. Morphological similarities between two animals have a genetic basis and are the result of a common evolutionary history. Thus, in grouping animals according to shared characteristics, von Linné grouped them according to their evolutionary relationships (figure 4.1). Ideally, members of the same taxonomic group are more closely related to each other than to members of different taxa.

Above the species level, there are no precise definitions of what constitutes a particular taxon. (The species concept was discussed in chapter 3.) Disagreements as to whether two species should be grouped into the same taxon or different taxa are common.

Nomenclature

Do you call certain freshwater crustaceans crawdads, crayfish, or crawfish? Do you call a common sparrow an English sparrow, a barn sparrow, or a house sparrow? (1) The binomial system of nomenclature brings order to a chaotic world of common names. There are two problems with common names. First, common names vary from country to country, and from region to region within a country. Some species have literally hundreds of different common names. Biology transcends regional and nationalistic boundaries, and so must the names of what biologists study. Second, many common names refer to taxonomic categories higher than the species level. A superficial examination will simply not distinguish most different kinds of pillbugs (class Crustacea, order Isopoda) or most different kinds of crayfish (class Crustacea, order Decapoda). A common name, even if one recognizes it, often does not specify a particular species.

The binomial system of nomenclature is universal, and one always knows what level of classification is involved in any description. No two kinds of animals are given the same binomial name, and every animal has only one correct name, as required by the *International Code of Zoological Nomenclature*. The confusion caused by common names, therefore, is avoided. When writing the scientific name of an animal, the genus begins with a capital letter, the species designation begins with a lowercase letter, and the entire scientific name is italicized or underlined because it is latinized. Thus, the scientific name of humans is written *Homo sapiens*, and when the genus is understood, the binomial name can be abbreviated *H. sapiens*.

Kingdoms of Life

In the 1960s, a system of classification that uses five kingdoms gained widespread acceptance (figure 4.2). It distinguishes between kingdoms based on cellular organization and mode of nutrition. According to this system, members of the kingdom **Monera** are the bacteria and the cyanobacteria. They are distinguished from all other organisms by being prokaryotic. (Prokaryotic organisms lack a membrane-bound nucleus and other membranous organelles.) Members of the kingdom **Protista** are eukaryotic and consist of single cells, or colonies of cells. This kingdom includes *Amoeba, Paramecium*, and many others. Members of the kingdom **Plantae** are eukaryotic, multicellular, and photosynthetic. Plants are characterized by walled cells and are usually nonmotile. Members of the kingdom **Fungi** are also eukaryotic and multicellular. Like plants, they have walled cells and are usually nonmotile. Fungi are distinguished from plants by mode of nutrition. They digest organic matter extracellularly and absorb the breakdown products. Members of the kingdom **Animalia** are eukaryotic, multicellular, and feed by ingesting other organisms or parts of other organisms. Their cells lack walls and they are usually motile.

In recent years, the five-kingdom classification system has been challenged by new information regarding the evolutionary relationships among the monerans and protists. Much of this information comes from molecular studies of ribosomal RNA from bacteria and protists. One system of classification that has the support of many microbiologists divides organisms into three domains (a taxonomic level above the kingdom level). All eukaryotic organisms are members of one domain and the eubacteria and archaebacteria make up the two other domains.

This textbook is primarily devoted to the animals. The next chapter, however, covers the animallike protists (protozoa). Its inclusion is a part of a tradition that originated with an old two-kingdom classification system. (2) Animallike protists (e.g., *Amoeba, Paramecium*) were once considered a phylum (Protozoa) in the animal kingdom, and general zoology courses usually include them.

Animal Systematics

The goal of animal systematics is to arrange animals into groups that reflect evolutionary relationships. Ideally, these groups should include a single ancestral species and all of its descendants. Such a group is called a **monophyletic group.** In searching out monophyletic groups, taxonomists look for attributes of animals, referred to as characters, that indicate relatedness. A character is virtually anything that has a genetic basis and that can be measured—from an anatomical feature to a sequence of nitrogenous bases in DNA or RNA. **Polyphyletic groups** that sometimes result from systematic studies reflect insufficient knowledge regarding the particular group because it is impossible for a species to evolve through two separate evolutionary pathways.

As in any human endeavor, disagreements have arisen in animal systematics. These disagreements revolve around methods of investigation and whether or not data may be used in describing distant evolutionary relationships. Three contemporary schools of systematics exist: evolutionary systematics, numerical taxonomy, and phylogenetic systematics (cladistics).

Evolutionary systematics is the oldest of the three approaches to systematics. It is sometimes called the "traditional approach," although it has certainly changed since the beginnings of animal systematics. A basic assumption of evolutionary systematists is that organisms closely related to an ancestor will resemble that ancestor more closely than they resemble distantly related organisms. Two kinds of similarities between organisms are recognized (*see also the discussion of homology and analogy in chapter 1*). Homologies are resemblances that result from common ancestry and are useful in classifying animals. An example is the similarity in the arrangement of bones in the wing of a bird and the arm of a human (*see figure 1.7*). Analogies are resemblances that result from organisms adapting under similar evolutionary pressures. The latter process is sometimes called convergent evolution. Analogies do not reflect common ancestry and are not used in animal taxonomy. The similarity between the wings of birds and insects is an analogy.

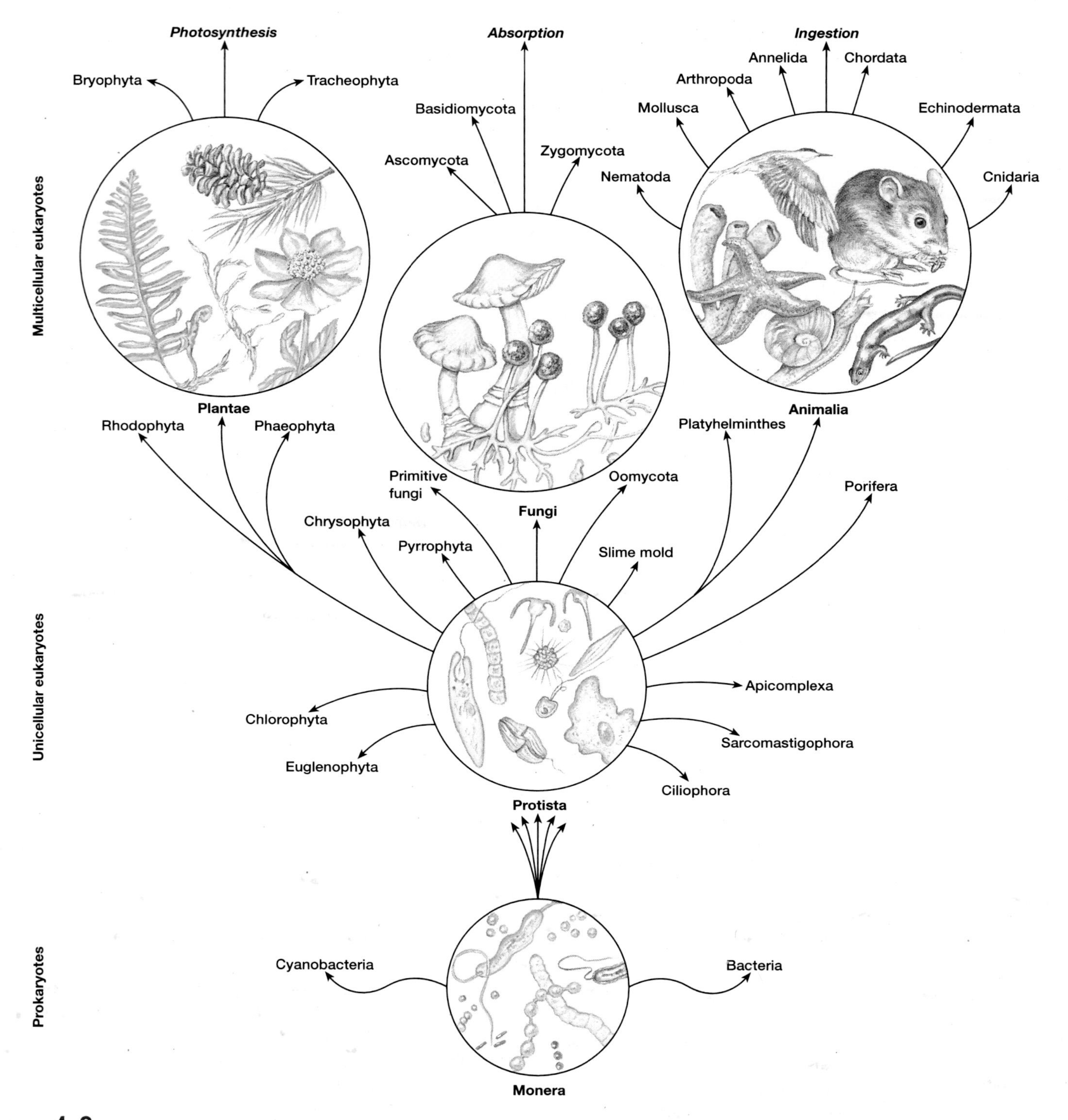

Figure 4.2
The Kingdoms of Life. In 1969 Robert H. Whittaker described a five-kingdom classification system that is widely used today.

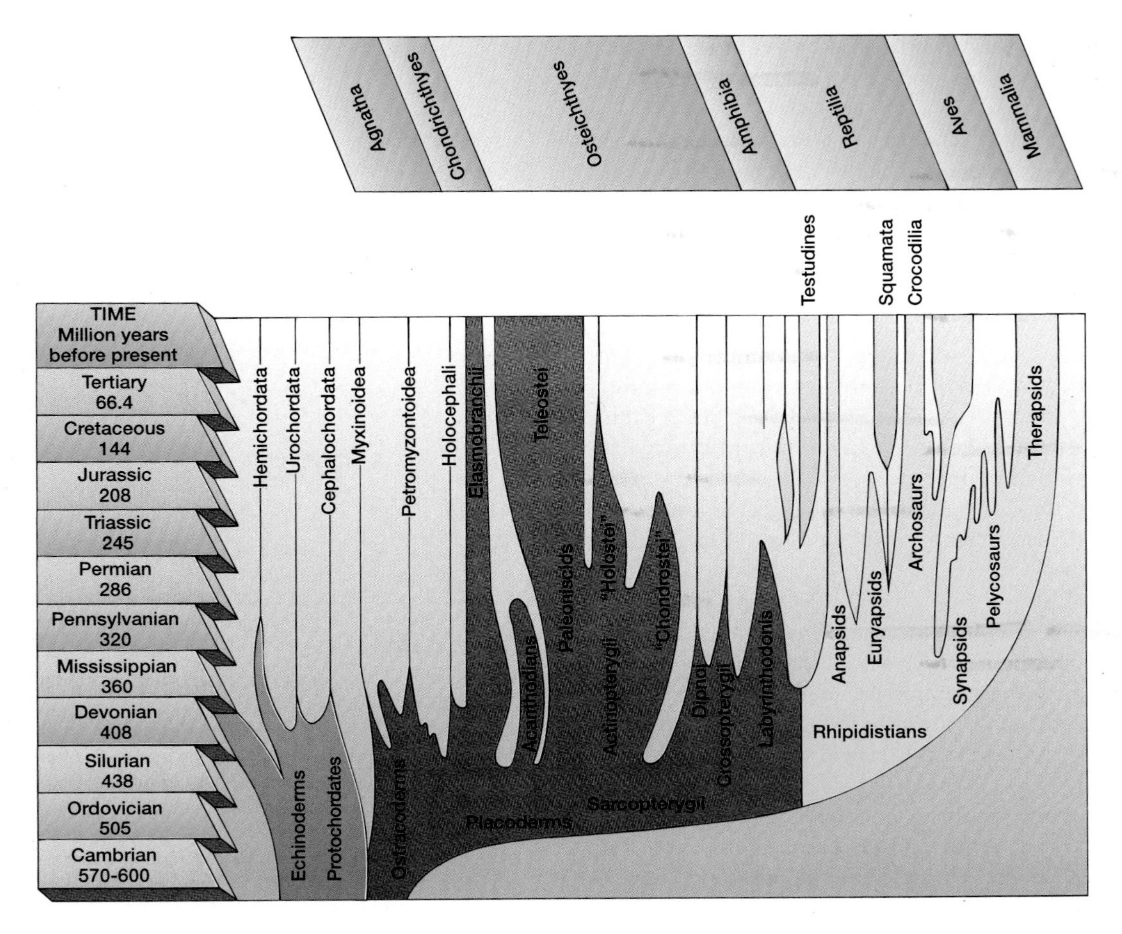

FIGURE 4.3

A Phylogenetic Tree Showing Vertebrate Phylogeny. A phylogenetic tree derived from evolutionary systematics depicts the degree of divergence since branching from a common ancestor, which is indicated by the time periods on the vertical axis. The width of the branches indicates the number of recognized genera for a given time period. Note that this diagram shows the birds (Aves) as being closely related to the reptiles (Reptilia), and both groups as having class-level status.

Evolutionary systematists often portray the results of their work on phylogenetic trees, where organisms are ranked according to their evolutionary relationships. Figure 4.3 is a phylogenetic tree showing vertebrate evolutionary relationships. In addition to depicting ancestry, time scales and relative abundance of animal groups are shown. These diagrams reflect judgments made about rates of evolution and the relative importance of certain key characters (e.g., feathers in birds).

Numerical taxonomy emerged during the 1950s and 1960s and represents the opposite end of the spectrum from evolutionary systematics. The founders of numerical taxonomy believed that the criteria for grouping taxa had become too arbitrary and vague. They tried to make taxonomy more objective. Numerical taxonomists use mathematical models and computer-aided techniques to group samples of organisms according to overall similarity. There is no attempt to distinguish between homologies and analogies. Numerical taxonomists admit that analogies exist. They contend, however, that it is sometimes impossible to tell one from the other and analogies will be overshadowed by the numerous homologies used in data analysis. A second major difference between evolutionary systematics and numerical taxonomy is that numerical taxonomists limit discussion of evolutionary relationships to closely related taxa. Numerical taxonomy is the least popular of the three taxonomic schools; however, computer programs developed for use by numerical taxonomists are used by all taxonomists.

Phylogenetic systematics (cladistics) is a third approach to animal systematics. The goal of cladistics is similar to that described for evolutionary systematics—the generation of hypotheses of genealogical relationships among monophyletic groups of organisms. Cladists contend, however, that their methods are more scientific than those of evolutionary systematists because cladists' methods are more open to analysis and testing.

As do evolutionary systematists, cladists differentiate between homologies and analogies. Cladists contend, however, that homologies that are of recent origin are most useful in phylogenetic studies. Characters that are shared by all members of a group are referred to as **symplesiomorphies** (Gr. *sym*, together + *plesio*, near + *morphe*, form). These characters are homologies that may indicate a shared ancestry, but they are useless in describing relationships within the group. **Synapomorphies** (Gr. *syn*, together + *apo*, away + *morphe*, form), on the other hand, are characters that have arisen within the group since it diverged from a common ancestor. Synapomorphies are also called shared, derived characters and are more useful in cladistic analysis because they give information regarding degrees of relatedness. The identification of synapomorphies is a fundamental task for cladists.

In order to illustrate the difference between shared, ancestral characteristics (symplesiomorphies) and shared, derived characteristics (synapomorphies), we can consider the presence of hair and mammary glands in members of the class Mammalia. These characters are present in all mammals but are not in any other animal group. Hair and mammary glands indicate a common ancestry, but they cannot be used to distinguish between different groups of mammals because they are present in all mammals. On the other hand, a characteristic like teeth arrangement can be used as a synapomorphy because that characteristic is variable within the class. It is important to realize that a synapomorphy at one level of taxonomy may be a symplesiomorphy at a lower taxonomic level. Within the mammals, for example, all members of the class Rodentia (the rodents) have a characteristic arrangement of teeth. The arrangement of teeth, therefore, cannot be used to distinguish different rodents and is a common, ancestral character (a symplesiomorphy) for members of the order. Figure 4.4 illustrates the concepts of symplesiomorphy and synapomorphy in a hypothetical lineage.

Cladograms are evolutionary diagrams that depict a sequence in the origin of unique, derived characteristics. Cladograms are interpreted as a family tree depicting a hypothesis regarding monophyletic lineages. New data in the form of newly investigated characters, or reinterpretation of old data, are used to test the hypothesis described by the cladogram. Figure 4.5 is a cladogram depicting the evolutionary relationships among the reptiles, birds, and mammals.

Cladistics has become very widely accepted among zoologists. This acceptance has resulted in some nontraditional interpretations of animal phylogeny. One example of different interpretations derived through evolutionary systematics and cladistics can be seen by comparing figures 4.3 and 4.5. The birds have been assigned class-level status (Aves) by generations of taxonomists. Reptiles also have had class-level status (Reptilia). Cladistic analysis has shown, however, that birds are more closely tied by common ancestry to the alligators and crocodiles than to any other group. According to this interpretation, birds and crocodiles would be assigned by cladists to a group that reflects this close common ancestry. Birds would become a subgroup within a larger group that included both birds and reptiles. Crocodiles would be depicted more closely related to the birds than they would be to snakes and lizards. Traditional evolutionary systematists maintain that the traditional interpretation is still correct because it takes into account the greater importance of key characteristics of birds (e.g., feathers and endothermy) that make the group unique. Cladists support their position by making the point that the designation of "key characteristics" involves value judgments that cannot be tested.

As debates between cladists and evolutionary systematists continue, our knowledge of evolutionary relationships among animals will become more complete. Debates like these are the fuel that force scientists to examine and reexamine old assumptions. Animal systematics is certain to be a lively and exciting field in future years.

The chapters that follow are a survey of the animal kingdom. The organization of these chapters reflects the traditional taxonomy that makes most zoologists comfortable. Cladograms

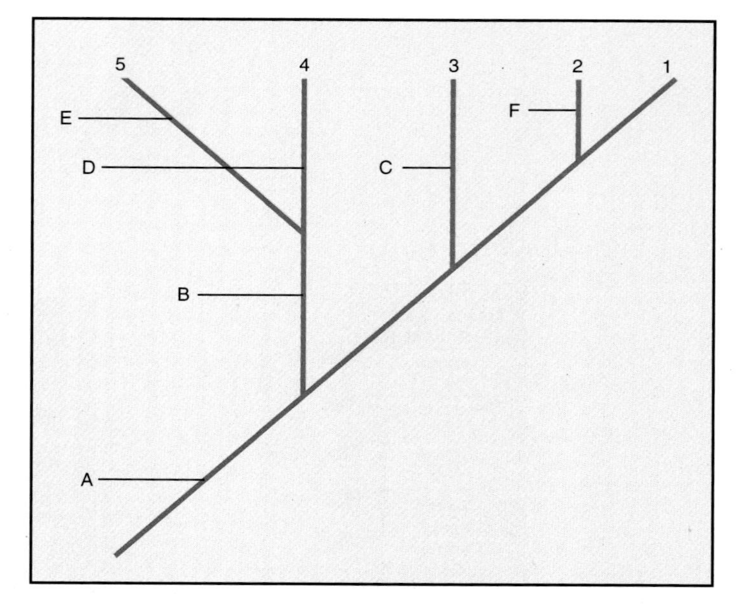

FIGURE 4.4

Interpreting Cladograms. This hypothetical cladogram shows five taxa (1–5) and the characters (A–F) used in deriving the taxonomic relationships. Character A is a symplesiomorphy (shared, ancestral characteristic) because it is shared by members of all five taxa. Because it is present in all taxa, character A cannot be used to distinguish members of this monophyletic lineage from each other. Character B is a synapomorphy (derived, ancestral character) because it is present in taxa 4 and 5 and can be used to distinguish these taxa from 1–3. Character B, however, is on the common branch giving rise to taxa 4 and 5. Character B is, therefore, symplesiomorphic for those two taxa. Characters D and E can be used to distinguish members of taxa 4 and 5.

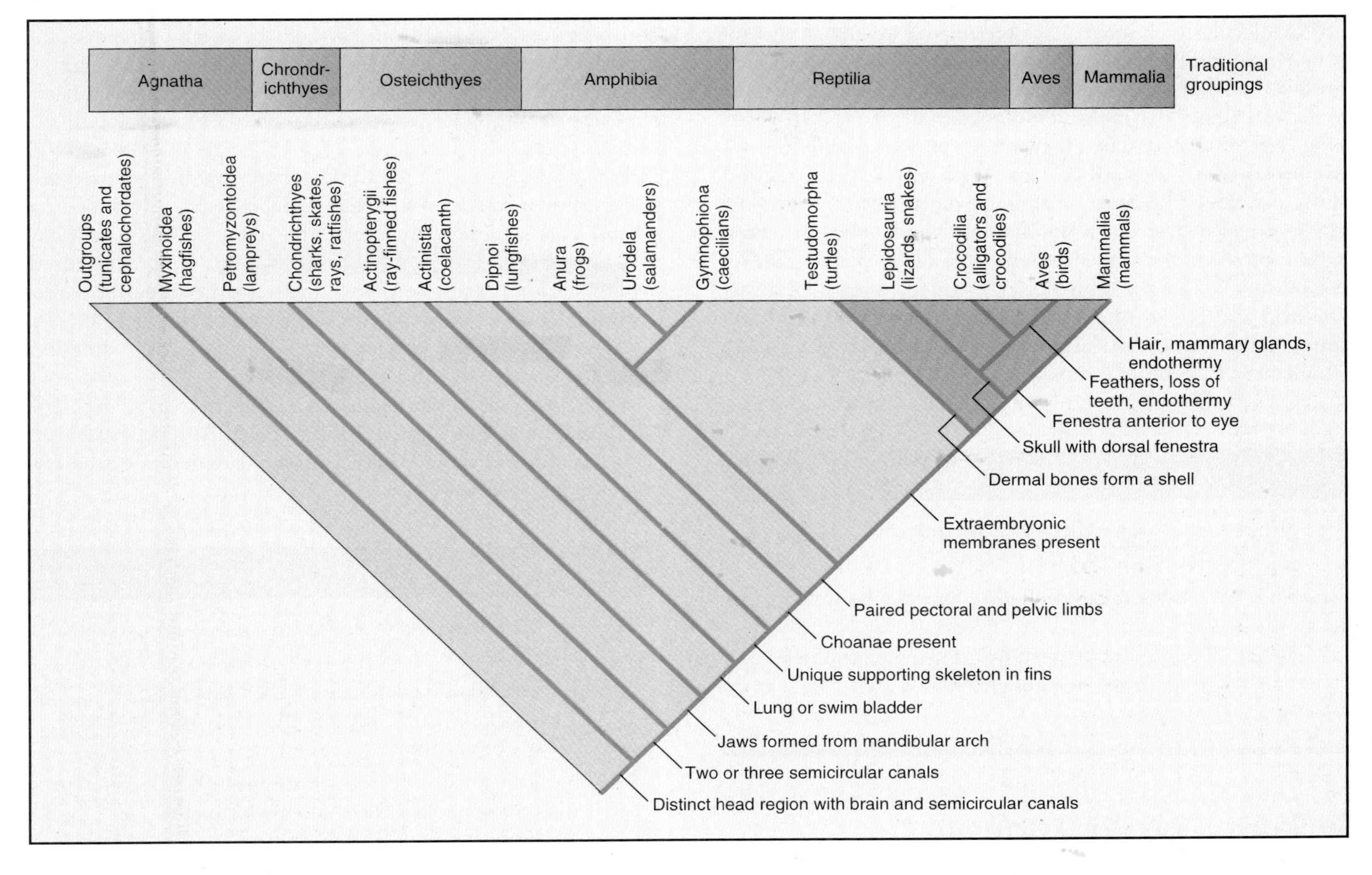

Figure 4.5

A Cladogram Showing Vertebrate Phylogeny. A cladogram is constructed by identifying points at which two groups diverged. Animals that share a branching point are included in the same taxon. Notice that timescales are not given or implied. The relative abundance of taxa is also not shown. Notice that this diagram shows the birds and crocodilians sharing a common branch, and that these two groups are more closely related to each other than either is to any other group of animals.

are usually included in the "Further Phylogenetic Considerations" at the end of most chapters, and any different interpretations of animal phylogeny implicit in these cladograms are discussed.

Molecular Approaches to Animal Systematics

In recent years, molecular biological techniques have provided important information for taxonomic studies. The relatedness of animals is reflected in the gene products (proteins) animals produce and in the genes themselves (the sequence of nitrogenous bases in DNA). Related animals have DNA derived from a common ancestor. Genes and proteins of related animals, therefore, are more similar than genes and proteins from distantly related animals. By comparing the sequence of amino acids in proteins, or the sequence of nitrogenous bases in DNA or RNA, and assuming a relatively constant mutation rate (referred to as a molecular clock), taxonomists can estimate the time elapsed since divergence from a common ancestor. Sequencing the nuclear DNA and the mitochondrial DNA of animals has become commonplace. Mitochondrial DNA is useful in taxonomic studies because mitochondria have their own genetic systems and are inherited cytoplasmically. That is, mitochondria are transmitted from parent to offspring through the egg cytoplasm and can be used to trace maternal lineages. Using mitochondrial DNA involves relatively small quantities of DNA that change at a relatively constant rate. The sequencing of ribosomal RNA has been used extensively in studying taxonomic relationships within the protists. Ribosomal RNA is ideal because it makes up an essential organelle (the ribosome) in all organisms, and ribosomal RNA's structure changes slowly with time.

Molecular techniques have provided a wealth of new information useful to animal taxonomists. These techniques, however, are not a panacea that will replace traditional taxonomic methods. The molecular clocks that are used to determine rates of evolutionary change have provided important information that helps fill in time gaps in the fossil record. Molecular clocks, however, apparently run at different rates depending on whether one is looking at the sequence of amino acids in proteins, the sequence of bases in DNA from organelles like mitochondria, the sequence of bases in nuclear DNA, or data from different evolutionary lineages. Molecular and traditional methods of investigation will probably always be used to complement each other in taxonomic studies.

Stop and Ask Yourself

1. What does it mean when one says that our classification system is hierarchical?
2. Why are common names inadequate for scientific purposes?
3. What is a synapomorphy? Why are synapomorphies essential for the work of cladists?
4. What is a cladogram?

Evolutionary Relationships and Tree Diagrams

Although evolutionary tree diagrams can help one appreciate evolutionary relationships and timescales, they are often a source of misunderstanding. Evolutionary tree diagrams often illustrate relationships among levels of classification above the species (*see figure 4.5*). 3 Depicting phyla or classes as ancestral is misleading because evolution occurs in species groups (populations), not at higher taxonomic levels. Also, when phyla or classes are depicted as ancestral, one should remember that modern representatives of these "ancestral phyla" have had just as long an evolutionary history as animals in other taxonomic groups that may have descended from the common ancestor. All modern representatives of any group of animals should be visualized at the tips of a "tree branch," and they may be very different from ancestral species. We use modern representatives to help visualize general characteristics of an ancestral species, but never to specify details of the ancestor's structure, function, or ecology.

In addition to these problems of interpretation, evolutionary trees often imply a ladderlike progression of increasing complexity. This is misleading because evolution has often resulted in reduced complexity and body forms that are evolutionary failures. In many cases, evolution has led to extinction, not evolutionary progression. Further, the common representation of a phylogeny as an inverted cone, or a tree with a narrow trunk and many higher branches, is often misleading. This implies that evolution is a continuous process of increasing diversification. The fossil records show that this is often wrong. There are, for example, 20 to 30 groups of echinoderms (sea stars and their relatives) in the fossil record and only 5 modern groups. This evolutionary lineage, like many others, underwent very rapid initial evolutionary diversification. After the initial diversification, extinction—not further diversification—was the rule. Contemporary paleontologist Stephen J. Gould uses the term contingency to refer to rapid evolutionary explosion followed by a high likelihood of extinction.

In spite of these problems, tree diagrams persist in scientific literature and are used in this textbook. As long as we keep their limitations in mind, they can help us visualize evolutionary relationships.

Stop and Ask Yourself

5. Why is it misleading to depict taxonomic groups higher than species on evolutionary tree diagrams?
6. Are the animals depicted along the trunk of an evolutionary tree diagram those that we see today? Why or why not?
7. What is meant by the term contingency? How can it be used to help us understand evolutionary relationships and tree diagrams?

Patterns of Organization

One of the most strikingly ordered series of changes in evolution is reflected in body plans in the animal kingdom and the protists. We can look at evolutionary changes in animal body plans and see what might be likened to a road map through a mountain range. What is most easily depicted are the starting and ending points and a few of the "attractions" along the route. What one cannot see from this perspective are the torturous curves and grades that must be navigated, and the extra miles that must be traveled, as one tries uncharted back roads. Unlike what we depict on a grand scale, evolutionary changes do not always mean "progress" and increased complexity. Evolution frequently results in backtracking, in experiments that fail, and in inefficient or useless structures. Evolution results in frequent dead ends, even though the route to that dead end may be filled with grandeur. The account that follows is a look at patterns of animal organization. As far as evolutionary pathways are concerned, one should view this account as an inexplicit road map through the animal kingdom. On a grand scale, it can be viewed as portraying evolutionary trends, but it should never be thought of as depicting an evolutionary sequence.

TABLE 4.2 ANIMAL SYMMETRY

TERM	MEANING
Asymmetry	The arrangement of body parts without a central axis or point (e.g., the sponges).
Bilateral symmetry	The arrangement of body parts such that a single plane passing dorsoventrally through the longitudinal axis divides the animal into right and left mirror images (e.g., the vertebrates).
Radial symmetry	The arrangement of body parts such that any plane passing through the oral-aboral axis divides the animal into mirror images (e.g., the cnidarians). Radial symmetry can be modified by the arrangement of some structures in pairs, or other combinations, around the central axis (e.g., biradial symmetry in the ctenophorans and some anthozoans, and pentaradial symmetry in the echinoderms).

SYMMETRY

The bodies of animals and protists are organized into almost infinitely diverse forms. Within this diversity, however, certain patterns of organization can be described. The concept of symmetry is fundamental to understanding animal organization. **Symmetry** describes how the parts of an animal are arranged around a point or an axis (table 4.2).

Asymmetry, which is the absence of a central point or axis around which body parts are equally distributed, is characteristic of most protists and many sponges (figure 4.6). Asymmetry cannot be said to be an adaptation to anything or advantageous to an organism. Asymmetrical organisms do not develop complex communication, sensory, or locomotor functions. It is clear, however, that protists and animals whose bodies consist of aggregates of cells have flourished.

A sea anemone can move along a substrate, but only very slowly (figure 4.7). How is it to gather food? How does it detect and protect itself from predators? For this animal, a blind side would leave it vulnerable to attack and cause it to miss many meals. The sea anemone, as is the case for most sedentary animals, has sensory and feeding structures uniformly distributed around its body. (4) Sea anemones do not have distinct head and tail ends. Instead, one point of reference is the end of the animal that possesses the mouth (the oral end), and a second point of reference is the end opposite the mouth (the aboral end). Animals such as the sea anemone are said to be radially symmetrical. **Radial symmetry** is the arrangement of body parts such that any plane passing through the central oral-aboral axis divides the animal into mirror images. Radial symmetry is often modified by the arrangement of some structures in pairs, or in other combinations, around the central oral-aboral axis. The paired arrangement of some structures in radially symmetrical animals is referred to as biradial symmetry. The arrangement of structures in fives around a radial animal is referred to as pentaradial symmetry.

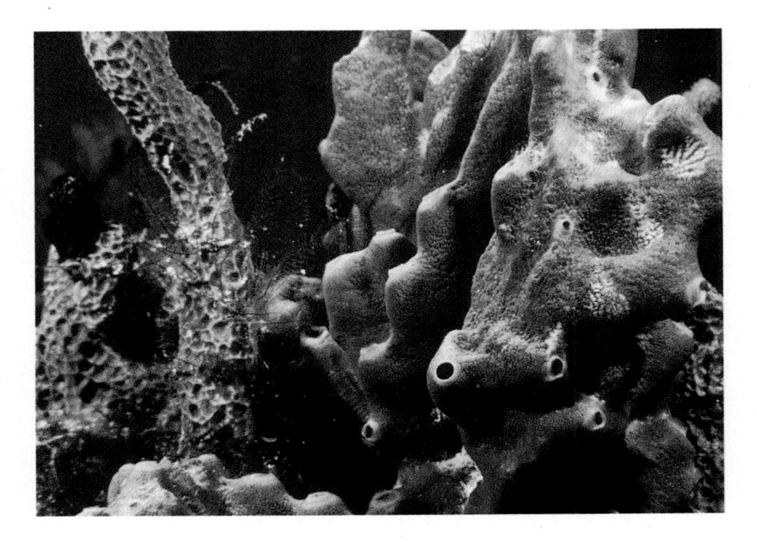

FIGURE 4.6
Asymmetry. Sponges display a cell-aggregate organization, and as seen in this brown volcano sponge (*Hemectyon ferox*), many are asymmetrical.

FIGURE 4.7
Radial Symmetry. Radially symmetrical animals, such as this tube coral polyp (*Tubastraea*), can be divided into equal halves by planes that pass through the oral/aboral axis. Sea anemones have their radial symmetry modified by certain arrangements of internal structures.

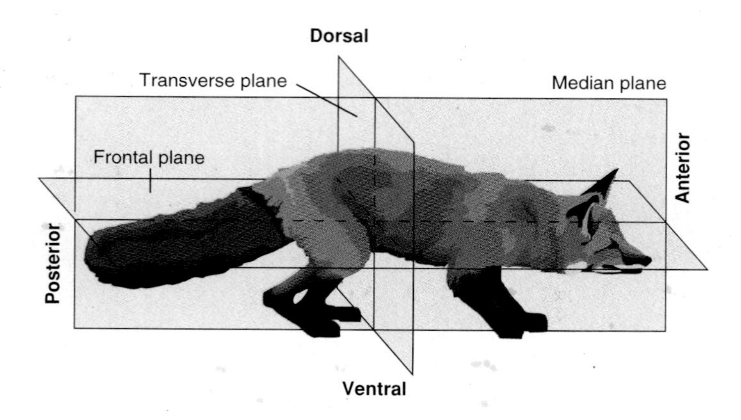

FIGURE **4.8**

Bilateral Symmetry. Planes and terms of direction that are useful in locating parts of a bilateral animal are indicated. A bilaterally symmetrical animal, such as this fox, has only one plane of symmetry. An imaginary median plane is the only plane through which the animal could be cut to yield mirror-image halves.

Although the sensory, feeding, and locomotor structures found in radially symmetrical animals could never be called "simple," one never sees structures comparable to the complex sensory, locomotor, and feeding structures found in many other animals. The evolution of such structures in radially symmetrical animals would require repeated distribution of very specialized structures around the animal.

Bilateral symmetry is the arrangement of body parts such that there is only a single plane, passing dorsoventrally through the longitudinal axis of an animal, that divides the animal into right and left mirror images (figure 4.8). Bilateral symmetry is characteristic of active, crawling, or swimming animals. Because bilateral animals move primarily in one direction, one end of the animal is continually encountering the environment. The end that meets the environment is usually where complex sensory, nervous, and feeding structures evolve and develop. These developments result in the formation of a distinct head, and are called **cephalization** (Gr. *kephale*, head). The anterior end of an animal is where cephalization occurs. Posterior is opposite anterior; it is the animal's tail end. Other important terms of direction and terms describing body planes and sections are applied to bilateral animals. These terms are used for locating body parts relative to a point of reference or an imaginary plane passing through the body (tables 4.2 and 4.3; figure 4.8).

Other Patterns of Organization

In addition to body symmetry, there are other recognizable patterns of animal organization. In a broad context, these patterns of organization may reflect evolutionary trends. As explained earlier, however, one should not view these trends as exact sequences in animal evolution.

TABLE 4.3 Terms of Direction

Term	Description
Aboral	The end opposite the mouth
Oral	The end containing the mouth
Anterior	The head end; usually the end of a bilateral animal that meets its environment
Posterior	The tail end
Caudal	Toward the tail
Cephalic	Toward the head
Distal	Away from the point of attachment of a structure on the body (e.g., the toes are distal to the knee)
Proximal	Toward the point of attachment of a structure on the body (e.g., the hip is proximal to the knee)
Dorsal	The back of an animal; usually the upper surface; synonymous with posterior for animals that walk upright
Ventral	The belly of an animal; usually the lower surface; synonymous with anterior for animals that walk upright
Inferior	Below a point of reference (e.g., the mouth is inferior to the nose in humans)
Superior	Above a point of reference (e.g., the neck is superior to the chest)
Lateral	Away from the plane that divides a bilateral animal into mirror images
Medial (median)	On or near the plane that divides a bilateral animal into mirror images

The Unicellular (Cytoplasmic) Level of Organization

Organisms whose bodies consist of single cells or cellular aggregates display the unicellular level of organization. Unicellular body plans are characteristic of the Protista. Some zoologists prefer to use the designation "cytoplasmic" to emphasize the fact that all living functions are carried out within the confines of a single plasma membrane. It is a mistake to consider unicellular organization "simple." All unicellular organisms must provide for the functions of locomotion, food acquisition, digestion, water and ion regulation, sensory perception, and reproduction in a single cell.

Cellular aggregates (colonies) consist of loose associations of cells in which there is little interdependence, cooperation, or coordination of function—therefore, cellular aggregates cannot be considered tissues. In spite of the absence of interdependence, some division of labor is found in these organisms. Some cells may be specialized for reproductive, nutritive, or structural functions.

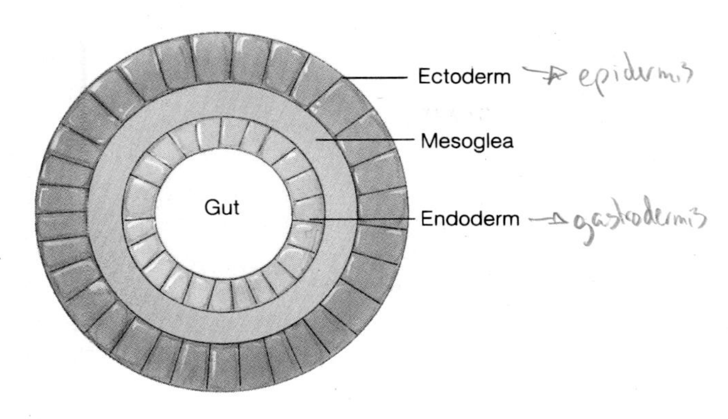

FIGURE 4.9

The Diploblastic Body Plan. Diploblastic animals have tissues derived from ectoderm and endoderm. Between these two layers is a noncellular mesoglea.

Diploblastic Organization

Cells are organized into tissues in most animal phyla. **Diploblastic** (Gr. *dis*, twice + *blaste*, to sprout) organization is the simplest, tissue-level organization (figure 4.9). Body parts are organized into layers that are derived from two tissue layers of the embryo. Ectoderm gives rise to the epidermis, the outer layer of the body wall. Endoderm gives rise to the gastrodermis, the tissue that lines the gut cavity. Between the epidermis and the gastrodermis is a noncellular layer called mesoglea. In some diploblastic organisms, cells occur in the mesoglea, but they are always derived from ectoderm or endoderm.

The cells in each tissue layer are functionally interdependent. The gastrodermis consists of nutritive (digestive) and muscular cells, and the epidermis contains epithelial and muscular cells. You may already be familiar with the feeding movements of *Hydra* or the swimming movements of a jellyfish. These kinds of functions are only possible when groups of cells cooperate, showing tissue-level organization.

Triploblastic Organization

Animals described in chapters 7 to 19 are **triploblastic** (Gr. *treis*, three + *blaste*, to sprout). That is, their tissues are derived from three embryological layers. As with diploblastic animals, ectoderm forms the outer layer of the body wall, and endoderm lines the gut. In addition to these two layers, a third embryological layer is sandwiched between the ectoderm and endoderm. This layer is mesoderm, which gives rise to supportive, contractile, and blood cells. Most triploblastic animals have an organ system level of organization. Tissues are organized together to form excretory, nervous, digestive, reproductive, circulatory, and other systems. Triploblastic animals are usually bilaterally symmetrical (or have evolved from bilateral ancestors) and are relatively active animals.

Triploblastic animals are organized into several subgroups based on the presence or absence of a body cavity and, for those that possess one, the kind of body cavity present. A body cavity is a fluid-filled cavity in which the internal organs can be suspended and separated from the body wall. (5) Body cavities are advantageous because they

1. Provide more room for organ development.
2. Provide more surface area for diffusion of gases, nutrients, and wastes into and out of organs.
3. Provide an area for storage.
4. Often act as hydrostatic skeletons.
5. Provide a vehicle for eliminating wastes and reproductive products from the body.
6. Facilitate increased body size.

Of these, the hydrostatic skeleton deserves further comment. Body cavity fluids give support while allowing the body to remain flexible. Hydrostatic skeletons can be illustrated with a water-filled balloon, which is rigid yet flexible. Because the water in the balloon is incompressible, squeezing one end causes the balloon to lengthen. Compressing both ends causes the middle of the balloon to become fatter. In a similar fashion, body-wall muscles, acting on coelomic fluid, are responsible for movement and shape changes in many animals.

The Triploblastic Acoelomate Pattern

Triploblastic animals whose mesodermally derived tissues form a relatively solid mass of cells between ectodermally and endodermally derived tissues are referred to as being **acoelomate** (Gr. *a*, without + *kilos*, hollow) (figure 4.10*a*). Some cells between the ectoderm and endoderm of acoelomate animals are loosely organized cells called parenchyma. Parenchymal cells are not specialized for a particular function.

The Triploblastic Pseudocoelomate Pattern

A **pseudocoelom** (Gr. *pseudes*, false) is a body cavity not entirely lined by mesoderm (figure 4.10*b*). No muscular or connective tissues are associated with the gut tract, no mesodermal sheet covers the inner surface of the body wall, and no membranes suspend organs in the body cavity. Embryologically, the pseudocoelom is derived from the blastocoel of the embryo.

The Triploblastic Coelomate Pattern

A **coelom** is a body cavity that is completely surrounded by mesoderm (figure 4.10*c*). The inner body wall is lined by a thin mesodermal sheet, the peritoneum, and visceral organs are

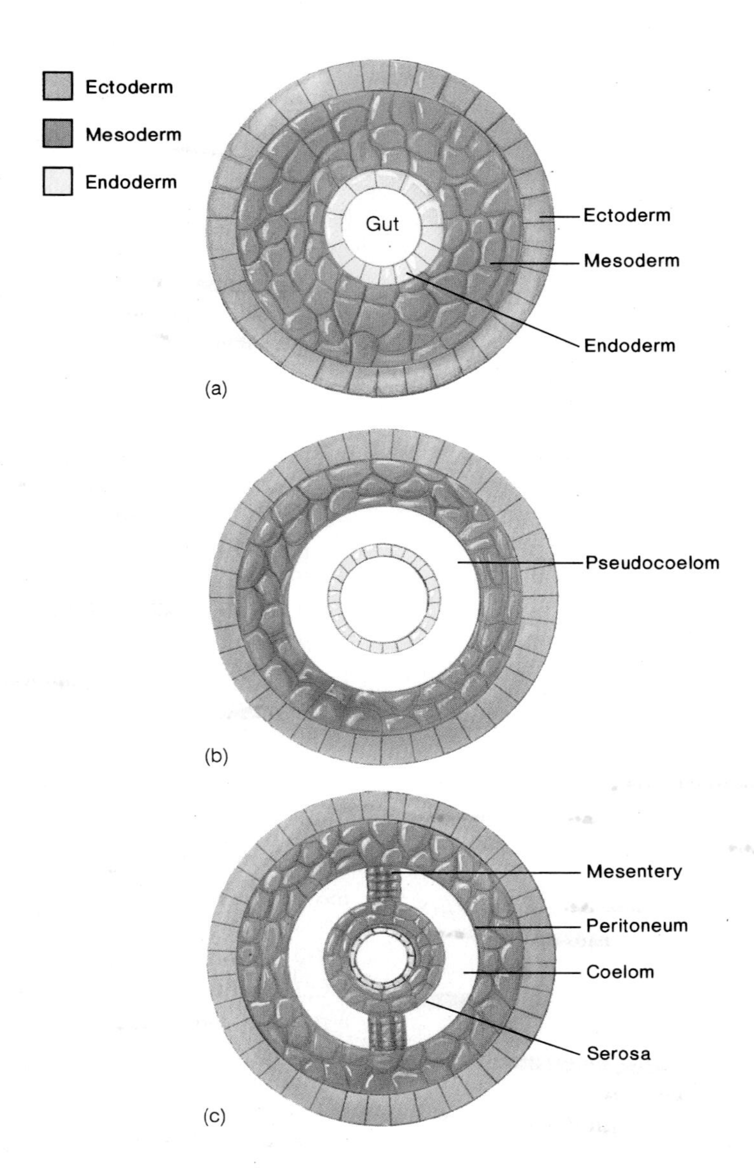

FIGURE 4.10

Triploblastic Body Plans. Triploblastic animals have tissues derived from ectoderm, mesoderm, and endoderm. (*a*) The triploblastic acoelomate pattern. (*b*) The triploblastic pseudocoelomate pattern. Note that there is no mesodermal lining on the gut track. (*c*) The triploblastic coelomate pattern. The coelom is completely surrounded by mesodermally derived tissues.

lined on the outside by a mesodermal sheet, the serosa. The peritoneum and the serosa are continuous, and suspend visceral structures in the body cavity. These suspending sheets are called mesenteries. Having mesodermally derived tissues, such as muscle and connective tissue, associated with internal organs enhances the function of virtually all internal body systems. In the chapters that follow, you will find many variations on the triploblastic coelomate pattern.

Stop and Ask Yourself

8. How would you distinguish between radial symmetry and bilateral symmetry? Why is radial symmetry advantageous for sedentary animals?
9. Why does cephalization usually accompany bilateral symmetry?
10. What kinds of tissues form from the mesoderm?
11. What advantages do body cavities confer to coelomate animals?

Summary

1. Systematics is the study of the evolutionary history and classification of organisms. The binomial system of classification originated with von Linné and is used throughout the world in classifying organisms.
2. Organisms are classified into very broad categories called kingdoms. The five-kingdom classification system used in recent years is being challenged as new information regarding evolutionary relationships among the monerans and protists is discovered.
3. There are three modern approaches to systematics. They are evolutionary systematics, numerical taxonomy, and phylogenetic systematics (cladistics). Systematists believe that the ultimate goal of systematics is to establish evolutionary relationships in monophyletic groups. Evolutionary systematists use homologies and rank the importance of different characteristics in establishing evolutionary relationships. These taxonomists take into consideration differing rates of evolution in taxonomic groups. Phylogenetic systematists (cladists) look for shared, derived characteristics that can be used to investigate evolutionary relationships. Cladists do not attempt to weigh the importance of different characteristics. Wide acceptance of cladistic methods has resulted in some nontraditional taxonomic groupings of animals.
4. Evolutionary tree diagrams are useful for depicting evolutionary relationships, but their limitations must be understood.
5. The bodies of animals are organized into almost infinitely diverse forms. Within this diversity, however, certain patterns of organization can be described. Symmetry describes how the parts of an animal are arranged around a point or an axis.
6. Other patterns of organization reflect how cells are associated together into tissues, and how tissues are organized into organs and organ systems.

Selected Key Terms

bilateral symmetry (*p.* 58)
class (*p.* 50)
coelom (*p.* 59)
family (*p.* 50)
genus (*p.* 50)
kingdom (*p.* 50)
order (*p.* 50)
phylum (*p.* 50)
radial symmetry (*p.* 57)
species (*p.* 50)

Critical Thinking Questions

1. In one sense, our classification system above the species level is artificial. In another sense, however, it is real. Explain this paradox.
2. Give proper scientific names to 10 hypothetical animal species. Assume that you have four different genera represented in your group of 10. Be sure your format for writing scientific names is correct.
3. Describe hypothetical synapomorphies that would result in an assemblage of one class, two orders, and three families (in addition to the four genera and 10 species).
4. Construct a cladogram, similar to that shown in figure 4.5, using your hypothetical animals from questions 2 and 3. Make drawings of your animals.
5. Describe the usefulness of evolutionary tree diagrams in zoological studies. Describe two problems associated with their use.

chapter 5

Animallike Protists: The Protozoa

Outline

Concepts

1. The kingdom Protista is a polyphyletic group with origins in ancestral members of the kingdom Monera.
2. Protozoa display unicellular or colonial organization. All functions must be carried out within the confines of a single plasma membrane.
3. According to the most widely accepted classification scheme, there are seven protozoan phyla: Sarcomastigophora, Labyrinthomorpha, Apicomplexa, Microspora, Acetospora, Myxozoa, and Ciliophora.
4. Certain members of the above phyla have had, and continue to have, important influences on human health and welfare.

Would You Like to Know:

1. whether or not all protozoan phyla can be traced back to a single moneran ancestor? (*p. 64*)
2. whether some protozoa should be considered single cells, multicellular organisms, or both? (*p. 65*)
3. what causes "red tides"? (*p. 68*)
4. what protozoan causes dysentery in humans? (*p. 72*)
5. what protozoan has caused more deaths in armies throughout history than actual combat? (*p. 74*)
6. why sandboxes should always be kept covered when not in use? (*p. 75*)
7. what group of protozoa has members that are considered the most complex of all protists? (*p. 76*)

These and other useful questions will be answered in this chapter.

This chapter contains evolutionary concepts, which are set off in this font.

Evolutionary Perspective

Where are your "roots"? Although most people are content to go back into their family tree a few hundred years, scientists look back millions of years to the origin of all life-forms. The fossil record indicates that virtually all protist and animal phyla living today were present during the Cambrian period, about 550 million years ago (*see table 1.3*). Unfortunately, there is little fossil evidence of the evolutionary pathways that gave rise to these phyla. Instead, evidence is gathered from examining the structure and function of living species. The "evolutionary perspective" in chapters 5 to 19 presents hypotheses regarding the origin of protist and animal phyla. These hypotheses seem reasonable to most zoologists; however, they are untestable, and alternative interpretations can be found in the scientific literature.

Ancient members of the kingdom Monera were the first living organisms on this planet. The Monera gave rise to the kingdom Protista (also called Protoctista) about 1.5 billion years ago. The endosymbiont hypothesis is one of a number of explanations of how this could have occurred. (1) Most scientists agree that the protists probably arose from more than one ancestral moneran group. Depending on the classification system used, between 7 and 45 phyla of protists are recognized today. These phyla represent numerous evolutionary lineages. When groups of organisms are believed to have had separate origins, they are said to be **polyphyletic** (Gr. *polys,* many + *phylon,* race). Some protists are plantlike because they are primarily autotrophic (they produce their own food). Others are animallike because they are primarily heterotrophic (they feed on other organisms). This chapter covers seven phyla of protists commonly called the protozoa (figures 5.1, 5.2).

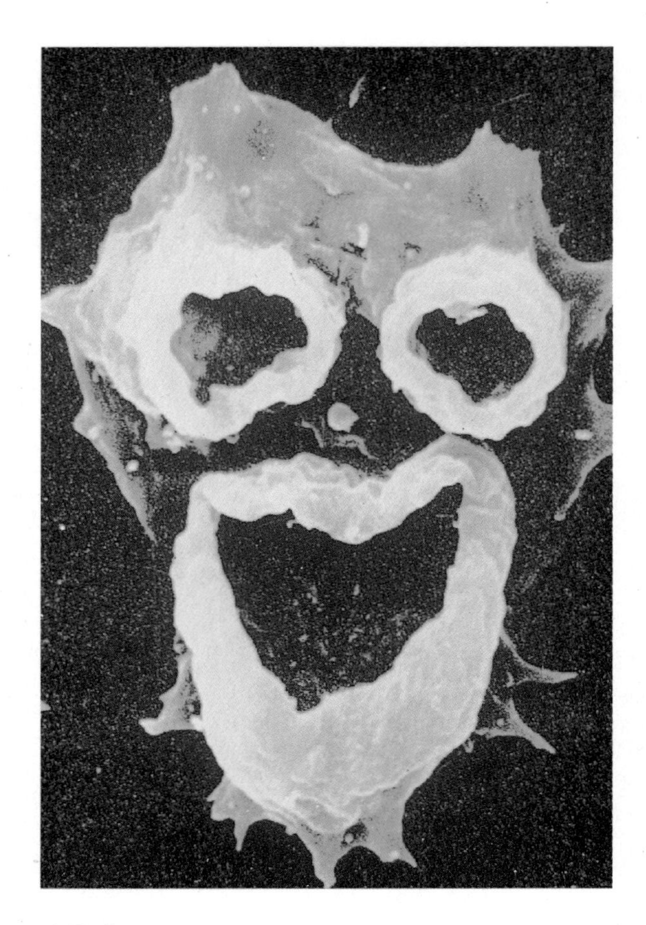

Figure 5.1
A Protozoan. Scanning electron micrograph of *Naegleria fowleri,* the cause of primary amebic meningoencephalitis in humans. The facelike structures are used for attack and engulfment of food sources (× 600).

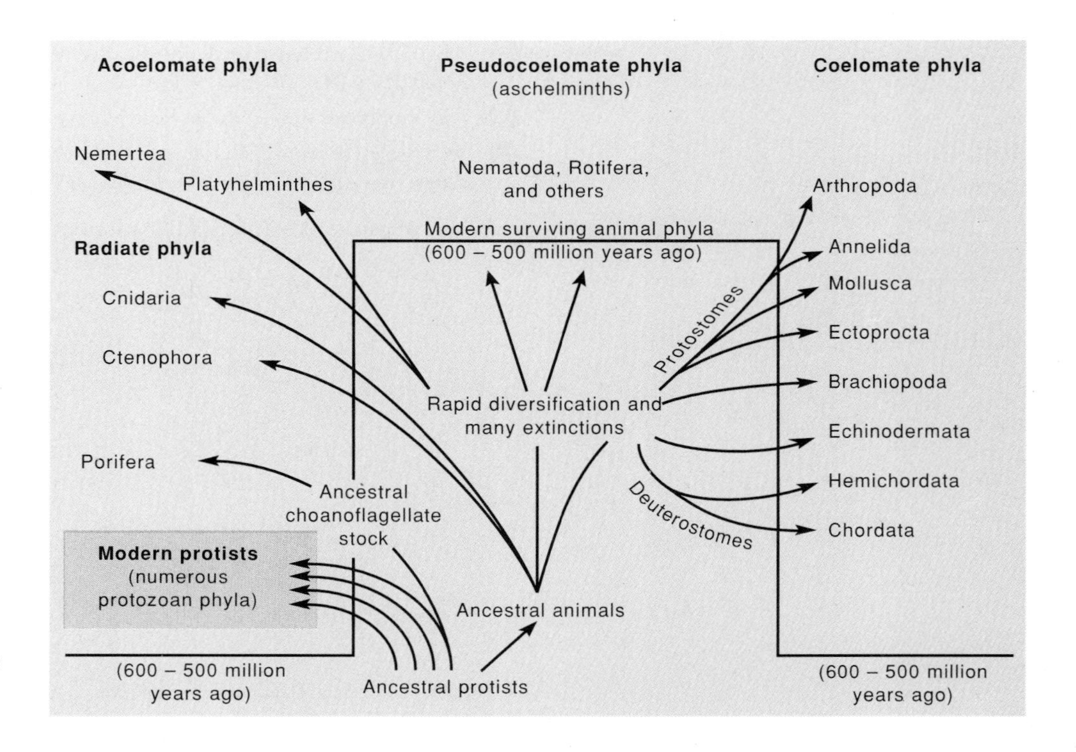

Figure 5.2
Animallike Protists: The Protozoa. A generalized evolutionary tree depicting the major events and possible lines of descent for the protozoa (shaded in orange).

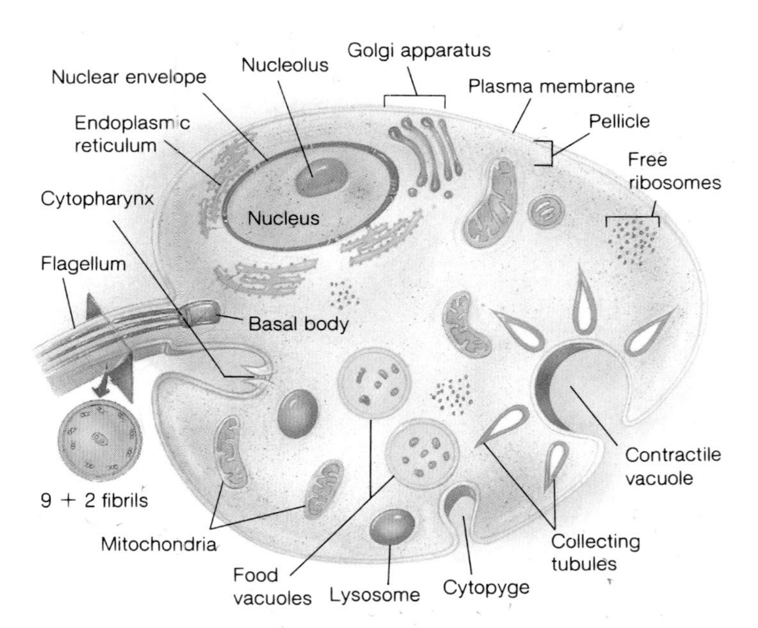

FIGURE 5.3

A Protozoan Protist. This drawing of a stylized protozoan with a flagellum illustrates the basic protozoan morphology. *From:* "A LIFE OF INVERTEBRATES" © *1979 W. D. Russell-Hunter.*

LIFE WITHIN A SINGLE PLASMA MEMBRANE

Protozoa (Gr. *proto,* first + *zoa,* animal) display unicellular (cytoplasmic) organization. Their single-celled organization does not necessarily imply that they are simple organisms. (2) Often, they are more complex than any particular cell in higher organisms. In some protozoan phyla, individuals may group together to form colonies, which are associations of individuals that are not dependent on one another for most functions. Protozoan colonies, however, can become very complex, with some individuals becoming so specialized that it becomes difficult to tell whether one is observing a colony or a multicellular organism.

MAINTAINING HOMEOSTASIS

Specific functions are carried out in protozoa by organelles that are similar to the organelles of other eukaryotic cells (figure 5.3). Some protozoan organelles, however, reflect specializations for unicellular life-styles.

The plasma membrane of many protozoa is underlaid by a regular arrangement of microtubules. Together, they are called the **pellicle.** The pellicle is rigid enough to maintain the shape of the protozoan, but it is also flexible.

The cytoplasm of a protozoan is differentiated into two regions. The portion of the cytoplasm just beneath the pellicle is called **ectoplasm** (Gr. *ectos,* outside + *plasma,* to form). It is relatively clear and firm. The inner cytoplasm is called **endoplasm** (Gr. *endon,* within). It is usually granular and more fluid. The conversion of cytoplasm between these two states is important in one kind of protozoan locomotion and is discussed later.

Most marine protozoa have solute concentrations similar to that of their environments. Freshwater protozoa, however, must regulate the water and solute concentrations of their cytoplasm. Water enters freshwater protozoa by osmosis because of higher solute concentrations in the protozoan than in the environment. This excess water is removed by **contractile vacuoles** (figure 5.3). In some protozoa, contractile vacuoles are formed by coalescence of smaller vacuoles. In others, the vacuoles are permanent organelles that are filled by collecting tubules that radiate into the cytoplasm. The contraction of microfilaments has been implicated in the emptying of contractile vacuoles.

Most protozoa ingest other organisms or products of other organisms—either by absorbing dissolved nutrients by active transport or by ingesting whole or particulate food through endocytosis. In some protozoa, food may be ingested in a specialized region analogous to a mouth, called the **cytopharynx.** Digestion and transport of food occurs in **food vacuoles** that form during endocytosis. Digestion is mediated by enzymes and acidity changes. Food vacuoles fuse with enzyme-containing lysosomes and circulate through the cytoplasm, distributing the products of digestion. After digestion is completed, the vacuoles are called **egestion vacuoles.** They release their contents by exocytosis, sometimes at a specialized region of the plasma membrane or pellicle called the **cytopyge.**

Because protozoa are small, they have a large surface area in proportion to their volume. This high surface area-to-volume ratio facilitates two other maintenance functions. Gas exchange involves acquiring quantities of oxygen needed for cellular respiration and eliminating the carbon dioxide that is produced as a by-product. Excretion is the elimination of the nitrogenous by-products of protein metabolism. The principal nitrogenous waste in protozoa is ammonia. Both gas exchange and excretion occur by diffusion across the plasma membrane.

REPRODUCTION

Both asexual and sexual reproduction occur among the protozoa. One of the simplest and most common forms of asexual reproduction is **binary fission.** In binary fission, mitosis produces two nuclei that are distributed into two similar-sized individuals when the cytoplasm divides. During cytokinesis, some organelles are also duplicated to ensure that each new protozoan will possess the needed organelles to begin life. Depending on the group of protozoa, cytokinesis may be longitudinal or transverse (figures 5.4 and 5.5).

Other forms of asexual reproduction are common. During **budding,** mitosis is followed by the incorporation of one nucleus into a cytoplasmic mass that is much smaller than the parent cell. **Multiple fission** or **schizogony** (Gr. *schizein,* to split) occurs when a large number of daughter cells are formed from

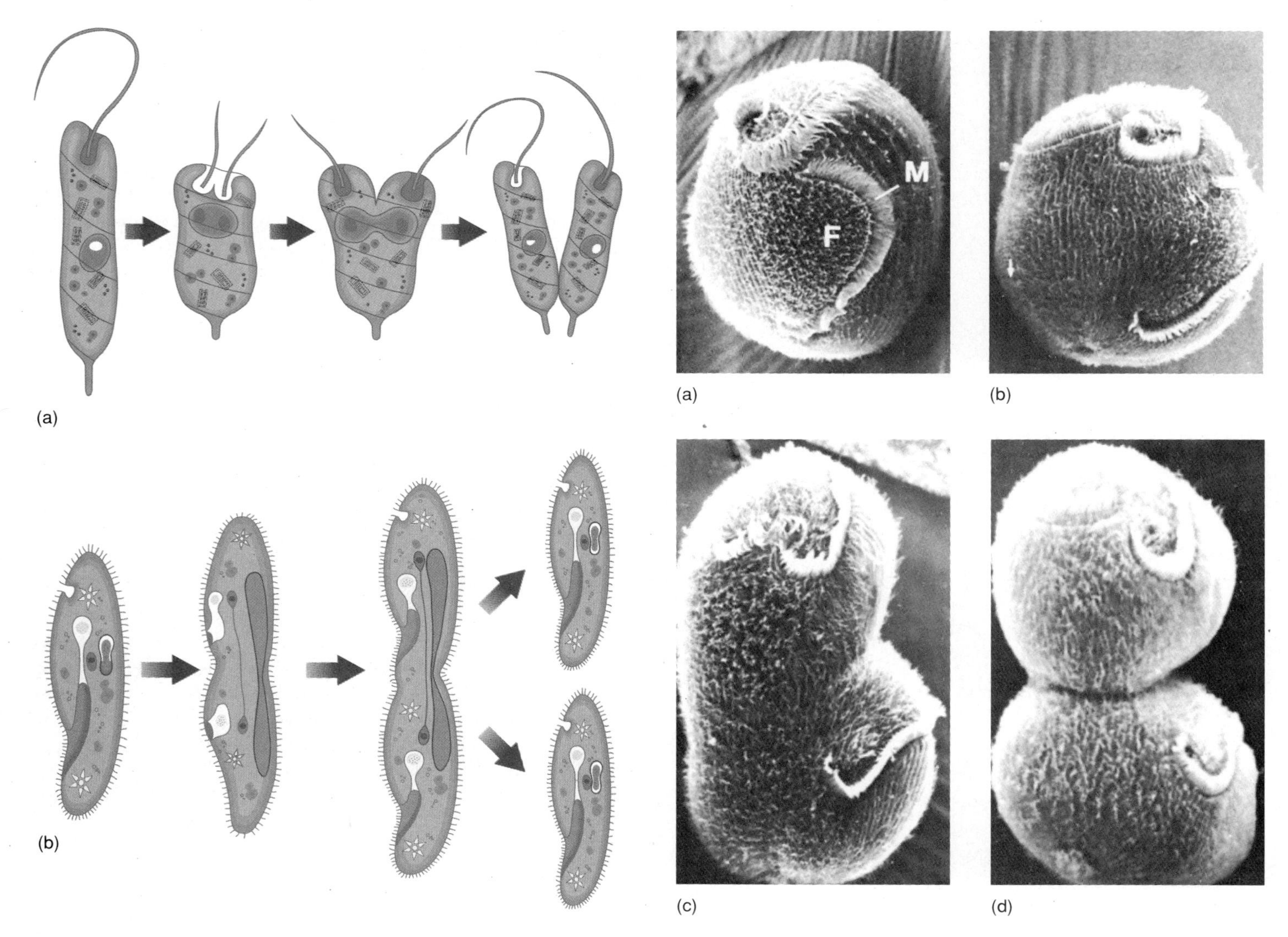

Figure 5.4
Asexual Reproduction in Protozoa. Binary fission begins with mitosis. Cytoplasmic division (cytokinesis) divides the organelles between the two cells and results in two similarly sized protozoa. Binary fission is (*a*) longitudinal in some protozoa (e.g., mastigophorans) and (*b*) transverse in other protozoa (e.g., ciliates).

Figure 5.5
Binary Fission of the Ciliate, *Stentor coeruleus*. Fission includes the division of some surface features (*a*,*b*), in this case cilia modified into a bandlike structure called a membranelle (M). F designates the frontal field, and the beginning of a fission furrow is shown by the arrow in (*b*). Fission is completed by division of the cytoplasm (*c*,*d*). (Scanning electron micrographs × 250.)

the division of a single protozoan. Schizogony begins with multiple mitotic divisions in a mature individual. When a certain number of nuclei have been produced, cytoplasmic division results in the separation of each nucleus into a new cell.

Sexual reproduction involves the formation of gametes and the subsequent fusion of gametes to form a zygote. In most protozoa, the sexually mature individual is haploid. Gametes are produced by mitosis, and meiosis occurs following union of gametes. Ciliated protozoa are an exception to this pattern. Specialized forms of sexual reproduction will be covered as individual protozoan groups are discussed.

Symbiotic Life-Styles

Symbiotic life-styles are important for many protozoa. **Symbiosis** (Gr. *syn*, with + *bios*, life) occurs when one organism lives in an intimate association with another. For many protozoa, these interactions involve a form of symbiosis called **parasitism,** in which one organism lives in or on a second organism, called a **host.** The host is harmed but usually survives, at least long enough for the parasite to complete one or more life cycles.

The relationships between a parasite and its host(s) are often complex. Some parasites have life histories involving multiple hosts. The **definitive host** is the host that harbors the sexual stages of the parasite. The sexual stages may produce offspring that enter another host. This second host is called an **intermediate host,** and asexual reproduction occurs here. More than one intermediate host and more than one immature stage may be involved in some life cycles. For the life cycle to be completed, the final, asexual stage must have access to a definitive host.

Other kinds of symbiosis involve relationships that do not harm the host. **Commensalism** is a symbiotic relationship in which one member of the relationship benefits, and the second member is neither benefited nor harmed. **Mutualism** is a symbiotic relationship in which both species benefit.

Protozoan Taxonomy

Zoologists who specialize in the study of protozoa are called **protozoologists.** Most protozoologists now regard the Protozoa as a subkingdom, consisting of seven separate phyla within the kingdom Protista. These are the phylum Sarcomastigophora, consisting of flagellates and amoebae with a single type of nucleus; the phyla Labyrinthomorpha, Apicomplexa, Microspora, Ascetospora, and Myxozoa, consisting of either saprozoic or parasitic species; and the phylum Ciliophora, containing ciliated protozoa with two types of nuclei. The classification of this subkingdom into phyla is based primarily on types of nuclei, mode of reproduction, and mechanism of locomotion (table 5.1). These taxa are further discussed in subsequent sections of this chapter.

Stop and Ask Yourself

1. Ancestral members of what kingdom of organisms gave rise to the protozoa?
2. What is the function of the following structures: pellicle? contractile vacuoles? cytopharynx? egestion vacuoles?
3. What are three forms of asexual reproduction in protozoa?
4. Why are gametes produced by mitosis rather than by meiosis in most protozoa?

Table 5.1 Summary of Protozoan Classification

Phylum Sarcomastigophora (sar′ko-mas-ti-gof′o-rah)
Protozoa that possess flagella, pseudopodia, or both for locomotion and feeding; single type of nucleus.

Subphylum Mastigophora (mas-ti-gof′o-rah)
One or more flagella used for locomotion; autotrophic, heterotrophic, or saprozoic.

Class Phytomastigophorea (fi′to-mas-ti-go-for′ah)
Chloroplasts usually present; mainly autotrophic, some heterotrophic. *Euglena*, *Volvox*, *Chlamydomonas*.

Class Zoomastigophorea (zo′o-mas-tigo-for′ah)
Lack chloroplasts; heterotrophic or saprozoic. *Trypanosoma*, *Trichonympha*, *Trichomonas*, *Giardia*, *Leishmania*.

Subphylum Sarcodina (sar′ko-din″ah)
Pseudopodia for movement and food gathering; naked or with shell or test; mostly free living.

Superclass Rhizopoda (ri-zop′o-dah)
Lobopodia, filopodia, reticulopodia, or no distinct pseudopodia. *Amoeba*, *Entamoeba*, *Naegleria*, *Arcella*, *Difflugia*; forminiferans (*Gumbelina*).

Superclass Actinopoda (ak′ti-nop″o-dah)
Spherical, planktonic; axopodia supported by microtubules; includes marine radiolarians with siliceous tests and freshwater heliozoans (*Actinophrys*).

Subphylum Opalinata (op′ah-li-not′ah)
Cylindrical; covered with cilia. *Opalina*, *Zelleriella*.

Phylum Labyrinthomorpha (la′brinth-o-morp′ha)
Trophic stage as ectoplasmic network with spindle-shaped or spherical, nonamoeboid cells; saprozoic and parasitic on algae and seagrass; mostly marine and estuarine. *Labyrinthula*.

Phylum Apicomplexa (a′pi-kom-plex′ah)
Parasitic with an apical complex used for penetrating host cells; cilia and flagella lacking, except in certain reproductive stages. The gregarines (*Monocystis*), coccidians (*Eimeria*, *Isospora*, *Sarcocystis*, *Toxoplasma*), *Plasmodium*.

Phylum Microspora (mi-cro-spor′ah)
Unicellular spores; intracellular parasites in nearly all major animal groups. The microsporeans (*Nosema*).

Phylum Acetospora (ah-seat-o-spor′ah)
Multicellular spore; all parasitic in invertebrates. The acetosporans (*Paramyxa*, *Haplosporidium*).

Phylum Myxozoa (myx-o-zoo-ah)
Spores of multicellular origin; all parasitic. The myxozoans (*Myxosoma*).

Phylum Ciliophora (sil-i-of′or-ah)
Protozoa with simple or compound cilia at some stage in the life history; heterotrophs with a well-developed cytostome and feeding organelles; at least one macronucleus and one micronucleus present. *Paramecium*, *Stentor*, *Euplotes*, *Vorticella*, *Balantidium*.

Phylum Sarcomastigophora

With over 18,000 described species, Sarcomastigophora (sar′ko-mas-ti-gof″o-rah) (Gr. *sarko*, fleshy + *mastigo*, whip + *phoros*, to bear) is the largest protozoan phylum. Characteristics of the phylum Sarcomastigophora include the following:

1. Unicellular or colonial
2. Locomotion by flagella, pseudopodia, or both
3. Autotrophic, saprozoic, or heterotrophic
4. Single type of nucleus
5. Sexual reproduction usually occurs

Subphylum Mastigophora: Flagellar Locomotion

Members of the subphylum Mastigophora (mas-ti-gof′o-rah) use flagella in locomotion. Movements of flagella may be

Figure 5.6

Class Phytomastigophorea: Dinoflagellates. Scanning electron micrograph of two species of dinoflagellates, *Peridinium* in the upper left and *Ceratium* in the lower half. The transverse grooves in the center of each dinoflagellate are the location for one of the two flagella (scanning electron micrograph × 800).

two-dimensional, whiplike movements or helical movements that result in the protozoan being pushed or pulled through its aquatic medium.

Class Phytomastigophorea

The subphylum Mastigophora is divided into two classes. Members of the class Phytomastigophorea (fi′to-mas-ti-go-for′ah) (Gr. *phytos*, plant) possess chlorophyll and one or two flagella. Phytomastigophoreans are responsible for producing a large portion of the food in marine food webs. Much of the oxygen in our atmosphere also comes from photosynthesis by these marine organisms.

Marine phytomastigophoreans include the dinoflagellates (figure 5.6). Dinoflagellates have one flagellum that wraps around the organism in a transverse groove. The primary action of this flagellum causes the organism to spin on its axis. A second flagellum is a trailing flagellum that pushes the organism forward. In addition to chlorophyll, many dinoflagellates contain xanthophyll pigments, which give them a golden-brown color. At times, dinoflagellates become so numerous that they color the water. One genus, *Gymnodinium*, has representatives that produce toxins. (3) Periodic "blooms" of these organisms are called "red tides" and result in fish kills along the continental shelves. Human deaths may result from consuming tainted molluscs or fish. The Bible reports that the first plague Moses visited upon the Egyptians was a blood-red tide that killed fish and fouled water. Indeed, the Red Sea is probably named after these toxic dinoflagellate blooms.

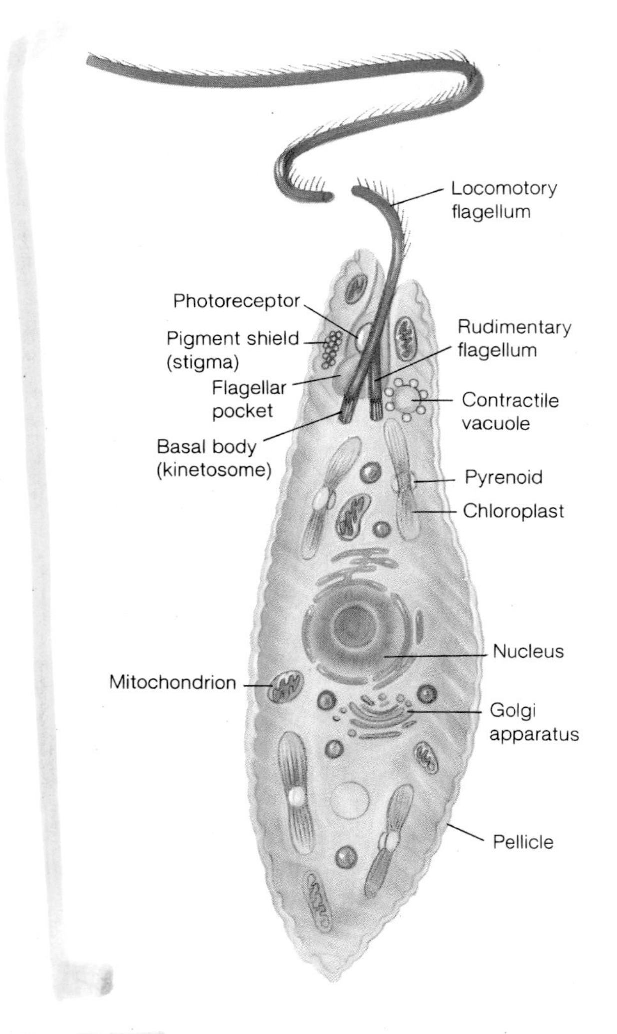

Figure 5.7

Phytomastigophorean Anatomy: The Structure of *Euglena*. Note the large, well-organized chloroplasts. The photoreceptor allows the organism to swim toward light.

Euglena is a phytomastigophorean found in fresh water (figure 5.7). Each chloroplast has a **pyrenoid,** which synthesizes and stores polysaccharides. If cultured in the dark, euglenoids feed by absorption and lose their green color. Some euglenoids (e.g., *Peranema*) lack chloroplasts and are always heterotrophic.

Euglena orients toward light of certain intensities. A pigment shield **(stigma)** covers a photoreceptor at the base of the flagellum. The stigma permits light to strike the photoreceptor from only one direction, allowing *Euglena* to orient and move in relation to a light source.

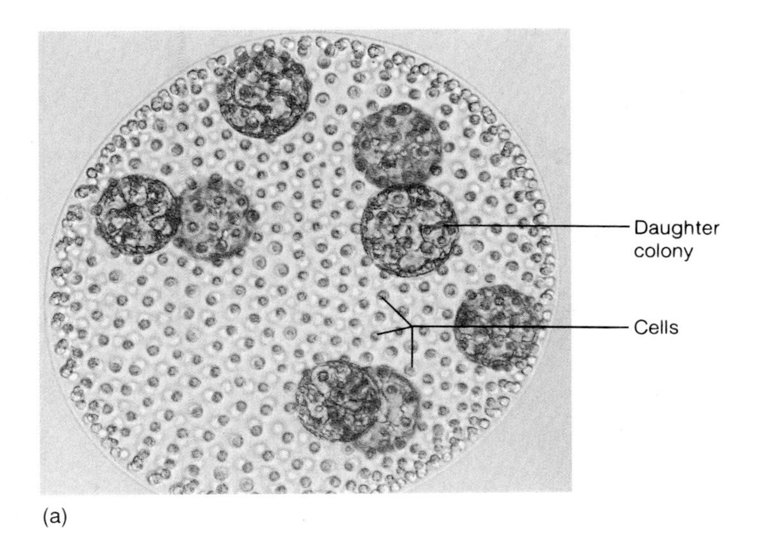

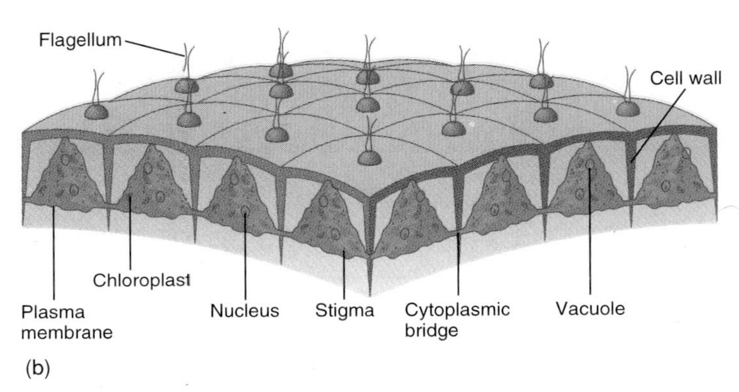

FIGURE 5.8
Class Phytomastigophorea: *Volvox,* a Colonial Flagellate. (*a*) A *Volvox* colony showing asexually produced daughter colonies (light micrograph × 400). (*b*) An enlargement of a portion of the colony wall.

Euglenoid flagellates are haploid and reproduce by longitudinal binary fission (*see figure* 5.4a). Sexual reproduction in these species is unknown.

Volvox is a colonial flagellate consisting of up to 50,000 cells embedded in a spherical, gelatinous matrix (figure 5.8*a*). Cells possess two flagella, which cause the colony to roll and turn gracefully through the water (figure 5.8*b*).

Although most cells of *Volvox* are relatively unspecialized, reproduction is dependent upon certain specialized cells. Asexual reproduction occurs in the spring and summer when certain cells withdraw to the watery interior of the parental colony and form daughter colonies. When the parental colony dies and ruptures, daughter colonies are released.

Sexual reproduction in *Volvox* occurs during autumn. Some species are **dioecious** (having separate sexes), other species are **monoecious** (having both sexes in the same colony). In autumn, specialized cells differentiate into macrogametes or microgametes. Macrogametes are large, filled with nutrient reserves, and nonmotile. Microgametes form as a packet of flagellated cells that leaves the parental colony and swims to a colony containing macrogametes. The packet then breaks apart and syngamy occurs between macro- and microgametes. The zygote, an overwintering stage, secretes a resistant wall around itself and is released when the parental colony dies. Because the parental colony consists of haploid cells, the zygote must undergo meiosis to reduce the chromosome number from the diploid zygotic condition. One of the products of meiosis then undergoes repeated mitotic divisions to form a colony consisting of just a few cells. The other products of meiosis degenerate. This colony is released from the protective zygotic capsule in the spring.

CLASS ZOOMASTIGOPHOREA

Members of the class Zoomastigophorea (zo′o-mas-ti-go-for′ah) (Gr. *zoion,* animal) lack chloroplasts and are heterotrophic. Some members of this class are important parasites of humans (box 5.1).

One of the most important species of zoomastigophoreans is *Trypanosoma brucei.* This species is divided into three subspecies: *T. b. brucei, T. b. gambiense,* and *T. b. rhodesiense.* The first of these three subspecies is a parasite of nonhuman mammals of Africa. The latter two cause sleeping sickness in humans. Tsetse flies (*Glossina* spp.) are intermediate hosts and vectors of all three subspecies. When a tsetse fly bites an infected human or mammal, parasites are picked up with the meal of blood. Trypanosomes multiply asexually in the gut of the fly for about 10 days, then migrate to the salivary glands. While in the fly, the trypanosomes are transformed, in 15 to 35 days, through a number of body forms. When the infected tsetse fly bites another vertebrate host, the parasites travel with salivary secretions into the blood of a new definitive host. The parasites multiply asexually in the new host and are again transformed through a number of body forms. Parasites may live in the blood, lymph, spleen, central nervous system, and cerebrospinal fluid (figure 5.9*a,b*).

When trypanosomes enter the central nervous system, they cause general apathy, mental dullness, and lack of coordination. "Sleepiness" develops and the infected individual may fall asleep during normal daytime activities. Death results from any combination of the above symptoms, from heart failure, malnutrition, and other weakened conditions. If detected early, sleeping sickness is curable. However, if an infection has advanced to the central nervous system, recovery is unlikely.

BOX 5.1 Giardiasis: "Backpacker's Disease" in the Rocky Mountains

Many of the pristine rivers, streams, and lakes in the Rocky Mountains now harbor the mastigophorean *Giardia lamblia* (figure 1). *Giardia* can inhabit the intestine of humans, and wild and domestic animals. In the intestine, the parasites multiply by binary fission and thickly carpet the intestinal wall. Because the parasites are so numerous, a person or animal can shed millions of *Giardia* cysts in the feces. If a human or animal infected by the parasite defecates close to a body of water, the durable cysts can infect someone else who drinks the water even 2 to 3 months later. Dogs, cattle, beavers, deer, bear and other *Giardia* hosts add to the reservoir of contaminated feces and have helped to create the current situation.

In animals and humans, *Giardia* causes the disease **giardiasis,** commonly called "backpacker's disease." Giardiasis is exceedingly unpleasant and usually involves severe diarrhea and painful intestinal cramps that can last for days or weeks. Antiparasitic drugs can quickly cure the disease. However, the best advice is to avoid giardiasis in the first place by not drinking water from wilderness streams or lakes.

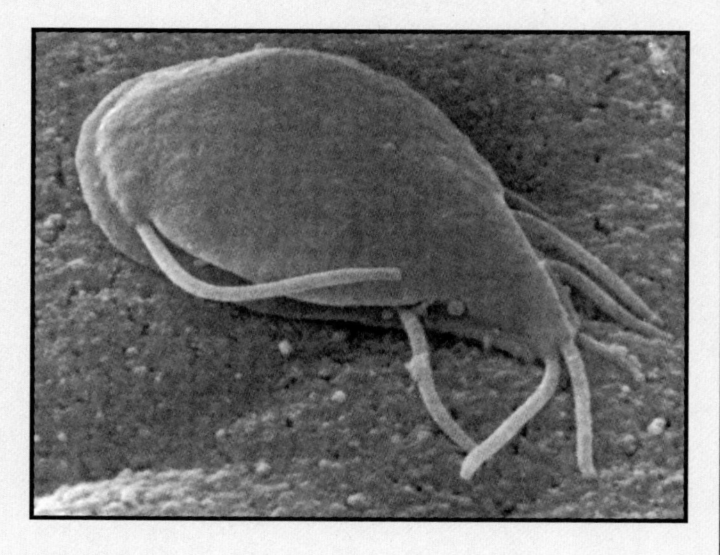

FIGURE 1 **Scanning Electron Micrograph of *Giardia lamblia* Adhering to the Wall of the Intestine.** The organism is 12 to 15 μm long.

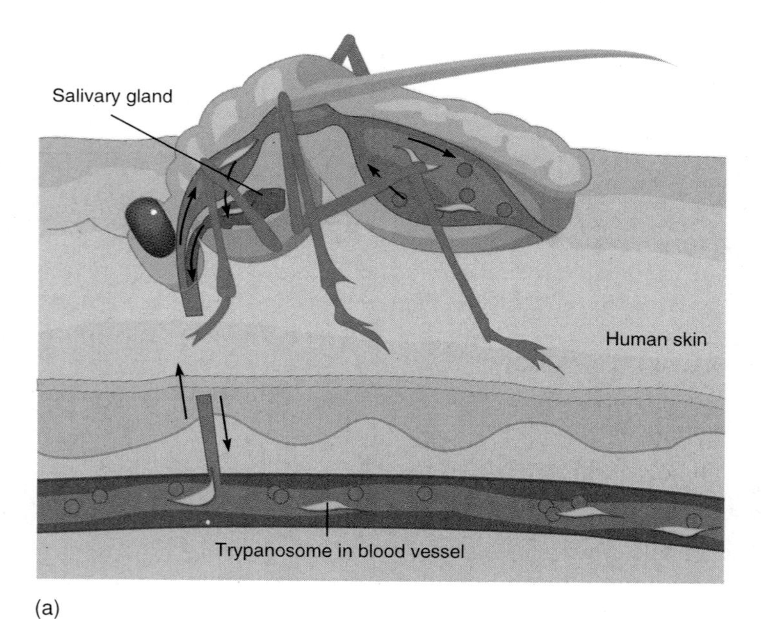

(a)

(b)

FIGURE 5.9

Class Zoomastigophorea: The Life Cycle of *Trypanosoma brucei*. (*a*) When a tsetse fly feeds on a vertebrate host, trypanosomes enter the vertebrate's circulatory system with the fly's saliva. Trypanosomes multiply in the circulatory and lymphatic systems by binary fission. When another tsetse fly bites this vertebrate host again, trypanosomes move into the gut of the fly, and undergo binary fission. Trypanosomes then migrate to the fly's salivary glands, where they are available to infect a new host. (*b*) Light micrograph showing trypanosomes among red blood cells (× 1,200).

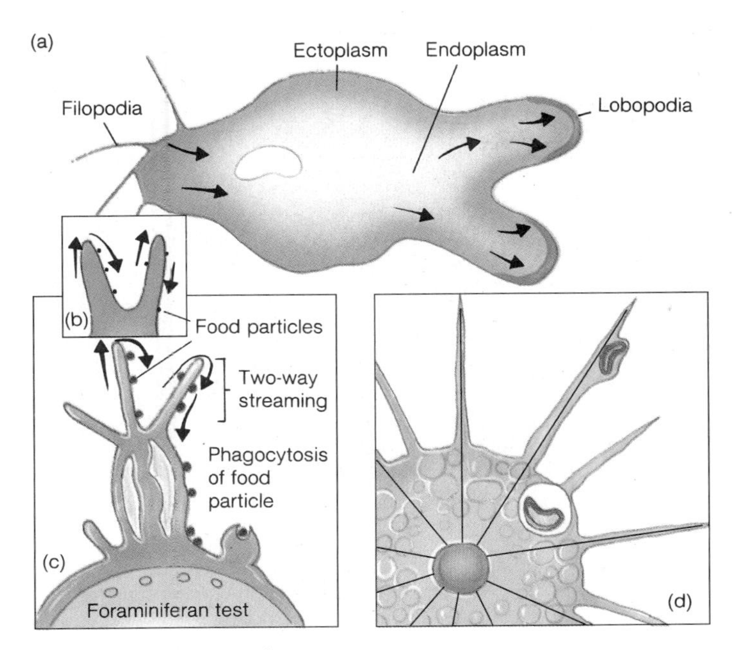

FIGURE 5.10
Variations in Pseudopodia. (*a*) Lobopodia of *Amoeba* contain both ectoplasm and endoplasm and are used for locomotion and engulfing food. (*b*) Filopodia of a shelled amoeba contain ectoplasm only and provide constant two-way streaming that delivers food particles to this protozoan in a conveyor-belt fashion. (*c*) Reticulopodia are similar to filopodia except that they branch and rejoin to form a netlike series of cell extensions. They occur in foraminiferans such as *Globigerina*. (*d*) Axopodia on the surface of a heliozoan such as *Actinosphaerium* deliver food to the central cytoplasm.

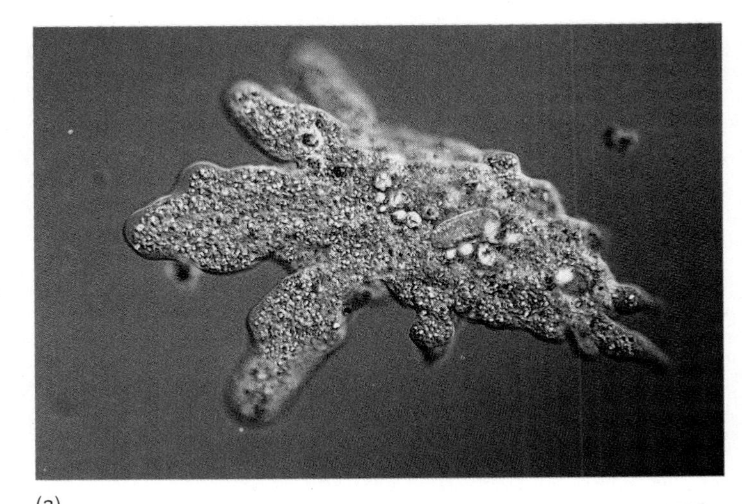
(a)

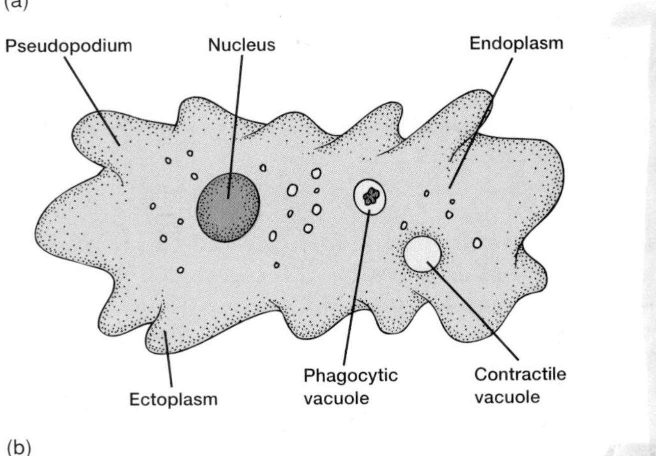

(b)

FIGURE 5.11
Subphylum Sarcodina: Class Lobosea. (*a*) Light micrograph of *Amoeba proteus* showing blunt lobopodia (× 160). (*b*) Drawing showing the anatomy of *Amoeba proteus*.

SUBPHYLUM SARCODINA: PSEUDOPODIA AND AMOEBOID LOCOMOTION

Members of the subphylum Sarcodina (sar′ko-din″ah) are the amoebae (singular amoeba). When feeding and moving, they form temporary cell extensions called **pseudopodia** (s., pseudopodium) (Gr. *pseudes*, false + *podion*, little foot). Pseudopodia exist in a variety of forms. **Lobopodia** (Gr. *lobos*, lobe; s., lobopodium) are broad cell processes containing ectoplasm and endoplasm and are used for locomotion and engulfing food (figure 5.10*a*). **Filopodia** (L. *filum*, thread; s., filopodium) contain ectoplasm only and provide a constant two-way streaming that delivers food in a conveyor-belt fashion (figure 5.10*b*). **Reticulopodia** (L. *reticulatus*, netlike; s., reticulopodium) are similar to filopodia, except that they branch and rejoin to form a netlike series of cell extensions (figure 5.10*c*). **Axopodia** (L. *axis*, axle; s., axopodium) are thin, filamentous, and supported by a central axis of microtubules. The cytoplasm covering the central axis is adhesive and movable. Food caught on axopodia can be delivered to the central cytoplasm of the amoeba (figure 5.10*d*).

CLASS LOBOSEA

The most familiar amoebae belong to the class Lobosea (lo-bo′sah) (Gr. *lobos*, lobe) and the genus *Amoeba* (figure 5.11). These amoebae are naked (they have no test or shell) and are normally found on substrates in shallow water of freshwater ponds, lakes, and slow-moving streams, where they feed on other protists and bacteria. Food is engulfed by phagocytosis, a process that involves the cytoplasmic changes described earlier for amoeboid locomotion. In the process, food is incorporated into food vacuoles. Binary fission occurs when an amoeba reaches a certain size limit. As with other amoebae, no sexual reproduction is known to occur.

Other members of the class Lobosea possess a test or shell. **Tests** are protective structures secreted by the cytoplasm. They may be calcareous (made of calcium carbonate),

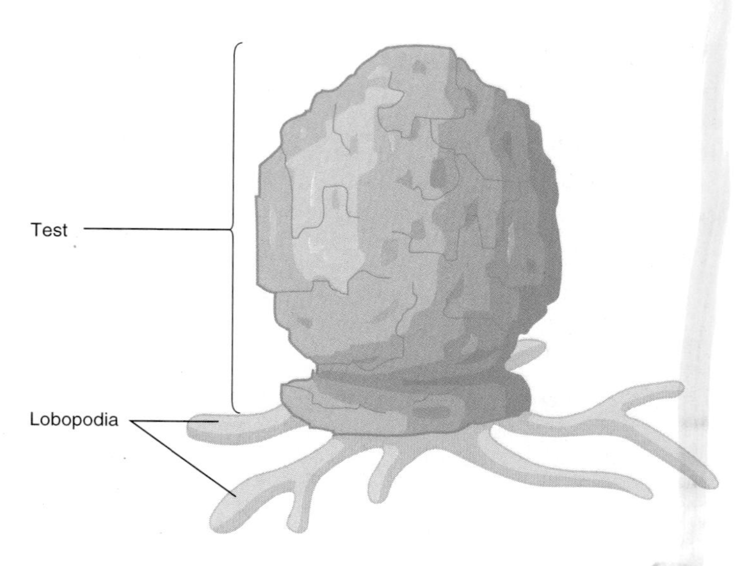

FIGURE 5.12
Subphylum Sarcodina. *Difflugia oblongata*, a common freshwater, shelled amoeba. The test is composed of cemented mineral particles.

FIGURE 5.13
Subphylum Sarcodina: Foraminiferan Test (*Polystomella*). As this foraminiferan grows, it secretes new, larger chambers that remain attached to older chambers, making this protozoan reminiscent of a tiny snail (light micrograph × 63).

proteinaceous (made of protein), siliceous (made of silica [SiO_2]), or chitinous (made of chitin—a polysaccharide). Other tests may be composed of sand or other debris cemented into a secreted matrix. Usually one or more openings in the test allow pseudopodia to be extruded. *Arcella* is a common freshwater, shelled amoeba. It has a brown, proteinaceous test that is flattened on one side and domed on the other. Pseudopodia project from an opening on the flattened side. *Difflugia* is another common freshwater, shelled amoeba (figure 5.12). Its test is vase shaped and is composed of mineral particles embedded in a secreted matrix.

Most amoebae are symbiotic; a few are pathogenic. For example, (4) *Entamoeba histolytica* causes one form of dysentery in humans. Dysentery is marked by inflammation and ulceration of the lower intestinal tract, accompanied by a debilitating diarrhea that includes blood and mucus. Amoebic dysentery is a worldwide problem that plagues humans in crowded, unsanitary conditions.

A significant problem in the control of *Entamoeba histolytica* is the fact that an individual can be infected and contagious without experiencing symptoms of the disease. Amoebae live in the folds of the intestinal wall, feeding on starch and mucoid secretions. Amoebae are passed from one host to another in the form of cysts that are transmitted by fecal contamination of food or water. Amoebae leave the cysts after a host ingests contaminated food or water and take up existence in the intestinal wall.

Foraminiferans, Heliozoans, and Radiolarians

Foraminiferans (commonly called forams) are primarily a marine group of amoebae. Foraminiferans possess reticulopodia and secrete an exoskeleton called a test that is mostly calcium carbonate. As foraminiferans grow, they secrete new, larger chambers that remain attached to the older chambers (figure 5.13). Test enlargement follows a symmetrical pattern that may result in a straight chain of chambers or spiral arrangement that resembles a snail shell. Many of these tests may reach relatively large sizes; for example, "Mermaid's pennies," found in Australia, may be several centimeters in diameter.

Foraminiferan tests have left an abundant fossil record that began in the Cambrian period. Foram tests make up a large component of marine sediments, and the accumulation of foram tests on the floor of primeval oceans has resulted in our limestone and chalk deposits. The white cliffs of Dover are one example of a foraminiferan-chalk deposit. Oil geologists use fossilized forams to identify geologic strata when taking exploratory cores.

Heliozoans are freshwater amoebae that are either planktonic or live attached by a stalk to some substrate. (The plankton of a body of water consists of those organisms that float freely in the water.) Heliozoans are either naked or enclosed within a test that contains openings for axopodia (figure 5.14*a*).

(a)

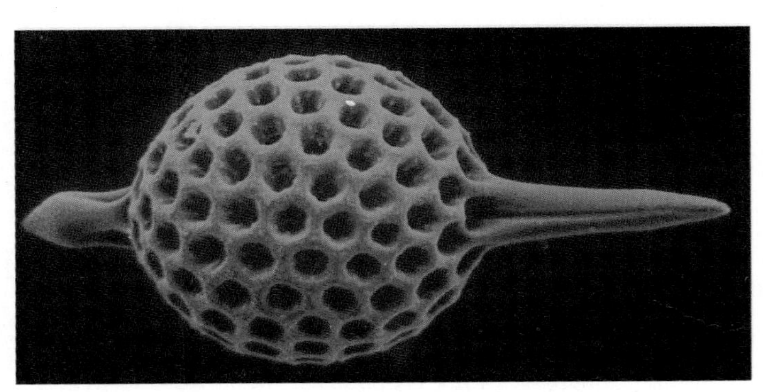

(b)

FIGURE 5.14

Subphylum Sarcodina: Heliozoan and Radiolarian Tests.
(*a*) *Actinosphaerium sol* has a spherical body covered with fine, long axopodia made up of numerous microtubules and surrounded by streaming cytoplasm. Following phagocytosis by the axopodium, waves of cytoplasmic movement carry trapped food particles into the main body of this protozoan (light micrograph × 450). (*b*) The radiolarian *Spaerostylus* is typically spherical with a highly sculptured test (light micrograph × 480).

Radiolarians are planktonic, marine amoebae. They are relatively large; some colonial forms may reach several centimeters in diameter. They possess a test (usually siliceous) of either long, movable spines and needles or a highly sculptured and ornamented lattice (figure 5.14*b*). When radiolarians die, their tests drift to the ocean floor. Some of the oldest known fossils of eukaryotic organisms are radiolarians.

PHYLUM LABYRINTHOMORPHA

The very small phylum Labyrinthomorpha (la′brinth-o-morp′ha) consists of protozoa that have spindle-shaped nonamoeboid vegetative cells. In some genera, amoeboid cells move within a network of mucous tracks using a typical gliding motion. Most members are marine and either saprozoic or parasitic on algae or seagrass. Several years ago, *Labyrinthula* killed most of the "eel grass" on the Atlantic coast, starving many ducks that feed on the grass.

Stop and Ask Yourself

5. How and when does sexual reproduction occur in *Volvox?*
6. How do trypanosomes gain entry to their human hosts?
7. What mechanism explains amoeboid locomotion?
8. How is *Entamoeba histolytica* transmitted among humans? *Giardia lamblia?*

PHYLUM APICOMPLEXA

Members of the phylum Apicomplexa (a″pi-kom-plex′ah) (L. *apex*, point + *com*, together, + *plexus*, interweaving) are all parasites. Characteristics of the phylum include the following:

1. Apical complex that aids in the penetration of host cells
2. Single type of nucleus
3. Cilia and flagella lacking, except in certain reproductive stages
4. Life cycles typically include asexual (schizogony, sporogony) and sexual (gametogony) phases

CLASS SPOROZOEA

The most important species in this phylum are members of the class Sporozoea (spor′o-zo″e). The class name derives from the fact that most sporozoeans produce a resistant spore or oocyst following sexual reproduction. Some members of this class, including *Plasmodium* and coccidians, cause a variety of diseases in domestic animals and humans.

Although there is considerable variability in life cycles of sporozoeans, certain generalizations are possible. Many are intracellular parasites, and their life cycles may be divided into three phases. **Schizogony** is multiple fission of an asexual stage in host cells, resulting in the formation of many more (usually asexual) individuals, called merozoites, that leave the host cell and infect many other cells.

Some of the merozoites produced (as described above) undergo **gametogony,** which begins the sexual phase of the life cycle. In so doing, the parasite forms either microgametocytes or macrogametocytes. Microgametocytes undergo multiple fission to produce biflagellate microgametes that emerge from the infected host cell. The macrogametocyte develops directly into a single macrogamete. Fertilization of the macrogamete by a microgamete produces a zygote that becomes enclosed and is called an oocyst.

The zygote undergoes meiosis, and the resulting cells divide repeatedly by mitosis. This process, called **sporogony,** produces many rodlike sporozoites in the oocyst. Sporozoites infect the cells of a new host when the oocyst is ingested and digested by the new host, or sporozoites are otherwise introduced (e.g., by a mosquito bite).

One sporozoean genus, *Plasmodium*, causes malaria and has been responsible for more human suffering than most other diseases. Accounts of the disease go back as far as 1550 B.C. (5) Malaria was a significant contributor to the failure of the Crusades during the medieval era, and it has contributed more to the devastation of armies than has actual combat. Recently (since the early 1970s), there has been a resurgence of malaria throughout the world. It is estimated that over 100 million humans annually contract the disease.

The life cycle of *Plasmodium* involves vertebrate and mosquito hosts (figure 5.15). Schizogony occurs first in liver cells and later in red blood cells, and gametogony also occurs in red blood cells. Gametocytes are taken into a mosquito during a meal of blood and subsequently fuse. The zygote penetrates the gut of the mosquito and is transformed into an oocyst. Sporogony forms haploid sporozoites that may enter a new host when the mosquito bites the host.

The symptoms of malaria recur periodically and are called paroxysms (box 5.2). Chills and fever are correlated with the maturation of parasites, the rupture of red blood cells, and the release of toxic metabolites.

Four species of *Plasmodium* are the most important human malarial species. *P. vivax* causes malaria in which the paroxysms recur every 48 hours. This species occurs in temperate regions and has been nearly eradicated in many parts of the world. *P. falciparum* causes the most virulent form of malaria in humans. Paroxysms occur more irregularly than in the other species. It was once worldwide, but is now mainly tropical and subtropical in distribution. It remains one of the greatest killers of humanity, especially in Africa. *P. malariae* is worldwide in distribution and causes malaria with paroxysms that recur every 72 hours. *P. ovale* is the rarest of the four human malarial species, and is primarily tropical in distribution.

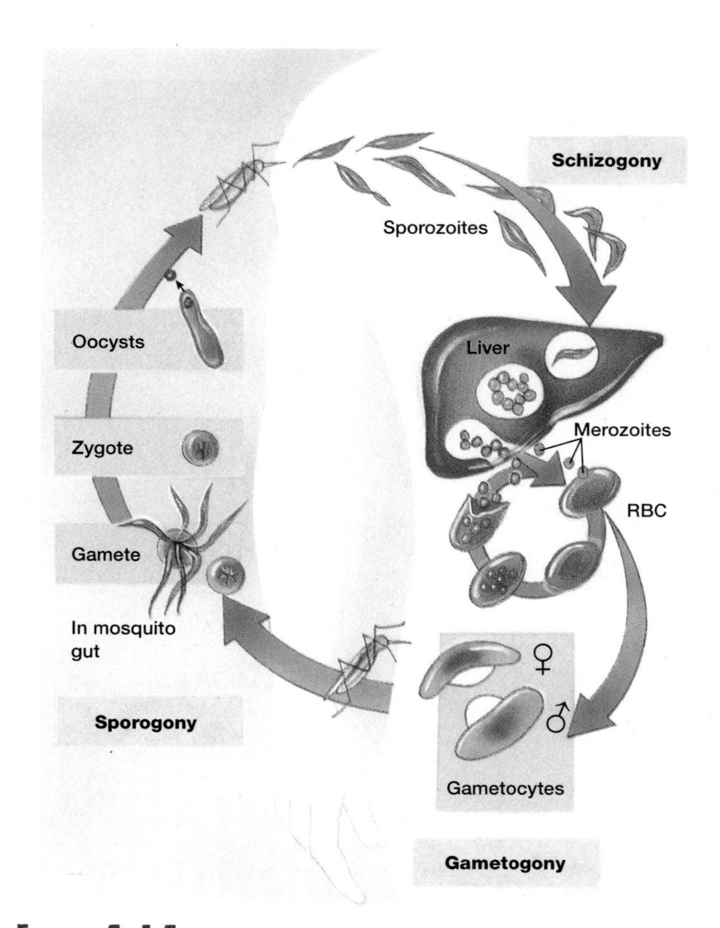

Figure 5.15

Phylum Apicomplexa: The Life Cycle of *Plasmodium*. Schizogony occurs in liver cells and, later, in the red blood cells of humans. Gametogony occurs in red blood cells. Microgametes and macrogametes are taken into a mosquito during a blood meal and fuse to form zygotes. Zygotes penetrate the gut of the mosquito and form the oocysts. Meiosis and sporogony form many haploid sporozoites that may enter a new host when the mosquito bites the host.

Other members of the class Sporozoea also cause important diseases. Coccidiosis is primarily a disease of poultry, sheep, cattle, and rabbits. Two genera, *Isospora* and *Eimeria*, are particularly important parasites of poultry. Yearly losses to the poultry industry in the United States from coccidiosis have approached 35 million dollars. Another coccidian, *Cryptosporidium*, has become more well-known with the advent of AIDS. It causes chronic diarrhea. Toxoplasmosis is a disease of mammals, including humans, and birds. Sexual reproduction of *Toxoplasma* occurs primarily in cats. Infections occur when oocysts are ingested with food contaminated by cat feces, or when meat containing encysted sporozoites is eaten raw or poorly cooked. Most infections in humans are asymptomatic and once infected an effective immunity develops. However, if a woman is infected near the time of pregnancy, or during pregnancy, congenital toxoplasmosis may develop in a fetus. Congenital toxoplasmosis is a major cause of stillbirths and spontaneous abortions. Fetuses that survive frequently show signs of mental retardation and epileptic seizures. There is no cure for congenital toxoplasmosis.

BOX 5.2 Malaria Control—A Glimmer of Hope

The following quotation is from *The Lake Regions of Central Africa* by Sir Richard Burton (1821–1890). Sir Richard Burton was an adventurer whose visits to the Far East and Africa brought him into contact with the greatest killer of humanity, *Plasmodium falciparum*.

> *The approach of malignant fever is very insidious. An attack begins mostly with an ordinary chill, attended by no unusual or marked symptoms. Sometimes the patient has had a light chill a day or two before this, which he has neglected. Sometimes he has felt slightly unwell for ten or fourteen days; has complained of loss of appetite and general weakness; but as these symptoms are not very marked, they are very apt to be overlooked, especially with newcomers.*
>
> *The real attack may begin with a chill or with a fever, but its effects are, in either case, at once evident in a peculiarly yellow skin and haggard countenance. In fever there is profuse perspiration, a rush of blood to the head, high and irregular pulse, and general prostration. Sometimes the body is hot, but dry. Thirst is urgent, but the stomach rejects whatever is drunk.*
>
> *If the paroxysm [sudden attack] of fever returns, it is with renewed force, and the third attack is commonly fatal. Before death the patient becomes insensible; there is violent vomiting, which is, in fact, regurgitation of ingesta, mixed with green and yellow fluid. Immediately after the chill, and even before this has passed off, the urine becomes dark red or black. The pulse is very irregular, the breathing slow and finally the patient sinks away into a state of coma, and dies without a struggle.*

Malaria is still a fact of life—especially in Africa. Nearly 300 million people are afflicted annually, and 1/3 of them die. The fight against this disease has centered on two fronts: elimination of the mosquito species known to carry *Plasmodium* parasites and treatment of infected persons with antimalarial drugs. In the fight against mosquitoes, DDT (dichloro-diphenyl-trichloroethane) was employed successfully in the 1950s and 1960s. Its use greatly reduced the incidence of malaria by the middle 1960s. Mosquito-control programs, however, began to lose their effectiveness, largely because mosquitoes developed resistance to DDT. DDT has also been found to be a highly persistent pesticide. It retains its toxicity for many years and can build up to lethal levels in aquatic and terrestrial environments. Its use became less effective and more expensive—both in economic and environmental terms. Other pesticides are now being used, but mosquito resistance is becoming a problem with these pesticides too.

Even more serious is the worldwide development of resistance by *Plasmodium* parasites to antimalarial drugs, such as chloroquine.

As bleak as this picture sounds, scientists and health officials are optimistic that malaria will eventually be conquered. Recent advancements in molecular biology and immunology are being employed in the development of antimalarial vaccines. Some vaccines are now being tested. Researchers have found that the complexity of the disease presents special problems in the development of vaccines. Not only do each of the *Plasmodium* species require a separate vaccine, but each stage of the life cycle of a single species requires a separate immunological component in a vaccine.

Toxoplasmosis also ranks high amongst the opportunistic diseases afflicting AIDS patients. (6) Steps can be taken to avoid infections by *Toxoplasma*. Precautions include keeping stray and pet cats away from children's sandboxes; using sandbox covers; and awareness, on the part of couples considering having children, of the potential dangers of eating raw or very rare pork, lamb, and beef.

Phylum Microspora

Members of the phylum Microspora (mi-cro-spor′ah), commonly called microsporidia, are small obligatory intracellular parasites. Included in this phylum are several species that parasitize beneficial insects. *Nosema bombicus* parasitizes silkworms (figure 5.16) causing the disease **pebrine,** and *N. apis* causes serious dysentery (foul brood) in honeybees. There has been an increased interest in these parasites because of their possible role as biological control agents for insect pests. For example, *N. locustae* has been approved and registered with the United States Environmental Protection Agency (EPA) for use in residual control of rangeland grasshoppers. Recently, four microsporidian genera have been implicated in secondary infections of immunosuppressed and AIDS patients.

Phylum Acetospora

Acetospora (ah-seat-o-spor′ah) is a relatively small phylum that consists exclusively of obligatory extracellular parasites characterized by spores lacking polar caps or polar filaments. The acetosporeans (e.g., *Haplosporidium*) primarily are parasitic in the cells, tissues, and body cavities of molluscs.

Phylum Myxozoa

The phylum Myxozoa (myx-o-zoo-ah), commonly called myxosporeans, are all obligatory extracellular parasites in freshwater and marine fish. They have a resistant spore with one to six coiled polar filaments. The most economically important myxosporean is *Myxosoma cerebralis*, which infects the nervous system and auditory organ of trout and salmon causing whirling or tumbling disease.

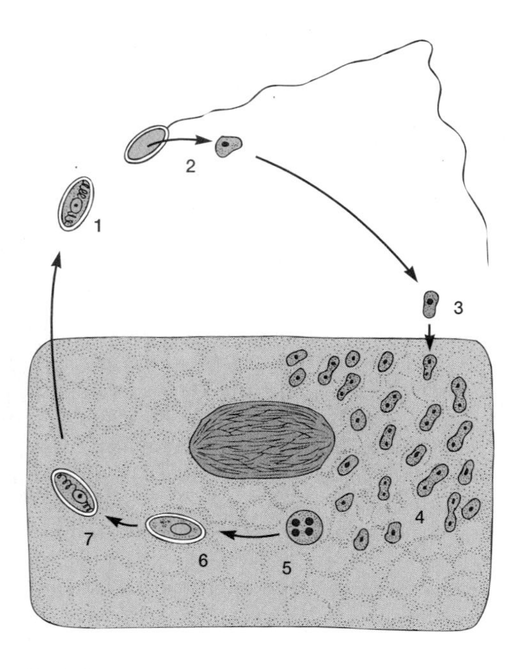

FIGURE 5.16

Phylum Microspora. Drawing of the Microsporean *Nosema bombicus*, which Is Fatal to Silkworms. The cell of a silkworm infected with *N. bombicus* is shown here. (*1*) A typical spore with one coiled filament: (*2*) When ingested, it extrudes the filament, which is used in locomotion. (*3*) The parasite enters an epithelial cell in the intestine of the silkworm and (*4*) divides many times to form small amoebae that eventually fill the cell and kill it. During this phase, some of the amoebae with four nuclei become spores (*5,6,7*). Silkworms are infected by eating leaves contaminated by the feces of infected worms.

Stop and Ask Yourself

9. What is an apical complex? What is it used for?
10. What is schizogony?
11. Members of which protozoan genus have caused the most human suffering since antiquity?
12. What are the three stages in the life cycle of *Plasmodium?*
13. How are infections of *Toxoplasma* acquired?
14. What are the dangers of toxoplasmosis for a fetus?
15. What are the four most important sporozoean parasites and the diseases they cause?
16. What is one economically important disease caused by a microsporidian? What group of animals do myxosporeans usually parasitize?

PHYLUM CILIOPHORA

7 The phylum Ciliophora (sil-i-of′or-ah) includes some of the most complex protozoa (*see table 5.1*). Ciliates are widely distributed in freshwater and marine environments. A few ciliates are symbiotic. Characteristics of the phylum Ciliophora include the following:

1. Cilia for locomotion and for the generation of feeding currents in water
2. Relatively rigid pellicle and more or less fixed shape
3. Distinct cytostome (mouth) structure
4. Dimorphic nuclei, typically a larger macronucleus and one or more smaller micronuclei

CILIA AND OTHER PELLICULAR STRUCTURES

Cilia are generally similar to flagella, except that they are much shorter, more numerous, and widely distributed over the surface of the protozoan (figure 5.17). Ciliary movements are coordinated, so that ciliary waves pass over the surface of the ciliate. Many ciliates can reverse the direction of ciliary beating and the direction of cell movement.

Basal bodies (kinetosomes) of adjacent cilia are interconnected with an elaborate network of fibers that are believed to anchor the cilia and give shape to the organism.

The evolution of some ciliates has resulted in the specialization of cilia. Cilia may cover the outer surface of the protozoan. They may be joined together to form **cirri,** which are used in movement. Alternatively, cilia may be lost from large regions of a ciliate.

Trichocysts are pellicular structures primarily used for attachment. They are rodlike or oval organelles oriented perpendicular to the plasma membrane. In *Paramecium,* they have a "golf tee" appearance. Trichocysts can be discharged from the pellicle and, after discharge, they remain connected to the body by a sticky thread (figure 5.18).

NUTRITION

Some ciliates, such as *Paramecium,* have a ciliated oral groove along one side of the body (*see figure 5.17*). Cilia of the oral groove sweep small, organic particles toward the cytopharynx where a food vacuole is formed. When a food vacuole reaches an upper size limit, it breaks free and circulates through the endoplasm.

Some free-living ciliates prey upon other protists or small animals. Prey capture is usually a case of fortuitous contact. *Didinium* is a ciliate that feeds principally on *Paramecium,* a prey item that is bigger than itself. The *Didinium* forms a temporary opening that can greatly enlarge to consume its prey (figure 5.19).

Suctorians are ciliates that live attached to their substrate by a stalk. They possess tentacles to which prey adhere. Their prey, often ciliates or amoebae, are paralyzed by secretions of the tentacles. The tentacles digest an opening in the pellicle of the prey, and prey cytoplasm is drawn into the suctorian through tiny channels in the tentacle. The mechanism for this probably involves tentacular microtubules (figure 5.20).

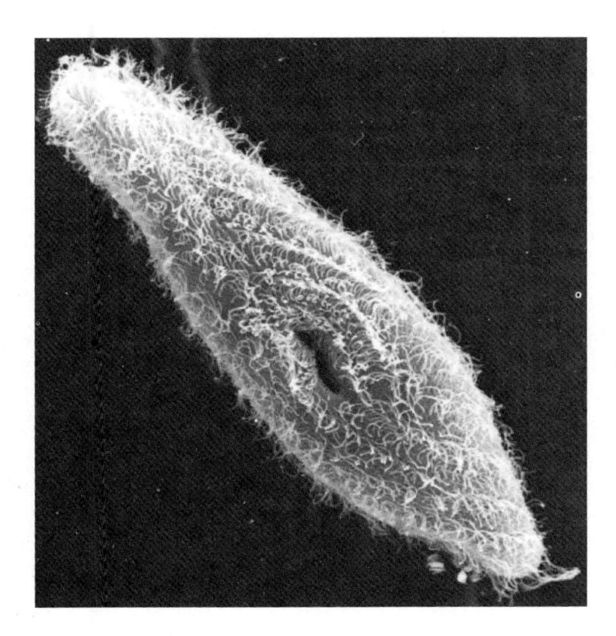

(a)

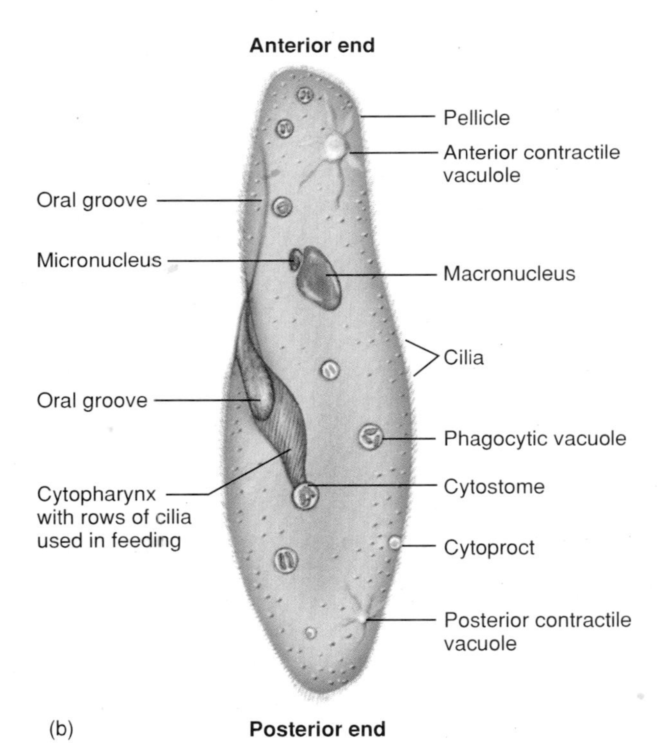

(b)

FIGURE 5.17

Phylum Ciliophora. (*a*) Scanning electron micrograph of the ciliate, *Paramecium sonneborn*. This paramecium is 40 µm in length. Note the oral groove near the middle of the body that leads into the cytopharynx (× 1,600). (*b*) The structure of a typical ciliate such as *Paramecium*.

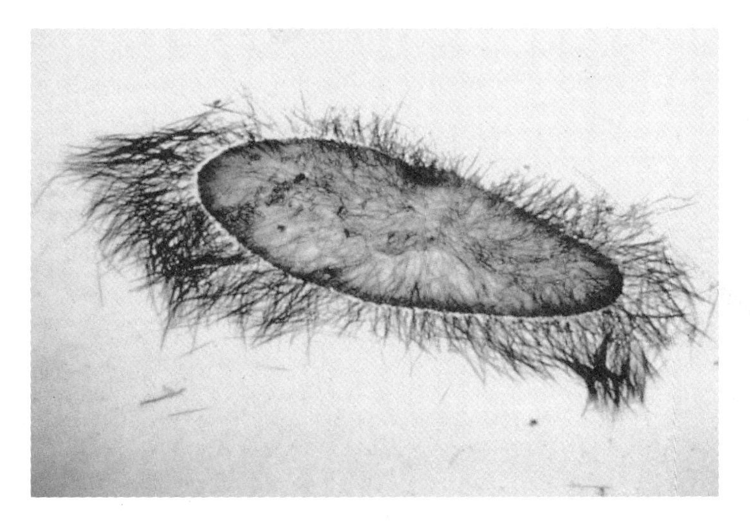

FIGURE 5.18

The Discharged Trichocysts of *Paramecium*. Each trichocyst is able to produce a long, sticky proteinaceous thread when discharged. This thread connects the trichocyst in the body of the protozoan (light micrograph × 150).

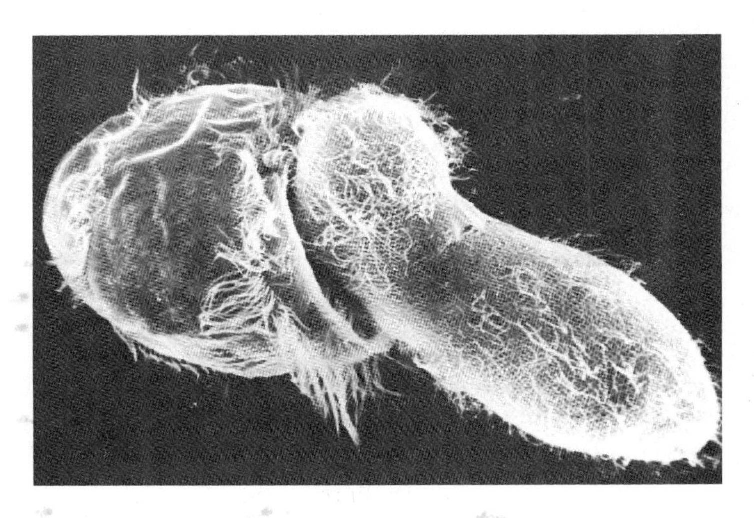

FIGURE 5.19

A Single-Celled Hunter and Its Prey. The juglike *Didinium* (left) is shown swallowing a slipper-shaped *Paramecium* (right) (scanning electron micrograph × 550).

GENETIC CONTROL AND REPRODUCTION

Ciliates have two kinds of nuclei. A large, polyploid **macronucleus** regulates the daily metabolic activities. One or more smaller **micronuclei** serve as the genetic reserve of the cell.

Asexual reproduction of ciliates occurs by transverse binary fission and occasionally by budding. Budding occurs in suctorians

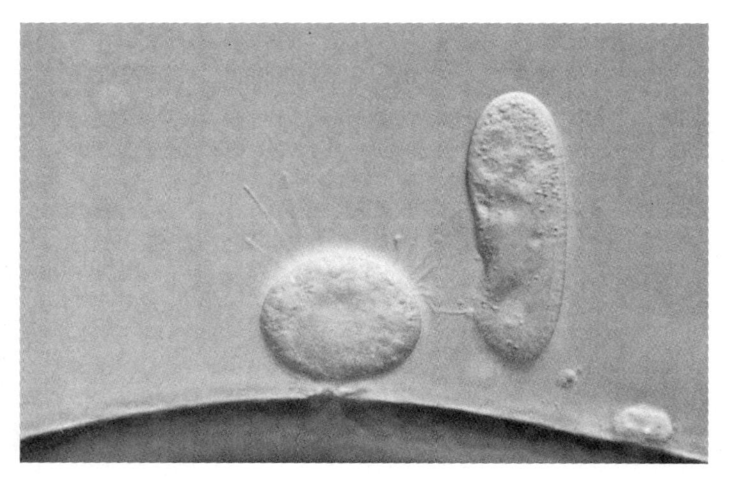

Figure 5.20

Suctorian (*Tokophrya* spp.) Feeding. A ciliate (right) is held by the knobbed tip of a tentacle. Tentacles discharge enzymes that immobilize the prey and dissolve the pellicle. Pellicles of tentacle and prey fuse, and the tentacle enlarges and invaginates to form a feeding channel. Prey cytoplasm is moved down the feeding channel and incorporated into endocytic vacuoles at the bottom of the tentacle (light micrograph × 181).

and results in the formation of ciliated, free-swimming organisms that attach to the substrate and take the form of the adult.

Sexual reproduction of ciliates occurs by **conjugation** (figure 5.21). The partners involved are called conjugants. Many species of ciliates have numerous mating types, not all of which are compatible with one another. The initial contact between individuals apparently occurs randomly, and adhesion is facilitated by sticky secretions of the pellicle. Fusion of ciliate plasma membranes occurs and lasts for several hours.

The macronucleus does not participate in the genetic exchange that follows. Instead, the macronucleus breaks up during or after micronuclear events and is reformed from micronuclei of the daughter ciliates.

After separation, the exconjugants undergo a series of nuclear divisions to restore the nuclear characteristics of the particular species, including the formation of a macronucleus from one or more micronuclei. These events are accompanied by cytoplasmic divisions which form daughter cells.

Symbiotic Ciliates

Most ciliates are free living; however, some are commensalistic or mutualistic and a few are parasitic. *Balantidium coli* is an important parasitic ciliate that lives in the large intestines of humans, pigs, and other mammals. At times, it is a ciliary feeder; at other times, it produces proteolytic enzymes that digest host epithelium, causing a flask-shaped ulcer. (Its pathology resembles that of *Entamoeba histolytica*.) *B. coli* is passed from one host to another in cysts that are formed as feces begin to dehydrate in the large intestine. Fecal contamination of food or water is the most common form of transmission. It is potentially worldwide in distribution, but is most common in the Philippines.

Large numbers of different species of ciliates also inhabit the rumen of many ungulates (hoofed animals). These ciliates contribute to the digestive processes of their hosts.

Further Phylogenetic Considerations

The origin of protozoa probably took place about 1.5 billion years ago. Although there are over 30,000 known fossil species, they are of little use in investigations of the origin and evolution of the various protozoan groups. Only protozoa with hard parts (tests) have left much of a fossil record, and only the foraminiferans and radiolarians have well-established fossil records in Precambrian rocks. Recent evidence from the study of base sequences in ribosomal RNA (figure 5.22) indicates that each of the seven protozoan phyla probably had separate origins, and that each is sufficiently different from the others to warrant elevating all of these groups to phylum status as has been done in this chapter. Additional modifications to the present scheme of protozoan classification are continually being proposed as the results of new ultrastructural and molecular studies are published. For example, in 1993, T. Cavalier-Smith proposed that the protozoa be elevated to kingdom status with 18 phyla. The acceptance of this new classification by protozoologists, however, remains to be determined.

Stop and Ask Yourself

17. What are four characteristics of the phylum Ciliophora?
18. What are trichocysts used for in ciliates?
19. What occurs during conjugation in *Paramecium*?
20. Why is it difficult to find protozoan fossils?

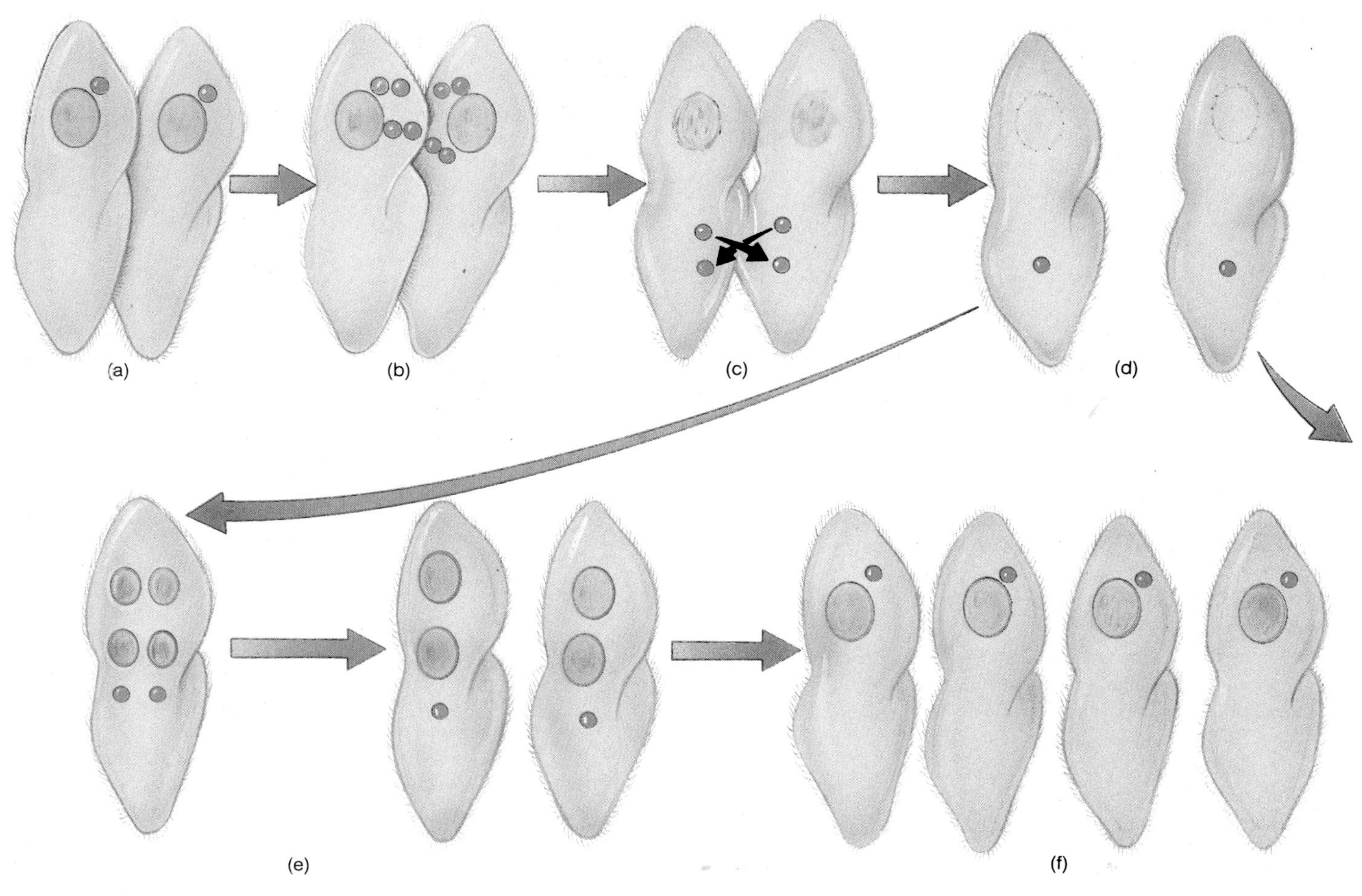

FIGURE 5.21

Conjugation in *Paramecium*. (*a*) Random contact brings individuals of opposite mating types together. (*b*) Meiosis results in four haploid pronuclei. (*c*) Three pronuclei and the macronucleus degenerate. Mitosis and mutual exchange of pronuclei is followed by fusion of pronuclei. (*d–f*) Separation of conjugants is followed by nuclear divisions that restore nuclear characteristics of the species. These events may be accompanied by cytoplasmic divisions.

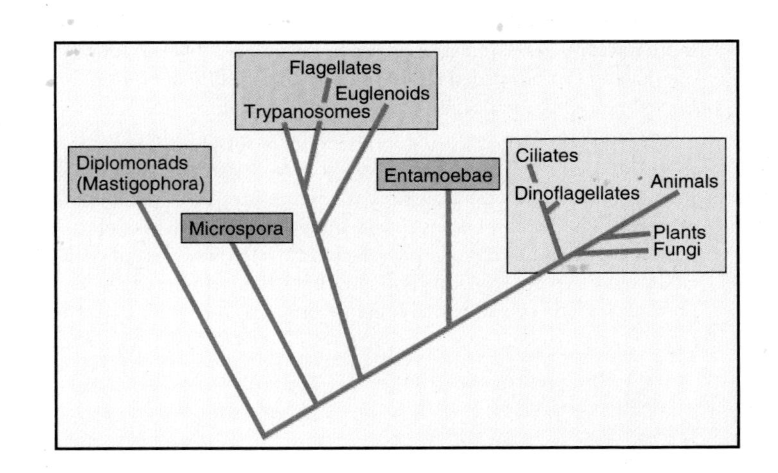

FIGURE 5.22

A Cladogram Showing the Evolutionary Relationships of Protozoa and Other Eukaryotes Based on 18S Ribosomal RNA Sequence Comparisons. This cladogram suggests that evolution along the nuclear line of descent was not a continuous process, but instead occurred in major epochs (an epoch is a particular period of time marked by distinctive features and events). Five major evolutionary radiations (colored boxes) are apparent for the protozoa. As shown, it is clear that the Mastigophora (e.g., *Giardia*) and Microspora (e.g., *Nosema*) are modern relatives of the earliest major eukaryotic cell lines. Following the development of these protozoa, the other groups of protozoa radiated off the nuclear line of descent.

Summary

1. The kingdom Protista is a polyphyletic group that arose about 1.5 billion years ago from the Monera. The evolutionary pathways leading to modern protozoa are uncertain.
2. Protozoa are both single cells and entire organisms. Many of their functions are carried out by organelles specialized for the unicellular life-style.
3. Many protozoa live in symbiotic relationships with other organisms, often in a host-parasite relationship.
4. Members of the phylum Sarcomastigophora possess pseudopodia and/or one or more flagella.
5. Members of the class Phytomastigophorea are photosynthetic and include the genera *Euglena* and *Volvox*. Members of the class Zoomastigophorea are heterotrophic and include *Trypanosoma*, which causes sleeping sickness.
6. Amoebae use pseudopodia for feeding and locomotion.
7. Members of the subphylum Sarcodina include the freshwater genera *Amoeba*, *Arcella*, and *Difflugia*, and the symbiotic genus *Entamoeba*. Foraminiferans and radiolarians are common marine amoebae.
8. Members of the phylum Apicomplexa are all parasites. The phylum includes *Plasmodium* and *Toxoplasma*, which cause malaria and toxoplasmosis, respectively. Many apicomplexans have a three-part life cycle involving schizogony, gametogony, and sporogony.
9. The phylum Microspora consists of very small protozoa that are intracellular parasites of every major animal group. They are transmitted from one host to the next as a spore, the form from which the group obtains its name.
10. The phylum Acetospora contains protozoa that produce spores lacking polar capsules. These protozoa are primarily parasitic in molluscs.
11. The phylum Myxozoa consists entirely of parasitic species, usually found in fishes. The spore is characterized by one to six polar filaments.
12. The phylum Ciliophora contains some of the most complex of all protozoa. Its members possess cilia, a macronucleus, and one or more micronuclei. Cilia are coordinated by mechanical coupling of cilia and can be specialized for different kinds of locomotion. Sexual reproduction occurs in ciliates by conjugation. Diploid ciliates undergo meiosis of the micronuclei to produce haploid pronuclei that are exchanged between two conjugants.
13. Precise evolutionary relationships are difficult to determine for the protozoa. The fossil record is sparse, and what does exist is not particularly helpful in deducing relationships. However, ribosomal RNA sequence comparisons indicate that each of the seven protozoan phyla probably had separate origins.

Selected Key Terms

ectoplasm (*p.* 65)
endoplasm (*p.* 65)
giardiasis (*p.* 70)
macronucleus (*p.* 77)
micronuclei (*p.* 77)
multiple fission (schizogony) (*p.* 65)
pellicle (*p.* 65)
protozoa (*p.* 65)
protozoologists (*p.* 67)
trichocysts (*p.* 76)

Critical Thinking Questions

1. If it is impossible to know for certain the evolutionary pathways that gave rise to protozoa and animal phyla, do you think it is worth constructing hypotheses about those relationships? Why or why not?
2. In what ways are protozoa similar to animal cells? In what ways are they different?
3. If sexual reproduction is unknown in *Euglena*, how do you think this lineage of organisms has survived through evolutionary time? (Recall that sexual reproduction provides the genetic variability that allows species to adapt to environmental changes.)
4. The use of DDT has been greatly curtailed for ecological reasons. In the past, it has proven to be the greatest malaria deterrent known throughout the world. Many organizations would like to see this form of mosquito control resumed. Do you agree or disagree? Explain your reasoning.
5. If you were traveling out of this country and you were concerned about contracting amoebic dysentery, what steps could you take to prevent acquiring the disease? How would the precautions differ if you were going to a country where malaria was a problem?

chapter 6

MULTICELLULAR AND TISSUE LEVELS OF ORGANIZATION

Outline

Concepts

1. How multicellularity originated in animals, and whether the animal kingdom is monophyletic, diphyletic, or polyphyletic, are largely unknown.
2. Animals whose bodies consist of aggregations of cells, but whose cells do not form tissues, are found in the phylum Porifera as well as some lesser known phyla.
3. Animals that show diploblastic, tissue-level organization are found in the phyla Cnidaria and Ctenophora.
4. Members of the phylum Cnidaria are important in zoological research because of their relatively simple organization and their contribution to coral reefs.

Would You Like to Know:

1. how multicellularity could have arisen in the animal kingdom? (*p.* 82)
2. how natural sponges are prepared for use in cleaning applications? (*p.* 87)
3. what value the intricate branching canal systems are to a sponge? (*p.* 87)
4. whether cells of a sponge body can communicate with one another? (*p.* 88)
5. how soft-bodied cnidarians support themselves? (*p*.92)
6. why one should avoid touching blue, gas-filled floats washed up on beaches of temperate and tropical waters? (*p.* 95)
7. which jellyfish should be avoided when swimming in coastal waters? (*p.* 95)
8. what organisms are responsible for the formation of coral reefs? (*p.* 99)

These and other useful questions will be answered in this chapter.

This chapter contains evolutionary concepts, which are set off in this font.

Evolutionary Perspective

Animals with multicellular and tissue levels of organization have captured the interest of scientists and laypersons alike. A description of some members of the phylum Cnidaria, for example, could fuel a science fiction writer's imagination.

> From a distance I was never threatened, in fact I was infatuated with its beauty. A large, inviting, bright blue float lured me closer. As I swam nearer I could see that hidden from my previous view was an infrastructure of tentacles, some of which dangled nearly nine meters below the water's surface! The creature seemed to consist of many individuals and I wondered whether or not each individual was the same kind of being because, when I looked closely, I counted eight different body forms!
>
> I was drawn closer and the true nature of this creature was painfully revealed. The beauty of the gas-filled float hid some of the most hideous weaponry imaginable. When I brushed against those silky tentacles I experienced the most excruciating pain. Had it not been for my life vest, I would have drowned. Indeed, for some time, I wished that had been my fate.

This fictitious account is not far from reality for swimmers of tropical waters who have come into contact with *Physalia physalis*, the Portuguese man-of-war (figure 6.1). In organisms such as *Physalia physalis*, cells are grouped together, specialized for various functions, and are interdependent. This chapter covers three animal phyla with multicellular organization that varies from a loose association of cells to cells organized into two distinct tissue layers. These phyla include the Porifera, Cnidaria, and Ctenophora (figure 6.2).

Figure 6.1
***Physalia physalis*, the Portuguese Man-of-War.** The bluish float is about 12 cm long, and the nematocyst-laden tentacles can be up to 9 m long. Nematocysts are lethal to small vertebrates and are dangerous to humans. Note the fish that has been captured by the tentacles. Digestion will eventually leave only the fish's skeletal remains.

Origins of Multicellularity

Multicellular life has been a part of the earth's history for approximately 550 million years. Although this seems a very long period of time, it represents only 10 percent of the earth's geological history. Multicellular life arose very quickly—in the 100 million years prior to the Precambrian/Cambrian boundary. What occurred during this 100 million years is often viewed as an evolutionary explosion. These evolutionary events resulted not only in the appearance of all of the 20 to 30 animal phyla recognized today, but also another 15 to 20 animal groups that are now extinct. Following this initial evolutionary explosion, most of the history of multicellular life has been one of extinction.

The evolutionary events leading to multicellularity are shrouded in mystery. (1) Many zoologists believe that multicellularity could have arisen as dividing cells remained together, in the fashion of many colonial protists. Although there are a number of variations of this hypothesis, they are all treated here as the **colonial hypothesis** (figure 6.3*a*).

A second proposed mechanism is called the **syncytial hypothesis** (figure 6.3*b*). A syncytium is a large, multinucleate cell. The formation of plasma membranes in the cytoplasm of a syncytial protist could have formed a small, multicellular organism. These hypotheses are supported by the fact that colonial and syncytial organization occurs in some protist phyla.

Animal Origins

A fundamental question concerning animal origins is whether animals are monophyletic (derived from a single ancestor), diphyletic (derived from two ancestors), or polyphyletic (derived from many ancestors). The view that animals are polyphyletic is attractive to a growing number of zoologists. The nearly simultaneous

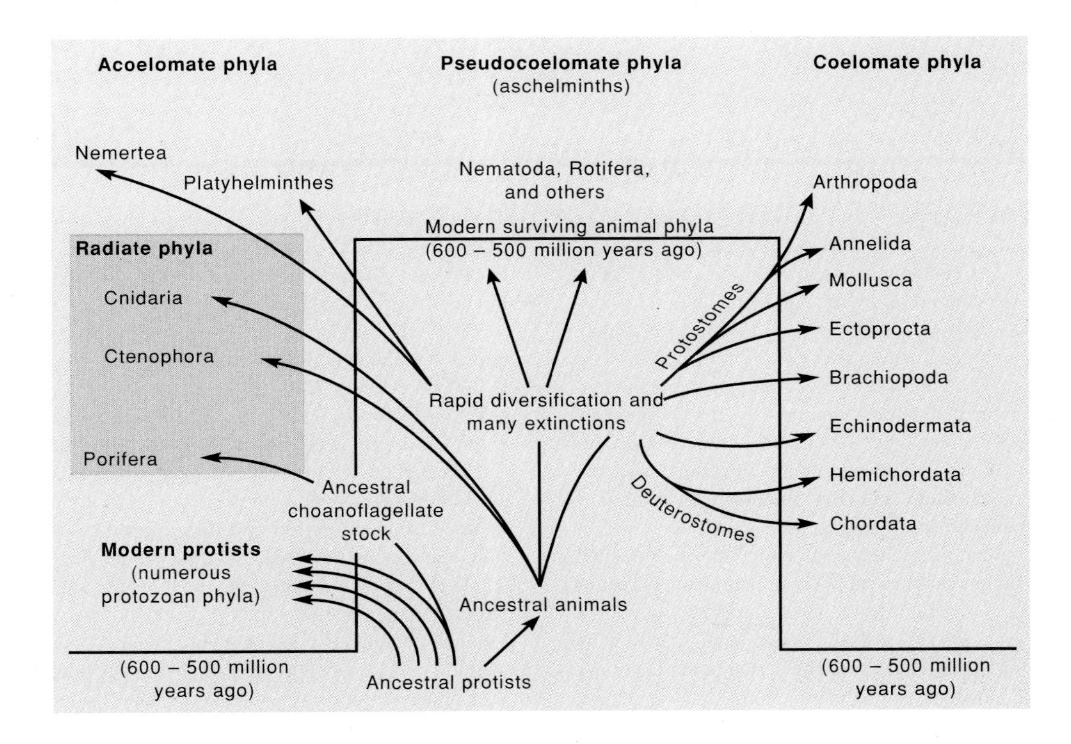

Figure 6.2

Evolutionary Relationships of the Poriferans and the Radiate Phyla. Members of the phylum Porifera are derived from ancestral protozoan stocks independently of other animal phyla. The radiate animals (shaded in orange) include members of the phyla Cnidaria and Ctenophora. This figure shows a diphyletic origin of the animal kingdom in which sponges are depicted as arising from the protists separate from other animals. Other interpretations of sponge origins are discussed in the text.

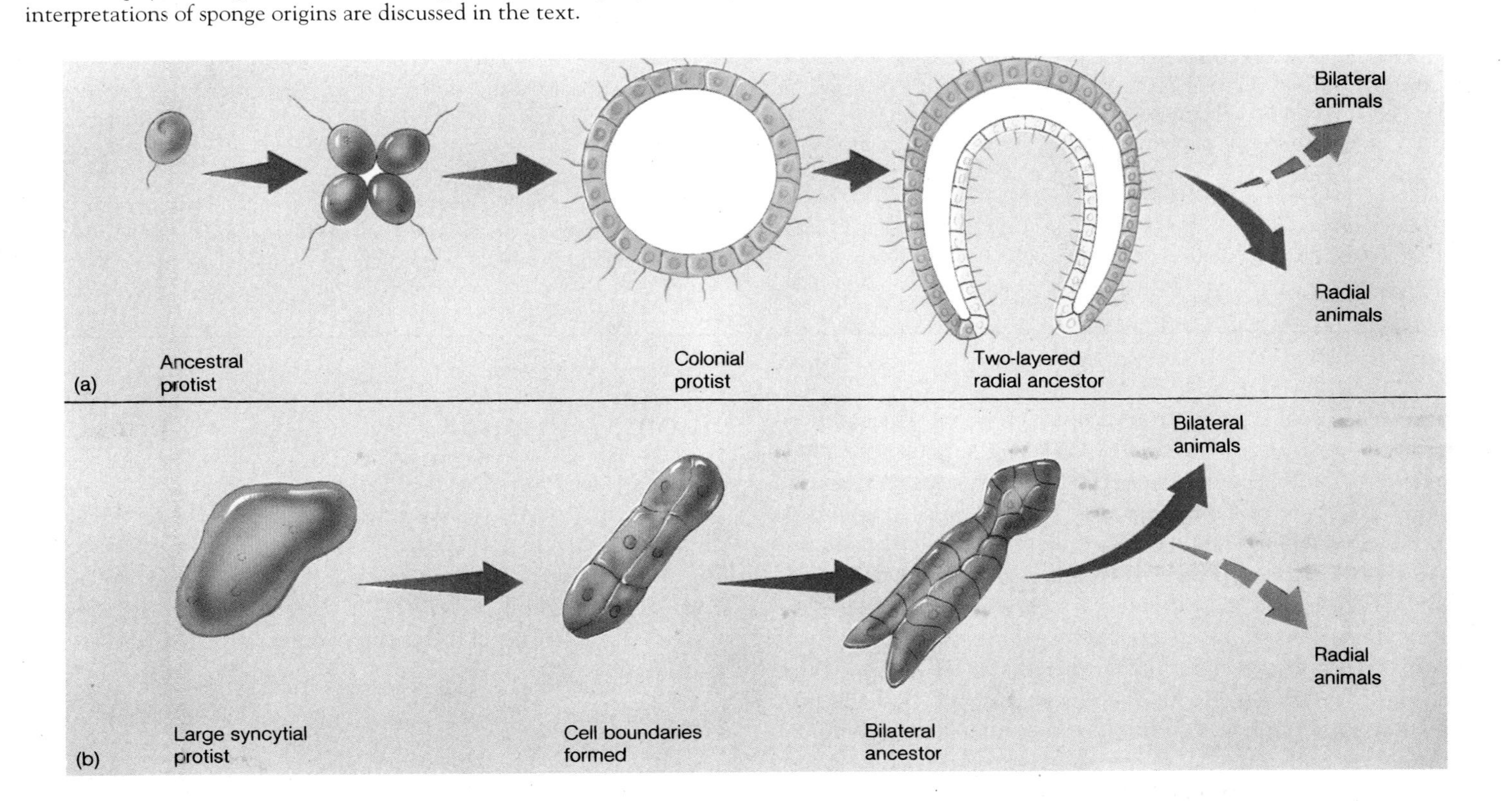

Figure 6.3

Two Hypotheses Regarding the Origin of Multicellularity. (*a*) The colonial hypothesis. Multicellularity may have arisen when cells produced by a dividing protist remained together. Invagination of cells could have formed a second cell layer. This hypothesis is supported by colonial organization of some Sacromastigophora. (The colonial protist and the two-layered radial ancestor are shown in sectional views.) (*b*) The syncytial hypothesis. Multicellularity could have arisen when plasma membranes were formed within the cytoplasm of a large, multinucleate protist. This hypothesis is supported by multinucleate, bilateral ciliates.

BOX 6.1 Animal Origins—The Cambrian Explosion

The geological timescale (*see table 1.3*) is marked by significant geological and biological events including the origin of the earth about 4.6 billion years ago, the origin of life about 3.5 billion years ago (*see box 2.1*), the origin of eukaryotic life-forms about 1.5 billion years ago, and the origin of animals about 0.6 billion years ago. The latter event marks the beginning of the Cambrian period. The origin of animals came relatively late in the history of the earth—only about 10% of the earth's history is marked by the presence of animals. During a geologically brief 100 million-year period, all modern animal phyla (along with other animals that are now extinct) evolved. This rapid origin and diversification of animals is often referred to as "the Cambrian explosion."

Important questions about this explosion have been asked ever since the time of Charles Darwin. Why did it occur so late in the history of the earth? The origin of multicellularity would seem a relatively simple step in comparison to the origin of life itself. Why are there no fossil records that document the series of evolutionary changes that occurred during the evolution of the animal phyla? Why did the evolution of animal life occur so quickly? Paleontologists continue to search the fossil records for answers to these kinds of questions.

One interpretation regarding the absence of fossils during this important 100 million-year period is that early animals were soft-bodied and simply did not fossilize. Fossilization of soft-bodied animals is less likely than fossilization of hard-bodied animals, but it does occur. Conditions that promote fossilization of soft-bodied animals include very rapid covering by sediments that creates an anoxic environment that discourages decomposition. In fact, fossil beds containing soft-bodied animals have been known for many years.

The Ediacara fossil formation, which contains the oldest known animal fossils, is made up exclusively of soft-bodied forms. Although it is named after a site in Australia, the Ediacara formation is worldwide in distribution and dates into Precambrian times. This 700 million-year-old formation gives us few clues to the origin of modern animals, however, because it is believed to represent an evolutionary experiment that failed. It contains no ancestors of modern animal phyla.

A slightly younger fossil formation containing animal remains is called the Tommotian formation—named after a locale in Russia. It dates to the very early Cambrian period and it also contains only soft-bodied forms. At one time, the animals present in these fossil beds were assigned to various modern phyla, including the Porifera and Cnidaria, but most paleontologists now agree that all Tommotian fossils represent unique body forms that arose early in the Cambrian period and disappeared before the end of the period, leaving no descendants in the modern animal phyla.

A third fossil formation containing soft-bodied animals provides evidence of the results of the Cambrian explosion. This fossil formation, called the Burgess Shale, is in Yoho National Park in the Canadian Rocky Mountains of British Columbia. Shortly after the Cambrian explosion, mud slides rapidly buried thousands of marine animals under conditions that favored fossilization. These fossil beds provide evidence of virtually all of the 31 phyla described in this textbook plus about 20 other animal body forms. These animal body forms are so different from any modern animals that they cannot be assigned to any modern phyla (figure 1). These unassignable animals include a large swimming predator called *Anomalocaris* and a soft-bodied detritus-eating or algae-eating animal called *Wiwaxia*. Not only are there unique body forms in the Burgess Shale, but there are also fossils of many extinct representatives of modern phyla. For example, a well-known Burgess Shale animal called *Sidneyia* is a representative of a previously unknown group of arthropods (insects, spiders, mites, crabs).

There are many lessons that have been, and will be, learned from fossil formations like the Burgess Shale. One of these is that evolution cannot always be thought of as a slow progression. The story of the Cambrian explosion involves very rapid evolutionary

appearance of all animal phyla in fossils from the Precambrian/Cambrian boundary is difficult to explain if animals are monophyletic. If animals are polyphyletic, more than one explanation of the origin of multicellularity could be possible and more than one body form could be ancestral. Conversely, the impressive similarities in cellular organization in all animals support the view that all or most animals are derived from a single ancestor. For example, asters are formed during mitosis in most animals, certain cell junctions are similar in all animal cells, flagellated sperm are produced by most animals, and the proteins that accomplish movement are similar in most animal cells. These common features are difficult to explain, assuming polyphyletic origins. If one assumes one or two ancestral lineages, then only one or two hypotheses regarding the origin of multicellularity can be correct (box 6.1).

Phylum Porifera

The Porifera (po-rif′er-ah) (L. *porus,* pore + *fera,* to bear), or sponges, are mostly marine animals consisting of loosely organized cells (figure 6.4; table 6.1). There are about 9,000 species of sponges, which vary in size from less than a centimeter to a mass that would fill one's arms.

Characteristics of the phylum Porifera include the following:

1. Asymmetrical or radially symmetrical
2. Three cell types: pinacocytes, mesenchyme cells, and choanocytes
3. Central cavity, or a series of branching chambers, through which water is circulated during filter feeding
4. No tissues or organs

diversification. A remarkable diversity of forms is seen in the Burgess Shale, but this diversity did not last. After an initial diversification, the story of evolution involved the extinction of many unique animals. Why was this evolution so very rapid? No one really knows. Many zoologists believe it was because there were so many ecological niches available and virtually no competition from existing species. Will we ever know the evolutionary sequences involved in the Cambrian explosion? Perhaps another ancient fossil bed of soft-bodied animals from 600 million-year-old seas is waiting for discovery.

FIGURE 1 **The Burgess Shale.** An artist's reconstruction of the Burgess Shale. The Burgess Shale contained numerous unique forms of animal life as well as representatives of most animal phyla described in this textbook. A trilobite is shown on the lower left. Tall sponges are shown on the right and left in the foreground. *Sidneyia* is shown on the seafloor in the middle foreground and in the middle left.

Cell Types, Body Wall, and Skeletons

In spite of their relative simplicity, sponges are more than colonies of independent cells. As in all animals, sponge cells are specialized for particular functions. This organization is often referred to as division of labor.

Thin, flat cells, called **pinacocytes,** line the outer surface of a sponge. Pinacocytes may be mildly contractile, and their contraction may change the shape of some sponges. In a number of sponges, some pinacocytes are specialized into tubelike, contractile **porocytes,** which can regulate water circulation (figure 6.5*a*). Openings through porocytes are pathways for water moving through the body wall.

Just below the pinacocyte layer of a sponge is a jellylike layer referred to as the **mesohyl** (Gr. *meso,* middle + *hyl,* matter). Amoeboid cells are found moving about in the mesohyl and are specialized for reproduction, secreting skeletal elements, transporting food, storing food, and forming contractile rings around openings in the sponge wall.

Below the mesohyl and lining an inner chamber(s) are choanocytes, or collar cells. **Choanocytes** (Gr. *choane,* funnel + *cyte,* cell) are flagellated cells that have a collarlike ring of microvilli surrounding a flagellum. Microfilaments connect the microvilli, forming a netlike mesh within the collar. The flagellum creates water currents through the sponge, and the collar filters microscopic food particles from the water (figure 6.5*b*).

Figure 6.4
Phylum Porifera. Many sponges are brightly colored, commonly with hues of red, orange, green, or yellow. (*a*) *Verongia*. (*b*) *Axiomella*.

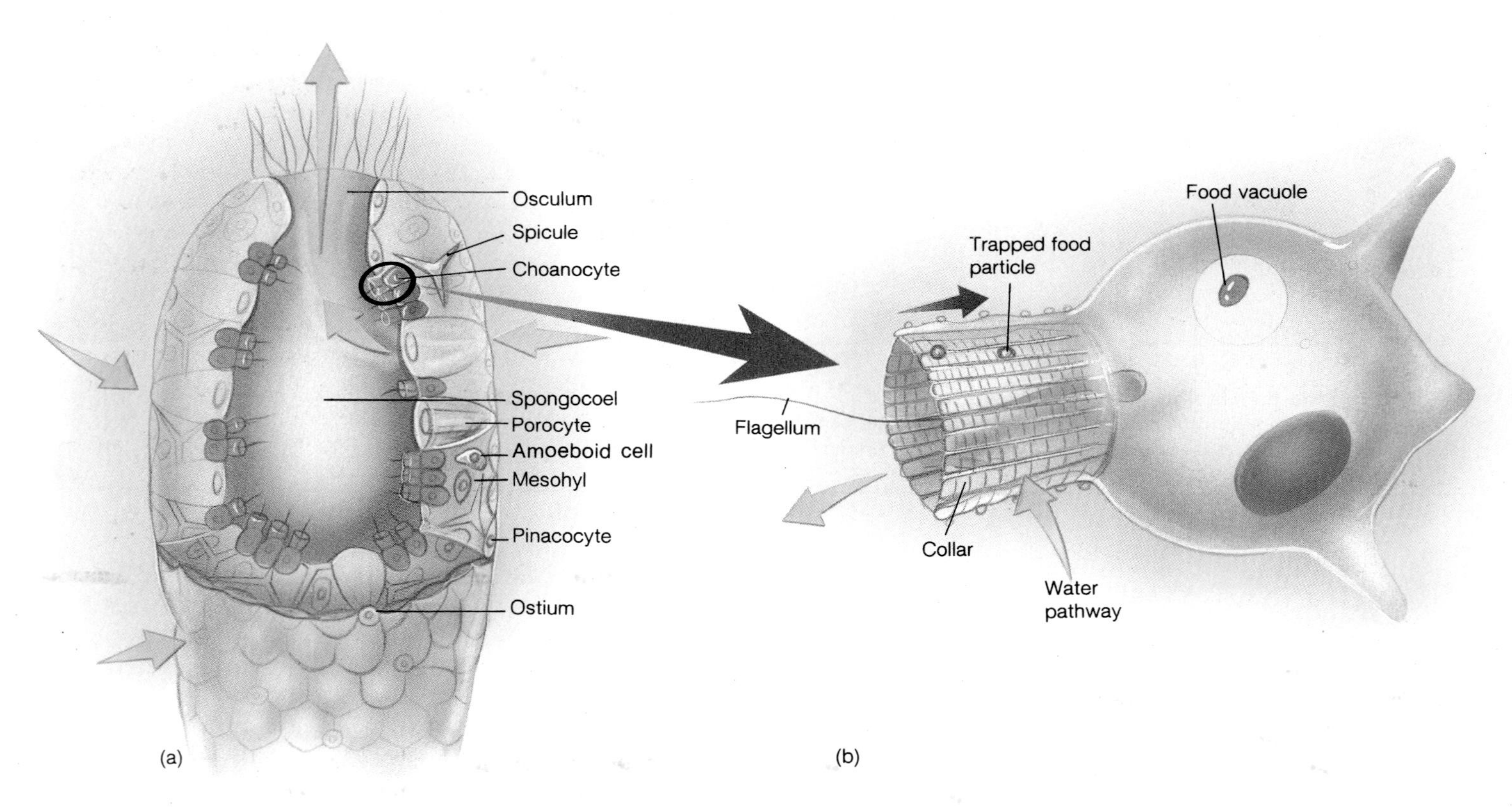

Figure 6.5
Morphology of a Simple Sponge. (*a*) In this example, pinacocytes form the outer body wall, and mesenchyme cells and spicules are found in the mesohyl. Ostia are formed by porocytes that extend through the body wall. (*b*) Choanocytes are cells with a flagellum surrounded by a collar of microvilli that traps food particles. Food is moved toward the base of the cell, where it is incorporated into a food vacuole and passed to amoeboid mesenchyme cells, where digestion takes place. Blue arrows show water flow patterns. The brown arrow shows the direction of movement of trapped food particles.

TABLE 6.1 CLASSIFICATION OF THE PORIFERA*

Phylum Porifera (po-rif'er-ah)
The animal phylum whose members are sessile and either asymmetrical or radially symmetrical; body organized around a system of water canals and chambers; cells not organized into tissues or organs.

Class Calcarea (kal-kar'ea)
Spicules composed of calcium carbonate; spicules needle shaped or with three or four rays; ascon, leucon, or scyon body forms; all marine. Calcareous sponges. *Grantia (Scypha)*, *Leucosolenia*.

Class Hexactinellida (hex-act'in-el'id-ah)
Spicules composed of silica and usually six rayed; spicules often fused into an intricate lattice; cup or vase shaped; sycon or leucon body form; found at 450 to 900 m depths in tropical West Indies and eastern Pacific. Glass sponges. *Euplectella* (Venus flower-basket).

Class Demospongiae (de-mo-spun'je-e)
Brilliantly colored sponges with needle-shaped or four-rayed siliceous spicules or spongin or both; leucon body form; up to 1 m in height and diameter. Includes one family of freshwater sponges, Spongillidae, and the bath sponges. *Cliona*, *Spongilla*.

*The class Sclerospongiae has been recently abandoned and its members assigned to Calcarea and Demospongiae.

The presence of choanocytes in sponges suggests an evolutionary link between the sponges and a group of protists called choanoflagellates. This link is discussed further at the end of this chapter.

Sponges are supported by a skeleton that may consist of microscopic needlelike spikes called **spicules.** Spicules are formed by ameboid cells, are made of calcium carbonate or silica, and may take on a variety of shapes (figure 6.6). ② Alternatively, the skeleton may be made of **spongin** (a fibrous protein made of collagen), which is dried, beaten, and washed until all cells are removed to produce a commercial sponge. The nature of the skeleton is an important characteristic in sponge taxonomy.

WATER CURRENTS AND BODY FORMS

The life of a sponge depends on the water currents created by choanocytes. Water currents bring food and oxygen to a sponge and carry away metabolic and digestive wastes. The way in which food filtration and circulation are accomplished is reflected in the body forms present in the phylum. Zoologists have described three sponge body forms.

The simplest and least common sponge body form is the **ascon** (figure 6.7*a*). Ascon sponges are vaselike. Ostia are the outer openings of porocytes and lead directly to a chamber called the spongocoel. Choanocytes line the spongocoel, and their flagellar movements draw water into the spongocoel through the ostia. Water exits the sponge through the osculum, which is a single, large opening at the top of the sponge.

FIGURE 6.6
Sponge Spicules. Photomicrograph of a variety of sponge spicules (× 150).

In the **sycon** body form, the sponge wall appears folded (figure 6.7*b*). Water enters a sycon sponge through openings called dermal pores. Dermal pores are the openings of invaginations of the body wall, called incurrent canals. Pores in the body wall connect incurrent canals to radial canals, and the radial canals lead to the spongocoel. Choanocytes line radial canals (rather than the spongocoel), and the beating of choanocyte flagella moves water from ostia, through incurrent and radial canals, to the spongocoel, and out the osculum.

Leucon sponges have an extensively branched canal system (figure 6.7*c*). Water enters the sponge through ostia and moves through branched incurrent canals, which lead to choanocyte chambers. Canals leading away from the chambers are called excurrent canals. Proliferation of chambers and canals has resulted in the absence of a spongocoel, and often multiple exit points (oscula) for water leaving the sponge.

③ In complex sponges, an increased surface area for choanocytes results in large volumes of water being moved through the sponge and greater filtering capabilities. Although the evolutionary pathways in the phylum are complex and incompletely described, most pathways have resulted in the leuconoid body form.

MAINTENANCE FUNCTIONS

Sponges feed on particles that are in the 0.1 to 50 μm size range. Their food consists of bacteria, microscopic algae, protists, and other suspended organic matter. Recent investigations have discovered that a few sponges are carnivorous. These deep-water sponges (*Asbestopluma*) can capture small crustaceans using spicule-covered filaments. The prey are slowly drawn into the sponge and consumed. Large populations of sponges play an important role in reducing turbidity of coastal waters. A single leuconoid sponge, 1 cm in diameter and 10 cm high, can filter in excess of 20 liters of water every day!

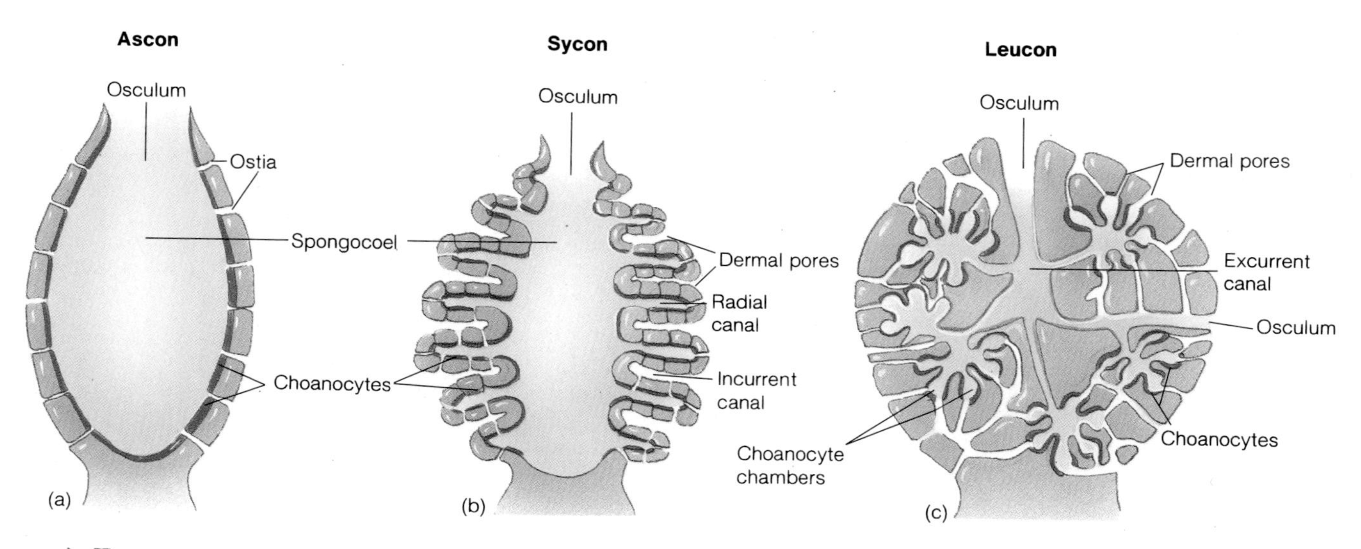

Figure 6.7
Sponge Body Forms. (*a*) An asconoid sponge. Choanocytes line the spongocoel in ascon sponges. (*b*) A syconoid sponge. The body wall of sycon sponges appears folded. Choanocytes line radial canals that open into the spongocoel. (*c*) A leuconoid sponge. The proliferation of canals and chambers has resulted in the loss of the spongocoel as a distinct chamber. Multiple oscula are frequently present.

Choanocytes filter small, suspended food particles. Water passes through their collar near the base of the cell then moves into a sponge chamber at the open end of the collar. Suspended food is trapped on the collar and moved along microvilli to the base of the collar, where it is incorporated into a food vacuole (*see figure* 6.5b). Digestion begins in the food vacuole by lysosomal enzymes and pH changes. Partially digested food is passed to amoeboid cells, which distribute it to other cells.

Filtration is not the only way that sponges feed. Larger food particles (up to 50 μm in size) may be phagocytized by pinacocytes lining incurrent canals. Nutrients dissolved in seawater may also be absorbed by active transport.

Because of extensive canal systems and the circulation of large volumes of water through sponges, all sponge cells are in close contact with water. Thus, the loss of nitrogenous wastes (principally ammonia) and gas exchange occur by diffusion.

Sponges do not have nerve cells to coordinate body functions. Most reactions are the result of individual cells responding to a stimulus. For example, water circulation through some sponges is at a minimum at sunrise and at a maximum just before sunset because light inhibits the constriction of porocytes and other cells surrounding ostia, keeping incurrent canals open. 4 Other reactions, however, suggest some communication between cells. For example, the rate of water circulation through a sponge can drop suddenly without any apparent external cause. This reaction can be due only to choanocytes ceasing activities more or less simultaneously and implies some form of internal communication. The nature of this communication is unknown. Chemical messages transmitted by amoeboid cells and ion movement over cell surfaces are possible control mechanisms.

Reproduction

Most sponges are monoecious (both sexes occur in the same individual) but do not usually undergo self-fertilization because they produce eggs and sperm at different times. Certain choanocytes lose their collars and flagella and undergo meiosis to form flagellated sperm. Other choanocytes (and ameboid cells in some sponges) probably undergo meiosis to form eggs. Eggs are retained in the mesohyl of the parent. Sperm cells exit one sponge through the osculum and enter another sponge with the incurrent water. Sperm are trapped by choanocytes and incorporated into a vacuole. The choanocytes lose their collar and flagellum, become ameboid, and transport sperm to the eggs.

In most sponges, early development occurs in the mesohyl. Cleavage of a zygote results in the formation of a flagellated larval stage. (A **larva** is an immature stage that may undergo a dramatic change in structure before attaining the adult body form.) The larva breaks free and is carried out of the parent sponge by water currents. After no more than 2 days of a free-swimming existence, the larva settles to the substrate and begins development of the adult body form (figure 6.8*a*,*b*).

Asexual reproduction of freshwater and some marine sponges involves the formation of resistant capsules containing masses of amoeboid cells. These capsules, called **gemmules,** are released when the parent sponge dies in the winter and can survive both freezing and drying (figure 6.8*c*). When favorable conditions return in the spring, amoeboid cells stream out of a tiny opening, called the micropyle, and organize themselves into a sponge.

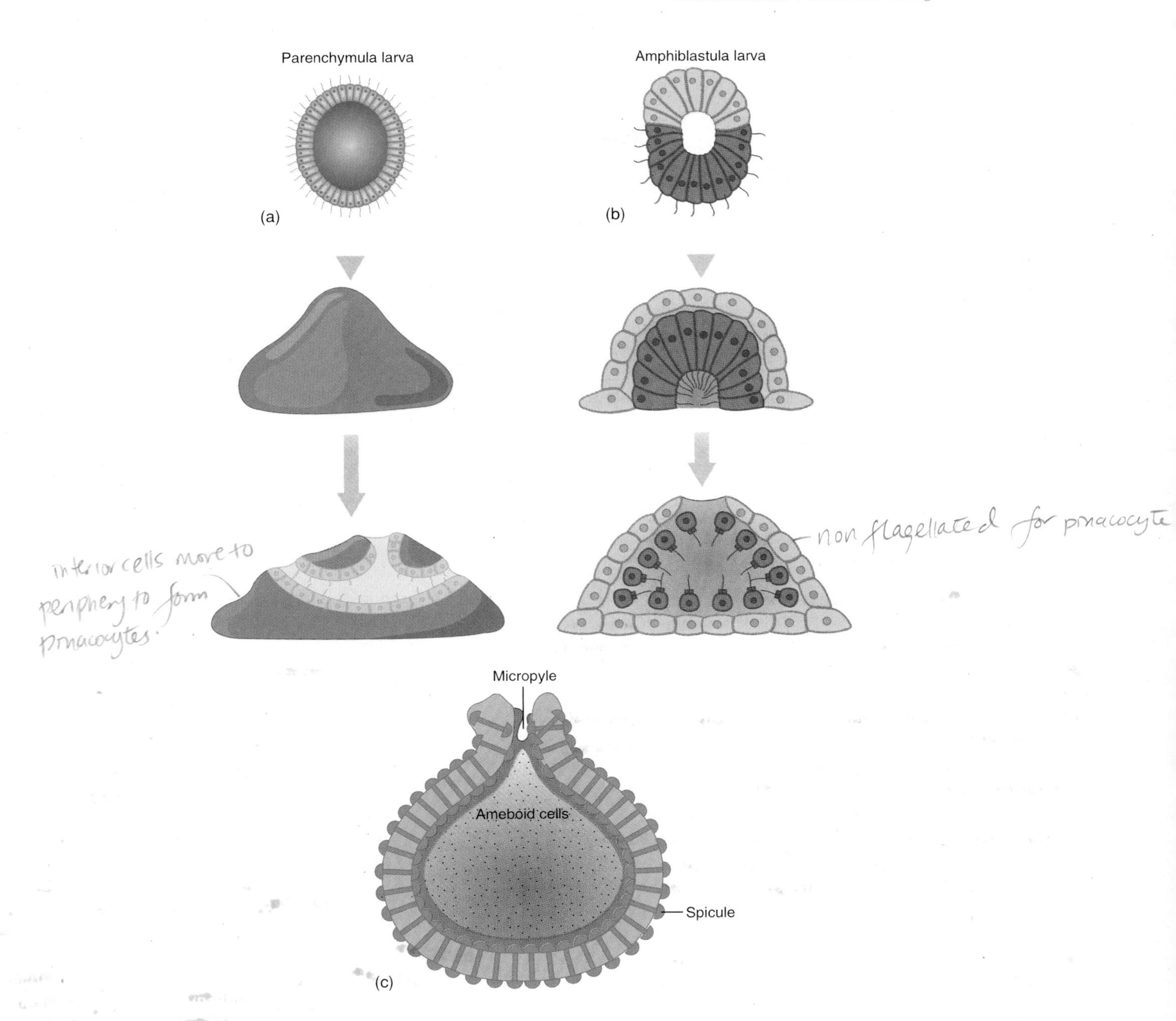

Figure 6.8

Development of Sponge Larval Stages. (*a*) Most sponges have a parenchymula larva. Flagellated cells cover most of the outer surface of the larva. After the larva settles and attaches, the outer cells lose their flagella, move to the interior, and form choanocytes. Interior cells move to the periphery and form pinacocytes. (*b*) Some sponges have an amphiblastula larva, which is hollow and has $\frac{1}{2}$ of the larva composed of flagellated cells. On settling, the flagellated cells invaginate into the interior of the embryo and will form choanocytes. Nonflagellated cells overgrow the choanocytes and form the pinacocytes. (*c*) Gemmules are resistant capsules containing masses of amoeboid cells. Gemmules are released when a parent sponge dies (e.g., in the winter) and amoeboid cells form a new sponge when favorable conditions return.

Some sponges possess remarkable powers of regeneration. Portions of a sponge that are cut or broken from one individual will regenerate new individuals.

Stop and Ask Yourself

1. How do the "colonial" and "syncytial" hypotheses account for the origin of multicellularity?
2. What evidence supports the idea that the kingdom Animalia is polyphyletic?
3. What are three kinds of cells found in the Porifera and what are their functions?
4. What is the path of water circulating through an ascon sponge? Through a sycon sponge? Through a leucon sponge?

Phylum Cnidaria (Coelenterata)

Members of the phylum Cnidaria (ni-dar'e-ah) (Gr. *knide*, nettle) possess radial or biradial symmetry. Biradial symmetry is a modification of radial symmetry in which a single plane, passing through a central axis, divides the animal into mirror images. It results from the presence of a single or paired structure in a basically radial animal and differs from bilateral symmetry in that there is no distinction between dorsal and ventral surfaces. Radially symmetrical animals have no anterior or posterior ends. Thus, terms of direction are based on the position of the mouth opening. Recall that the end of the animal that contains the mouth is the oral end, and the opposite end is the aboral end. Radial symmetry is advantageous for sedentary animals because sensory receptors are evenly distributed around the body. These organisms can respond to stimuli that come from all directions.

The Cnidaria include over 9,000 species, are mostly marine, and are very important in coral reef ecosystems (table 6.2).

Characteristics of the phylum Cnidaria include the following:

1. Radial or biradial symmetry
2. Diploblastic, tissue-level organization
3. Gelatinous mesoglea located between the epidermal and gastrodermal tissue layers
4. Gastrovascular cavity
5. Nervous system in the form of a nerve net
6. Specialized cells, called cnidocytes, used in defense, feeding, and attachment

TABLE 6.2 Classification of the Cnidaria

Phylum Cnidaria (ni-dar'e-ah)
Radial or biradial symmetry, diploblastic organization, a gastrovascular cavity, and cnidocytes.

Class Hydrozoa (hi'dro-zo"ah)
Cnidocytes present in the epidermis; gametes produced epidermally and always released to the outside of the body; no wandering mesenchyme cells in mesoglea; medusae usually with a velum; many polyps colonial; mostly marine with some freshwater species. *Hydra, Obelia, Gonionemus, Physalia.*

Class Scyphozoa (si'fo-zo"ah)
Medusa prominent in the life history; polyp small; gametes gastrodermal in origin and released into the gastrovascular cavity; cnidocytes present in the gastrodermis as well as epidermis; medusa lacks a velum; mesoglea with wandering mesenchyme cells of epidermal origin; marine. *Aurelia.*

Class Cubozoa (ku'bo-zo"ah)
Medusa prominent in life history; polyp small; gametes gastrodermal in origin; medusa cuboidal in shape with tentacles that hang from each corner of the bell; marine. *Chironex.*

Class Anthozoa (an'tho-zo"ah)
Colonial or solitary polyps; medusae absent; cnidocytes present in the gastrodermis; gametes gastrodermal in origin; gastrovascular cavity divided by mesenteries that bear nematocysts; internal biradial or bilateral symmetry present; mesoglea with wandering mesenchyme cells; marine. Anemones and corals. *Metridium.*

The Body Wall and Nematocysts

Cnidarians possess diploblastic, tissue-level organization (*see figure* 4.9). Cells are organized into tissues that carry out specific functions, and all cells are derived from two embryological layers. The ectoderm of the embryo gives rise to an outer layer of the body wall, called the **epidermis,** and the inner layer of the body wall, called the **gastrodermis,** is derived from endoderm (figure 6.9). Cells of the epidermis and gastrodermis are differentiated into a number of cell types that function in protection, food gathering, coordination, movement, digestion, and absorption. Between the epidermis and gastrodermis is a jellylike layer called **mesoglea.** Cells are present in the middle layer of some cnidarians, but they have their origin in either the epidermis or the gastrodermis.

One kind of cell is characteristic of the phylum. Epidermal and/or gastrodermal cells called **cnidocytes** produce structures

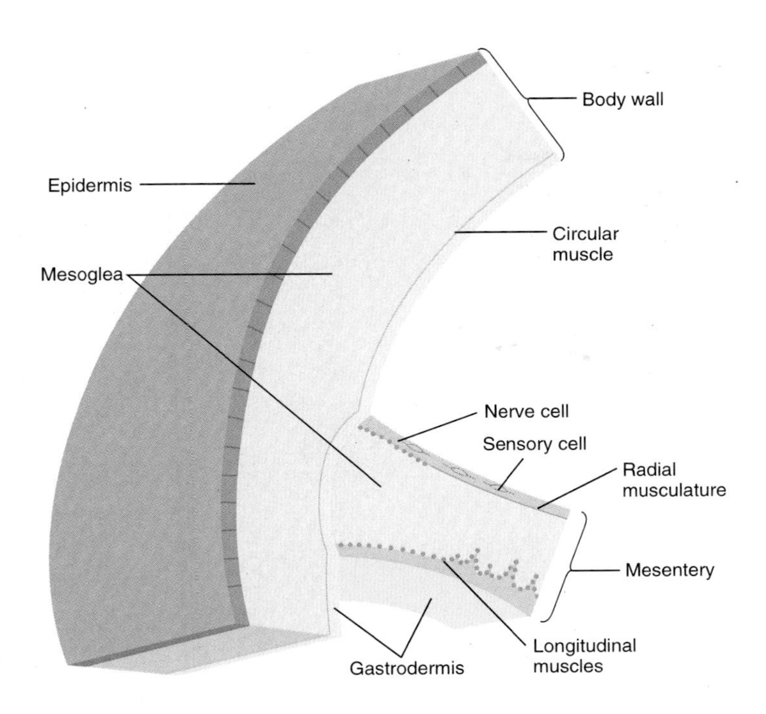

FIGURE 6.9

The Body Wall of a Cnidarian (Class Anthozoa). Cnidarians are diploblastic (two tissue layers). The epidermis is derived embryologically from ectoderm and the gastrodermis is derived embryologically from endoderm. Between these layers is mesoglea. *Source: After Bullock and Horridge.*

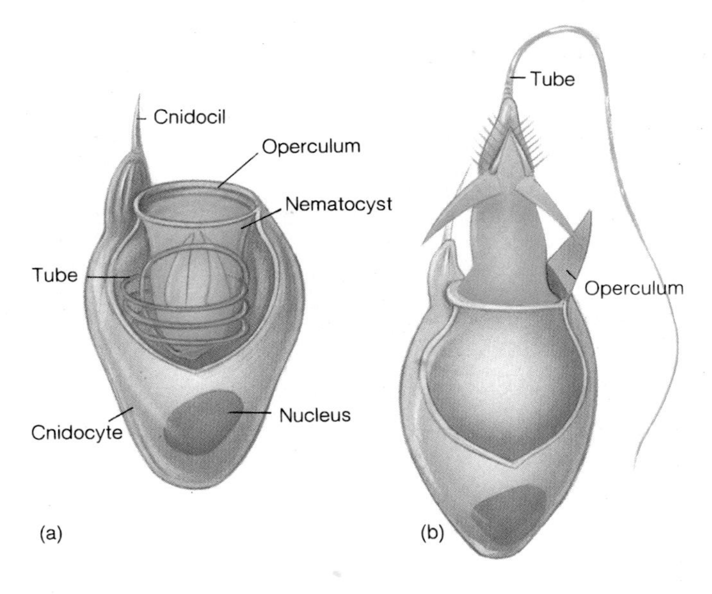

FIGURE 6.10

Cnidocyte Structure and Nematocyst Discharge. (*a*) A nematocyst develops in a capsule in the cnidocyte. The capsule is capped at its outer margin by an operculum (lid) that is displaced upon discharge of the nematocyst. The triggerlike cnidocil is responsible for nematocyst discharge. (*b*) A discharged nematocyst. When the cnidocil is stimulated, there is a rapid (osmotic) influx of water, causing the nematocyst to evert, first near its base, and then progressively along the tube from base to tip. The tube revolves at enormous speeds as the nematocyst is discharged. In nematocysts that are armed with barbs, the advancing tip of the tube is aided in its penetration of the prey as barbs spring forward from the interior of the tube and then flick backward along the outside of the tube.

called nematocysts, which are used for attachment, defense, and feeding. A **nematocyst** is a fluid-filled capsule enclosing a coiled, hollow tube (figure 6.10). The capsule is capped at one end by a lidlike operculum. The cnidocyte has a modified cilium, called a cnidocil. When the cnidocil is stimulated, the operculum is forced open and the coiled tube is discharged—as one would evert a sweater sleeve that had been turned inside out.

Nearly 30 kinds of nematocysts have been described. Nematocysts used in food gathering and defense may discharge a long tube armed with spines that penetrates the prey. The spines have hollow tips that discharge paralyzing toxins. Other nematocysts contain unarmed tubes that wrap around prey or a substrate. Still other nematocysts have sticky secretions that help the animal anchor itself. Six or more kinds of nematocysts may be present in one individual.

ALTERNATION OF GENERATIONS

Most cnidarians possess two body forms in their life histories (figure 6.11). The **polyp** is usually asexual and sessile. It is attached to a substrate at the aboral end, has a cylindrical body, called the column, and a mouth surrounded by food-gathering tentacles. The **medusa** (plural medusae) is dioecious and free swimming. It is shaped like an inverted bowl and has tentacles dangling from its margins. The mouth opening is centrally located facing downward, and the medusa swims by gentle pulsations of the body wall. The mesoglea is more abundant in a medusa than in a polyp, giving the former a jellylike consistency.

MAINTENANCE FUNCTIONS

The gastrodermis of all cnidarians lines a blind-ending cavity, called the **gastrovascular cavity.** This cavity functions in digestion, the exchange of respiratory gases and metabolic wastes, and for discharge of gametes. Food, digestive wastes, and reproductive stages enter and leave the gastrovascular cavity through the mouth.

The food of most cnidarians consists of very small crustaceans. Nematocysts entangle and paralyze prey, contractile cells in the tentacles cause the tentacles to shorten, and food is drawn

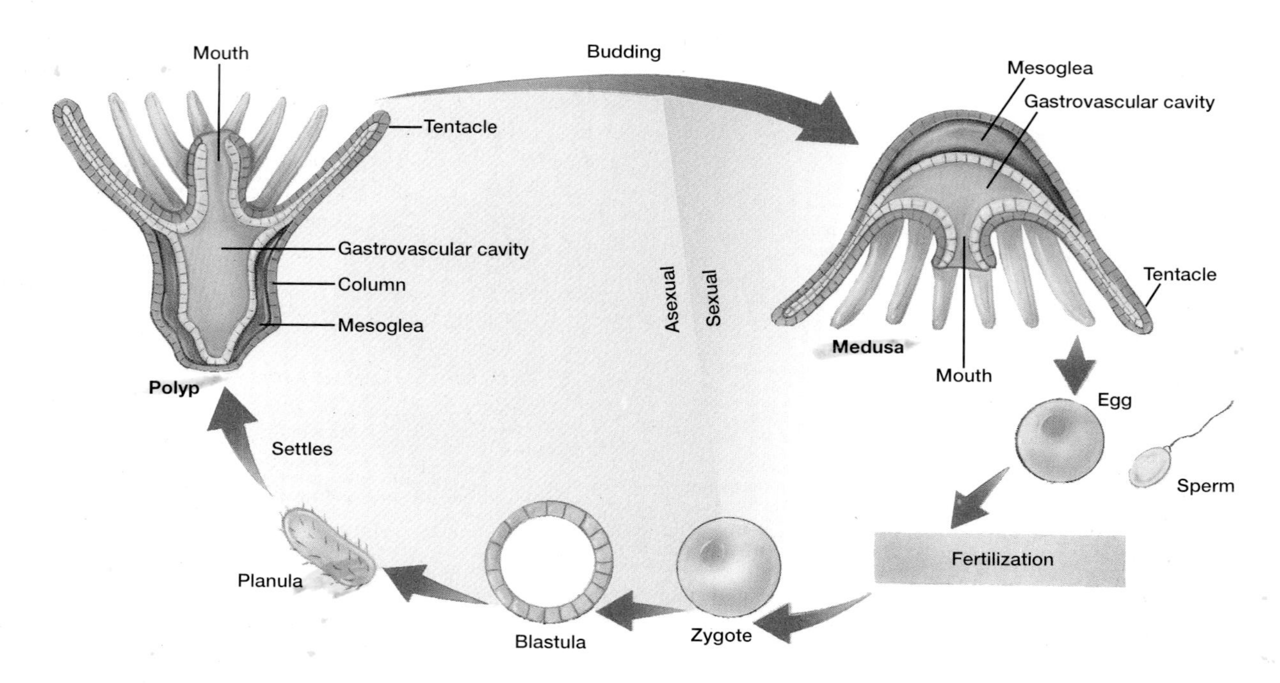

Figure 6.11

A Generalized Cnidarian Life Cycle. This figures show alternation between medusa and polyp body forms. Dioecious medusae produce gametes that may be shed into the water, where fertilization takes place. Early development forms a ciliated planula larva. After a brief free swimming existence, the planula settles to the substrate and forms a polyp. Budding of the polyp produces additional polyps and medusa buds. Medusae break free of the polyp and swim away. The polyp or medusa stage of many species is either lost or reduced and the sexual and asexual stages have been incorporated into one body form.

toward the mouth. As food enters the gastrovascular cavity, gastrodermal gland cells secrete lubricating mucus and enzymes, which reduce food to a soupy broth. Certain gastrodermal cells, called nutritive-muscular cells phagocytize partially digested food and incorporate it into food vacuoles, where digestion is completed. Nutritive-muscular cells also have circularly oriented contractile fibers that help move materials into or out of the gastrovascular cavity by peristaltic contractions. During peristalsis, ringlike contractions move along the body wall, pushing contents of the gastrovascular cavity ahead of them, expelling undigested material through the mouth.

5 Cnidarians derive most of their support from the buoyancy of water around them. In addition, they have a hydrostatic skeleton to aid in support and movement. A **hydrostatic skeleton** is water or body fluids confined in a cavity of the body and against which contractile elements of the body wall act. In the Cnidaria, the water-filled gastrovascular cavity acts as a hydrostatic skeleton. Certain cells of the body wall, called epithelio-muscular cells, are contractile and aid in movement. When a polyp closes its mouth (to prevent water from escaping) and contracts longitudinal epithelio-muscular cells on one side of the body, the polyp bends toward that side. If these cells contract while the mouth is open, water escapes from the gastrovascular cavity, and the polyp collapses. Contraction of circular epithelio-muscular cells causes constriction of a part of the body and, if the mouth is closed, water in the gastrovascular cavity is compressed, and the polyp elongates.

Polyps use a variety of forms of locomotion. They may move by somersaulting from base to tentacles and from tentacles to base again, or move in an inchworm fashion, using their base and tentacles as points of attachment. Polyps may also glide very slowly along a substrate while attached at their base or walk on their tentacles.

Medusae move by swimming and floating. Most horizontal movements are from being passively carried by water currents and wind. Vertical movements are the result of swimming. Contractions of circular and radial epitheliomuscular cells cause rhythmic pulsations of the bell and drive water from beneath the bell, propelling the medusa through the water.

Cnidarian nerve cells have been of interest to zoologists for many years because they may be the most primitive nervous elements in the animal kingdom. By studying these cells, zoologists may gain insight into the evolution of animal nervous systems. Nerve cells are located below the epidermis, near the mesoglea, and interconnect to form a two-dimensional nerve net. This net conducts nerve impulses around the body in response to a localized stimulus. The extent to which a nerve impulse spreads over the body depends on the strength of a stimulus. For example, a weak stimulus applied to a polyp's tentacle may cause the tentacle to be retracted. A strong stimulus at the same point may cause the entire polyp to withdraw from the stimulus.

Sensory structures of cnidarians are distributed throughout the body and include receptors for perceiving touch and certain chemicals. More specialized receptors are located at specific sites on a polyp or medusa.

Because cnidarians have large surface area to volume ratios, all cells are a short distance from the body surface, and oxygen, carbon dioxide, and nitrogenous wastes can be exchanged by diffusion.

Reproduction

Most cnidarians are dioecious. Sperm and eggs may be released into the gastrovascular cavity or to the outside of the body. In some instances, eggs are retained in the parent until after fertilization.

A blastula forms early in development, and migration of surface cells to the interior fills the embryo with cells that will eventually form the gastrodermis. The embryo elongates to form a ciliated, free-swimming larva, called a **planula.** The planula attaches to a substrate, interior cells split to form the gastrovascular cavity, and a young polyp develops (*see figure 6.11*).

Medusae are nearly always formed by budding from the body wall of a polyp, and polyps may form other polyps by budding. Buds may detach from the polyp or they may remain attached to the parent to contribute to a colony of individuals. Variations on this general pattern will be discussed in the survey of cnidarian classes that follows.

Stop and Ask Yourself

5. What are the three layers of the cnidarian body wall?
6. What is a nematocyst? What are several functions of nematocysts?
7. How would you characterize the nervous organization of cnidarians?
8. What is a hydrostatic skeleton? How does this function in cnidarians?

Class Hydrozoa

Hydrozoans (hi′dro-zo″anz) are small, relatively common cnidarians. The vast majority are marine, but this is the one cnidarian class with freshwater representatives. Most hydrozoans have life cycles that display alternation of generations; however, in some the medusa stage is lost, while in others the polyp stage is very small.

Hydrozoans can be distinguished from other cnidarians by three features (*see table 6.2*). Nematocysts are only in the epidermis; gametes are epidermal and released to the outside of the body rather than into the gastrovascular cavity; and the mesoglea never contains amoeboid mesenchyme cells.

Most hydrozoans have colonial polyp forms, some of which may be specialized for feeding, producing medusae by budding, or defending the colony. In *Obelia*, a common marine cnidarian, the planula develops into a feeding polyp, called a **gastrozooid** (gas′tra-zo′oid) (figure 6.12). The gastrozooid has tentacles, feeds on microscopic organisms in the water, and secretes a skeleton of protein and chitin called the perisarc, around itself.

Growth of an *Obelia* colony results from budding of the original gastrozooid. Rootlike processes grow into and horizontally along the substrate. They anchor the colony and give rise to branch colonies. The entire colony has a continuous gastrovascular cavity, body wall, and perisarc, and is a few centimeters high. Gastrozooids are the most common type of polyp in the colony; however, as an *Obelia* colony grows, gonozooids are produced. A **gonozooid** (gon′o-zo′oid) is a reproductive polyp that produces medusae by budding. *Obelia*'s small medusae are formed on a stalklike structure of the gonozooid. When medusae mature, they break free of the stalk and swim out an opening at the end of the gonozooid. Medusae reproduce sexually to give rise to more colonies of polyps.

Gonionemus (figure 6.13*a*) is a hydrozoan in which the medusa stage predominates. It lives in shallow marine waters, where it is often found clinging to seaweeds by adhesive pads on its tentacles. The biology of *Gonionemus* is typical of most hydrozoan medusae. The margin of the *Gonionemus* medusa projects inward to form a shelflike lip, called the velum. A velum is found on most hydrozoan medusae but is absent in all other cnidarian classes. The velum concentrates water expelled from beneath the medusa to a smaller outlet, creating a jet-propulsion system. The mouth is at the end of a tubelike **manubrium** that hangs from the medusa's oral surface. The gastrovascular cavity leads from the inside of the manubrium into four radial canals that extend to the margin of the medusa. Radial canals are connected at the margin of the medusa by an encircling ring canal.

In addition to a nerve net, *Gonionemus* has a concentration of nerve cells, called a nerve ring, that encircles the margin of the medusa. The nerve ring coordinates swimming movements. Embedded in the mesoglea around the margin of the medusa are sensory structures called statocysts (figure 6.13*b*). A **statocyst** consists of a small sac surrounding a calcium carbonate concretion called a statolith. When *Gonionemus* tilts, the statolith moves in response to the pull of gravity, and nerve impulses are initiated, which may change the animal's swimming behavior.

Gonads of *Gonionemus* medusae hang from the oral surface, below the radial canals. *Gonionemus* is dioecious, and the

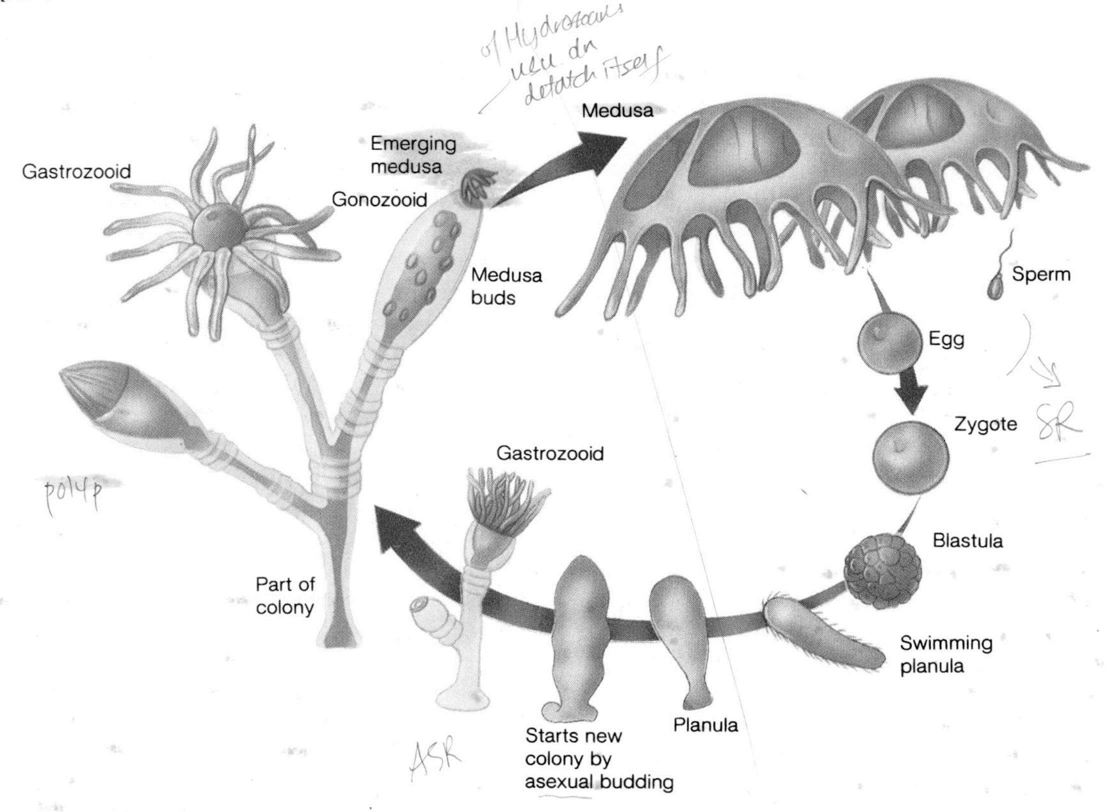

Figure 6.12

***Obelia* Structure and Life Cycle.** This hydrozoan displays alternation between polyp and medusa stages. Unlike *Obelia*, the majority of colonial hydrozoans have medusae that remain attached to the parental colony and gametes or larval stages are released from the medusa through the gonozooid. The medusa often degenerate and may be little more than gonadal specializations in the gonozooid.

(a)

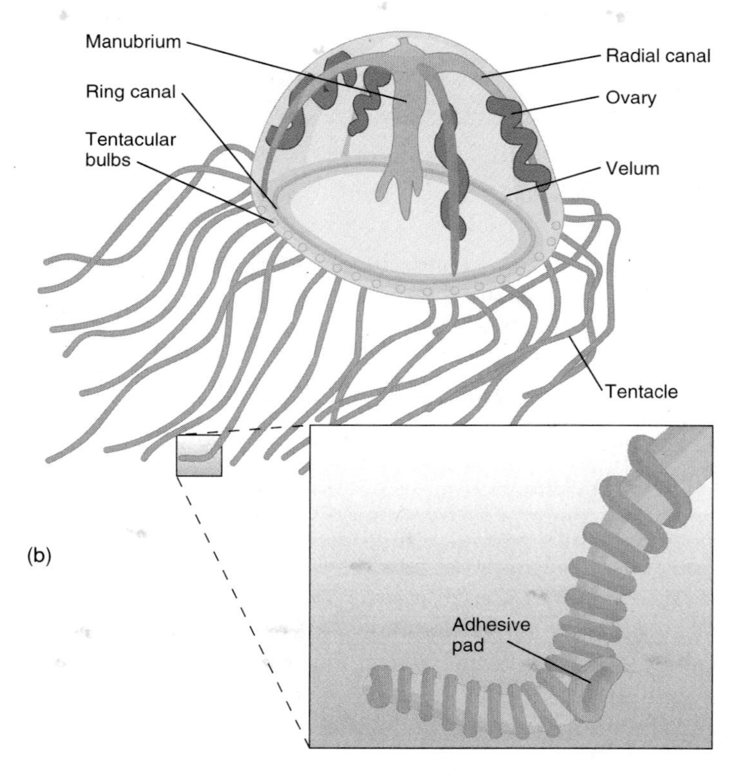

(b)

Figure 6.13

A Hydrozoan Medusa. (*a*) A *Gonionemus* medusa. (*b*) The structure of *Gonionemus*.

gametes are shed directly into seawater. A planula larva develops and attaches to the substrate, eventually forming a polyp (about 5 mm tall). The polyp reproduces by budding to make more polyps and medusae.

Hydra is a common freshwater hydrozoan that is found hanging from the underside of floating plants in clean streams and ponds. *Hydra* lacks a medusa stage and reproduces both asexually by budding from the side of the polyp and sexually. Hydras are somewhat unusual hydrozoans, because sexual reproduction occurs in the polyp stage. Testes are conical elevations of the body surface that form from mitosis of certain epidermal cells called interstitial cells. Sperm form by meiosis in the testes. Mature sperm exit the testes through temporary openings. Ovaries also form from interstitial cells. One large egg is formed per ovary. During egg formation, yolk is incorporated into the egg cell from gastrodermal cells. As ovarian cells disintegrate, the egg is left attached to the body wall by a thin stalk of tissue. After fertilization and early development, epithelial cells lay down a resistant chitinous shell. The embryo drops from the parent, overwinters, hatches in the spring, and develops into an adult.

Large oceanic hydrozoans belong to the order Siphonophora. These colonies are associations of numerous polypoid and medusoid individuals. Some polyps, called dactylozooids, possess a single, long (up to 9 m) tentacle armed with cnidocytes that are used in capturing prey. Other polyps are specialized for digesting prey. Various medusoid individuals form swimming bells, sac floats, oil floats, leaflike defensive structures, and gonads.

Physalia, commonly called the Portuguese man-of-war, is a very large, colonial siphonophore. It lacks swimming capabilities and moves at the mercy of wind and waves. 6 Its cnidocyte-laden dactylozooids are lethal to small vertebrates and dangerous to humans.

Class Scyphozoa

Members of the class Scyphozoa (si′fo-zo″ah) are all marine and are called "true jellyfish" because the dominant stage in their life history is the medusa (figure 6.14). Unlike hydrozoan medusae, scyphozoan medusae lack a velum, the mesoglea contains amoeboid mesenchyme cells, cnidocytes occur in the gastrodermis as well as the epidermis, and gametes are gastrodermal in origin (table 6.2).

Many scyphozoans are harmless to humans; others can deliver unpleasant and even dangerous stings. 7 For example, *Mastigias quinquecirrha*, the so-called stinging nettle, is a common Atlantic scyphozoan whose populations increase in late summer and become hazardous to swimmers (figure 6.14*a*). A rule of thumb for swimmers is to avoid helmet-shaped jellyfish with long tentacles and fleshy lobes hanging from the oral surface.

Aurelia is a common scyphozoan in both Pacific and Atlantic coastal waters of North America (figure 6.14*b*). The margin of its medusa has a fringe of short tentacles and is divided by notches. The mouth of *Aurelia* leads to a stomach with four gastric pouches, which contain cnidocyte-laden gastric filaments. Radial canals lead from gastric pouches to the margin of the bell. In *Aurelia*, but not all scyphozoans, the canal system is extensively branched and leads to a ring canal around the margin of the medusa. Gastrodermal cells of all scyphozoans possess cilia for the continuous circulation of seawater and partially digested food.

(a)

(b)

Figure 6.14

Representative Scyphozoans. (*a*) *Mastigias quinquecirrha*. (*b*) *Aurelia*.

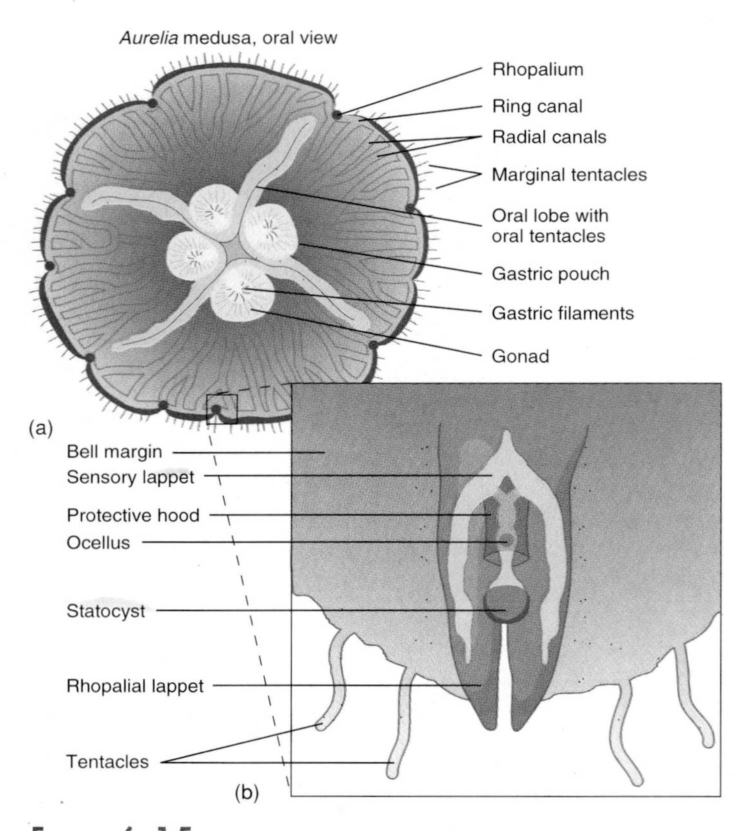

Figure 6.15

The Structure of a Scyphozoan Medusa. (*a*) Internal structure of *Aurelia*. (*b*) A section through a rhopalium of *Aurelia*. Each rhopalium consists of two sensory (olfactory) lappets, a statocyst, and a photoreceptor called an ocellus. *(b) Source: After L. H. Hyman,* Biology of the Invertebrates, *copyright 1940 McGraw-Hill Publishing Co.*

Aurelia is a plankton feeder. At rest, it sinks slowly in the water, and microscopic animals are trapped in mucus on its epidermal surfaces. This food is carried by cilia to the margin of the medusa. Four fleshy lobes, called oral lobes, hang from the manubrium and are used to scrape food from the margin of the medusa (figure 6.15*a*). Cilia on the oral lobes carry food to the mouth.

In addition to sensory receptors located on the epidermis, *Aurelia* has eight specialized structures, called rhopalia, located in the notches at the margin of the medusa. Each **rhopalium** (figure 6.15*b*) consists of two sensory pits (presumed to be olfactory), lappets (probably touch receptors), and a statocyst. Photoreceptors, called ocelli, are also associated with rhopalia. *Aurelia* displays a distinct negative phototaxis, coming to the surface at twilight and descending to greater depths during bright daylight.

Scyphozoans are dioecious. *Aurelia's* eight gonads are located in gastric pouches, two per pouch. Gametes are released into the gastric pouches. Sperm swim through the mouth to the outside of the medusa. In some scyphozoans, eggs are fertilized in the female's gastric pouches, and early development occurs there. In *Aurelia,* eggs lodge in the oral lobes, where fertilization and development to the planula stage occurs.

The planula develops into a polyp called a **scyphistoma** (figure 6.16). The scyphistoma lives a year or more, during which time budding produces miniature medusae, called **ephyrae.** Repeated budding of the scyphistoma results in ephyrae being stacked on the polyp—as one might pile saucers on top of one another. After ephyrae are released, they gradually attain the adult form.

Class Cubozoa

The class Cubozoa (ku′bo-zo″ah) was formerly classified as an order in the Scyphozoa. The medusa is cuboidal, and tentacles hang from each of its corners. Polyps are very small and, for some species, polyps are unknown. Cubozoans are very active swimmers and feeders in warm tropical waters. Some possess dangerous nematocysts (figure 6.17).

Class Anthozoa

Members of the class Anthozoa (an′tho-zo″ah) are colonial or solitary and lack medusae. They include anemones and stony and soft corals. Anthozoans are all marine and are found at all depths.

Anthozoan polyps differ from hydrozoan polyps in three respects. (1) The mouth of an anthozoan leads to a pharynx, which is an invagination of the body wall that leads into the gastrovascular cavity. (2) The gastrovascular cavity is divided into sections by mesenteries (membranes) that bear cnidocytes and gonads on their free edges. (3) The mesoglea contains amoeboid mesenchyme cells (table 6.2).

Externally, anthozoans appear to show perfect radial symmetry. Internally, the mesenteries and other structures convey biradial symmetry to members of this class.

Sea anemones are solitary, frequently large, and colorful (figure 6.18*a*). Some attach to solid substrates, some burrow in soft substrates, and some live in symbiotic relationships (figure 6.18*b*). The polyp attaches to its substrate by a pedal disk (figure 6.19). An oral disk contains the mouth and hollow, oral tentacles. At one or both ends of the slitlike mouth is a siphonoglyph, which is a ciliated tract that moves water into the gastrovascular cavity to maintain the hydrostatic skeleton.

Mesenteries are arranged in pairs. Some attach at the body wall at their outer margin and to the pharynx along their inner margin. Other mesenteries attach to the body wall, but are free along their entire inner margin. Openings in mesenteries near the oral disk permit circulation of water between compartments set off by the mesenteries. The free lower edges of the mesenteries form a trilobed mesenterial filament. Mesenterial filaments bear cnidocytes, cilia that aid in water circulation, gland cells that secrete digestive enzymes, and absorptive cells that absorb products of digestion. Threadlike acontia at

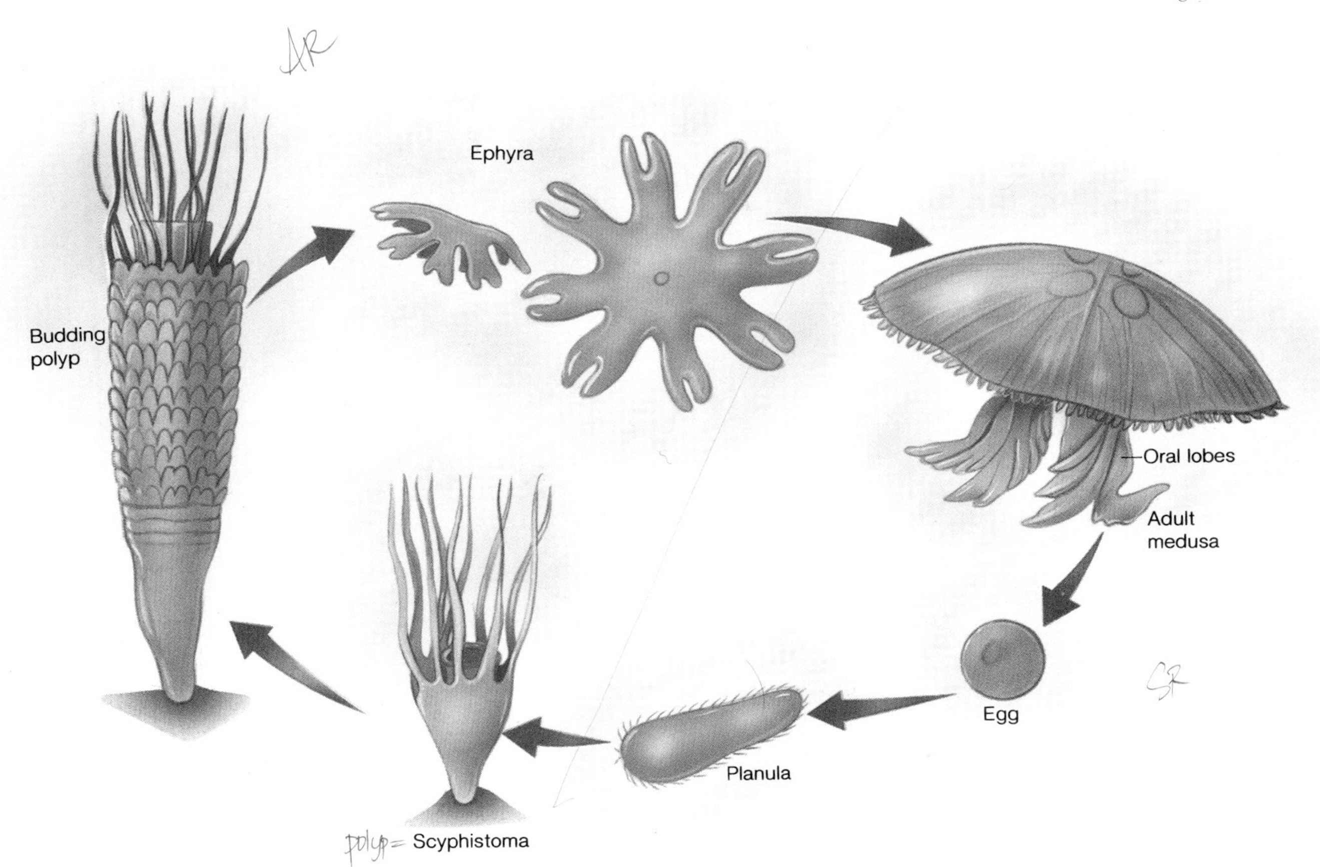

Figure 6.16

Aurelia Life History. *Aurelia* is dioecious and, like all scyphozoans, the medusa predominates in the life history of the organism. The planula develops into a polyp called a scyphistoma, which produces young medusae, or ephyrae, by budding.

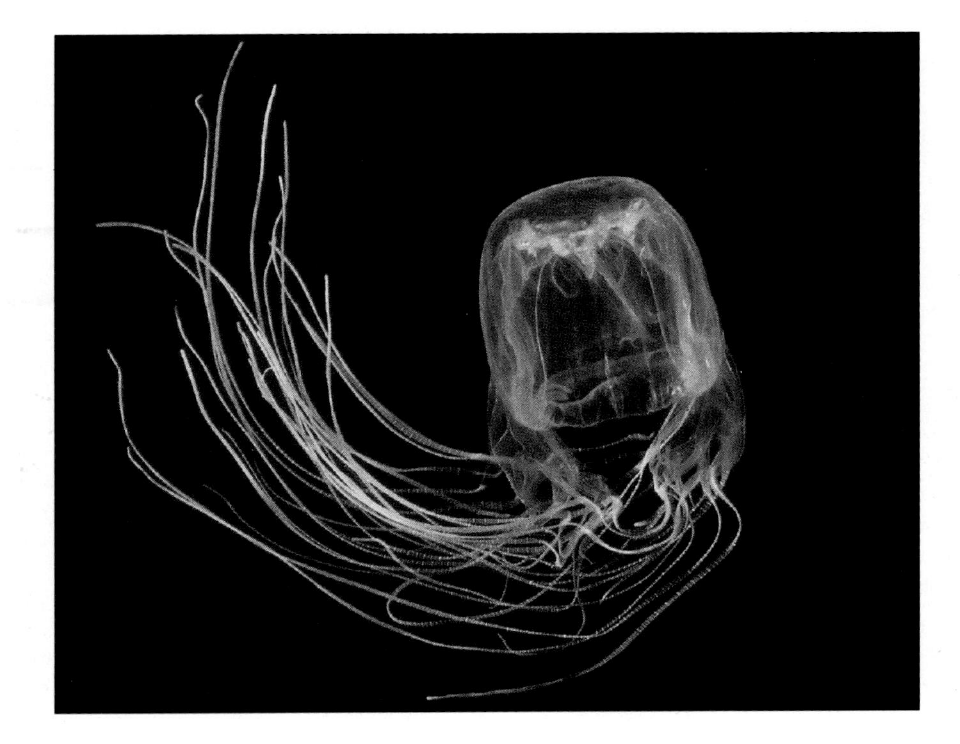

Figure 6.17

Class Cubozoa. The sea wasp, *Chironex fleckeri*. The medusa is cuboidal in shape; note the tentacles that hang from the corners of the bell. *Chironex fleckeri* has caused more human suffering and death off Australian coasts than the Portuguese man-of-war has in any of its home waters. Death from heart failure and shock is not likely unless one is repeatedly stung.

(a)

(b)

Figure 6.18

Representative Sea Anemones. (*a*) Giant sea anemone (*Anthopleura xanthogrammica*). (*b*) This sea anemone lives in a mutualistic relationship with a hermit crab (*Petrochirus diobenes*). Hermit crabs lack a heavily armored exoskeleton over much of their bodies and seek refuge in empty snail shells. When this crab outgrows its present home it will take its anemone with it to a new snail shell. This anemone, riding on the shell of the hermit crab, has a degree of mobility that is unusual for other anemones. The crab, in turn, is protected from predators by the anemone's nematocysts.

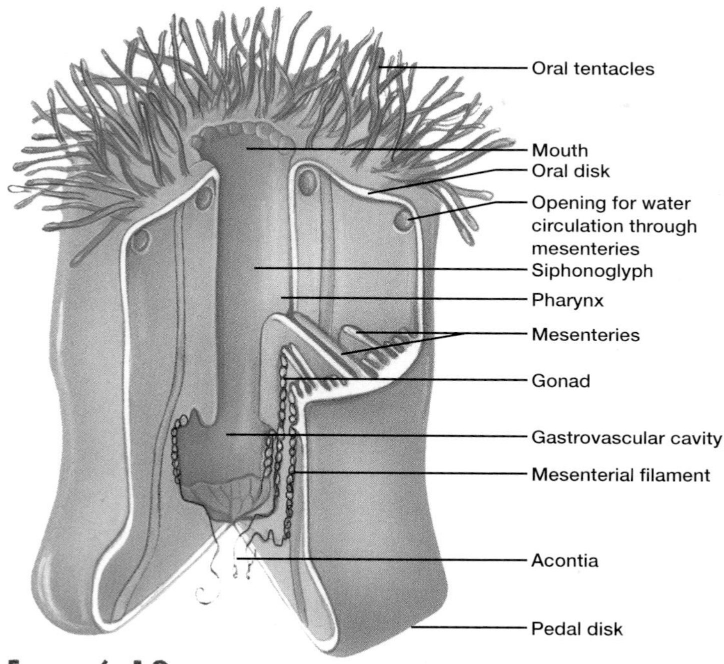

Figure 6.19

Class Anthozoa. The structure of the anemone, *Metridium*.

the ends of mesenterial filaments bear cnidocytes. Acontia are used in subduing live prey in the gastrovascular cavity and can be extruded through small openings in the body wall or through the mouth when an anemone is threatened.

Muscle fibers are largely gastrodermal. Longitudinal muscle bands are restricted to the mesenteries. Circular muscles are in the gastrodermis of the column. When threatened, anemones contract their longitudinal fibers, allowing water to escape from the gastrovascular cavity. This action causes the oral end of the column to fold over the oral disk, and the anemone appears to collapse. The reestablishment of the hydrostatic skeleton depends on gradual uptake of water into the gastrovascular cavity via the siphonoglyphs.

Anemones are capable of limited locomotion. Movement is accomplished by gliding on their pedal disk, crawling on their side, and walking on their tentacles. When disturbed, some "swim" by thrashing their body or tentacles. Some anemones float using a gas bubble held within folds of the pedal disk.

Anemones feed on invertebrates and fishes. Tentacles capture prey and draw it toward the mouth. Radial muscle fibers in the mesenteries open the mouth to receive the food.

Both sexual and asexual reproduction are shown by anemones. In asexual reproduction, a piece of pedal disk may break away from the polyp and grow into a new individual in a process called pedal laceration. Alternatively, longitudinal or

BOX 6.2 Coral Reefs

Not all corals build coral reefs. Those that do not are called soft corals and often live at great depths in cold seawater. Stony corals are reef-building species. Coral reefs are built as calcium carbonate exoskeletons of one generation of stony corals are secreted on the exoskeletons of preceding generations. It requires millions of years for massive reefs, such as those found in warm, shallow waters of the Indian Ocean, the south Pacific Ocean, and the Caribbean Sea to develop (figure 1). Reef formation requires constantly warm (20°C), shallow water (less than 90 m), and constant salinity near 3.5%.

Most reef-building activities are the result of stony corals living in a mutualistic relationship (*see chapter 5*) with a group of dinoflagellate protists called **zooxanthellae** (*see figure 6.20*). Stony corals depend on photosynthetic activities of zooxanthellae as a principal source of carbohydrates. Predatory activities serve mainly as a source of protein for polyps. Zooxanthellae also promote exceptionally high rates of calcium deposition. As zooxanthellae carry on photosynthesis, they remove CO_2 from the environment of the polyp. Associated pH changes induce the precipitation of dissolved $CaCO_3$ as aragonite (coral limestone). It is thought that the 90 m depth limit for reef building corresponds to the limits to which sufficient light penetrates to support dinoflagellate photosynthesis.

Certain algae, called **coralline algae,** live outside the coral organisms and create their own calcium carbonate masses. These algae contribute to the reef by cementing together larger coral formations.

Reefs can extend hundreds of meters below the ocean's surface; however, only the upper and outer layer includes coral animals and algae. Most of the reef formation consists of exoskeletons of previous generations of stony corals. (The depth of the reef mass is evidence of changing oceanic levels during glacial periods and of the subsidence of the ocean floor.) In addition to the outer layer of photosynthetic and cnidarian life-forms, the reef supports a host of other organisms, including fishes, molluscs, arthropods, echinoderms, soft corals, and sponges. The exceptionally high productivity of reef communities depends on the ability of reef organisms to recycle nutrients rather than to lose them to the ocean floor.

FIGURE 1 **Coral Reefs.** A fringing reef surrounding an island off the eastern coast of Australia. Note the human-made shipping channel cut through the reef.

There are three types of coral reefs. (1) Fringing reefs are built up from the sea bottom so close to a shoreline that no navigable channel exists between the shoreline and the reef. This reef formation frequently creates a narrow, shallow lagoon between the reef and the shore. Surging water creates frequent breaks and irregular channels through these reefs. (2) Barrier reefs are separated from shore by wide, deep channels. The Great Barrier Reef of Australia is 1,700 km long with a channel 20 to 50 m deep and up to 48 km wide. (The Great Barrier Reef actually consists of a number of different reef forms, including barrier reefs.) (3) Atolls are circular reefs that enclose a lagoon in the open ocean. One hypothesis regarding their origin, first described by Charles Darwin, is that atolls were built up around islands that later sank.

transverse fission may divide one individual into two, with missing parts being regenerated. Unlike other cnidarians, anemones may be either monoecious or dioecious. In monoecious species, male gametes mature earlier than female gametes so that self-fertilization does not occur. This is called **protandry** (Gr. *protos*, first + *andros*, male). Gonads occur in longitudinal bands behind mesenterial filaments. Fertilization may be external or within the gastrovascular cavity. Cleavage results in the formation of a planula, which develops into a ciliated larva that settles to the substrate, attaches, and eventually forms the adult.

Other anthozoans are corals. (8) Stony corals are responsible for the formation of coral reefs (box 6.2) and, except for lacking siphonoglyphs, are similar to the anemones. Their common name derives from a cuplike calcium carbonate exoskeleton secreted around their base and the lower portion of

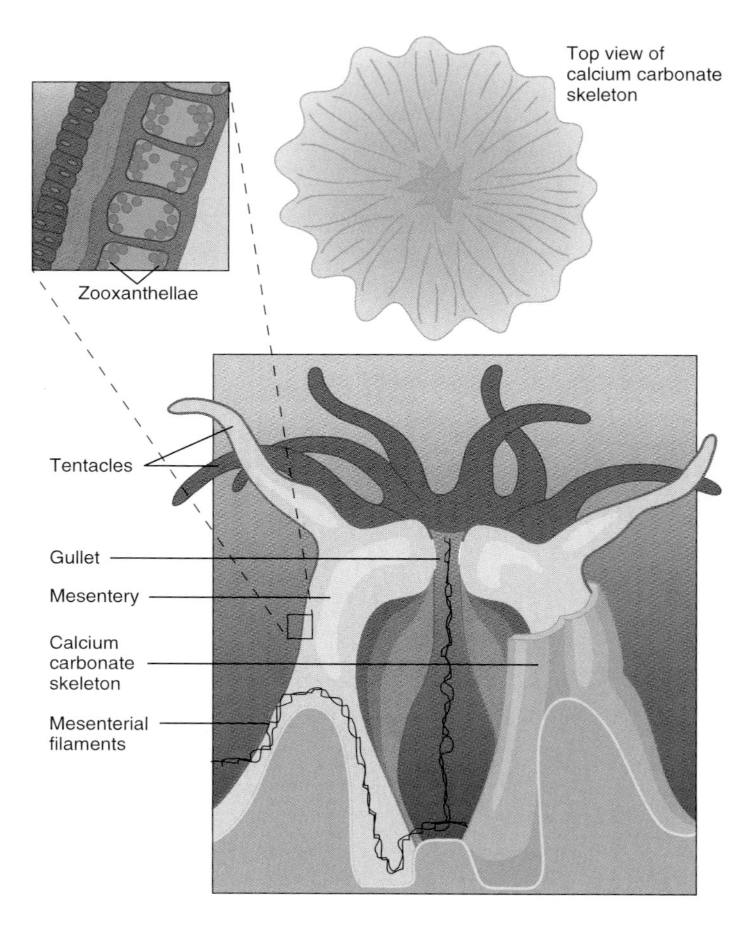

FIGURE 6.20
Class Anthozoa. A stony coral polyp in its calcium carbonate skeleton (longitudinal section).

their column by epithelial cells (figure 6.20). When threatened, polyps retract into their protective exoskeletons. Sexual reproduction is similar to that of anemones, and asexual budding produces other members of the colony.

The colorful octacorallian corals are common in warm waters. They have eight pinnate (featherlike) tentacles, eight mesenteries, and one siphonoglyph. The body walls of members of a colony are connected, and mesenchyme cells secrete an internal skeleton of protein or calcium carbonate. Sea fans, sea pens, sea whips, red corals, and organ-pipe corals are members of this group (figure 6.21).

Stop and Ask Yourself

9. What hydrozoan has well-defined alternation of generations? What hydrozoan has reduced alternation of generations?
10. What kinds of sensory structures occur in cnidarians?
11. How do the following fit into the life history of a scyphozoan: scyphistoma? ephyra? planula?
12. What is protandry? How does it apply to the Anthozoa?

(a)

(b)

FIGURE 6.21
Representative Octacorallian Corals. (*a*) Fleshy sea pen (*Ptilosaurus gurneyi*). (*b*) Purple sea fan (*Gorgonia ventalina*).

PHYLUM CTENOPHORA

Animals in the phylum Ctenophora (ti-nof'er-ah) (Gr. *kteno,* comb + *phoros,* to bear) are called sea walnuts or comb jellies (table 6.3). There are approximately 90 described species, all of which are marine (figure 6.22*a*). Most ctenophorans have a spherical form, although several groups are flattened and/or elongate.

TABLE 6.3 CLASSIFICATION OF THE CTENOPHORA

Phylum Ctenophora (ti-nof′er-ah)
The animal phylum whose members are biradially symmetrical, diploblastic, usually ellipsoid or spherical in shape, possess colloblasts, and have meridionally arranged comb rows.

Class Tentaculata (ten-tak′u-lata)
With tentacles that may or may not be associated with sheaths, into which the tentacles can be retracted. *Pleurobranchia*.

Class Nuda (nuda)
Without tentacles; flattened; a highly branched gastrovascular cavity. *Beroe*.

Characteristics of the phylum Ctenophora include the following:

1. Diploblastic, tissue-level organization
2. Biradial symmetry
3. Gelatinous mesoglea located between the epidermal and gastrodermal tissue layers
4. Gastrovascular cavity
5. Nervous system in the form of a nerve net
6. Adhesive structures called colloblasts
7. Eight rows of ciliary bands, called comb rows, that are used in locomotion

Pleurobranchia has a spherical or ovoid, transparent body about 2 cm in diameter. It occurs in the colder waters of the Atlantic and Pacific Oceans (figure 6.22*b*). *Pleurobranchia*, like most ctenophorans, has eight meridional bands of cilia, called **comb rows,** that run between the oral and aboral poles. Comb rows are locomotor structures that are coordinated through a statocyst at the aboral pole. *Pleurobranchia* normally swims with its aboral pole oriented downward. Tilting is detected by the statocyst, and the comb rows adjust the animal's orientation. Two long, branched tentacles arise from pouches near the aboral pole. Tentacles possess contractile fibers that retract the tentacles and adhesive cells, called **colloblasts,** which are used for prey capture (figure 6.22*c*).

Ingestion occurs as the tentacles wipe the prey across the mouth. The mouth leads to a branched gastrovascular canal system. Some canals are blind; however, two small, anal canals open to the outside near the apical sense organ. Thus, unlike the cnidarians, ctenophores have an anal opening. Some undigested wastes are eliminated through these canals, and some are probably also eliminated through the mouth (*see figure* 6.22b).

Pleurobranchia is monoecious, as are all ctenophores. Two bandlike gonads are associated with the gastrodermis. One of these is an ovary and the other a testis. Gametes are shed through the mouth, fertilization is external, and a slightly flattened larva develops.

(a)

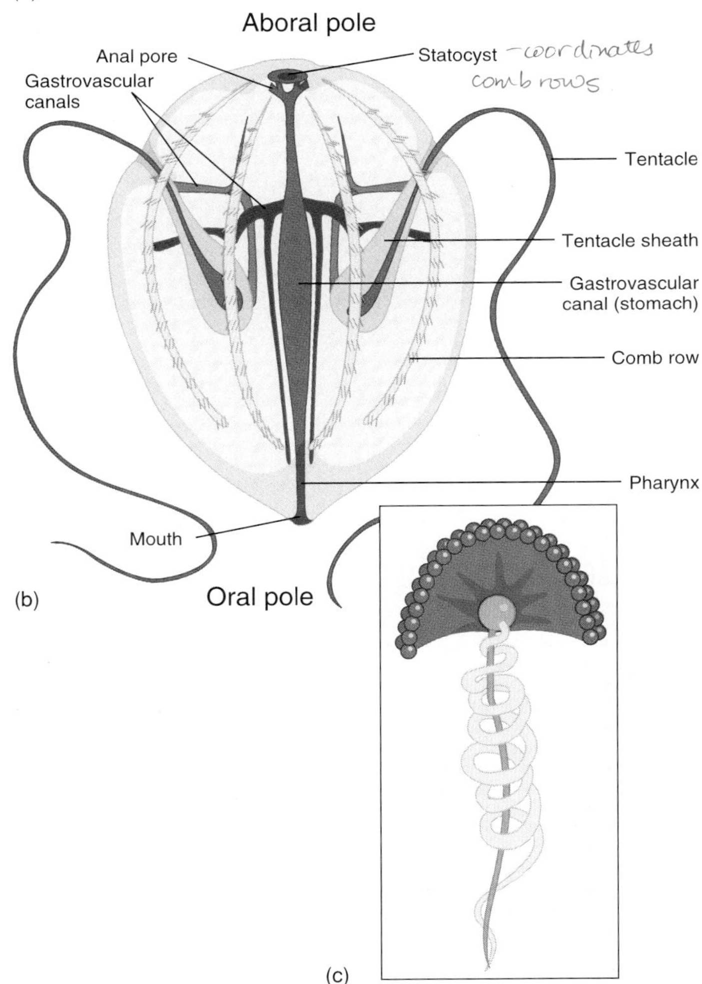

FIGURE 6.22
Phylum Ctenophora. (*a*) The ctenophore *Mnemiopsis*. Ctenophorans are well known for their bioluminescence. Light-producing cells are located in the walls of their digestive canals, which are located beneath comb rows. (*b*) The structure of *Pleurobranchia*. The animal usually swims with the oral end forward or upward. (c) Colloblasts consist of a hemispherical sticky head that is connected to the core of the tentacle by a straight filament. A contractile spiral filament coils around the straight filament. Straight and spiral filaments prevent struggling prey from escaping.

Further Phylogenetic Considerations

The evolutionary position of the phyla covered in this chapter is subject to debate. If the animal kingdom is polyphyletic, then all phyla could have had separate origins, although scientists who believe in multiple origins agree that the number of independent origins is probably small. Some zoologists believe it to be at least diphyletic, with the Porifera being derived separately from all other phyla. The similarity of poriferan choanocytes and choanoflagellate protists suggests evolutionary ties between these groups. Many other zoologists believe that the sponges have a common, although remote, ancestry with other animals. The presence of amoeboid and flagellated cells in sponges and higher animals is used to support this view. One thing that nearly everyone agrees upon, however, is that the Porifera are evolutionary "dead ends." They gave rise to no other animal phyla.

If two origins are assumed, the origin of the nonporiferan lineage is also debated. One interpretation is that the ancestral animal was derived from a radially symmetrical ancestor, which in turn may have been derived from a colonial flagellate similar in form to *Volvox* (*see figure 5.8*). If this is true, then the radiate phyla (Cnidaria and Ctenophora) could be closely related to that ancestral group. Other zoologists contend that bilateral symmetry is the ancestral body form, and a bilateral ancestor gave rise to both the radiate phyla and bilateral phyla. In this interpretation, the radiate phyla are further removed from the base of the evolutionary tree.

The probable evolutionary relationships of the cnidarian classes are shown in figure 6.23. The classical interpretation is that primitive Hydrozoa were the ancestral radial animals, and that the medusoid body form is the primitive body form. This ancestry is suggested by the fact that the medusa is the adult body form in the Hydrozoa, and that the medusa is the only body form present in some trachyline hydrozoans. The polyp may have evolved secondarily as a larval stage. In the evolution of the other three orders, septa appeared—dividing the gastrovascular cavity—and gonads became endodermal in origin. The Scyphozoa and Cubozoa are distinguished from the Anthozoa by the evolutionary reduction of the polyp stage in the former classes and the loss of the medusa stage in the latter class. The Scyphozoa and Cubozoa are distinguished from each other by budding of the polyp (Scyphozoa) and the cuboidal shape of the medusa (Cubozoa).

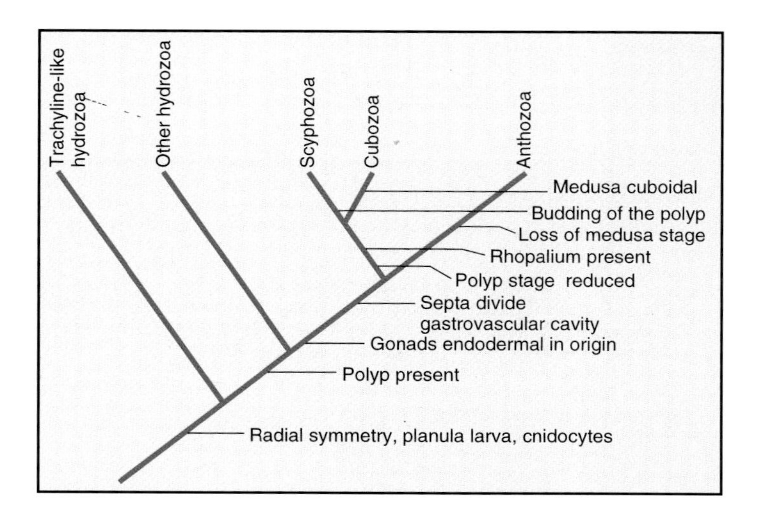

Figure 6.23

A Cladogram Showing Cnidarian Taxonomy. Selected synapomorphic characters are shown. Hydrozoans are believed by most zoologists to be ancestral to other cnidarians.

The relationships of the Ctenophora to any other group of animals are uncertain. The Ctenophora and Cnidaria share important characteristics such as radial (biradial) symmetry, diploblastic organization, nerve nets, and gastrovascular cavities. In spite of these similarities, differences in adult body forms and embryological development make it very difficult to derive the Ctenophora from any group of cnidarians. Relationships between the Cnidaria and Ctenophora are probably very distant.

Stop and Ask Yourself

13. What characteristics are shared by the Ctenophora and the Cnidaria?
14. What are colloblasts?
15. How are a statocyst and comb rows used by ctenophores to maintain an upright position in the water?
16. The sponges are considered evolutionary "dead ends." Why?

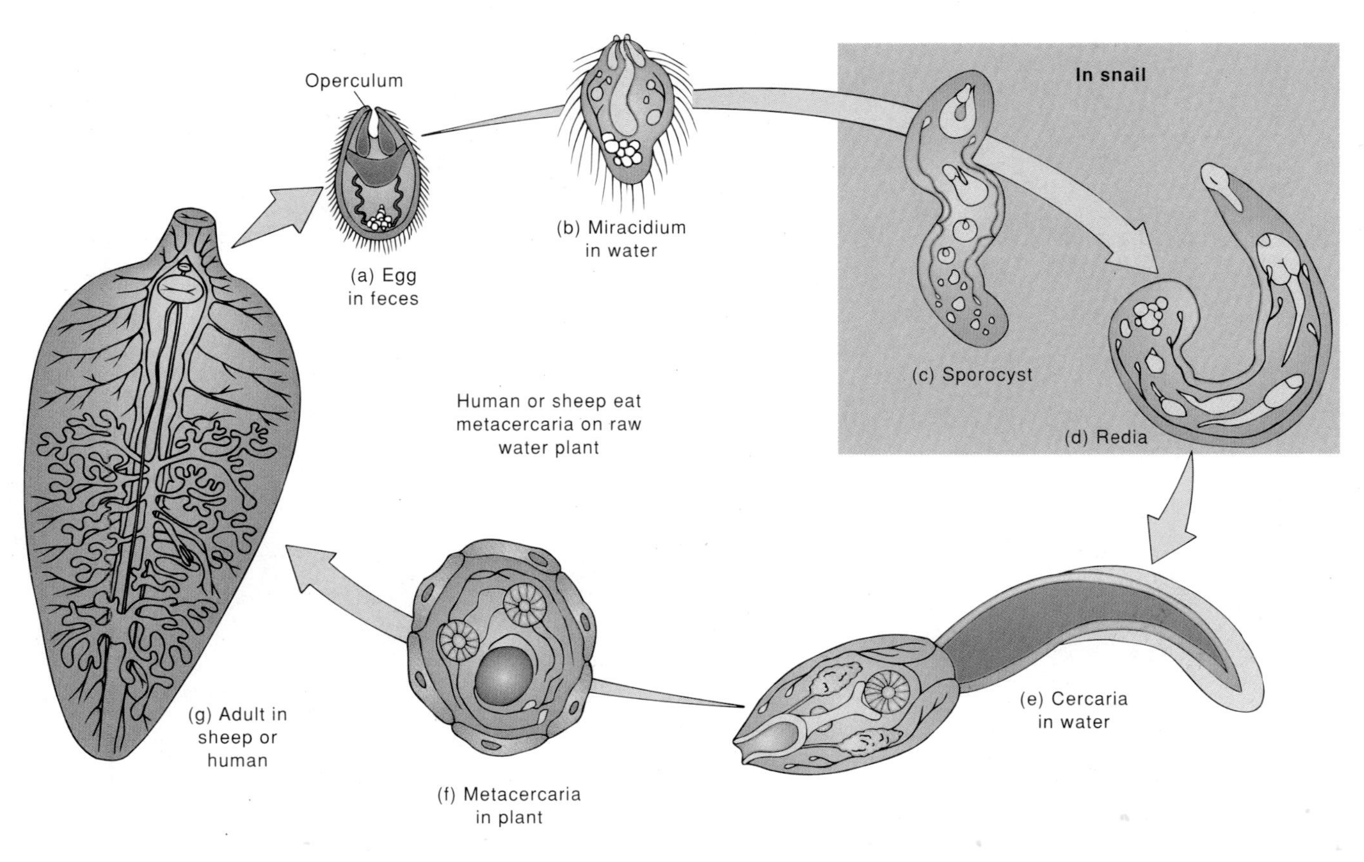

FIGURE 7.14
Class Trematoda: Subclass Digenea. The life cycle of the digenetic trematode, *Fasciola hepatica* (the common liver fluke).

(figure 7.14*b*). The miracidium swims until it finds a suitable first intermediate host (a snail) to which it is chemically attracted. The miracidium penetrates the snail, loses its cilia, and develops into a **sporocyst** (figure 7.14*c*). (Alternately, the miracidium may remain in the egg and hatch after being eaten by a snail.) Sporocysts are baglike and contain embryonic cells that develop into either **daughter sporocysts,** or **rediae** (s., redia) (figure 7.14*d*). At this point in the life cycle, asexual reproduction first occurs. From a single miracidium, hundreds of daughter sporocysts, and in turn, hundreds of rediae can form by asexual reproduction. Embryonic cells in each daughter sporocyst or redia produce hundreds of the next larval stage, called **cercariae** (s., cercaria) (figure 7.14*e*). (This phenomenon of producing many cercariae is called polyembryony. It enhances greatly the chances that one cercaria will further the life cycle.) A cercaria has a digestive tract, suckers, and a tail. Cercariae leave the snail and swim freely until they encounter a second intermediate or final host, which may be a vertebrate or invertebrate. The cercaria penetrates this host and encysts as a **metacercaria** (pl., metacercariae) (figure 7.14*f*). When the second intermediate host is eaten by the definitive host, the metacercaria excysts and develops into an adult (figure 7.14*g*).

Some Important Trematode Parasites of Humans

The Chinese liver fluke, *Clonorchis sinensis*, is a common parasite of humans in the Orient, where over 50 million people are infected. The adult lives in the bile ducts of the liver, where it feeds on epithelial tissue and blood (figure 7.15*a*). Embryonated eggs are released by the adults into the common bile duct, make their way to the intestine, and are eliminated with feces (figure 7.15*b*). The miracidia are released when a snail ingests the eggs. Following sporocyst and redial stages, cercariae emerge into the water. If the cercaria contacts a fish (the second intermediate host), it penetrates the epidermis of the fish, loses its tail, and encysts. The metacercaria develops into an adult in a human who eats raw or poorly cooked fish, a delicacy in the Orient and gaining in popularity in the Western world.

5 *Fasciola hepatica* is called the sheep liver fluke (*see figure 7.14*a) because it is common in sheep-raising areas and uses sheep or humans as its definitive host. The adults live in the bile duct of the liver. Eggs pass via the common bile duct to the intestine, from which they are eliminated. When eggs are

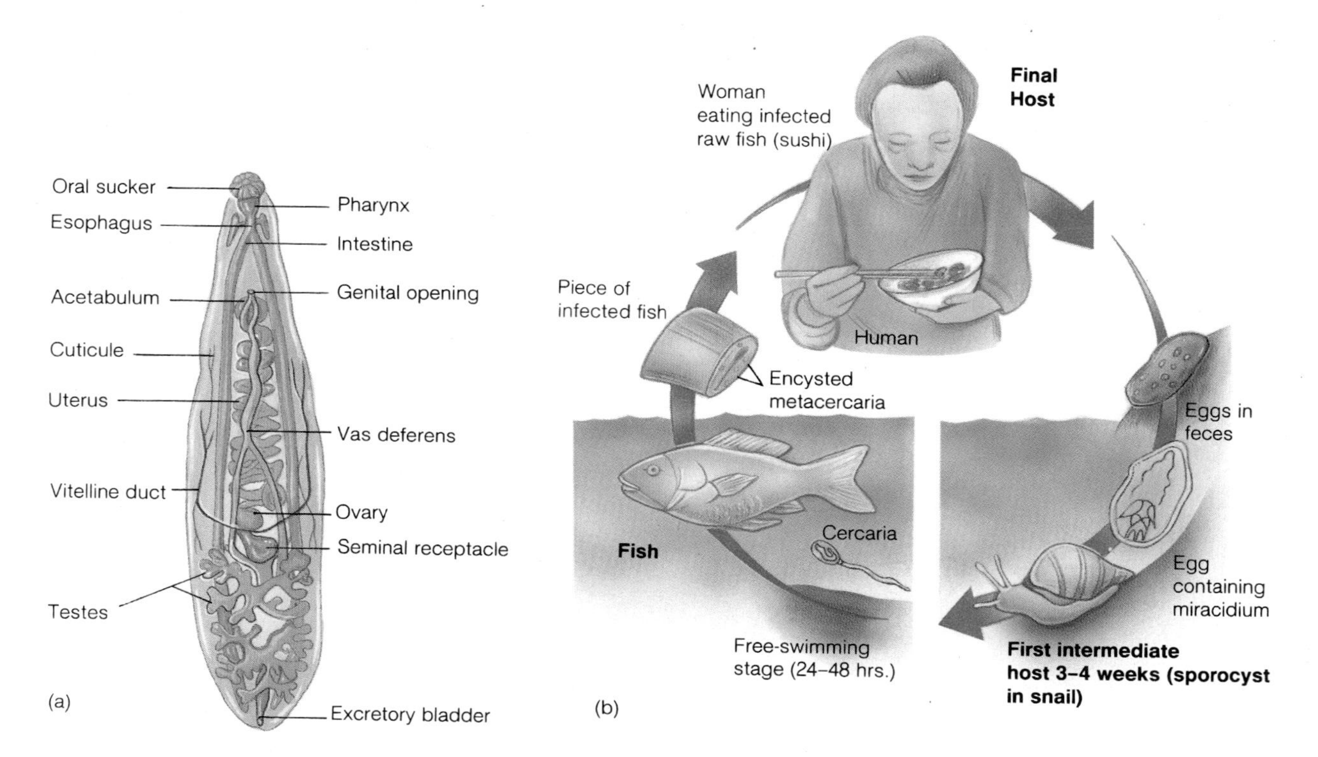

Figure 7.15
The Chinese Liver Fluke. *Clonorchis sinensis*. (*a*) Dorsal view and (*b*) life cycle.

deposited in fresh water, they hatch, and the miracidia must locate the proper species of snail. If a snail is found, miracidia penetrate the snail's soft tissue and develop into sporocysts that develop into rediae and give rise to cercariae. After the cercariae emerge from the snail, they encyst on aquatic vegetation. Sheep or other animals become infected when they graze on the aquatic vegetation. Humans may become infected with *Fasciola hepatica* by eating a freshwater plant called watercress that contains the encysted metacercaria.

Schistosomes are blood flukes that are of vast medical significance. The impact these flukes have had on history is second only to that of *Plasmodium* (malaria; *see box 5.2*). They infect over 200 million people throughout the world. Infections are most common in Africa (*Schistosoma haematobium* and *S. mansoni*), South and Central America (*S. mansoni*), and Southeast Asia (*S. japonicum*). The adult dioecious worms live in the bloodstream of humans (figure 7.16*a*). The male fluke is shorter and thicker than the female, and the sides of the male's body are curved under to form a canal along the ventral surface ("schistosoma" means "split body"). The female fluke is long and slender, and is carried in the canal of the male (figure 7.16*b*). Copulation is continuous, and the female produces thousands of eggs. Each egg contains a spine that aids it in moving through host tissue until it is eliminated in either the feces or urine (figure 7.16*c*). Unlike other flukes, schistosome eggs lack an operculum. The miracidium escapes through a slit that develops in the egg when the egg reaches fresh water (figure 7.16*d*). The miracidium seeks a snail (figure 7.16*e*), penetrates it, develops into a sporocyst (figure 7.16*f*), daughter sporocysts (figure 7.16*f*), and then forked-tailed cercariae (figure 7.16*f*). There is no redial generation. The cercariae leave the snail and penetrate the skin of a human (figure 7.16*g*). Penetration is aided by anterior digestive glands that secrete digestive enzymes. Once in a human, the cercariae lose their tails and develop into adults in the intestinal veins, skipping the metacercaria stage (box 7.1).

Stop and Ask Yourself

4. What is unique about the trematode tegument?
5. What is the function of an opisthaptor?
6. What is the major difference between the life cycle of a monogenean and that of a digenean fluke?
7. How can each of the following be described: miracidium? sporocyst? redia? cercaria? metacercaria? operculum?

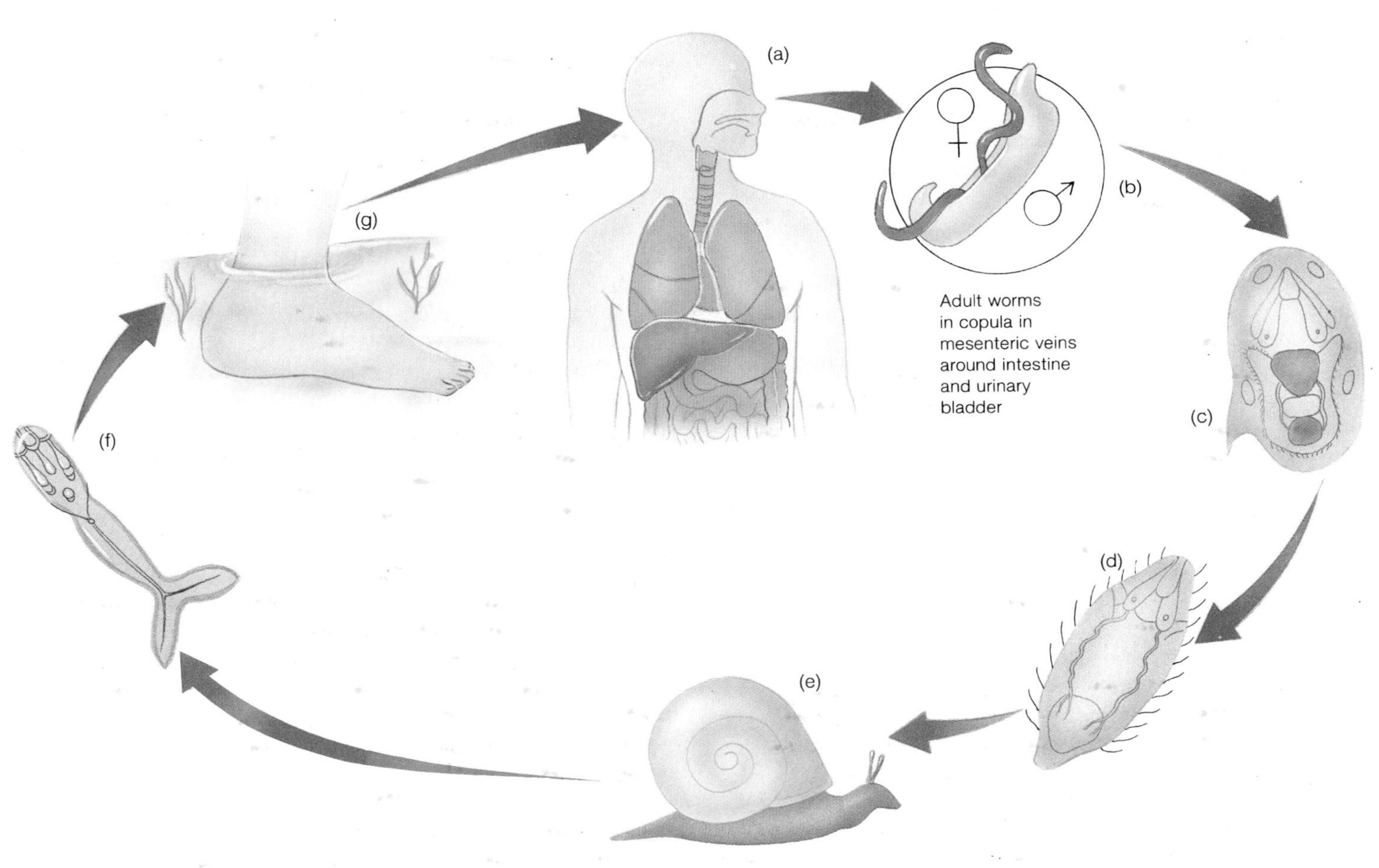

FIGURE 7.16

A Representative Life Cycle of a Schistosome Fluke. The cycle begins in a human (*a*) when the female fluke lays eggs (*b,c*) in the thin-walled, small vessels of the small intestine (*S. mansoni* and *S. japonicum*) or urinary bladder (*S. haematobium*). Secretions from the eggs weaken the walls, and the blood vessels rupture, releasing eggs into the intestinal lumen or urinary bladder. From there the eggs leave the body. If they reach fresh water, the eggs hatch into ciliated, free-swimming larvae called miracidia (*d*). A miracidium burrows into the tissues of an aquatic snail (*e*), losing its cilia in the process, and develops into a sporocyst, then daughter sporocysts. Eventually forked-tailed larvae (cercariae) are produced (*f*). After the cercariae leave the snail, they actively swim about. If they encounter human skin (*g*), they attach to it and release tissue-degrading enzymes. The larvae enter the body and migrate to the circulatory system, where they mature. They end up at the intestinal vessels, where sexual reproduction takes place and the cycle begins anew.

BOX 7.1 Swimmer's Itch

An interesting phase of schistosome biology concerns cercarial dermatitis or "swimmer's itch." Cercariae of species that normally infect birds (waterfowl and shorebirds) and other vertebrates, especially certain mammals (e.g., muskrats and mink), attempt to penetrate the skin of humans, and in doing so, sensitize the area of attack, resulting in itchy rashes. Because humans are not suitable definitive hosts for these cercariae, the flukes do not enter the bloodstream and mature, but perish after penetrating the skin. One of the most common causative agents of marine swimmer's itch on both the east and west coasts of North America is a blood parasite of sea gulls, the cercariae of which develop in a common mudflat snail.

Class Cestoidea: The Tapeworms

The most highly specialized class of flatworms are members of the class Cestoidea (ses-toid′ da) (Gr. *kestos*, girdle + *eidos*, form), commonly called either tapeworms or cestodes. All of the approximately 3,500 species are endoparasites that usually reside in the digestive system of vertebrates. Their color is often white with shades of yellow or gray. Adult tapeworms range in size from 1 mm to 15 m in length.

Tapeworms are characterized by two unique adaptations to their parasitic way of life. 6 (1) Tapeworms lack a mouth and digestive tract in all of their life-cycle stages; they absorb nutrients directly across their body wall. (2) Most adult tapeworms consist of a long series of repeating units called **proglottids.** Each proglottid contains a complete set of reproductive structures.

As with most endoparasites, adult tapeworms live in a very stable environment. The intestinal tract of a vertebrate has very few environmental variations that would require the development of great anatomical or physiological complexity in any single tapeworm body system. Homeostasis (internal constancy) of a tapeworm is maintained by the physiology of the tapeworm's host. In adapting to such a specialized environment, tapeworms have lost some of the structures believed to have been present in ancestral turbellarians. Tapeworms are, therefore, a good example of the fact that evolution does not always result in greater complexity.

Subclass Cestodaria

Representatives of the subclass Cestodaria are all endoparasites in the intestine and coelom of primitive fishes. About 15 species have been identified. They possess some digenetic trematode features in that only one set of both reproductive systems is present in each animal, some bear suckers, and their bodies are not divided into proglottids as in other cestodes. Yet the complete absence of a digestive system, the presence of larval stages similar to those of cestodes, and the presence of parenchymal muscle cells, which are not present in any other platyhelminth, all suggest strong phylogenetic affinities with other cestodes.

Subclass Eucestoda

Almost all of the cestodes belong to the subclass Eucestoda and are called true tapeworms. They represent the ultimate degree of specialization of any parasitic animal. The body is divided into three regions (figure 7.17*a*). At one end is a holdfast structure called the **scolex** that contains circular or leaflike suckers and sometimes a rostellum of hooks (figure 7.17*b*). It is via the scolex that the tapeworm firmly anchors itself into the intestinal wall of its definitive vertebrate host. No mouth is present.

Posteriorly, the scolex narrows to form the neck. Transverse constrictions in the neck give rise to the third body region, the **strobila** (Gr. *strobilus*, a linear series; pl., strobilae). The strobila consists of a series of linearly arranged proglottids, which function primarily as reproductive units. As a tapeworm grows, new proglottids are added in the neck region, and older proglottids are gradually pushed posteriorly. As they move posteriorly, proglottids mature and begin producing eggs. Thus, anterior proglottids are said to be immature, those in the midregion of the strobila are mature, and those at the posterior end that have accumulated eggs are gravid (L. *gravida*, heavy, loaded, pregnant).

The outer body wall of tapeworms consists of a tegument similar in structure to that of trematodes (*see figure 7.12*). It plays a vital role in the absorption of nutrients because tapeworms have no digestive system. The tegument even absorbs some of the host's own enzymes to facilitate digestion.

With the exception of the reproductive systems, the body systems of tapeworms are reduced in structural complexity. The nervous system consists of only a pair of lateral nerve cords that arise from a nerve mass in the scolex and extend the length of the strobila. A protonephridial system also runs the length of the tapeworm (*see figure 7.6*).

Tapeworms are monoecious and most of their physiology is devoted to producing large numbers of eggs. Each proglottid contains a complete set of male and female reproductive organs (figure 7.17*a*). Numerous testes are scattered throughout the proglottid and deliver sperm via a duct system to a copulatory organ called a cirrus. The cirrus opens through a genital pore, which is an opening shared with the female system. The male system of a proglottid matures before the female system, so that copulation usually occurs with another mature proglottid of the same tapeworm or with another tapeworm in the same host. As previously mentioned, the avoidance of self-fertilization leads to hybrid vigor.

Eggs in each proglottid are produced in a single pair of ovaries. Sperm stored in a seminal receptacle fertilize eggs as they move through the oviduct. After passing the vitelline (yolk) gland to pick up yolk cells, the eggs move into the ootype, which is an expanded region of the oviduct that shapes the capsules around the eggs. The ootype is surrounded by Mehlis' gland, which aids in the formation of the egg capsule. Most tapeworms have a blind-ending uterus where eggs accumulate (figure 7.17*a*). As eggs accumulate, the reproductive organs degenerate; thus, gravid proglottids can be thought of as "bags of eggs." Eggs are released when gravid proglottids break free from the end of the tapeworm and pass from the host with the feces. In a few tapeworms, the uterus opens to the outside of the worm, and eggs are released into the intestine of the host. Because the proglottids are not continuously lost in these worms, the adult tapeworms usually become very long.

Some Important Tapeworm Parasites of Humans

One medically important tapeworm of humans is the beef tapeworm, *Taeniarhynchus saginatus* (figure 7.18). Adults live in the small intestine and may reach lengths of 3 m. About 80,000 eggs per proglottid are released as proglottids break free of the

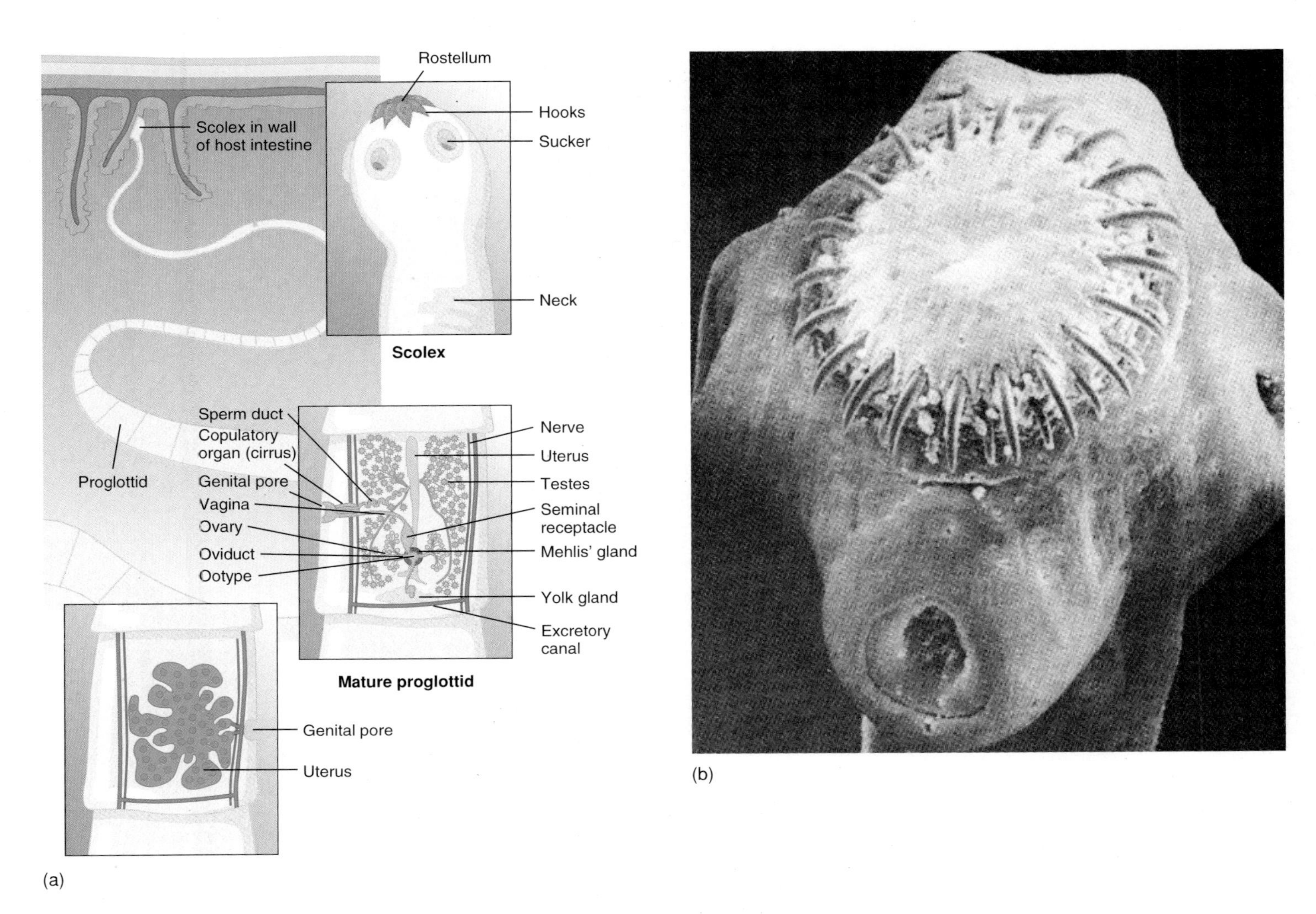

Figure 7.17

Class Cestoidea: A Tapeworm. (*a*) Diagram showing the scolex, neck, and proglottids of the pork tapeworm, *Taenia solium*. Included is a detailed view of a mature proglottid with a complete set of male and female reproductive structures. (*b*) A scanning electron micrograph (×100) of the scolex of the cestode *Taenia solium*. Notice the rostellum with two circles of hooks.

adult worm. As an egg develops, it forms a six-hooked (hexacanth) larva called the **onchosphere.** As cattle (the intermediate host) graze on pastures contaminated with human feces, the oncospheres (or proglottids) are ingested. Digestive enzymes of the cattle free the oncospheres, and the larvae use their hooks to bore through the intestinal wall into the bloodstream. The bloodstream carries the larvae to skeletal muscles, where they encyst and form a fluid-filled bladder called a **cysticercus** (pl., cysticerci) or **bladder worm.** When a human eats infected meat (termed "measly beef") that is raw or improperly cooked, the cysticercus is released from the meat, the scolex attaches to the human intestinal wall, and the tapeworm matures.

A closely related tapeworm, *Taenia solium* (the pork tapeworm), has a life cycle similar to *Taeniarhynchus saginatus*, except that the intermediate host is the pig. The strobila has been reported as being 10 m long, but 2 to 3 m is more common. The pathology is more serious in the human than in the pig. Oncospheres are frequently released from gravid proglottids before the proglottids have had a chance to leave the small intestine of the human host. When these larvae hatch, they move through the intestinal wall, enter the bloodstream and are distributed throughout the body where they eventually encyst in human tissue as cysticerci. The disease that results is called **cysticercosis** and can be fatal if the cysticerci encyst in the brain.

The broad fish tapeworm, *Diphyllobothrium latum*, is relatively common in the northern parts of North America and in the Great Lakes area of the United States. This tapeworm has a scolex with two longitudinal grooves (bothria; s., bothrium; figure 7.19) that act as holdfast structures. The adult worm may attain a length of 10 m and shed up to a million eggs a day. Many proglottids release eggs through uterine pores. When eggs are deposited in fresh water, they hatch,

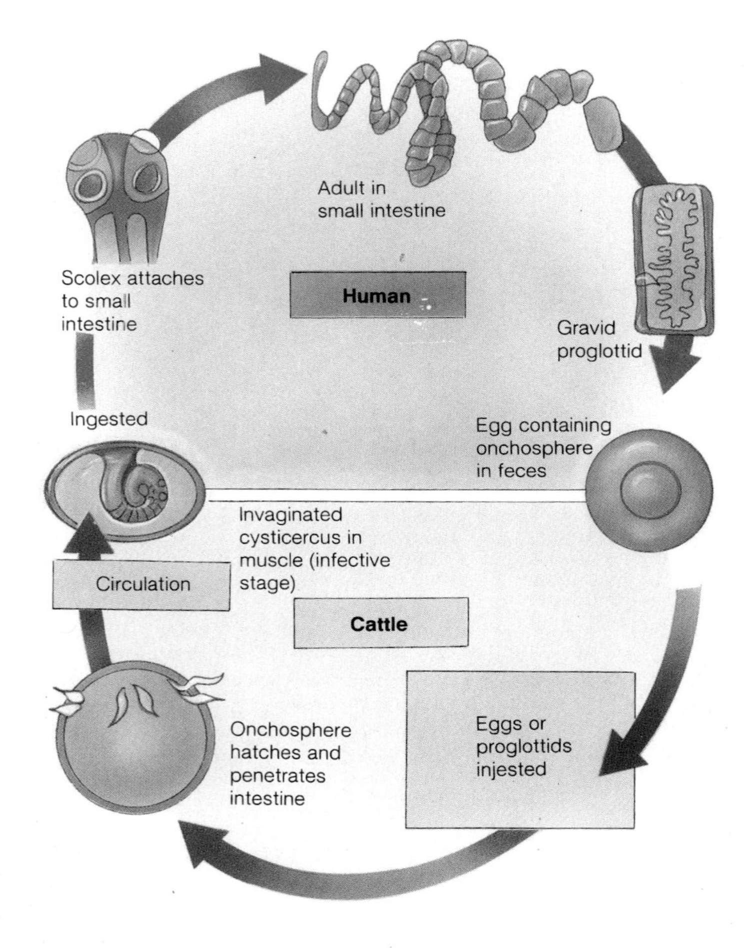

Figure 7.18

The Life Cycle of the Beef Tapeworm, *Taeniarhynchus saginatus*. *Source: Redrawn from Centers for Disease Control, Atlanta.*

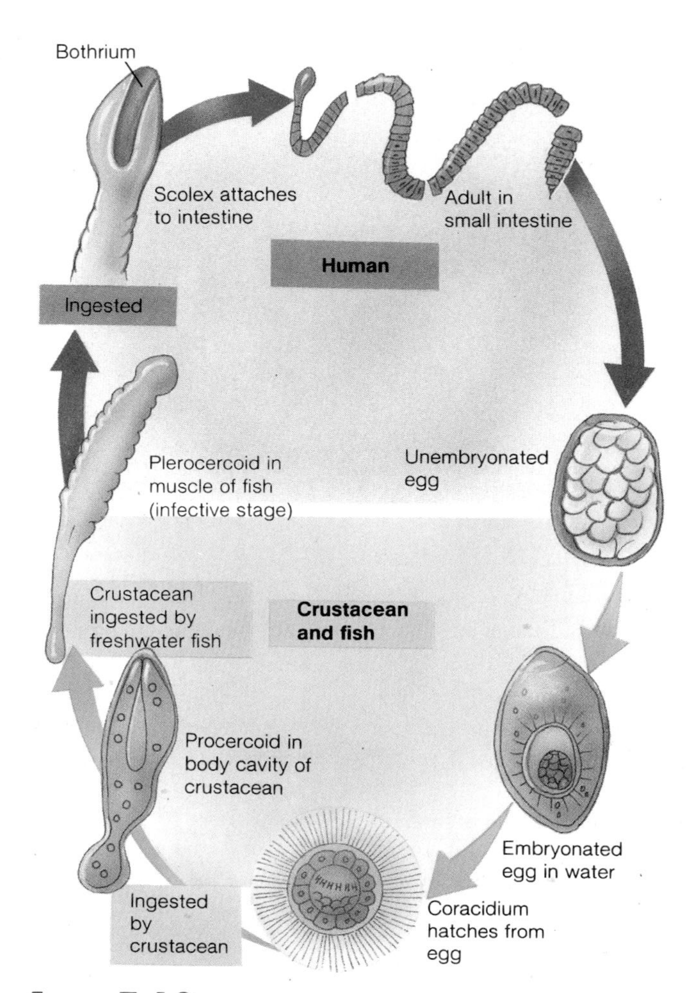

Figure 7.19

The Life Cycle of the Broad Fish Tapeworm, *Diphyllobothrium latum*. *Source: Redrawn from Centers for Disease Control, Atlanta.*

and ciliated larvae called **coracidia** (s., coracidium) emerge. These coracidia swim about until they are ingested by small crustaceans called copepods. The larvae shed their ciliated coats in the copepods and develop into **procercoid larvae.** When copepods are eaten by a fish, the procercoids burrow into the muscle of the fish and become **plerocercoid larvae.** Larger fishes that eat smaller fishes become similarly infected with plerocercoids. When infected, raw, or poorly cooked fishes are eaten by a human (or other carnivore), the plerocercoids attach to the small intestine and grow into adult worms.

Stop and Ask Yourself

8. What is a proglottid? A scolex? A strobila?
9. How do tapeworms obtain nutrients?
10. What is the life cycle of a typical tapeworm?
11. What is each of the following: coracidium? cysticercus? procercoid larva? plerocercoid larva?

Phylum Nemertea

Most of the approximately 900 species of nemerteans (nem-er′te-ans) (Gr. *Nemertes*, a Mediterranean sea nymph; the daughter of Nereus and Doris) are elongate, flattened worms found in marine mud and sand. Due to the presence of a long proboscis, nemerteans are commonly called proboscis worms. Adult worms range in size from a few millimeters to several centimeters in length. Most nemerteans are pale yellow, orange, green, or red. Characteristics of the phylum Nemertea include the following:

1. Triploblastic, acoelomate, bilaterally symmetrical unsegmented worms possessing a ciliated epidermis containing mucous glands
2. Complete digestive tract with an anus

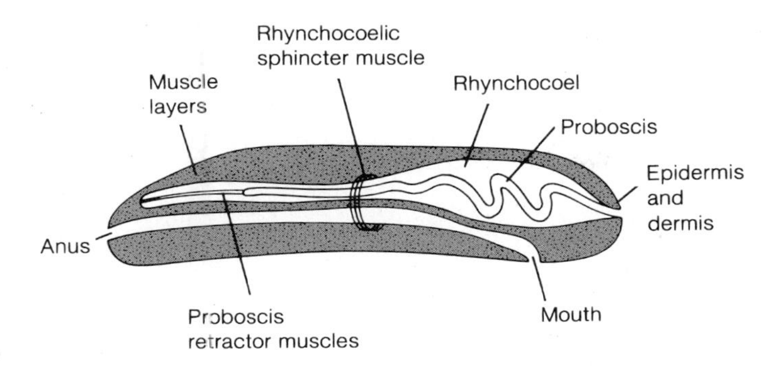

FIGURE 7.20

Phylum Nemertea. Diagram of a nemertean; longitudinal section, showing the tubular gut and proboscis. *Source: Modified from Turbeville and Ruppert, 1983, Zoomorphology, 103:103, Copyright 1983, Springer-Verlag, Heidelberg, Germany.*

3. Protonephridia
4. Cerebral ganglion, longitudinal nerve cords, and transverse commissures
5. Closed circulatory system
6. Body musculature organized into two or three layers

The most distinctive feature of nemerteans is a long proboscis that is held in a sheath called a **rhynchocoel** (figure 7.20). The proboscis may be tipped with a barb called a stylet. Carnivorous species use the proboscis to capture annelid (segmented worms) and crustacean prey.

Unlike the platyhelminths, nemerteans have a complete one-way digestive tract. They have a mouth for ingesting food and an anus for eliminating digestive wastes. This characteristic enables mechanical breakdown of food, digestion, absorption, and feces formation to proceed sequentially from an anterior to posterior direction—a major evolutionary innovation found in all higher bilateral animals.

Another major innovation found in all higher animals evolved first in the nemerteans—a circulatory system consisting of two lateral blood vessels and often, tributary vessels that branch from lateral vessels. However, no heart is present, and contractions of the walls of the large vessels help to propel blood along. Blood does not circulate but simply moves forward and backward through the longitudinal vessels. Blood cells are present in some species. This combination of blood vessels with their capacity to serve local tissues, and a one-way digestive system with its greater efficiency at processing nutrients, allows nemerteans to reach lengths much larger than most flatworms.

Nemerteans are dioecious. Male and female reproductive structures develop from parenchymal cells along each side of the body. External fertilization results in the formation of a helmet-shaped, ciliated **pilidium larva.** After a brief free-swimming existence, the larva develops into a young worm that settles to the substrate and begins feeding.

When they move, adult nemerteans glide on a trail of mucus. Cilia and peristaltic contractions of body muscles provide the propulsive forces.

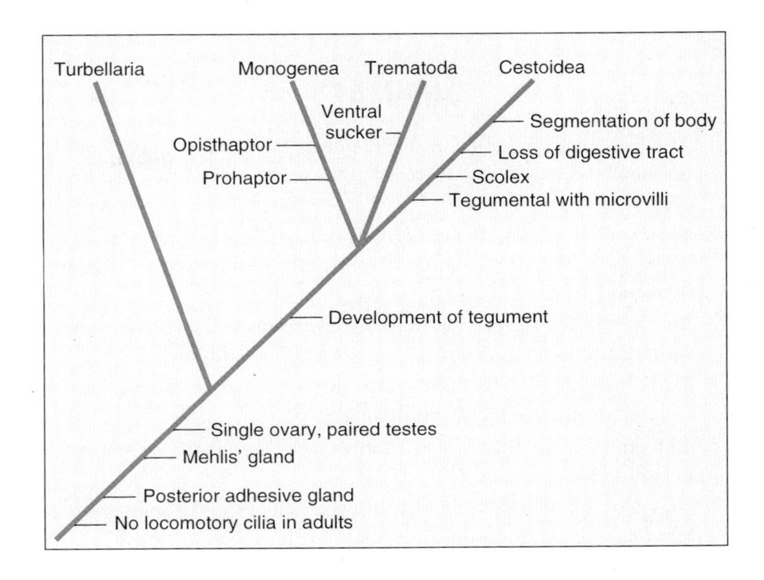

FIGURE 7.21

A Cladogram Showing Evolutionary Relationships between the Classes of Platyhelminthes. The absence of synapomorphies for the Turbellaria suggests that the ancestral platyhelminth was itself a turbellarian, and that some member of that class was ancestral to the three living parasitic groups. *Source: Data from D. R. Brooks, Journal of Parasitology, 1989, pp. 606-616.*

FURTHER PHYLOGENETIC CONSIDERATIONS

Zoologists who believe that the platyhelminth body form is central to animal evolution envision an ancestral flatworm similar to a turbellarian. Figure 7.21 illustrates a cladogram emphasizing the uniqueness of the tegument as a synapomorphy (a shared, evolutionarily derived character that is used to describe common descent among two or more species) uniting the Monogenea, Trematoda, and Cestoidea.

More conclusive evidence exists linking the parasitic flatworms to ancient, free-living ancestors. The divergence between the free-living and parasitic ways of life probably occurred in the Cambrian period, 600 million years ago. The first flatworm parasites were probably associated with primitive molluscs, arthropods, and echinoderms. It must have been much later that they acquired the vertebrate hosts and complex life cycles that have been described in this chapter.

Stop and Ask Yourself

12. Why are the nemerteans called proboscis worms?
13. What is a rhynchocoel?
14. What major body system developed for the first time in the nemerteans?

Summary

1. The free-living platyhelminths, members of the class Turbellaria, are small, bilaterally symmetrical acoelomate animals with some cephalization.
2. Most turbellarians move entirely by cilia and are predators and scavengers. Digestion is initially extracellular and then intracellular.
3. Protonephridia are present in many flatworms and are involved in osmoregulation. A primitive brain and nerve cords are present.
4. Turbellarians are monoecious with the reproductive systems adapted for internal fertilization.
5. The mongenetic flukes (class Monogenea) are mostly ectoparasites of fishes.
6. The class Trematoda is divided into two subclasses (Aspidogastrea and Digenea) and most are external or internal parasites of vertebrates. A gut is present and most of these flukes are monoecious.
7. Cestodes, or tapeworms, are gut parasites of vertebrates. They are structurally more specialized than flukes, having a scolex with attachment organs, a neck region, and a strobila, which consists of a chain of segments (proglottids) budded off from the neck region. A gut is absent, and the reproductive system is repeated in each proglottid.
8. Nemerteans are similar to platyhelminths but can be much larger. They are predatory on other invertebrates, which they capture with a unique proboscis. They have a one-way digestive tract, a blood-vascular system, and the sexes are separate.

Selected Key Terms

acetabulum (*p. 114*)
cercariae (*p. 115*)
coracidia (*p. 120*)
cysticercus (*p. 119*)
daughter sporocysts (*p. 115*)
metacercaria (*p. 115*)
miracidium (*p. 114*)
onchosphere (*p. 119*)
oncomiracidium (*p. 113*)
plerocercoid larvae (*p. 120*)
procercoid larvae (*p. 120*)
rediae (*p. 115*)
sporocyst (*p. 115*)

Critical Thinking Questions

1. Describe the morphological and developmental similarities and differences between nemerteans and turbellarians.
2. How do parasitic flatworms evade their host's immune system?
3. How would a zoologist go about documenting the complex life cycle of a digenetic trematode?
4. Describe some of the key features of acoelomate animals.
5. Consider the cestode body form and respond to the following statement: "Many very successful animals are anything but highly complex organisms. In fact, for some, evolution has meant a reduction in complexity." In what ways would this statement accurately describe the cestode body form? In what ways would it not be accurate?

chapter 8

The Pseudocoelomate Body Plan: Aschelminths

Outline

Concepts

1. Nine phyla are grouped together into the aschelminths: Gastrotricha, Rotifera, Kinorhyncha, Nematoda, Nematomorpha, Acanthocephala, Loricifera, Priapulida, and Entoprocta. Because most of these phyla have had a separate evolutionary history, this grouping is mostly one of convenience.
2. The major unifying aschelminth feature is a pseudocoelom. The pseudocoelom is a type of body cavity that develops from the blastocoel (the primitive cavity in the embryo) and is not fully lined by mesoderm, as in the true coelomates. In the pseudocoelomates, the muscles and other structures of the body wall and internal organs are in direct contact with fluid in the pseudocoelom (figure 8.1).
3. Other common aschelminth features include a complete digestive tract, a muscular pharynx, constant cell numbers (eutely), protonephridia, cuticle, and adhesive glands.

Would You Like to Know:

1. why worms molt? (*p. 125*)
2. how rotifers got their name? (*p. 126*)
3. in what animal the reproductive organ called a penis first developed? (*p. 128*)
4. what some of the most abundant animals on earth are? (*p. 130*)
5. why roundworms move in an undulating, wavelike fashion? (*p. 130*)
6. why all nematodes are round? (*p. 131*)
7. what the most common roundworm parasite in the United States is? (*p. 133*)
8. why you should not eat improperly cooked pork products? (*p. 135*)
9. what causes the disease elephantiasis? (*p. 135*)
10. what causes heartworm disease in dogs? (*p. 135*)
11. why horsehair worms were thought to arise from horses' tails? (*p. 135*)
12. how spiny-headed worms got their name? (*p. 136*)
13. what the most recently described animal phylum is? (*p. 137*)

These and other useful questions will be answered in this chapter.

This chapter contains evolutionary concepts, which are set off in this font.

Evolutionary Perspective

The nine different phyla that are grouped for convenience as the **aschelminths** (Gr. *askos*, bladder + *helmins*, worm) are very diverse animals. They have obscure phylogenetic affinities, and their fossil record is meager. Two hypotheses have been proposed for their phylogeny. The first hypothesis contends that the phyla are related based on the presence of the following structures: a pseudocoelom, a cuticle, a muscular pharynx, and adhesive glands. The second hypothesis contends that the various aschelminth phyla are not related to each other. The absence of any single unique feature found in all groups strongly suggests independent evolution of each phylum. The similarities among the living aschelminths may simply be the result of convergent evolution as these various animals adapted to similar environments.

The correct phylogeny may actually be something between the two hypotheses. All phyla are probably distantly related to each other based on the few anatomical and physiological features they share (figures 8.1, 8.2). True convergent evolution may have also produced some visible analogous similarities, but each phylum probably arose from a common acoelomate ancestor, and diverged very early in evolutionary history (figure 8.3). Such an ancestor might have been a primitive, ciliated, acoel turbellarian (*see figure 7.4a*), from which it would follow that the first ancestor was ciliated, acoelomate, marine, probably monoecious, and lacked a cuticle.

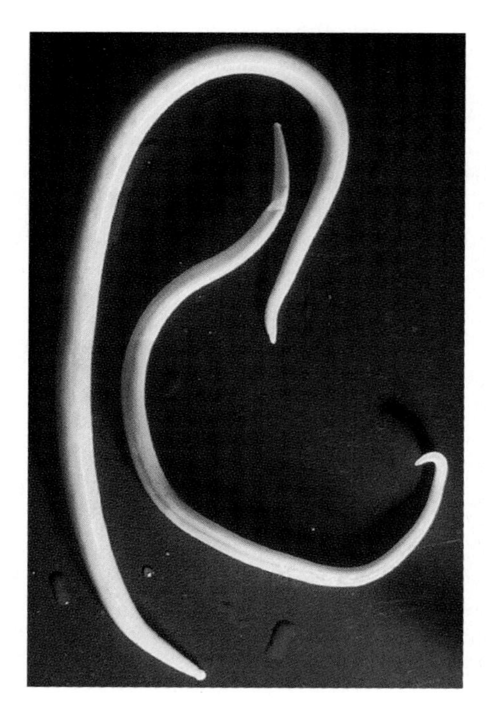

Figure 8.1

Roundworm Characteristics: A Fluid-Filled Body Cavity and a Complete Digestive System. *Ascaris lumbriocoides* inhabits the intestines of both pigs and humans. Male ascarid worms are smaller than females and have a curved posterior end.

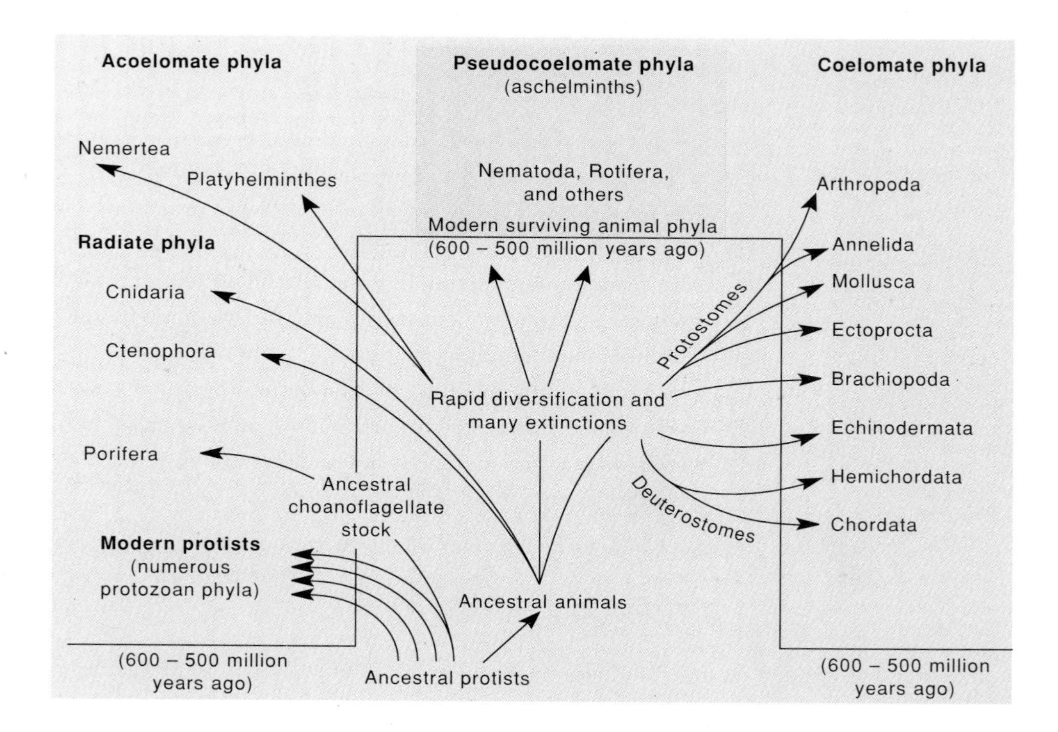

Figure 8.2

Pseudocoelomate Phyla. A generalized evolutionary tree depicting the major events and possible lines of descent for the pseudocoelomates (shaded in orange).

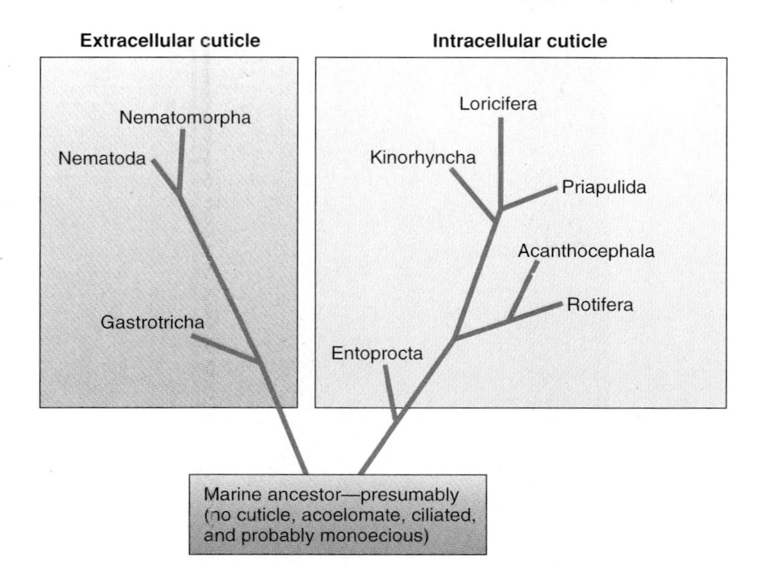

FIGURE 8.3
The Possible Phylogenetic Relationships among the Nine Aschelminth Phyla. Notice that the pseudocoelomate condition has probably been derived several times. That is to say, "aschelminths" are almost certainly a polyphyletic taxon. This may be seen in the structure of the aschelminth cuticle. For example, the cuticle of gastrotrichs, nematodes, and nematomorphs is extracellular, whereas the other members have an intracellular cuticle.

BOX 8.1 AN APPLICATION OF EUTELY

The science of aging is called **gerontology** (Gr. *gerontos,* old man). The eutelic characteristic of aschelminths makes them excellent research animals for studies on aging because (1) none of the cells of aschelminths are being continually renewed; (2) aschelminths are devoid of the capacity to repair cells; and (3) a specific number of cells is present and their exact lineage is known. In a eutelic animal, cellular longevity appears to be a simple, measurable parameter, and the onset of aging can be easily studied. Some of the characteristics of aging that have been studied in these animals include the progressive disorganization of muscle and nerve cells, mitochondrial degeneration, decrease in cellular motility, accumulation of age pigments, and increase in specific gravity.

GENERAL CHARACTERISTICS

The aschelminths are the first assemblage of animals to possess a distinct body cavity, but one that lacks the peritoneal linings and membranes called mesenteries that are found in more advanced animals. As a result, the various internal (visceral) organs lie free in the cavity. Such a cavity is called a pseudocoelom or **pseudocoel** (*see figure 4.10*b), and the animals are called **pseudocoelomates.** The pseudocoelom is often fluid filled or may contain a gelatinous substance with mesenchyme cells, serves as a cavity for circulation, aids in digestion, and acts as an internal hydrostatic skeleton that functions in locomotion.

Most aschelminths (the acanthocephalans and nematomorphs are exceptions) have a complete tubular digestive tract that extends from an anterior mouth to a posterior anus. This tube-within-a-tube body plan was first encountered in the nemerteans (*see figure 7.20*) and is characteristic of almost all other higher animals. It permits, for the first time, the mechanical breakdown of food, digestion, absorption, and feces formation to proceed sequentially and continually from an anterior to posterior direction—an evolutionary advancement over the blind-ending digestive system. Most aschelminths also have a specialized muscular pharynx that is adapted for feeding.

Many aschelminths show **eutely** (Gr. *euteia,* thrift), a condition in which the number of cells (or nuclei in syncytia) are constant both for the entire animal and for each given organ, in all the animals of that species (box 8.1). For example, the number of body (somatic) cells in all adult *Caenorhabditis elegans* nematodes is 959, and the number of cells in the pharynx of every worm in the species is precisely 80.

Most aschelminths are microscopic, although some grow to a length of over a meter. They are bilaterally symmetrical, unsegmented, triploblastic, and cylindrical in cross section. An osmoregulatory system of protonephridia (*see figure 7.6*) is found in most aschelminths. This system is best developed in freshwater forms in which osmotic problems are the greatest. No separate blood or gas exchange systems are present. There is some cephalization, with the anterior end containing a primitive brain, sensory organs, and a mouth. The vast majority of aschelminths are dioecious. Both reproductive systems are relatively uncomplicated; life cycles are usually simple, except in parasites. Cilia are generally absent from the external surface, but a thin, tough external cuticle is present for the first time in any animal group. The **cuticle** (L. *cutis,* skin) may bear spines, scales, or other forms of ornamentation that protect the animal and are useful to taxonomists. (1) Some aschelminths shed this cuticle in a process called **molting or ecdysis** (Gr. *ekdysis,* to strip off) in order to grow. Beneath the cuticle is a syncytial epidermis that actively secretes the cuticle. Several longitudinal muscle layers lie beneath the epidermis.

Most aschelminths are freshwater animals; only a few are found in marine environments. The nematomorphs, acanthocephalans, and many nematodes are parasitic. The remainder of this assemblage is mostly free living; some rotifers are colonial.

Phylum Gastrotricha

The gastrotrichs (gas-tro-tri'ks) (Gr. *gastros*, stomach + *trichos*, hair) are members of a small phylum of about 500 free-living marine and freshwater species that inhabit the space between bottom sediments. They range from 0.01 to 4 mm in length. Gastrotrichs move over the substrate using cilia on their ventral surface. The phylum contains a single class divided into two orders.

The dorsal cuticle often contains scales, bristles, or spines, and a forked tail is often present (figure 8.4). A syncytial epidermis is found beneath the cuticle. Sensory structures include tufts of long cilia and bristles on the rounded head. The nervous system includes a brain and a pair of lateral nerve trunks. The digestive system is a straight tube with a mouth, a muscular pharynx, a stomach-intestine, and an anus. Microorganisms and organic detritus from the bottom sediment and water are ingested by the action of the pumping pharynx. Digestion is mostly extracellular. Adhesive glands in the forked tail secrete materials that anchor the animal to solid objects. Paired protonephridia occur in freshwater species, rarely in marine ones. Gastrotrich protonephridia, however, are morphologically different from those found in the acoelomates. Each protonephridium possesses a single flagellum instead of the cilia found in flame cells.

Most of the marine species reproduce sexually and are hermaphroditic. Most of the freshwater species reproduce asexually by parthenogenesis; the females can lay two kinds of unfertilized eggs. Thin-shelled eggs hatch into females during favorable environmental conditions, whereas thick-shelled, resting eggs can withstand unfavorable conditions for long periods before hatching into females. There is no larval stage: development is direct, and the juveniles have the same form as the adults.

Phylum Rotifera

2 The rotifers (ro-tif'ers) (L. *rota*, wheel + *fera*, to bear) derive their name from the characteristic ciliated organ, the **corona** (Gr. *krowe*, crown), located around lobes on the head of these animals (figure 8.5*a*). The cilia of the corona do not beat in synchrony; instead, each cilium is at a slightly earlier stage in the beat cycle than the next cilium in the sequence. A wave of beating cilia thus appears to pass around the periphery of the ciliated lobes and gives the impression of a pair of spinning wheels. (Interestingly, the rotifers were first called "wheel animalicules.")

Rotifers are small animals (0.1 to 3 mm in length) that are abundant in most freshwater habitats; a few (less than 10%) are marine. There are about 2,000 species in three classes (table 8.1). The body is composed of approximately 1,000 cells, and the organs are eutelic. Rotifers are usually solitary free-swimming animals, although some colonial forms are known. Others occur interstitially in sediments.

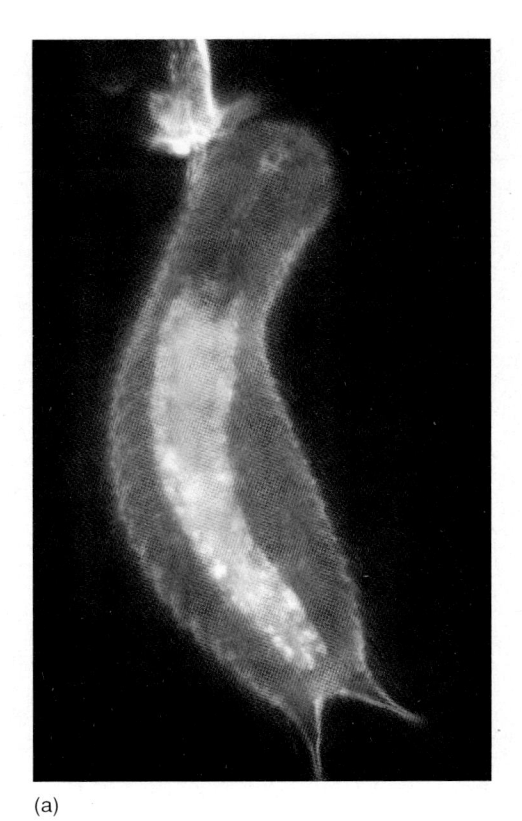

(a)

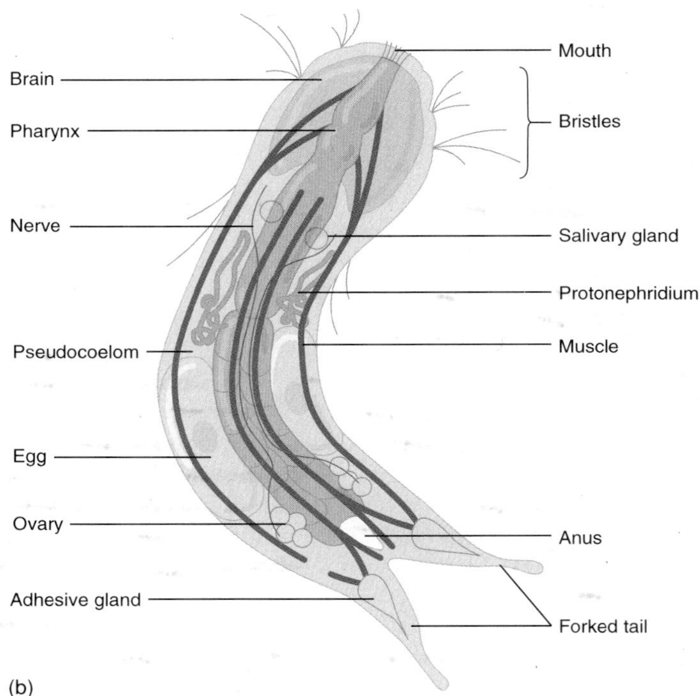

Figure 8.4
Phylum Gastrotricha. (*a*) Light micrograph of a gastrotrich *Chaetonotus*. Notice the transparent dorsoventrally flattened body. (*b*) Illustration of the internal anatomy of a freshwater gastrotrich.

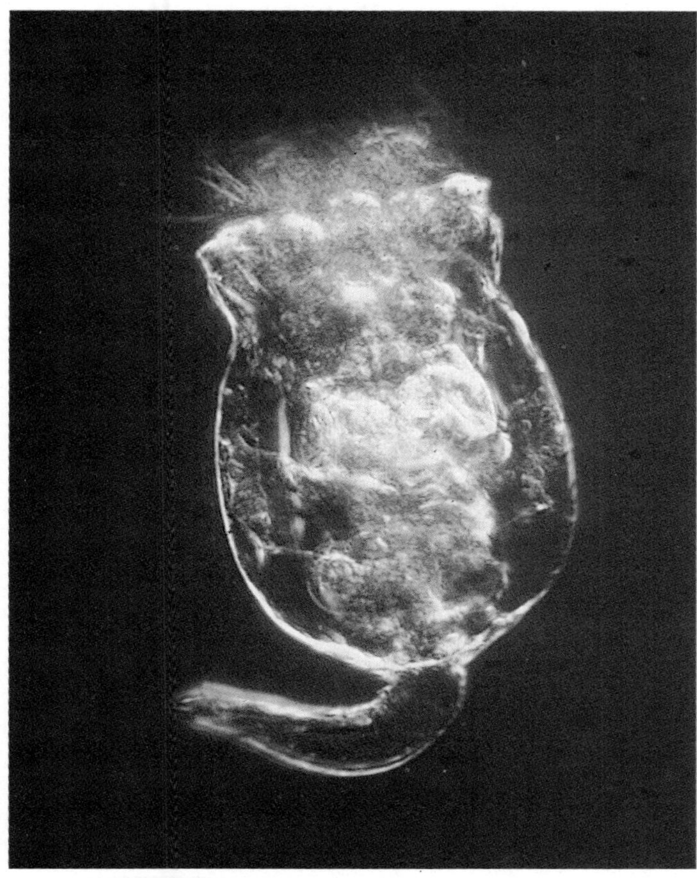

(a)

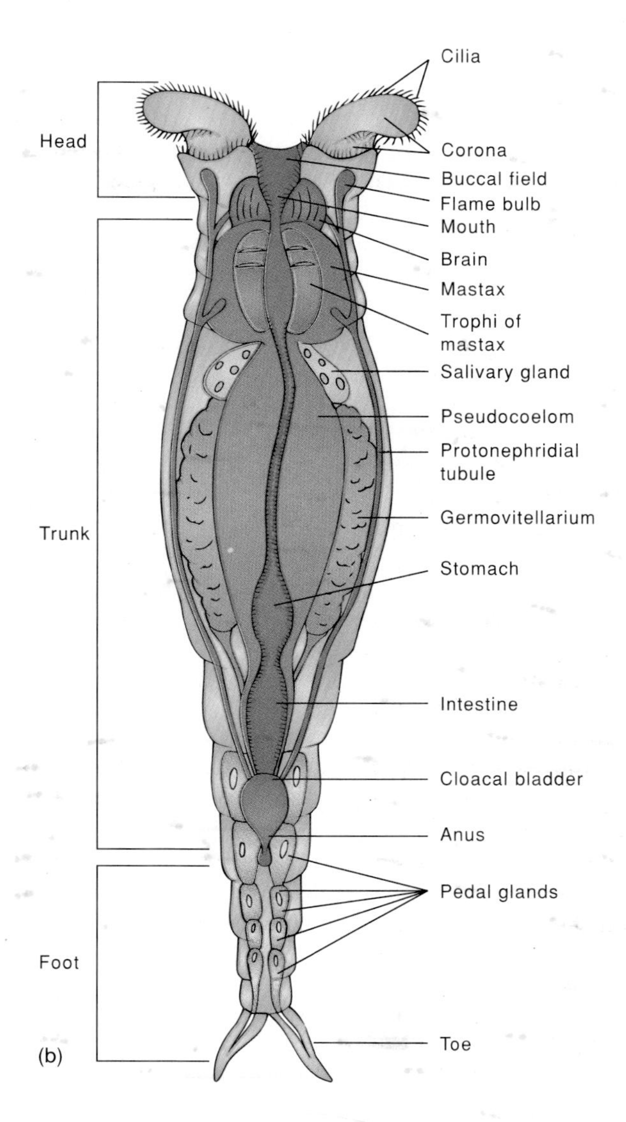

FIGURE 8.5
Phylum Rotifera. (*a*) Light micrograph of a rotifer, *Brachionus* (×150). (*b*) Illustration of the internal anatomy of a typical rotifer, *Philodina*.

TABLE 8.1 CLASSIFICATION OF THE ROTIFERA

Phylum Rotifera (ro-tif′e-ra)
A cilated corona surrounding a mouth; muscular pharynx (mastax) present with jawlike features; nonchitinous cuticle; parthenogenesis is common; both freshwater and marine species.

Class Seisonidea (sy′son-id′ea)
A single genus of marine rotifers that are commensals of crustaceans; large and elongate body with reduced corona. *Seison*.

Class Bdelloidea (Digonota) (del-oid′e-a)
Anterior end retractile and bearing two trochal disks; mastax adapted for grinding; paired ovaries; cylindrical body; males absent. *Adineta, Philodina, Rotaria*.

Class Monogononta (mon′o-go-no′ta)
Rotifers with one ovary; mastax not designed for grinding; produce mictic and amictic eggs. Males appear only sporadically. *Conochilus, Collotheca, Notommata*.

Characteristics of the phylum Rotifera include the following:

1. Triploblastic, bilateral, unsegmented, pseudocoelomate
2. Complete digestive system, regionally specialized
3. Anterior end often has a ciliated organ called a corona
4. Posterior end with toes and adhesive glands
5. Well-developed cuticle
6. Protonephridia
7. Males generally reduced in number or absent; parthenogenesis common

EXTERNAL FEATURES

The external surface of a rotifer is covered by an epidermally secreted cuticle. In many species, the cuticle is thickened to form an encasement called a **lorica** (L. *corselet*, a loose-fitting case). The cuticle or lorica provides protection and is the main supportive element, although fluid in the pseudocoelom also provides hydrostatic support. The epidermis is syncytial; that is, there are no cell membranes between nuclei.

The head contains the corona, mouth, sensory organs, and brain (figure 8.5*b*). The corona surrounds a large ciliated area called the buccal field. The trunk is the largest part of a rotifer and is elongate and saclike. The anus occurs dorsally on the posterior trunk. The posterior, narrow portion is called the foot. The terminal portion of the foot usually bears one or two toes. At the base of the toes are many pedal glands whose ducts open on the toes. Secretions from these glands aid in temporary attachment of the foot to a substratum.

Feeding and the Digestive System

Most rotifers feed on small microorganisms and suspended organic material. The coronal cilia create a current of water that brings food particles to the mouth. The pharynx contains a unique structure called the **mastax** (Gr. *jaws*). The mastax is a muscular organ in which food is ground. The inner walls of the mastax contain several sets of jaws called *trophi* (figure 8.5*b*). The trophi vary in morphological detail and are used by taxonomists to distinguish species.

From the mastax, food passes through a short, ciliated esophagus to the ciliated stomach. Salivary and digestive glands secrete digestive enzymes into the pharynx and stomach. The complete extracellular digestion of food and its absorption occur in the stomach. In some species, a short ciliated intestine extends posteriorly and becomes a cloacal bladder, which receives water from the protonephridia and eggs from the ovaries, as well as digestive waste. The cloacal bladder opens to the outside via an anus at the junction of the foot with the trunk.

Other Organ Systems

All visceral organs lie in a pseudocoelom that is filled with fluid and interconnecting amoeboid cells. Osmoregulation is accomplished by protonephridia that empty into the cloacal bladder. Rotifers, like other pseudocoelomates, exchange gases and dispose of nitrogenous wastes across body surfaces. The nervous system is composed of two lateral nerves and a bilobed, ganglionic brain that is located on the dorsal surface of the mastax. Sensory structures include numerous ciliary clusters and sensory bristles concentrated on either one or more short antennae or the corona. One to five photosensitive eyespots may be found on the head.

Reproduction and Development

Some rotifers reproduce sexually, although several types of parthenogenesis occur in most species. Smaller males appear only sporadically in one class (Monogononta) and no males are known in another class (Bdelloidea). In the class Seisonidea, fully developed males and females are equally common in the population. Most rotifers have a single ovary and an attached syncytial vitellarium, which produces yolk that is incorporated into the eggs. The ovary and vitellarium often fuse to form a single germovitellarium (figure 8.5*b*). After fertilization, each egg travels through a short oviduct to the cloacal bladder and out its opening.

In males, the mouth, cloacal bladder, and other digestive organs are either degenerate or absent. A single testis produces sperm that travel through a ciliated vas deferens to the gonopore. (3) Male rotifers typically have an eversible penis that injects sperm, like a hypodermic needle, into the pseudocoelom of the female (hypodermic impregnation).

In one class (Seisonidea), the females produce haploid eggs that must be fertilized to develop into either males or females. In another class (Bdelloidea), all females are parthenogenetic and produce diploid eggs that hatch into diploid females. In the third class (Monogononta), two different types of eggs are produced (figure 8.6). **Amictic eggs** (Gr. *a*, without + *miktos*, mixed or blended; thin-shelled summer eggs) are produced by mitosis, are diploid, cannot be fertilized, and develop directly into amictic females. Thin-shelled, **mictic** (Gr. *miktos*, mixed or blended) **eggs** are haploid. If the mictic egg is not fertilized, it develops parthenogenetically into a male; if fertilized, mictic eggs secrete a thick, heavy shell and become dormant or resting winter eggs. Dormant eggs always hatch with melting snows and spring rains into amictic females which begin a first amictic cycle, building up large populations quickly. By early summer, some females have begun to produce mictic eggs, males appear, and dormant eggs are produced. Another amictic cycle, as well as the production of more dormant eggs, occurs before the yearly cycle is over. Dormant eggs are often dispersed by winds or birds, accounting for the unique distribution patterns of many rotifers. Most females lay either amictic or mictic eggs, but not both. Apparently, during oocyte development, the physiological condition of the female determines whether her eggs will be amictic or mictic.

Stop and Ask Yourself

1. How do the cilia of a rotifer beat?
2. How can the anatomy of a rotifer be described?
3. How do rotifers feed? Reproduce?
4. What is the difference between a mictic and an amictic rotifer egg?

Phylum Kinorhyncha

Kinorhynchs (kin'o-rink's) are small (less than 1 mm long), elongate bilaterally symmetrical worms found exclusively in marine environments, where they live in mud and sand. Because they have no external cilia or locomotor appendages, they simply burrow through the mud and sand with their snout.

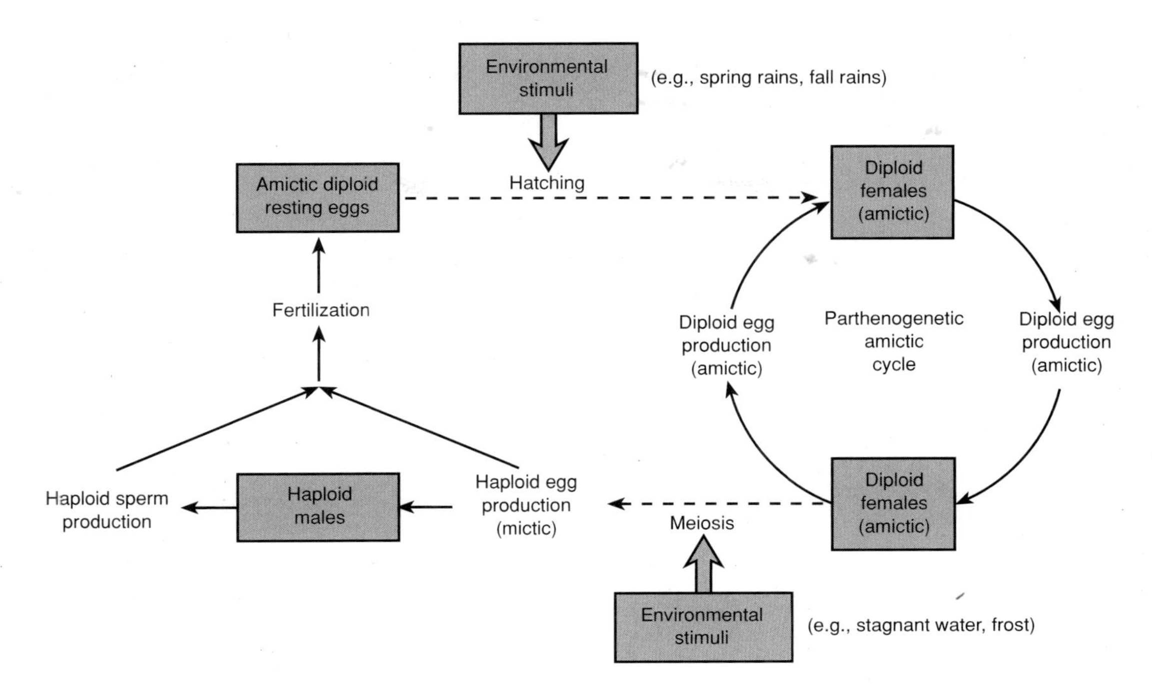

Figure 8.6

The Life Cycle of a Monogonont Rotifer. Dormant, diploid, resting eggs hatch in response to environmental stimuli (e.g., melting snows and spring rains) to begin a first amictic cycle. Other environmental stimuli (e.g., population density, stagnating water) later stimulate the production of haploid mictic eggs that lead to the production of dormant eggs that carry the species through the summer (e.g., when the pond dries up). With autumn rains, there is a second amictic cycle. Frost stimulates the production of mictic eggs again and the eventual dormant resting eggs that allow the population of rotifers to overwinter.

In fact, the phylum takes its name (Kinorhyncha, Gr. *kinein*, motion + *rhynchos*, snout) from this method of locomotion. The phylum Kinorhyncha contains about 150 known species.

The body surface of a kinorhynch is devoid of cilia and is composed of 13 or 14 definite units called **zonites** (figure 8.7). The head, represented by zonite 1, bears the mouth, an oral cone, and spines. The neck, represented by zonite 2, contains spines called **scalids** and plates called **placids.** The head can be retracted into the neck. The trunk consists of the remaining 11 or 12 zonites and terminates with the anus. Each trunk zonite bears a pair of lateral spines and one dorsal spine.

The body wall consists of a cuticle, epidermis, and two pairs of muscles: a dorsolateral and ventrolateral pair. The pseudocoelom is large and contains amoeboid cells.

A complete digestive system is present, consisting of a mouth, buccal cavity, muscular pharynx, esophagus, stomach-intestine (where digestion and absorption take place), and anus. Most kinorhynchs feed on diatoms, algae, and organic matter.

A pair of protonephridia is located in zonite 11. The nervous system consists of a brain and single ventral nerve cord with a ganglion (a mass of nerve cells) in each zonite. Eyespots and sensory bristles are found in some species. Kinorhynchs are dioecious with paired gonads. The male gonopore is surrounded by several spines that may be used in copulation. The young

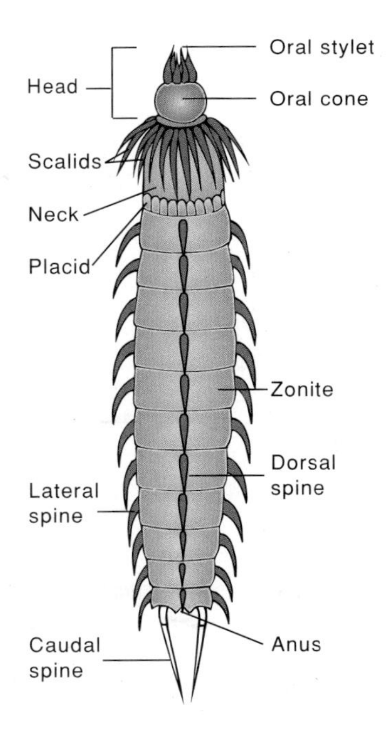

Figure 8.7

Phylum Kinorhyncha. An illustration of the external anatomy of an adult kinorhynch (dorsal view).

BOX 8.2 The Ecology of Soil Nematodes

Nematodes are incredibly abundant and diverse in soils. Some are parasitic on the roots of various plants, where they do considerable damage. Their large reproductive potential is generated at the expense of the plants on whose tissues they feed. The result is millions of dollars worth of damage to garden, truck farm, woody, and ornamental plants.

The vast majority of soil nematodes, however, are free living. They move between soil particles and are very important in the ecology of the soil. For example, some nematodes are voracious feeders on soil bacteria and fungi; they help control the populations of these microorganisms. Other nematodes feed, in turn, on the microbial feeders and also play an important role in biological control. Still other nematodes are eminently important in the entire process of decomposition. Many of these species are omnivorous or saprophytic (eat decomposing organic matter). The abundant soil nematodes are fed on by soil arthropods, some fungi, and earthworms. Overall, nematodes are essential to the energy flow and nutrient cycling in soil ecosystems.

hatch into larvae that do not have all of the zonites. As the larvae grow and molt, the adult morphology appears. Once adulthood is attained, molting no longer occurs.

Phylum Nematoda

(4) Nematodes (nem-a-to′des) (Gr. *nematos*, thread) or roundworms are some of the most abundant animals on earth—there may be some 5 billion in every acre of fertile garden soil. Zoologists estimate that there are anywhere from 10,000 to 500,000 roundworm species, feeding on every conceivable source of organic matter—from rotting substances to the living tissues of other invertebrates, vertebrates, and plants (box 8.2). They range in size from microscopic to several meters long. Many nematodes are parasites of plants or animals; others are free living in marine, freshwater, or soil habitats. Some nematodes play an important role in recycling nutrients in soils and bottom sediments.

Except in their sensory structures, nematodes lack cilia; a characteristic they share with arthropods. Also in common with some arthropods, the sperm of nematodes is amoeboid. Two classes of nematodes are recognized (table 8.2).

Characteristics of the phylum Nematoda include the following:

1. Triploblastic, bilateral, vermiform (resembling a worm in shape; long and slender), unsegmented, pseudocoelomate
2. Body round in cross section and covered by a layered cuticle; growth in juveniles usually accompanied by molting

TABLE 8.2 Classification of the Nematodes

Phylum Nematoda (nem-a-to′da)
Nematodes, or roundworms.

Class Secernentea (Phasmidea) (ses-er-nen′te-a)
Paired glandular or sensory structures called phasmids in the tail region; similar pair of structures (amphids) poorly developed in anterior end; excretory system present; both free-living and parasitic species. *Ascaris*, *Enterobius*, *Rhabditis*, *Turbatrix*, *Necator*, *Wuchereria*.

Class Adenophorea (Aphasmidia) (a-den″o-for′e-a)
Phasmids absent; most free living, but some parasitic species occur. *Dioctophyme*, *Trichinella*, *Trichuris*.

3. Complete digestive tract; mouth usually surrounded by lips bearing sense organs
4. Most with unique excretory system comprised of one or two renette cells or a set of collecting tubules
5. Body wall has only longitudinal muscles

External Features

A typical nematode body is slender, elongate, cylindrical, and tapered at both ends (figure 8.8*a*,*b*). Much of the success of nematodes is due to their outer, noncellular, collagenous cuticle (figure 8.8*c*) that is continuous with the foregut, hindgut, sense organs, and parts of the female reproductive system. The cuticle may be either smooth or contain spines, bristles, papillae (small, nipplelike projections), warts, or ridges, all of which are of taxonomic significance. Three primary layers make up the cuticle: cortical, matrix, and basal. The cuticle functions to maintain internal hydrostatic pressure, provide mechanical protection, and aid in resisting digestion by the host in parasitic species. The cuticle is usually molted four times during maturation.

Beneath the cuticle is the epidermis, or hypodermis, which surrounds the pseudocoelom (figure 8.8*d*). The epidermis may be syncytial, and its nuclei are usually located in the four epidermal cords (one dorsal, one ventral, and two lateral) that project inward. The longitudinal muscles are the principal means of locomotion in nematodes. (5) Contraction of these muscles results in undulatory waves that pass from the anterior to posterior end of the animal, creating characteristic thrashing movements. Nematodes lack circular muscles and therefore cannot crawl as do worms with more complex musculature.

Some nematodes have lips surrounding the mouth, and some species bear spines or teeth on or near the lips. In others, the lips have disappeared. Some roundworms have head shields that afford protection. Sensory organs include amphids, phasmids, or ocelli. **Amphids** are anterior depressions in the cuticle that contain modified cilia and function in chemoreception. **Phasmids** are located near the anus and also function in chemoreception. The presence or absence of these

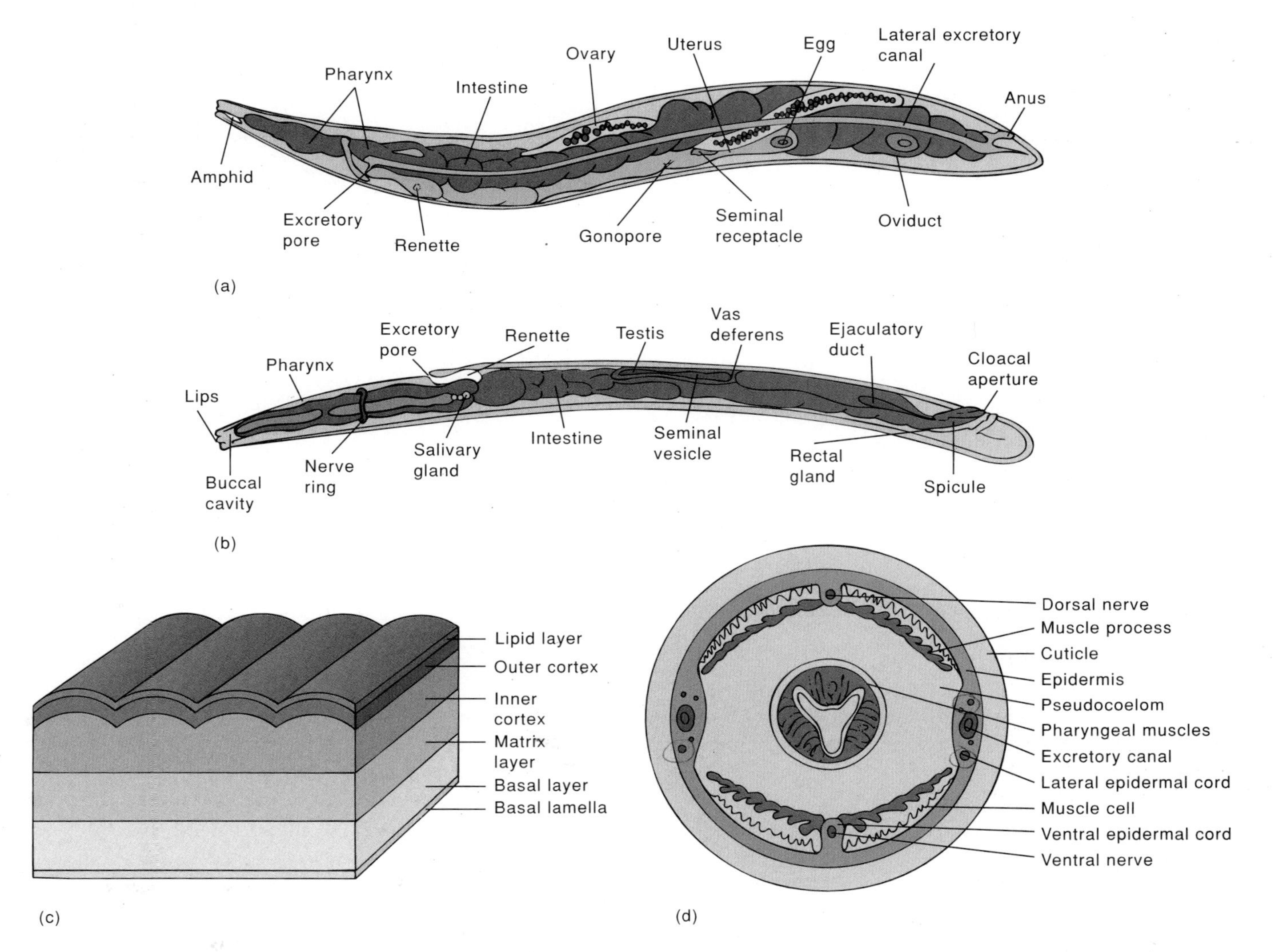

FIGURE 8.8

Phylum Nematoda. (*a*) Internal anatomical features of a female and (*b*) male *Rhabditis*. (*c*) A section through a nematode cuticle showing the various layers. (*d*) A cross section through the region of the muscular pharynx of a nematode. The hydrostatic pressure in the pseudocoelom acts to maintain the rounded body shape of a nematode and also to collapse the intestine, which aids in food and waste material moving from the mouth to the anus.

organs determines the taxonomic class (*see table 8.2*) to which nematodes belong. Paired ocelli (eyes) are present in aquatic nematodes.

INTERNAL FEATURES

The nematode pseudocoelom is a spacious, fluid-filled cavity that contains the visceral organs and forms a hydrostatic skeleton. 6 All nematodes are round because of the equal outward force generated in all directions by the body muscles contracting against the pseudocoelomic fluid (figure 8.8*d*).

FEEDING AND THE DIGESTIVE SYSTEM

Depending on the environment, nematodes are capable of feeding on a wide variety of foods—they may be carnivores, herbivores, omnivores, saprobes that consume decomposing organisms, or parasitic species that feed on blood and tissue fluids of their hosts.

Nematodes have a complete digestive system consisting of a mouth, which may have teeth, jaws, or stylets; buccal cavity; muscular pharynx; long tubular intestine where digestion and absorption occur; short rectum; and anus. Hydrostatic

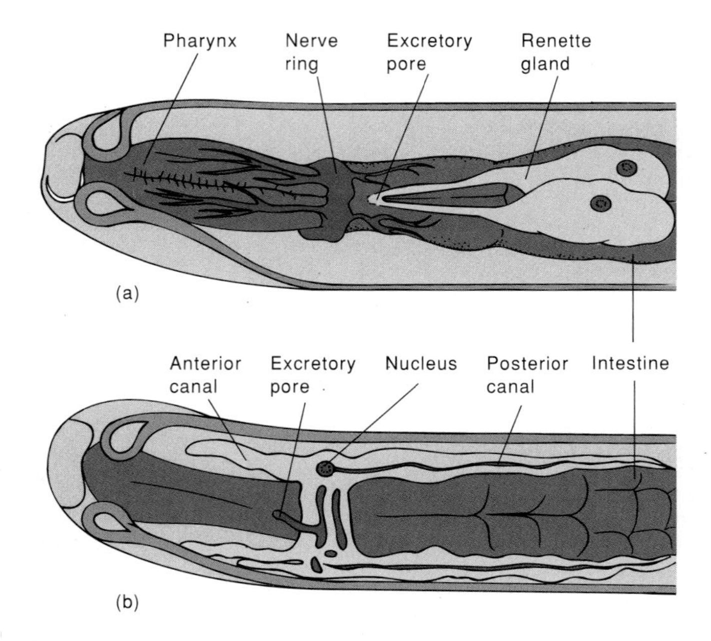

Figure 8.9
Nematode Excretory Systems. (*a*) Glandular, as found in *Rhabditis*. (*b*) Tubular, as found in *Ascaris*.

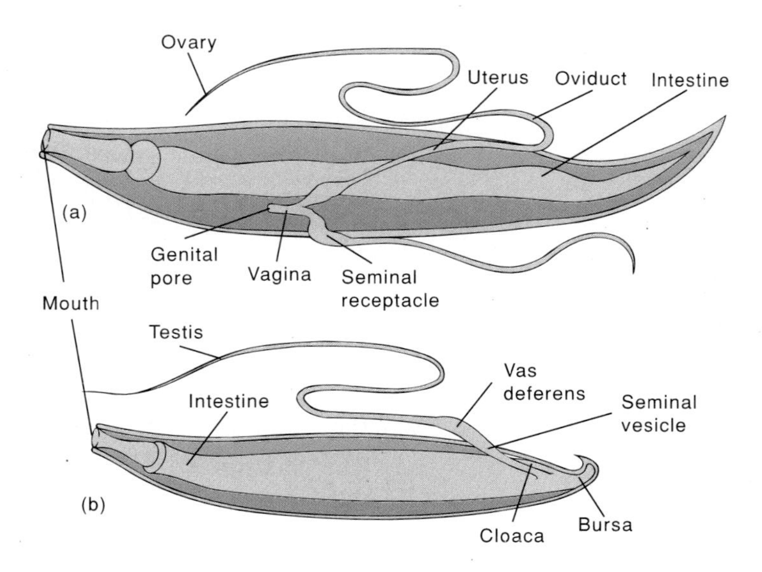

Figure 8.10
Nematode Reproductive Systems. Illustrations showing the reproductive systems of a (*a*) female and (*b*) male nematode such as *Ascaris*. The sizes of the reproductive systems are exaggerated to show details.

pressure in the pseudocoelom and the pumping action of the pharynx are responsible for the passage of food through the alimentary canal.

Other Organ Systems

Nematodes accomplish osmoregulation and excretion of nitrogenous waste products (ammonia, urea) with two unique systems. The glandular system (figure 8.9*a*) is found in aquatic species, and consists of ventral gland cells, called **renettes,** that are located posterior to the pharynx. Each gland absorbs waste material from the pseudodocoelom and empties it to the outside through an excretory pore. Parasitic nematodes have a more advanced system, called the tubular system (figure 8.9*b*), that develops from the renette system. In this system, the renettes unite to form a large canal, which opens to the outside via an excretory pore.

The nervous system consists of an anterior nerve ring (*see figure* 8.8b). Nerves extend anteriorly and posteriorly; many connect to each other via commissures. Certain neuroendocrine secretions are involved in growth, molting, cuticle formation, and metamorphosis.

Reproduction and Development

Most nematodes are dioecious and dimorphic, with the males being smaller than the females. The long, coiled gonads lie free in the pseudocoelom.

The female system consists of a pair of convoluted ovaries (figure 8.10*a*). Each ovary is continuous with an oviduct whose proximal end is swollen to form a seminal receptacle. Each oviduct becomes a tubular uterus; the two uteri unite to form a vagina that opens to the outside through a genital pore.

The male system consists of a single testis, which is continuous with a vas deferens that eventually expands into a seminal vesicle (figure 8.10*b*). The seminal vesicle connects to the cloaca. Males are commonly armed with a posterior flap of tissue called a bursa. The bursa aids the male in the transfer of sperm to the female genital pore during copulation.

After copulation, each fertilized egg is moved to the gonopore by hydrostatic forces in the pseudocoelom (*see figure* 8.8d). The number of eggs produced varies with the species; some nematodes produce only several hundred, whereas others may produce 200,000 daily. Some nematodes give birth to larvae (ovoviviparity). The development and hatching of the eggs are influenced by external factors, such as temperature and moisture. Hatching produces a larva (also referred to by some parasitologists as a juvenile) that has most adult structures. The larva (juvenile) undergoes four molts, although in some species, the first one or two molts may take place before the eggs hatch.

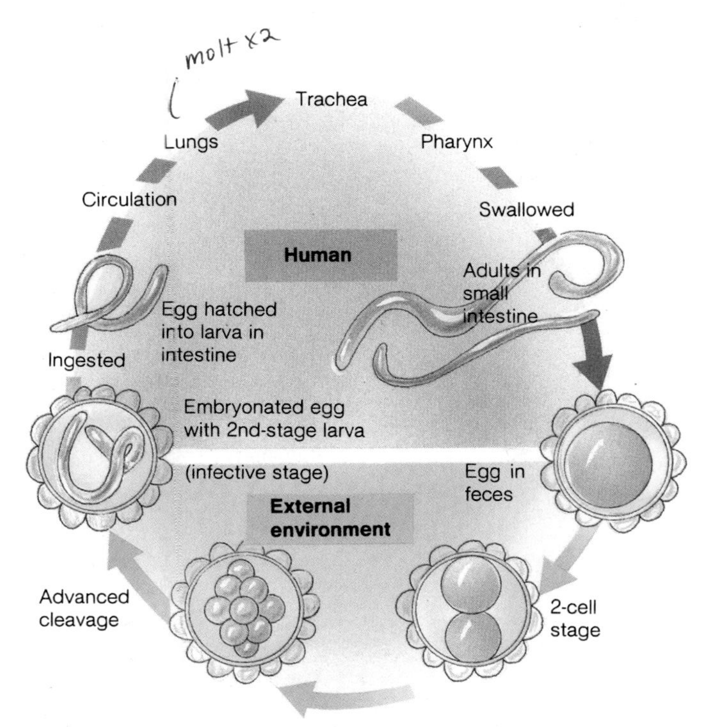

FIGURE 8.11
The Life Cycle of *Ascaris lumbricoides*. (See text for details.) *Source: Redrawn from Centers for Disease Control, Atlanta.*

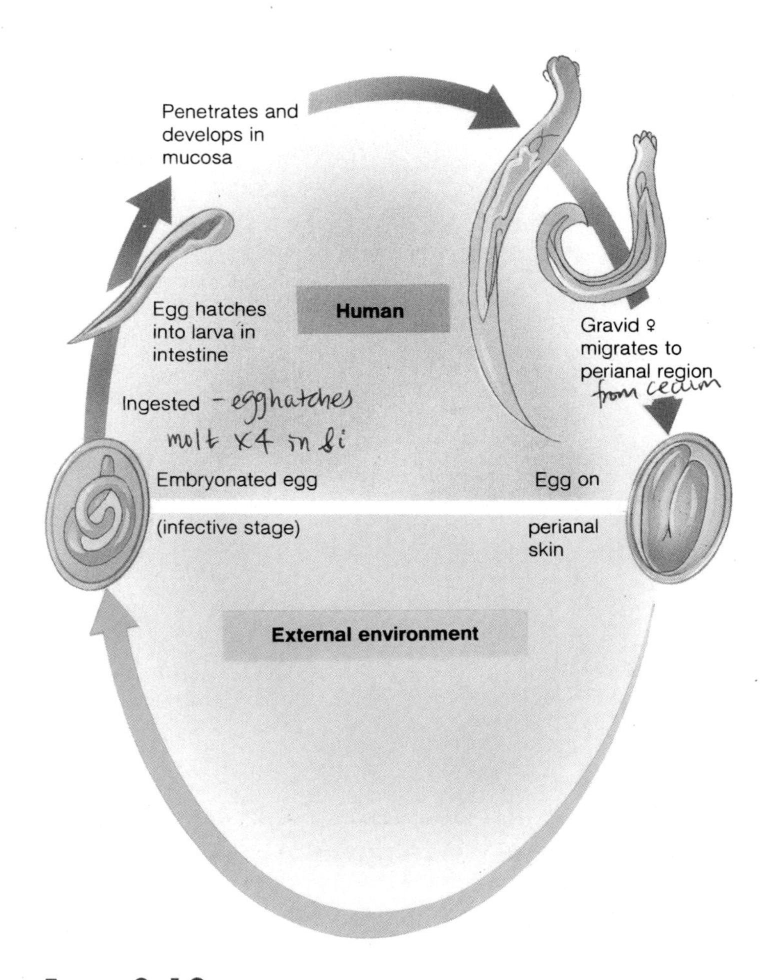

FIGURE 8.12
The Life Cycle of *Enterobius vermicularis*. (See text for details.) *Source: Redrawn from Centers for Disease Control, Atlanta.*

SOME IMPORTANT NEMATODE PARASITES OF HUMANS

Parasitic nematodes possess a number of evolutionary adaptations to their way of life. These include a high reproductive potential, life cycles that make transmission from one host to another more likely, an enzyme-resistant cuticle, resistant eggs, and encysted larvae. The life cycles of nematodes are not as complicated as those of cestodes or trematodes because only one host is usually involved. Discussions of the life cycles of five important human parasites follow.

Ascaris lumbricoides: The Giant Intestinal Roundworm of Humans

It has been estimated that as many as 800 million people throughout the world may be infected with *Ascaris lumbricoides*. Adult *Ascaris* (Gr. *askaris*, intestinal worm) live in the small intestine of humans (*see figure 8.1*). Large numbers of eggs are produced and exit with the feces (figure 8.11). A first-stage larva develops rapidly in the egg, molts, and matures into a second-stage larva, the infective stage. When a human ingests embryonated eggs, they hatch in the intestine, the larvae penetrate the intestinal wall, and are carried via the circulation to the lungs. They molt twice in the lungs, migrate up the trachea, and are swallowed. The worms attain sexual maturity in the intestine, mate, and begin egg production.

Enterobius vermicularis: The Human Pinworm

7 Pinworms (*Enterobius*; Gr. *enteron*, intestine + *bios*, life) are the most common roundworm parasites in the United States. Adult *Enterobius vermicularis* are located in the lower region of the large intestine. At night, gravid females migrate out of the cecum to the perianal area, where they deposit eggs containing a first-stage larva (figure 8.12). When humans ingest these eggs, the eggs hatch, the larvae molt four times in the small intestine and migrate to the large intestine. Adults mate, and females begin egg production in a short period of time.

Necator americanus: The New World Hookworm

The New World or American hookworm, *Necator americanus* (L. *necator*, killer), can be found in the southern United States. The adults live in the small intestine, where they hold onto the intestinal wall with teeth and feed on blood and tissue fluids

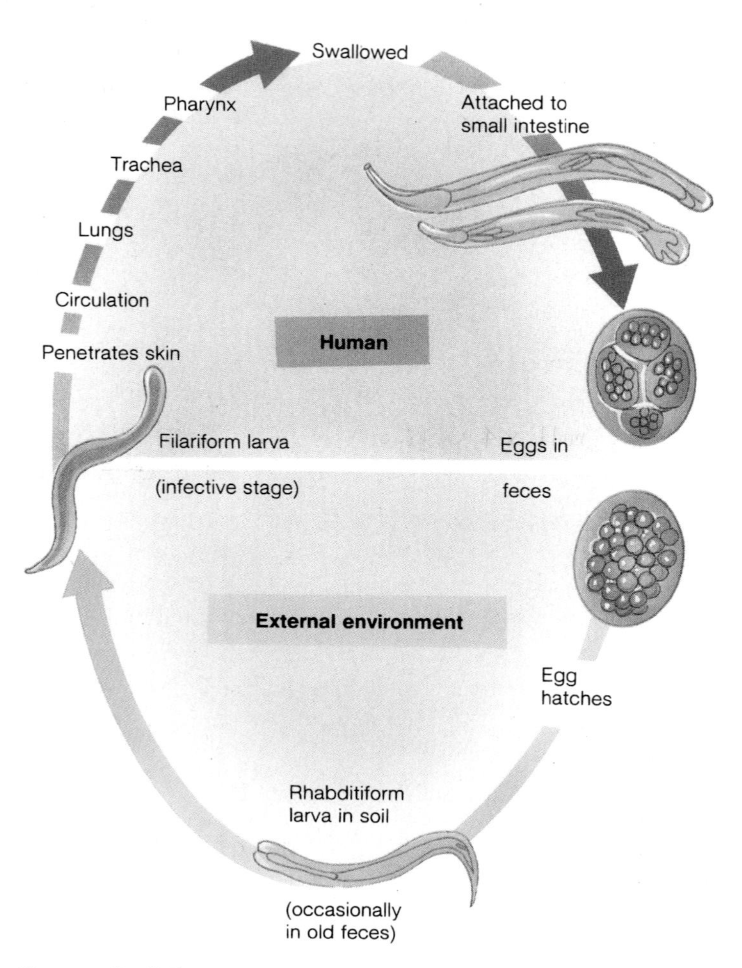

FIGURE 8.13
The Life Cycle of *Necator americanus*. (See text for details.) *Source: Redrawn from Centers for Disease Control, Atlanta.*

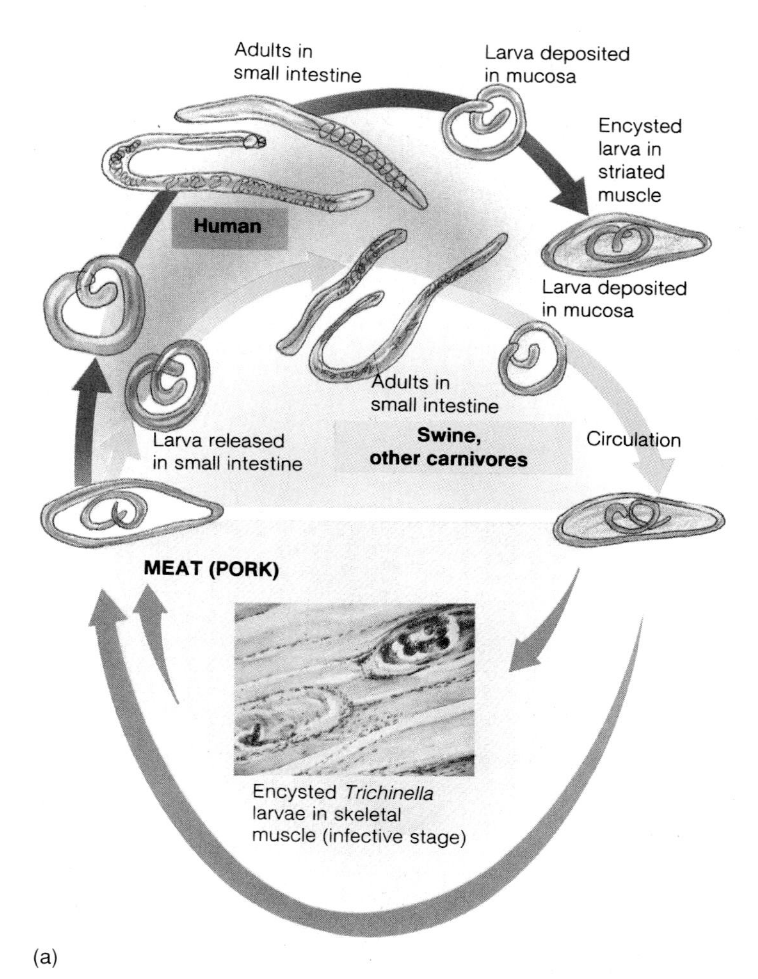

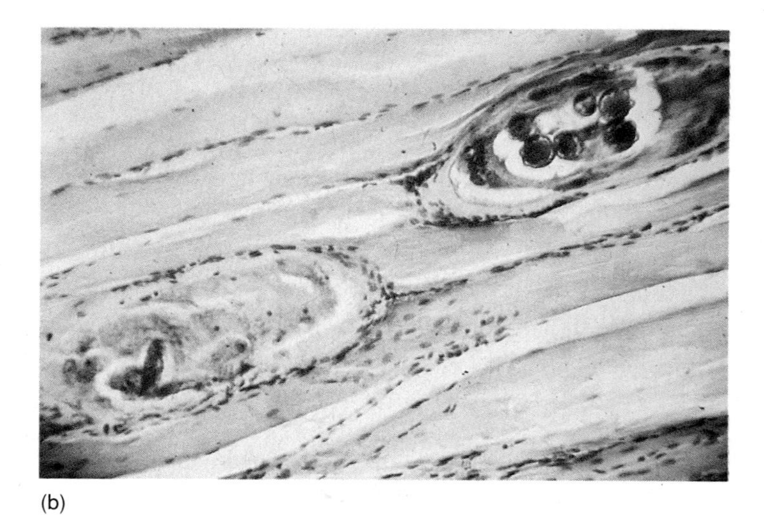

FIGURE 8.14
The Life Cycle of *Trichinella spiralis*. (*a*) See text for details. (*b*) An enlargement of the insert in (*a*) showing two encysted larvae in skeletal muscle (light micrograph ×450). *(a) Source: Redrawn from Centers for Disease Control, Atlanta.*

(figure 8.13). Individual females may produce as many as 10,000 eggs daily, which pass out of the body in the feces. The eggs hatch on warm, moist soil and release a small rhabditiform larva. It molts and becomes the infective filariform larva. Humans become infected when the filariform larva penetrates the skin, usually on the foot. (Outside defecation and subsequent walking barefoot through the immediate area maintains the life cycle in humans.) After the larva burrows through the skin, it reaches the circulatory system. The rest of its life cycle is similar to that of *Ascaris* (*see figure 8.11*).

Trichinella spiralis: The Porkworm

Adult *Trichinella* (Gr. *trichinos*, hair) *spiralis* live in the mucosa of the small intestine of humans and other carnivores (e.g., the pig). In the intestine, adult females give birth to young larvae that then enter the circulatory system and are carried to skeletal (striated) muscles of the same host (figure 8.14). The young

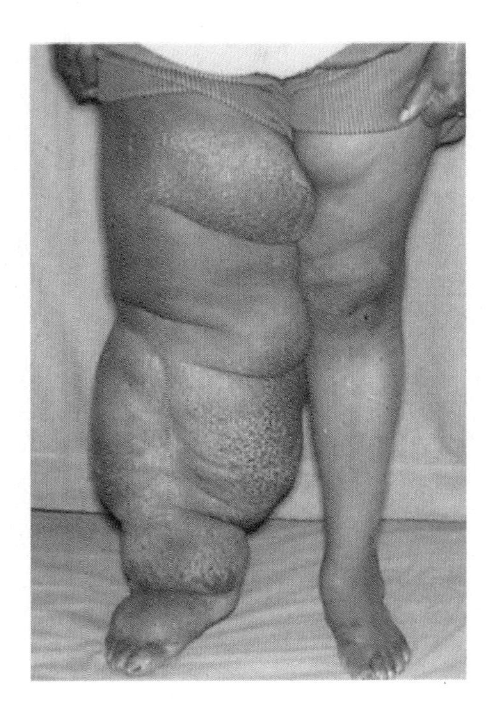

Figure 8.15
Elephantiasis. An Example of Elephantiasis Caused by the Filarial Worm, *Wuchereria bancrofti*. It takes years for this condition to become as shown in this figure.

larvae encyst in the skeletal muscles and remain infective for many years. The disease caused by this nematode is called **trichinosis.** Infective meat (muscle) must be ingested by another host to continue the life cycle. (8) Humans most often become infected by eating improperly cooked pork products. Once ingested, the larvae excyst in the stomach and make their way to the small intestine, where they molt four times and develop into adults.

Wuchereria spp.: The Filarial Worms

In tropical countries, over 250 million humans are infected with filarial (L. *filium*, thread) worms. Two examples of human filarial worms are *Wuchereria bancrofti* and *W. malayi*. These elongate, threadlike nematodes live in the lymphatic system, where they block the vessels. Because lymphatic vessels return tissue fluids to the circulatory system, when the filiarial nematodes block these vessels, fluids tend to accumulate in peripheral tissues. (9) This fluid accumulation causes the enlargement of various appendages, a condition called **elephantiasis** (figure 8.15).

In the lymphatic vessels, adults copulate and produce larvae called **microfilariae** (figure 8.16). The microfilariae are released into the bloodstream of the human host and migrate to the peripheral circulation at night. When a mosquito feeds on a human, it ingests the microfilariae. The microfilariae migrate to the mosquito's thoracic muscles, where they molt twice and become infective. When the mosquito takes another meal of blood, the infective third-stage larvae are injected into the blood of the human host through the mosquito's proboscis. The final two molts take place as the larvae enter the lymphatic vessels.

A filarial worm prevalent in the United States is *Dirofilaria immitis*, a parasite of dogs. (10) Since the adult worms live in the heart and large arteries of the lungs, the infection is called **heartworm disease.** Once established, these filarial worms are difficult to eliminate, and can be fatal; prevention with heartworm medicine is thus advocated for all dogs.

Stop and Ask Yourself

5. What is the function of the nematode cuticle?
6. What are some internal structures of nematodes? Some external structures?
7. What are the two types of excretory systems found in nematodes?
8. What are the life cycles of the following human parasites: *Ascaris? Enterobius? Necator? Trichinella? Wuchereria?*

Phylum Nematomorpha

Nematomorphs (nem′a-to-mor′phs) (Gr. *nema*, thread + *morphe*, form) are a small group (about 250 species) of elongate worms commonly called either **horsehair worms,** or **Gordian worms.** (11) The hairlike nature of these worms is so striking that they were formerly thought to arise spontaneously from the hairs of a horse's tail in drinking troughs or other stock-watering places. The adults are free living, but the juveniles are all parasitic in arthropods. They have a worldwide distribution and can be found in both running and standing water.

The body of a nematomorph is extremely long and threadlike and has no distinct head (figure 8.17). The body wall has a thick cuticle, a cellular epidermis, longitudinal cords, and a muscle layer of longitudinal fibers. The nervous system contains an anterior nerve ring and a vental cord.

Nematomorphs have separate sexes; two long gonads extend the length of the body. After copulation, the eggs are deposited in water. When an egg hatches, a small larva that has a protrusible proboscis armed with spines emerges. Terminal stylets are also present on the proboscis. The larva must quickly enter an arthropod (e.g., a beetle, cockroach) host, either by penetrating the host or being eaten. Lacking a digestive system, the larva feeds by absorbing material directly across its body wall. Once mature, the worm leaves its host only when the arthropod is near water. Sexual maturity is attained during the free-living adult phase of the life cycle.

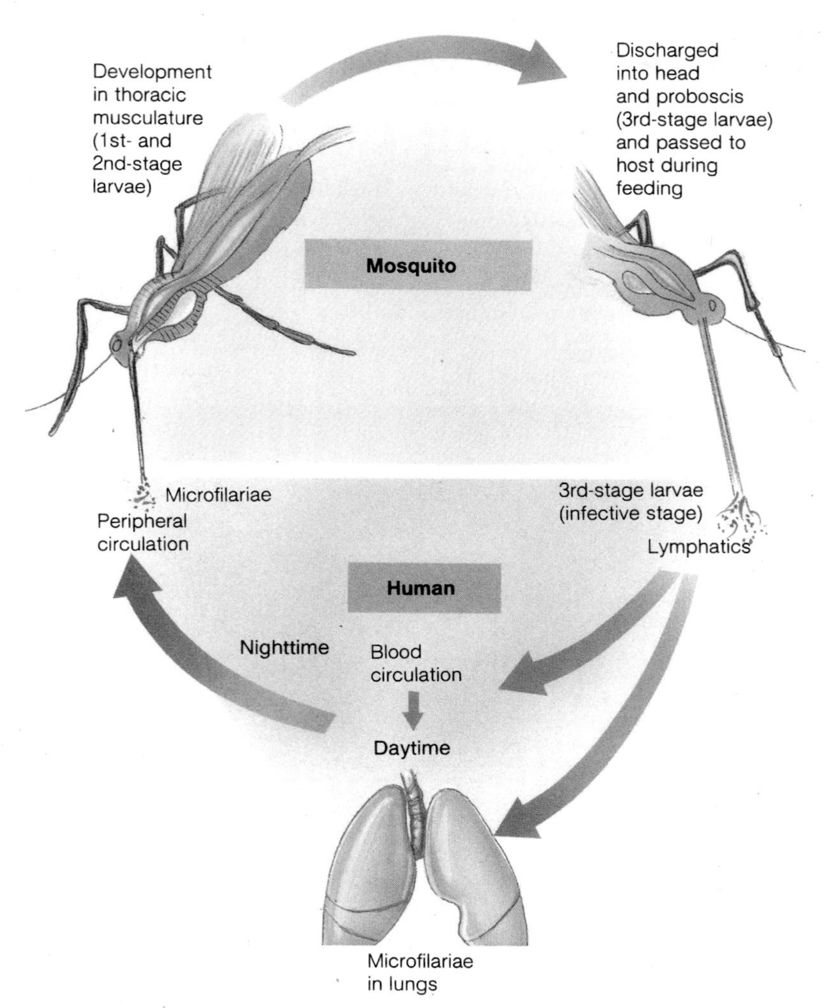

Figure 8.16

The Life Cycle of *Wuchereria* spp. (See text for details.) *Source: Redrawn from Centers for Disease Control, Atlanta.*

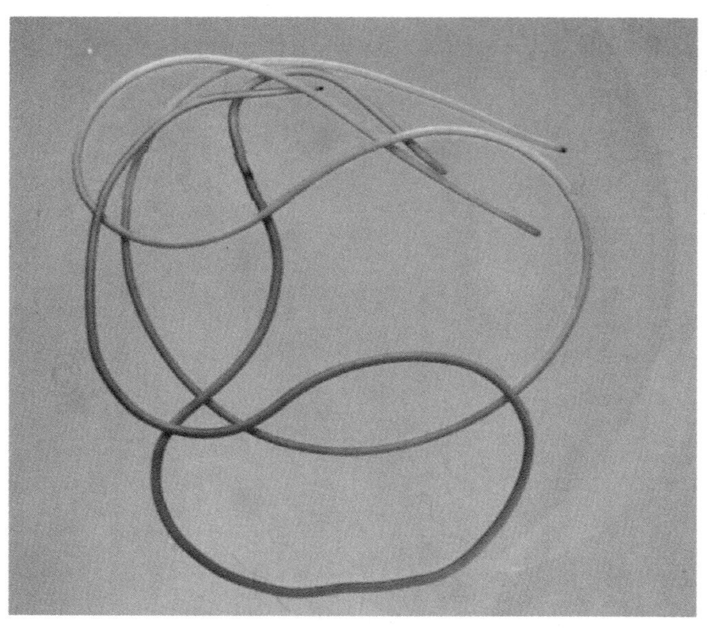

Figure 8.17

Phylum Nematomorpha. Photomicrograph of two adult worms. As illustrated, these worms tend to twist and turn upon themselves, giving the appearance of complicated knots, thus the name "Gordian worms." (Legend has it that King Gordius of Phrygia tied a formidable knot—the Gordian knot—and declared that whoever might undo it would be the ruler of all Asia. No one could accomplish this until Alexander the Great cut through it with his broadsword.)

Phylum Acanthocephala

Adult acanthocephalans (a-kan′tho-sef′a-lans) (Gr. *akantha*, spine or thorn + *kephale*, head) are endoparasites in the intestinal tract of vertebrates (especially fishes). Two hosts are required to complete the life cycle. The juveniles are parasites of crustaceans and insects. Acanthocephalans are generally small (less than 40 mm long), although one important species that occurs in pigs, *Macracanthorhynchus hirudinaceus*, can be up to 80 cm long. The body of the adult is elongate and composed of a short anterior proboscis, a neck region, and a trunk (figure 8.18*a*). (12) The proboscis is covered with recurved spines (figure 8.18*b*); hence the name "spiny-headed worms." The retractible proboscis provides the means of attachment in the host's intestine. Females are always larger than males and there are about 1,000 species.

The body wall of acanthocephalans is covered by a living syncytial tegument that has been adapted to the parasitic way of life. A glycocalyx consisting of mucopolysaccharides and glycoproteins covers the tegument and protects against host enzymes and immune defenses. No digestive system is present; food is absorbed directly through the tegument from the host by specific membrane transport mechanisms and pinocytosis. Protonephridia may be present. The nervous system is composed of a ventral, anterior ganglionic mass from which anterior and posterior nerves arise. Sensory organs are poorly developed.

The sexes are separate, and the male has a protrusible penis. Fertilization is internal, and development of eggs takes place in the pseudocoelom. The biotic potential of certain acanthocephalans is great; for example, a gravid female *Macroacanthorhynchus hirudinaceus* may contain up to 10 million embryonated eggs. The eggs pass out of the host with the feces and must be eaten by certain insects (e.g., cockroaches or grubs), or by aquatic crustaceans (e.g., amphipods, isopods, ostracods). Once in the invertebrate, the larva emerges from the egg and is now called an **acanthor.** It burrows through the gut wall and lodges in the hemocoel, where it develops into an **acanthella** and eventually into a **cystacanth.** When the intermediate host is eaten by a mammal, fish, or bird, the cystacanth excysts and attaches to the intestinal wall with its spiny proboscis.

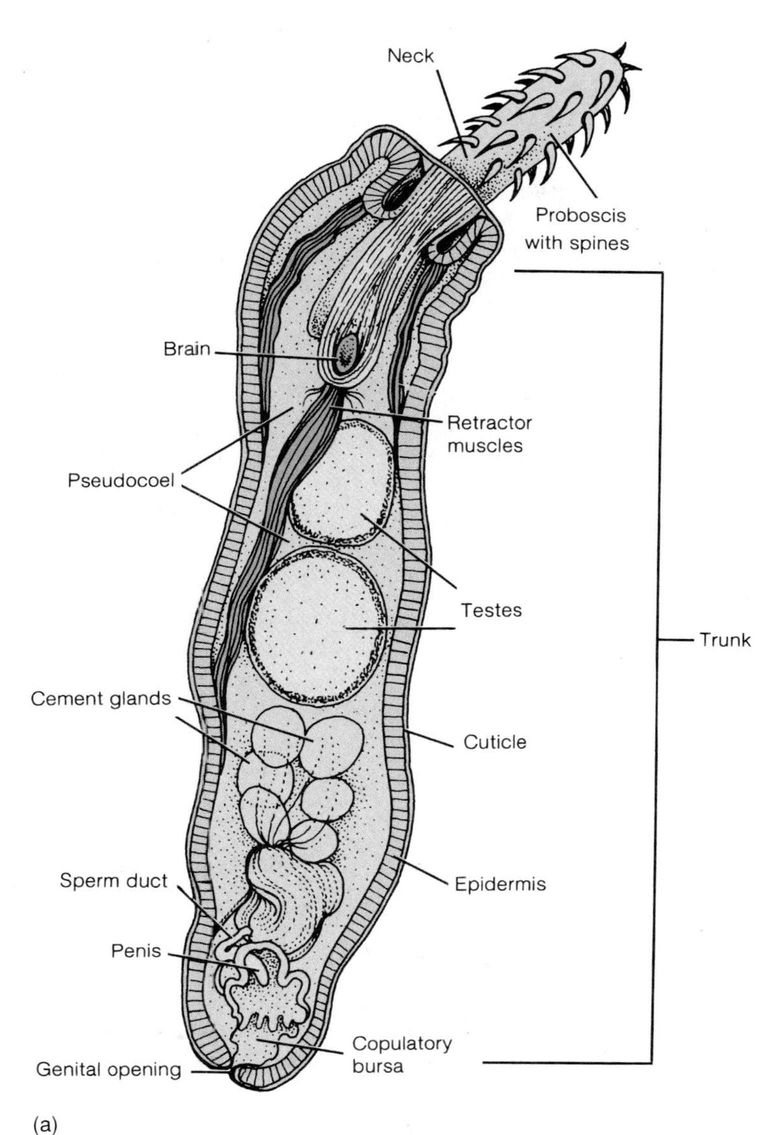

Figure 8.18

Phylum Acanthocephala. (*a*) Drawing of an adult male, dorsal view. (*b*) Light micrograph of the proboscis of a spiny-headed worm (×50).

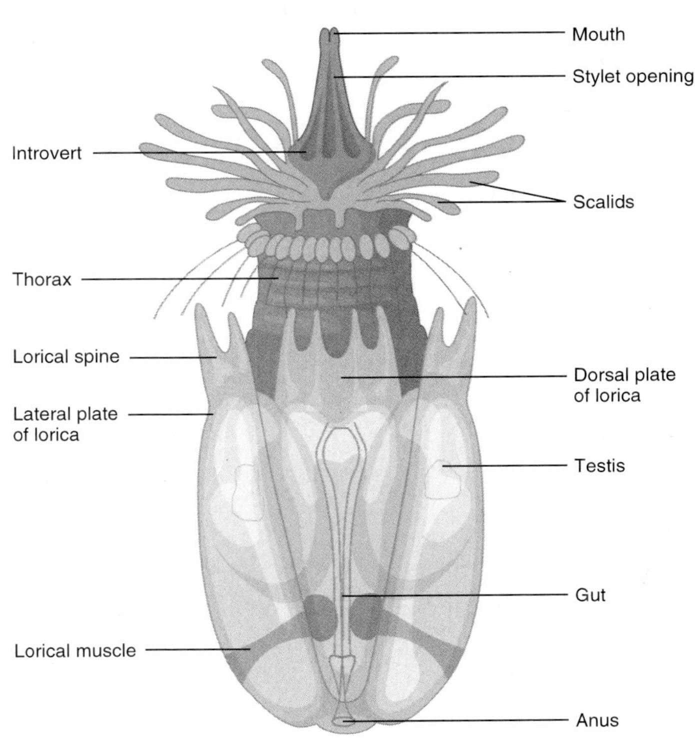

Figure 8.19

Phylum Loricifera. The anatomy of an adult male *Nanaloricus* (dorsal view).

Stop and Ask Yourself

9. Why are nematomorphs called horsehair worms or Gordian worms?
10. What is the life cycle of a nematomorph?
11. What is unique about the tegument of an acanthocephalan?
12. What is the life cycle of an acanthocephalan?

Phylum Loricifera

13 The phylum Loricifera (lor′a-sif-er-a) (L. *lorica*, clothed in armor + *fero*, to bear) is the most recently described animal phylum. Its first members were identified and named in 1983. Loriciferans occur in spaces between marine gravel. A characteristic species is *Nanaloricus mysticus*. It is a small, bilaterally symmetrical worm that has a spiny head called an **introvert,** a thorax, and an abdomen surrounded by a lorica (figure 8.19). Both the introvert and thorax can be retracted into the anterior end of the lorica. The introvert bears eight oral stylets that surround the mouth. The lorical cuticle is periodically molted. A pseudocoelom is present and contains a short digestive system, brain, and several ganglia. Loriciferans are dioecious with paired gonads. About 14 species have been formally described.

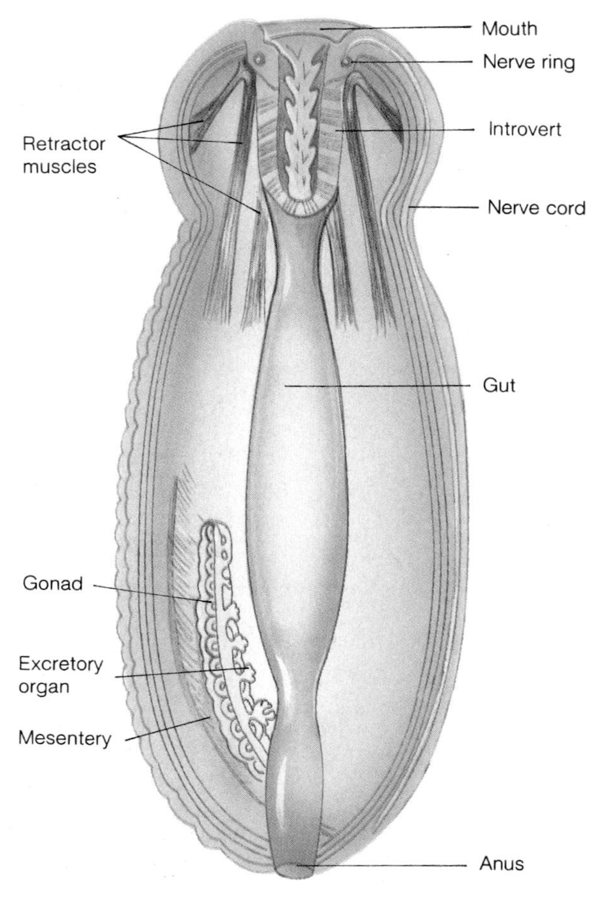

Figure 8.20
Phylum Priapulida. The internal anatomy of the priapulid, *Priapulus*. In this drawing, the introvert is shown withdrawn into the body.

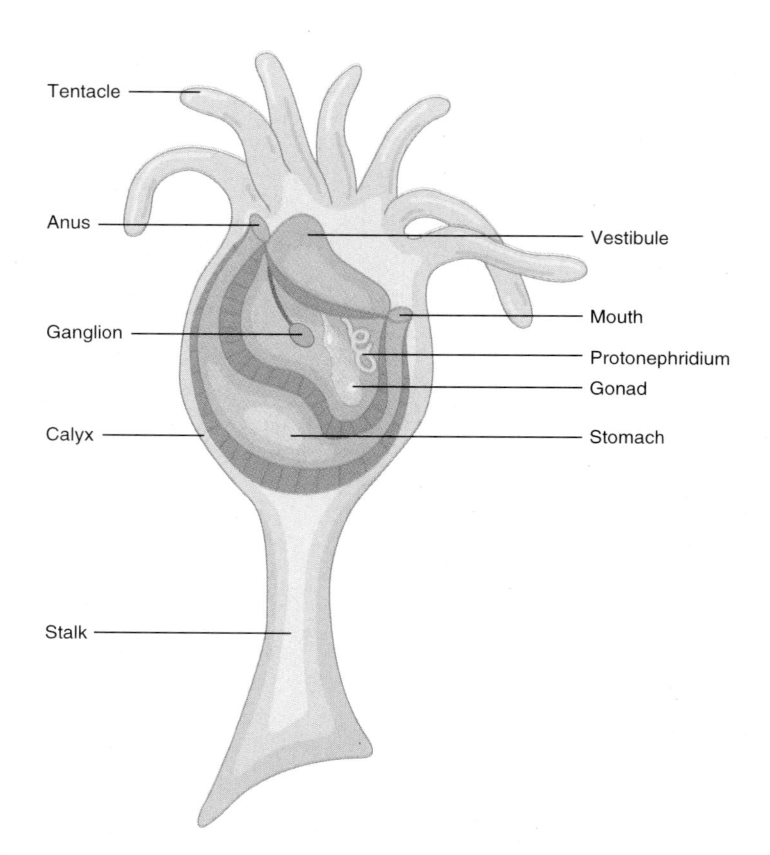

Figure 8.21
Phylum Entoprocta. Some morphological features of a typical entoproct.

Phylum Priapulida

The priapulids (pri′a-pyu-lids) (Gr. *priapos*, phallus + *ida*, pleural suffix; from *Priapos*, the Greek god of reproduction, symbolized by the penis) are a small group (only 16 species) of marine worms found in cold waters. They live buried in the mud and sand of the seafloor, where they feed on small annelids and other invertebrates.

The body (figure 8.20) is cylindrical in cross section, and ranges in length from 2 mm to about 8 cm. The anterior part of the body is an introvert (proboscis), which can be drawn into the longer, posterior trunk. The introvert functions in burrowing and is surrounded by spines. The muscular body is covered with a thin cuticle that bears spines and the trunk bears superficial annuli. A straight digestive tract is suspended in a large pseudocoelom that acts as a hydrostatic skeleton. In some species, the pseudocoelom contains amoeboid cells that probably function in gas transport. The nervous system consists of a nerve ring around the pharynx and a single midventral nerve cord. The sexes are separate but not superficially distinguishable. A pair of gonads is suspended in the pseudocoelom and shares a common duct with the protonephridia. The duct opens near the anus, and gametes are shed into the sea. Fertilization is external, and the eggs eventually sink to the bottom, where the larvae develop into adults. The cuticle is repeatedly molted throughout life. The most commonly encountered species is *Priapulus caudatus*.

Phylum Entoprocta

The entoprocts (en′to-prok′s) (Gr. *entos*, within + *proktos*, anus) comprise a small phylum of about 100 species of sedentary marine filter feeders. They are either solitary or colonial and are found in coastal waters. One group is commensalistic on the body surface of various invertebrates. Most entoprocts are microscopic in size. Entoprocts may form large, matlike colonies on rocks. An individual entoproct consists of a muscular stalk bearing a cup-shaped **calyx** with a crown of ciliated tentacles (figure 8.21). The stalk is surrounded by a chitinous cuticle and may bear an attachment disk with adhesive glands. The pseudocoel is filled with loose connective tissue. Entoprocts are filter feeders, and when they feed, the cilia on the tentacles convey food into the mouth. The digestive tract forms a U-shaped gut located in the calyx. Also found in the calyx is a pair of protonephridial tubules that open through a single pore into the mouth. The nervous system consists of a small, central ganglion and radiating nerves. Exchange of gases occurs across the body surface.

Entoprocts reproduce by asexual budding and also sexually. Most entoprocts are hermaphroditic, but usually eggs and sperm are produced at different times in one animal. Sperm are released freely into the water, and fertilization occurs internally. Embryos develop in a brood chamber, from which free-swimming larvae are released. Eventually, the larvae settle to the substrata and develop into adults.

Further Phylogenetic Considerations

The aschelminths are clearly a diverse assemblage of animals. Despite the common occurrence of a cuticle, pseudocoelom, a muscular pharynx, and adhesive glands, there are no distinctive features that occur in every phyla.

The gastrotrichs show some distant relationships to the acoelomates. For example, many gastrotrichs lack a body cavity, are monoecious, are small, and their ventral cilia may have been derived from the same ancestral sources as those of the turbellarian flatworms.

The rotifers also have certain features in common with the acoelomates. The protonephridia of rotifers closely resemble those of some freshwater turbellarians, and it is generally believed that rotifers originated in freshwater habitats. Both flatworms and rotifers have separate ovaries and vitellaria. Rotifers probably had their origins from the earliest acoelomates and may have had a common bilateral, metazoan ancestor.

The kinorhynchs, acanthocephalans, loriciferans, and priapulids all have a spiny anterior end that can be retracted; thus, they are probably related. Loriciferans and kinorhynchs appear to be most closely related.

The affinities of the nematodes to other phyla are vague. No other living group is believed to be closely related to these worms. Nematodes probably evolved in freshwater habitats and then colonized the oceans and soils. The ancestral nematodes may have been sessile, attached at the posterior end, with the anterior end protruding upward into the water. The nematode cuticle, feeding structures, and food habits probably preadapted these worms for parasitism. In fact, free-living species could become parasitic without substantial anatomical or physiological changes.

Nematomorphs may be more closely related to nematodes than to any other group by virtue of both groups being cylindrical in shape, having a cuticle, dioecious, and sexually dimorphic. However, because the larval form of some nematomorphs has a resemblance to the Priapulida, the exact affinity to the nematodes is questionable.

The phylogenetic position of the entoprocts is still controversial. Some zoologists consider entoprocts more closely related to a phylum of coelomates called the Ectoprocta. However, because entoprocts have a pseudocoelom and protonephridia similar to flatworms and rotifers, they are discussed with the aschelminths.

Stop and Ask Yourself

13. What are the characteristics of a typical loriciferan?
14. How can a typical priapulan be described?
15. How can a typical entoproct be described?
16. What evidence is there that the individual aschelminth phyla exhibit diverse phylogenetic relationships?

Summary

1. Nine phyla are grouped together as the aschelminths. Most have a well-defined pseudocoelom, a constant number of body cells or nuclei (eutely), protonephridia, and a complete digestive system with a well-developed pharynx. No organs are developed for gas exchange or circulation. The body is covered with a cuticle that may be molted. Only longitudinal muscles are often present in the body wall.
2. The phylogenetic affinities among the nine phyla and with other phyla are uncertain.
3. Gastrotrichs are microscopic, aquatic animals with a head, neck, and trunk. Numerous adhesive glands are present. The group is generally hermaphroditic, although males are rare and female parthenogenesis is common in freshwater species.
4. The majority of rotifers inhabit fresh water. The head of these animals bears a unique ciliated corona used for locomotion and food capture. Males are smaller than females and unknown in some species. Females may develop parthenogenetically.
5. Kinorhynchs are minute worms living in marine habitats. Their bodies are comprised of 13 zonites, which have cuticular scales, plates, and spines.
6. Nematodes live in aquatic and terrestrial environments; many are parasitic and of medical and agricultural importance. They are all elongate, slender, and circular in cross section. Two sexes are present.
7. Nematomorpha are threadlike and free living in fresh water. They lack a digestive system.
8. Acanthocephalans are also known as spiny-headed worms because of their spiny proboscis. All are endoparasites in vertebrates.
9. The phylum Loricifera was described in 1983. These microscopic animals have a spiny head and thorax and are found in gravel in marine environments.
10. The phylum Priapulida contains only 15 known species of cucumber-shaped, wormlike animals that live buried in the bottom sand and mud in marine habitats.
11. The phylum Entoprocta contains about 100 species of sessile or sedentary filter feeders; most are hermaphroditic.

Selected Key Terms

amictic eggs (*p. 128*)
aschelminths (*p. 124*)
corona (*p. 126*)
cuticle (*p. 125*)
eutely (*p. 125*)
mastax (*p. 128*)
mictic eggs (*p. 128*)
zonites (*p. 129*)

Critical Thinking Questions

1. Discuss the limitations placed on shape changes in nematodes by the structure of the body wall.
2. What characteristics set the Nematomorpha apart from the Nematoda? What characteristics do the Nematomorpha share with the Nematoda?
3. In what respects are the kinorhynchs like nematodes? Like rotifers?
4. How are gastrotrichs related to the rotifers?
5. What are some environmental factors that appear to trigger the production of mictic females in monogonont rotifers?

chapter 9

Molluscan Success

Outline

Concepts

1. Triploblastic animals are often assembled into two groups, the protostomes and deuterostomes, based on certain developmental features.
2. Molluscs have a coelom, as well as a head-foot, visceral mass, mantle, and a mantle cavity. Most also have a radula.
3. Members of the class Gastropoda are the snails and slugs. They include the only terrestrial molluscs. Their bodies are modified by torsion and shell coiling.
4. Clams, oysters, mussels, and scallops are members of the class Bivalvia. They are all aquatic filter feeders and are often found burrowing in soft substrates or attached to hard substrates.
5. The class Cephalopoda includes the octopuses, squids, cuttlefish, and nautili. They are the most complex of all invertebrates and are adapted for predatory life-styles.
6. Other molluscs include members of the classes Scaphopoda (tooth shells), Monoplacophora, Aplacophora (solenogasters), and Polyplacophora (chitons). Members of these classes are all marine.
7. The exact relationship of molluscs to other animal phyla is debated. Specializations of molluscs have obscured evolutionary relationships among molluscan classes.

Would You Like to Know:

1. what characteristics can be used to identify molluscs? (*p. 145*)
2. how coiling of a snail's shell affects the arrangement of the snail's internal organs? (*p. 148*)
3. how a snail extends its tentacles? (*p. 148*)
4. how a pearl is formed? (*p. 150*)
5. how a clam feeds? (*p. 150*)
6. how some bivalves burrow through limestone and coral? (*p. 153*)
7. what invertebrate preys on whales? (*p. 155*)
8. what molluscs have the most advanced nervous and sensory functions among invertebrates? (*p. 155*)

These and other useful questions will be answered in this chapter.

This chapter contains evolutionary concepts, which are set off in this font.

Evolutionary Perspective

Octopuses, squids, and cuttlefish (the cephalopods) may be considered some of the invertebrate world's most adept predators. Predatory life-styles have resulted in the evolution of large brains (by invertebrate standards), complex sensory structures (by any standards), rapid locomotion, grasping tentacles, and tearing mouthparts. In spite of these adaptations, cephalopods rarely make up a major component of any community. Once numbering about 9,000 species, the class Cephalopoda now includes only about 550 species (figure 9.1).

Zoologists do not know why the cephalopods have declined so dramatically. Cephalopods may have been outcompeted by vertebrates, because the vertebrates were also making their appearance in prehistoric seas, and some vertebrates acquired active, predatory life-styles. Alternatively, the cephalopods may have declined simply because of random evolutionary events.

The same has not been the case for all molluscs. This group has, as a whole, been very successful. If success is measured by numbers of species, the molluscs can be considered twice as successful as vertebrates! The vast majority of its nearly 100,000 living species belong to two classes: Gastropoda, the snails and slugs; and Bivalvia, the clams and their close relatives.

Molluscs are triploblastic, as are all the remaining animals covered in this textbook. In addition, they are the first animals described in this textbook that possess a coelom, although the coelom of molluscs is only a small cavity (the pericardial cavity) surrounding the heart and gonads. A coelom is a body cavity that arises in mesoderm and is lined by a sheet of mesoderm called the peritoneum (*see figure 4.10c*).

Figure 9.1
Phylum Mollusca. The phylum Mollusca includes nearly 100,000 living species including members of the class Cephalopoda—some of the invertebrate world's most adept predators. A member of the genus *Octopus* is shown here.

Relationships to Other Animals

Comparative embryology is the study of similarities and differences in early development of animals. Events in embryology of animals may be similar because of shared ancestry; however, similarities in development can also reflect adaptations of distantly related, or unrelated, species to similar environments. Comparative embryologists, therefore, have a difficult task of sorting homologous developmental sequences from analogous developmental sequences (*see chapter 1*). This difficulty is well illustrated in the attempt to determine the phylogenetic relationships of the molluscs to other phyla.

The phyla described in the following chapters are divided into two large groups, and many of the reasons for this separation stem from comparative embryology. Although there are exceptions to the following generalizations, most zoologists are convinced that these two groups are true evolutionary assemblages (figure 9.2).

Protostomes include animals in the phyla Mollusca, Annelida, Arthropoda, and others. The developmental characteristics that unite these phyla are shown in figure 9.3*a–d*. One characteristic is the pattern of the early cleavages of the zygote. In spiral cleavage, the mitotic spindle is oriented obliquely to the animal/vegetal axis of a cell. For example, division of the four-celled embryo produces an eight-celled embryo in which the upper tier of cells is twisted out of line with the lower cells. A second characteristic that is common to many protostomes is that early cleavage is determinate, meaning that the fate of cells is established very early in development. If blastomeres of a two- or four-celled embryo are separated, none will develop into a complete organism. A third characteristic is reflected in the term "protostome" (Gr. *protos*, first + *stoma*, mouth). The blastopore, which forms during an embryonic process called gastrulation, usually remains open and forms the mouth. Other attributes of many protostomes include a top-shaped larva called a **trochophore larva** and a pattern of coelom and mesoderm formation, called schizocoelous, in which the mesoderm splits to form the coelom.

The other group, the **deuterostomes**, includes animals in the phyla Echinodermata, Hemichordata, Chordata, and others. The developmental characteristics that unite these phyla are shown in figure 9.3*e–h*. Radial cleavage occurs when the mitotic spindle is oriented perpendicularly to the animal/vegetal axis of a cell and results in blastomeres oriented directly over one another. Cleavage is indeterminate, meaning that the fate of blastomeres is determined late in embryology and, if blastomeres are separated, they can develop into entire individuals. In deuterostomes (Gr. *deutero*, second + *stoma*, mouth), the blastopore either forms the anus, or the blastopore closes, and the anus forms in the region of the blastopore. Mesoderm and coelom formation also differ from that of protostomes. The

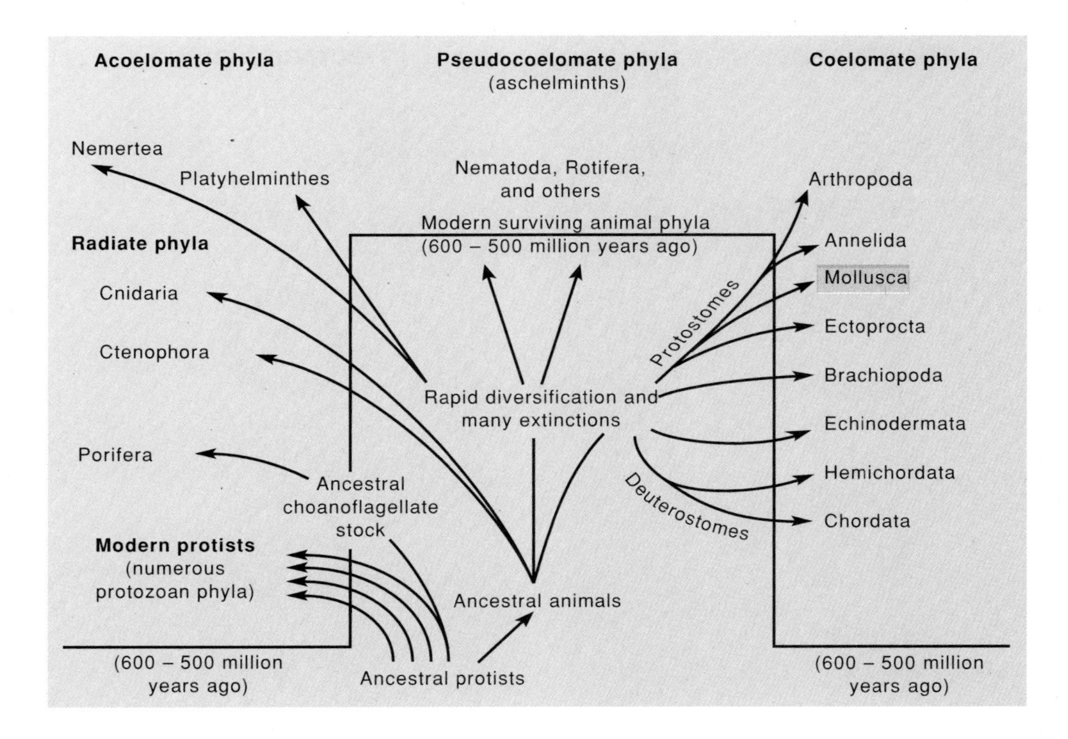

FIGURE 9.2

Evolutionary Relationships of the Molluscs. Molluscs (shaded in orange) share certain developmental characteristics with annelids, arthropods, ectoprocts, and brachiopods. These groups are placed into a larger assemblage called the protostomes. These developmental characteristics are discussed in this chapter.

mesoderm pinches off as pouches of the gut, and the coelom forms inside the mesodermal pouches. This method of coelom formation is called enterocoelous. A kidney-bean-shaped larva, called a dipleurula, is often represented as being characteristic of deuterostomes. There is, however, no single kind of deuterostome larval stage.

ORIGIN OF THE COELOM

There are a number of hypotheses regarding the origin of the coelom. These hypotheses are important because they influence how one pictures the evolutionary relationships among triploblastic phyla.

The schizocoel hypothesis (Gr. *schizen,* to split + *koilos,* hollow) is patterned after the method of mesoderm development and coelom formation in many protostomes (figure 9.3*a*), in which all mesoderm is derived from a particular ectodermal cell of the blastula. Mesoderm derived from this cell fills the area between ectoderm and endoderm. The coelom arises from a splitting of this mesoderm. If the coelom formed in this way during evolution, mesodermally derived tissues would have preceded the coelom, implying that a triploblastic, acoelomate (flatworm) body form could be the forerunner of the coelomate body form (*see figure* 4.10a).

The enterocoel hypothesis (Gr. *enteron,* gut + *koilos,* hollow) suggests that the coelom may have arisen as outpocketings of a primitive gut tract. This hypothesis is patterned after the method of coelom formation in deuterostomes (other than vertebrates; *figure* 9.3b). The implication of this hypothesis is that mesoderm and the coelom formed from the gut of a diploblastic animal. If this is true, the triploblastic, acoelomate body form would have been secondarily derived by mesoderm filling the body cavity of a coelomate animal.

Unfortunately zoologists may never know which, if either, of these hypotheses is accurate. Some zoologists believe that the coelom may have arisen more than once in different evolutionary lineages, in which case, more than one explanation could be correct.

MOLLUSCAN CHARACTERISTICS

Molluscs range in size and body form from the giant squid, measuring 18 m in length, to the smallest garden slug, less than 1 cm long. In spite of this diversity, the phylum Mollusca (mol-lus'kah) (L. *molluscus,* soft) is not difficult to characterize (table 9.1).

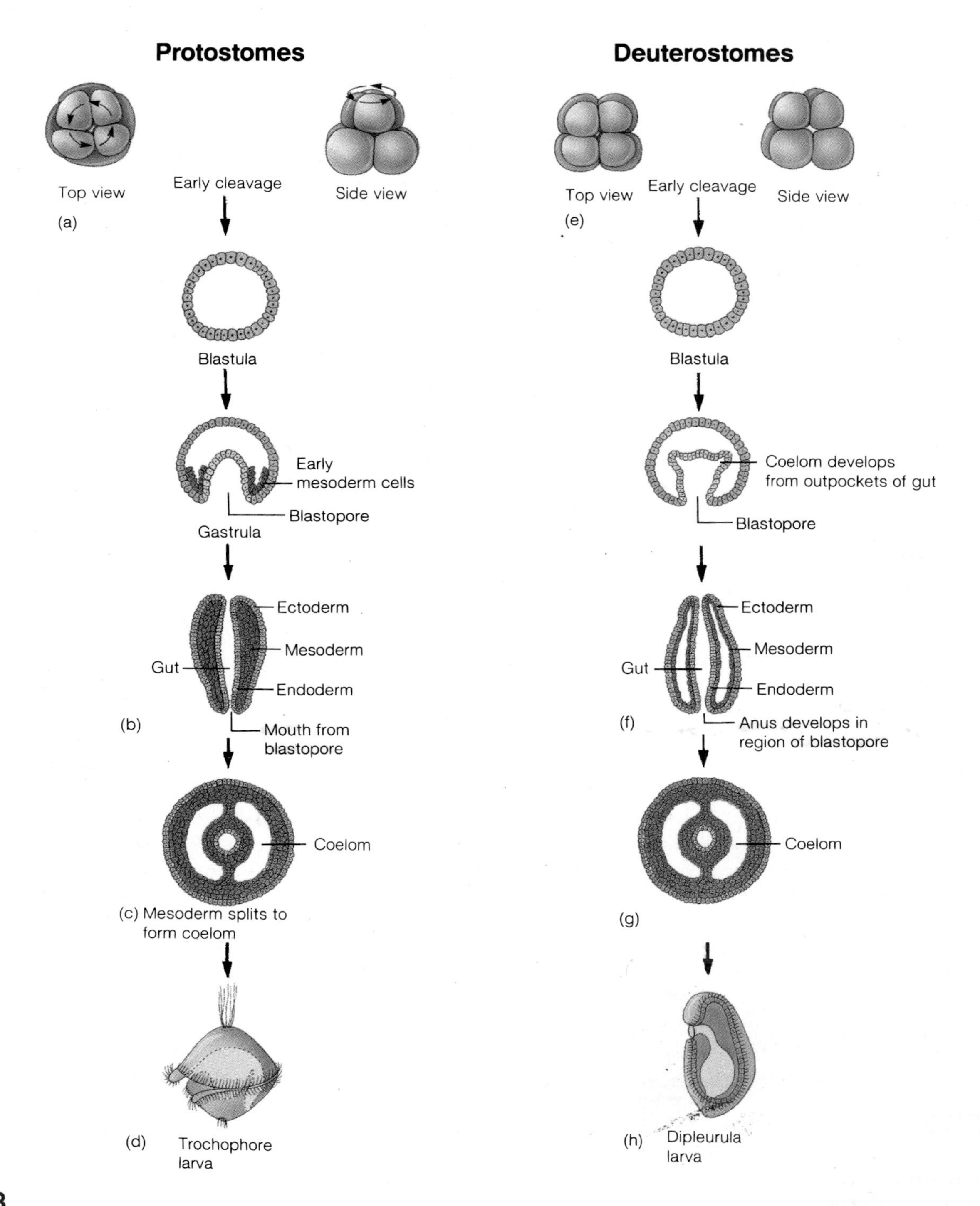

Figure 9.3

Developmental Characteristics of Protostomes and Deuterostomes. Protostomes are characterized by spiral and determinate cleavage (*a*), a mouth that forms from an embryonic blastopore (*b*), schizocoelous coelom formation (*c*), and a trochophore larva (*d*). Deuterostomes are characterized by radial and indeterminate cleavage (*e*), an anus that forms in the region of the embryonic blastopore (*f*), and enterocoelous coelom formation (*g*). A dipleurula larva is often represented as being characteristic of deuterostomes. Although a dipleurula is present in some echinoderms, there is no characteristic deuterostome larval stage (*h*).

TABLE 9.1 CLASSIFICATION OF THE MOLLUSCA

Phylum Mollusca (mol-lus'kah)
The coelomate animal phylum whose members possess a head-foot, visceral mass, mantle, and mantle cavity. Most molluscs also possess a radula and a calcareous shell.

Class Caudofoveata (kaw'do-fo've-a'ta)
Wormlike molluscs with a cylindrical, shell-less body and scalelike calcareous spicules; lack eyes, tentacles, statocysts, crystalline style, foot, and nephridia. Deep-water, marine burrowers. *Chaetoderma*.

Class Aplacophora (a'pla-kof"o-rah)
Shell, mantle, and foot lacking; wormlike; head poorly developed; burrowing molluscs. Marine. *Neomenia*.

Class Polyplacophora (pol'e-pla-kof'o-rah)
Elongate, dorsoventrally flattened; head reduced in size; shell consisting of eight dorsal plates. Marine, on rocky intertidal substrates. *Chiton*.

Class Monoplacophora (mon'o-pla-kof"o-rah)
Molluscs with a single arched shell; foot broad and flat; certain structures serially repeated. Marine. *Neopilina*.

Class Scaphopoda (ska-fop'o-dah)
Body enclosed in a tubular shell that is open at both ends; tentacles used for deposit feeding. No head. Marine. *Dentalium*.

Class Bivalvia (bi"val've-ah)
Body enclosed in a shell consisting of two valves, hinged dorsally; no head or radula; wedge-shaped foot. Marine and fresh water. *Anodonta, Mytilus, Venus*.

Class Gastropoda (gas-trop'o-dah)
Shell, when present, usually coiled; body symmetry distorted by torsion; some monoecious species. Marine, fresh water, terrestrial. *Nerita, Orthaliculus, Helix*.

Class Cephalopoda (sef'ah-lop'o-dah)
Foot modified into a circle of tentacles and a siphon; shell reduced or absent; head in line with the elongate visceral mass. Marine. *Octopus, Loligo, Sepia, Nautilus*.

This taxonomic listing reflects a phylogenetic sequence. The discussions that follow, however, begin with molluscs that are familiar to most students.

Characteristics of the phylum Mollusca include the following:

1. Body of two parts: head-foot and visceral mass
2. Mantle that secretes a calcareous shell and covers the visceral mass
3. Mantle cavity functions in excretion, gas exchange, elimination of digestive wastes, and release of reproductive products
4. Bilateral symmetry
5. Protostome characteristics including trochophore larvae, spiral cleavage, and schizocoelous coelom formation
6. Coelom reduced to cavities surrounding the heart, nephridia, and gonads
7. Open circulatory system in all but one class (Cephalopoda)
8. Radula usually present and used in scraping food

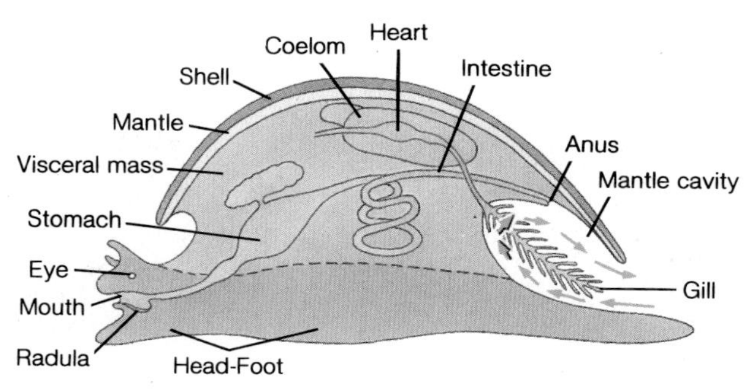

FIGURE 9.4
Molluscan Body Organization. All molluscs possess three features unique to the phylum. The head-foot is a muscular structure usually used for locomotion and sensory perception. The visceral mass contains organs of digestion, circulation, reproduction, and excretion. The mantle is a sheet of tissue that enfolds the rest of the body and secretes the shell. Arrows indicate the flow of water through the mantle cavity.

1 The body of a mollusc is divided into two main regions—the head-foot and the visceral mass (figure 9.4). The **head-foot** is elongate with an anterior head, containing the mouth and certain nervous and sensory structures, and an elongate foot, used for attachment and locomotion. A **visceral mass** contains the organs of digestion, circulation, reproduction, and excretion and is attached at the dorsal aspect of the head-foot.

The **mantle** of a mollusc usually attaches to the visceral mass, enfolds most of the body, and may secrete a shell that overlies the mantle. The shell of a mollusc is secreted in three layers (figure 9.5). The outer layer of the shell is called the periostracum. This protein layer is secreted by mantle cells at the outer margin of the mantle. The middle layer of the shell, called the prismatic layer, is the thickest of the three layers and consists of calcium carbonate mixed with organic materials. It is also secreted by cells at the outer margin of the mantle. The inner layer of the shell, the nacreous layer, forms from thin sheets of calcium carbonate alternating with organic matter. The nacreous layer is secreted by cells along the entire epithelial border of the mantle. Secretion of nacre causes the shell to grow in thickness.

Between the mantle and the foot is a space called the **mantle cavity.** The mantle cavity opens to the outside and functions in gas exchange, excretion, elimination of digestive wastes, and release of reproductive products.

The mouth of most molluscs possesses a rasping structure, called a **radula,** which consists of a chitinous belt and rows of posteriorly curved teeth (figure 9.6). The radula overlies a fleshy, tonguelike structure supported by a cartilaginous

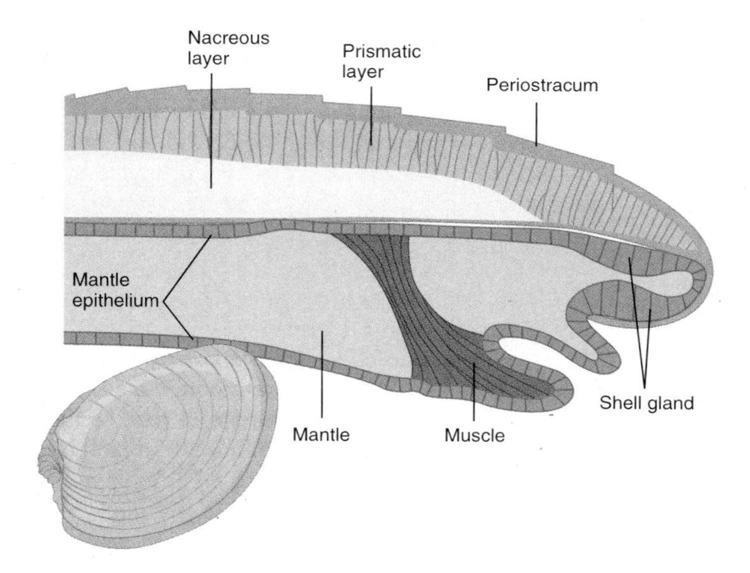

Figure 9.5

The Molluscan Shell and Mantle. A transverse section of a bivalve shell and mantle shows the three layers of the shell and the portions of the mantle responsible for secretion of the shell.

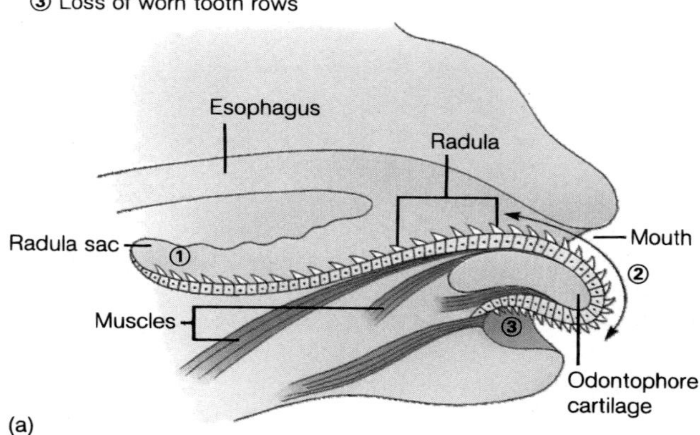

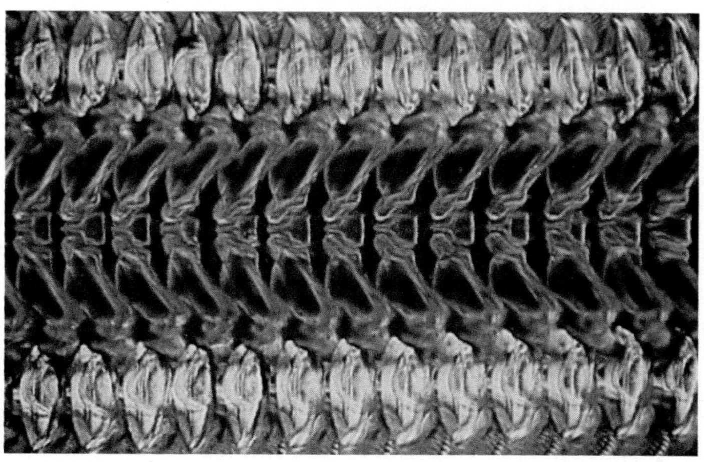

Figure 9.6

Radular Structure. (*a*) The radular apparatus lies over the cartilaginous odontophore. Muscles attached to the radula move the radula back and forth over the odontophore. (*b*) Micrograph of radular teeth arrangement of the marine snail, *Nerita.* Tooth structure is an important taxonomic characteristic for zoologists who study molluscs. (*a*) *From:* "A LIFE OF INVERTEBRATES" © *1979 W. D. Russel-Hunter.*

odontophore. Muscles associated with the odontophore permit the radula to be protruded from the mouth. Muscles associated with the radula move the radula back and forth over the odontophore. Food is scraped from a substrate and passed posteriorly to the digestive tract.

Stop and Ask Yourself

1. What characteristics unite the protostomes? The deuterostomes?
2. Why does the schizocoel hypothesis of the origin of the coelom suggest that coelomate animals evolved from triploblastic animals?
3. What are the three regions of the body of a generalized mollusc? What are the functions of these three regions?
4. What are the functions of the mantle cavity of a mollusc?

Class Gastropoda

The class Gastropoda (gas-trop′o-dah) (Gr. *gaster*, gut + *podos*, foot) includes the snails, limpets, and slugs. With over 35,000 living species (*see table 9.1*), Gastropoda is the largest and most varied molluscan class. Its members occupy a wide variety of marine, freshwater, and terrestrial habitats. Most people give gastropods little thought unless they encounter *Helix pomatia* (escargot) in a French restaurant or are pestered by garden slugs and snails. One important impact of gastropods on the lives of many humans is that gastropods are intermediate hosts for some medically important trematode parasites of humans (*see chapter 7*).

Torsion

One of the most significant modifications of the molluscan body form in the gastropods occurs early in gastropod development. **Torsion** is a 180°, counterclockwise twisting of the visceral mass, the mantle, and the mantle cavity. After torsion, the gills, anus, and openings from the excretory and reproductive systems are positioned just behind the head and

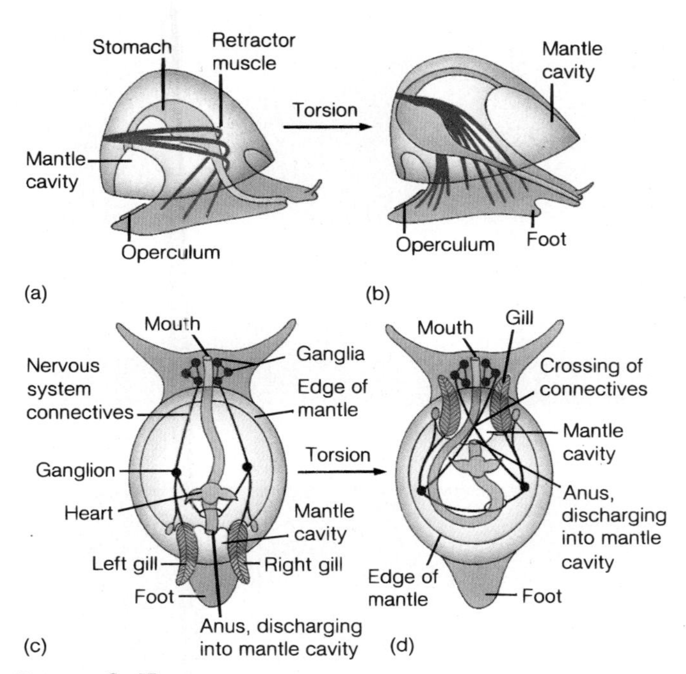

FIGURE 9.7

Torsion in Gastropods. (*a*) A pretorsion gastropod larva. Note the posterior opening of the mantle cavity and the untwisted digestive tract. (*b*) After torsion, the digestive tract is looped, and the mantle cavity opens near the head. The foot is drawn into the shell last, and the operculum closes the shell opening. (*c*) A hypothetical (adult) ancestor shows the arrangement of internal organs prior to torsion. (*d*) Modern adult gastropods have an anterior opening of the mantle cavity and the looped digestive tract. *Redrawn from L. Hyman,* The Invertebrates, *Volume VI. Copyright © 1967 McGraw-Hill, Inc. Used by permission.*

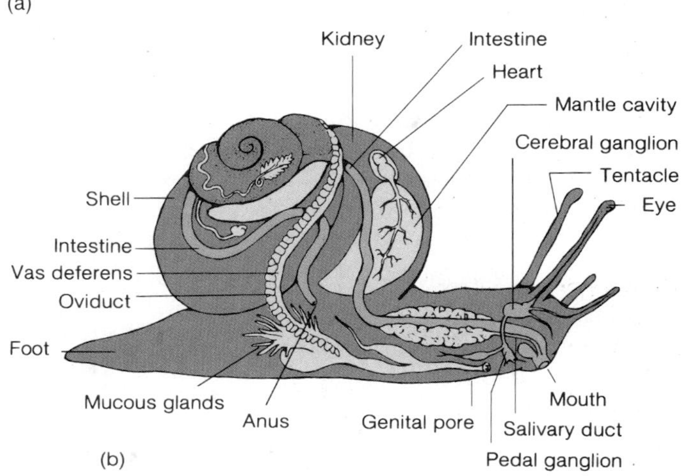

FIGURE 9.8

Gastropod Structure. (*a*) A land (pulmonate) gastropod (*Orthaliculus*). (*b*) Internal structure of a generalized gastropod.

nerve cords, and the digestive tract is twisted into a U shape (figure 9.7).

The adaptive significance of torsion is speculative; however, three advantages are plausible. First, without torsion, withdrawal into the shell would proceed with the foot entering first, and the more vulnerable head entering last. With torsion, the head enters the shell first and is followed by the foot, leaving the head less exposed to potential predators. In some snails, protection is enhanced by a proteinaceous covering, called an **operculum,** on the dorsal, posterior margin of the foot. When the foot is drawn into the mantle cavity, the operculum closes the opening of the shell, thus preventing desiccation when the snail is in drying habitats. Another advantage of torsion concerns an anterior opening of the mantle cavity that allows clean water from in front of the snail to enter the mantle cavity, rather than water contaminated with silt stirred up by the snail's crawling. An additional advantage to torsion derives from the twist in the mantle's sensory organs around to the head region. This positioning allows a snail greater sensitivity to stimuli coming from the direction in which it moves.

Note in figure 9.7*d* that after torsion the anus and nephridia empty dorsal to the head and create potential fouling problems. However, a number of evolutionary adaptations seem to circumvent this problem. Various modifications allow water and the wastes it carries to exit the mantle cavity through notches or openings in the mantle and shell posterior to the head. Some gastropods undergo detorsion, in which the embryo undergoes a full 180° torsion and then untwists approximately 90°. The mantle cavity thus opens on the right side of the body, behind the head.

SHELL COILING

The earliest fossil gastropods possessed a shell that was coiled in one plane. This arrangement is not common in later fossils, probably because growth resulted in an increasingly cumbersome shell. (Some modern snails, however, have secondarily returned to this shell form.)

Most modern snail shells are asymmetrically coiled into a more compact form, with successive coils or whorls slightly larger than, and ventral to, the preceding whorl (figure 9.8*a*).

2 This pattern leaves less room on one side of the visceral mass for certain organs, and it is thought that organs that are now single were probably paired ancestrally. This asymmetrical arrangement of internal organs will be described further when particular body systems are described.

Locomotion

Nearly all gastropods have a flattened foot that is often ciliated, covered with gland cells, and used to creep across the substrate (figure 9.8*b*). The smallest gastropods use cilia to propel themselves over a mucous trail. Larger gastropods use waves of muscular contraction that move over the foot. The foot of some gastropods is modified for clinging, as in abalones and limpets, or for swimming, as in sea butterflies and sea hares.

Feeding and Digestion

Most gastropods feed by scraping algae or other small, attached organisms from their substrate. Others are herbivores that feed on larger plants, scavengers, parasites, or predators (figure 9.9).

The anterior portion of the digestive tract may be modified into an extensible proboscis, which contains the radula. This structure is important for some predatory snails that must extract animal flesh from hard to reach areas. The digestive tract of gastropods, like that of most molluscs, is ciliated. Food is trapped in mucous strings and incorporated into a rotating mucoid mass called the **protostyle,** which extends to the stomach and is rotated by cilia. Enzymes and acid are released into the stomach from a digestive gland located in the visceral mass, and food trapped on the protostyle is freed and digested. Wastes are formed into fecal pellets in the intestine.

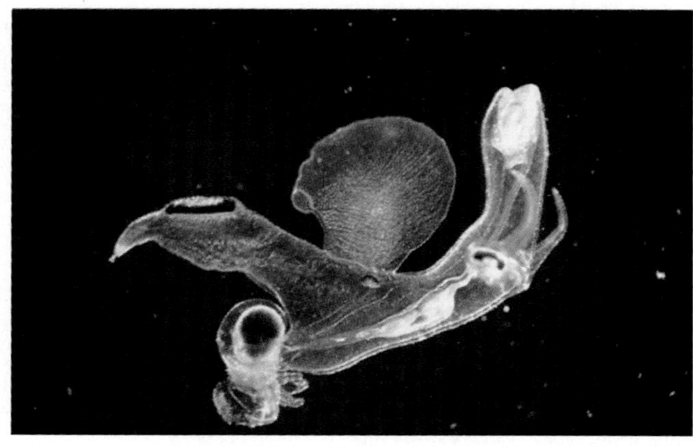

(a)

(b)

Figure 9.9

Variations in the Gastropod Body Form. (*a*) This heteropod (*Carinaria*) is a predator that swims upside down in the open ocean. Its body is nearly transparent. The head is at the left, and the shell is below and to the right. (*b*) Colorful nudibranchs have no shell or mantle cavity. The projections on the dorsal surface are used in gas exchange. In some nudibranchs, the dorsal projections are armed with nematocysts that the nudibranchs use for protection. Nudibranchs prey on sessile animals, such as soft corals and sponges.

Other Maintenance Functions

Gas exchange always involves the mantle cavity. Primitive gastropods had two gills; modern gastropods have lost one gill because of coiling. Some gastropods have a rolled extension of the mantle, called a **siphon,** that serves as an inhalant tube. Species that burrow extend the siphon to the surface of the substrate to bring in water. Gills are lost or reduced in land snails (pulmonates), but these snails have a richly vascular mantle that is used for gas exchange between blood and air. Contractions of the mantle help circulate air and water through the mantle cavity.

Gastropods, as do most molluscs, have an **open circulatory system.** During part of its circuit around the body, blood leaves the vessels and directly bathes cells in tissue spaces called sinuses. Molluscs typically have a heart consisting of a single, muscular ventricle and two auricles. Most gastropods have lost one member of the pair of auricles because of coiling.

In addition to transporting nutrients, wastes, and gases, the blood of molluscs acts as a hydraulic skeleton. A **hydraulic skeleton** uses blood confined to tissue spaces for support. A mollusc uses its hydraulic skeleton to extend body structures by contracting muscles distant from the extending structure. 3 For example, snails have sensory tentacles on their heads, and if the tentacle is touched, it can be rapidly withdrawn by retractor muscles. However, no antagonistic muscles exist to extend the tentacle. Extension is accomplished more slowly by contracting distant muscles to squeeze blood into the tentacle from adjacent blood sinuses.

The nervous system of primitive gastropods is characterized by six ganglia located in the head-foot and visceral mass (*see figure* 9.8b). The nerves that link these ganglia are twisted by torsion. The evolution of the gastropod nervous system has resulted in the untwisting of nerves and the concentration of nervous tissues into fewer, larger ganglia, especially in the head.

Gastropods have well-developed sensory structures. Eyes may be at the base or at the end of tentacles, they may be simple pits of photoreceptor cells, or consist of a lens and cornea. Statocysts are in the foot. Osphradia are chemoreceptors in the anterior wall of the mantle cavity that detect sediment and chemicals in inhalant water or air. The osphradia of predatory gastropods help detect prey.

Primitive gastropods possessed two nephridia. In modern species, the right nephridium has disappeared, probably because of shell coiling. The nephridium consists of a sac with highly folded walls and is connected to the reduced coelom, the pericardial cavity. Excretory wastes are derived largely from fluids filtered and secreted into the coelom from the blood. The nephridium modifies this waste by selectively reabsorbing certain ions and organic molecules. The nephridium opens to the mantle cavity or, in land snails, on the right side of the body adjacent to the mantle cavity and anal opening. Ammonia is the primary excretory product for aquatic species, because they have access to water in which toxic ammonia is diluted. Terrestrial snails must convert ammonia to a less toxic form—uric acid. Because uric acid is relatively insoluble in water and less toxic, it can be excreted in a semisolid form, which helps conserve water.

Reproduction and Development

Many marine snails are dioecious. Gonads lie in spirals of the visceral mass (*see figure* 9.8b). Ducts discharge gametes into the sea, where external fertilization occurs.

Many other snails are monoecious, and internal, cross-fertilization is the rule. Copulation may result in mutual sperm transfer, or one snail may act as the male and the other as the female. A penis has evolved from a fold of the body wall, and portions of the female reproductive tract have become glandular and secrete a protective jelly or capsule around the fertilized egg. Some monoecious snails are protandric in that testes develop first, and after they degenerate, ovaries mature.

Eggs are shed singly or in masses for external fertilization. Internally fertilized eggs are deposited in gelatinous strings or masses. Eggs of terrestrial snails are large and yolky. These are deposited in moist environments, such as forest-floor leaf litter, and may be encapsulated by a calcareous shell. In marine gastropods, spiral cleavage results in a free-swimming trochophore larva that develops into another free-swimming larva with foot, eyes, tentacles, and shell, called a **veliger larva.** Sometimes, the trochophore is suppressed, and the veliger is the primary larva. Torsion occurs during the veliger stage, followed by settling and metamorphosis to the adult.

Class Bivalvia

With close to 30,000 species, the class Bivalvia (bi″val′ve-ah) (L. *bis*, twice + *valva*, leaf) is the second largest molluscan class. This class includes the clams, oysters, mussels, and scallops (*see table* 9.1). These laterally compressed animals are covered by a sheetlike mantle and a shell consisting of two valves (hence the class name). Many bivalves are edible, and some form pearls. Because most bivalves are filter feeders, another value is in the removal of bacteria from polluted water.

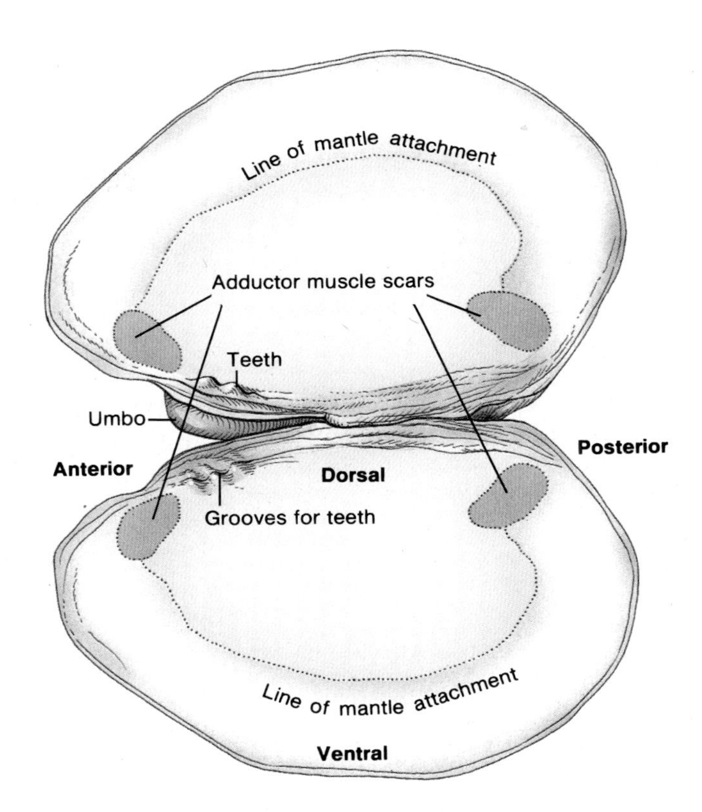

Figure 9.10
Inside View of a Bivalve Shell. The umbo is the oldest part of the bivalve shell. As the bivalve grows, the mantle lays down more shell in concentric lines of growth.

Shell and Associated Structures

The two convex halves of the shell are called **valves.** Along the dorsal margin of the shell is a proteinaceous hinge and a series of tongue-and-groove modifications of the shell, called teeth, that prevent the valves from twisting (figure 9.10). The oldest part of the shell is the **umbo,** a swollen area near the anterior margin of the shell. Although bivalves appear to have two shells, embryologically the shell forms as a single structure. The shell is continuous along its dorsal margin, but the mantle, in the region of the hinge, secretes relatively greater quantities of protein and relatively little calcium carbonate. The result is an elastic hinge ligament. The elasticity of the hinge ligament opens the valves when certain muscles are relaxed.

Adductor muscles are located at either end of the dorsal half of the shell and are responsible for closing the shell. Anyone who has tried to force apart the valves of a bivalve mollusc knows the effectiveness of these muscles. This is important for bivalves, because the primary defense of most bivalves against

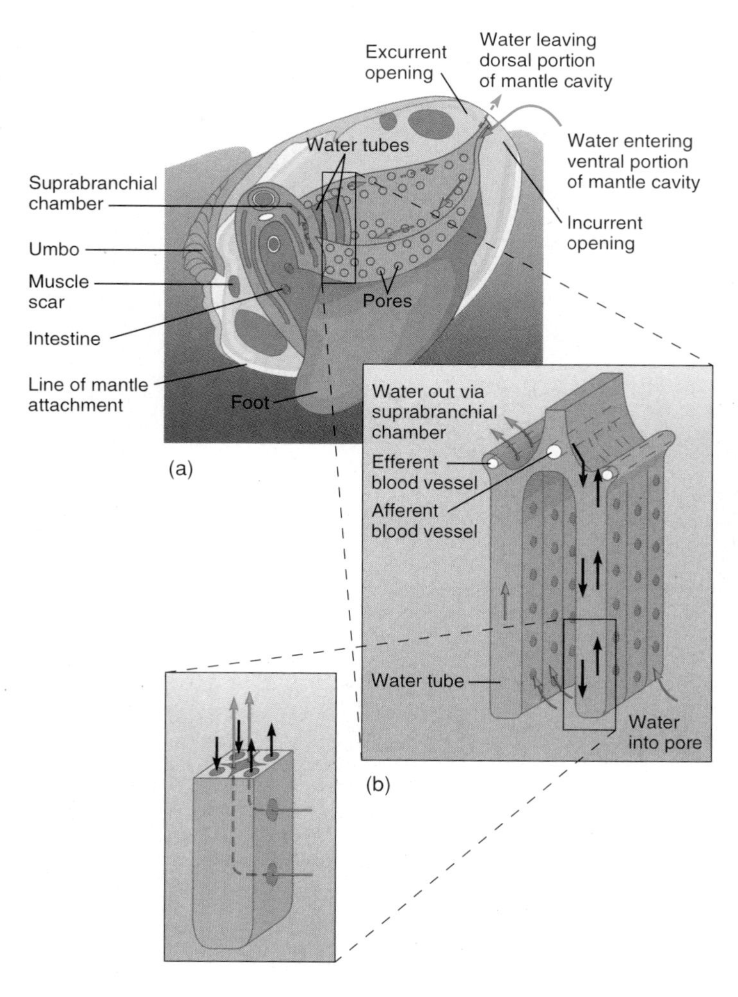

FIGURE 9.11

The Lamellibranch Gill of a Bivalve. (*a*) Blue arrows indicate incurrent and excurrent water currents. Food is filtered as water enters water tubes through pores in the gills. (*b*) Cross section through a portion of a gill. Water passing through a water tube passes in close proximity to blood. Gas exchange occurs between water and blood in the water tubes. Blue arrows show the path of water. Black arrows show the path of blood.

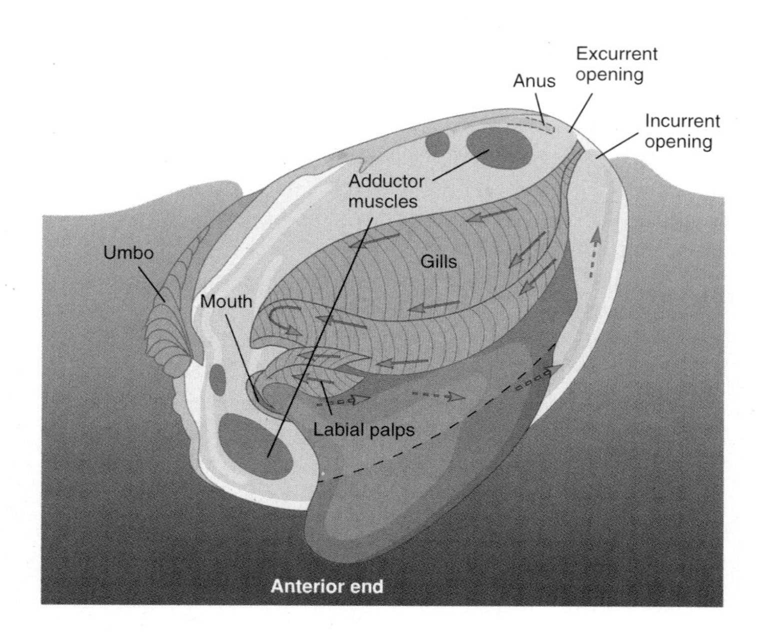

FIGURE 9.12

Bivalve Feeding. Solid blue arrows show the path of food particles after being filtered by the gills. Dashed blue arrows show the path of particles rejected by the gills and the labial palps.

predatory sea stars is to tenaciously refuse to open their shell. In chapter 13, you will see how sea stars are adapted to meet this defense strategy.

The mantle of bivalves is attached to the shell around the adductor muscles and near the margin of the shell. 4 If a sand grain or a parasite becomes lodged between the shell and the mantle, the mantle secretes nacre around the irritant, gradually forming a pearl. Highest quality pearls are formed by the Pacific pearl oysters *Pinctada margaritifera* and *Pinctada mertensi*.

GAS EXCHANGE, FILTER FEEDING, AND DIGESTION

The adaptation of bivalves to sedentary, filter-feeding life-styles includes the loss of the head and radula and, except for a few bivalves, the gills have become greatly expanded and covered with cilia. Gills form folded sheets (lamellae), with one end attached to the foot and the other end attached to the mantle. The mantle cavity ventral to the gills is the inhalant region, and the cavity dorsal to the gills is the exhalant region (figure 9.11*a*). Cilia move water into the mantle cavity through an incurrent opening of the mantle. Sometimes this opening is at the end of a siphon, which is an extension of the mantle. A bivalve buried in the substrate can extend its siphon to the surface and still feed and exchange gases. Water moves from the mantle cavity into small pores in the surface of the gills, and from there, into vertical channels in the gills, called water tubes. In moving through water tubes, blood and water are in close proximity, and gases are exchanged by diffusion (figure 9.11*b*). Water exits the bivalve through a part of the mantle cavity at the dorsal aspect of the gills called the suprabranchial chamber, and through an excurrent opening in the mantle (figure 9.11*a*).

5 Cilia covering the gills not only create water currents, but they also filter suspended food from the water, transport food toward the mouth, and sort filtered particles (figure 9.12). Cilia covering leaflike **labial palps,** located on either side of the mouth, also sort filtered particles. Small particles are carried by cilia into the mouth, and larger particles are moved to the edges of the palps and gills. This rejected material, called pseudofeces, falls, or is thrown, onto the mantle and is transported posteriorly by a ciliary tract on the mantle. Pseudofeces are washed from the mantle cavity by water rushing out when the valves are forcefully closed.

The digestive tract of bivalves is similar to that of other molluscs (figure 9.13*a*). Food entering the esophagus is entangled in a mucoid food string, which extends to the stomach and is rotated by cilia lining the digestive tract. A consolidated mucoid

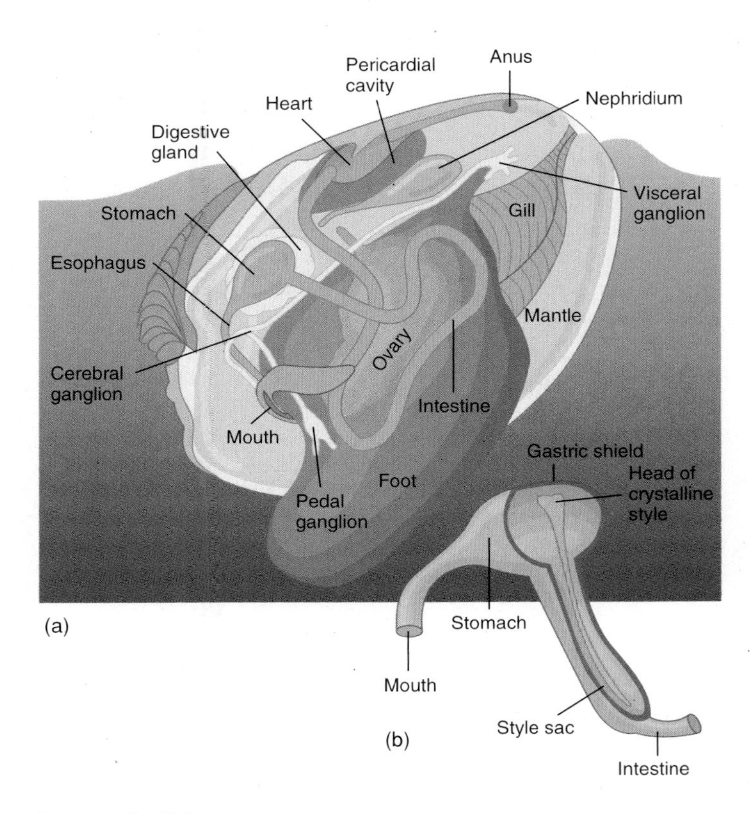

FIGURE 9.13
Bivalve Structure. (*a*) Internal structure of a bivalve. (*b*) The bivalve stomach showing the crystalline style and associated structures.

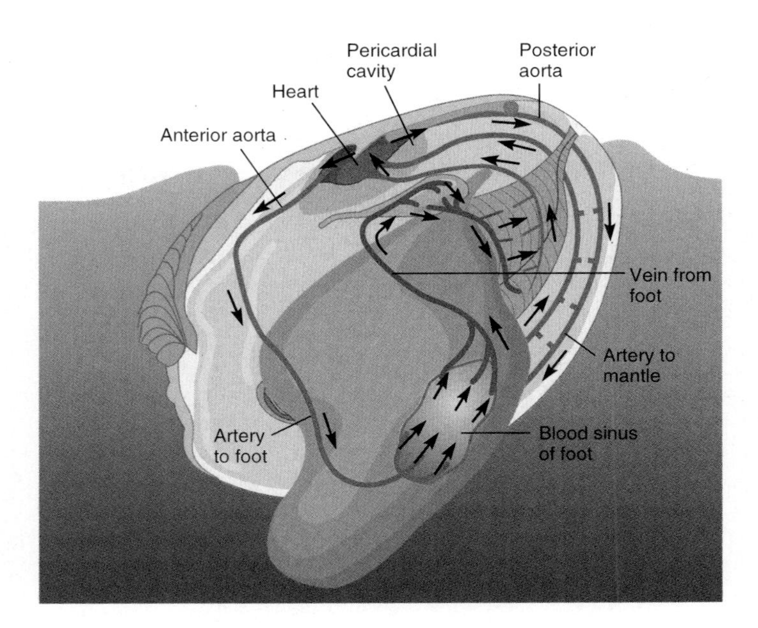

FIGURE 9.14
Bivalve Circulation. Blood flows from the single ventricle to tissue sinuses through anterior and posterior aortae. Blood from tissue sinuses flows to the nephridia, to the gills, and then enters the auricles of the heart. In all bivalves, the mantle is an additional site for oxygenation. In some bivalves, a separate aorta delivers blood to the mantle. This blood returns directly to the heart. The ventricle of bivalves is always folded around the intestine. Thus, the pericardial cavity (the coelom) encloses the heart and a portion of the digestive tract.

mass, the **crystalline style,** projects into the stomach from a diverticulum, called the style sac (figure 9.13*b*). Enzymes for carbohydrate and fat digestion are incorporated into the crystalline style. Cilia of the style sac rotate the style against a chitinized **gastric shield.** Abrasion of the style against the gastric shield and acidic conditions in the stomach dislodge enzymes. As the crystalline style rotates, the mucoid food string winds around the crystalline style, and is pulled farther into the stomach from the esophagus. Food particles in the food string are dislodged by this action and the acidic pH in the stomach. Further sorting separates fine particles from undigestible coarse materials. The latter are sent on to the intestine. Partially digested food from the stomach enters a digestive gland, where intracellular digestion occurs. Undigested wastes in the digestive gland are carried back to the stomach and then to the intestine by cilia. The intestine empties through the anus near the excurrent opening, and feces are carried away by excurrent water.

Other Maintenance Functions

In bivalves, blood flows from the heart to tissue sinuses, nephridia, gills, and back to the heart (figure 9.14). The mantle is an additional site for oxygenation. In some bivalves, a separate aorta delivers blood directly to the mantle. Two nephridia are located below the pericardial cavity (the coelom). Their duct system connects to the coelom at one end and opens at nephridiopores in the anterior region of the suprabranchial chamber.

The nervous system of bivalves consists of three pairs of interconnected ganglia associated with the esophagus, the foot, and the posterior adductor muscle. The margin of the mantle is the principal sense organ. It always has sensory cells, and it may have sensory tentacles and photoreceptors. In some species, photoreceptors are in the form of complex eyes with a lens and a cornea (e.g., scallops). Other receptors include statocysts near the pedal ganglion and an osphradium in the mantle, beneath the posterior adductor muscle.

Reproduction and Development

Most bivalves are dioecious. A few are monoecious, and some of these species are protandric. Gonads are located in the visceral mass, where they surround the looped intestine. Ducts of these gonads open directly to the mantle cavity or by the nephridiopore to the mantle cavity.

External fertilization occurs in most bivalves. Gametes exit through the suprabranchial chamber of the mantle cavity and the exhalant opening. Development proceeds through trochophore and veliger stages (figure 9.15*a*,*b*). When the veliger settles, the adult form is assumed.

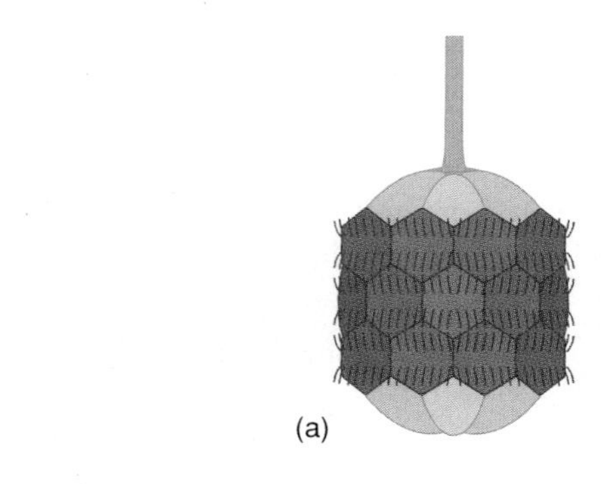

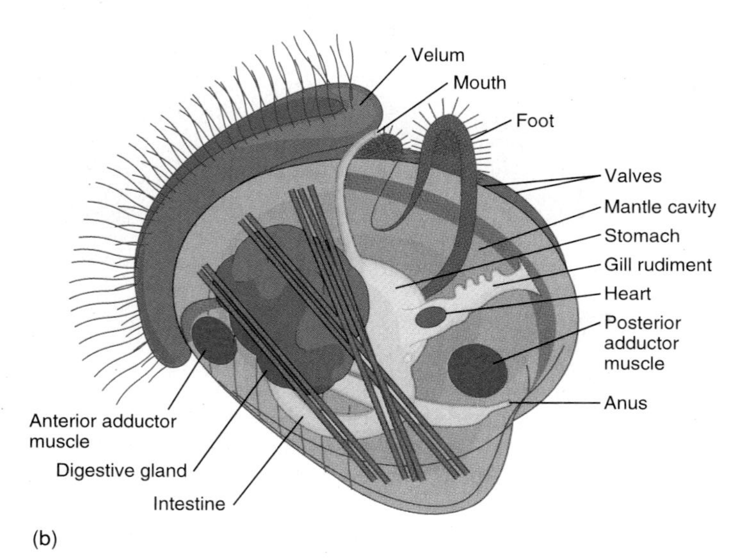

Figure 9.15

Larval Stages of Bivalves. (*a*) A trochophore larva of *Yoldia limatula*. (*b*) A veliger of an oyster. (*c*) A glochidium of a freshwater clam. Note the tooth used to attach to fish gills.

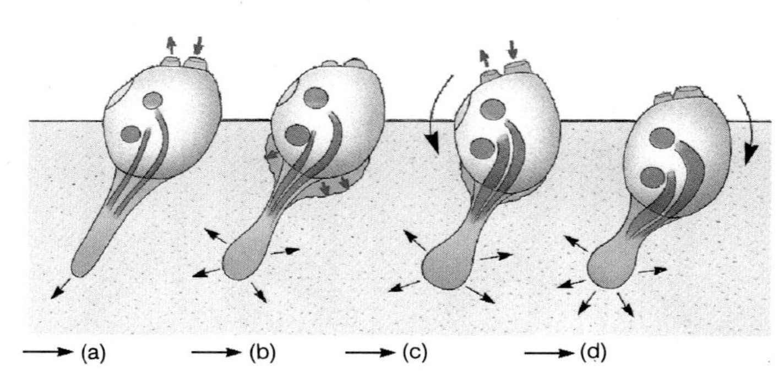

Figure 9.16

Bivalve Burrowing. (*a*) Adductor muscles relax, and the valves push against the substrate to anchor the bivalve. The foot is extended by hydraulic pressure and pedal protractor muscles, which extend from each side of the foot to each valve. Circular and transverse muscles contract to form the foot into a narrow, probing structure. Blue arrows show movement of water into and out of the mantle cavity. (*b*) The tip of the foot is dilated with blood to form a second anchor, and the valves are closed. Water from the mantle cavity washes substrate away from the bivalve when the valves close (lower blue arrows). (*c*,*d*) Pedal retractor muscles contract and pull the bivalve downward. The process is repeated as necessary. *From:* "A LIFE OF INVERTEBRATES" © *1979 W. D. Russell-Hunter.*

The largest families of freshwater bivalves brood their young. Fertilization occurs in the mantle cavity by sperm brought in with inhalant water. Some brood their young in maternal gills through reduced trochophore and veliger stages. Young clams are shed from the gills. Others brood their young to a modified veliger stage called a **glochidium,** which is parasitic on fishes (figure 9.15*c*). These larvae possess two tiny valves, and some species have toothlike hooks. Larvae exit through the exhalant aperture and sink to the substrate. Most of these will die. If a fish contacts a glochidium, however, the larva attaches to the gills, fins, or another body part and begins to feed on host tissue. The fish forms a cyst around the larva. After a period of larval development, during which it begins acquiring its adult structures, the miniature clam falls from its host and takes up its filter-feeding life-style. The glochidium acts as a dispersal stage for an otherwise sedentary animal.

Bivalve Diversity

Bivalves are found in nearly all aquatic habitats. They may live completely or partially buried in sand or mud, attached to solid substrates, or burrowed into submerged wood, coral, or limestone (box 9.1).

The mantle margins of burrowing bivalves are frequently fused to form distinct openings in the mantle cavity (siphons). This fusion helps to direct the water washed from the mantle cavity during burrowing and helps keep sediment from accumulating in the mantle cavity (figure 9.16).

BOX 9.1 THE ZEBRA MUSSEL—ANOTHER BIOLOGICAL INVASION

About 100 years ago, a bird fancier released a few starlings in New York City's Central Park. Today the starling is the most common bird in the United States. In 1866 the gypsy moth was transported from Europe to the New England states. It proliferated at the expense of North American forests. Today, these European invaders are joined by another, the zebra mussel (*Dreissena polymorpha*). This invasion, like the others before it, has been very costly, both economically and ecologically for much of North America.

The zebra mussel is actually one of about 120 exotic invaders of the Great Lakes. The invasion began in 1985 or 1986 when larval mussels were picked up in freshwater ports of Europe when cargo ships filled their ballast tanks with fresh water. The larvae were released when ballast tanks were emptied into the Great Lakes. Within three or four years the mussel spread into Lakes Erie, Ontario, Huron, and southern Lake Michigan. By June 1991, records of the mussels were being made in the Illinois River and the mussel now threatens much of the Mississippi River drainage basin.

Many of the problems associated with the zebra mussels are a result of their high reproductive potential. A single female may release 40,000 eggs. Their veliger larval stages may drift in the plankton for up to five weeks and be carried long distances by water currents. When larvae settle on a hard substrate, they attach by tough byssal threads. They grow to a length of about 2 cm, and densities of 200 individuals/m^2 are common (figure 1).

Economic problems associated with zebra mussels result from their settling on, and clogging, water intake pipes. Detroit Edison officials reported 700,000 mussels/m^2 on a single water intake screen. In December 1988, mussels and ice shards blocked water intake to the Detroit Edison plant, which resulted in power outages throughout Detroit. It cost the company $250,000 to restore electricity to the city. Detroit Edison officials spent six million dollars on a new intake system that they hope will reduce the fouling problems. It is estimated that throughout the Great Lakes, two billion dollars will be spent cleaning and refitting pipes in Great Lakes port cities through the 1990s.

Zebra mussels also threaten the ecology of freshwater ecosystems. They are very efficient filter feeders and are expected to disrupt freshwater food webs. As larvae settle and encrust hard substrates, they may disrupt the spawning ground of game fish such as walleyed pike. There is particular concern for the Mississippi River drainage basin. The Mississippi River and its tributaries contain the highest diversity of clams in the world. Some of these clams are endangered species. In other places, where native clams have been displaced by the zebra mussel, the valves of a native clam make an excellent substrate for the attachment of the zebra mussel. The native clam can be so densely covered that feeding is impossible.

FIGURE 1 The zebra mussel (*Dreissena polymorpha*) invaded U.S. fresh waters and is threatening native bivalves and other freshwater species.

While research efforts are under way to monitor the spread of the zebra mussel and to search for its "Achilles' heel," a larger question looms in the background: "Are more invaders on the way?" One study of the ballast water of 55 cargo ships revealed that 17 species of animals were still alive in each ship by the time they arrived in fresh waters of North America. Estimates of the number of individuals alive per species ranged between 10,000 and eight billion! A relatively simple, partial solution to the problem of ballast-water invaders has been to require ships to dump ballast water from foreign, freshwater ports into the open ocean. This ballast water is then replaced with seawater. Seawater kills most freshwater organisms, and freshwater kills most marine organisms.

Some surface-dwelling bivalves are attached to the substrate either by proteinaceous strands called byssal threads, which are secreted by a gland in the foot, or by cementation to the substrate. The former method is used by the common marine mussel *Mytilus*, and the latter by oysters.

Boring bivalves live beneath the surface of limestone, clay, coral, wood, and other substrates. (6) Boring begins after the larvae settle to the substrate, and it occurs by mechanical abrasion of the substrate by the anterior margin of the valves. Physical abrasion is sometimes accompanied by acidic secretions from the mantle margin that dissolve limestone. As the bivalve grows, portions of the burrow recently bored are larger in diameter than other, usually external, portions of the burrow. Thus, the bivalve is often imprisoned in its rocky burrow.

(a)

(b)

FIGURE 9.17
Class Cephalopoda. (*a*) A chambered nautilus. (*b*) An octopus.

Stop and Ask Yourself

5. What are three advantages conferred to gastropods by torsion?
6. What is a hydraulic skeleton?
7. How do bivalves discourage predatory sea stars?
8. How do bivalves feed?

CLASS CEPHALOPODA

The class Cephalopoda (sef′ah-lop′o-dah) (L. *cephalic*, head + Gr. *podos*, foot) includes the octopuses, squids, cuttlefish, and nautili (figure 9.17; table 9.1). They are the most complex molluscs and, in many ways, the most complex invertebrates. The anterior portion of their foot has been modified into a circle of tentacles or arms that are used for prey capture, attachment, locomotion, and copulation (figure 9.18). The foot is also incorporated into a funnel that is associated with the mantle cavity and used for jetlike locomotion. The molluscan body plan is further modified in that the cephalopod head is in line with the visceral mass. Cephalopods have a highly muscular mantle that encloses all of the body except the head and tentacles. The mantle acts as a pump for bringing large quantities of water into the mantle cavity.

SHELL

Ancestral cephalopods probably had a conical shell. The only living cephalopods that possess an external shell are the nautili (*see figure* 9.17a). They have a coiled shell that is subdivided by septa. As the nautilus grows, it moves forward, secreting new shell around itself and leaving an empty septum behind. Only

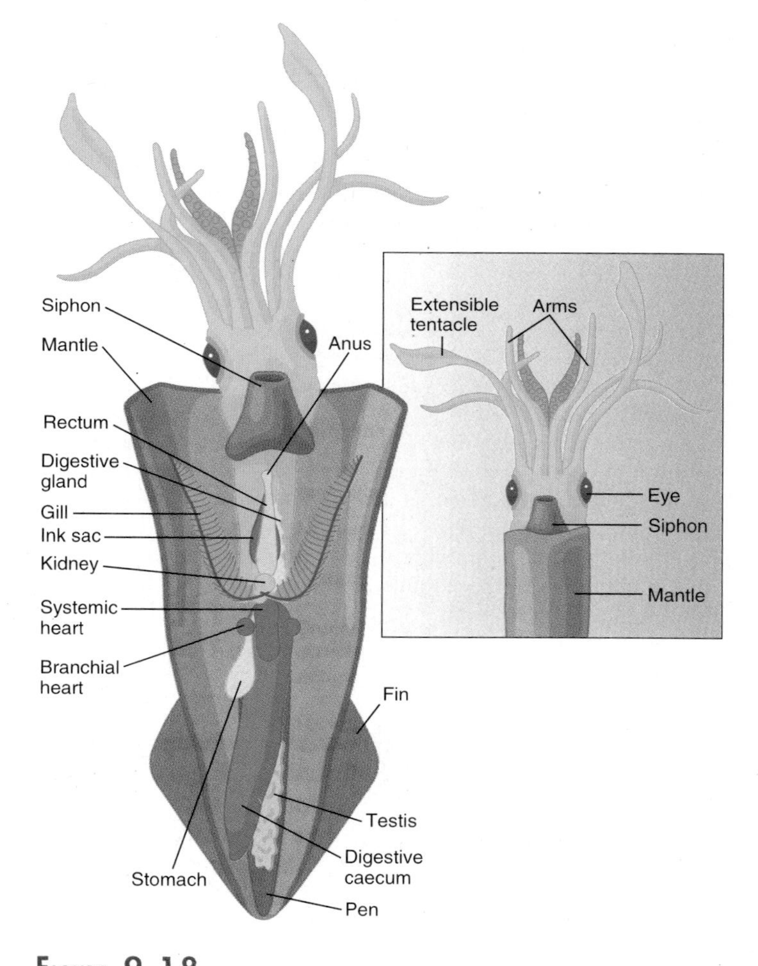

FIGURE 9.18
Internal Structure of the Squid, *Loligo*. The shell of most cephalopods is reduced or absent, and the foot is modified into a circle of tentacles and/or arms that encircle the head and a funnel. The inset shows the undissected anatomy of the squid.

the last chamber is occupied. These chambers are fluid filled when they are formed. Septa are perforated by a cord of tissue called a siphuncle, which absorbs fluids by osmosis and replaces them with metabolic gases. The amount of gas in the chambers is regulated to alter the buoyancy of the animal.

In all other cephalopods, the shell is reduced or absent. In cuttlefish, the shell is internal and laid down in thin layers, leaving small, gas-filled spaces that increase buoyancy. Cuttlefish shell, called cuttlebone, has been used to make powder for polishing and has been fed to pet birds to supplement their diet with calcium. The shell of a squid is reduced to an internal, chitinous structure called the pen. In addition to this reduced shell, squid also have cartilaginous plates in the mantle wall, neck, and head that support the mantle and protect the brain. The shell is absent in octopuses.

Locomotion

As predators, cephalopods depend upon their ability to move quickly using a jet-propulsion system. The mantle of cephalopods contains radial and circular muscles. When circular muscles contract, they decrease the volume of the mantle cavity and close collarlike valves to prevent water from moving out of the mantle cavity between the head and the mantle wall. Water is thus forced out of a narrow funnel. Muscles attached to the funnel control the direction of the animal's movement. Radial mantle muscles bring water into the mantle cavity by increasing the volume of the mantle cavity. Posterior fins act as stabilizers in squid and also aid in propulsion and steering in cuttlefish. "Flying squid" (family Onycoteuthidae) have been clocked at speeds of 20 knots (about 20 mph). Octopuses are more sedentary animals. They may use jet propulsion in an escape response, but normally they crawl over the substrate using their tentacles.

Feeding and Digestion

Cephalopod prey are located by sight and are captured with tentacles, which bear adhesive cups. In squid, these cups are reinforced with tough protein around their margins and sometimes possess small hooks (figure 9.19).

Jaws and a radula are present in all cephalopods. The jaws are powerful, beaklike structures used for tearing food, and the radula rasps food, forcing it into the mouth cavity.

Cuttlefish and nautili feed on small invertebrates on the ocean floor. Octopuses are nocturnal hunters and feed on snails, fish, and crustaceans. Octopuses have salivary glands that are used to inject venom into their prey. Squid feed on fishes and shrimp, which they kill by biting across the back of the head. (7) Giant squid even prey upon the young of sperm whales—just as adult sperm whales prey upon the young of giant squid.

The digestive tract of cephalopods is muscular, and peristalsis (coordinated muscular waves) replaces ciliary action in moving food. Most digestion occurs in a stomach and a large cecum. Digestion is primarily extracellular with large digestive glands supplying enzymes. An intestine ends at the anus, near the funnel, and wastes are carried out of the mantle cavity with exhalant water.

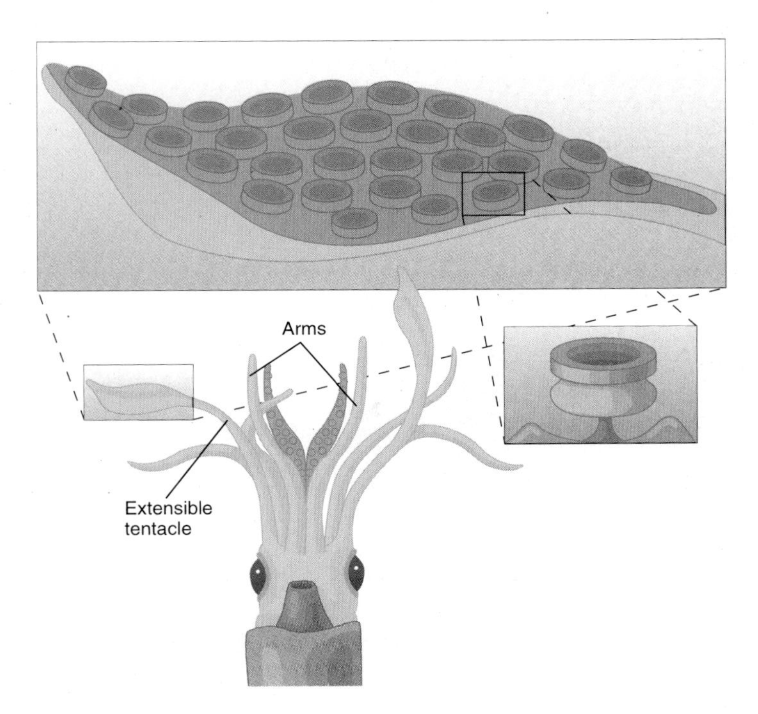

Figure 9.19
A Cephalopod Tentacle. Suction cups are used for prey capture and as holdfast structures.

Other Maintenance Functions

Cephalopods, unlike other molluscs, have a **closed circulatory system.** Blood is confined to vessels throughout its circuit around the body. Capillary beds connect arteries and veins, and exchanges of gases, nutrients, and metabolic wastes occur across capillary walls. In addition to having a heart consisting of two auricles and one ventricle, cephalopods have contractile arteries and structures called branchial hearts. The latter are located at the base of each gill and help move blood through the gill. These modifications increase blood pressure and the rate of blood flow—necessary for active animals with relatively high metabolic rates. Large quantities of water circulate over the gills at all times.

Greater excretory efficiency is achieved in the cephalopods because of the closed circulatory system. A close association of blood vessels with nephridia allows both filtration and secretion of wastes directly from the blood into the excretory system.

(8) The cephalopod nervous system is unparalleled by any other invertebrate. Cephalopod brains are large, and their evolution is directly related to cephalopod predatory habits and dexterity.

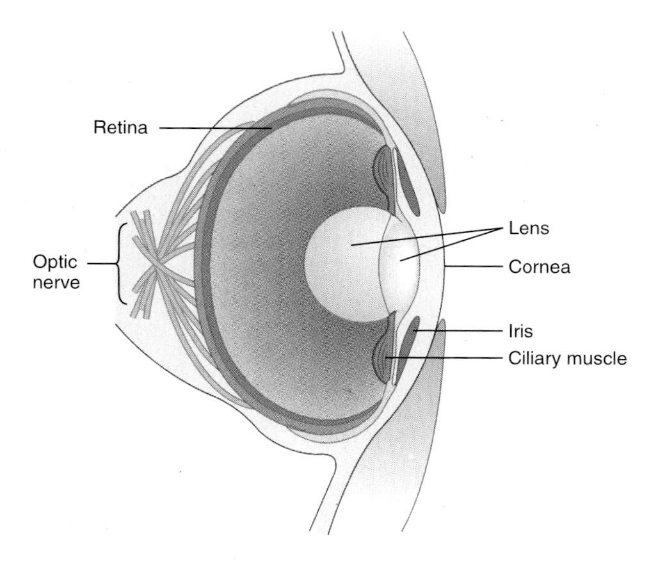

FIGURE 9.20

The Cephalopod Eye. The eye is immovable in a supportive and protective socket of cartilages. It contains a rigid, spherical lens. An iris in front of the lens forms a slitlike pupil that can open and close in response to varying light conditions. Note that the optic nerve comes off the back of the retina.

The brain is formed by a fusion of ganglia. It has large areas devoted to controlling muscle contraction (e.g., swimming movements and sucker closing), sensory perception, and functions such as memory and decision making. Research on cephalopod brains has provided insight into human brain functions.

The eyes of octopuses, cuttlefish, and squids are similar in structure to vertebrate eyes (figure 9.20). (This similarity is an excellent example of convergent evolution.) In contrast to the vertebrate eye, nerve cells leave the eye from the outside of the eyeball, so that no blind spot exists. Like many aquatic vertebrates, cephalopods focus by moving the lens back and forth. Cephalopods can form images, distinguish shapes, and discriminate some colors. The nautiloid eye is less complex. It lacks a lens, and the interior is open to seawater, thus it acts as a pinhole camera.

Cephalopod statocysts respond to gravity and acceleration and are located in cartilages next to the brain. Osphradia are present only in *Nautilus*. Tactile receptors and additional chemoreceptors are widely distributed over the body.

Cephalopods have pigment cells called **chromatophores.** Tiny muscles attach to these pigment cells and, when these muscles contract, the chromatophores quickly expand and change the color of the animal. Color changes, in combination with ink discharge, function in alarm responses. In defensive displays, color changes may spread in waves over the body to form large, flickering patterns. Color changes may also help cephalopods to blend with their background. The cuttlefish, *Sepia,* can even make a remarkably good impression of a checkerboard background. Color changes are also involved with courtship displays. Some species combine chromatophore displays with bioluminescence.

All cephalopods possess an ink gland that opens just behind the anus. Ink is a brown or black fluid containing melanin and other chemicals. Discharged ink confuses a predator, allowing the cephalopod to escape. For example, *Sepiola* reacts to danger by darkening itself with chromatophore expansion prior to releasing ink. After ink discharge, *Sepiola* changes to a lighter color again to assist its escape.

Reproduction and Development

Cephalopods are dioecious. Gonads are located in the dorsal portion of the visceral mass. The male reproductive tract consists of testes and structures for encasing sperm in packets called **spermatophores.** The female reproductive tract produces large, yolky eggs and is modified with glands that secrete gellike cases around eggs. These cases frequently harden on exposure to seawater.

One tentacle of male cephalopods, called the **hectocotylus,** is modified for spermatophore transfer. In *Loligo* and *Sepia,* the hectocotylus has several rows of smaller suckers capable of picking up spermatophores. During copulation, male and female tentacles intertwine, and the male removes spermatophores from his mantle cavity. The male inserts his hectocotylus into the mantle cavity of the female and deposits a spermatophore near the opening to the oviduct. Spermatophores have an ejaculatory mechanism that frees sperm from the baseball-bat-shaped capsule. Eggs are fertilized as they leave the oviduct and deposited singly or in stringlike masses. They are usually attached to some substrate, such as the ceiling of an octopus' den. Octopuses tend eggs during development by cleaning them of debris with their arms and squirts of water.

Development of cephalopods occurs in the confines of the egg membranes, and the hatchlings are miniatures of adults. Young are never cared for after hatching.

Stop and Ask Yourself

9. How does the jet-propulsion system of a squid work?
10. In what way is the mechanism of movement of food in the cephalopod digestive tract different from that in other molluscs?
11. What is a closed circulatory system?
12. What is a chromatophore? A spermatophore?

Class Polyplacophora

The class Polyplacophora (pol′e-pla-kof″o-rah) (Gr. *polys,* many + *plak,* plate + *phoros,* to bear) contains the chitons. Chitons are common inhabitants of shallow marine waters, wherever hard substrates occur. Chitons were used for food by early Native Americans. They have a fishy flavor but are tough to chew, and difficult to collect.

Chitons are easy to recognize. They have a reduced head, a flattened foot, and a shell that is divided into eight articulating

(a)

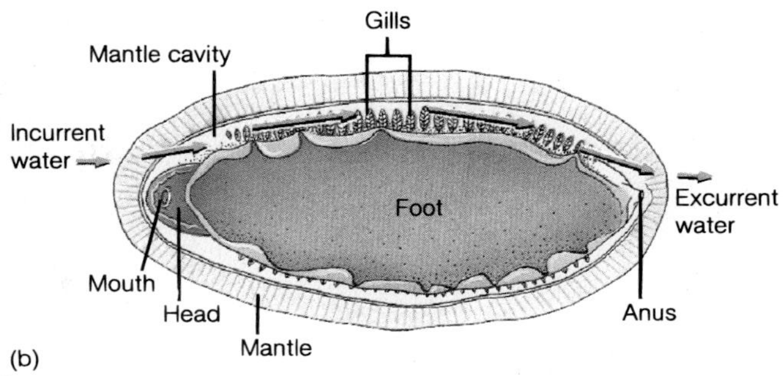

(b)

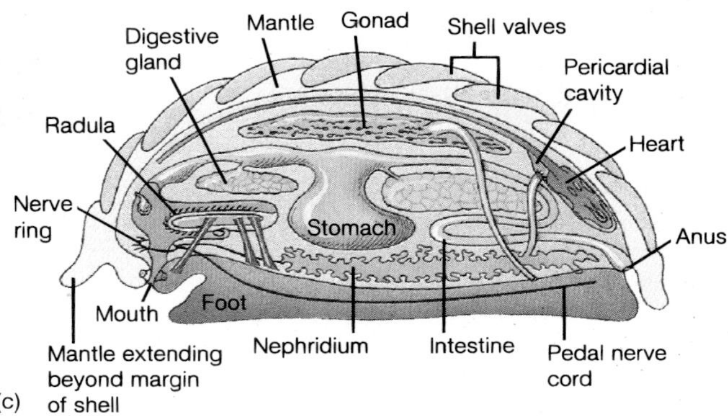

(c)

FIGURE 9.21
Class Polyplacophora. (*a*) Dorsal view of a chiton. Note the shell consisting of eight valves and the mantle extending beyond the margins of the shell. (*b*) A ventral view of a chiton. The mantle cavity is the region between the mantle and the foot. Arrows show the path of water moving across gills in the mantle cavity. (*c*) Internal structure.

dorsal valves (figure 9.21*a*). The broad foot is covered by a muscular mantle that extends beyond the margins of the shell and foot (figure 9.21*b*). The mantle cavity is restricted to the space between the margin of the mantle and the foot. Chitons crawl over their substrate in a manner similar to gastropods. The muscular foot promotes secure attachment to a substrate, which allows chitons to withstand strong waves and tidal currents. When chitons are disturbed, the edges of the mantle are applied tightly to the substrate, and contraction of foot muscles raises the middle of the foot, creating a powerful vacuum that holds the chiton to its substrate. Articulations in the shell allow chitons to roll into a ball when dislodged from the substrate.

A linear series of gills are located in the mantle cavity on each side of the foot. Currents of water, created by cilia on the gills, enter below the anterior mantle margins and exit posteriorly. Openings of the digestive, excretory, and reproductive tracts are located near the exhalant area of the mantle cavity, and products of these systems are carried away with exhalant water.

Most chitons feed on attached algae. A chemoreceptor, the subradular organ, is extended from the mouth to detect food, which the radula rasps from the substrate. Food is trapped in mucus and enters the esophagus by ciliary action. Extracellular digestion occurs in the stomach, and wastes are sent on to the intestine (figure 9.21*c*).

The nervous system is ladderlike, with four anteroposterior nerve cords and numerous transverse nerves. A nerve ring encircles the esophagus. Sensory structures include osphradia, tactile receptors on the mantle margin, chemoreceptors near the mouth, and statocysts in the foot. In some chitons, photoreceptors dot the surface of the shell.

Sexes are separate in chitons. External fertilization and development result in a swimming trochophore that settles and metamorphoses into an adult without passing through a veliger stage.

CLASS SCAPHOPODA

Members of the class Scaphopoda (ska-fop′o-dah) (Gr. *skaphe*, boat + *podos*, foot) are called tooth shells or tusk shells. There are over 300 species, and all are burrowing marine animals that inhabit moderate depths. Their most distinctive characteristic is a conical shell that is open at both ends. The head and foot project from the wider end of the shell, and the rest of the body, including the mantle, is greatly elongate and extends the length of the shell (figure 9.22). Scaphopods live mostly buried in the substrate with head and foot oriented down and with the apex of the shell projecting into the water above. Incurrent and excurrent water enters and leaves the mantle cavity through the opening at the apex of the shell. Functional gills are absent, and gas exchange occurs across mantle folds. Scaphopods have a radula and tentacles, which they use in feeding on foraminiferans. Sexes are separate, and trochophore and veliger larvae are produced.

CLASS MONOPLACOPHORA

Members of the class Monoplacophora (mon′o-pla-kof″o-rah) (Gr. *monos*, one + *plak*, plate + *phoros* to bear) possess an undivided, arched shell; a broad, flat foot; and serially repeated pairs of gills and foot retractor muscles. They are dioecious;

FIGURE 9.22
Class Scaphopoda. This conical shell is open at both ends. In its living state, the animal would be mostly buried with the apex of the shell projecting into the water.

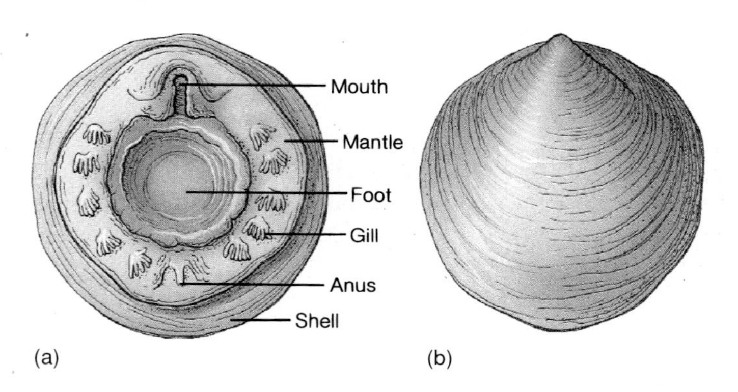

FIGURE 9.23
Class Monoplacophora. (*a*) Ventral and (*b*) dorsal views of *Neopilina*.

however, nothing is known of their embryology. This group of molluscs was known only from fossils until 1952, when a limpetlike monoplacophoran, named *Neopilina*, was dredged up from a depth of 3,520 m off the Pacific coast of Costa Rica (figure 9.23).

CLASS CAUDOFOVEATA

Members of the class Caudofoveata (kaw′do-fo′ve-a′ta) (L. *cauda*, tail + *fovea*, small pit) are wormlike molluscs that range in size from 2 mm to 14 cm and live in vertical burrows on the deep-sea floor. They possess scalelike spicules on the body wall and lack the following typical molluscan characteristics: shell, crystalline style, statocysts, foot, and nephridia. Approximately 70 species have been described but little is known of their ecology.

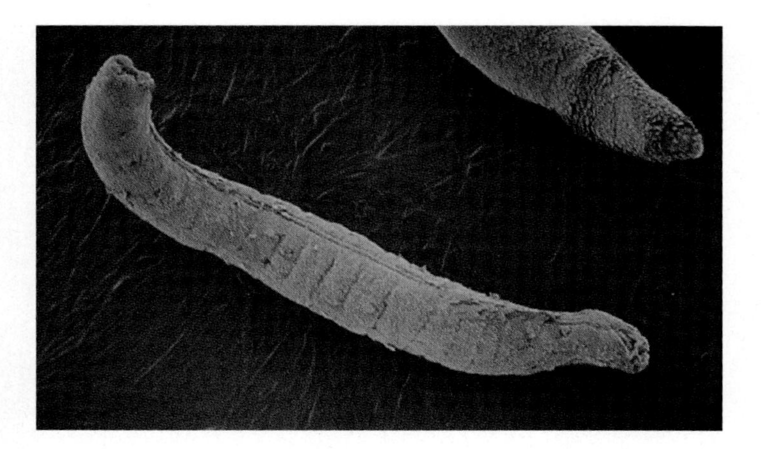

FIGURE 9.24
Class Aplacophora. A scanning electron micrograph of the solenogaster, *Meiomenia*. The body is covered by flattened, spinelike calcareous spicules. The ventral groove shown here may be formed from a rolling of the mantle margins. *Meiomenia* is approximately 2 mm long.

CLASS APLACOPHORA

Members of the class Aplacophora (a′pla-kof″o-rah) (Gr. *a*, without + *plak*, plate + *phoros*, to bear) are called solenogasters (figure 9.24). There are about 250 species of these cylindrical molluscs that lack a shell and crawl on their ventral surface. Their nervous system is ladderlike and reminiscent of the flatworm body form, causing some to suggest that this group may be closely related to the ancestral molluscan stock. One small group of aplacophorans contains burrowing species that feed on microorganisms and detritus and possess a radula and nephridia. The majority of these molluscs, however, lack nephridia and a radula, are mostly surface dwellers on corals and other substrates, and are carnivores, frequently feeding on cnidarian polyps.

FURTHER PHYLOGENETIC CONSIDERATIONS

Fossil records of molluscan classes indicate that the phylum is over 500 million years old. Although molluscs have protostome affinities, zoologists do not know the exact relationship of this phylum to other animal phyla. The discovery of *Neopilina* (class Monoplacophora) in 1952 seemed to revolutionize ideas regarding the position of the molluscs in the animal phyla. The most striking feature of *Neopilina* was a segmental arrangement of gills, excretory structures, and nervous system. Because annelids and arthropods (*see chapters 11 and 12*) also have a segmental arrangement of body parts, monoplacophorans were considered a "missing link" between other molluscs and the annelid-arthropod evolutionary line. This link was further supported by the fact that molluscs, annelids, and arthropods share certain protostome characteristics (*see figure 9.3a*), and chitons also show a repetition of some body parts.

Most zoologists now agree that the segmentation seen in some molluscs is very different from that of annelids and arthro-

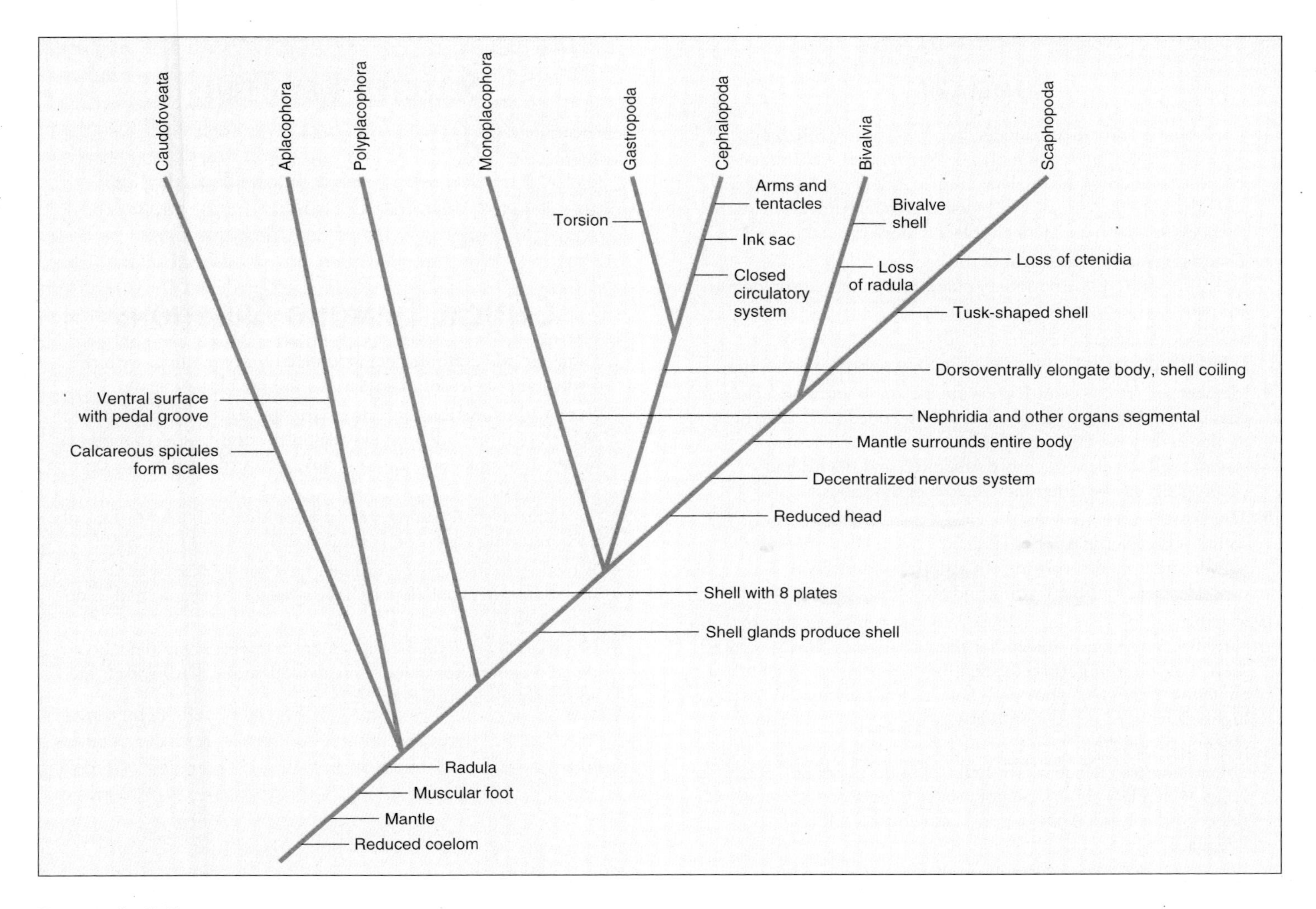

FIGURE 9.25
Molluscan Phylogeny. A cladogram showing possible evolutionary relationships among the molluscs.

pods. Although information on the development of the serially repeating structures in *Neopilina* is not available, no serially repeating structure in any other mollusc develops in an annelid-arthropod fashion. Segmentation is probably not an ancestral molluscan characteristic. Many zoologists now believe that molluscs diverged from ancient triploblastic stocks independently of any other phylum. Other zoologists maintain that, in spite of the absence of annelidlike segmentation in molluscs, protostomate affinities still tie the molluscs to the annelid-arthropod line. Whichever is the case, the relationship of the molluscs to other animal phyla is distant.

The diversity of body forms and life-styles in the phylum Mollusca is an excellent example of adaptive radiation. Molluscs probably began in Precambrian times as slow-moving, marine, bottom dwellers. The development of unique molluscan features allowed them to diversify relatively quickly. By the end of the Cambrian period, some were filter feeders, some were burrowers, and others were swimming predators. Later, some molluscs became terrestrial and invaded many habitats, from tropical rain forests to arid deserts.

One interpretation of molluscan phylogeny is shown in figure 9.25. The lack of a shell in the classes Caudofoveata and Aplacophora is thought to be a primitive character. All other molluscs have a shell or are derived from shelled ancestors. The multipart shell distinguishes the Polyplacophora from other classes. Other selected synapomorphies, discussed earlier in this chapter, are noted in the cladogram. There are, of course, other interpretations of molluscan phylogeny. The extensive adaptive radiation of this phylum has made higher taxonomic relationships very difficult to discern.

Stop and Ask Yourself

13. What molluscs have eight dorsal, articulating plates?
14. How would you characterize members of the class Scaphopoda?
15. What mollusc has serially repeated pairs of gills and an undivided, arched shell?
16. What is the significance of *Neopilina* in hypotheses concerning molluscan phylogeny?

Summary

1. Triploblastic animals are divided into two groups, which may represent evolutionary assemblages. Protostomes, or spiralians, include members of the phyla Mollusca, Annelida, Arthropoda, and others. Deuterostomes, or radialians, include members of the phyla Echinodermata, Hemichordata, Chordata, and others.
2. Theories regarding the origin of the coelom influence how zoologists interpret evolutionary relationships among triploblastic animals.
3. Molluscs are characterized by a head-foot, a visceral mass, a mantle, and a mantle cavity. Most molluscs also have a radula.
4. Members of the class Gastropoda are the snails and slugs. They are characterized by torsion and often have a coiled shell. Like most molluscs, they use cilia for feeding, have an open circulatory system, well-developed sensory structures, and nephridia. Gastropods may be either monoecious or dioecious.
5. The class Bivalvia includes the clams, oysters, mussels, and scallops. Bivalves lack a head and are covered by a sheetlike mantle and a shell consisting of two valves. Most bivalves use expanded gills for filter feeding and most are dioecious.
6. Members of the class Cephalopoda are the octopuses, squids, cuttlefish, and nautili. Except for the nautili, they have a reduced shell. The anterior portion of their foot has been modified into a circle of tentacles. Cephalopods have a closed circulatory system, highly developed nervous and sensory systems, and they are efficient predators.
7. Other molluscs include tooth shells (class Scaphopoda), *Neopilina* (class Monoplacophora), caudofoveates (class Caudofoveata), solenogasters (class Aplacophora), and chitons (class Polyplacophora).
8. Some zoologists believe that the molluscs are derived from the annelid-arthropod lineage. Others believe that they arose from triploblastic stocks independently of any other phylum. Adaptive radiation in the molluscs has resulted in the invasion of most ecosystems of the earth.

Selected Key Terms

deuterostomes (*p. 142*)
glochidium (*p. 152*)
head-foot (*p. 145*)
mantle (*p. 145*)
protostomes (*p. 142*)
radula (*p. 145*)
torsion (*p. 146*)
trochophore larva (*p. 142*)
veliger larva (*p. 149*)
visceral mass (*p. 145*)

Critical Thinking Questions

1. Compare and contrast hydraulic skeletons of molluscs with the hydrostatic skeletons of cnidarians and pseudocoelomates.
2. Review the functions of body cavities presented in chapter 16. Which of those functions would apply to the coelom of molluscs? What additional function(s), if any, are carried out by the coelom of molluscs?
3. Torsion and shell coiling are often confused by students. Describe each and their effects on gastropod structure and function.
4. Why do you think that nautiloids have retained their external shell, but other cephalopods have a reduced internal shell or no shell at all?
5. Bivalves are often used as indicators of environmental quality. Based on your knowledge of bivalves, describe why bivalves are good animals for this purpose.

ANNELIDA: THE METAMERIC BODY FORM

Outline

Concepts

1. Members of the phylum Annelida are the segmented worms. The relationships of annelids to lower triploblastic animals are debated.
2. Metamerism has important influences on virtually every aspect of annelid structure and function.
3. Members of the class Polychaeta are annelids that have become adapted to a variety of marine habitats. Some live in or on marine substrates; others live in burrows or are free swimming. They are characterized by the presence of parapodia and numerous, long setae.
4. Members of the class Oligochaeta are found in freshwater and terrestrial habitats. They lack parapodia and have fewer, short setae.
5. The class Hirudinea contains the leeches. They are predators in freshwater, marine, and terrestrial environments. Body-wall musculature and the coelom are modified from the pattern found in the other annelid classes. These differences influence locomotor and other functions of leeches.
6. The ancient polychaetes are probably the ancestral stock from which modern polychaetes, oligochaetes, and leeches evolved.

Would You Like to Know:

1. what worm is the basis of a great communal feast? (*p. 162*)
2. what metamerism is? (*p. 162*)
3. how the fan of a fanworm is used? (*p. 166*)
4. why the blood of some marine worms is green? (*p. 166*)
5. why swarming of epitokes is advantageous for some marine worms? (*p. 169*)
6. why an earthworm is so difficult to extract from its burrow? (*p. 171*)
7. why leeches should be referred to as predators rather than parasites? (*p. 173*)
8. why earthworms are found in soil around deciduous vegetation? (*p. 176*)

These and other useful questions will be answered in this chapter.

This chapter contains evolutionary concepts, which are set off in this font.

Evolutionary Perspective

At the time of the November full moon, on islands near Samoa in the South Pacific, natives rush about preparing for one of their biggest yearly feasts. In just one week, the sea will yield a harvest that can be scooped up in nets and buckets. Worms by the millions transform the ocean into what one writer called a "vermicelli soup!" Celebrants gorge themselves on worms that have been cooked or wrapped in breadfruit leaves. (1) The Samoan palolo worm (*Eunice viridis*) spends its entire adult life in coral burrows at the sea bottom. Each November, 1 week after the full moon, this worm emerges from its burrow, and specialized body segments devoted to sexual reproduction break free and float to the surface, while the rest of the worm is safe on the ocean floor. The surface water is discolored as gonads release their countless eggs and sperm. The natives' feast is shortlived, however; these reproductive swarms last only 2 days and do not recur for another year.

The Samoan palolo worm is a member of the phylum Annelida (ah-nel'i-dah) (*L. annellus*, ring). Other members of this phylum include countless marine worms (figure 10.1), the soil-building earthworms, and predatory leeches (table 10.1).

Characteristics of the phylum Annelida include the following:

1. Body metameric, bilaterally symmetrical, and wormlike
2. Protostome characteristics include spiral cleavage, trochophore larvae (when larvae are present), and schizocoelous coelom formation
3. Paired, epidermal setae
4. Closed circulatory system
5. Dorsal suprapharyngeal ganglia and ventral nerve cord(s) with ganglia
6. Metanephridia (usually) or protonephridia

Figure 10.1
Phylum Annelida. The phylum Annelida includes about 9,000 species of segmented worms. Most of these are marine members of the class Polychaeta. The fanworm (*Sriro branchus*) is shown here. The fan of this tube-dwelling polychaete is specialized for feeding and gas exchange.

Relationships to Other Animals

Annelids are protostomes (*see figure* 9.3). Protostome characteristics, such as spiral cleavage, a mouth derived from an embryonic blastopore, schizocoelous coelom formation, and trochophore larvae are present in most members of the phylum (figure 10.2). (Certain exceptions will be discussed later.) The origin of this diverse phylum, like that of most other phyla, occurred at least as early as Precambrian times, more than 600 million years ago. Unfortunately, there is little evidence documenting the evolutionary pathways that resulted in the first annelids.

There are a number of hypotheses that account for annelid origins. These hypotheses are tied into hypotheses regarding the origin of the coelom (*see chapter* 9). If one assumes a schizocoelous origin of the coelom, as many zoologists believe, then the annelids evolved from ancient flatworm stock. On the other hand, if an enterocoelous coelom origin is correct, then annelids evolved from ancient diploblastic animals, and the triploblastic, acoelomate body may have been derived from a triploblastic, coelomate ancestor. The recent discovery of a worm, *Lobatocerebrum*, which shares annelid and flatworm characteristics, has lent support to the enterocoelous origin hypothesis. *Lobatocerebrum* is classified as an annelid based on the presence of certain segmentally arranged excretory organs, an annelidlike body covering, a complete digestive tract, and an annelidlike nervous system. As do flatworms, however, it has a ciliated epidermis and is acoelomate. Some zoologists believe that *Lobatocerebrum* illustrates how the triploblastic, acoelomate design could have been derived from the annelid lineage.

Metamerism and Tagmatization

When one looks at an earthworm, one of the first characteristics noticed is the organization of the body into a series of ringlike segments. What is not externally obvious, however, is that the body is divided internally as well. (2) Segmental

TABLE 10.1

CLASSIFICATION OF THE PHYLUM ANNELIDA

Phylum Annelida (ah-nel'i-dah)
The phylum of triploblastic, coelomate animals whose members are metameric (segmented), elongate, and cylindrical or oval in cross section. Annelids have a complete digestive tract; paired, epidermal setae; and a ventral nerve cord. The phylum is divided into three classes.

Class Polychaeta (pol"e-ket'ah)
The largest annelid class; mostly marine; head with eyes and tentacles; parapodia bear numerous setae; monoecious or dioecious; development frequently involves a trochophore larval stage. *Nereis, Arenicola, Sabella.*

Class Oligochaeta (ol"i-go-ket'ah)
Few setae and no parapodia; no distinct head; monoecious with direct development; primarily fresh water or terrestrial. *Lumbricus, Tubifex.*

Class Hirudinea (hi'roo-din"eah)
Leeches; bodies with 34 segments; each segment subdivided into annuli; anterior and posterior suckers present; parapodia absent; setae reduced or absent. Fresh water, marine, and terrestrial. *Hirudo.*

arrangement of body parts in an animal is called **metamerism** (Gr. *meta*, after + *mere*, part).

Metamerism has profound influences on virtually every aspect of annelid structure and function, such as the anatomical arrangement of organs that are coincidentally associated with metamerism. For example, the compartmentalization of the body has resulted in each segment having its own excretory, nervous, and circulatory structures. Two related functions are probably the primary adaptive features of metamerism: flexible support and efficient locomotion. These functions depend on the metameric arrangement of the coelom and can be understood by examining the development of the coelom and the arrangement of body-wall muscles.

During embryonic development, the body cavity of annelids arises by a segmental splitting of a solid mass of mesoderm that occupies the region between ectoderm and endoderm on either side of the embryonic gut tract. Enlargement of each cavity forms a double-membraned septum on the anterior and posterior margin of each coelomic space and dorsal and ventral mesenteries associated with the digestive tract (figure 10.3).

Muscles also develop from the mesodermal layers associated with each segment. A layer of circular muscles lies below the epidermis, and a layer of longitudinal muscles, just below

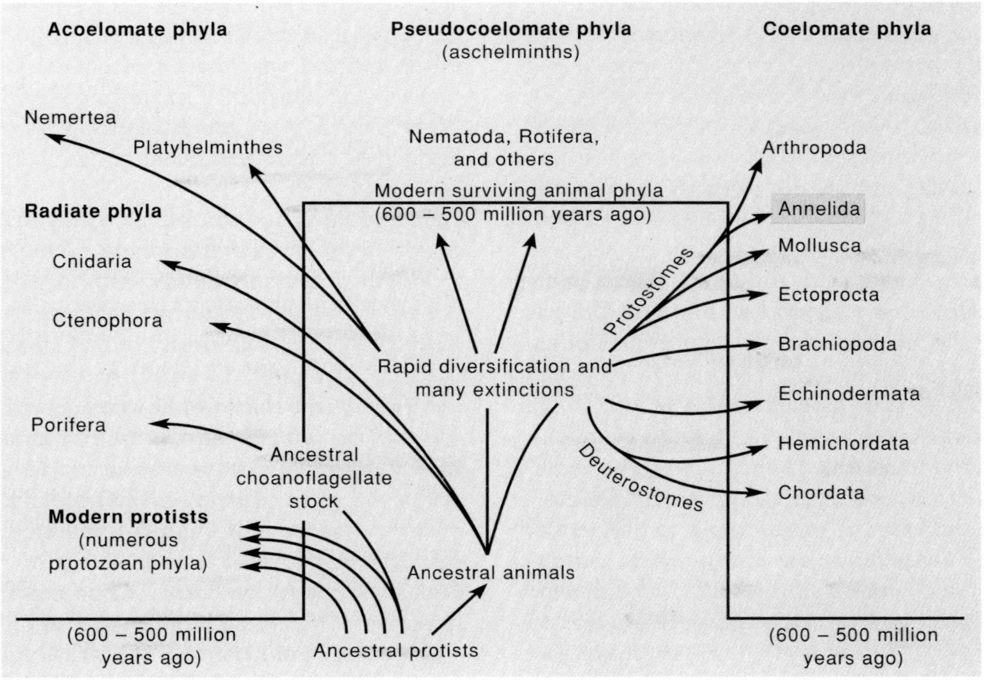

FIGURE 10.2
Evolutionary Relationships of the Annelida. Annelids (shaded in orange) are protostomes with close evolutionary ties to the arthropods.

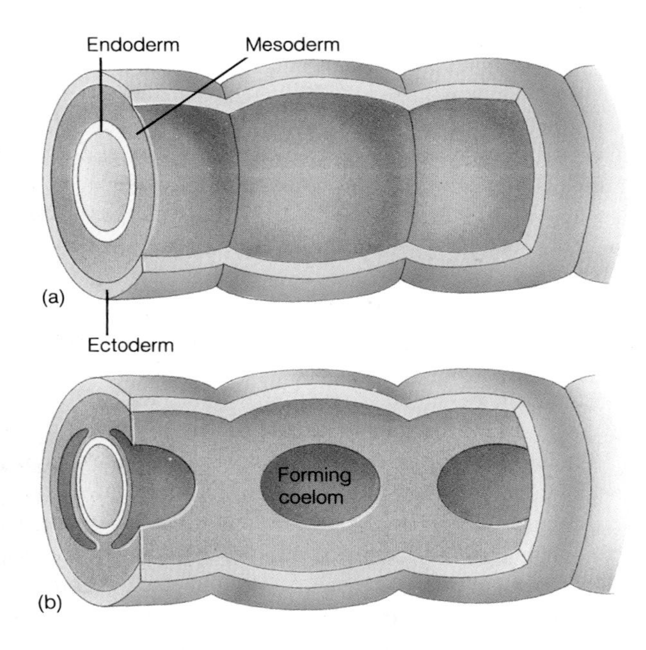

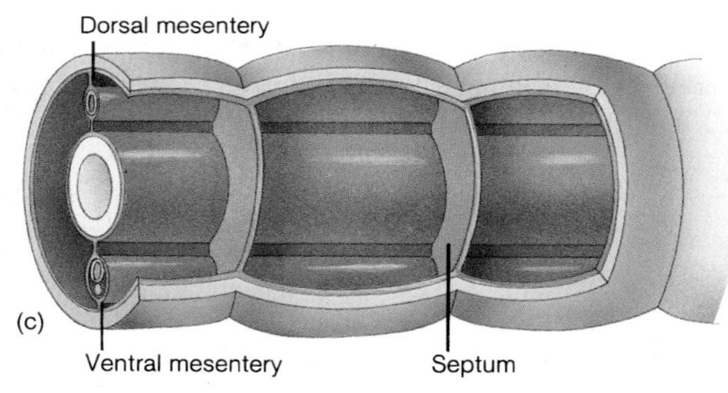

FIGURE 10.3

Development of Metameric, Coelomic Spaces in Annelids. (*a*) A solid mesodermal mass separates ectoderm and endoderm in early embryological stages. (*b*) Two cavities in each segment form from a splitting of the mesoderm on each side of the endoderm (schizocoelous coelom formation). (*c*) These cavities spread in all directions. Enlargement of the coelomic sacs leaves a thin layer of mesoderm applied against the outer body wall (the parietal peritoneum) and the gut tract (the visceral peritoneum), and it forms dorsal and ventral mesenteries. Anterior and posterior expansion of the coelom in adjacent segments results in the formation of the double-membraned septum that separates annelid metameres.

the circular muscles, runs between the septa that separate each segment. In addition, some polychaetes have oblique muscles and the leeches have dorsoventral muscles.

One advantage of the segmental arrangement of coelomic spaces and muscles is the creation of hydrostatic compartments, which makes possible a variety of advantageous locomotor and supportive functions not possible in nonmetameric animals that utilize a hydrostatic skeleton. Each segment can be controlled independently of distant segments, and muscles can act as antagonistic pairs within a segment. The constant volume of coelomic fluid provides a hydrostatic skeleton against which

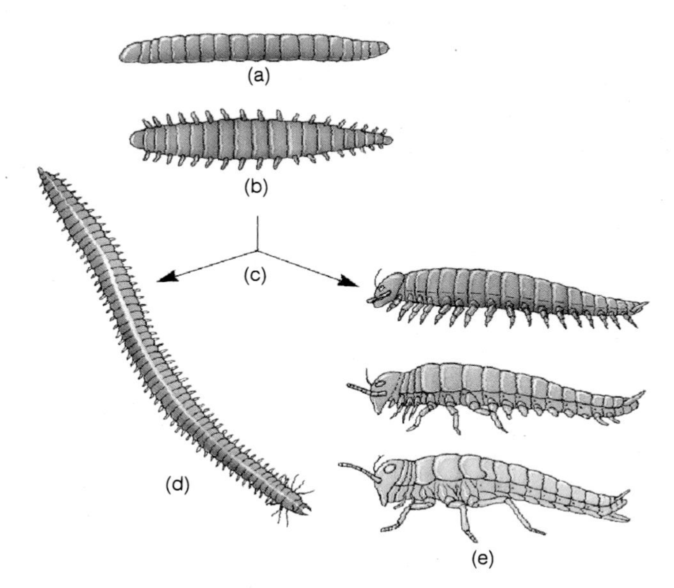

FIGURE 10.4

A Possible Origin of Annelids and Arthropods. A possible sequence in the evolution of the annelid/arthropod line from a hypothetical wormlike ancestor. (*a*) A wormlike prototype. (*b*) Paired, metameric appendages develop. (*c*) Divergence of the annelid and arthropod lines. (*d*) Paired appendages develop into parapodia of ancestral polychaetes. (*e*) Extensive tagmatization results in specializations characteristic of the arthropods. A head is a sensory and feeding tagma, a thorax is a locomotor tagma, and an abdomen contains visceral organs.

muscles operate. Resultant localized changes in the shape of groups of segments provide the basis for swimming, crawling, and burrowing.

A second advantage of metamerism is that it lessens the impact of injury. If one or a few segments are injured, adjacent segments, set off from injured segments by septa, may be able to maintain nearly normal functions, which increases the likelihood that the worm, or at least a part of it, will survive the trauma.

A third advantage of metamerism is that it permits the modification of certain regions of the body for specialized functions, such as feeding, locomotion, and reproduction. The specialization of body regions in a metameric animal is called **tagmatization** (Gr. *tagma*, arrangement). Although it is best developed in the arthropods, some annelids also display tagmatization. (The arthropods include animals such as insects, spiders, mites, ticks, and crayfish.)

Virtually all zoologists agree that, because of similarities in the development of metamerism in the two groups, annelids and arthropods are closely related. Other common features include triploblastic coelomate organization, bilateral symmetry, a complete digestive tract, and a ventral nerve cord. As usual, there is little fossil evidence documenting ancestral pathways that led from a common ancestor to the earliest representatives of these two phyla. Zoologists are confident that the annelids and arthropods evolved from a marine, wormlike, bilateral ancestor that possessed metameric design. Figure 10.4 depicts a sequence of evolutionary changes that may have given rise to these two phyla.

Figure 10.5
Class Polychaeta. External structure of *Nereis virens*. Note the numerous parapodia.

Stop and Ask Yourself

1. What are two hypotheses regarding the origin of the phylum Annelida?
2. What is metamerism? What is tagmatization?
3. What are three advantages of metameric organization?
4. What other phylum is closely related to the annelids?

Class Polychaeta

Members of the class Polychaeta (pol″-e-ket′ah) (Gr. *polys*, many + *chaite*, hair) are mostly marine, and are usually between 5 and 10 cm long (*see table 10.1*). With more than 5,300 species, Polychaeta is the largest of the annelid classes. Polychaetes have become adapted to a variety of habitats. Many live on the ocean floor, under rocks and shells, and within the crevices of coral reefs. Other polychaetes are burrowers and move through their substrate by peristaltic contractions of the body wall. A bucket of intertidal sand normally yields vast numbers and an amazing variety of these burrowing annelids. Other polychaetes construct tubes of cemented sand grains or secreted organic materials. Mucus-lined tubes serve as protective retreats and feeding stations.

External Structure and Locomotion

In addition to metamerism, the most distinctive feature of polychaetes is the presence of lateral extensions called **parapodia** (Gr. *para*, beside + *podion*, little foot) (figure 10.5). Parapodia are supported by chitinous rods, and numerous setae project from the parapodia, giving them their class name. **Setae** (L. *saeta*, bristle) are bristles that are secreted from invaginations of the distal ends of parapodia. They aid locomotion by digging into the substrate and are also used to hold a worm in its burrow or tube.

The **prostomium** (Gr. *pro*, before + *stoma*, mouth) of a polychaete is a lobe that projects dorsally and anteriorly to the mouth and contains numerous sensory structures, including eyes, antennae, palps, and ciliated pits or grooves, called nuchal organs. The first body segment, the **peristomium** (Gr. *peri*, around), surrounds the mouth and bears sensory tentacles or cirri.

The epidermis of polychaetes consists of a single layer of columnar cells that secrete a protective, nonliving **cuticle.** Some polychaetes have epidermal glands that secrete luminescent compounds.

Various species of polychaetes are capable of walking, fast crawling, or swimming. To do so, the longitudinal muscles on one side of the body act antagonistically to the longitudinal muscles on the other side of the body so that undulatory waves move along the length of the body from the posterior end toward the head. The propulsive force is the result of parapodia and setae acting against the substrate or water. Parapodia on opposite sides of the body are out of phase with one another. When longitudinal muscles on one side of a segment contract, the parapodial muscles on that side also contract, stiffening the parapodium and protruding the setae for the power stroke (figure 10.6*a*). As a polychaete changes from a slow crawl to swimming, the period and amplitude of undulatory waves increase (figure 10.6*b*).

Burrowing polychaetes push their way through sand and mud by contractions of the body wall or by eating their way through the substrate. In the latter, organic matter in the substrate is digested and absorbed and undigestible materials are eliminated via the anus.

Feeding and the Digestive System

The digestive tract of polychaetes is a straight tube and is suspended in the body cavity by mesenteries and septa. The anterior region of the digestive tract is modified into a proboscis, which can be everted through the mouth by special protractor muscles and coelomic pressure. Retractor muscles bring the proboscis back into the peristomium. In some, when the proboscis is everted, paired jaws are opened and may be used for seizing prey. Predatory polychaetes may not leave their burrow or coral crevice. When prey approaches a burrow entrance, the anterior portion of the worm is quickly extended, the proboscis is everted, and the prey is pulled back into the burrow. Some polychaetes have poison glands at the base of the jaw. Other polychaetes are herbivores and scavengers and use jaws for tearing food. Deposit-feeding polychaetes (e.g., *Arenicola*, the lugworm) extract organic matter from the marine sediments they ingest. The digestive tract consists of a pharynx that

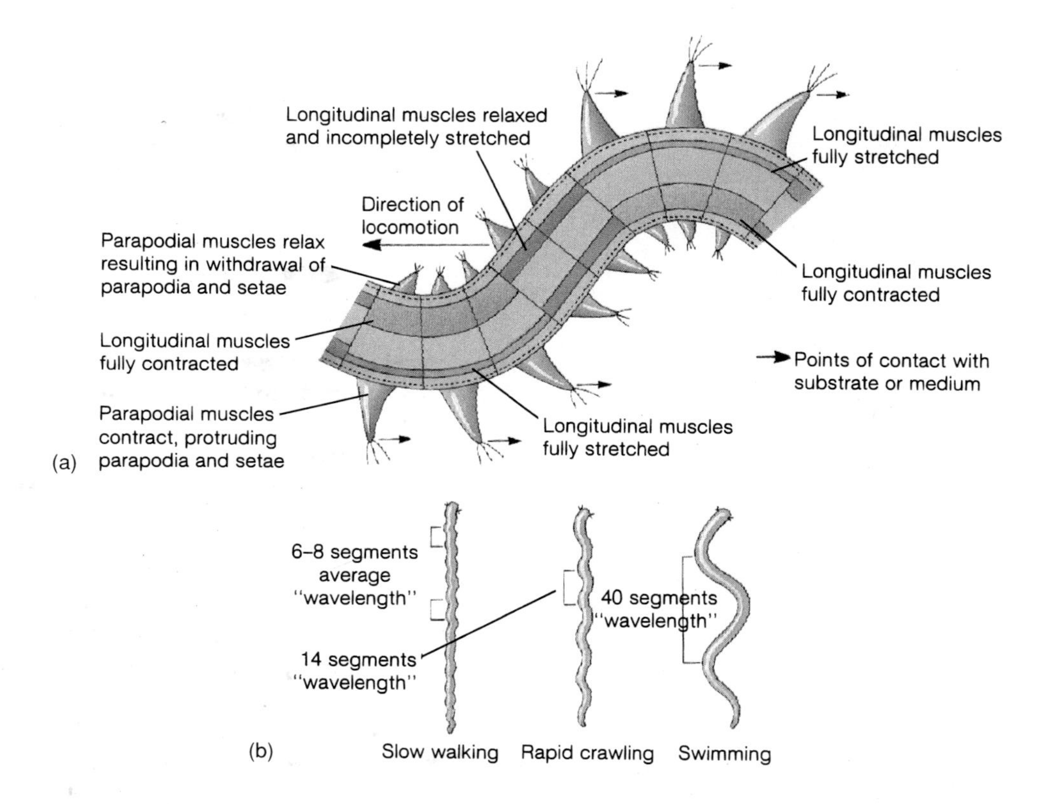

Figure 10.6

Polychaete Locomotion. (*a*) Dorsal view of a primitive polychaete showing the antagonism of longitudinal muscles on opposite sides of the body and the resultant protrusion and movement of parapodia. (*b*) Both the period and amplitude of locomotor waves increase as a polychaete changes from a "slow walk" to a swimming mode. *From:* "A LIFE OF INVERTEBRATES" © *1979 W. D. Russell-Hunter.*

when everted, forms the proboscis; a storage sac, called a crop; a grinding gizzard; and a long, straight intestine. These are similar to digestive organs of earthworms (*see figure 10.13*). Organic matter is digested extracellularly, and the inorganic particles are passed through the intestine and released as "castings."

Many sedentary and tube-dwelling polychaetes are filter feeders. They usually lack a proboscis but possess other specialized feeding structures. Some tube dwellers, called fanworms, possess radioles that form a funnel-shaped fan. (3) Cilia on the radioles circulate water through the fan, trapping food particles. Trapped particles are carried along a food groove at the axis of the radiole. During transport, a sorting mechanism rejects the largest particles and transports the finest particles to the mouth. Another filter feeder, *Chaetopterus,* lives in a U-shaped tube and secretes a mucous bag that collects food particles, which may be as small as 1 micrometer (μm). The parapodia of segments 14 through 16 are modified into fans that create filtration currents. When full, the entire mucous bag is ingested.

Elimination of digestive waste products can be a problem for tube-dwelling polychaetes. Those that live in tubes that open at both ends simply have wastes carried away by water circulating through the tube. Those that live in tubes that open at one end must turn around in the tube to defecate, or they may use ciliary tracts along the body wall to carry feces to the tube opening.

Polychaetes that inhabit substrates rich in dissolved organic matter can absorb as much as 20 to 40% of their energy requirements across their body wall as sugars and other organic compounds. This method of feeding occurs in other animal phyla too, but rarely accounts for more than 1% of their energy needs.

Gas Exchange and Circulation

Respiratory gases of most annelids simply diffuse across the body wall, and parapodia increase the surface area for these exchanges. In many polychaetes, the surface area for gas exchange is further increased by parapodial gills.

The circulatory system of polychaetes is a closed system. Oxygen is usually carried in combination with molecules called respiratory pigments, which are usually dissolved in the plasma rather than contained in blood cells. (4) Blood may be

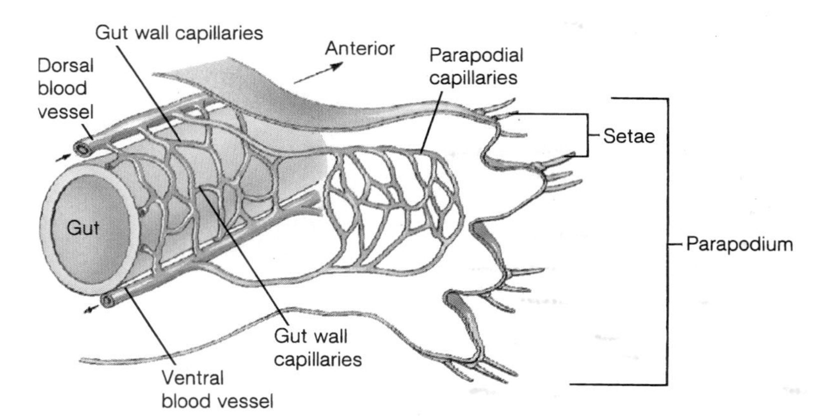

FIGURE 10.7
The Circulatory System of a Polychaete. In the closed circulatory system shown here, blood passes posterior to anterior in the dorsal vessel and anterior to posterior in the ventral vessel. Dorsal and ventral vessels are interconnected by capillary beds.

colorless, green, or red, depending on the presence and/or type of respiratory pigment.

Contractile elements of polychaete circulatory systems consist of a dorsal aorta that lies just above the digestive tract and propels blood from rear to front, and a ventral aorta that lies ventral to the digestive tract and propels blood from front to rear. Running between these two vessels are two or three sets of segmental vessels that receive blood from the ventral aorta and break into capillary beds in the gut and body wall. Capillaries coalesce again into segmental vessels that deliver blood to the dorsal aorta (figure 10.7).

NERVOUS AND SENSORY FUNCTIONS

Nervous systems are similar in all three classes of annelids. The annelid nervous system includes a pair of suprapharyngeal ganglia, which are connected to a pair of subpharyngeal ganglia by circumpharyngeal connectives that run dorsoventrally along either side of the pharynx. A double ventral nerve cord runs the length of the worm along the ventral margin of each coelomic space, and there is a paired segmental ganglion in each segment. The double ventral nerve cord and paired segmental ganglia may be fused to varying extents in different taxonomic groups. Lateral nerves emerge from each segmental ganglion, supplying the body-wall musculature and other structures of that segment (figure 10.8*a*).

Segmental ganglia are responsible for coordinating swimming and crawling movements in isolated segments. (Anyone who has used portions of worms as live fish bait can confirm that the head end—with the pharyngeal ganglia—is not necessary for coordinated movements.) Each segment acts separately from, but is closely coordinated with, neighboring segments. The subpharyngeal ganglia help mediate locomotor functions requiring coordination of distant segments. The suprapharyngeal ganglia probably control motor and sensory functions involved with feeding, and sensory functions associated with forward locomotion.

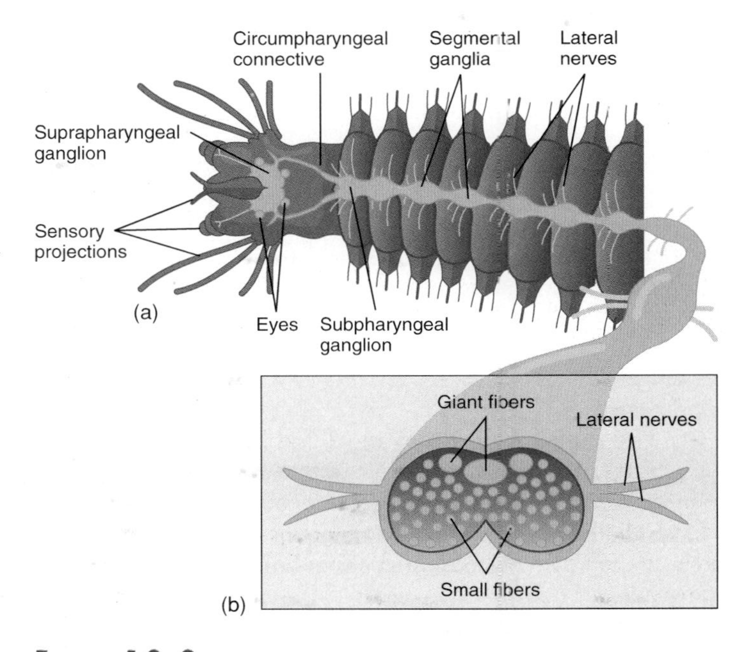

FIGURE 10.8
The Nervous System of a Polychaete. (*a*) Suprapharyngeal and subpharyngeal ganglia are linked by connectives. Segmental ganglia and lateral nerves occur along the length of the worm. (*b*) Cross section of the ventral nerve cord showing giant fibers.

In addition to small-diameter fibers that help coordinate locomotion, the ventral nerve cord also contains giant fibers (figure 10.8*b*). Annelid giant fibers are involved with escape reactions. For example, a harsh stimulus at one end of a worm, such as a fishhook, causes a very rapid withdrawal from the stimulus. Giant fibers are approximately 50 μm in diameter and conduct nerve impulses at 30 m/second (as opposed to 0.5 m/second in the smaller, 4 μm diameter annelid fibers).

Polychaetes have various sensory structures. Two to four pairs of eyes occur on the surface of the prostomium. They vary in complexity from a simple cup of receptor cells to structures made up of a cornea, lens, and vitreous body. Most polychaetes react negatively to increased light intensities. Fanworms, however, react negatively to decreasing light intensities. If shadows cross them, fanworms retreat into their tubes. This response is believed to help protect fanworms from passing predators. Nuchal organs are pairs of ciliated sensory pits or slits in the head region. They are innervated by nerves from the suprapharyngeal ganglia and are thought to be chemoreceptors for food detection. Statocysts are found in the head region of polychaetes, and the body wall is covered by ciliated tubercles, ridges, and bands, all of which contain receptors for tactile senses.

Excretion

Annelids excrete ammonia, and because ammonia diffuses readily into the water, most nitrogen excretion probably occurs across the body wall. Excretory organs of annelids are more active in regulating water and ion balances, although these abilities are limited. Most marine polychaetes, if presented with extremely diluted seawater, cannot survive the osmotic influx of water and the loss of ions that results. The evolution of efficient osmoregulatory abilities has allowed only a few polychaetes to invade fresh water.

The excretory organs of annelids, like those of many invertebrates, are called nephridia. Two types of nephridia are found in annelids. A protonephridium consists of a tubule with a closed bulb at one end and a connection to the outside of the body at the other end. Protonephridia have a tuft of flagella in their bulbular end that drives fluids through the tubule (figure 10.9*a*; *see also figure 7.6*). Some primitive polychaetes possess paired, segmentally arranged protonephridia that have their bulbular end projecting through the anterior septum into an adjacent segment and the opposite end opening through the body wall at a nephridiopore.

Most polychaetes possess a second kind of nephridium, called a metanephridium. A **metanephridium** consists of an open, ciliated funnel, called a nephrostome, that projects through an anterior septum into the coelom of an adjacent segment. At the opposite end, a tubule opens through the body wall at a nephridiopore or occasionally through the intestine (figure 10.9*b,c*). There is usually one pair of metanephridia per segment, and tubules may be extensively coiled, with one portion dilated into a bladder. A capillary bed is usually associated with the tubule of a metanephridium for active transport of ions between the blood and the nephridium (figure 10.9*d*).

Some polychaetes also have chloragogen tissue associated with the digestive tract. This tissue functions in amino acid metabolism in all annelids and will be described further in a later section.

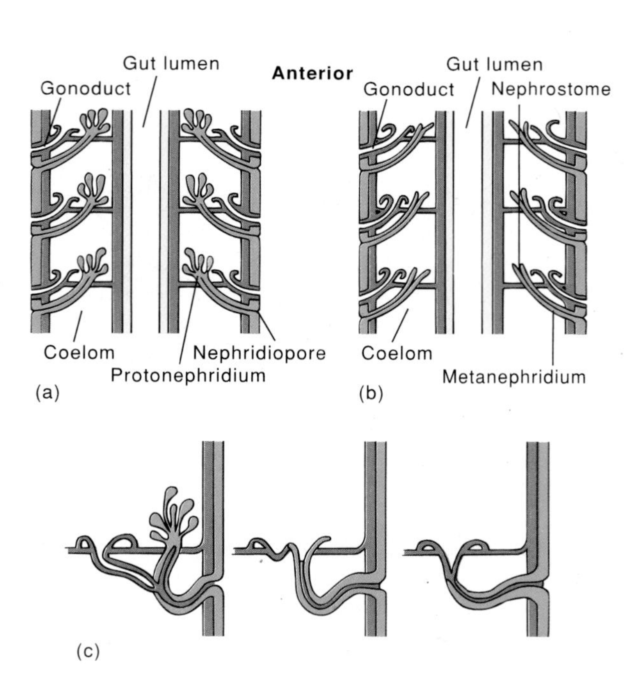

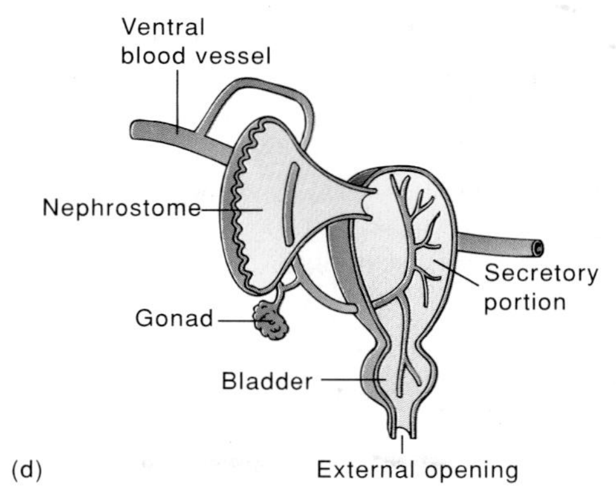

Figure 10.9

Annelid Nephridia. (*a*) The protonephridium. The bulbular ends of this nephridium contain a tuft of flagella that drives wastes to the outside of the body. In primitive polychaetes, a gonoduct (coelomoduct) carries reproductive products to the outside of the body. (*b*) The metanephridium. An open ciliated funnel (the nephrostome) drives wastes to the outside of the body. (*c*) In modern annelids, the gonoduct and the nephridial tubules undergo varying degrees of fusion. (*d*) Nephridia of modern annelids are closely associated with capillary beds for secretion, and nephridial tubules may have an enlarged bladder. *From:* "A LIFE OF INVERTEBRATES" © *1979 W. D. Russell-Hunter.*

Regeneration, Reproduction, and Development

All polychaetes have remarkable powers of regeneration. They can replace lost parts, and some species have break points that allow worms to sever themselves when grabbed by a predator. Lost segments are later regenerated.

Some polychaetes reproduce asexually by budding or by transverse fission; however, sexual reproduction is much more common. Most polychaetes are dioecious. Gonads develop as masses of gametes and project from the coelomic peritoneum. Primitively, gonads occur in every body segment, but most polychaetes have gonads limited to specific segments. Gametes are shed into the coelom where they mature. Mature female worms are often packed with eggs. Gametes may exit worms by entering nephrostomes of metanephridia and exiting through the nephridiopore, or they may be released, in some polychaetes, after the worm ruptures. In these cases, the adult soon dies. Only a few polychaetes have separate gonoducts, a condition that is believed to be primitive (*see figure 10.9a–c*).

Fertilization is external in most polychaetes, although copulation occurs in a few species. One of the most unique copulatory habits has been reported in *Platynereis megalops* from Woods Hole, Massachusetts. Toward the end of their lives, male and female worms cease feeding, and their intestinal tracts begin to degenerate. At this time, gametes have accumulated in the body cavity. During sperm transfer, male and female worms coil together, and the male inserts his anus into the mouth of the female. Because the digestive tracts of the worms have degenerated, sperm are transferred directly from the male's coelom to the egg-filled coelom of the female. This method ensures fertilization of most eggs, and after fertilization is accomplished, the female sheds eggs from her anus. Both worms die soon after copulation.

Epitoky is the formation of a reproductive individual (an epitoke) that differs from the nonreproductive form of the species (an atoke). Frequently, an epitoke has a body that is modified into two body regions. Anterior segments carry on normal maintenance functions, and posterior segments are enlarged and filled with gametes. The epitoke may have modified parapodia for more efficient swimming.

At the beginning of this chapter, there was an account of the reproductive swarming habits of *Eunice viridis* (the Samoan palolo worm), and one culture's response to those swarms. Similar swarming occurs in other species, usually in response to changing light intensities and lunar periods. The Atlantic palolo worm, for example, swarms at dawn during the first and third quarters of the July lunar cycle.

5 Swarming of epitokes is believed to accomplish at least three things. First, because nonreproductive individuals remain safe below the surface waters, predators cannot devastate an entire population. Second, external fertilization requires that individuals become reproductively active at the same time and in close proximity to one another. Swarming ensures that large numbers of individuals will be in the right place at the proper time. Third, swarming of vast numbers of individuals for brief periods provides a banquet for predators. However, because vast numbers of prey are available for only short periods during the year, predator populations cannot increase beyond the limits of their normal diets. Therefore, predators can dine gluttonously and still leave epitokes that will yield the next generation of animals.

Spiral cleavage of fertilized eggs may result in planktonic trochophore larvae that bud segments anterior to the anus. Larvae eventually settle to the substrate (figure 10.10). As growth proceeds, newer segments continue to be added posteriorly. Thus, the anterior end of a polychaete is the oldest end. Many other polychaetes lack a trochophore and display direct development or metamorphosis from another larval stage.

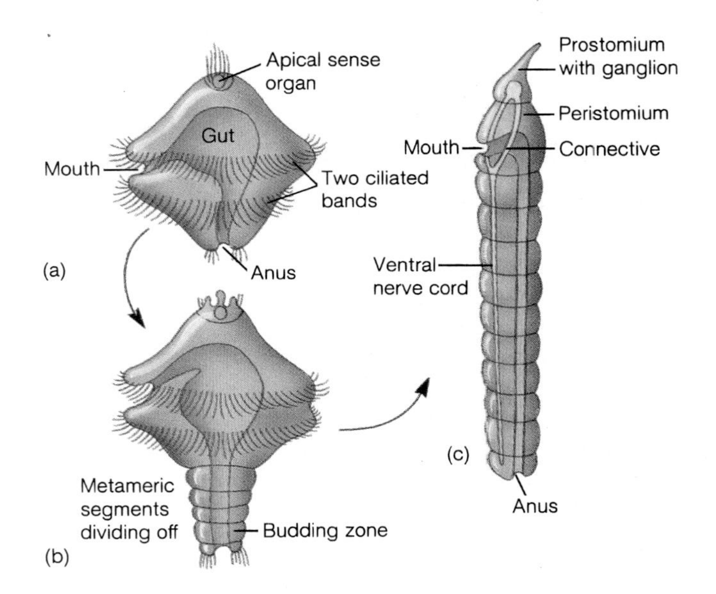

Figure 10.10

Polychaete Development. (*a*) A trochophore. (*b*) A later planktonic larva showing the development of body segments. As more segments develop, the larva will settle to the substrate. (*c*) A juvenile worm.
From: "A LIFE OF INVERTEBRATES" © *1979 W. D. Russell-Hunter.*

Stop and Ask Yourself

5. How is the body wall of a polychaete involved with crawling movement?
6. What is the function of giant nerve fibers in polychaetes?
7. What is epitoky? Why is it advantageous for some polychaetes?

BOX 10.1 Soil Conditioning by Earthworms

Earthworms have had an inestimable impact on the development of our planet's soil. For the past 100 million years earthworms have evolved with deciduous vegetation. Leaf fall and the death of land plants have provided a massive food source for earthworms and other soil-inhabiting organisms. The excrement, death, and decay of these soil organisms build the organic constituents of our soils, and burrowing by earthworms aerates the soil and improves drainage.

In typical grassland and woodland soils, earthworms reach populations of hundreds of animals per square meter. They often dominate the invertebrate biomass (the total mass of invertebrate animals) in a region. Earthworms ingest soil as they feed on organic matter and as they burrow through the soil. Charles Darwin estimated that 15 tons of soil per acre per year passed through earthworm bodies. Recent, more accurate estimates of earthworm populations, give even more impressive tillage figures of nearly 40 tons per acre per year!

Earthworms are surface feeders that emerge at night to feed on leaf fragments and other plant debris. Some of these fragments are ingested immediately, others are carried into burrows. Of the plant matter ingested by earthworms, less than 10% is incorporated into worm tissues. The rest passes through the digestive tract, is incorporated into castings (fecal material and soil), and released deeper in the soil. Earthworm castings are also rich in ammonia, which is a form of nitrogen usable by some plants.

Earthworms function as vegetation shredders in our soils. Shredding vegetation and incorporating it into fecal material increases the surface area of plant matter by several orders of magnitude and hastens its eventual decomposition by bacteria and fungi. One study demonstrated the role of shredders in the breakdown of plant litter by soil animals. Leaf-filled nylon bags were buried in the soil. Some bags had a 0.5 mm mesh size, which excluded all earthworms and other large invertebrates. Other bags had a mesh size of 7 mm, which allowed all invertebrates to enter. The rate of breakdown of leaf litter was reduced by approximately two-thirds in the small mesh bags.

Class Oligochaeta

The class Oligochaeta (ol″i-go-ket′ah) has over 3,000 species that are found throughout the world in freshwater and terrestrial habitats (*see table 10.1*). A few oligochaetes are estuarine, and some are marine. Aquatic species live in shallow water, where they burrow in mud and debris. Terrestrial species are found in soils with high organic content, and these species rarely leave their burrows. In hot, dry weather they may retreat to depths of 3 m below the surface. Soil-conditioning habits of earthworms are well known (box 10.1). *Lumbricus terrestris* is commonly used in zoology laboratories because of its large size. It was introduced to the United States from northern Europe and has flourished. Common native species like *Eisenia foetida* and various species of *Allolobophora* are smaller.

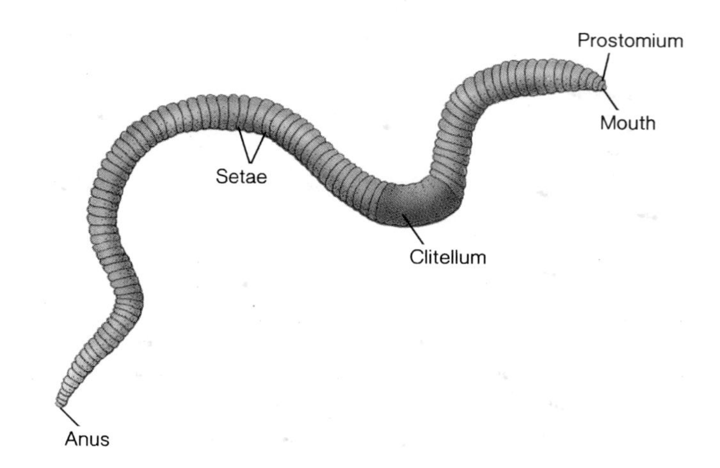

Figure 10.11
Class Annelida. External structures of the earthworm, *Lumbricus terrestris*.

External Structure and Locomotion

Oligochaetes (Gr. *oligos*, few + *chaite*, hair) have setae, but fewer than are found in polychaetes (thus the derivation of the class name). Oligochaetes lack parapodia, because parapodia and long setae would interfere with their burrowing life-styles, although they do have short setae on their integument. The prostomium consists of a small lobe or cone in front of the mouth and lacks sensory appendages. A series of segments in the anterior half of an oligochaete is usually swollen into a girdlelike structure called the **clitellum** that is used for mucous secretion during copulation and cocoon formation (figure 10.11). As in the polychaetes, the body is covered by a nonliving, secreted cuticle.

Oligochaete locomotion involves the antagonism of circular and longitudinal muscles in groups of segments. Neurally controlled waves of contraction move from rear to front.

Segments bulge and setae protrude when longitudinal muscles are contracted, providing points of contact with the burrow wall. In front of each region of longitudinal muscle contraction, circular muscles contract, causing the setae to retract, and the segments to elongate and push forward. Contraction of longitudinal muscles in segments behind a bulging region causes those segments to be pulled forward. Thus, segments move forward relative to the burrow as waves of muscle contraction move anteriorly on the worm (figure 10.12).

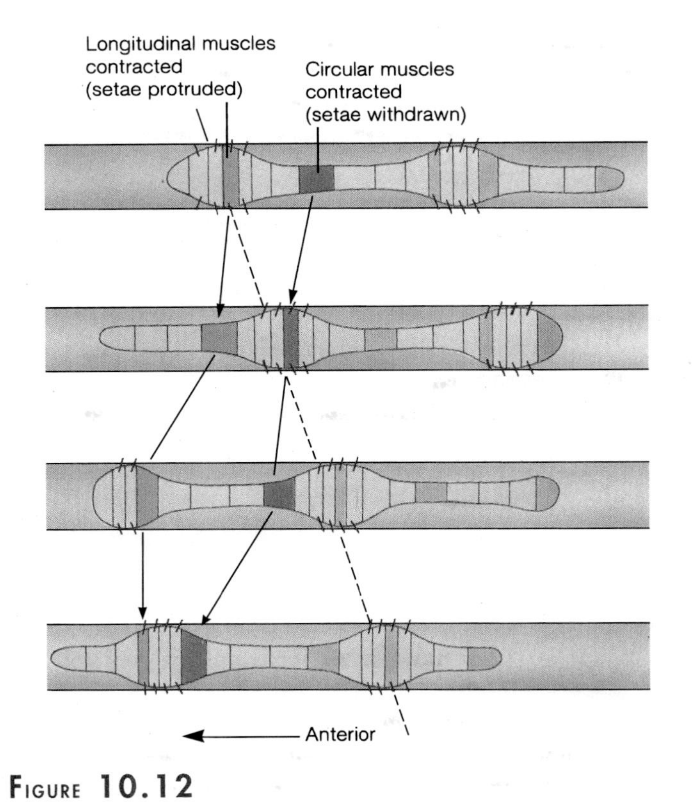

FIGURE 10.12
Earthworm Locomotion. Arrows designate activity in specific segments of the body, and the broken lines indicate regions of contact with the substrate. *From:* "A LIFE OF INVERTEBRATES" © *1979 W. D. Russell-Hunter.*

Burrowing is the result of coelomic hydrostatic pressure being transmitted toward the prostomium. As an earthworm pushes its way through the soil, it uses expanded posterior segments and protracted setae to anchor itself to its burrow wall. (6) (Any child pursuing fishing worms experiences the effectiveness of this anchor when trying to extract a worm from its burrow.) Contraction of circular muscles transforms the prostomium into a conical wedge, 1 mm in diameter at its tip. Contraction of body-wall muscles generates coelomic pressure that forces the prostomium through the soil. During burrowing, earthworms swallow considerable quantities of soil.

Feeding and the Digestive System

Oligochaetes are scavengers and feed primarily on fallen and decaying vegetation, which they drag into their burrows at night. The digestive tract of oligochaetes is tubular and straight (figure 10.13). The mouth leads to a muscular pharynx. In the earthworm, pharyngeal muscles attach to the body wall. The pharynx acts as a pump for ingesting food. The mouth pushes against food and the pharynx pumps the food into the esophagus. The esophagus is narrow and tubular and frequently is expanded to form a stomach, crop, or gizzard; the latter two are common in terrestrial species. A crop is a thin-walled storage structure, and a gizzard is a muscular, cuticle-lined grinding structure. Calciferous glands are evaginations of the esophageal wall that rid the body of excess calcium absorbed from food. Calciferous glands also have an important function in regulating the pH of body

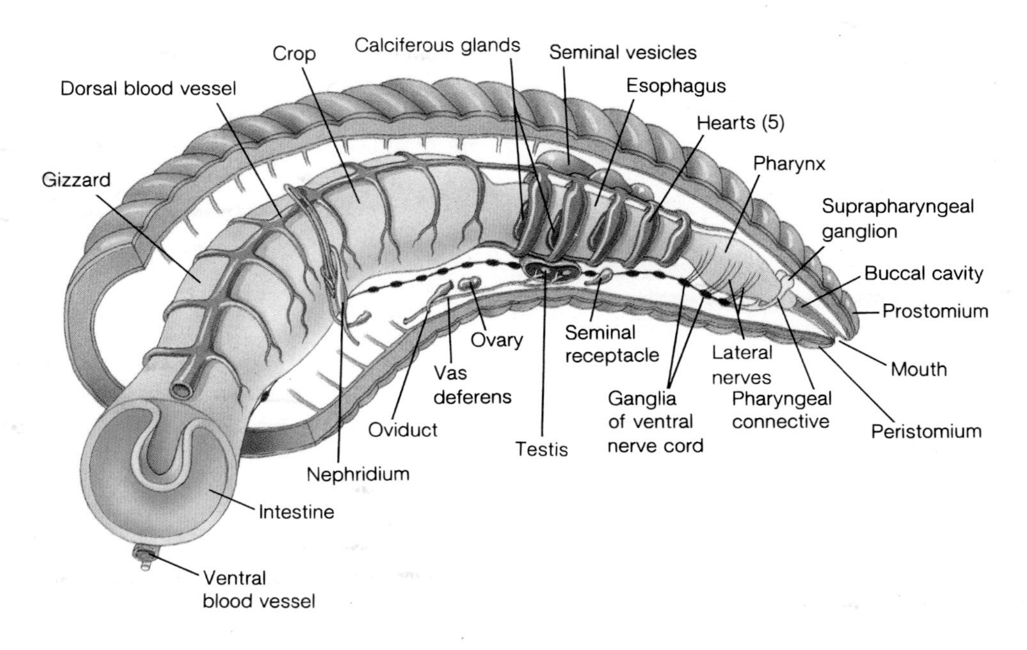

FIGURE 10.13
Earthworm Structure. This diagram shows a lateral view of the internal structures in the anterior ⅓ of an earthworm.

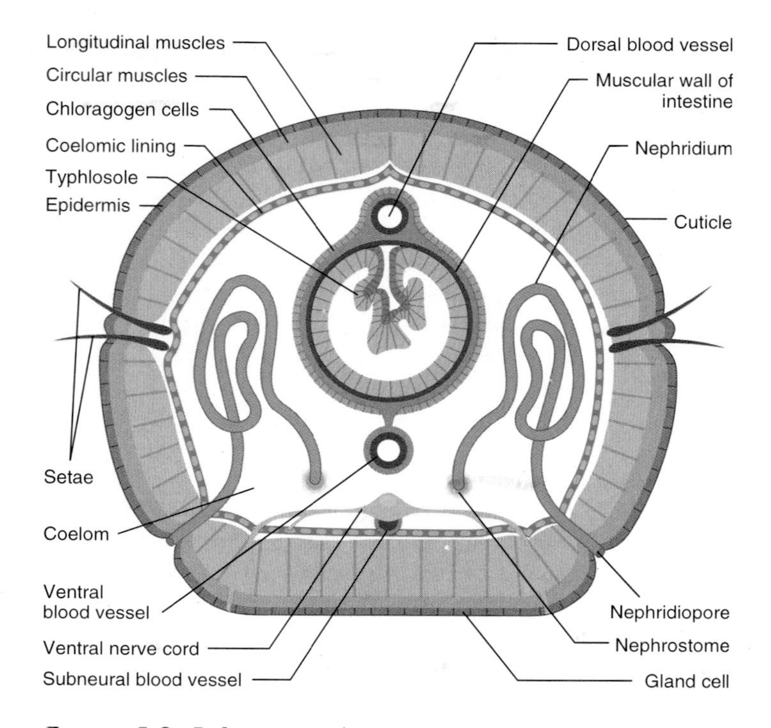

Figure 10.14

Earthworm Cross Section. The nephrostomes shown here would actually be associated with the next anterior segment.

fluids. The intestine is a straight tube and is the principal site of digestion and absorption. The surface area of the intestine is increased substantially by a dorsal fold of the lumenal epithelium called the typhlosole (figure 10.14). The intestine ends at the anus.

Gas Exchange and Circulation

Both respiratory and circulatory functions are as described for polychaetes. Some segmental vessels are expanded and may be contractile. In the earthworm, for example, expanded segmental vessels surrounding the esophagus propel blood between dorsal and ventral blood vessels and anteriorly in the ventral vessel toward the mouth. Even though these are sometimes called "hearts," the main propulsive structures are the dorsal and the ventral vessels (*see figure 10.13*). Branches from the ventral vessel supply the intestine and body wall.

Nervous and Sensory Functions

The ventral nerve cords and all ganglia of oligochaetes have undergone a high degree of fusion. Other aspects of nervous structure and function are essentially the same as those described earlier for polychaetes. As with polychaetes, giant fibers mediate escape responses. An escape response results from the stimulation of either the anterior or the posterior end of a worm. An impulse conducted to the opposite end of the worm initiates the formation of an anchor, and longitudinal muscles contract to quickly pull the worm away from the stimulus.

Oligochaetes lack well-developed eyes, which should not be surprising given their subterranean life-style. It is not unusual for animals living in perpetual darkness to be without well-developed eyes. Other oligochaetes have simple pigment-cup ocelli, and all have a "dermal light sense" that arises from photoreceptor cells scattered over the dorsal and lateral surfaces of the body. Scattered photoreceptor cells mediate a negative phototaxis in strong light (evidenced by movement away from the light source) and a positive phototaxis in weak light (evidenced by movement toward the light source).

Oligochaetes are sensitive to a wide variety of chemical and mechanical stimuli. Receptors for these stimuli are scattered over the surface of the body, especially around the prostomium.

Excretion

Oligochaetes use metanephridia for excretion and for ion and water regulation. As with polychaetes, funnels of metanephridia are associated with the segment just anterior to the segment containing the tubule and the nephridiopore. Nitrogenous wastes include ammonia and urea. Oligochaetes excrete copious amounts of very dilute urine, although they retain vital ions, which is very important for organisms living in environments where water is plentiful but essential ions are limited.

Oligochaetes (as well as other annelids) possess chloragogen tissue that surrounds the dorsal blood vessel and lies over the dorsal surface of the intestine (*see figure 10.14*). **Chloragogen tissue** acts similarly to the vertebrate liver. It is a site of amino acid metabolism. Deamination of amino acids, and the conversion of ammonia to urea occurs there. Chloragogen tissue also converts excess carbohydrates into energy-storage molecules of glycogen and fat.

Reproduction and Development

All oligochaetes are monoecious, and mutual sperm exchange occurs during copulation. One or two pairs of testes and one pair of ovaries are located on the anterior septum of certain anterior segments. Both the sperm ducts and the oviducts have ciliated funnels at their proximal ends to draw gametes into their respective tubes.

Testes are closely associated with three pairs of **seminal vesicles,** which are sites for maturation and storage of sperm prior to their release. **Seminal receptacles** receive sperm during copulation. A pair of very small ovisacs, associated with oviducts, are sites for the maturation and storage of eggs prior to egg release (figure 10.15).

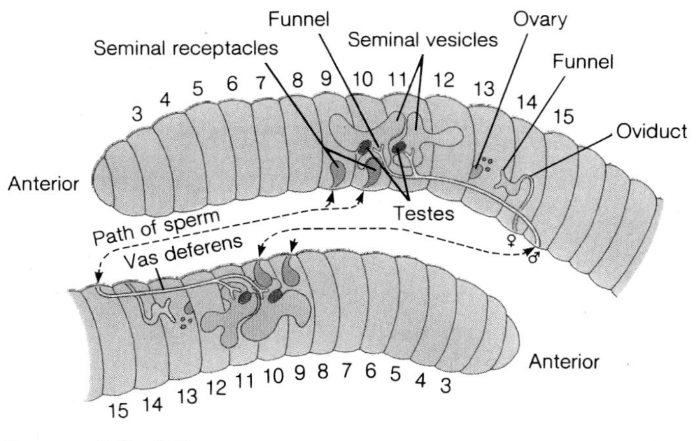

Figure 10.15
Earthworm Reproduction. Mating earthworms showing arrangements of reproductive structures and the path taken by sperm during sperm exchange (shown by arrows).

During copulation of *Lumbricus*, two worms line up facing in opposite directions, with the ventral surfaces of their anterior ends in contact with each other. This orientation lines up the clitellum of one worm with the genital segments of the other worm. Worms are held in place by a mucous sheath, secreted by the clitellum, that envelopes the anterior halves of both worms. Some species also have penile structures and genital setae that help maintain contact between worms. In *Lumbricus*, sperm are released from the sperm duct and travel along the external, ventral body wall in sperm grooves formed by the contraction of special muscles. Muscular contractions along this groove help propel sperm toward the openings of the seminal receptacles. In other oligochaetes, copulation results in the alignment of sperm duct and seminal receptacle openings and transfer of sperm is direct. Copulation lasts 2 to 3 hours, during which both worms give and receive sperm.

Following copulation, the clitellum forms a cocoon for the deposition of eggs and sperm. The cocoon consists of mucoid and chitinous materials that encircle the clitellum. A food reserve, albumen, is secreted into the cocoon by the clitellum, and the worm begins to back out of the cocoon. Eggs are deposited in the cocoon as the cocoon passes the openings to the oviducts, and sperm are released as the cocoon passes the openings to the seminal receptacles. Fertilization occurs in the cocoon and, as the worm continues backing out, the ends of the cocoon are sealed, and the cocoon is deposited in moist soil.

Spiral cleavage is modified, and no larva is formed. Hatching occurs in one to a few weeks, depending on the species, when young worms emerge from one end of the cocoon.

Asexual reproduction also occurs in freshwater oligochaetes. It consists of transverse division of a worm, followed by regeneration of missing segments.

Stop and Ask Yourself

8. How does an earthworm move across a substrate? How does an earthworm burrow?
9. What is the function of calciferous glands? The typhlosole? Chloragogen tissue?
10. How would you describe the method of sperm transfer and egg deposition in earthworms?
11. In what way is development of an oligochaete different from that of a polychaete?

Class Hirudinea

The class Hirudinea (hi′roo-din″eah) (L. *hirudin*, leech) contains approximately 500 species of leeches (*see table 10.1*). Most leeches are fresh water; others are marine, or completely terrestrial. Leeches prey on small invertebrates or feed on body fluids of vertebrates (box 10.2).

External Structure and Locomotion

Leeches lack parapodia and head appendages. Setae are absent in most leeches. In a few species, setae occur only on anterior segments. Leeches are dorsoventrally flattened and taper anteriorly. They have 34 segments, but the segments are difficult to distinguish externally because they have become secondarily divided. Several secondary divisions, called **annuli,** are in each true segment. Anterior and posterior segments are usually modified into suckers (figure 10.16).

Modifications of body-wall musculature and the coelom influence patterns of locomotion in the leeches. The musculature of leeches is more complex than that of other annelids. A layer of oblique muscles is present between the circular and longitudinal muscle layers. In addition, dorsoventral muscles are responsible for the typical leech flattening. The coelom of leeches has lost its metameric partitioning. Septa are lost, and the coelom has been invaded by connective tissue, resulting in a series of interconnecting sinuses.

These modifications have resulted in altered patterns of locomotion. Rather than being able to utilize independent coelomic compartments, the leech has a single hydrostatic cavity and uses it in a looping type of locomotion. The mechanics of this locomotion are described in figure 10.17. Leeches also swim using undulations of the body.

Feeding and the Digestive System

Many leeches feed on body fluids or the entire bodies of other invertebrates. Some feed on blood of vertebrates, including human blood. (7) Leeches are sometimes called parasites; however, the association between a leech and its host is relatively brief.

BOX 10.2 Leeches and Science

At first mention of the word leech, most thoughts turn to the medicinal leech, *Hirudo medicinalis* (figure 1), and the practice of bloodletting. In past centuries, various illnesses have been attributed to "bad blood," and the practice of bloodletting was common. Medicinal leeches were used in bloodletting because when they feed, they ingest seven to eight times their weight in blood—one of the biggest meals in the animal kingdom. After such a meal, the leech may not feed again for a year.

The use of leeches in bloodletting during the second century is documented in the writings of Galen, an early Greek physician. The practice remained very common through the early nineteenth century. The medicinal leech and medicine became almost synonymous. Physicians themselves were sometimes (respectfully) referred to as "leeches." During the late nineteenth and early twentieth century, the use of leeches in medicine declined. Now, however, the medicinal leech is again being used in medicine to remove excess blood from tissues after plastic surgery, or after the reattachment of amputated appendages. If the excess blood is not removed, blood accumulating in tissues of postoperative patients often slows the regrowth of capillary beds and can cause tissues to die.

Leeches are also being used to investigate the physiology of nervous systems. The fact that the nervous system of a leech is simpler than that of many other animals that show system-level organization, and the fact that the nerve cells of all animals share similar physiological properties, make leeches ideal subjects for research into animal nervous systems. Certain chemicals, such as serotonin, are secreted from leech nerve cells and have been found to help regulate all aspects of leech feeding—from finding prey to ingesting blood. Organisms such as leeches may hold the keys to discovering the roles of essentially identical chemicals in the nervous systems of other animals.

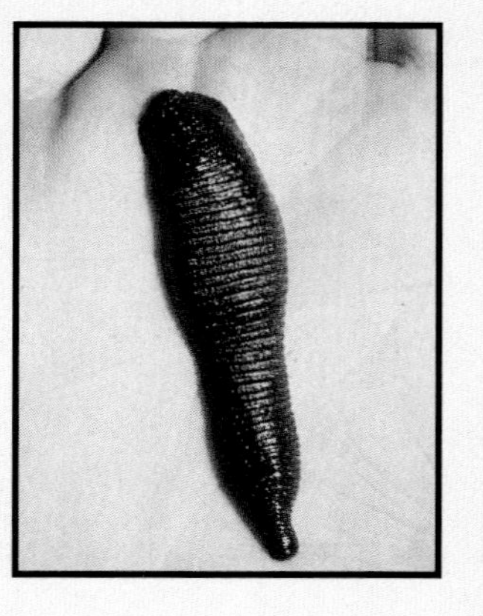

Figure 1 **The Medicinal Leech, *Hirudo medicinalis*.**

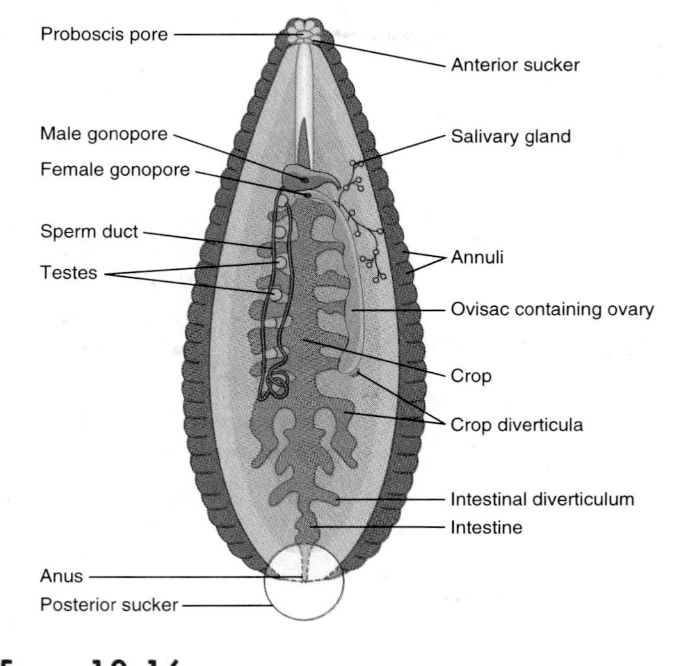

Figure 10.16
Internal Structure of a Leech. Each true segment is subdivided by annuli and the coelom is not subdivided by septa.

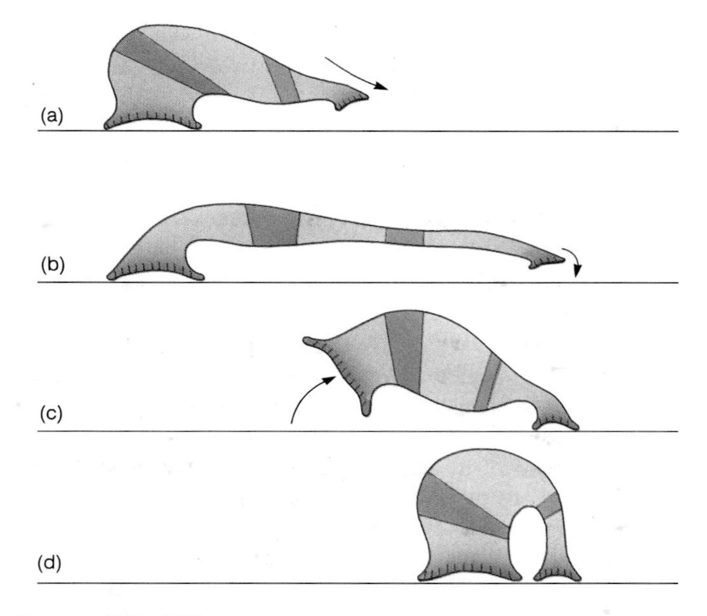

Figure 10.17
Leech Locomotion. (*a,b*) Attachment of the posterior sucker causes reflexive release of the anterior sucker, contraction of circular muscles, and relaxation of longitudinal muscles. This muscular activity compresses fluids in the single hydrostatic compartment, and the leech extends. (*c,d*) Attachment of the anterior sucker causes reflexive release of the posterior sucker, the relaxation of circular muscles, and the contraction of longitudinal muscles, causing body fluids to expand the diameter of the leech. The leech shortens, and the posterior sucker again attaches. *From:* "A LIFE OF INVERTEBRATES" © *1979 W. D. Russell-Hunter.*

Therefore, describing leeches as predatory is probably more accurate. Leeches are also not species specific, as are most parasites. (Leeches are, however, class specific. That is, a leech that preys upon a turtle may also prey on an alligator, but probably would not prey on a fish or a frog.)

The mouth of a leech opens in the middle of the anterior sucker. In some leeches, the anterior digestive tract is modified into a protrusible proboscis, lined inside and outside by a cuticle. In others, the mouth is armed with three chitinous jaws. While feeding, a leech attaches to its prey by the anterior sucker and either extends its proboscis into the prey or uses its jaws to slice through host tissues. Salivary glands secrete an anticoagulant called "hirudin" that prevents blood from clotting.

Behind the mouth is a muscular pharynx that pumps body fluids of the prey into the leech. The esophagus follows the pharynx and leads to a large stomach with lateral cecae. Most leeches ingest large quantities of blood or other body fluids and gorge their stomachs and lateral cecae, increasing their body mass 2 to 10 times. After engorgement, a leech can tolerate periods of fasting that may last for months. The digestive tract ends in a short intestine and anus (*see figure 10.16*).

Gas Exchange and Circulation

Gas exchange occurs across the body wall. The basic annelid circulatory pattern is retained in some leeches, but in most it is highly modified, and vessels are replaced by coelomic sinuses. Coelomic fluid has taken over the function of blood and, except in two orders, respiratory pigments are lacking.

Nervous and Sensory Functions

The nervous system of leeches is similar to that of other annelids. Ventral nerve cords are unfused, except at the ganglia. The suprapharyngeal and subpharyngeal ganglia and the pharyngeal connectives are all fused into a nerve ring that surrounds the pharynx. There is also a similar fusion of ganglia at the posterior end of the animal.

A variety of epidermal sense organs are widely scattered over the body. Most leeches have photoreceptor cells located in pigment cups (2 to 10) along the dorsal surface of the anterior segments. Normally, leeches are negatively phototactic, but when searching for food, the behavior of some leeches changes, and they become positively phototactic, which increases the likelihood of contacting prey that happens to pass by.

Hirudo medicinalis, the medicinal leech, has a well-developed temperature sense. This sense helps the leech detect the higher body temperature of its mammalian prey. Other leeches are attracted to extracts of prey tissues.

All leeches have sensory cells with terminal bristles in a row along the middle annulus of each segment. These sensory cells, called sensory papillae, are of uncertain function but are taxonomically important.

Excretion

Leeches have 10 to 17 pairs of metanephridia, 1 per segment in the middle segments of the body. Their metanephridia are highly modified and possess, in addition to the nephrostome and tubule, a capsule that is believed to be involved with the production of coelomic fluid. Chloragogen tissue is proliferated through the body cavity of most leeches.

Reproduction and Development

All leeches reproduce sexually and are monoecious. None are capable of asexual reproduction or regeneration. They have a single pair of ovaries and from four to many testes. Leeches have a clitellum that includes three body segments. The clitellum can be seen only in the spring when most leeches breed.

Sperm transfer and egg deposition usually occur in the same manner as described for oligochaetes. A penis assists the transfer of sperm between individuals. A few transfer sperm by expelling a spermatophore from one leech into the integument of another, a form of hypodermic impregnation. Special tissues within the integument connect to the ovaries by short ducts. Cocoons are deposited in the soil or are attached to underwater objects. There are no larval stages, and the offspring are mature by the following spring.

Further Phylogenetic Considerations

Although the origins of the phylum as a whole are speculative and somewhat controversial, the evolutionary relationships among members of the three annelid classes are as clear as those in any other phylum of animals. Polychaetes are the most primitive of the three annelid classes, as evidenced by basic metamerism, spiral cleavage, and trochophore larval stages in some species. (Some zoologists contend that the occurrence of the latter is too variable to be considered a part of the evidence of the ancestral status of the class.) Adaptive radiation of polychaetes has resulted in the great diversity of modern polychaetes.

Some members of the ancient annelid stock invaded fresh waters, which required the ability to regulate the salt and water content of body fluids. It was from this group that the oligochaetes probably evolved. Initially, oligochaetes were strictly fresh water, and many remain in that habitat; however, during the Cretaceous period, approximately 100 million years ago, oligochaetes invaded

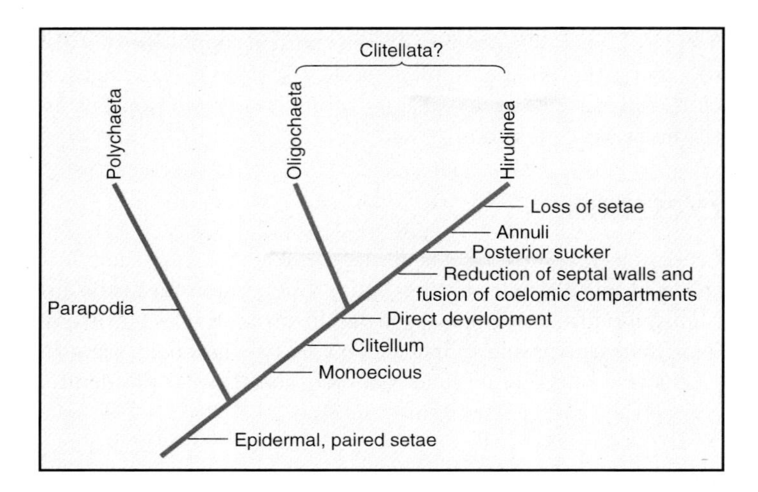

FIGURE 10.18

Annelid Phylogeny. A cladogram illustrating the evolutionary relationships of the three annelid classes. The ancestors of the annelids and arthropods were metameric coelomate animals in the protostome lineage. Paired epidermal setae are diagnostic of the phylum. The polychaetes were the first modern annelids to be derived from the ancestral annelids. Some zoologists believe that these animals should be grouped into a single class, Clitellata. (Note the question mark in the cladogram.) The oligochaetes and leeches were derived from a second major lineage of annelids. Note that the oligochaetes are distinguished from the leeches by the absence of derived characters.

moist, terrestrial environments. This period saw the climax of the giant land reptiles, but, more importantly, it was a time of proliferation of flowering land plants. (8) The reliance of modern earthworms on deciduous vegetation can be traced back to the exploitation of this food source by their ancestors. As described earlier, terrestrial oligochaetes deserve a large share of the credit for developing the soils of this planet (*see box 10.1*). A few oligochaetes have secondarily invaded marine environments.

Some of those early freshwater oligochaetes also gave rise to the Hirudinea. As with the oligochaetes, some leeches colonized marine habitats from fresh water.

Figure 22.18 is a cladogram showing the evolutionary relationships among the three annelid classes. The polychaetes are depicted as the first modern annelids. The oligochaetes and leeches are shown as sharing a second major lineage. There are no known derived characteristics (synapomorphies) unique to the oligochaetes. Instead, the oligochaetes are defined by the absence of leech characteristics. This fact supports the idea that the oligochaetes are ancestral to the leeches, and it also leads some taxonomists to believe that the oligochaetes and leeches should be combined into a single class—Clitellata.

Stop and Ask Yourself

12. How is basic annelid metamerism modified in leeches?
13. How are the body-wall musculature and the body cavity of a leech used in locomotion?
14. Why is it more accurate to call leeches predators than parasites?
15. What are the phylogenetic relationships of the annelid classes?

Summary

1. The origin of the Annelida is largely unknown. A diagnostic characteristic of the annelids is metamerism.
2. Metamerism allows efficient utilization of separate coelomic compartments as a hydrostatic skeleton for support and movement. Metamerism also lessens the impact of injury and makes tagmatization possible.
3. Members of the class Polychaeta are mostly marine and possess parapodia with numerous setae. Locomotion of polychaetes involves the antagonism of longitudinal muscles on opposite sides of the body, which creates undulatory waves along the body wall and causes parapodia to act against the substrate.
4. Polychaetes may be predators, herbivores, scavengers, or filter feeders.
5. The nervous system of polychaetes usually consists of a pair of suprapharyngeal ganglia, subpharyngeal ganglia, and double ventral nerve cords that run the length of the worm.
6. Polychaetes have a closed circulatory system. Oxygen is carried by respiratory pigments dissolved in blood plasma.
7. Either protonephridia or metanephridia are used in excretion in polychaetes.
8. Most polychaetes are dioecious, and gonads develop from coelomic epithelium. Fertilization is usually external. Epitoky occurs in some polychaetes.
9. Development of polychaetes usually results in a planktonic trochophore larva that buds off segments near the anus.
10. The class Oligochaeta includes primarily freshwater and terrestrial annelids. Oligochaetes possess few setae and they lack a head and parapodia.
11. Oligochaetes are scavengers that feed on dead and decaying vegetation. Their digestive tract is tubular, straight, and frequently has modifications for storing and grinding food, and increasing the surface area for secretion and absorption.
12. Oligochaetes possess metanephridia. Chloragogen tissue is a site for the formation of urea from protein metabolism and synthesis and storage of glycogen and fat.
13. Oligochaetes are monoecious and exchange sperm during copulation.
14. Members of the class Hirudinea are the leeches. Complex arrangements of body-wall muscles and the loss of septa influence patterns of locomotion.
15. Leeches are predatory and feed on body fluids, the entire bodies of other invertebrates, and the blood of vertebrates.
16. Leeches are monoecious, and reproduction and development occur as in oligochaetes.
17. Ancestral annelids gave rise to three major classes. The earliest lineage gave rise to modern polychaetes. Some members of this ancient annelid stock invaded fresh water and gave rise to early freshwater oligochaetes. These oligochaetes gave rise to modern freshwater oligochaetes, terrestrial oligochaetes, and leeches.

Selected Key Terms

chloragogen tissue (*p. 172*)
clitellum (*p. 170*)
epitoky (*p. 169*)
metamerism (*p. 163*)
metanephridium (*p. 168*)
parapodia (*p. 165*)
peristomium (*p. 165*)
prostomium (*p. 165*)
tagmatization (*p. 164*)

Critical Thinking Questions

1. What evidence is there that would link the annelids and arthropods in the same evolutionary line?
2. Distinguish between a protonephridium and a metanephridium. Name a class of annelids whose members may have protonephridia. What other phylum have we studied whose members also had protonephridia? Do you think that metanephridia would be more useful for a coelomate or an acoelomate animal? Explain.
3. In what annelid groups does one see the loss of septa between coelomic compartments? What advantages does this loss give each group?
4. What are the differences in the structure of nephridia that one might expect to see in freshwater and marine annelids?
5. Very few polychaetes have invaded fresh water. Can you think of a reasonable explanation for this?

endpaper TWO

Some Lesser Known Invertebrates: Possible Annelid Relatives

Three Phyla of Likely Annelid Relatives: The Echiura, Pogonophora, and Sipuncula

The coelomate phyla Echiura, Pogonophora, and Sipuncula comprise fewer than 600 protostome species that may be annelid relatives. They probably branched off from various points along the annelid-arthropod lineage.

Phylum Echiura: The Spoon Worms

The echiurans (ek-ee-yur′ans) (Gr. *echis*, serpent + *oura*, tail) consist of about 130 species of marine animals that have a worldwide distribution. Echiurans usually live in shallow waters, where they either burrow in mud or sand, or live protected in rock crevices. The soft body is covered only by a thin cuticle. As a result, the animals keep to the safety of their burrows or crevices, even when feeding. An echiuran feeds by sweeping organic material into its spatula-shaped proboscis that contains a ciliated gutter (figure 1). The proboscis can be extended for a considerable distance, but it can never be retracted into the body. Echiurans are sometimes called spoon worms because of the spatulate nature of the proboscis. Individual echiurans are from 15 to 50 cm in length, but the extensible proboscis may increase their length up to 2 m.

All echiurans are dioecious, and sexual dimorphism is extreme in some species. The eggs or sperm do not complete their development in the single ventral gonad but are released into the coelom. After they mature, they are collected by special collecting organs and then released into the seawater, where fertilization occurs, giving rise to free-swimming trochophore larvae. The early development of echiurans is similar to that of annelids with spiral cleavage. However, later development diverges from the annelid pattern in that no segmentation occurs.

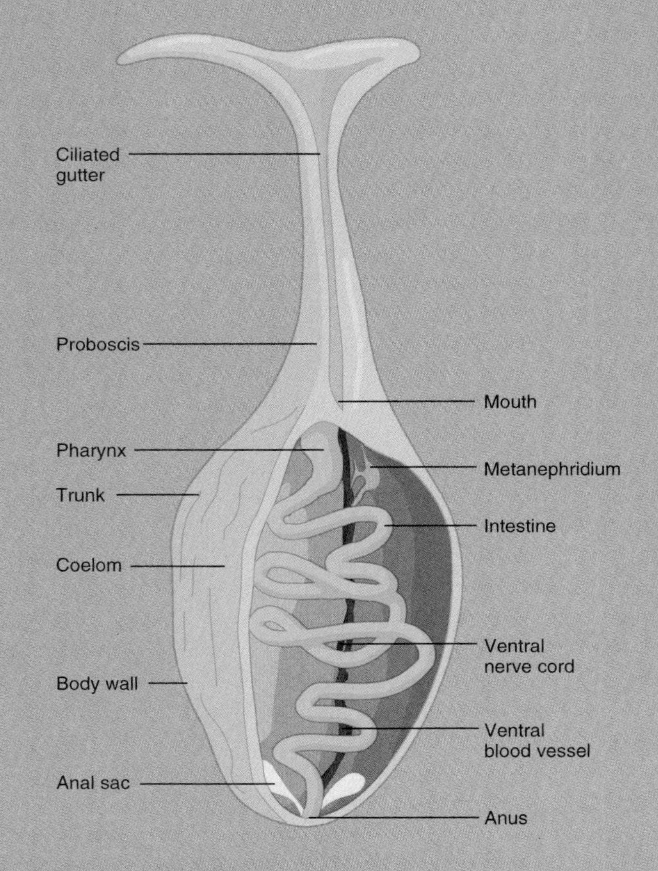

FIGURE 1 **Phylum Echiura.** Internal structure of an echiuran. The muscular body wall surrounds the large coelom, in which a long, coiled intestine is located. A simple closed circulatory system is present, as well as a ventral nerve cord that extends into the proboscis, several pairs of metanephridia, and a pair of anal sacs that empty into the anus at the end of the worm. The exchange of gases takes place through the body wall and proboscis.

FIGURE 2 **Phylum Pogonophora.** Giant red pogonophorans (*Riftia*) inside their tubes.

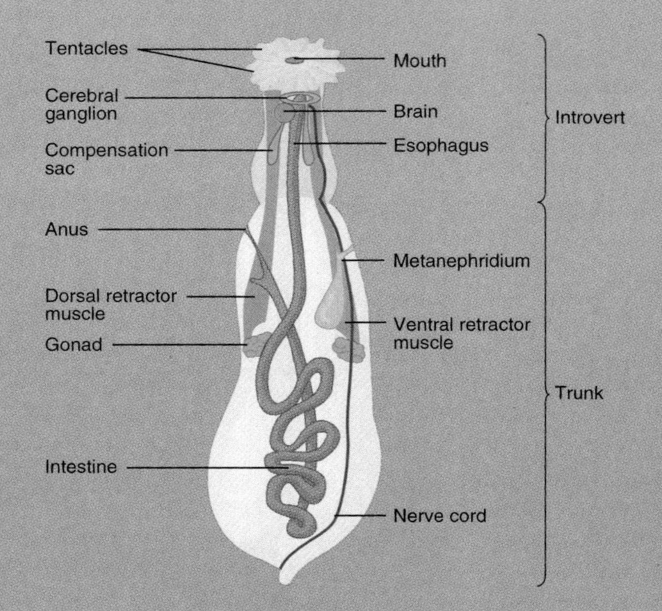

FIGURE 3 **Phylum Sipuncula.** Anatomy of a typical sipunculan. The body is composed of an anterior introvert and a posterior trunk. When the introvert is extended, the anterior portion, with its ciliated tentacles, surrounds the mouth. The long, U-Shaped intestine is arranged in a spiral coil. Anteriorly, the intestine ends at an anus that opens to the outside, near the introvert. A large pair of metanephridia is located in the anterior trunk. The anterior nervous system is annelidlike, with a supraesophageal brain and a ventral nerve cord that runs the length of the trunk.

Phylum Pogonophora: The Pogonophorans, or Beard Worms

The pogonophorans (po′go-nof′or-ans) (Gr. *pogon*, beard + *phora*, bearing) or beard worms are a group of about 120 species of tube-dwelling marine worms distributed throughout the world's oceans, especially along the continental slopes. They are named for the thick tuft of white or reddish tentacles (*see box 1.1*).

The slender, delicate body is protected in a secreted chitinous tube consisting of a series of rings, to which the worm adds as it grows (figure 2). The tubes are embedded in soft marine sediments in cold, deep (over 100 m), nutrient-poor waters. They range in length from about 10 cm to over 2 m.

Pogonophorans have no mouth or digestive tract. Nutrient uptake is via the outer cuticle and from the endosymbiotic bacteria that they harbor in the posterior part (trophosome) of the body. These bacteria are able to fix carbon dioxide into organic compounds that both the host and symbiont can use.

Very little is known about pogonophoran reproduction and development. In general, there are separate sexes, and sperm are packaged into spermatophores before being released by a male. The mechanism of fertilization is unknown. After fertilization, a solid blastula develops following radial cleavage.

Phylum Sipuncula: The Sipunculans, or Peanut Worms

The sipunculans (sigh-pun′kyu-lans) (Gr. L. *sipunculus*, little siphon) or peanut worms (because of their peanut shape when disturbed) consist of about 350 species of unsegmented, coelomate, burrowing worms found in oceans throughout the world. These worms live in mud, sand, or any protected retreat. Their burrows may be mucus lined, but sipunculans do not construct true tubes as do pogonophorans. They range in length from about 2 mm to 75 cm (figure 3).

Sipunculans are dioecious. Gonads are attached to the coelomic wall and liberate their gametes into the coelom. After maturity, the gametes escape into the seawater via the metanephridia. Fertilization is external, cleavage is spiral, and development is either direct (no larva) or it may produce a free-swimming trochophore larva. The larva eventually settles to the bottom and grows into an adult. In a few species, asexual reproduction can also occur by transverse fission—the posterior part of the parent constricts to give rise to a new individual.

chapter 11

THE ARTHROPODS: BLUEPRINT FOR SUCCESS

Outline

Concepts

1. Arthropods have been successful in almost all habitats on the earth. Some ancient arthropods were the first animals to live most of their lives in terrestrial environments.
2. Metamerism with tagmatization, a chitinous exoskeleton, and metamorphosis have contributed to the success of arthropods.
3. Members of the subphylum Trilobitomorpha are extinct arthropods that were a dominant form of life in the oceans between 345 and 600 million years ago.
4. Members of the subphylum Chelicerata have a body divided into two regions and have chelicerae. The class Merostomata contains the horseshoe crabs. The class Arachnida contains the spiders, mites, ticks, and scorpions. Some ancient arachnids were among the earliest terrestrial arthropods, and modern arachnids have numerous adaptations for terrestrial life. The class Pycnogonida contains the sea spiders.
5. Animals in the subphylum Crustacea have biramous appendages and two pairs of antennae. The class Branchiopoda includes the fairy shrimp, brine shrimp, and water fleas. The class Malacostraca includes the crabs, lobsters, crayfish, and shrimp. The classes Copepoda and Cirrepedia include the copepods and barnacles, respectively.

Would You Like to Know:

1. what the most abundant animal is? (*p. 182*)
2. how an arthropod grows within the confines of a rigid exoskeleton? (*p. 184*)
3. how arthropods were preadapted for terrestrialism? (*p. 188*)
4. why some spiders go ballooning? (*p. 190*)
5. what two spiders found in the United States are dangerous to humans? (*p. 190*)
6. what causes the bite of a chigger to itch so badly? (*p. 192*)
7. what mite lives in the hair follicles of most readers of this textbook? (*p. 192*)
8. what crustaceans colonize the hulls of ships? (*p. 198*)

These and other useful questions will be answered in this chapter.

This chapter contains evolutionary concepts, which are set off in this font.

Evolutionary Perspective

What animal species has the greatest number of individuals? 1 The answer can only be an educated guess; however, many zoologists would argue that one of the many species of small (1 to 2 mm) crustaceans, called copepods, that drift in the open oceans must have this honor. Copepods have been very successful, feeding on the vast photosynthetic production of the open oceans (figure 11.1). After only 20 minutes of towing a plankton net behind a slowly moving boat (at the right location and during the right time of year), one can collect over 3 million copepods—enough to solidly pack a 2 gallon pail! Copepods are food for fish, such as herring, sardines, mackerel, as well as for whale sharks and the largest mammals, the blue whale and its relatives. Humans benefit from copepod production by eating fish that feed on copepods. (Unfortunately, we use a small fraction of the total food energy in these animals. In spite of ½ of the earth's inhabitants lacking protein in their diet, humans process into fish meal most of the herring and sardines caught, which is then fed to poultry and hogs. In eating the poultry and hogs, we lose over 99% of the original energy present in the copepods!)

Copepods are one of many groups of animals belonging to the phylum Arthropoda (ar′thra-po′dah) (Gr. *arthro*, joint + *podos*, foot). Crayfish, lobsters, spiders, mites, scorpions, and insects are also arthropods. About 1 million species of arthropods have been described, and recent studies estimate that there may be 30 to 50 million undescribed species. In this chapter and chapter 12, you will discover the many ways in which some arthropods are considered among the most successful of all animals.

Characteristics of members of the phylum Arthropoda include the following:

1. Metamerism modified by the specialization of body regions for specific functions (tagmatization)
2. Chitinous exoskeleton provides support, protection, and is modified to form sensory structures
3. Paired, jointed appendages
4. Growth accompanied by ecdysis or molting
5. Ventral nervous system
6. Coelom reduced to cavities surrounding gonads and sometimes excretory organs
7. Open circulatory system in which blood is released into tissue spaces (hemocoel) derived from the blastocoel
8. Complete digestive tract
9. Metamorphosis often present; reduces competition between immature and adult stages

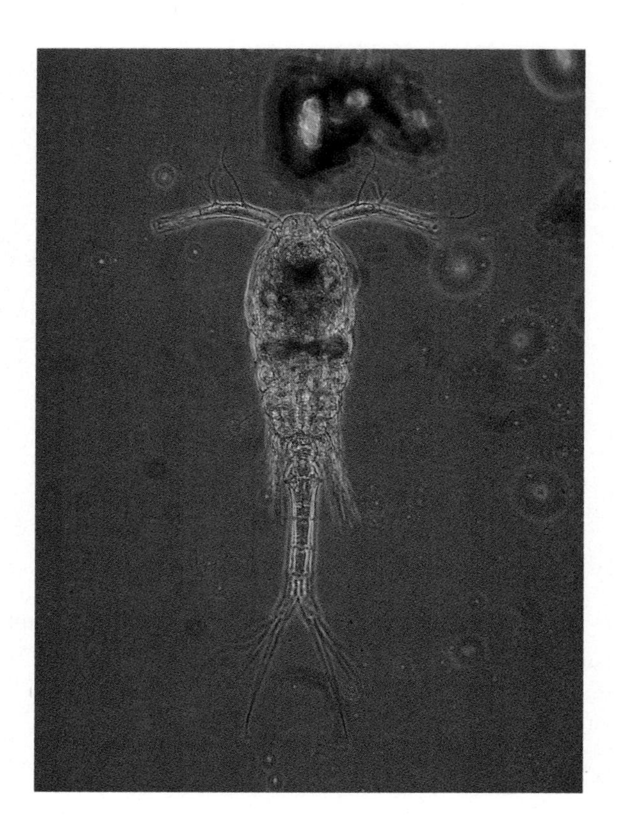

Figure 11.1
The Most Abundant Animal? Copepods, such as *Calanus*, are extremely abundant in the oceans of the world and form important links in oceanic food webs.

Classification and Relationships to Other Animals

As discussed in chapter 10, arthropods and annelids are closely related to each other. Shared protostome characteristics, such as the development of the mouth from the blastopore and schizocoelous coelom formation, as well as other common characteristics, such as the presence of a paired ventral nerve cord and metamerism, give evidence of a common ancestry (figure 11.2).

Zoologists, however, disagree about the evolutionary relationships among the arthropods. Many zoologists believe that it is not one phylum, but three. These ideas are discussed at the end of chapter 12. The arthropods are treated in this textbook as members of a single phylum. Living arthropods are divided into three subphyla: Chelicerata, Crustacea, and Uniramia. All members of a fourth subphylum, Trilobita, are extinct (table 11.1). Trilobita, Chelicerata, and Crustacea are discussed in this chapter and the Uniramia are discussed in chapter 12.

Metamerism and Tagmatization

Three aspects of arthropod biology have been important in contributing to their success. One of these is metamerism. Metamerism of arthropods is most evident externally, because the arthropod body is often composed of a series of similar segments, each bearing a pair of appendages (*see figure 10.4e*). Internally, however, the body cavity of arthropods is not divided by septa, and most organ systems are not metamerically arranged. The reason for the loss of internal metamerism is speculative; however, the presence of

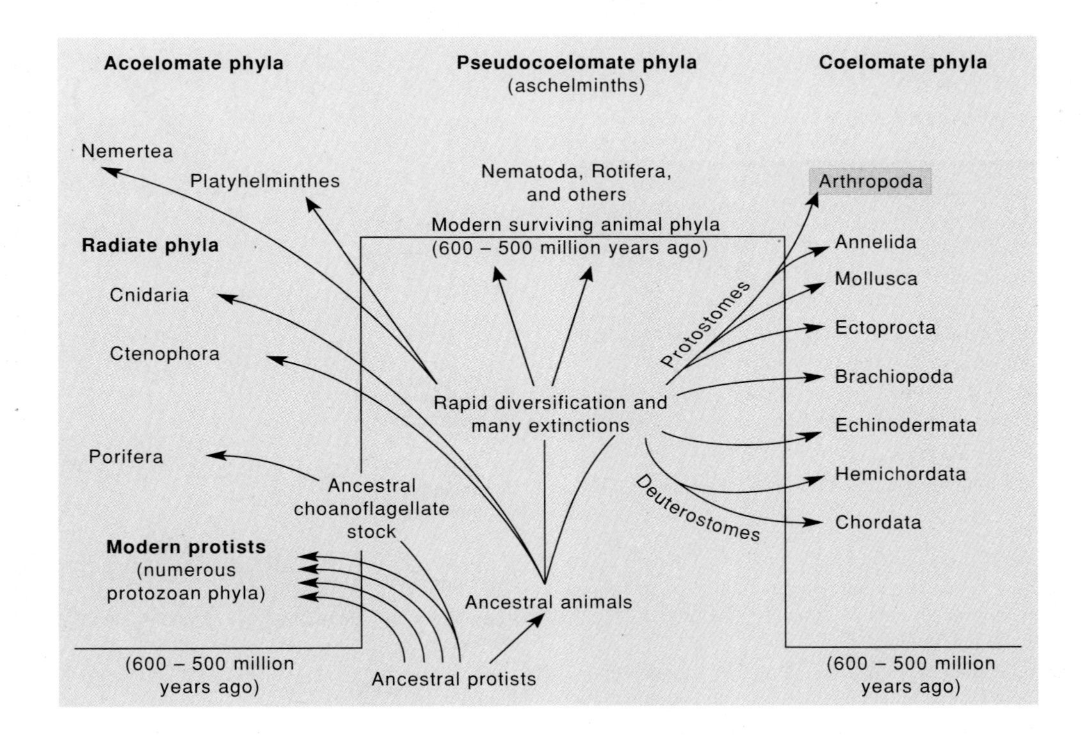

Figure 11.2

Evolutionary Relationships of the Arthropods. Arthropods (shaded in orange) are protostomes with close evolutionary ties to the annelids. This is shown by the presence of a paired ventral nerve cord and metamerism in both groups.

metamerically arranged hydrostatic compartments would be of little value in the support or locomotion of animals enclosed by an external skeleton (discussed below).

As discussed in chapter 10, metamerism permits the specialization of regions of the body for specific functions. This regional specialization is called tagmatization. In arthropods, body regions, called tagmata (s., tagma), are specialized for feeding and sensory perception, locomotion, and visceral functions.

The Exoskeleton

Arthropods are enclosed by an external, jointed skeleton, called an **exoskeleton** or **cuticle.** The exoskeleton is often cited as the major reason for arthropod success. It provides structural support, protection, impermeable surfaces for prevention of water loss, and a system of levers for muscle attachment and movement.

The exoskeleton covers all body surfaces and invaginations of the body wall, such as the anterior and posterior portions of the gut tract. It is nonliving and is secreted by a single layer of epidermal cells (figure 11.3). The epidermal layer is sometimes called the hypodermis because, unlike other epidermal tissues, it is covered on the outside by exoskeleton rather than being directly exposed to air or water.

The exoskeleton is composed of two layers. The epicuticle is the outermost layer. Because it is made of a waxy lipoprotein, it is impermeable to water and serves as a barrier to microorganisms and pesticides. The bulk of the exoskeleton is below the epicuticle and is called the procuticle. (In crustaceans, the procuticle is sometimes called the endocuticle.) The procuticle is composed of **chitin,** a tough, leathery polysaccharide and several kinds of proteins. Hardening of the procuticle is accomplished through a process called sclerotization and sometimes by impregnation with calcium carbonate. Sclerotization is a tanning process in which layers of protein are chemically cross-linked with one another—hardening and darkening the exoskeleton. In insects and most other arthropods, this bonding occurs in the outer portion of the procuticle. Hardening of the exoskeleton of crustaceans is accomplished by sclerotization and by the deposition of calcium carbonate in the middle regions of the procuticle. Some proteins give the exoskeleton resiliency. When the exoskeleton is distorted, energy is stored. Stored energy can be used in activities such as flapping wings and jumping. The inner portion of the procuticle is not hardened.

Hardening in the procuticle provides armorlike protection for arthropods, but it also necessitates a variety of adaptations to allow arthropods to live and grow within their confines. Invaginations of the exoskeleton form firm ridges and bars for muscle attachment. Another modification of

TABLE 11.1 CLASSIFICATION OF THE PHYLUM ARTHROPODA

Phylum Arthropoda (ar′thra-po′dah)
Animals that show metamerism with tagmatization, a jointed exoskeleton, and a ventral nervous system.

Subphylum Trilobitomorpha (tri″lo-bit′o-mor′fah)
Marine, all extinct; lived from Cambrian to Carboniferous periods; bodies are divided into three longitudinal lobes; head, thorax, and abdomen present; one pair of antennae and biramous appendages.

Subphylum Chelicerata (ke-lis″e-ra′tah)
Body usually divided into prosoma and opisthosoma; first pair of appendages piercing or pincerlike (chelicerae) and used for feeding.

Class Merostomata (mer′o-sto′mah-tah)
Marine, with book gills on opisthosoma. Two subclasses: Eurypterida, a group of extinct arthropods called giant water scorpions, and Xiphosura, the horseshoe crabs. *Limulus*.

Class Arachnida (ah-rak′ni-dah)
Mostly terrestrial, with book lungs, tracheae, or both; usually four pairs of walking legs in adults. Spiders, scorpions, ticks, mites, harvestmen, and others.

Class Pycnogonida (pik′no-gon″i-dah)
Reduced abdomen; no special respiratory or excretory structures; four to six pairs of walking legs; common in all oceans. Sea spiders.

Subphylum Crustacea (krus-tas′eah)
Most aquatic, head with two pairs of antennae, one pair of mandibles, and two pairs of maxillae; biramous appendages.

Class Remipedia (ri-mi-pe′de-ah)
A single species of cave-dwelling crustaceans from the Bahamas; body with approximately 30 segments that bear uniform, biramous appendages.

Class Cephalocarida (sef′ah-lo-kar′i-dah)
Small (3 mm) marine crustaceans with uniform, leaflike, triramous appendages.

Class Branchiopoda (brang′ke-o-pod′ah)
Flattened, leaflike appendages used in respiration, filter feeding, and locomotion; found mostly in fresh water. Fairy shrimp, brine shrimp, clam shrimp, water fleas.

Class Malacostraca (mal-ah-kos′trah-kah)
Appendages may be modified for crawling, feeding, swimming. Lobsters, crayfish, crabs, shrimp, isopods (terrestrial).

Class Copepoda (ko′pepod′ah)
Maxillipeds modified for feeding. Copepods.

Class Cirripedia (sir′i-ped′eah)
Sessile as adults, marine, and enclosed by calcium carbonate valves. Barnacles.

Subphylum Uniramia (yoo′ne-ram′eah)
Head with one pair of antennae and usually one pair of mandibles; all appendages uniramous. Insects and their relatives. See chapter 24.

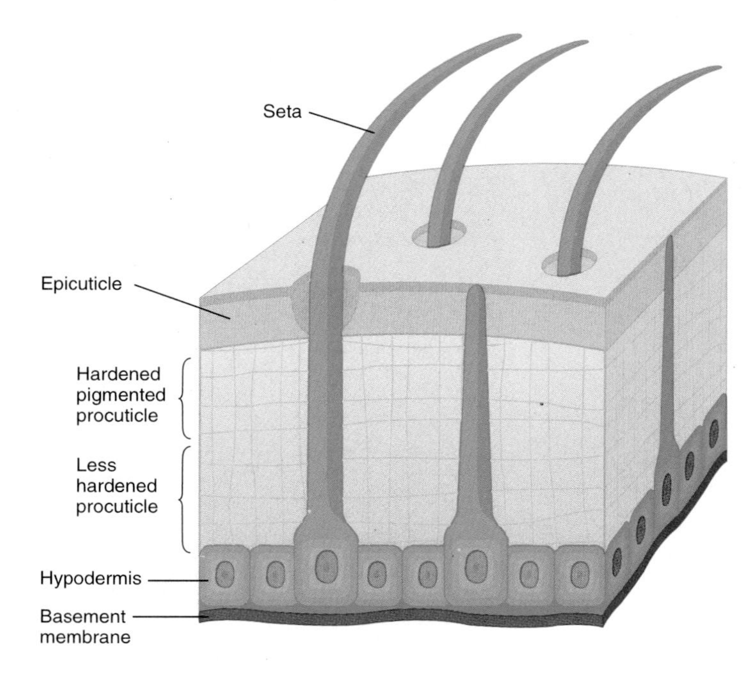

FIGURE 11.3
The Arthropod Exoskeleton. The epicuticle is made of a waxy lipoprotein and is impermeable to water. The outer layer of the procuticle is hardened by calcium carbonate deposition and/or sclerotization. Chitin, a tough, leathery polysaccharide, and several kinds of proteins make up the bulk of the procuticle. The entire exoskeleton is secreted by the hypodermis.

the exoskeleton is the formation of joints. A flexible membrane, called an articular membrane, is present in regions where the procuticle is thinner and less hardened (figure 11.4). Other modifications of the exoskeleton include sensory receptors, called sensilla, that are in the form of pegs, bristles, and lenses, and modifications of the exoskeleton that permit gas exchange.

2 Growth of an arthropod would be virtually impossible unless the exoskeleton were periodically shed; such as in the molting process called **ecdysis** (Gr. *ekdysis*, getting out). Ecdysis is divided into four stages: (1) enzymes, secreted from hypodermal glands, begin digesting the old endocuticle. This digestion separates the hypodermis and the exoskeleton (figure 11.5*a,b*); (2) digestion of the endocuticle is followed by the secretion of new procuticle and epicuticle (figure 11.5*c,d*); (3) the old exoskeleton is split open along predetermined ecdysal lines when the animal stretches by air or water intake. Additional epicuticle is secreted through pores in the procuticle (figure 11.5*e*); (4) finally, the new exoskeleton is hardened by deposition of calcium carbonate and/or sclerotization (figure 11.5*f*). During the few hours or days of the hardening process, the arthropod is vulnerable to predators and remains hidden. All of these changes are controlled by the nervous and endocrine systems; the controls will be discussed in more detail later.

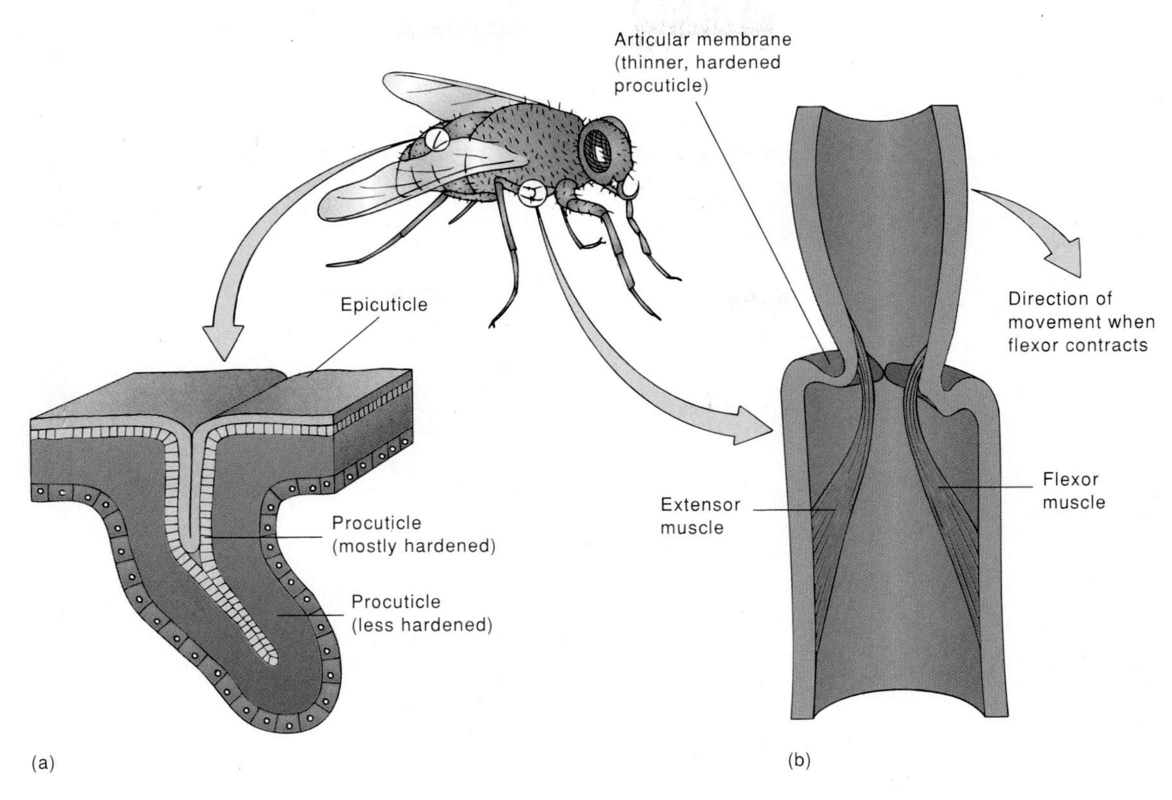

Figure 11.4

Modifications of the Exoskeleton. (*a*) Invaginations of the exoskeleton result in firm ridges and bars when the procuticle in the region of the invagination remains thick and hard. These are used as muscle attachment sites. (*b*) Regions where the procuticle is thinned are flexible and form membranes and joints. *From:* "A LIFE OF INVERTEBRATES" © *1979 W. D. Russell-Hunter.*

Metamorphosis

A third characteristic that has contributed to arthropod success is a reduction of competition between adults and immature stages because of metamorphosis. Metamorphosis is a radical change in body form and physiology that occurs as an immature stage, usually called a larva, becomes an adult. The evolution of arthropods has resulted in an increasing divergence of body forms, behaviors, and habitats between immature and adult stages. Adult crabs, for example, are usually found prowling the sandy bottoms of their marine habitats for live prey or decaying organic matter, whereas larval crabs live and feed in the plankton. Similarly, the caterpillar that feeds on leafy vegetation eventually develops into a nectar-feeding adult butterfly or moth. Having different adult and immature stages means that they will not compete with each other for food or living space. In some arthropod and other animal groups, larvae also serve as the dispersal stage.

Subphylum Trilobitomorpha

Members of the subphylum Trilobitomorpha (tri″lo-bit′o-mor′fah) (Gr. *tri*, three + *lobos*, lobes) were a dominant form of life in the oceans from the Cambrian period (600 million years ago) to the Carboniferous period (345 million years ago). They crawled along the substrate feeding on annelids, molluscs, and decaying organic matter. The body of trilobites was oval, flattened, and divided into three longitudinal regions (figure 11.6). All body segments articulated so the trilobite could roll into a ball to protect its soft ventral surface. Most fossilized trilobites are found in this position. Trilobite appendages consist of two lobes. The inner lobe served as a walking leg, and the outer lobe bore spikes or teeth that may have been used in digging or swimming or as gills in gas exchange. Because they possessed two lobes or rami, these appendages are called **biramous** (L. *bi*, twice + *ramus*, branch) **appendages.**

Stop and Ask Yourself

1. What are the four subphyla of arthropods?
2. What three aspects of structure and function have been important in arthropod success?
3. How does metamorphosis reduce competition between adult and immature forms?
4. In what way is the name "trilobite" descriptive of that group of animals?

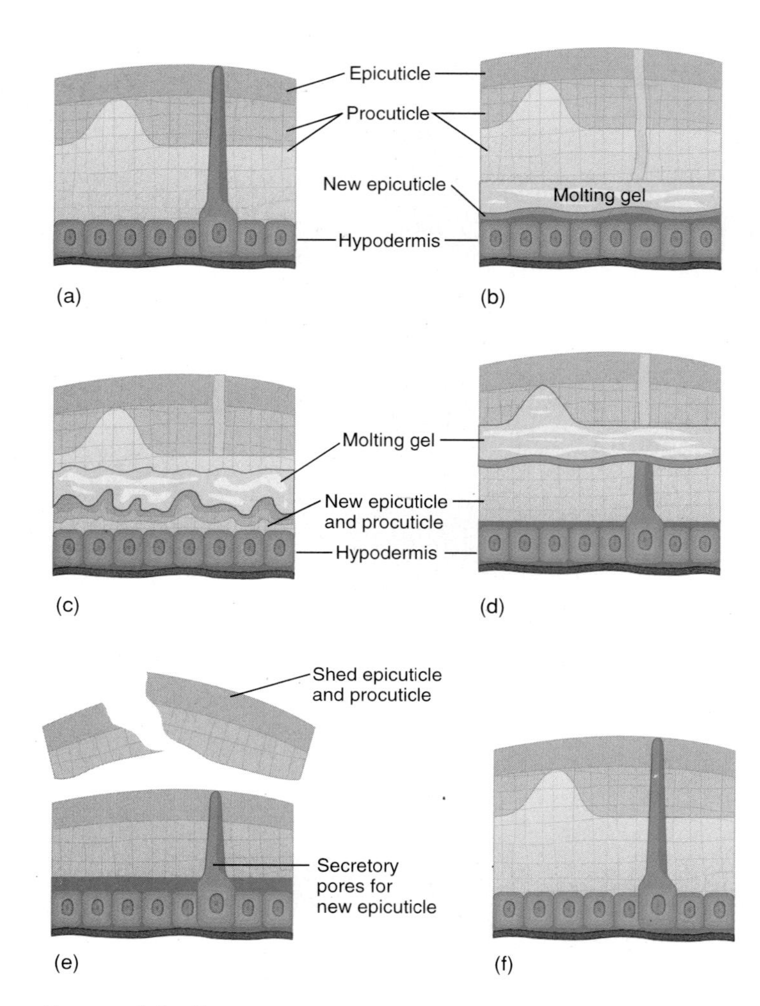

FIGURE 11.5

Events of Ecdysis. (*a,b*) During preecdysis, the hypodermis detaches from the exoskeleton and the space between the old exoskeleton and the hypodermis is filled with a fluid called molting gel. (*c,d*) The hypodermis begins secreting a new epicuticle and a new procuticle is formed as the old procuticle is digested. The products of digestion are incorporated into the new procuticle. Note that the new epicuticle and procuticle are wrinkled beneath the old exoskeleton to allow for increased body size after ecdysis. (*e*) Ecdysis occurs when the animal swallows air or water, and the exoskeleton splits along predetermined ecdysal lines. The animal pulls out of the old exoskeleton. (*f*) After ecdysis, the new exoskeleton hardens by calcium carbonate deposition, and/or sclerotization, and pigments are deposited in the outer layers of the procuticle. Additional material is added to the epicuticle.

SUBPHYLUM CHELICERATA

One arthropod lineage, the subphylum Chelicerata (ke-lis″e-ra′tah) (Gr. *chele*, claw + *ata*, plural suffix) includes familiar animals, such as spiders, mites, and ticks, and less familiar animals, such as horseshoe crabs and sea spiders. These animals have two tagmata. The **prosoma** or **cephalothorax** is a sensory, feeding, and locomotor tagma. It usually bears eyes, but unlike other arthropods, never has antennae. Paired appendages attach to the prosoma. The first pair, called **chelicerae,** are often pincerlike or chelate and are most often used in feeding. They may also be specialized as hollow fangs or for a variety of other functions. The second pair, called **pedipalps,** are usually sensory but may also be used in feeding, locomotion, or reproduction. Pedipalps are followed by paired walking legs. Posterior to the prosoma is the **opisthosoma,** which contains digestive, reproductive, excretory, and respiratory organs.

FIGURE 11.6

Trilobite Structure. The body of a trilobite was divided into three longitudinal sections (thus the subphylum name). It was also divided into three tagmata. A head, or cephalon, bore a pair of antennae and eyes. The trunk, or thorax, bore appendages used in swimming or walking. A series of posterior segments formed the pygidium, or tail.

CLASS MEROSTOMATA

Members of the class Merostomata (mer′o-sto′mah-tah) are divided into two subclasses. The Xiphosura are the horseshoe crabs, and the Eurypterida are the giant water scorpions (figure 11.7). The latter are extinct, having lived from the Cambrian period (600 million years ago) to the Permian period (280 million years ago).

There are only four species of horseshoe crabs living today; one species, *Limulus polyphemus*, is widely distributed in the Atlantic Ocean and the Gulf of Mexico (figure 11.8*a*). Horseshoe crabs scavenge sandy and muddy substrates for annelids, small molluscs, and other invertebrates. Their body form has remained virtually unchanged for over 200 million years, and they were cited in chapter 3 as an example of stabilizing selection.

The cephalothorax of horseshoe crabs is covered with a hard, horseshoe-shaped carapace. The chelicerae, pedipalps, and first three pairs of walking legs are chelate and are used for walking and food handling. The last pair of appendages has leaflike plates at their tips and are used for locomotion and digging (figure 11.8*b*).

The opisthosoma of a horseshoe crab includes a long, unsegmented telson. If a horseshoe crab is flipped over by wave action, it arches its opisthosoma dorsally, causing the animal to

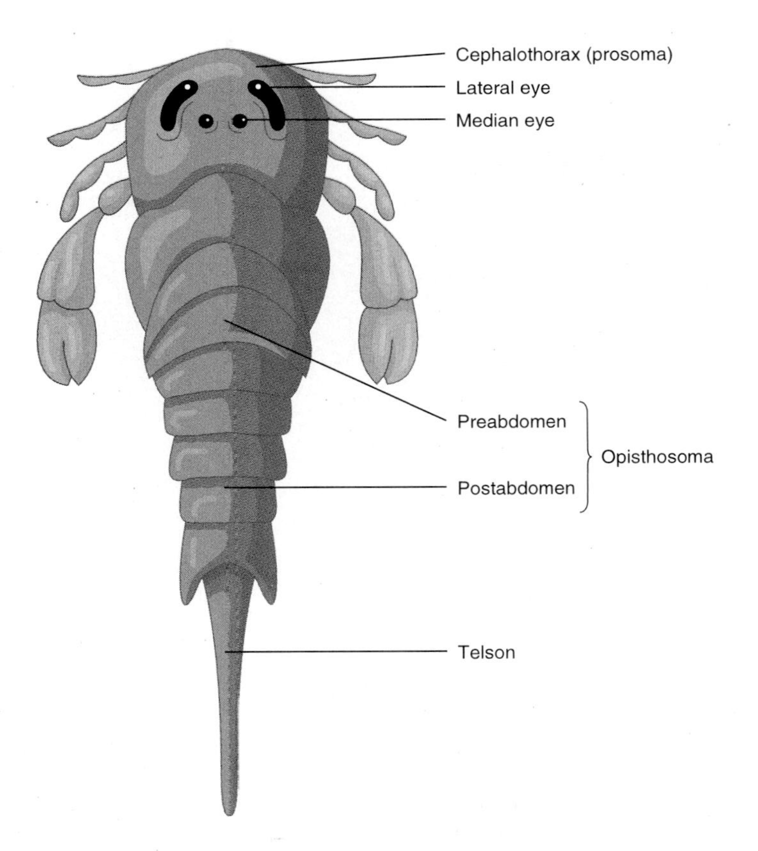

FIGURE 11.7
Class Merostomata. A eurypterid, *Euripterus remipes*.

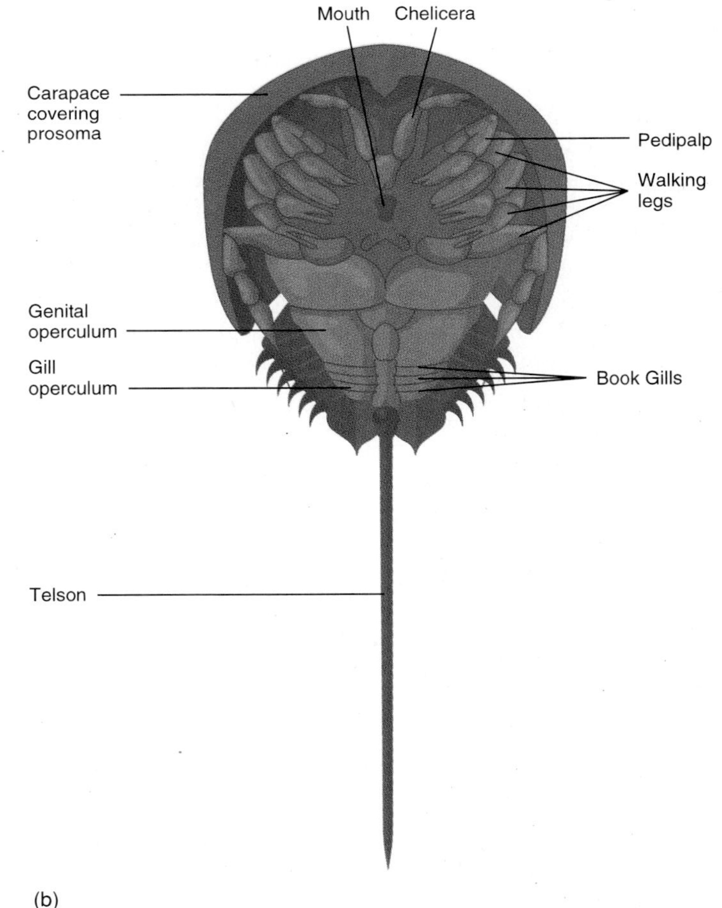

FIGURE 11.8
Class Merostomata. (*a*) Dorsal view of the horseshoe crab, *Limulus polyphemus*. (*b*) Ventral view.

roll to its side and flip right side up again. The first pair of opisthosomal appendages cover genital pores and are called genital opercula. The remaining five pairs of appendages are **book gills.** The name is derived from the resemblance of these platelike gills to the pages of a closed book. Gas exchange between the blood and water occurs as blood circulates through the book gills. Horseshoe crabs have an open circulatory system, as do all arthropods. Blood circulation in horseshoe crabs is similar to that described later in this chapter for arachnids and crustaceans.

Horseshoe crabs are dioecious. During reproductive periods, males and females congregate in intertidal areas. The male mounts the female and grasps her with his pedipalps. The female excavates shallow depressions in the sand, and eggs are fertilized by the male as they are shed from the female into depressions. Fertilized eggs are covered with sand and develop unattended.

CLASS ARACHNIDA

Members of the class Arachnida (ah-rak′ni-dah) (Gr. *arachne*, spider) are some of the most misrepresented members of the animal kingdom. Their reputation as fearsome and grotesque creatures is vastly exaggerated. The majority of spiders, mites, ticks, scorpions, and related forms are either harmless or very beneficial to humans.

Most zoologists believe that arachnids arose from the eurypterids and were very early terrestrial inhabitants. The earliest fossils of aquatic scorpions date back to the Silurian period (405 to 425 million years ago), fossils of terrestrial scorpions date from the Devonian period (350 to 400 million years ago), and fossils of all other arachnid groups are present by the Carboniferous period (280 to 345 million years ago).

3 Water conservation is a major concern for any terrestrial organism, and ancestral arachnids were preadapted for terrestrialism by their relatively impermeable exoskeleton. **Preadaptation** occurs when a structure present in members of a species proves useful in promoting reproductive success when an individual encounters new environmental situations. Later adaptations included evolution of efficient excretory structures, internal surfaces for gas exchange, appendages modified for locomotion on land, and greater deposition of wax in the epicuticle.

Form and Function

Most arachnids are carnivores. Small arthropods are usually held by the chelicerae while enzymes from the gut tract pour over the prey. Partially digested food is then taken into the mouth. The gut tract of arachnids is divided into three regions. The anterior portion is called the foregut, and the posterior portion is called the hindgut. Both develop as infoldings of the body wall and are lined with cuticle. A portion of the foregut is frequently modified into a pumping pharynx, and the hindgut is frequently a site of water reabsorption. The midgut is between the foregut and hindgut. It is noncuticular and lined with secretory and absorptive cells. Lateral diverticula increase the area available for absorption and storage.

Arachnids use coxal glands and/or malpighian tubules for excreting nitrogenous wastes. **Coxal glands** are paired, thin-walled, spherical sacs bathed in the blood of body sinuses. Nitrogenous wastes are absorbed across the wall of the sacs, transported into a long, convoluted tubule, and excreted through excretory pores at the base of the posterior appendages. Arachnids that are adapted to dry environments possess blind-ending diverticula of the gut tract that arise at the juncture of the midgut and hindgut. These tubules are called **malpighian tubules.** They absorb waste materials from the blood, and empty them into the gut tract. Excretory wastes are then eliminated with digestive wastes. The major excretory product of arachnids is uric acid, which is advantageous for terrestrial animals because it is excreted as a semisolid with little water loss.

Gas exchange also occurs with minimal water loss because arachnids have few exposed respiratory surfaces. Some arachnids possess structures, called **book lungs,** that are assumed to be modifications of the book gills found in the Merostomata. Book lungs are paired invaginations of the ventral body wall that are folded into a series of leaflike lamellae (figure 11.9). Air enters the book lung through a slitlike opening and circulates between lamellae. Diffusion of respiratory gases occurs between the blood moving among the lamellae

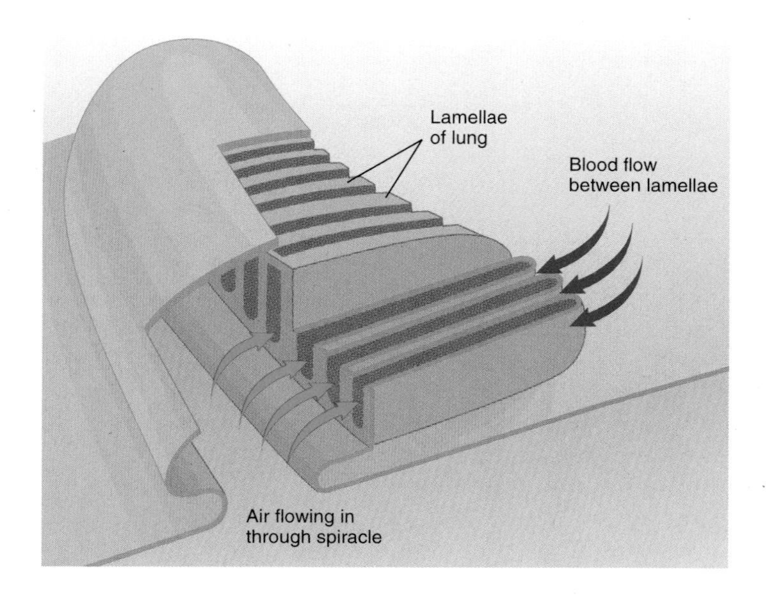

FIGURE 11.9

An Arachnid Book Lung. Air and blood moving on opposite sides of a lamella of the lung exchange respiratory gases by diffusion. Figure 11.12 shows the location of book lungs in spiders.

and the air in the lung chamber. Other arachnids possess a series of branched, chitin-lined tubules that deliver air directly to body tissues. These tubule systems, called **tracheae** (s., trachea), open to the outside through openings called **spiracles** that are located along the ventral or lateral aspects of the abdomen. (Tracheae are also present in insects but had a separate evolutionary origin. Aspects of their physiology will be described in chapter 12.)

The circulatory system of arachnids, like that of most arthropods, is an open system in which blood is pumped by a dorsal contractile vessel (usually called the dorsal aorta or "heart") and is released into tissue spaces. In arthropods, the coelom is reduced to cavities surrounding the gonads and sometimes the coxal glands. Large tissue spaces, or sinuses, are derived from the blastocoel and are called the **hemocoel.** Blood bathes the tissues and then returns to the dorsal aorta through openings in the aorta called ostia. Arachnid blood contains the dissolved respiratory pigment hemocyanin and has amoeboid cells that aid in clotting and body defenses.

The nervous system of all arthropods is ventral and, in ancestral arthropods, must have been laid out in a pattern similar to that of the annelids (*see figure 10.8a*). With the exception of scorpions, the nervous system of arachnids is greatly concentrated by fusion of ganglia.

The body of an arachnid is supplied with a variety of sensory structures. Most mechanoreceptors and chemoreceptors are modifications of the exoskeleton, such as projections, pores, and slits together with sensory and accessory cells. Collectively, these receptors are called **sensilla.** For example, setae are hairlike, cuticular modifications that may be set into membranous sockets. Displacement of a seta initiates a nerve im-

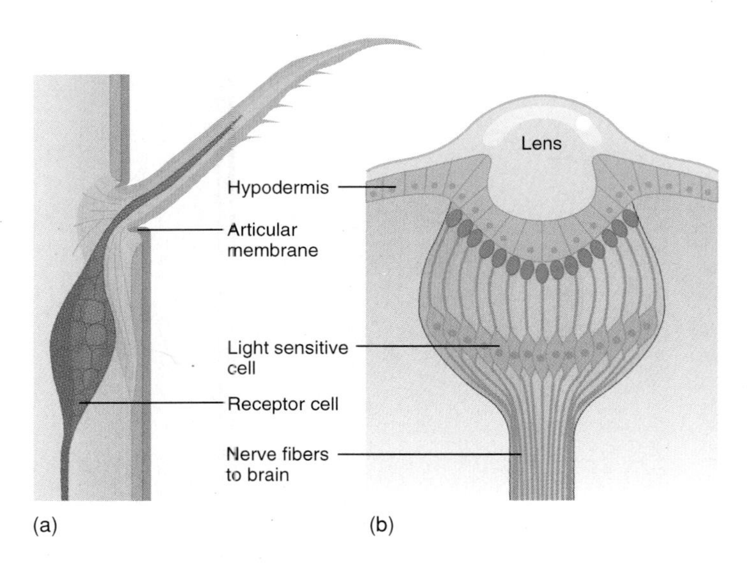

FIGURE 11.10

Arthropod Seta and Eye (Ocellus). (*a*) A seta is a hairlike modification of the cuticle set in a membranous socket. Displacement of the seta initiates a nerve impulse in a receptor cell (sensillum) associated with the base of the seta. (*b*) The lens of this spider eye is a thickened, transparent modification of the cuticle. Below the lens and hypodermis are light-sensitive sensillae that contain pigments that convert light energy into nerve impulses.

pulse in an associated nerve cell (figure 11.10*a*). Vibration receptors are very important to some arachnids. Spiders that use webs to capture prey, for example, determine both the size of the insect and its position on the web by the vibrations the insect makes while struggling to free itself. The chemical sense of arachnids is comparable to taste and smell in vertebrates. Small pores in the exoskeleton are frequently associated with peglike, or other, modifications of the exoskeleton, and allow chemicals to stimulate nerve cells. Arachnids possess one or more pairs of eyes (figure 11.10*b*). These eyes are used primarily for detecting movement and changes in light intensity. The eyes of some hunting spiders probably form images.

Arachnids are dioecious. Paired genital openings are on the ventral side of the second abdominal segment. Sperm transfer is usually indirect. The male often packages sperm in a spermatophore, after which it is transferred to the female. Courtship rituals confirm that individuals are of the same species, attract a female to the spermatophore, and position the female to receive the spermatophore. In some taxa (e.g., spiders), copulation occurs, and sperm transfer is accomplished via a modified pedipalp of the male. Development is direct, and the young hatch from eggs as miniature adults. Many arachnids tend their developing eggs and young during and after development.

Order Scorpionida

Members of the order Scorpionida (skor″pe-ah-ni′dah) are the scorpions (figure 11.11*a*). They are common from tropical to warm temperate climates. Scorpions are secretive and nocturnal, spending most of the daylight hours hidden under logs and stones.

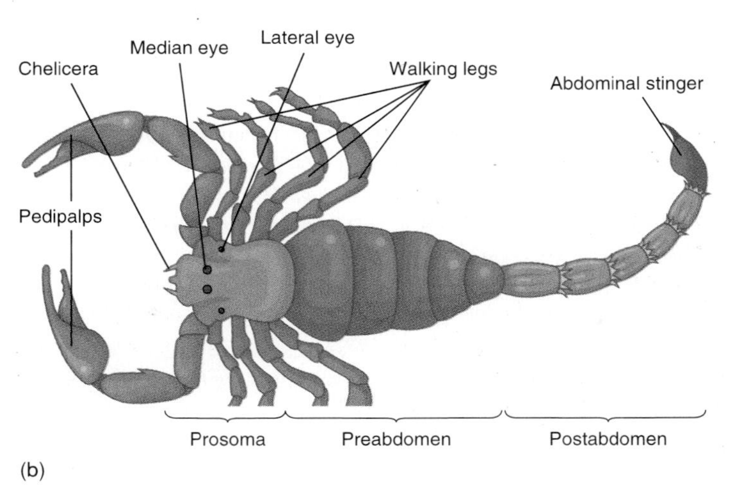

FIGURE 11.11

Order Scorpionida. (*a*) A scorpion captures its prey using chelate pedipalps. Venom from its sting paralyzes the prey prior to feeding. This Kenyan scorpion (*Pardinus*) is shown eating a cricket. (*b*) External anatomy of a scorpion.

Scorpions have small chelicerae that project anteriorly from the front of the carapace (figure 11.11*b*). A pair of enlarged, chelate pedipalps are posterior to the chelicerae. The opisthoma is divided. An anterior preabdomen contains the slitlike openings to book lungs, comblike tactile and chemical receptors called pectines, and genital openings. The postabdomen (commonly called the tail) is narrower than the preabdomen and is curved dorsally and anteriorly over the body when aroused. At the tip of the postabdomen is a sting. The sting has a bulbular base that contains venom-producing glands and a hollow, sharp, barbed point. Smooth muscles eject venom during stinging. Only a few scorpions have venom that is highly toxic to humans. Species in the genera *Androctonus* (northern Africa) and *Centuroides* (Mexico, Arizona, and New Mexico) have been responsible for human deaths. Other scorpions from the southern and southwestern areas of North America give stings comparable to wasp stings.

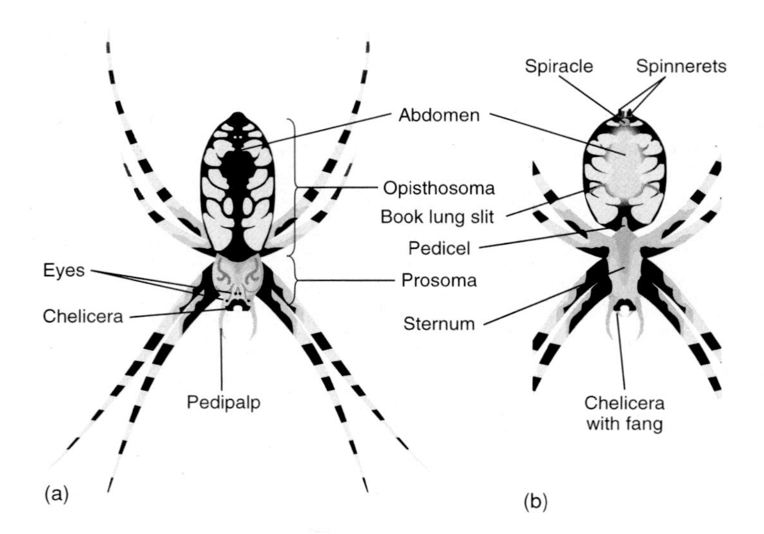

Figure 11.12
External Structure of a Spider. (*a*) Dorsal view. (*b*) Ventral view.
Sources: (a) After Sherman and Sherman. (b) After the Kastons.

Prior to reproduction, male and female scorpions have a period of courtship that lasts from five minutes to several hours. Male and female scorpions face each other and extend their abdomens high into the air. The male seizes the female with his pedipalps, and they repeatedly walk backward and then forward. The lower portion of the male reproductive tract forms a spermatophore that is deposited on the ground. During courtship, the male positions the female so that the genital opening on her abdomen is positioned over the spermatophore. Downward pressure of the female's abdomen on a triggerlike structure of the spermatophore causes sperm to be released into the female's genital chamber.

Most arthropods are **oviparous;** females lay eggs that develop outside the body. Many scorpions and some arthropods are **ovoviviparous;** internal development occurs, although large, yolky eggs provide all the nourishment for development. Some scorpions, however, are **viviparous** meaning the embryos are nourished by nutrients provided by the mother. Eggs develop in diverticula of the ovary that are closely associated with diverticula of the digestive tract. Nutrients pass from the digestive tract diverticula to the developing embryos. Development requires up to 1.5 years and 20 to 40 young are brooded. After birth, the young crawl onto the mother's back, where they remain for up to 1 month.

Order Araneae

With about 34,000 species, the order Araneae (ah-ran′a-e) is the largest group of arachnids (figure 11.12). The prosoma of spiders bears chelicerae with poison glands and fangs. Pedipalps are leglike, and in males, are modified for sperm transfer. There are usually eight eyes on the dorsal, anterior margin of the carapace.

The prosoma is attached to the opisthosoma by a slender, waistlike pedicel. The abdomen is swollen or elongate and contains openings to the reproductive tract, book lungs, and tracheae. It also has six to eight conical projections, called spinnerets, that are associated with silk glands. The protein that forms silk is emitted as a liquid, but hardens as it is drawn out. Several kinds of silk are formed, each with its own use. In addition to being formed into webs for capturing prey, (figure 11.13) silk is used to line retreats, to lay a safety line that is fastened to the substrate to interrupt a fall, and to wrap eggs into a case for development. Silk lines produced by young spiders are caught by air currents and serve as a dispersal mechanism. (4) Silk lines have been known to carry spiders at great altitudes for hundreds of miles. This behavior is called ballooning.

Figure 11.13
Order Araneae. Some of the most beautiful and intricate spider webs are produced by members of the family Araneidae, the orb weavers. Many species are relatively large, like this garden spider—*Argiope*. A web is not a permanent construction. When webs become wet with rain or dew, or when they age, they lose their stickiness. The entire web, or at least the spiraled portion, is eaten and replaced.

Most spiders feed on insects and other arthropods. A few (e.g., tarantulas or "bird spiders") feed on small vertebrates. Once captured in webs or by hunting, prey are paralyzed by the spider's bite and sometimes wrapped in silk. Enzymes are introduced through a puncture in the body wall, and predigested products are sucked into the spider's digestive tract by a pumping stomach. The venom of most spiders is harmless to humans. (5) Black widow spiders (*Lactrodectus*) and brown recluse spiders (*Loxosceles*) are exceptions since their venom is toxic to humans (box 11.1).

BOX 11.1 A Fearsome Twosome

The fearsome reputation of spiders is vastly exaggerated. All spiders have poison glands and fangs (modifications of chelicerae) that are used in immobilizing prey. The venom of most spiders, however, is not dangerous to humans. There are only about 20 species of spiders whose bite is considered dangerous to humans. Of these dangerous species, only two are found in North America.

The black widow spider (*Latrodectus mactans*) is found throughout the United States and southern Canada (figure 1*a*). The black widow's venom is a neurotoxin. The bite of a black widow is not particularly painful, but symptoms of the bite can be very severe. Symptoms include abdominal and leg pain, high cerebrospinal fluid pressure, nausea, muscular spasms, and respiratory paralysis. Since an antivenom is available, human deaths are rare.

The brown recluse spider (*Loxosceles reclusa*) is common in the midwestern United States (figure 1*b*). The venom of a brown recluse is a hemolytic toxin. The effects of a brown recluse's bite are initially confined to the site of the bite and consist of localized tissue death and ulceration. The ulceration, however, quickly spreads to adjacent tissues and creates a large ulcer that heals very slowly. Human deaths from brown recluse bites are likewise also rare since an antivenom is available.

Of these two species of spiders, one is probably more likely to encounter the brown recluse. The black widow is usually found in the wild—under rocks and logs, in brush piles, and in other natural areas. Its distinctive marking—a ventral red hourglass pattern on a shiny black body—is very easy to recognize. The brown recluse, on the other hand, lives closely with humans and domestic animals. It is nocturnal, secretive, and may go unnoticed. During the winter months, the brown recluse is relatively inactive. Unlike the black widow, the brown recluse is a drab-looking spider. It is brown, but if one looks closely, a distinctive violin-shaped mark on the dorsal aspect of the prosoma is clearly visible. An encounter with the brown recluse is fairly likely. In the midwestern United States, most human dwellings have had or will have brown recluse spiders in residence. Controlling these spiders by spraying with pesticides is possible; but because of their reclusive habits and inactive periods, the sprays will not always reach their intended targets.

The best advice concerning these two species of spiders is to learn to recognize them, and to shake out shoes, boots, and other articles of clothing that have not been recently worn. These spiders, like all of their relatives, have exaggerated reputations—but the key to a peaceful coexistence is to respect the spiders' venomous bites.

(a)

(b)

FIGURE 1 **Two Venomous Spiders.** (*a*) A black widow spider (*Lactrodectus mactans*) is recognized by its shiny black body with a red hourglass pattern on the ventral surface of its opisthosoma. (*b*) A brown recluse spider (*Loxosceles reclusa*) is recognized by the dark brown, violin-shaped mark on the dorsal aspect of its prosoma.

Mating of spiders involves complex behaviors that include chemical, tactile, and/or visual signals. Chemicals called pheromones are deposited by a female on her web or on her body to attract a male. (Pheromones are chemicals released into the environment by one individual to create a behavioral change in another member of the same species.) A male may attract a female by plucking the strands of a female's web. The pattern of plucking is species specific and helps identify and locate a potential mate. The tips of a male's pedipalps possess a bulblike reservoir with an ejaculatory duct and a penislike structure called an embolus. Prior to mating, the male fills the reservoir of his pedipalps by depositing sperm on a small web and then collecting sperm with his pedipalps. During mating, a pedipalp is engorged with blood, the embolus is inserted into the female's reproductive opening, and sperm are discharged. The female deposits up to 3,000 eggs in a silken egg case, which is sealed and attached to webbing, placed in a retreat, or carried about by the female.

Order Opiliones

Members of the order Opiliones (o′pi-le″on-es) are the harvestmen or daddy longlegs. The prosoma is broadly joined to the opisthosoma, and thus the body appears ovoid. Legs are very long and slender. Many harvestmen are omnivores (they feed on a variety of plant and animal material), and others are strictly predators. Prey are seized by pedipalps and ingested as described for other arachnids. Digestion is both external and internal. Sperm transfer is direct, as males have a penislike structure. Females have a tubular ovipositor that is projected from a sheath at the time of egg laying. Hundreds of eggs are deposited in damp locations on the ground.

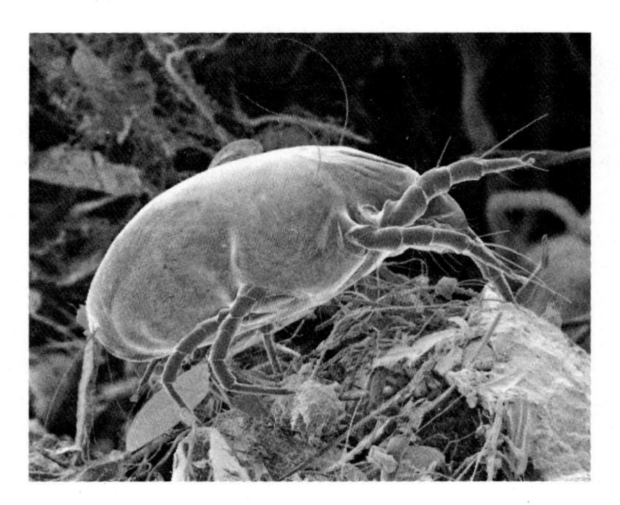

Figure 11.14

Order Acarina. *Dermatophagoides farinae* is common in homes and grain storage areas. It is believed to be a major cause of dust allergies (×200).

Order Acarina

Members of the order Acarina (ak′ar-i″nah) are the mites and ticks. Many are ectoparasites (parasites on the outside of the body) on humans and domestic animals. Others are free living in both terrestrial and aquatic habitats. Of all arachnids, acarines have had the greatest impact on human health and welfare.

Mites are 1 mm or less in length. The prosoma and opisthosoma are fused and covered by a single carapace. Mouthparts are carried on an anterior projection called the capitulum. Chelicerae and pedipalps are variously modified for piercing, biting, anchoring, and sucking, and adults have four pairs of walking legs.

Free-living mites may be herbivores or scavengers. Herbivorous mites, such as spider mites, cause damage to ornamental and agricultural plants. Scavenging mites are among the most common animals in soil and in leaf litter. These mites include some pest species that feed on flour, dried fruit, hay, cheese, and animal fur (figure 11.14).

Parasitic mites usually do not remain permanently attached to their hosts, but feed for a few hours or days and then drop to the ground. One mite, the notorious chigger or red bug (*Trombicula*), is a parasite during one of its larval stages on all groups of terrestrial vertebrates. (6) Host skin is enzymatically broken down and sucked by a larva, causing local inflammation and intense itching at the site of the bite. The chigger larva drops from the host and then molts to the next immature stage, called a nymph. Nymphs eventually molt to adults, and both nymphs and adults feed on insect eggs.

A few mites are permanent ectoparasites. (7) The follicle mite, *Demodex folliculorum*, is very common (but harmless) in hair follicles of most of the readers of this text. Itch mites cause scabies in humans and other animals. *Sarcoptes scabei* is the human itch mite. It tunnels in the epidermis of human skin, where females lay about 20 eggs each day. Secretions of the mites cause skin irritation, and infections are acquired by contact with an infected individual.

Ticks are ectoparasites during their entire life history. They may be up to 3 cm in length, but are otherwise similar to mites. Hooked mouthparts are used to attach to their hosts and to feed on blood. The female ticks, whose bodies are less sclerotized than those of males, expand when engorged with blood. Copulation occurs on the host, and after feeding, females drop to the ground to lay eggs. Eggs hatch into six-legged immatures called seed ticks. Immatures feed on host blood and drop to the ground for each molt. Some ticks transmit diseases to humans and domestic animals. For example, *Dermacentor andersoni* transmits the bacteria that cause Rocky Mountain spotted fever and tularemia, and *Ixodes scapularis* transmits the bacterium that causes Lyme disease (figure 11.15). Other orders of arachnids include whip scorpions, whip spiders, pseudoscorpions, and others.

Class Pycnogonida

Members of the class Pycnogonida (pik′no-gon″i-dah) are the sea spiders. All are marine and are most common in cold waters (figure 11.16). Pycnogonids live on the ocean floor and are frequently found feeding on cnidarian polyps and ectoprocts. Some sea spiders feed by sucking up prey tissues through a proboscis. Others tear at prey with their chelicerae.

Pycnogonids are dioecious. Gonads are U-shaped, and branches of the gonads extend into each leg. Gonopores are located on one of the pairs of legs. Eggs are released by the female, and as the male fertilizes the eggs, they are cemented into spherical masses and attached to a pair of elongate appendages of their male, called ovigers, where they are brooded until hatching.

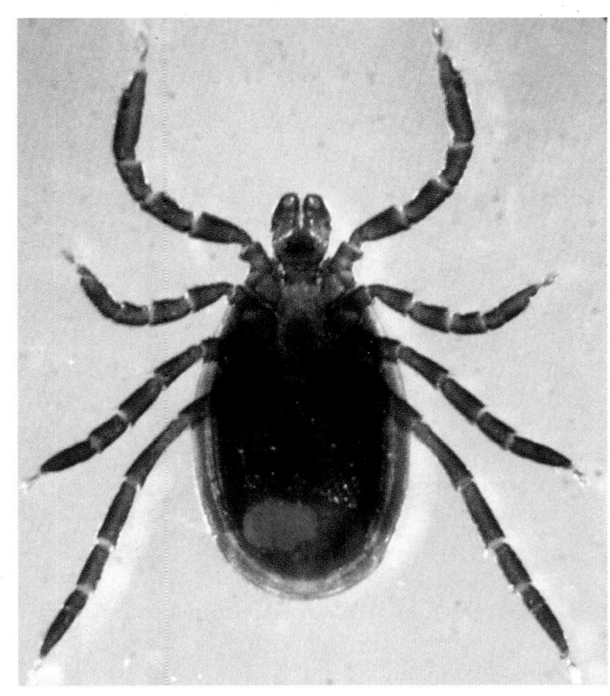

(a)

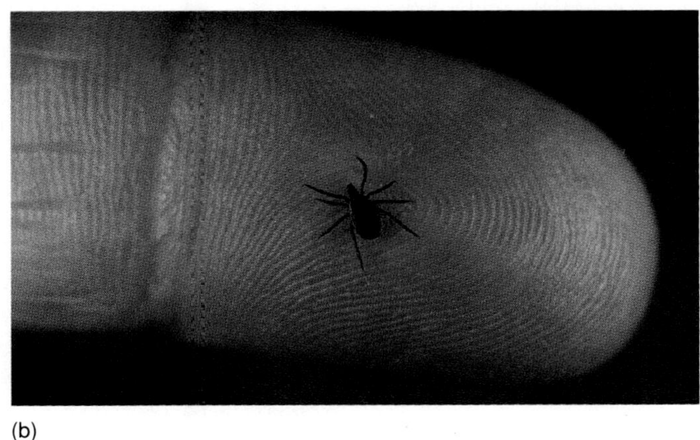

(b)

FIGURE 11.15
Order Acarina. *Ixodes scapularis*, the tick that transmits the bacterium that causes Lyme disease is shown here (*a*). The adult is about the size of a sesame seed (*b*), and the nymph is the size of a poppy seed. People walking in tick-infested regions should examine themselves regularly and remove any ticks found on their skin, because ticks can transmit diseases, such as Rocky Mountain spotted fever, tularemia, and Lyme disease.

Stop and Ask Yourself

5. How are the members of the subphylum Chelicerata characterized?
6. What are book lungs? What animals have them?
7. What characteristic of arachnids preadapted them for terrestrial environments?
8. What are three ways that spiders use silk?

FIGURE 11.16
Class Pycnogonida. Sea spiders, such as *Pycnogonum*, are often found in intertidal regions feeding on cnidarian polyps.

SUBPHYLUM CRUSTACEA

Some of the members of subphylum Crustacea (krus-tas′eah) (L. *crustaceus*, hard shelled), such as crayfish, shrimp, lobsters, and crabs, are familiar to nearly everyone. In addition, there are many lesser-known, but very common, taxa. These include copepods, cladocerans, fairy shrimp, isopods, amphipods, and barnacles. Except for some isopods and crabs, crustaceans are all aquatic.

Crustaceans differ from other living arthropods in two ways. They have two pairs of antennae, whereas all other arthropods have one pair or none. In addition, crustaceans possess biramous appendages, each of which consists of a basal segment, called the **protopodite,** with two rami (distal processes that give the appendage a Y shape) attached. The medial ramus is the **endopodite** and the lateral ramus is the **exopodite** (figure 11.17). A similar condition was described for the trilobites and may be evidence that the trilobites were ancestral to the crustaceans.

CLASS MALACOSTRACA

Malacostraca (mal-ah-kos′trah-kah) (Gr. *malakos*, soft + *ostreion*, shell) is the largest class of crustaceans. It includes crabs, lobsters, crayfish, shrimp, mysids, shrimplike krill, isopods, and amphipods.

The order Decapoda (dek-i-pod′ah) is the largest order of crustaceans and includes shrimp, crayfish, lobsters, and crabs. Shrimp have a laterally compressed, muscular abdomen and pleopods that are used for swimming. Lobsters, crabs, and crayfish are adapted to crawling on the surface of the substrate (figure 11.18). The abdomen of crabs is greatly reduced and is held flexed beneath the cephalothorax.

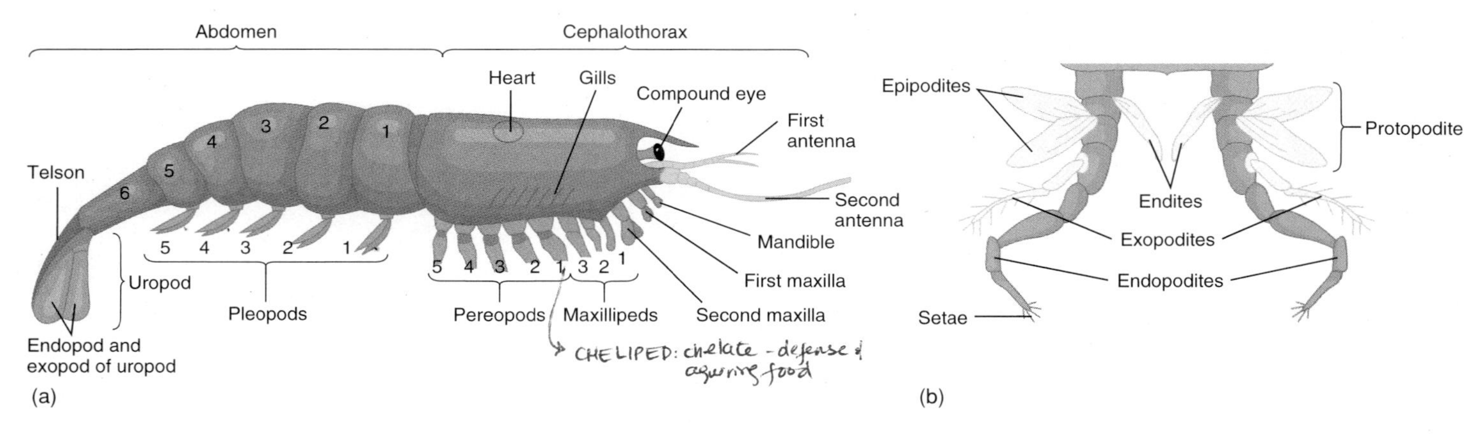

FIGURE 11.17
The Crustacean Body Form. (*a*) The external anatomy of a crustacean. (*b*) A pair of appendages showing the generalized biramous structure. A protopodite attaches to the body wall. An exopodite (a lateral ramus) and an endopodite (a medial ramus) attach at the end of the protopodite. In modern crustaceans, both the distribution of appendages along the length of the body and the structure of appendages are modified for specialized functions.

FIGURE 11.18
Order Decapoda. The lobsters, shrimp, crayfish, and crabs comprise the largest crustacean order. The lobster, *Homarus americanus*, is shown here.

Crayfish are often used to illustrate general crustacean structure and function. They are convenient to study because of their relative abundance and large size. The body of a crayfish is divided into two regions. A cephalothorax is derived from developmental fusion of a sensory and feeding tagma (the head) with a locomotor tagma (the thorax). The exoskeleton of the cephalothorax extends laterally and ventrally to form a shieldlike carapace. The abdomen is posterior to the cephalothorax, has locomotor and visceral functions, and in crayfish, takes the form of a muscular "tail."

Paired appendages are present in both body regions (figure 11.19). The first two pairs of cephalothoracic appendages are the first and second antennae. The third through the fifth pairs of appendages are associated with the mouth. During crustacean evolution, the third pair of appendages became modified into chewing or grinding structures called **mandibles.** The fourth and fifth pairs of appendages, called **maxillae,** are used for food handling. The second maxilla bears a gill and a thin, bladelike structure, called a scaphognathite (gill bailer), used to circulate water over the gills. The sixth through the eighth cephalothoracic appendages are called maxillipeds and are derived from the thoracic tagma. They are accessory sensory and food handling appendages. Each also bears a gill. Appendages 9 to 13 are thoracic appendages called periopods (walking legs). The first periopod, known as the cheliped, is enlarged and chelate (pincherlike) and used in defense and capturing food. All but the last pair of appendages of the abdomen are called pleopods (swimmerets) and used for swimming. In females, developing eggs are attached to pleopods, and the embryos are brooded until after hatching. In males, the first two pairs of pleopods are modified into gonopods (claspers) that are used for sperm transfer during copulation. The abdomen ends in a median extension called the telson. The telson bears the anus and is flanked on either side by flattened, biramous appendages of the last segment, called uropods. The telson and uropods make an effective flipperlike structure used in swimming and in escape responses.

All crustacean appendages, except the first antennae, have presumably evolved from an ancestral biramous form, as evidenced by their embryological development, in which they arise as simple two-branched structures. (First antennae develop as uniramous appendages and later acquire the branched form. The crayfish and their close relatives are unique in having branched first antennae.) Structures, such as the biramous appendages of a crayfish, whose form is based upon a common ancestral pattern and that have similar development in the segments of an animal, are said to be **serially homologous.**

Crayfish prey upon other invertebrates, eat plant matter, and scavenge dead and dying animals. The foregut includes an enlarged stomach, part of which is specialized for grinding. A digestive gland secretes digestive enzymes and absorbs products of digestion. The midgut extends from the stomach and is often called the intestine. A short hindgut ends in an anus and is important in water and salt regulation (figure 11.20*a*).

As described above, the gills of a crayfish are attached to the bases of some cephalothoracic appendages. Gills are located

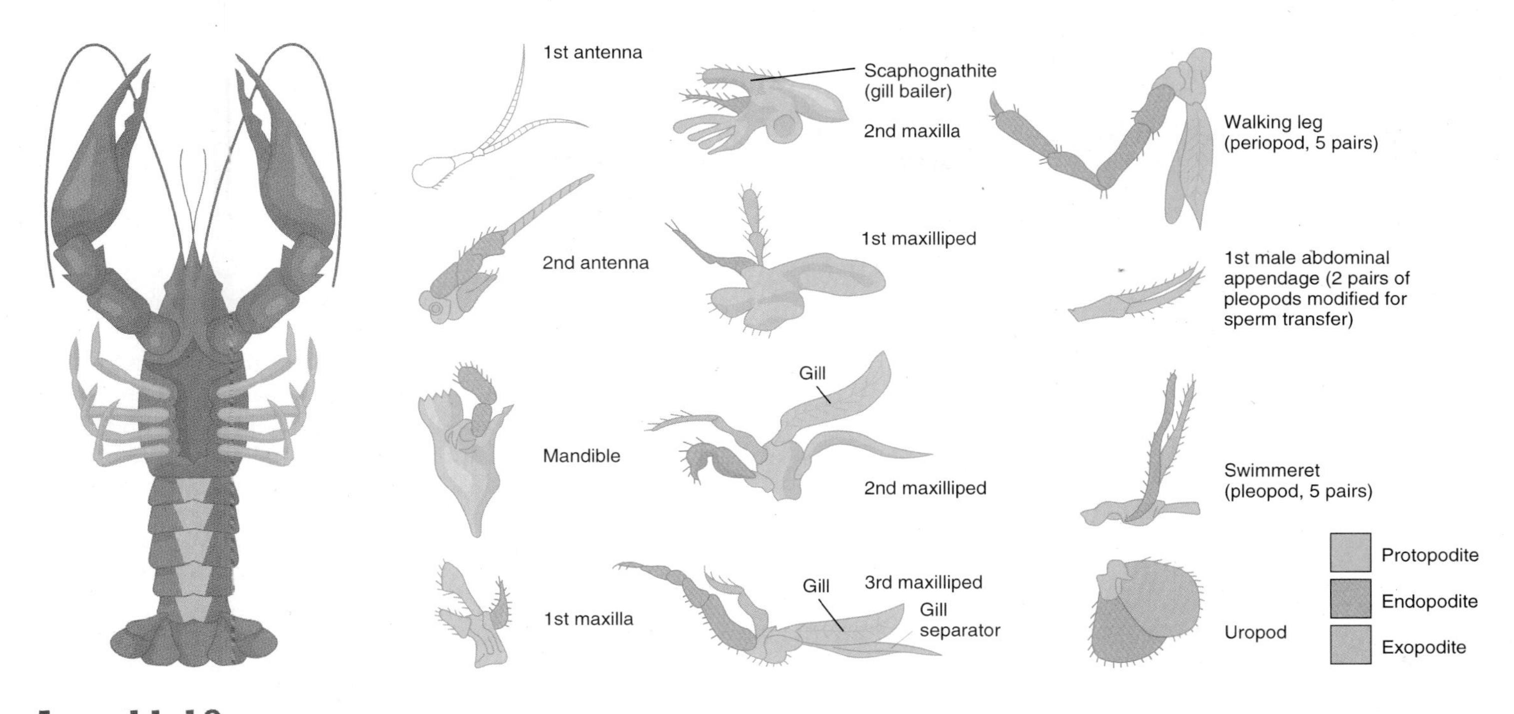

FIGURE 11.19

Crayfish Appendages. Ventral view of a crayfish. Appendages removed and arranged in sequence. Homologies regarding the structure of appendages are color coded. The origin and homology of the first antennae are uncertain.

in a branchial (gill) chamber, the space between the carapace and the lateral body wall (figure 11.20*b*). Water is driven anteriorly through the branchial chamber by the beating of the scaphognathite of the second maxilla. Oxygen and carbon dioxide are exchanged between blood and water across the gill surfaces, and oxygen is carried in blood plasma by a respiratory pigment, hemocyanin.

Circulation in crayfish is similar to that of most arthropods. Dorsal, anterior, and posterior arteries lead away from a muscular heart. Branches of these vessels empty into sinuses of the hemocoel. Blood returning to the heart collects in a ventral sinus and enters the gills before returning to the pericardial sinus, which surrounds the heart (figure 11.20*b*).

Crustacean nervous systems show trends similar to those in annelids and arachnids. Primitively, the ventral nervous system is ladderlike. Higher crustaceans show a tendency toward centralization and cephalization. In crayfish, there are supraesophageal and subesophageal ganglia that receive sensory input from receptors in the head and control the head appendages. There is a fusion of the ventral nerves and segmental ganglia, and giant neurons in the ventral nerve cord function in escape responses (figure 11.20*a*). When nerve impulses are conducted posteriorly along giant nerve fibers of a crayfish, powerful abdominal flexor muscles of the abdomen contract alternately with weaker extensor muscles, causing the abdomen to flex (the propulsive stroke) and then extend (the recovery stroke). The telson and uropods form a paddlelike "tail" that propels the crayfish posteriorly.

In addition to antennae, the sensory structures of crayfish include compound eyes, simple eyes, statocysts, chemoreceptors, proprioceptors, and tactile setae. Chemical receptors are widely distributed over the appendages and the head. Many of the setae covering the mouthparts and antennae are chemoreceptors that are used in sampling food and detecting pheromones. A single pair of statocysts is located at the bases of the first antennae. A statocyst is a pitlike invagination of the exoskeleton that contains setae and a group of cemented sand grains called a statolith. Movements of the crayfish cause the statolith to move and displace setae. Statocysts provide information regarding movement, orientation with respect to the pull of gravity, and vibrations of the substrate. Because the statocyst is cuticular, it is replaced with each molt. Sand is incorporated into the statocyst when the crustacean is buried in sand. Other receptors involved with equilibrium, balance, and position senses are tactile receptors on the appendages and at joints. When a crustacean is crawling or resting, stretch receptors at the joints are stimulated. Tilting is detected by changing patterns of stimulation. These widely distributed receptors are very important to most crustaceans, because many lack statocysts.

Crayfish have compound eyes that are mounted on movable eyestalks. The lens system consists of 25 to 14,000 individual receptors called ommatidia. Compound eyes also occur in insects, and their physiology is discussed in chapter 12. Larval crustaceans have a single, median photoreceptor consisting of a few sensilla. These simple eyes, called ocelli, allow larval crustaceans to orient

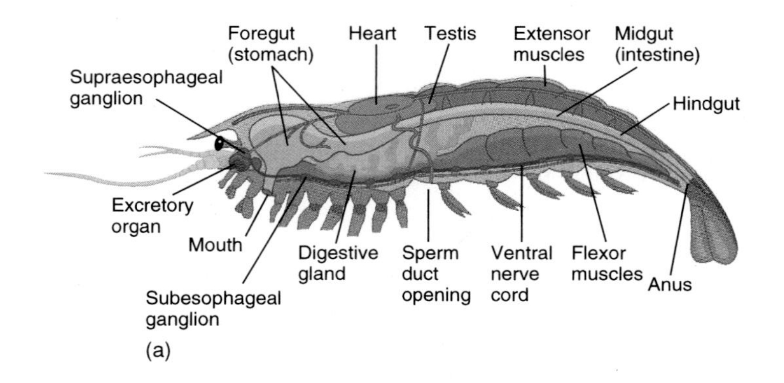

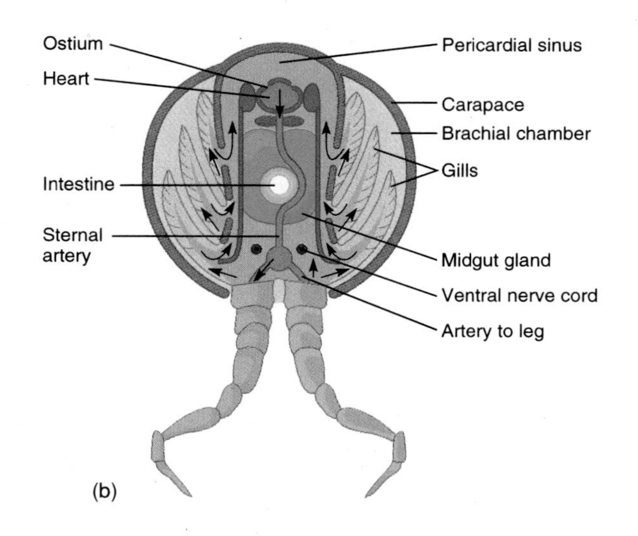

Figure 11.20

Internal Structure of a Crayfish. (*a*) Lateral view of a male. In the female, the ovary is located in the same place as the testis of the male, but the gonoducts open at the base of the third periopods. (*b*) Cross section of the thorax in the region of the heart. In this diagram, gills are shown attached higher on the body wall than they actually occur in order to show the path of blood flow (arrows) through them.

toward or away from the light but do not form images. Many larvae are planktonic and use their ocelli to orient toward surface waters.

The endocrine system of a crayfish controls functions such as ecdysis, sex determination, and color change. Endocrine glands release chemicals called hormones into the blood, where they circulate and cause responses at certain target tissues. In crustaceans, endocrine functions are closely tied to nervous functions. Nervous tissues that produce and release hormones are called neurosecretory tissues. X-organs are neurosecretory tissues located in eyestalks of crayfish. Associated with each X-organ is a sinus gland that accumulates and releases the secretions of the X-organ. Other glands, called Y-organs, are not directly associated with nervous tissues. They are located near the bases of the maxillae. Both the X-organ and the Y-organ control ecdysis. The X-organ produces molt-inhibiting hormone, and the sinus gland releases it. The target of this hormone is the Y-organ. As long as molt-inhibiting hormone is present, the Y-organ is inactive. Under certain conditions, molt-inhibiting hormone release is prevented and the Y-organ releases ecdysone hormone, leading to molting. (These "certain conditions" are often complex and species specific. They include factors such as nutritional state, temperature, and photoperiod.) Other hormones that facilitate molting have also been described, including, among others, a molt-accelerating factor.

Another endocrine function is mediated by androgenic glands, located in the cephalothorax of males. (Females possess rudiments of these glands during development, but the glands never mature.) Normally, androgenic hormone(s) promotes the development of testes and male characteristics, such as gonopods. Removal of androgenic glands from males results in the development of female sex characteristics, and if androgenic glands are experimentally implanted into a female, she will develop testes and gonopods.

Many other crustacean functions are probably regulated by hormones. Some that have been investigated include the development of brooding structures of females in response to ovarian hormones, the seasonal regulation of ovarian functions, the regulation of heart rate by eyestalk hormones, and the regulation of body color changes by eyestalk hormones.

The excretory organs of crayfish are called antennal glands (green glands) because they are located at the bases of the second antennae and are green in living crayfish. In other crustaceans, they are called maxillary glands because they are located at the base of the second maxillae. In spite of their name, they are not glands. They are structurally similar to the coxal glands of arachnids, and they presumably had a common evolutionary origin. Excretory products are formed by filtration of blood. Ions, sugars, and amino acids are reabsorbed in the tubule before the diluted urine is excreted. As with most aquatic animals, ammonia is the primary excretory product. However, crayfish do not rely solely on the antennal glands to excrete ammonia. Diffusion of ammonia across thin parts of the exoskeleton is very important. Even though it is toxic, ammonia is water soluble and rapidly diluted by water. All freshwater crustaceans face a continual influx of fresh water and loss of ions. Thus, the elimination of excess water and the reabsorption of ions become extremely important functions. Gill surfaces are also important in ammonia excretion and water and ion regulation (osmoregulation).

Crayfish, and all other crustaceans except the barnacles, are dioecious. Gonads are located in the dorsal portion of the thorax, and gonoducts open at the base of the third (females) or fifth (males) periopods. Mating occurs just after a female has molted. The male turns the female onto her back and deposits nonflagellated sperm near the openings of the female's gonoducts. Fertilization occurs after copulation, as the eggs are shed. The eggs are sticky and become securely fastened to the female's pleopods. Fanning movements of the pleopods over the eggs keeps them aerated. The development of crayfish embryos

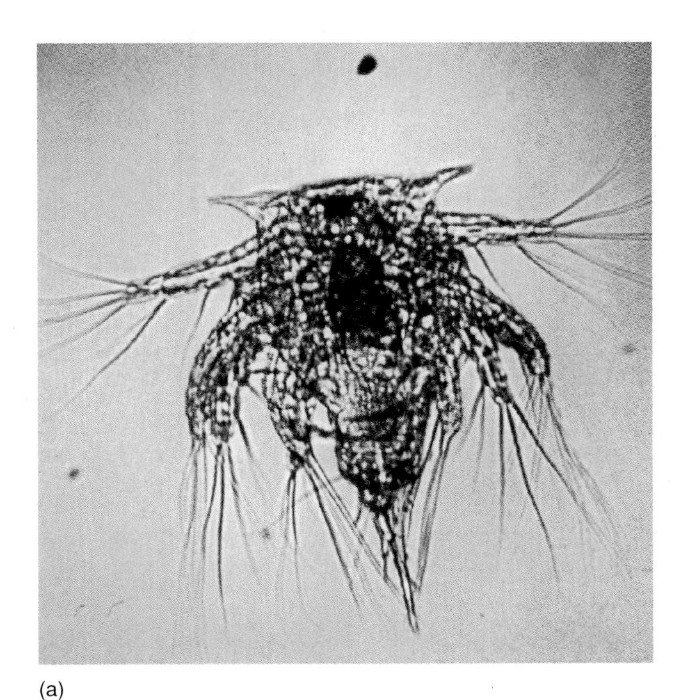

(a)

(b)

Figure 11.21
Crustacean Larvae. (*a*) Nauplius larva of a barnacle. (*b*) Zoea larvae of a crab.

(a)

(b)

Figure 11.22
Orders Isopoda and Amphipoda. (*a*) Some isopods roll into a ball when disturbed or threatened with drying—thus, the name "pillbug." (*b*) This amphipod (*Orchestoidea californiana*) spends some time out of the water hopping along beach sands—thus, the name "beachhopper."

is direct, with young hatching as miniature adults. In many other crustaceans, a planktonic, free-swimming larva called a nauplius is present (figure 11.21*a*). In some, the nauplius develops into a miniature adult. In crabs and their relatives, a second larval stage called a zoea is present (figure 11.21*b*). When all adult features are present, except sexual maturity, the immature is called the postlarva.

Two other orders of malacostracans have members that are encountered by many students. Members of the order Isopoda (i′so′pod′ah) include "pillbugs." Isopods are dorsoventrally flattened, and may be either aquatic or terrestrial, and scavenge decaying plant and animal material. Some have become modified for clinging to and feeding on other animals. Terrestrial isopods live under rocks and logs and in leaf litter (figure 11.22*a*). Members of the order Amphipoda (am-fi-pod′ah) have a laterally compressed body that gives them a shrimplike appearance. Amphipods move by crawling or swimming on their sides along the substrate. Some species are modified for burrowing, climbing, or jumping (figure 11.22*b*). Amphipods are scavengers, and a few species are parasites.

Class Branchiopoda

Members of the class Branchiopoda (bran′ke-o-pod′ah) (Gr. *branchio*, gill + *podos*, foot) are primarily found in fresh water. All branchiopods possess flattened, leaflike appendages that are used in respiration, filter feeding, and locomotion.

Fairy shrimp and brine shrimp comprise the order Anostraca (an-ost′ra-kah). Fairy shrimp are usually found in temporary ponds formed by spring thaws and rains. Eggs are brooded, and when the female dies, and the temporary pond begins to dry, the embryos become dormant in a resistant capsule. Embryos lay on the forest floor until the pond fills again the following spring, at which time they hatch into nauplius larvae. Dispersal may occur if embryos are carried to other locations by animals, wind, or water currents. Their short and uncertain life cycle is an adaptation to living in ponds that dry up. The vulnerability of these slowly swimming and defenseless crustaceans probably explains why they live primarily in temporary ponds, a habitat that contains few larger predators. Brine shrimp also form resistant embryos. They live in salt lakes and ponds (e.g., the Great Salt Lake in Utah).

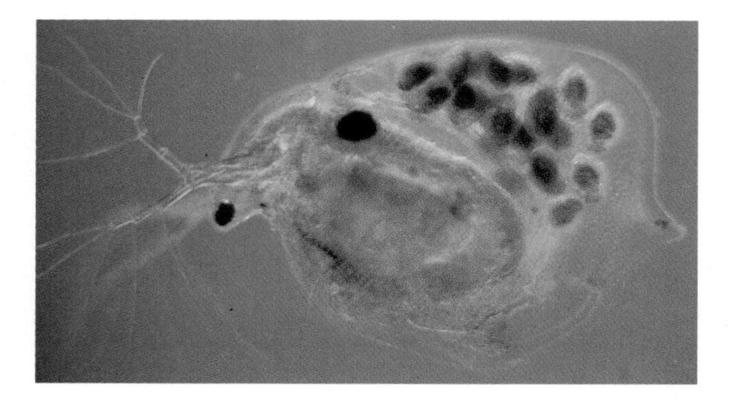

Figure 11.23
Class Branchiopoda. This cladoceran water flea (*Vetulus*) is carrying young under its carapace.

Members of the order Cladocera (kla-dos′er-ah) are called water fleas (figure 11.23). Their bodies are covered by a large carapace, and they swim using their second antennae, which they repeatedly thrust downward to create a jerky, upward locomotion. Females reproduce parthenogenetically (without fertilization) in spring and summer and can rapidly populate a pond or lake. Eggs are brooded in an egg case beneath the carapace. At the next molt, the egg case is released and either floats or sinks to the bottom of the pond or lake. In response to decreasing temperature, changing photoperiod, or decreasing food supply, females produce eggs that develop parthenogenetically into males. Sexual reproduction then occurs and produces resistant "winter eggs" that overwinter and hatch in the spring.

Class Copepoda

Members of the class Copepoda (ko′pe-pod′ah) (Gr. *kope,* oar + *podos,* foot) include some of the most abundant crustaceans. There are both marine and freshwater species. Copepods possess a cylindrical body and a median ocellus that develops in the nauplius stage and persists into the adult stage. The first antennae (and the thoracic appendages in some) are modified for swimming, and the abdomen is free of appendages. Most copepods are planktonic and use their second maxillae for filter feeding. Their importance in marine food webs was noted in the introduction to this chapter. A few copepods live on the substrate, a few are predatory, and others are commensals or parasites of marine invertebrates, fishes, or marine mammals.

Class Cirripedia

Members of the class Cirripedia (sir′i-ped′eah), the barnacles, are sessile and highly modified as adults. They are exclusively marine and include about 1,000 species. Most barnacles are monoecious (figure 11.24*a*). The planktonic nauplius of barnacles is followed by a planktonic larval stage, called a cypris larva, which has a bivalved carapace. Cypris larvae attach to the substrate by their first antennae and metamorphose to adults. In the process of metamorphosis, the abdomen is reduced, and the gut tract becomes U-shaped. Thoracic appendages are modified for filtering and moving food into the mouth. The larval carapace is covered by calcareous plates in the adult stage.

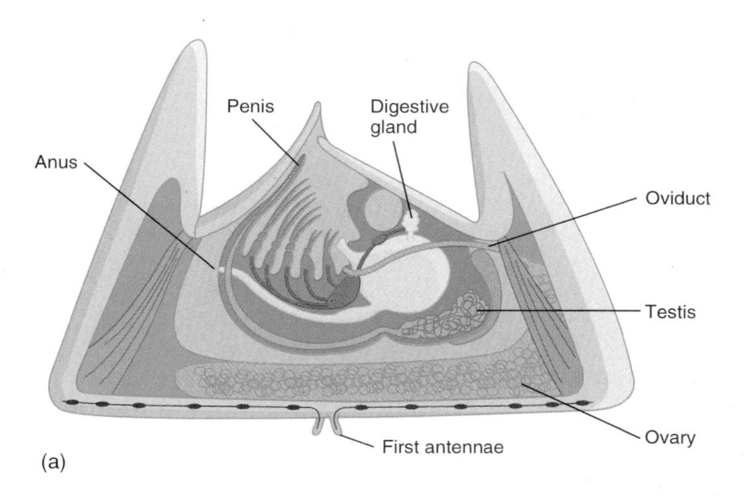

Figure 11.24
Class Cirripedia. (*a*) Internal structure of a stalkless (acorn) barnacle. (*b*) Stalked (gooseneck) barnacles (*Lepas*).

8 Barnacles attach to a variety of substrates, including rock outcroppings, the bottom of ships, whales, and other animals. Some barnacles attach to their substrate by a stalk (figure 11.24*b*). Others are nonstalked and are called acorn

BOX 11.2 *SACCULINA*: A HIGHLY MODIFIED PARASITE

Barnacles of the order Rhizocephala are parasites. Many are very similar to free-living barnacles but others, such as *Sacculina*, are some of the most highly modified of all animal parasites. Not only do adults not look like barnacles, they are difficult to recognize as animals and more closely resemble a fungus. Larval stages, however, disclose the true identity of this parasite.

The life cycle of *Sacculina* begins when a cypris larva attaches by its first antennae to a seta on a limb of a crab. The larva moves to a membranous area and bores through the crab exoskeleton. Once inside the body, the larva loses its exoskeleton, and dedifferentiated cells move through the blood to the midgut. The parasite then grows throughout the hemocoel and branches into a myceliumlike mass of parasite tissue. When the crab molts, a brood sac containing the parasite's eggs is formed in the flexed abdomen of the crab—in the same position that the female crabs normally brood their own young.

Early research indicated that all crabs parasitized by *Sacculina* were apparently females. It was later discovered that in fact males were parasitized and transformed into females by the parasite! It was once thought that the destruction of the testes caused the sex change, but it is now known that the parasite destroys the androgenic gland. Just as experimental removal of the androgenic glands transforms males to females, so does parasitization.

The crab cares for the parasite's brood sac as if it were its own. Fertilization occurs when a male cypris larva introduces sperm-forming tissue into the parasite's brood sac. Nauplius larvae are released from the brood sac, and they metamorphose into cypris larvae. Parasitism prevents further molting by the crab, results in sterility, and usually causes the crab's death.

barnacles. Barnacles that colonize the bottom of ships reduce both ship speed and fuel efficiency. Much time, effort, and money has been devoted to research on how to keep ships free of barnacles.

Some barnacles have become highly modified parasites (box 11.2). **The evolution of parasitism in barnacles is probably a logical consequence of living attached to other animals.**

FURTHER PHYLOGENETIC CONSIDERATIONS

After studying this chapter, it should be clear that the arthropods have been very successful. This is evidenced by diverse body forms and life-styles of copepods, crabs, lobsters, crayfish, and barnacles, which are an example of adaptive radiation. Few aquatic environments are without some crustaceans.

The subphylum Chelicerata is a very important group of animals from an evolutionary standpoint even though they are less numerous in terms of numbers of species and individuals than are many of the crustacean groups. **Their arthropod exoskeleton and the evolution of excretory and respiratory systems that minimize water loss, resulted in ancestral members of this subphylum becoming some of the first terrestrial animals.** Chelicerates, however, are not the only terrestrial arthropods. In terms of numbers of species and numbers of individuals, chelicerates are dwarfed in terrestrial environments by the fourth arthropod lineage—the insects and their relatives. This lineage and the evolutionary relationships within the entire phylum are the subject of the next chapter.

Stop and Ask Yourself

9. What evidence supports the hypothesis that trilobites were ancestral to the crustaceans?
10. What structures of a crayfish are serially homologous? What does this mean?
11. What are neurosecretory tissues? How are they involved in regulating crustacean metamorphosis?
12. What crustaceans belong to the class Branchiopoda and reproduce by parthenogenesis?

Summary

1. Arthropods and annelids are closely related animals. Living arthropods are divided into three subphyla: Chelicerata, Crustacea, and Uniramia. All members of a fourth subphylum, Trilobitomorpha, are extinct.
2. Arthropods have three distinctive characteristics: they are metameric and display tagmatization, they possess an exoskeleton, and many undergo metamorphosis during development.
3. Members of the extinct subphylum Trilobitomorpha had bodies that were oval and flattened and consisted of three tagmata and three longitudinal lobes. Appendages were biramous.
4. The subphylum Chelicerata has members whose bodies are divided into a prosoma and an opisthosoma. They also possess a pair of feeding appendages called chelicerae.
5. The horseshoe crabs and the giant water scorpions belong to the class Merostomata.
6. The class Arachnida includes spiders, mites, ticks, scorpions, and others. Their exoskeleton partially preadapted the arachnids for their terrestrial habitats.
7. The sea spiders are the only members of the class Pycnogonida.
8. The subphylum Crustacea contains animals characterized by two pairs of antennae and biramous appendages. All crustaceans, except for some isopods, are primarily aquatic.
9. Members of the class Branchiopoda have flattened, leaflike appendages. Examples are fairy shrimp, brine shrimp, and water fleas.
10. The class Malacostraca includes the crabs, lobsters, crayfish, shrimp, isopods, and amphipods. This is the largest crustacean class in terms of numbers of species and contains the largest crustaceans.
11. Members of the class Copepoda include the copepods.
12. The class Cirrepedia contains the barnacles, which are sessile filter feeders.

Selected Key Terms

biramous appendages (*p. 185*)
chelicerae (*p. 186*)
ecdysis (*p. 184*)
exoskeleton or cuticle (*p. 183*)
hemocoel (*p. 188*)
mandibles (*p. 194*)
oviparous (*p. 190*)
ovoviviparous (*p. 190*)
serially homologous (*p. 194*)
viviparous (*p. 190*)

Critical Thinking Questions

1. What is tagmatization, and why is it advantageous for metameric animals?
2. In spite of being an armorlike covering, the exoskeleton permits movement and growth. Explain how this is accomplished.
3. Why is the arthropod exoskeleton often cited as the major reason for arthropod success?
4. Explain why excretory and respiratory systems of ancestral arachnids probably preadapted these organisms for terrestrial habitats.
5. Barnacles are obviously very successful arthropods. What factors do you think are responsible for the evolution of their highly modified body form?

chapter

The Hexapods and Myriapods: Terrestrial Triumphs

Outline

Concepts

1. Flight, along with other arthropod characteristics, has resulted in insects becoming the most abundant and diverse group of terrestrial animals.
2. The myriapods include members of the classes Diplopoda, Chilopoda, Pauropoda, and Symphyla.
3. Members of the class Hexapoda (Insecta) are characterized by three pairs of legs, and they usually have wings.
4. Adaptations for living on land are reflected in many aspects of insect structure and function.
5. Insects have important effects on human health and welfare.
6. Some zoologists believe that the arthropods should be divided into three phyla: the Chelicerata, Crustacea, and Uniramia.

Would You Like to Know:

1. why insects have been so successful on land? (*p. 202*)
2. what animals were the first to invade terrestrial habitats? (*p. 203*)
3. how fast the wings of a midge beat? (*p. 207*)
4. how far a human could leap if one could jump, relative to body size, the same distance a flea can jump? (*p. 207*)
5. what an insect sees? (*p. 209*)
6. how insects communicate over long distances? (*p. 211*)
7. how the social organization in a honeybee hive is regulated? (*p. 214*)

These and other useful questions will be answered in this chapter.

This chapter contains evolutionary concepts, which are set off in this font.

Evolutionary Perspective

By almost any criterion, the insects have been enormously successful. There are approximately 750,000 described species, and some zoologists estimate that there may be as many as 30 million species of insects! Most of the undescribed species are found in tropical rain forests. The described species of insects comprise ¾ of all living species. Obviously, the total number of described and undescribed insects dwarfs all other kinds of living organisms. Although there are numerous freshwater and parasitic species, the success of insects has largely been due to their ability to exploit terrestrial habitats (figure 12.1).

During the late Silurian and early Devonian periods (about 400 million years ago), terrestrial environments were largely uninhabited by animals. Low-growing herbaceous plants and the first forests were beginning to flourish, and enough ozone had accumulated in the upper atmosphere to filter ultraviolet radiation from the sun. Animals with adaptations that permitted life on land had a wealth of photosynthetic production available, and unlike in marine habitats, had little competition with other animals for resources. However, the problems associated with terrestrial life were substantial. Support and movement outside of a watery environment were difficult on land, as were water, ion, and temperature regulation.

What factors have permitted the insect dominance of terrestrial habitats? (1) A number of factors contributed to their success. The exoskeleton preadapted the insects for life on land. Not only is the exoskeleton supportive, but also the evolution of a waxy epicuticle enhanced the exoskeleton's water-conserving properties. The evolution of flight has also played a big role in insect success. The ability to fly has allowed insects to use widely scattered food resources, to invade new habitats, and to escape unfavorable environments. These factors—along with desiccation resistant eggs, metamorphosis, high reproductive potential, and diversification of mouthparts and feeding habits—have permitted insects to become the dominant class of organisms on the earth.

The insects make up one of five classes in the subphylum Uniramia (table 12.1; figure 12.2). The four noninsect classes

Figure 12.1

Class Hexopoda. Insects were early inhabitants of terrestrial environments. The exoskeleton and the evolution of flight have contributed to the enormous success of this group of arthropods.

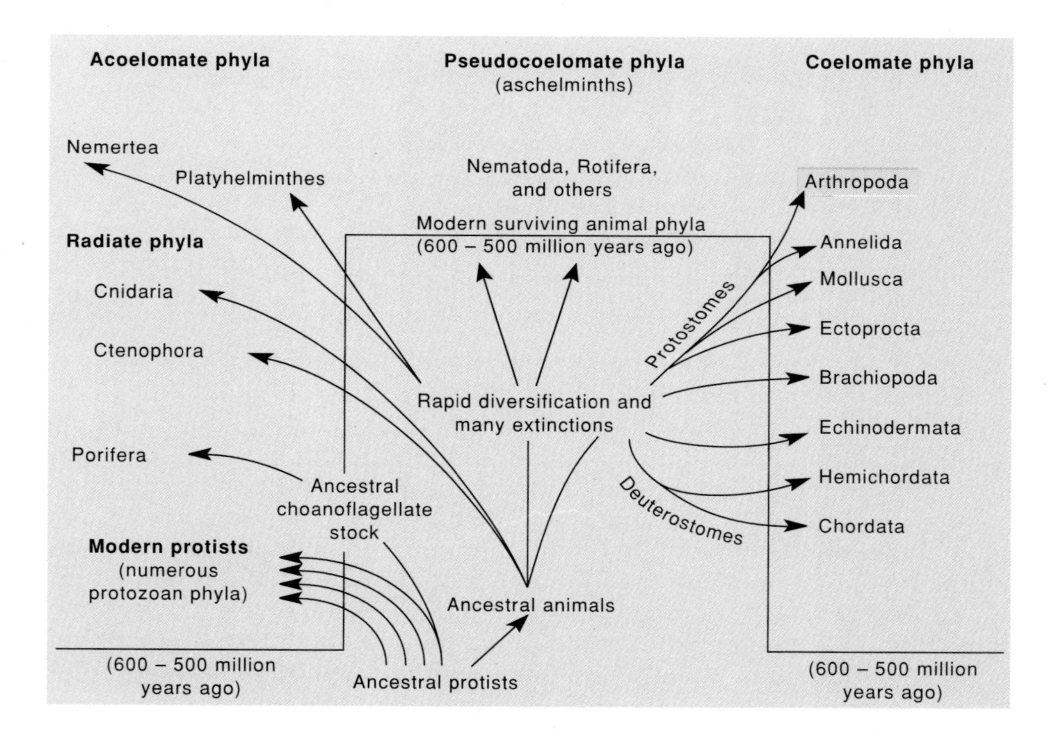

Figure 12.2

Evolutionary Relationships of the Arthropods. The arthropods (shaded in orange) are protostomes that are closely related to the annelids. Members of the subphylum Uniramia are the subject of this chapter.

(a)

(b)

Figure 12.3
The Myriapods. (*a*) A woodland millipede (*Ophyiulus pilosus*). (*b*) A centipede (*Scolopendra heros*).

(discussed next) are grouped into a convenient, nontaxonomic grouping called the **myriapods** (Gr. *myriad*, ten thousand + *podus*, foot).

Class Diplopoda

The class Diplopoda (dip'lah-pod'ah) (Gr. *diploos*, twofold + *podus*, foot) contains the millipedes. 2 Ancestors of this group appeared on land during the Devonian period and were among the first terrestrial animals. Millipedes have 11 to 100 trunk segments that have been derived from an embryological and evolutionary fusion of primitive metameres. An obvious result of this fusion is the occurrence of two pairs of appendages on each apparent trunk segment. Each segment is actually the fusion of two segments. Fusion is also reflected internally by two ganglia, two pairs of ostia, and two pairs of tracheal trunks per apparent segment. Most millipedes are round in cross section, although some are more flattened (figure 12.3*a*).

Millipedes are worldwide in distribution and are nearly always found in or under leaf litter, humus, or decaying logs. Their epicuticle does not contain much wax; therefore, their choice of habitat is important to prevent desiccation. Their many legs, simultaneously pushing against the substrate, help millipedes bulldoze through the habitat. Millipedes feed on decaying plant matter using their mandibles in a chewing or scraping fashion. A few millipedes have mouthparts modified for sucking plant juices.

Millipedes roll into a ball when faced with desiccation or when they are disturbed. Many also possess repugnatorial glands that produce hydrogen cyanide, which is repellant to other animals. Hydrogen cyanide is not synthesized and stored as hydrogen cyanide because it is very caustic and would destroy millipede tissues. Instead, a precursor compound and an enzyme are mixed as they are released from separate glandular compartments. Repellants increase the likelihood that the millipede will be dropped unharmed and decrease the chances that the same predator will try to feed on another millipede.

Sperm are transferred to the female millipede with modified trunk appendages of the male, called gonopods, or in spermatophores. Eggs are fertilized as they are laid, and hatch in several weeks. Immatures acquire more legs and segments with each molt until they reach adulthood.

Class Chilopoda

Members of the class Chilopoda (ki'lah-pod'ah) (Gr. *cheilos*, lip + *podus*, foot) are the centipedes. Most centipedes are nocturnal and spend their time scurrying about the surfaces of logs, rocks, or other forest-floor debris. As do millipedes, most centipedes lack a waxy epicuticle, and therefore are found in moist habitats. Their bodies are flattened in cross section, and they have a single pair of long legs on each of their 15 or more trunk segments. The last pair of legs is usually modified into long sensory appendages.

Centipedes are fast-moving predators. Food usually consists of small arthropods, earthworms, and snails; however, some feed on frogs and rodents (figure 12.3*b*). Poison claws (modified first trunk appendages called maxillipeds) are used to kill or immobilize prey. Maxillipeds, along with mouth appendages, hold the prey as mandibles chew and ingest the food. The venom of most centipedes is essentially harmless to humans, although many centipedes have bites that are comparable to wasp stings; a few human deaths have been reported from large, tropical species.

TABLE 12.1 CLASSIFICATION OF THE SUBPHYLUM UNIRAMIA*

Phylum Arthropoda (ar′thra-po′dah)
Animals with metamerism and tagmatization, a jointed exoskeleton, and a ventral nervous system.
Subphylum Uniramia (yoo′ne-ram′eah) (L. *unis*, one + *ramis*, branch)
Head with one pair of antennae and usually one pair of mandibles; all appendages uniramous.
Class Diplopoda (dip′le-pod′ah)
Two pairs of legs per apparent segment; body usually round in cross section. Millipedes.
Class Chilopoda (ki′le-pod′ah)
One pair of legs per segment; body oval in cross section; poison claws. Centipedes.
Class Pauropoda (por′e-pod′ah)
Small (0.5 to 2 mm), soft-bodied animals; 11 segments; nine pairs of legs; live in leaf mold. Pauropods.
Class Symphyla (sim-fi′lah)
Small (2 to 10 mm); long antennae; centipedelike; 10 to 12 pairs of legs; live in soil and leaf mold. Symphylans.
Class Hexapoda (hex′sah-pod′ah)**
Three pairs of legs; usually two pairs of wings; body with head, thorax, and abdomen; mandibulate mouthparts variously adapted. Insects.
Subclass Apterygota (ap-ter-i-go′tah)
Primitively wingless insects; pregenital abdominal appendages; ametabolous metamorphosis; indirect sperm transfer.
Order Collembola (col-lem′bo-lah)
Antennae with four to six segments; compound eyes absent; abdomen with six segments, most with springing appendage on fourth segment; inhabit soil and leaf litter. Springtails.
Order Protura (pro-tu′rah)
Minute, with cone-shaped head; antennae, compound eyes, and ocelli absent; abdominal appendages on first three segments; inhabit soil and leaf litter. Proturans.
Order Diplura (dip-lu′rah)
Head with many segmented antennae; compound eyes and ocelli absent; cerci multisegmented or forcepslike; inhabit soil and leaf litter. Diplurans.
Order Thysanura (thi-sa-nu′rah)
Tapering abdomen; flattened; scales on body; terminal cerci; long antennae. Silverfish.
Subclass Pterygota (ter-i-go′tah)
Insects descendant from winged ancestors. No pregenital abdominal appendages; direct sperm transfer.
Superorder Exopterygota (eks-op-ter-i-go′tah)
Paurometabolous (or hemimetabolous) metamorphosis; wings develop as external wing pads.
Order Ephemeroptera (e-fem-er-op′ter-ah)
Elongate, abdomen with two or three tail filaments; two pairs of membranous wings with many veins; forewings triangular; short bristlelike antennae. Mayflies.
Order Odonata (o-do-nat′ah)
Elongate, membranous wings with netlike venation; abdomen long and slender; compound eyes occupy most of head. Dragonflies, damselflies.
Order Phasmida (fas′mi-dah)
Body elongate and sticklike; wings reduced or absent; some tropical forms are flattened and leaflike. Walking sticks, leaf insects.
Order Orthoptera (or-thop′ter-ah)
Forewing long, narrow, and leathery; hindwing broad and membranous; chewing mouthparts. Grasshoppers, crickets, katydids.

*Selected orders of insects are described.
**Most entomologists now use the term "Hexapoda" as the inclusive class name. The term "Insecta" is used in a more restricted sense to refer to ectognathous hexapods (those whose mouthparts are more or less exposed). The common use of the term "insect" to refer to hexapods in general is followed in this textbook.

Centipede reproduction may involve courtship displays in which the male lays down a silk web using glands at the posterior tip of the body. A spermatophore is placed in the web and picked up by the female who introduces the spermatophore into her genital opening. Eggs are fertilized as they are laid. A female may brood and guard eggs by wrapping her body around the eggs, or they may be deposited in the soil. Young are similar to adults except that they have fewer legs and segments. Legs and segments are added with each molt.

CLASSES PAUROPODA AND SYMPHYLA

Members of the class Pauropoda (por′o-pod′ah) (Gr. *pauros*, small + *podus*, foot) are soft-bodied animals with 11 segments. These animals live in forest-floor litter, where they feed on fungi, humus, and other decaying organic matter. Their very small size and thin, moist exoskeleton allow gas exchange across the body surface and diffusion of nutrients and wastes in the body cavity.

Order Mantodea (man-to′deah)
Prothorax long; prothoracic legs long and armed with strong spines for grasping prey; predators. Mantids.
Order Blattaria (blat-tar′eah)
Body oval and flattened; head concealed from above by a shieldlike extension of the prothorax. Cockroaches.
Order Isoptera (i-sop′ter-ah)
Workers white and wingless; front and hind wings of reproductives of equal size; reproductives and some soldiers may be sclerotized; abdomen broadly joins thorax; social. Termites.
Order Dermaptera (der-map′ter-ah)
Elongate; chewing mouthparts; threadlike antennae; abdomen with unsegmented forcepslike cerci. Earwigs.
Order Phthiraptera (fthi-rap′ter-ah)
Small, wingless ectoparasites of birds and mammals; body dorsoventrally flattened; white. Sucking and chewing lice.
Order Hemiptera (hem-ip′ter-ah)
Proximal portion of forewing sclerotized, distal portion membranous; sucking mouthparts arise ventrally on anterior margin of head. True bugs.
Order Homoptera (ho-mop′ter-ah)
Wings entirely membranous; mouthparts arise ventrally on posterior margin of head (hypognathous). Cicadas, leafhoppers, aphids, whiteflies, scale insects.
Order Thysanoptera (thi-sa-nop′ter-ah)
Small bodied; sucking mouthparts; wings narrow and fringed with long setae; plant pests. Thrips.
Superorder Endopterygota (en-dop-ter-i-go′tah)
Holometabolous metamorphosis; wings develop internally during the pupal stage.
Order Neuroptera (neu-rop′ter-ah)
Wings membranous; hind wings held rooflike over body at rest. Lacewings, snakeflies, antlions, dobsonflies.
Order Coleoptera (ko-le-op′ter-ah)
Forewings sclerotized, forming covers (elytra) over the abdomen; hindwings membranous; chewing mouthparts; the largest insect order. Beetles.
Order Trichoptera (tri-kop′ter-ah)
Mothlike with setae-covered antennae; chewing mouthparts; wings covered with setae and held rooflike over abdomen at rest; larvae aquatic and often dwell in cases that they construct. Caddis flies.
Order Lepidoptera (lep-i-dop′ter-ah)
Wings broad and covered with scales; mouthparts formed into a sucking tube. Moths, butterflies.
Order Diptera (dip′ter-ah)
Mesothoracic wings well developed; metathoracic wings reduced to knoblike halteres; variously modified but never chewing mouthparts. Flies.
Order Siphonaptera (si-fon-ap′ter-ah)
Laterally flattened, sucking mouthparts; jumping legs; parasites of birds and mammals. Fleas.
Order Hymenoptera (hi-men-op′ter-ah)
Wings membranous with few veins; well-developed ovipositor, sometimes modified into a sting; mouthparts modified for biting and lapping; social and solitary species. Ants, bees, wasps.

Members of the class Symphyla (sim-fil′ah) (Gr. *sym*, same + *phyllos*, leaf) are small arthropods (2 to 10 mm in length) that occupy soil and leaf mold, superficially resemble centipedes, and are often called garden centipedes. They lack eyes and have 12 leg-bearing trunk segments. The posterior segment may have one pair of spinnerets or long sensory bristles. Symphylans normally feed on decaying vegetation; however, some species are pests of vegetables and flowers.

Stop and Ask Yourself

1. What probably accounts for the dominance of insects on land over other animals?
2. What classes of arthropods are collectively called the myriapods?
3. How would you characterize members of the following classes: Diplopoda? Chilopoda? Pauropoda? Symphyla?
4. How do millipedes discourage predators?

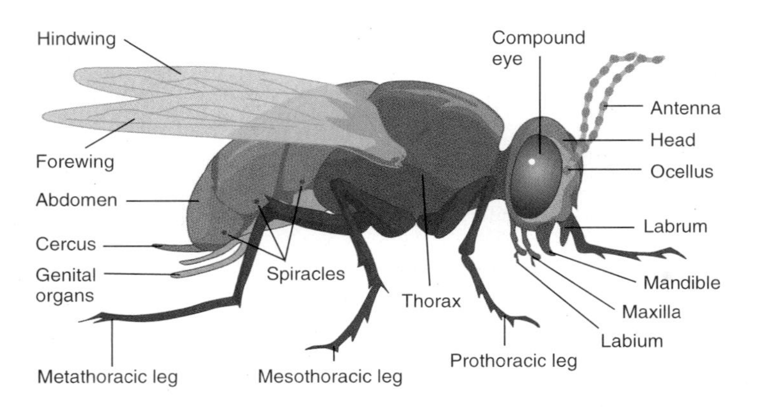

Figure 12.4

The External Structure of a Generalized Insect. Insects are characterized by a body divided into head, thorax, and abdomen; three pairs of legs; and two pairs of wings.

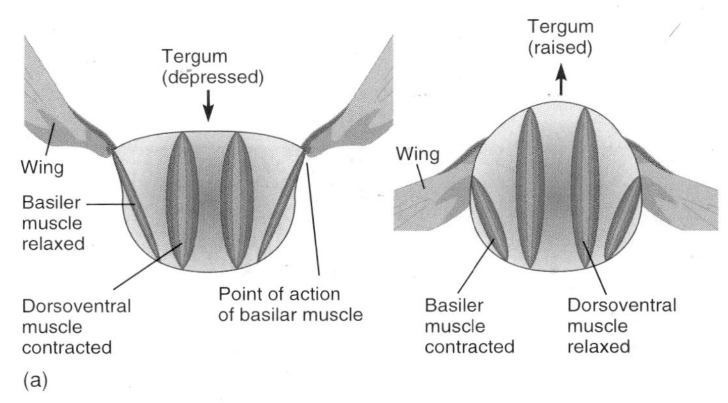

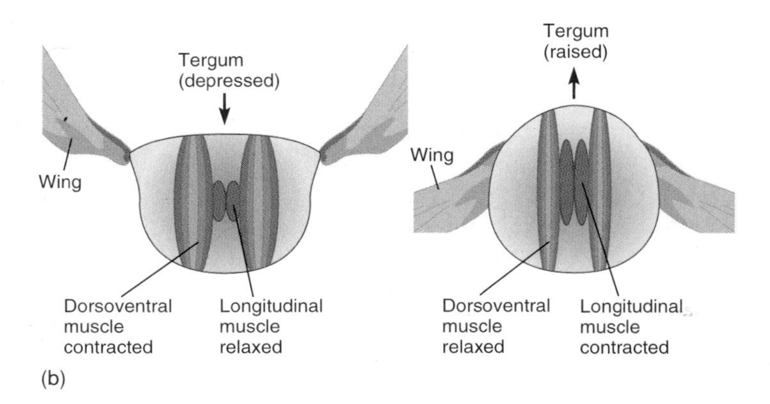

Figure 12.5

Insect Flight. (*a*) Muscle arrangements for the direct or synchronous flight mechanism. Note that muscles responsible for the downstroke attach at the base of the wings. (*b*) Muscle arrangements for an indirect or asynchronous flight mechanism. Wings move up and down as a result of muscles changing the shape of the thorax.

Class Hexapoda

Members of the class Hexapoda (Gr. *hexa*, six + *podus*, feet) are, in terms of numbers of species and individuals, the most successful land animals. Order-level classification varies depending on the authority consulted. One system is shown in table 12.1. In spite of obvious diversity, there are common features that make insects easy to recognize. Many insects have wings and one pair of antennae, and virtually all adults have three pairs of legs.

External Structure and Locomotion

The body of an insect is divided into three tagmata: head, thorax, and **abdomen** (figure 12.4). The head bears a single pair of antennae, mouthparts, compound eyes, and zero, two, or three ocelli. The thorax consists of three segments. They are, from anterior to posterior, the **prothorax,** the **mesothorax,** and the **metathorax.** One pair of legs attaches along the ventral margin of each thoracic segment, and a pair of wings, when present, attaches at the dorsolateral margin of the mesothorax and metathorax. Wings have thickened, hollow veins for increased strength. The thorax also contains two pairs of spiracles, which are openings to the tracheal system. Most insects have 10 or 11 abdominal segments, each of which has a lateral fold in the exoskeleton that allows the abdomen to expand when the insect has gorged itself or when it is full of mature eggs. Each abdominal segment has a pair of spiracles. Also present are genital structures used during copulation and egg deposition, and sensory structures called cerci. Gills are present on abdominal segments of certain immature aquatic insects.

Insect Flight

The great diversity of insects and insect habitats is accompanied by diversity in how insects move. From an evolutionary perspective, however, flight is the most important form of insect locomotion. Insects were the first animals to fly. One of the most popular hypotheses on the origin of flight states that wings may have evolved from rigid, lateral outgrowths of the thorax that probably served as protection for the legs or spiracles. Later, these fixed lobes could have been used in gliding from the top of tall plants to the forest floor. The ability of the wing to flap, tilt, and fold back over the body probably came later.

Another requirement for flight was the evolution of limited thermoregulatory abilities. Thermoregulation is the ability to maintain body temperatures at a level different from environmental temperatures. Achieving relatively high body temperatures, perhaps 25° C or greater, is needed for flight muscles to contract rapidly enough for flight.

Some insects use a **direct** or **synchronous flight** mechanism, in which a downward thrust of the wings results from the contraction of muscles acting on the bases of the wings. The upward thrust of the wings is accomplished by the contraction of muscles attaching dorsally and ventrally on the exoskeleton (figure 12.5*a*). The synchrony of direct flight mechanisms comes from the fact that each wingbeat must be preceded by a nerve impulse to the flight muscles.

Other insects use an **indirect** or **asynchronous flight** mechanism. Muscles act to change the shape of the exoskeleton for both upward and downward strokes of the wings. The upward thrust of the wing is produced by dorsoventral muscles pulling the dorsal exoskeleton downward. The downward thrust occurs when longitudinal muscles contract and cause the exoskeleton to arch upward (figure 12.5*b*). The power and velocity of these strokes are enhanced by the resilient properties of the exoskeleton. During a wingbeat, the thorax is deformed, and in the process, energy is stored in the exoskeleton. At a critical point midway into the downstroke, stored energy reaches a maximum, and at the same time, resistance to wing movement suddenly decreases. The wing then "clicks" through the rest of the cycle using energy stored in the exoskeleton. Asynchrony of this flight mechanism arises from the fact that there is no one-to-one correspondence between nerve impulses and wingbeats. (3) A single nerve impulse can result in approximately 50 cycles of the wing, and frequencies of 1,000 cycles per second (cps) have been recorded in some midges! The asynchrony between wingbeat and nerve impulses is dependent on flight muscles being stretched during the "click" of the thorax. The stretching of longitudinal flight muscles during the upward beat of the wing initiates the subsequent contraction of these muscles. Similarly, stretching during the downward beat of the wing initiates subsequent contraction of dorsoventral flight muscles. Indirect flight muscles are frequently called **fibrillar flight muscles.**

Simple flapping of wings is not enough for flight. The tilt of the wing must be controlled to provide lift and horizontal propulsion. In most insects, muscles that control wing tilting attach to sclerotized plates at the base of the wing.

Other Forms of Locomotion

Locomotion across the ground or other substrate is accomplished by walking, running, jumping, or swimming. When walking, insects have three or more legs on the ground at all times, creating a very stable stance. When running, fewer than three legs may be in contact with the ground. A fleeing cockroach (order Blattaria) reaches speeds of about 5 km/hour, although it seems much faster when trying to catch one. The apparent speed is the result of their small size and ability to quickly change directions. Jumping insects, such as grasshoppers (order Orthoptera), usually have long metathoracic legs in which leg musculature is enlarged to generate large, propulsive forces. Energy for a flea's (order Siphonaptera) jump is stored as elastic energy of the exoskeleton. Muscles that flex the legs distort the exoskeleton. A catch mechanism holds the legs in this "cocked" position until special muscles release the catches and allow the stored energy to quickly extend the legs. This action hurls the flea for distances that exceed 100 times its body length. (4) A comparable distance for a human long jumper would be the length of two football fields !

Nutrition and the Digestive System

The diversity of insect feeding habits parallels the diversity of insects themselves. Figure 12.6 shows the head and mouthparts of an insect such as a grasshopper or cockroach. An upper, liplike

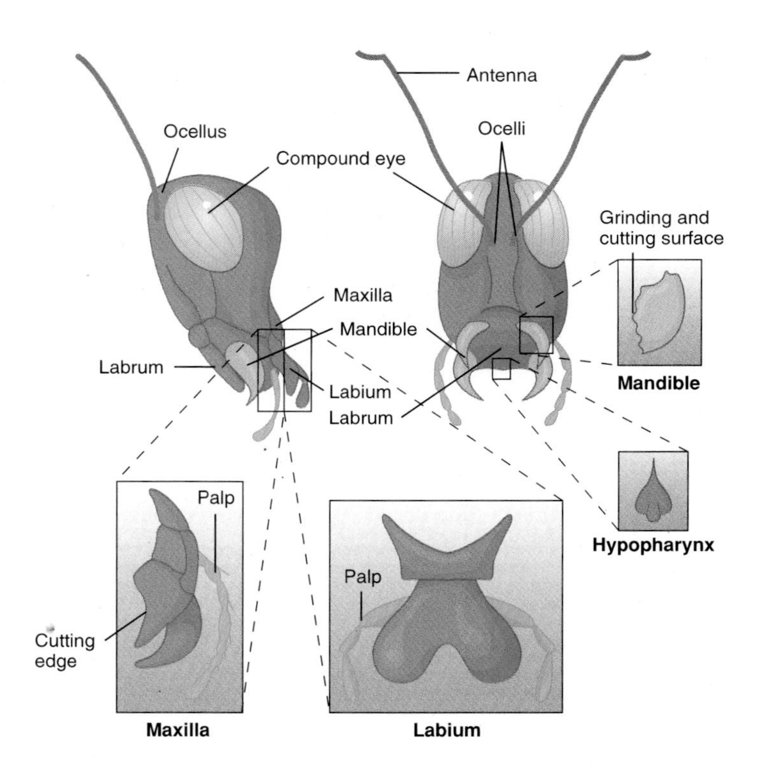

Figure 12.6
The Head and Mouthparts of a Grasshopper. All mouthparts except the labrum are derived from segmental appendages. The labrum is a sensory upper lip. The mandibles are heavily sclerotized and used for tearing and chewing. The maxillae have cutting edges and a sensory palp. The labium forms a sensory lower lip. The hypopharynx is a sensory, tonguelike structure.

structure is called the labrum. It is sensory, and unlike the remaining mouthparts, is not derived from segmental, paired appendages. Mandibles are sclerotized, chewing mouthparts. The maxillae often have cutting surfaces and bear a sensory palp. The **labium** is a sensory lower lip. All of these aid in food handling. Variations on this plan are specializations for sucking or siphoning plant or animal fluids (figure 12.7). The digestive tract, as in all arthropods, consists of a foregut, a midgut, and a hindgut (figure 12.8). Enlargements for storage and diverticula that secrete digestive enzymes are common.

Gas Exchange

Gas exchange with air requires a large surface area for the diffusion of gases. In terrestrial environments, these surfaces are also avenues for water loss. Respiratory water loss in insects, as in some arachnids, is reduced through the invagination of respiratory surfaces to form highly branched systems of chitin-lined tubes, called tracheae.

Tracheae open to the outside of the body through spiracles, which are usually provided with some kind of closure device to prevent excessive water loss. Spiracles lead to tracheal trunks that branch, eventually giving rise to smaller branches, the tracheoles. Tracheoles end intracellularly and

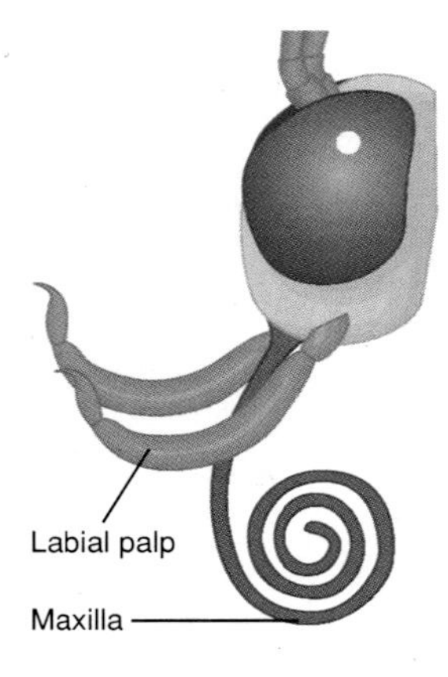

FIGURE 12.7
Specialization of Insect Mouthparts. The mouthparts of insects are often highly specialized for specific feeding habits. For example, the sucking mouthparts of a butterfly consist of modified maxillae that coil when not in use. Mandibles, labia, and the labrum are reduced in size. A portion of the anterior digestive tract is modified as a muscular pump for drawing liquids through the mouthparts.

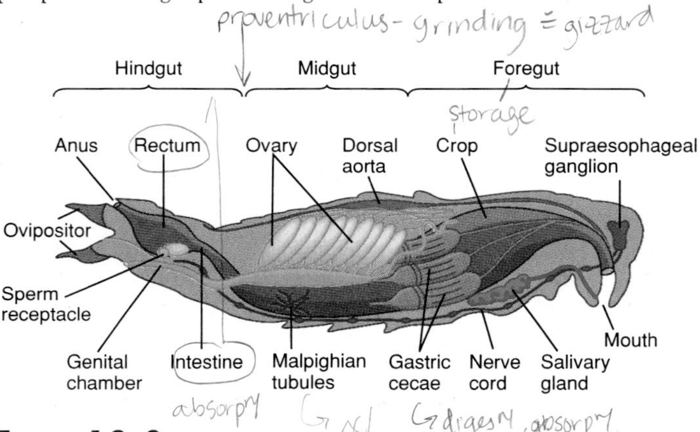

FIGURE 12.8
Internal Structure of a Generalized Insect. Salivary glands produce enzymes, but may be modified for the secretion of silk, anticoagulants, or pheromones. The crop is an enlargement of the foregut that is used for storing food. The proventriculus is a grinding and/or straining structure at the junction of the midgut and hindgut. Gastric cecae secrete digestive enzymes. The intestine and the rectum are modifications of the hindgut that function in absorbing water and the products of digestion.

are especially abundant in metabolically active tissues, such as flight muscles. No cells are more than 2 or 3 μm from a tracheole (figure 12.9).

Most insects have ventilating mechanisms that move air into and out of the tracheal system. For example, alternate compression and expansion of the larger tracheal trunks by contracting flight muscles ventilates the tracheae. In some insects, carbon dioxide produced by metabolically active cells is sequestered in the hemocoel as bicarbonate ions (HCO_3^-). As oxygen diffuses from the tracheae to the body tissues, and is not replaced by carbon dioxide, a vacuum is created that

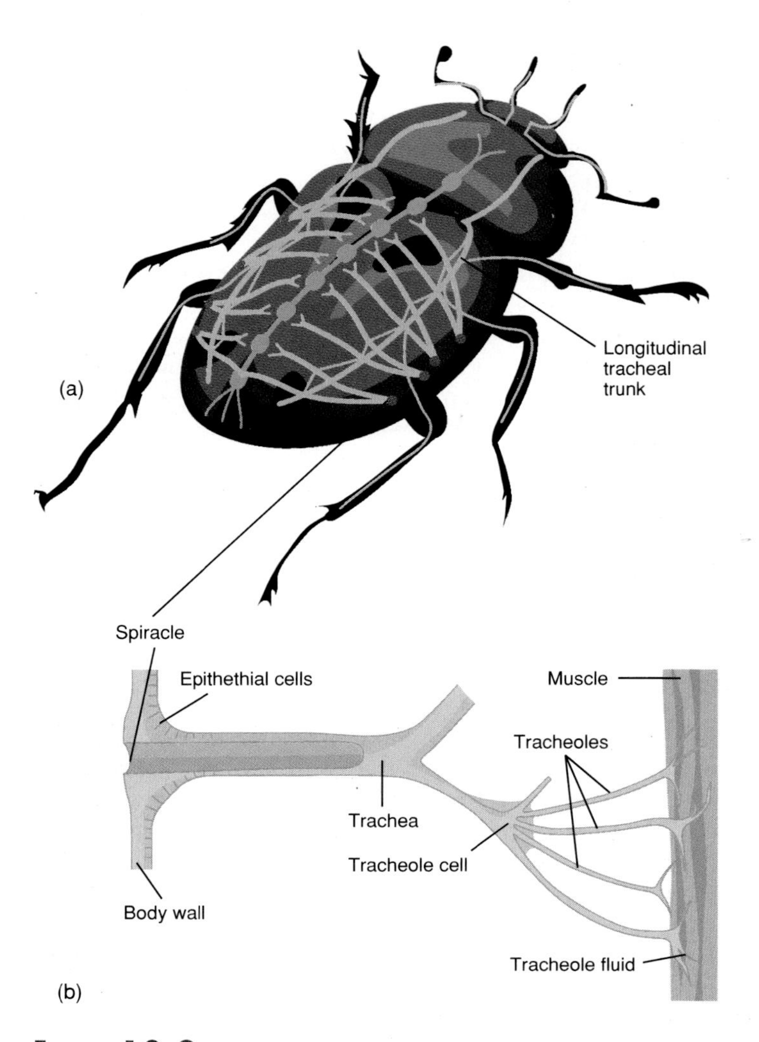

FIGURE 12.9
The Tracheal System of an Insect. (*a*) Major tracheal trunks. (*b*) Tracheoles end in cells, and the terminal portions of tracheoles are fluid filled.

draws more air into the spiracles. This process is called passive suction. Periodically, the sequestered bicarbonate ions are converted back into carbon dioxide, which escapes through the tracheal system. Other insects move air in and out of their tracheal systems by contracting abdominal muscles in a pumplike fashion.

CIRCULATION AND TEMPERATURE REGULATION

The circulatory system of insects is similar to that described for other arthropods, although the blood vessels are less well developed. Blood distributes nutrients, hormones, and wastes, and amoeboid blood cells participate in body defense and repair mechanisms. Blood is not important in gas transport.

As described earlier, thermoregulation is a requirement for flying insects. Virtually all insects warm themselves by basking in the sun or resting on warm surfaces. Because they use external heat sources in temperature regulation, insects are generally considered to be ectotherms. Other insects (e.g., some moths, alpine bumblebees, and beetles) can generate heat by rapid contraction of flight muscles, a process called shivering thermogenesis. Metabolic heat generated in this way can raise the temperature of thoracic muscles from near 0 to 30° C. Because some insects rely to a limited extent on metabolic heat sources, they have a variable body temperature and are sometimes called heterotherms. Insects are also able to cool themselves by seeking cool, moist habitats. Honeybees can cool a hive by beating their wings at the entrance of the hive, thus circulating cooler outside air through the hive.

Stop and Ask Yourself

5. How can you recognize an arthropod as being an insect?
6. Why was the evolution of thermoregulatory abilities an important step in the evolution of flight?
7. What is indirect or asynchronous flight?
8. What are the functions of the circulatory system of insects?

Nervous and Sensory Functions

The nervous system of insects is similar to the pattern described for annelids and other arthropods (*see figure 12.8*). The supraesophageal ganglion is associated with sensory structures of the head. It is joined by connectives to the subesophageal ganglion, which innervates the mouthparts and salivary glands and has a general excitatory influence on other body parts. Segmental ganglia of the thorax and abdomen undergo various degrees of fusion in different taxa. Insects also possess a well-developed visceral nervous system that innervates the gut tract, reproductive organs, and heart.

Research has demonstrated that insects are capable of some learning and have a memory. For example, bees (order Hymenoptera) instinctively recognize flowerlike objects by their shape and ability to absorb ultraviolet light, which makes the center of the flower appear dark. If a bee is rewarded with nectar and pollen, it learns the odor of the flower. Bees that feed once at artificially scented feeders choose that odor in 90% of subsequent feeding trials. Odor is a very reliable cue for bees because it is more constant than color and shape. The latter may be damaged by wind, rain, and herbivores.

Sense organs of insects are similar to those found in other arthropods, although they are usually specialized for functioning on land. Mechanoreceptors perceive physical displacement of the body or one of its parts. Setae are distributed over the mouthparts, antennae, and legs (*see figure 11.10*a). Displacement of setae may occur as a result of touch, air movements, and vibrations of the substrate. Stretch receptors at the joints, on other parts of the cuticle, and on muscles monitor posture and position.

Hearing is a mechanoreceptive sense in which airborne pressure waves displace certain receptors. All insects can respond to pressure waves with generally distributed setae; others have specialized receptors. For example, **Johnston's organs** are found in the base of the antennae of most insects, including mosquitoes and midges (order Diptera). Antennae of these insects are covered with long setae that vibrate when struck by certain frequencies of sound. The vibrating setae cause the antenna to move in its socket, stimulating sensory cells. Sound waves in the frequency range of 500 to 550 cycles per second (cps) attract and elicit mating behavior of male mosquitoes (*Aedes aegypti*). These waves are in the range of the sounds produced by the wings of females. **Tympanal (tympanic) organs** are found in the legs of crickets and katydids (order Orthoptera), in the abdomen of grasshoppers (order Orthoptera) and some moths (order Lepidoptera), and in the thorax of other moths. Tympanal organs consist of a thin, cuticular membrane covering a large air sac. The air sac acts as a resonating chamber. Just under the membrane are sensory cells that detect pressure waves. Grasshopper tympanal organs can detect sounds in the range of 1,000 to 50,000 cps. (The human ear can detect sounds between 20 and 20,000 cps.) Bilateral placement of tympanal organs allows insects to discriminate the direction and origin of a sound (box 12.1).

Chemoreception is used in feeding, selection of egg-laying sites, mate location, and in some insects, social organization. Chemoreceptors are usually abundant on the mouthparts, antennae, legs, and ovipositors and take the form of hairs, pegs, pits, and plates that have one or more pores leading to internal nerve endings. Chemicals diffuse through these pores and bind to and excite nerve endings.

All insects are capable of detecting light and may use light in orientation, navigation, feeding, or other functions. **Compound eyes** are well developed in most adult insects. They are similar in structure and function to those of other arthropods, although their possible homology (of common ancestry) with those of crustaceans, horseshoe crabs, and trilobites is debated. Compound eyes consist of a few to 28,000 receptors, called **ommatidia,** that are fused into a multifaceted eye. The outer surface of each ommatidium is a lens and is one **facet** of the eye. Below the lens is a **crystalline cone.** The lens and the crystalline cone serve as light-gathering structures. Certain cells of an ommatidium, called **retinula cells,** have special light-collecting areas, called the **rhabdom.** The cells of the rhabdom convert light energy into nerve impulses. **Pigment cells** surround the crystalline cone, and sometimes the rhabdom, and prevent the light that strikes one rhabdom from reflecting into an adjacent ommatidium (figure 12.10).

5 Although many insects form an image of sorts, the concept of an image has no real significance for most species.

BOX 12.1 Bat Echolocation and Moth Responses

Bats, some of the most successful nighttime hunters, feed on flying insects and rely on **echolocation** for finding their prey. During echolocation, bats emit ultrasonic sounds (sound at frequencies too high to be heard by humans) that reflect from flying insects back to the bats' unusually large external ears. Using this information, bats can determine the exact location of an insect and can even distinguish the kind of insect.

Noctuid moths, a prey item of these bats, possess an effective escape behavior. The tympanal organs of noctuid moths are sensitive to sounds in the 3,000 to 150,000 cps frequency range, which encompasses frequencies characteristic of a bat's cries. A weak stimulus from a bat a long distance away results in a moth flying away from the source of the sound. A stronger stimulus results in the moth's flight becoming very erratic. Often, strong stimuli result in the moth diving straight for the ground, a reflexive behavior advantageous for the moth, because sound echoing off the insect becomes indistinguishable from sound echoing off objects on the ground.

The stereotyped nature of a moth's response is explained by the structure and placement of its tympanal organs. They are located on either side of the metathoracic segment, and their bilateral placement helps a moth determine the direction of incoming sound. Sound arriving from a moth's right side strikes the right tympanal organ more strongly, because the moth's body shades the left tympanal organ from sound. Thus, the moth can determine the approximate location of the predator. In addition, each receptor consists of a cuticular membrane overlying an air-filled cavity. Nerve impulses are initiated when sound waves displace the cuticular membrane and one of two sensory cells associated with the inside of the membrane. One cell, called the A_1 cell, is stimulated by relatively low-energy sound waves. These sounds are made by a bat so far away that sounds echoing off the moth would be undetectable by the bat. In this situation, a moth turns away from the source of the sound. A second cell, called the A_2 cell, is stimulated by high-energy sound waves. These sounds are made by a bat near enough to detect the moth. The response of the moth to the activity of the A_2 cell is erratic flight and/or a dive toward the ground.

The apparent complexity of some insect behavior is deceptive. It may seem as if insects make conscious decisions in their actions; however, this is seldom the case. As with a noctuid moth's evasive responses to a bat's cries, most insect behavior patterns are reflexes programmed by specific interconnections of nerve cells.

The compound eye is better suited for detecting movement. Movement of a point of light less than 0.1° can be detected as light successively strikes adjacent ommatidia. For this reason, bees are attracted to flowers blowing in the wind, and predatory insects select moving prey. Compound eyes detect wavelengths of light that the human eye cannot detect, especially in the ultraviolet end of the spectrum. In some insects, compound eyes also detect polarized light, which may be used for navigation and orientation.

Ocelli consist of 500 to 1,000 receptor cells beneath a single cuticular lens (*see figure 11.10*b). Ocelli are sensitive to changes in light intensity and may be important in the regulation of daily rhythms.

Excretion

The primary insect excretory structures are the malpighian tubules and the rectum. Malpighian tubules end blindly in the hemocoel and open to the gut tract at the junction of the midgut and the hindgut. The inner surface of their cells is covered with microvilli. Various ions are actively transported into the tubules, and water passively follows. Uric acid is secreted into the tubules and then into the gut tract, as are amino acids and ions (figure 12.11). In the rectum, water, certain ions, and other materials are reabsorbed, and the uric acid is eliminated.

As described in chapter 11, excretion of uric acid is advantageous for terrestrial animals because it is accompanied by little water loss. There is, however, an evolutionary trade-off to consider. The conversion of primary nitrogenous wastes (ammonia) to uric acid is energetically costly. It has been estimated that nearly ½ of the food energy consumed by a terrestrial insect is used to process metabolic wastes! In aquatic insects, ammonia simply diffuses out of the body into the surrounding water.

Chemical Regulation

Many physiological functions of insects, such as cuticular sclerotization, osmoregulation, egg maturation, cellular metabolism, gut peristalsis, and heart rate, are controlled by the endocrine system. As in all arthropods, ecdysis is under neuroendocrine control. In insects, the subesophageal ganglion and two endocrine glands, the corpora allata and the prothoracic glands, control these activities.

Neurosecretory cells of the subesophageal ganglion manufacture ecdysiotropin. This hormone travels in neurosecretory cells to a structure called the corpora cardiaca. The corpora cardiaca then releases thoracotropic hormone, which stimulates the prothoracic gland to secrete ecdysone. Ecdysone initiates the reabsorption of the inner portions of the procuticle and the formation of the new exoskeleton. Other hormones are also involved

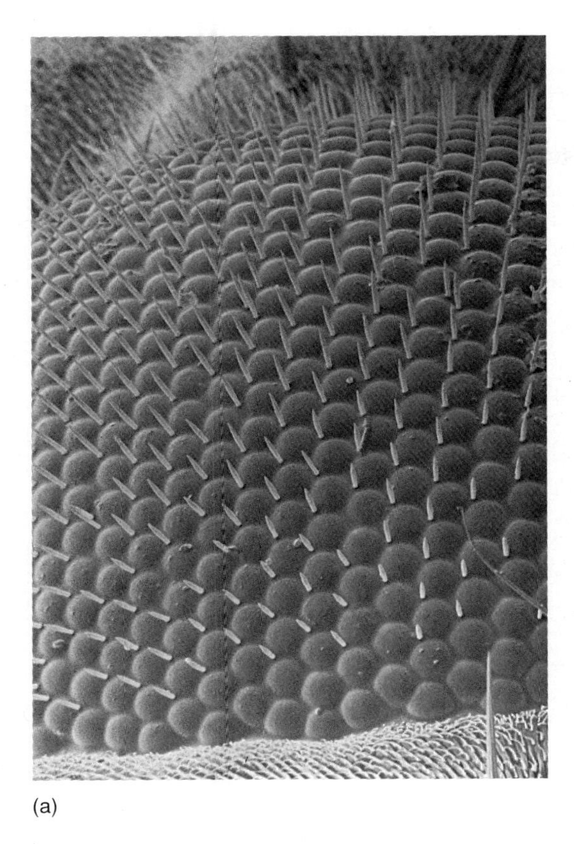

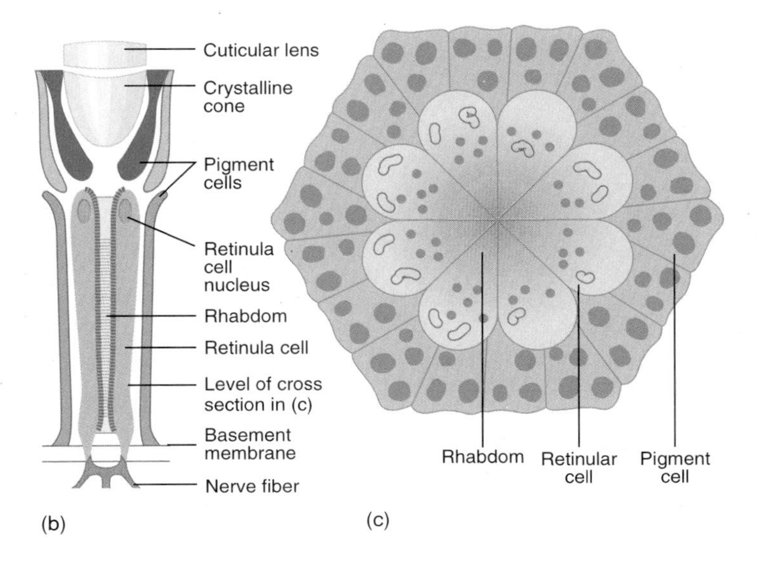

Figure 12.10

The Compound Eye of an Insect. (*a*) Scanning electron micrograph of the compound eye of *Drosophila* (×300). Each facet of the eye is the lens of a single sensory unit called an ommatidium. (*b*) The structure of an ommatidium. The lens and the crystalline cone serve as light-gathering structures. Retinula cells have light-gathering areas, called the rhabdom. Pigment cells prevent light in one ommatidium from reflecting into adjacent ommatidia. In insects that are active at night, the pigment cells are often migratory, and pigment can be concentrated around the crystalline cone. In these insects, low levels of light from widely scattered points can excite an ommatidium. (*c*) Cross section through the rhabdom region of an ommatidium.

in ecdysis. The recycling of materials absorbed from the procuticle, changes in metabolic rates, and pigment deposition are a few of probably many functions controlled by hormones.

In immature stages, the corpora allata produces and releases small amounts of juvenile hormone. The amount of juvenile hormone circulating in the hemocoel determines the nature of the next molt. Large concentrations of juvenile hormone result in a molt to a second immature stage. Intermediate concentrations of juvenile hormone result in a molt to a third immature stage. Low concentrations of juvenile hormone result in a molt to the adult stage. Decreases in the level of circulating juvenile hormone also lead to the degeneration of the prothoracic gland so that in most insects, no further molts occur once adulthood is reached. Interestingly, once the final molt has been accomplished, the level of juvenile hormone increases again, but now it promotes the development of accessory sexual organs, the synthesis of yolk, and the maturation of eggs.

Pheromones are chemicals released by an animal that cause behavioral or physiological changes in another member of the same species. Many different uses of pheromones by insects have been described (table 12.2). Pheromones are often so specific that the stereoisomer (chemical mirror image) of the pheromone may be ineffective in initiating a response. (6) They may be carried several kilometers by wind or water, and a few molecules falling on a chemoreceptor of another individual may be enough to elicit a response.

Stop and Ask Yourself

9. What evidence indicates that bees are capable of learning?
10. Compound eyes are particularly well suited for what visual function?
11. What are the endocrine functions that regulate ecdysis in insects?
12. What are pheromones? What are three kinds of functions they serve in insects?

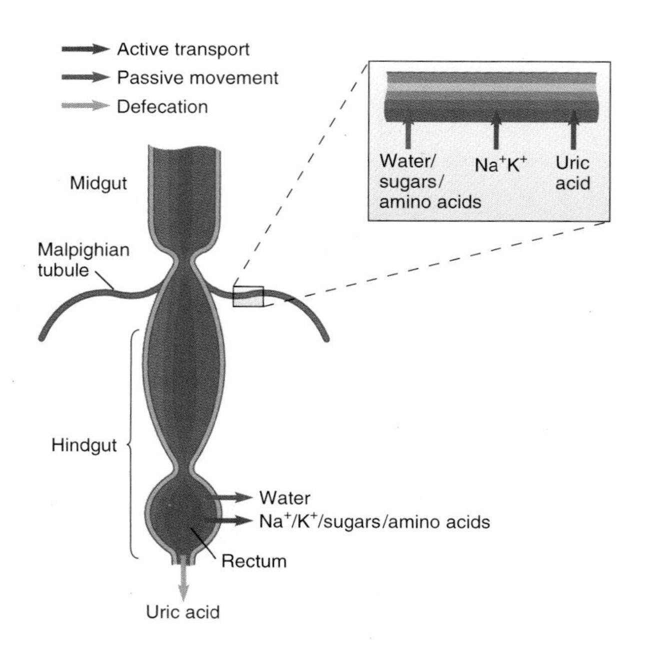

Figure 12.11

Insect Excretion. Malpighian tubules remove nitrogenous wastes from the hemocoel. Various ions are actively transported across the outer membrane of the tubule. Water follows these ions into the tubule and carries amino acids, sugars, and some nitrogenous wastes along passively. Some water, ions, and organic compounds are reabsorbed in the basal portion of the malpighian tubules and the hindgut; the rest are reabsorbed in the rectum. Uric acid moves into the hindgut and is excreted.

Table 12.2 Functions of Insect Pheromones

Functions of Insect Pheromones
Sex pheromones—Excite or attract members of the opposite sex; accelerate or retard sexual maturation. Example: female moths produce and release pheromones that attract males.
Caste regulating pheromones—Used by social insects to control the development of individuals in a colony. Example: the amount of "royal jelly" fed a female bee larva will determine whether the larva will become a worker or a queen.
Aggregation pheromones—Produced to attract individuals to feeding or mating sites. Example: certain bark beetles aggregate on pine trees during an attack on a tree.
Alarm pheromones—Warn other individuals of danger; may cause orientation toward the pheromone source and elicit a subsequent attack or flight from the source. Example: when one is stung by one bee, other bees in the area are alarmed and are likely to attack.
Trailing pheromones—Laid down by foraging insects to help other members of a colony identify the location and quantity of food found by one member of the colony. Example: ants can often be observed trailing on a pheromone path to and from a food source. The pheromone trail is reinforced each time an ant travels over it.

Reproduction and Development

One of the reasons for the success of insects is their high reproductive potential. Reproduction in terrestrial environments, however, has its risks. Temperature, moisture, and food supplies vary with the season. Internal fertilization requires highly evolved copulatory structures because gametes will dry quickly on exposure to air. In addition, mechanisms are required to bring males and females together at appropriate times.

Sexual maturity is regulated by complex interactions between internal and external environmental factors. Internal regulation includes interactions between endocrine glands (primarily the corpora allata) and reproductive organs. External regulating factors may include the quantity and quality of food. For example, the eggs of mosquitoes (order Diptera) do not mature until after the female takes a meal of blood, and the number of eggs produced is proportional to the quantity of blood ingested. The photoperiod (the relative length of daylight and darkness in a 24-hour period) is used by many insects for timing reproductive activities, because it can be used to anticipate seasonal changes. Population density, temperature, and humidity also influence reproductive activities.

A few insects, including silverfish (order Thysanura) and springtails (order Collembola) have indirect fertilization. The male deposits a spermatophore that is picked up later by the female. Most insects have a complex mating behavior that is used to locate and recognize a potential mate, to position a mate for copulation, or to pacify an aggressive mate. Mating behavior may involve the use of pheromones (moths, order Lepidoptera), visual signals (fireflies, order Coleoptera), and auditory signals (cicadas, order Homoptera; and grasshoppers, crickets, and katydids, order Orthoptera). Once other stimuli have brought the male and female together, tactile stimuli from the antennae and other appendages help position the insects for mating.

Sperm transfer is usually accomplished by abdominal copulatory appendages of the male, and sperm are stored in an outpocketing of the female reproductive tract, the sperm receptacle (*see figure 12.6*). Eggs are fertilized as they leave the female and are usually laid near the larval food supply. Females may use an **ovipositor** to deposit eggs in or on some substrate.

Insect Development and Metamorphosis

Insect evolution has resulted in the divergence of immature and adult body forms and habits. For insects in the superorder Endopterygota (*see table 12.1*), immature stages, called **larval instars,** have become a time of growth and accumulation of reserves for the transition to adulthood. The adult stage is associated with reproduction and dispersal. In these orders, there is a tendency for insects to spend a greater part of their lives in juvenile stages. The developmental patterns of insects reflect degrees of divergence between immatures and adults and are classified into three (or sometimes four) categories.

In insects that display **ametabolous** (Gr. *a*, without + *metabolos*, change) **metamorphosis,** the primary differences between adults and larvae are body size and sexual maturity.

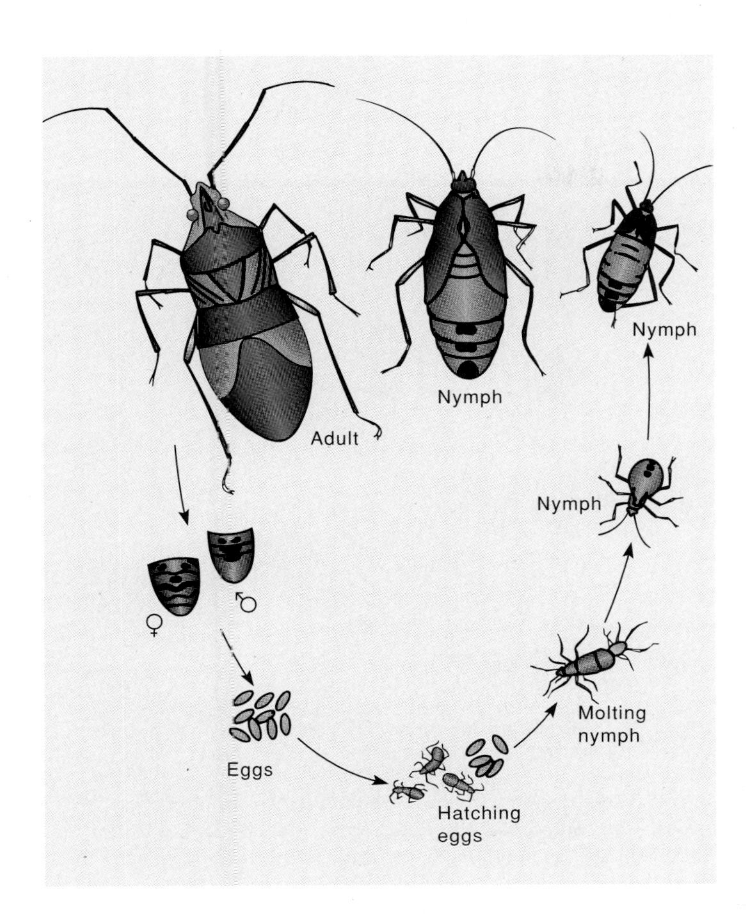

Figure 12.12

Paurometabolous Development of the Milkweed Bug, *Oncopeltus fasciatus* (Order Hemiptera). Eggs hatch into nymphs. Note the gradual increase in size of the nymphs and the development of external wing pads. In the adult stage, the wings are fully developed, and the insect is sexually mature.

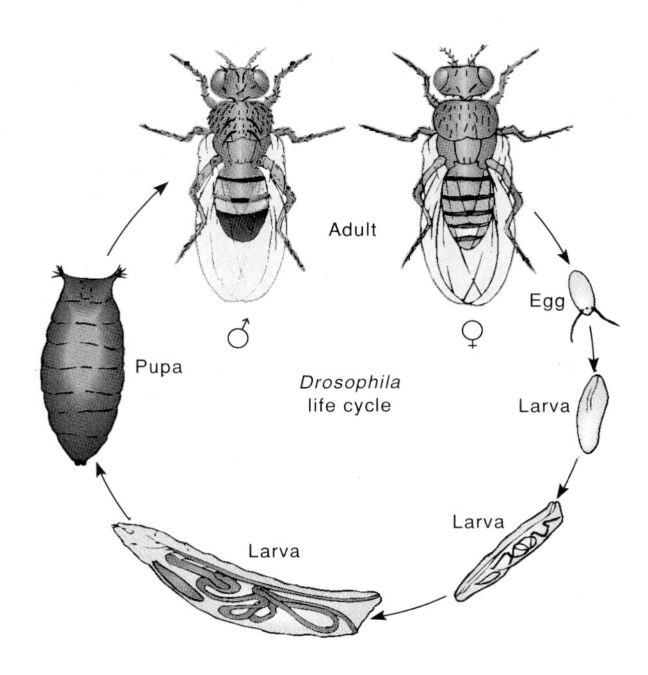

Figure 12.13

Holometabolous Development of the Fruit Fly, *Drosophila melanogaster* (Order Diptera). The egg hatches into a larva that is very different in form and habitat from the adult. After a certain number of larval instars, the insect pupates. During the pupal stage all characteristics of the adult are formed.

Both adults and larvae are wingless. The number of molts in the ametabolous development of a species is variable and, unlike most other insects, molting continues after sexual maturity is reached. Silverfish (order Thysanura) have ametabolous metamorphosis.

Paurometabolous (Gr. *pauros,* small) **metamorphosis** involves a species-specific number of molts between egg and adult stages, during which immatures gradually take on the adult form. The external development of wings (except in those insects, such as lice, that have secondarily lost wings), the attainment of adult body size and proportions, and the development of genitalia occur during this time. Immatures are called **nymphs.** Grasshoppers (order Orthoptera) and milkweed bugs (order Hemiptera) show paurometabolous metamorphosis (figure 12.12).

Some authors use an additional classification for insects that have a series of gradual changes in their development, but whose immature form is much different from the adult form usually due to the presence of gills (e.g., mayflies, order Ephemeroptera; dragonflies, order Odonata). This kind of development is called **hemimetabolous** (Gr. *hemi,* half) **metamorphosis,** and the immatures are aquatic and called **naiads** (L. *naiad,* water nymph).

In **holometabolous** (Gr. *holos,* whole) **metamorphosis,** immatures are called larvae because they are very different from the adult in body form, behavior, and habitat (figure 12.13). There is a species-specific number of larval instars, and the last larval molt forms the **pupa.** The pupa is a time of apparent inactivity but is actually a time of radical cellular change, during which all characteristics of the adult insect take form. The pupal stage may be enclosed in a protective case. A **cocoon** is constructed partially or entirely from silk by the last larval instar (e.g., moths, order Lepidoptera). The **chrysalis** (e.g., butterflies, order Lepidoptera) and **puparium** (e.g., flies, order Diptera) are the last larval exoskeletons and are retained through the pupal stage. Other insects (e.g., mosquitoes, order Diptera) have pupae that are unenclosed by a larval exoskeleton, and the pupa may be active. The final molt to the adult stage usually occurs within the cocoon, chrysalis, or puparium, and the adult then exits, frequently using its mandibles to open the cocoon or other enclosure. This final process is called emergence or eclosion.

Insect Behavior

Insects have many complex behavior patterns (*see box 12.1*). Most of these are innate (genetically programmed). For example, a newly emerged queen in a honeybee hive will search out and try to destroy other queen larvae and pupae in the hive.

BOX 12.2 Communication in Honeybees

The exploitation of food sources by honeybees has been studied for decades, but its study still offers important challenges for zoologists. One of these areas of research concerns the extent to which honeybees communicate the location of food to other bees. A worker bee that returns to a hive laden with nectar and pollen stimulates other experienced workers to leave the hive and visit productive pollen and nectar sources. Inexperienced workers are also recruited to leave the hive and search for nectar and pollen, but stronger stimuli are needed to elicit their searching behavior. In the darkness of the hive, the incoming bee performs what researchers have described as a round dance and a waggle dance. Throughout the dancing, other workers contact the dancing bee with their antennae and mouthparts, picking up the odors associated with pollen, nectar, and other objects in the vicinity of the incoming bee's food source. During the dance, the incoming bee moves first in a semicircle to the left, then in a straight course to the starting point. Next, she follows a semicircle to the right, and then another straight course to the starting point. During the linear parts of the dance, the abdomen of the bee moves in a waggling fashion (figure 1). These stimuli apparently encourage inexperienced workers to leave the hive and begin searching for food. As described in the text, their search relies heavily on olfaction, and workers tend to be attracted to pollen and nectar similar to that brought back to the hive by the dancing bee.

The round and waggle dances may also convey information on location of a food source. Biologists have found that information regarding the direction and distance of a food source from the hive are contained in the dance. The angle that the waggle dance makes with the vertical of the comb approximates the angle between the sun and the food source (figure 1). Similarly, the number of straight line runs per unit time, the duration of sounds made during the dance, and the number of waggles during the dance vary with the distance of the food source from the hive.

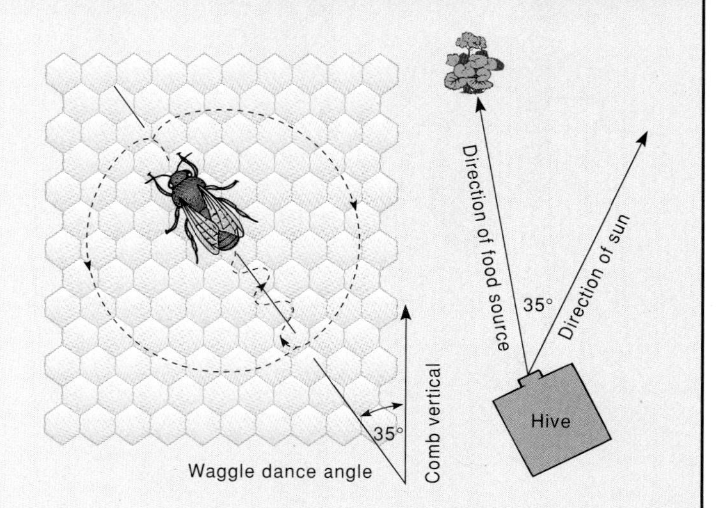

FIGURE 1 **Insect Communication.** The waggle dance of the honeybee. *From:* "A LIFE OF INVERTEBRATES" © 1979 *W. D. Russell-Hunter.*

These observations indicate that bees communicate information regarding distance, direction, and kind of food to other bees when returning from a foraging trip. Thus, the exploitation of pollen and nectar is a very efficient process and is one source of evidence of the highly evolved nature of the honeybee colony.

This behavior is innate because no experiences taught the potential queen that her survival in the hive required the death or dispersal of all other potential queens. Similarly, no experience taught her how queen-rearing cells differ from the cells containing worker larvae and pupae. Some insects are capable of learning and remembering and these abilities play important roles in insect behavior.

Social Insects

The evolution of social behavior has occurred in many insects, and is particularly evident in those insects that live in colonies (box 12.2). Usually, different members of the colony are specialized, often structurally as well as behaviorally, for performing different tasks. Social behavior is most highly evolved in the bees, wasps, and ants (order Hymenoptera) and termites (order Isoptera). Each kind of individual in an insect colony is called a **caste.** Often, three or four castes are present in a colony. Reproductive females are referred to as queens. Workers may be sterile males and females (termites) or sterile females (Hymenoptera) and are involved with support, protection, and maintenance of the colony. Their reproductive organs are often degenerate. Reproductive males inseminate the queen(s) and are called kings or drones. Soldiers are usually sterile and may possess large mandibles to defend the colony.

Honeybees (order Hymenoptera) have three of the above castes in their colonies (figure 12.14). A single queen lays all the eggs. Workers are female and they construct the comb out of wax that they produce. They also gather nectar and pollen, feed the queen and drones, care for the larvae, and guard and clean the hive. These tasks are divided among workers according to age. Younger workers take care of jobs around the hive, and older workers forage for nectar and pollen. Except for those that overwinter, workers live for about 1 month. Drones develop from unfertilized eggs, do not work, and are fed by workers until they leave the hive to attempt mating with a queen.

7 The honeybee caste system is controlled by a pheromone released by the queen. Workers lick and groom the queen and other workers. In so doing, they pick up and pass to

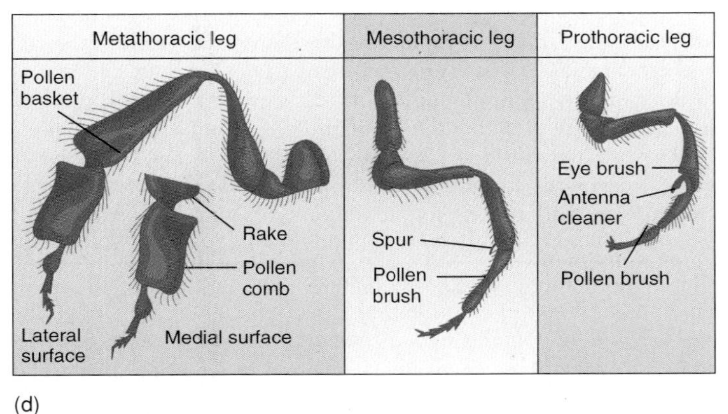

FIGURE 12.14
Honeybees (Order Hymenoptera). Honeybees have a social organization consisting of three castes. The castes are distinguished by overall body size, as well as the size of their eyes. (*a*) A worker bee. (*b*) A drone bee. (*c*) A queen bee. (*d*) The inner surface of metathoracic legs have setae, called the pollen comb, that remove pollen from the mesothoracic legs and the abdomen. Pollen is then compressed into a solid mass by being squeezed in a pollen press and moved to a pollen basket on the outer surface of the leg, where the pollen is carried. The mesothoracic legs are used for gathering pollen from body parts. The prothoracic legs of a worker bee are used to clean pollen from the antennae and body.

other workers a caste-regulating pheromone. This pheromone inhibits the rearing of new queens by workers. As the queen ages, or if she dies, the amount of caste-regulating pheromone in the hive decreases. As the pheromone decreases, workers begin to feed the food for queens ("royal jelly"), to several female larvae that are developing in the hive. This food contains chemicals that promote development of queen characteristics. The larvae that receive royal jelly develop into queens, and as they emerge, the new queens begin to eliminate each other until only one remains. The queen that remains goes on a mating flight and returns to the colony, where she will live for several years.

The evolution of social behavior with many individuals leaving no offspring and where individuals are sacrificed for the perpetuation of the colony has puzzled evolutionists for many years.

INSECTS AND HUMANS

Only about 0.5% of insect species adversely affect human health and welfare (box 12.3). Many others have provided valuable services throughout human history. Commercially valuable insect products, such as wax, honey, and silk have been utilized by humans for thousands of years. Insects are responsible for the pollination of approximately 65% of all plant species. Insects and flowering plants have coevolutionary relationships that directly benefit humans. The annual value of insect-pollinated crops is estimated at 19 billion dollars per year in the United States.

Insects also serve as agents of biological control. The classic example of the regulation of a potentially harmful insect by another insect is the control of cottony-cushion scale by vedalia (lady bird) beetles. The scale insect, *Icerya purchasi*, was introduced into California in the 1860s. Within 20 years, the citrus industry in California was virtually destroyed. The vedalia beetle (*Vedalia cardinalis*) was brought to the United States in 1888 and 1889 and cultured on trees infested with scale. In just a few years, the scale was under control, and the citrus industry began to recover.

There are many other beneficial insects. Soil-dwelling insects play important roles in aeration, drainage, and turnover of soil, and they promote decay processes. Other insects serve important roles in food webs. Insects are used in teaching and research and have contributed to advances in genetics, population ecology, and physiology. Insects have also given endless hours of pleasure to those who collect them and enjoy their beauty.

Some insects, however, are parasites and vectors of disease. Parasitic insects include head, body, and pubic lice (order Anoplura); bed bugs (order Hemiptera); and fleas (order Siphonaptera). Other insects transmit disease-causing microorganisms, nematodes, and flatworms. The impact of insect-transmitted diseases, such as malaria, yellow fever, bubonic plague, encephalitis, leishmaniasis, and typhus, has changed the course of history (*see box 5.2*).

Other insects are pests of domestic animals and plants. Some reduce the health of domestic animals and the quality of animal products. Insects feed on crops and transmit diseases of plants, such as Dutch elm disease, potato virus, and asters yellow. Annual lost revenue from insect damage to crops or insect-transmitted diseases in the United States is approximately 5 billion dollars.

FURTHER PHYLOGENETIC CONSIDERATIONS

A fundamental question regarding arthropod evolution concerns whether or not the arthropod taxa represent fundamentally different evolutionary lineages. Many zoologists believe that the living arthropods should be divided into three separate

BOX 12.3 "Killer Bees?"

The "African bee," *Apis mellifera scutellata,* is common in most parts of Africa. It is a small bee and is adapted to warm climates with extended dry seasons. In its home range, it has many enemies, including humans and birds. Aggressive behavior and frequent swarming have allowed colonies to escape predation and survive drought.

In 1956, a few *Apis mellifera scutellata* queens were imported to Brazil in hopes of breeding these queens with local stocks to create bees better adapted to tropical climates than were the local bees. (Bees of the Americas were imported from Europe in the 1600s.) A few of these queens escaped captivity and hybridized with local bees. (It is really not accurate to refer to the hybridized bees as "African bees," "Africanized bees," or "killer bees." Their reputation as "killers" has been exaggerated. Most authorities now refer to hybridized bees as Brazilian bees.) Many of the qualities that allowed the African bee to succeed in Africa allowed these Brazilian bees to spread 100 to 200 mi/year. Their frequent swarming and their ability to nest in relatively open shelters gives a distinct selective advantage over the bees imported from Europe many years earlier. By 1969, Brazilian bees had spread to Argentina; by 1973, to Venezuela; and by early 1980, they had crossed the Panama Canal. Mexico presented few barriers to their spread toward the southern United States, and they have now crossed the border into the southwestern United States. How far will Brazilian bees ultimately spread? An educated guess may be made by looking at the distribution of the African bee in Africa. African bees cannot overwinter outside of tropical regions. Similarly, the Brazilian bee will likely be limited in America to the warmer latitudes.

The primary implications of the spread of Brazilian bees has less to do with threats to human health than to the health of the beekeeping industry. The unpleasant temperament of these bees can be dealt with by wearing protective clothing. More formidable problems associated with the culture of these bees include their tendency to swarm and their lower productivity. Swarming for a beekeeper means the loss of bees and a lowered honey production. Frequent swarming makes profitable beekeeping almost impossible.

phyla: Chelicerata, Crustacea, and Uniramia. A polyphyletic origin of these groups implies convergent evolution of remarkably similar arthropodan features in all three (or at least two of three) phyla. Although there is evidence for dual origins of tracheae, mandibles, and compound eyes, convergence in all other arthropod traits seems unlikely to many zoologists.

This textbook assumes a monophyletic origin of the arthropods. Given this assumption, the phylogenetic relationships within the Arthropoda are also debated. Members of all four subphyla are present in the fossil record from the early Paleozoic era, and there currently are no known fossils of arthropods from Precambrian times. The fossil record, therefore, is of little help in discovering the evolutionary relationships among the arthropod subphyla. Zoologists must rely on comparative anatomy, comparative embryology, and molecular studies to investigate these relationships.

Central to the questions surrounding arthropod phylogeny are two important issues. One issue is whether or not the biramous limbs of crustaceans and trilobites are homologous. Homology of these appendages would imply that the trilobites were closely related to the crustaceans; therefore, many zoologists view the trilobites as an important ancestral group. It is possible to envision crustaceans, and possibly the arachnids, arising from the trilobites, but it is more difficult to envision a similar origin for the myriapods and insects. A second important issue is whether or not the mandibles of uniramians and crustaceans are homologous. Superficially, the mandibles of members of these groups are structurally similar and have similar functions. Muscle arrangements and methods of articulation, however, are different enough that many zoologists doubt that these appendages are homologous. Discussions of arthropod phylogeny also center around the origins and possible homologies of arthropod compound eyes, tracheal systems, and malpighian tubules.

Depending on how these issues have been interpreted, a number of hypotheses regarding the relationships of the arthropod subphyla have been presented. Two of these hypotheses are shown in figure 12.15. In figure 12.15*a,* the Uniramia and the Crustacea are depicted as being closely related. This hypothesis emphasizes possible homologies between the mandibles, compound eyes, and other structures of these two groups. In figure 12.15*b,* the Uniramia are shown diverging independently of the Chelicerata and Crustacea. The latter two subphyla are depicted as being closely related to the trilobites. This figure implies that the mandibles of the crustaceans and uniramians are not homologous and that the biramous appendages of the crustaceans and trilobites are homologous.

Questions regarding the evolutionary relationships within the arthropods are difficult to answer. These questions will probably remain unanswered until new fossils are discovered or data from molecular studies provide more information on ancestral arthropods.

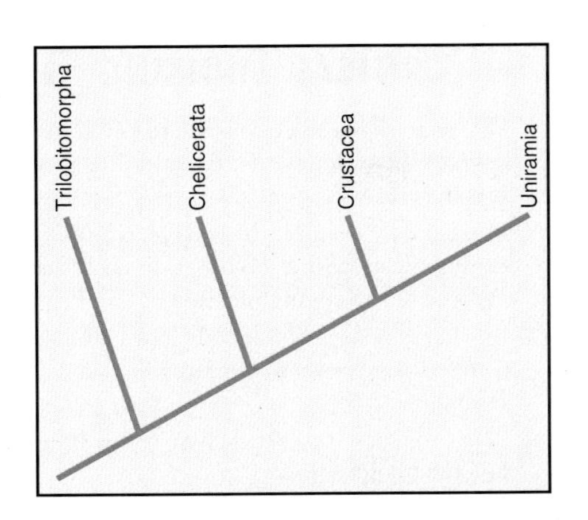

(b)
Trilobitomorpha
Crustacea
Chelicerata
Uniramia

FIGURE 12.15

Two Interpretations of Arthropod Phylogeny. A number of interpretations of arthropod phylogeny are possible. (*a*) Some zoologists think that the crustaceans and uniramians are closely related. Evidence for this hypothesis comes from possible homologies of mandibles, compound eyes, and other characters within the groups. (*b*) Other zoologists prefer to think of the uniramians as being more distantly related to other arthropods. Note that the trilobites are depicted here as an important ancestral group of arthropods. The many characters used in establishing these cladograms are not shown in these figures in order to simplify the presentation.

Stop and Ask Yourself

13. What is the sequence of changes that occurs in the life history of a holometabolous insect?
14. What are three castes in a honeybee hive and how is the caste system regulated?
15. What is the nature of the debate regarding arthropod phylogeny?

SUMMARY

1. During the Devonian period, insects began to exploit terrestrial environments. Flight, the exoskeleton, and metamorphosis are probably keys to insect success.
2. Myriapods include four classes of arthropods. Members of the class Diplopoda are the millipedes and are characterized by apparent segments bearing two pairs of legs. The centipedes are in the class Chilopoda. They are characterized by a single pair of legs on each of their 15 segments and a body that is flattened in cross section. The class Pauropoda contains soft-bodied animals that feed on fungi and decaying organic matter in forest-floor litter. Members of the class Symphyla are centipedelike arthropods that live in soil and leaf mold, where they feed on decaying vegetation.
3. Animals in the class Hexapoda are characterized by a head with one pair of antennae, compound eyes, and ocelli; a thorax with three pairs of legs and usually two pairs of wings; and an abdomen that is free of appendages except terminal sensory (cerci) and reproductive (ovipositor) structures.
4. Insect flight involves either a direct (synchronous) flight mechanism or an indirect (asynchronous) flight mechanism.
5. Mouthparts of insects are adapted for chewing, piercing, and/or sucking, and the gut tract may be modified for pumping, storage, digestion, and water conservation.
6. Gas exchange occurs through a tracheal system.
7. The insect nervous system is similar to that of other arthropods. Sensory structures include tympanal organs, compound eyes, and ocelli.
8. Malpighian tubules transport uric acid to the digestive tract. Conversion of nitrogenous wastes to uric acid conserves water but is energetically expensive.
9. Hormones regulate many insect functions, including ecdysis and metamorphosis. Pheromones are chemicals emitted by one individual that alter the behavior of another member of the same species.
10. Insect adaptations for reproduction on land include resistant eggs, external genitalia, and behavioral mechanisms that bring males and females together at appropriate times.
11. Metamorphosis of an insect may be ametabolous, paurometabolous, hemimetabolous, or holometabolous. Metamorphosis is controlled by neuroendocrine and endocrine secretions.
12. Insects show both innate and learned behavior.
13. Many insects are beneficial to humans, and a few are parasites and/or transmit diseases of humans or agricultural products. Others attack cultivated plants and stored products.
14. Whether the arthropods represent a monophyletic group or a polyphyletic group is a question that is still being debated.

Selected Key Terms

ametabolous metamorphosis (*p. 212*)
caste (*p. 214*)
direct or synchronous flight (*p. 206*)
hemimetabolous metamorphosis (*p. 213*)
holometabolous metamorphosis (*p. 213*)
indirect or asynchronous flight (*p. 206*)
larval instars (*p. 212*)
nymphs (*p. 213*)
paurometabolous metamorphosis (*p. 213*)
pupa (*p. 213*)

Critical Thinking Questions

1. What are the problems associated with living and reproducing in terrestrial environments? Explain how insects overcome these problems.
2. List as many examples as you can think of how insects communicate with each other. In each case, what is the form and purpose of the communication?
3. In what way does holometabolous metamorphosis reduce competition between immature and adult stages? Give specific examples.
4. What role does each stage play in the life history of holometabolous insects?
5. Some biologists think that the arthropods are a polyphyletic group. What does that mean? What would polyphyletic origins require in terms of the origin of the exoskeleton and its derivatives?

endpaper THREE

Some Lesser Known Invertebrates: Possible Arthropod Relatives

Three Phyla of Uncertain Affiliation: The Onychophora, Pentastomida, and Tardigrada

Animals in the phyla Onychophora, Pentastomida, and Tardigrada show a combination of arthropod and nonarthropod characteristics.

Phylum Onychophora: The Onychophorans, Velvet Worms or Walking Worms

The onychophorans (on-y-kof'o-rans) (Gr. *onyx,* claw + *pherein,* to bear), also known as velvet or walking worms, are free-living terrestrial animals that live in certain humid, tropical regions. heir ancestors have been considered an evolutionary transition between annelids and arthropods because of their many similarities to both phyla. These interesting worms may live up to 6 years. More than 100 species have been described, with *Peripatus* being the best-known genus.

Onychophorans usually come out at night and move by using their unjointed legs to crawl (figure 1). Most species are predaceous and feed on small invertebrates. In order to capture fast-moving prey, onychophorans secrete an adhesive slime (produced in their adhesive gland) from their oral papillae. Some species can eject a stream of slime with enough force to strike a prey animal 50 cm away. The slime hardens immediately, entangling the prey which is then masticated with the mandibles and sucked into the mouth.

The sexes are separate and fertilization is internal. Onychophorans are either oviparous or viviparous. The oviparous species lay large, yolky eggs, each enclosed in a shell, in moist places; cleavage is spiral. The viviparous species retain the embryos in the uterus.

(a)

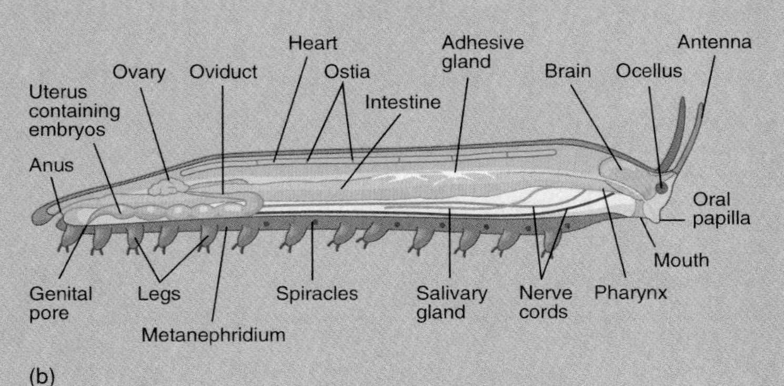

(b)

Figure 1 **Phylum Onychophora.** (*a*) *Peripatus.* (*b*) Lateral view of the internal anatomy of a female. The anterior end consists of two large antennae and ventral mouth. The mouth is surrounded by oral papillae and clawlike mandibles analogous to those of arthropods. The legs vary in number from 14 to 43 pairs; each leg has a pair of terminal claws. The entire surface of the body is covered by large and small tubercles that are covered by small scales and arranged in rings or bands.

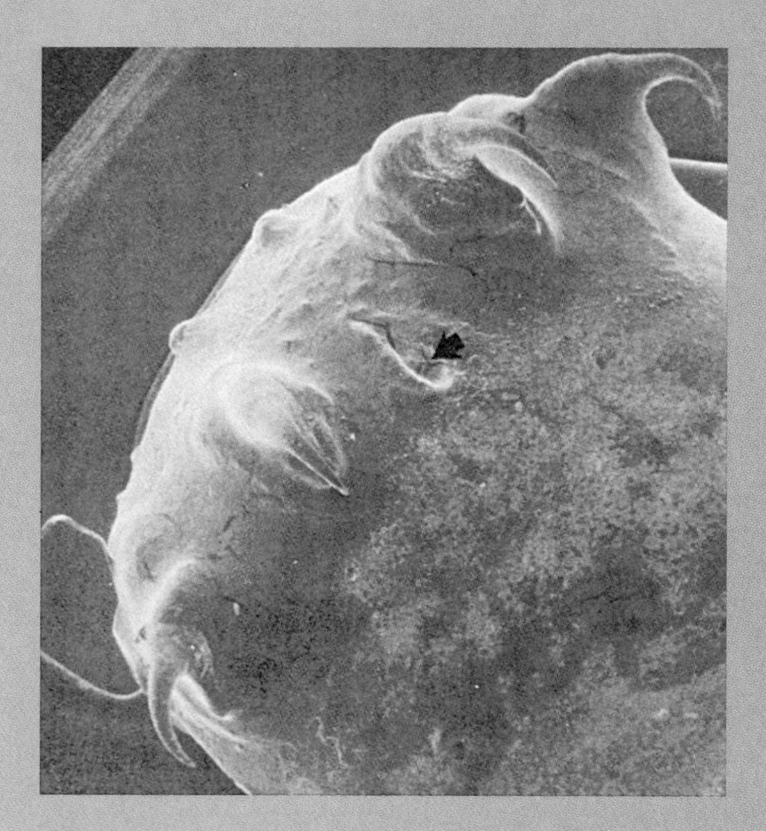

FIGURE 2 **Phylum Pentastomida.** Close-up view of the hooks used by pentastomids to attach to their hosts. Note the mouth between the middle hooks (arrow). When feeding, a pentastomid grasps the host tissue with its clawed legs, inflicting a wound. The mouth is applied to the wound, and blood and tissue fluid are sucked into the worm.

FIGURE 3 **Phylum Tardigrada.** Scanning electron micrograph of the tardigrade, *Macrobiotus tonolli*. The body is elongate, cylindrical, unsegmented, and has four pairs of unsegmented legs with claws. The entire body is covered with a proteinaceous cuticle that is periodically molted (×1,080).

PHYLUM PENTASTOMIDA: THE PENTASTOMIDS, OR TONGUE WORMS

The pentastomids (pent-ta-stom′ids) (Gr. *pente*, five + *stoma*, mouth) or tongue worms are all endoparasitic in the lungs or nasal passageways of carnivorous vertebrates (figure 2). Ninety percent of pentastomids' hosts are reptiles (e.g., snakes, lizards, crocodiles). There are about 90 species; two of the more well-known genera are *Linguatula* and *Raillietiella*.

Pentastomids are dioecious, and females are larger than males. Gonads are unpaired, and gonopores open to the outside. Male and female pentastomids mate in the final host. Following internal fertilization, females lay millions of shelled eggs, which pass out in the host's nasal secretions, saliva, or feces. If the eggs are eaten by one of a variety of vertebrate intermediate hosts, the larvae develop. The larva is characterized by having four to six arthropodlike jointed appendages. When the intermediate host is eaten by a final host, the larva is freed by the digestive enzymes and it migrates up the esophagus to the lungs, trachea, or nasal sinuses. Just as in the arthropods, pentastomids possess a brain and ventral nerve cord with ganglia and a hemocoel.

PHYLUM TARDIGRADA: THE TARDIGRADES, OR WATER BEARS

The tardigrades (tar-di-gray′ds) (L. *tardus*, slow + *gradus*, step) are commonly called water bears because of their body shape and legs and the way they lumber over aquatic vegetation (figure 3). These small animals (less than 1 mm in length) live in marine interstitial areas, in freshwater detritus, and in the water film on terrestrial lichens, liverworts, and mosses. There are about 500 species; the most common genera are *Echiniscus*, *Echiniscoides*, and *Macrobiotus*.

Tardigrades are dioecious with a single gonad dorsal to the midgut. A single oviduct or sperm duct empties into a gonopore. Fertilization is internal. Several dozen ornate eggs are laid by the female. After about 2 weeks, a juvenile hatches from the egg, molts, and develops into an adult. In some moss-dwelling species, males are rare or have never been observed, and parthenogenic reproduction presumably occurs.

Tardigrades (as well as nematodes and many rotifers) are able to enter a period of suspended animation termed cryptobiosis (Gr. *kryptos*, hidden + *bios*, life). This ability offers great survival benefit to these animals that live in habitats where conditions can suddenly become adverse. If a tardigrade begins to dry out (desiccate), it contracts into a shape that produces an ordered packing of organs and tissues to minimize mechanical damage caused by the desiccation. Overall metabolism slows. When rehydration occurs, the above events are reversed. Interestingly, repeated periods of cryptobiosis can extend a life span of approximately 1 year to 60 to 70 years.

chapter 13

THE ECHINODERMS

Outline

Concepts

1. Echinoderms are a part of the deuterostome evolutionary lineage. They are characterized by pentaradial symmetry, a calcium carbonate internal skeleton, and a water-vascular system.
2. Although there are many classes of extinct echinoderms, living echinoderms are divided into six classes. These are Asteroidea—the sea stars, Ophiuroidea—brittle stars and basket stars, Echinoidea—sea urchins and sand dollars, Holothuroidea—sea cucumbers, Crinoidea—sea lilies and feather stars, and Concentricycloidea—sea daisies.
3. Pentaradial symmetry of echinoderms probably developed during the evolution of sedentary life-styles, in which the water-vascular system was used in suspension feeding. Later, evolution resulted in some echinoderms becoming more mobile, and the water-vascular system came to be used primarily in locomotion.

Would You Like to Know:

1. what evidence links echinoderms and chordates to a common evolutionary pathway? (*p. 222*)
2. why most sea stars have five arms, rather than four or six? (*p. 223*)
3. how a sea star can open the shell of a clam? (*p. 225*)
4. how brittle stars use their snakelike arms? (*p. 227*)
5. what sea urchins use to chew through rock and coral? (*p. 230*)
6. how sea cucumbers "spill their insides" for a predator or a collector? (*p. 231*)
7. what the most recently discovered class of living echinoderms is? (*p. 233*)
8. what the original function of the water-vascular system was? (*p. 233*)

These and other useful questions will be answered in this chapter.

This chapter contains evolutionary concepts, which are set off in this font.

Evolutionary Perspective

If one could visit 400-million-year-old Paleozoic seas, one would see representatives of nearly every phylum studied in the previous eight chapters of this textbook. In addition, one would observe many representatives of the phylum Echinodermata (i-ki′na-dur″ma-tah) (Gr. *echinos*, spiny + *derma*, skin + *ata*, to bear). Many ancient echinoderms were attached to their substrate and probably lived as filter feeders—a feature found in only one class of modern echinoderms (figure 13.1). Today, we know this phylum by the relatively common sea stars, sea urchins, sand dollars, and sea cucumbers. In terms of numbers of species, echinoderms may seem to be a declining phylum. Studies of fossil records indicate that about 12 of 18 classes of echinoderms have become extinct. That does not mean, however, that living echinoderms are of minor importance. Members of three classes of echinoderms have flourished and often make up a major component of the biota of marine ecosystems (table 13.1).

Characteristics of the phylum Echinodermata include the following:

1. Calcareous endoskeleton in the form of ossicles that arise from mesodermal tissue
2. Adults with pentaradial symmetry and larvae with bilateral symmetry
3. Water-vascular system composed of water-filled canals used in locomotion, attachment, and/or feeding
4. Complete digestive tract that may be secondarily reduced
5. Hemal system derived from coelomic cavities
6. Nervous system consisting of a nerve net, nerve ring, and radial nerves

Figure 13.1
Phylum Echinodermata. This feather star (*Comanthina*) uses its highly branched arms in filter feeding. Although this probably reflects the original use of echinoderm appendages, most modern echinoderms use arms for locomotion, capturing prey, and scavenging the substrate for food.

Relationships to Other Animals

Most zoologists believe that echinoderms share a common ancestry with hemichordates and chordates. (1) Evidence of these evolutionary ties is seen in the deuterostome characteristics that they share (*see figure 9.3*): an anus that develops in the region of the blastopore, a coelom that forms from outpockets of the embryonic gut tract (vertebrate chordates are an exception), and radial, indeterminate cleavage. Unfortunately, no fossils have been discovered that document a common ancestor for these phyla or that demonstrate how the deuterostome lineage was derived from ancestral diploblastic or triploblastic stocks (figure 13.2).

Although adults are radially symmetrical, it is generally accepted that echinoderms evolved from bilaterally symmetrical ancestors. Evidence for this relationship includes bilaterally symmetrical echinoderm larval stages and extinct forms which were not radially symmetrical.

Table 13.1 Classification of the Phylum Echinodermata

Phylum Echinodermata (i-ki′na-dur″ma-tah).
The phylum of triploblastic, coelomate animals whose members are pentaradially symmetrical as adults, possess an endoskeleton covered by epithelium, and possess a water-vascular system. Pedicellaria often present.

Class Crinoidea (kri-noi′de-ah)
Free-living or attached by an aboral stalk of ossicles; flourished in the Paleozoic era; approximately 230 living species. Sea lilies, feather stars.

Class Asteroidea (as′te-roi″de-ah)
Rays not sharply set off from central disk; ambulacral grooves with tube feet; suction disks on tube feet; pedicellariae present. Sea stars.

Class Ophiuroidea (o-fe-u-roi″de-ah)
Arms sharply marked off from the central disk; tube feet without suction disks. Brittle stars.

Class Concentricycloidea (kon-sen′tri-si-kloi″de-ah)
Two concentric water-vascular rings encircle a disklike body; no digestive system; digest and absorb nutrients across their lower surface; internal brood pouches; no free-swimming larval stage. Sea daisies.

Class Echinoidea (ek′i-noi″de-ah)
Globular or disk shaped; no rays; movable spines; skeleton (test) of closely fitting plates. Sea urchins, sand dollars.

Class Holothuroidea (hol′o-thu-roi″de-ah)
No rays; elongate along the oral-aboral axis; microscopic ossicles embedded in a muscular body wall; circumoral tentacles. Sea cucumbers.

This listing reflects a phylogenetic sequence; however, the discussion that follows begins with the echinoderms that are familiar to most students.

Echinoderm Characteristics

There are approximately 7,000 species of living echinoderms. They are exclusively marine and occur at all depths in all oceans. Modern echinoderms have a form of radial symmetry, called **pentaradial symmetry**, in which body parts are arranged

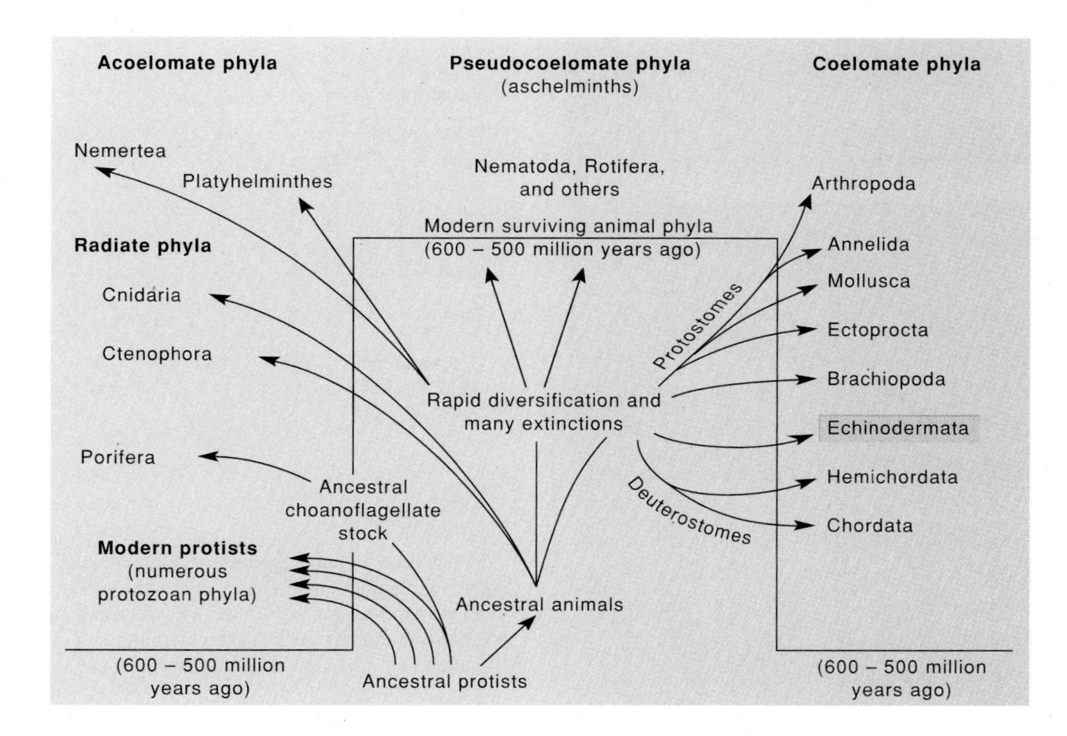

Figure 13.2

Evolutionary Relationships of the Echinoderms. The echinoderms (shaded in orange) diverged from the deuterostomate lineage at least 600 million years ago. Although modern echinoderms are pentaradially symmetrical, the earliest echinoderms were probably bilaterally symmetrical.

in fives, or a multiple of five, around an oral-aboral axis (figure 13.3*a*). Radial symmetry is adaptive for sedentary or slowly moving animals because it allows a uniform distribution of sensory, feeding, and other structures around the animal. Some modern mobile echinoderms, however, have secondarily returned to a basically bilateral form.

The skeleton of echinoderms consists of a series of calcium carbonate plates called ossicles. These plates are derived from mesoderm, held in place by connective tissues, and covered by an epidermal layer. If the epidermal layer is abraded away, the skeleton may be exposed in some body regions. The skeleton is frequently modified into fixed or articulated spines that project from the body surface.

2 The evolution of the skeleton may be responsible for the pentaradial body form of echinoderms. The joints between two skeletal plates represent a weak point in the skeleton (figure 13.3*b*). By not having weak joints directly opposite one another, the skeleton is made stronger than if the joints were arranged opposite each other.

The **water-vascular system** of echinoderms is a series of water-filled canals, and their extensions are called tube feet. It originates embryologically as a modification of the coelom and is ciliated internally. The water-vascular system includes a ring canal that surrounds the mouth (figure 13.4). The ring canal usually opens to the outside or to the body cavity through a stone canal and a sievelike plate, called the madreporite. The madreporite may serve as an inlet to replace water lost from the water-vascular system and may help equalize pressure differences between the water-vascular system and the outside. Tiedemann bodies are swellings that are often associated with the ring canal. They are believed to be the site for production of phagocytic cells, called coelomocytes, whose functions will be described later in this chapter. Polian vesicles are sacs that are also associated with the ring canal and function in fluid storage for the water-vascular system.

Five (or a multiple of five) radial canals branch from the ring canal. Radial canals are associated with arms of star-shaped echinoderms. In other echinoderms, they may be associated with the body wall and arch toward the aboral pole. Many lateral canals branch off each radial canal and end at the tube feet.

Tube feet are extensions of the canal system and usually emerge through openings in skeletal ossicles (*see figure 13.3*a). Internally, tube feet usually terminate in a bulblike, muscular ampulla. When an ampulla contracts, it forces water into a tube foot, which then extends. Valves prevent the backflow of water from the tube foot into the lateral canal. A tube foot often has a suction cup at its distal end. When the foot is extended and comes into contact with solid substrate, muscles of the suction cup contract and create a vacuum. In some taxa, the tube feet have a pointed or blunt distal end. These echinoderms may extend their tube feet into a soft substrate to secure contact during locomotion or to sift sediment during feeding.

(a)

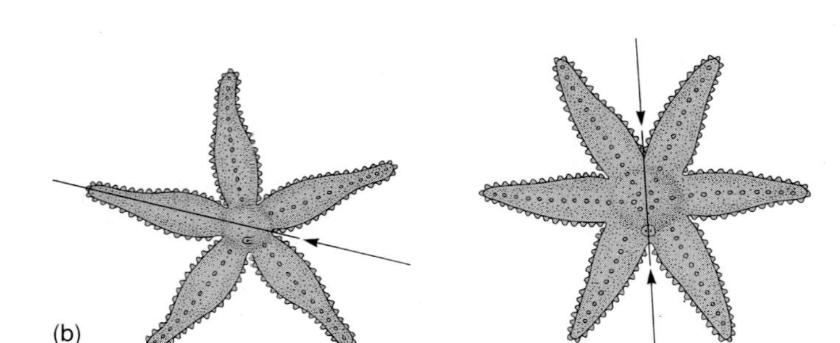

(b)

FIGURE 13.3

Pentaradial Symmetry. (*a*) Echinoderms possess a form of radial symmetry, called pentaradial symmetry, in which body parts are arranged in fives around an oral/aboral axis. Note the madreporite between the bases of the two arms in the foreground and the tube feet on the tips of the upturned arm. (*b*) Comparison of hypothetical penta- and hexaradial echinoderms. The five-part organization may be advantageous because joints between skeletal ossicles are never directly opposite one another, as they would be if an even number of parts were present. Having joints on opposite sides of the body in line with each other (arrows) could make the skeleton weaker.

The water-vascular system has other functions in addition to locomotion. As will be discussed at the end of this chapter, the original function of water-vascular systems was probably feeding, not locomotion. In addition, the soft membranes of the water-vascular system permit diffusion of respiratory gases and nitrogenous wastes across the body wall.

A **hemal system** consists of strands of tissue that encircle an echinoderm near the ring canal of the water-vascular system and run into each arm near the radial canals (*see figure 13.4*). The hemal system has been likened to a vestigial circulatory system; however, its function is largely unknown. It may aid in the transport of large molecules, hormones, or coelomocytes, which are cells that engulf and transport waste particles within the body.

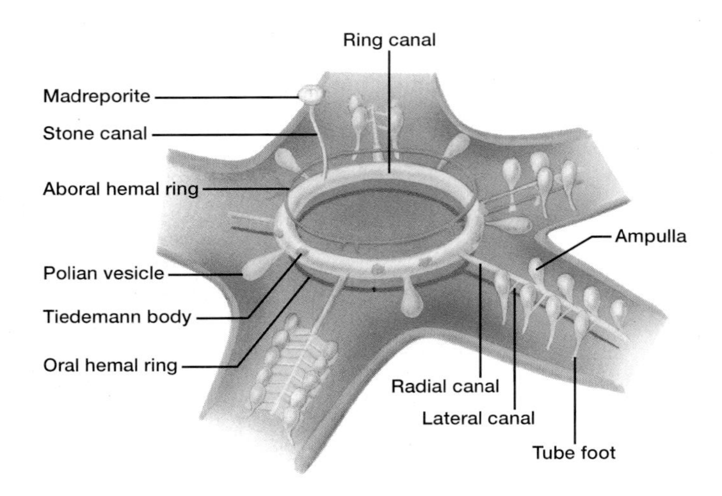

FIGURE 13.4

The Water-Vascular System of a Sea Star. The ring canal gives rise to radial canals that lead into each arm, a stone canal that ends at a madreporite on the aboral surface, and often has Polian vesicles and Tiedemann bodies associated with it.

CLASS ASTEROIDEA

The sea stars make up the class Asteroidea (as′te-roi″de-ah) (Gr. *aster*, star + *oeides*, in the form of) and include about 1,500 species (box 13.1). They are often found on hard substrata in marine environments, although some species are also found in sandy or muddy substrates. Sea stars may be brightly colored with red, orange, blue, or gray. *Asterias* is an orange sea star common along the Atlantic coast of North America and is frequently studied in introductory zoology laboratories.

Sea stars have five arms that radiate from a central disk. The oral opening, or mouth, is in the middle of one side of the central disk. It is normally oriented downward and is surrounded by movable oral spines. The aboral surface is roughened by movable and fixed spines that project from the skeleton. Thin folds of the body wall, called **dermal branchiae,** extend between ossicles and function in gas exchange (figure 13.5). In some sea stars, the aboral surface has numerous pincherlike structures called **pedicellariae,** which are used for cleaning the body surface of debris and for protection. Pedicellariae may be attached on a movable spine, or they may be immovably fused to skeletal ossicles.

An **ambulacral groove** runs the length of the oral surface of each arm and is formed by a series of ossicles in the arm. It houses the radial canal, and paired rows of tube feet protrude through the body wall on either side of the ambulacral groove. Tube feet of sea stars move in a stepping motion. Alternate extension, attachment, and contraction of tube feet move sea stars across their substrate. Tube feet are coordinated by the nervous system so that all feet move the sea star in the same direction; however, the tube feet do not move in unison. The

BOX 13.1 A Thorny Problem for Australia's Barrier Reef

The crown-of-thorns sea star (*Acanthaster planci*) is a common inhabitant of the South Pacific. Adults have a diameter of about 0.5 m and 13 to 16 arms (figure 1). Their common name is derived from their large venomous spines that can cause swelling, pain, and nausea in humans.

The crown-of-thorns sea star has become a problem in the waters off Australia and other South Pacific coasts. It feeds on coral polyps, and in one day, a single individual can extract polyps from 0.1 m^2 of reef. Over the last 20 years, the crown-of-thorns sea star has experienced a dramatic population increase. Thousands of sea stars have been observed moving slowly along a reef, leaving a white, almost sterile, limestone coral skeleton in their trail. They have seriously damaged over 500 km^2 of Australia's Great Barrier Reef.

One or more hypotheses may explain why there is a problem now, when these sea stars and coral polyps have coexisted for millions of years. Some believe that a part of the increase in sea stars may be due to the destruction of sea star predators, in particular, the giant triton gastropods. Tritons are valued for their beautiful shells, and their habitat has been disrupted by blasting to create shipping channels. Pesticides and other pollutants are also believed to be responsible for the destruction of predators of crown-of-thorns larvae. Another hypothesis suggests that the apparent increase in the sea star population may be a natural fluctuation in population size.

The reefs of the South Pacific are a source of economic wealth in the form of fisheries and tourism; therefore, the proliferation of crown-of-thorns sea stars has been the subject of intense study and control efforts. The Australian government has spent in excess of $3,000,000 and has not yet achieved satisfactory control. Control measures have included the injection of formaldehyde into adults by SCUBA divers, the erection of wire fences to divert populations away from reefs, and the use of computers to predict movements of colonies and local population explosions. One of the difficulties in these control efforts stems from the fact that sea star larvae are planktonic and are widely dispersed by oceanic currents.

FIGURE 1 **The Crown-of-Thorns Sea Star (*Acanthaster*).** This sea star is shown here feeding on coral (*Pocillopora*).

The original relationship between the crown-of-thorns sea star, coral polyps, and possibly sea star predators is an interesting example of a balanced predator-prey relationship. What is unfortunate is that humans often do not appreciate such balances until they have been altered.

suction disks of tube feet are effective attachment structures, allowing sea stars to maintain their position, or move from place to place, in spite of strong wave action.

Maintenance Functions

Sea stars feed on snails, bivalves, crustaceans, polychaetes, corals, detritus, and a variety of other food items. The mouth opens to a short esophagus and then to a large stomach that fills most of the coelom of the central disk. The stomach is divided into two regions. The larger, oral stomach, sometimes called the cardiac stomach, receives ingested food (figure 13.6). It joins the smaller, aboral stomach, sometimes called the pyloric stomach. The pyloric stomach gives rise to ducts that connect to secretory and absorptive structures called pyloric ceca. Two pyloric ceca extend into each arm. A short intestine leads to rectal ceca (uncertain functions) and to a nearly nonfunctional anus, which opens on the aboral surface of the central disk.

Some sea stars ingest whole prey, which are digested extracellularly within the stomach. Undigested material is expelled through the mouth. Many sea stars feed on bivalves by forcing the valves apart. (Anyone who has tried to pull apart the valves of a bivalve shell can appreciate that this is a remarkable accomplishment.) (3) When a sea star feeds on a bivalve, it wraps itself around the ventral margin of a bivalve. Tube feet attach to the outside of the shell, and the body-wall musculature forces the valves apart. (This is possible because the sea star changes tube feet when the muscles of engaged tube feet begin to tire.) When the valves are opened about 0.1 mm,

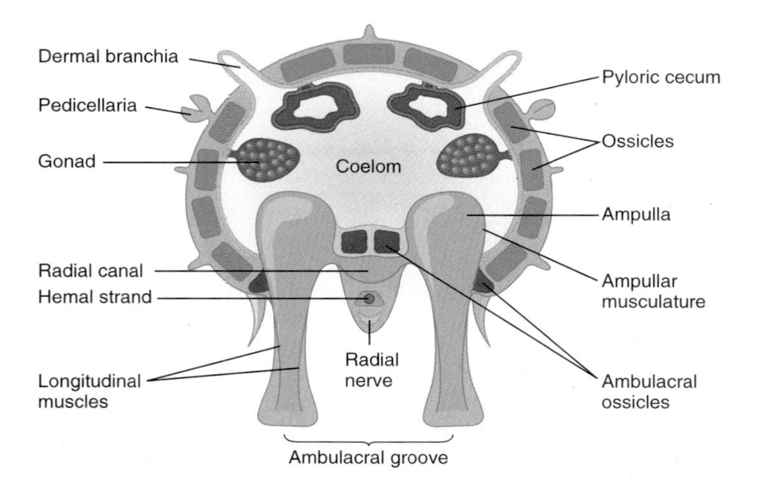

FIGURE 13.5
Body Wall and Internal Anatomy of a Sea Star. A cross section through one arm of a sea star shows the structures of the water-vascular system and the tube feet extending through the ambulacral groove.

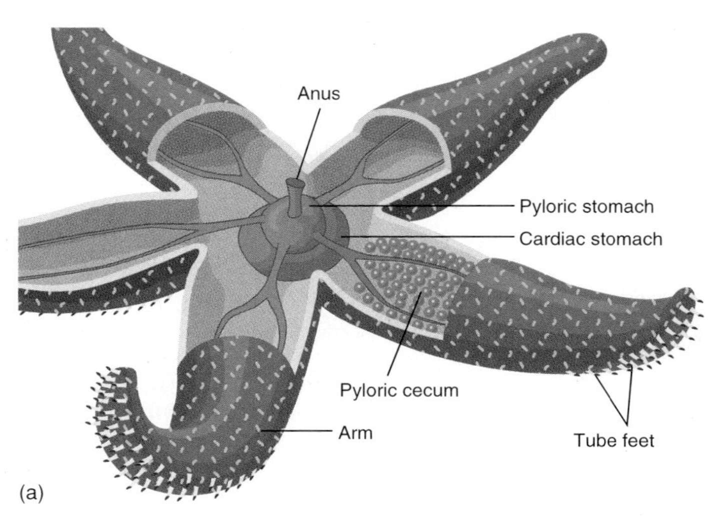

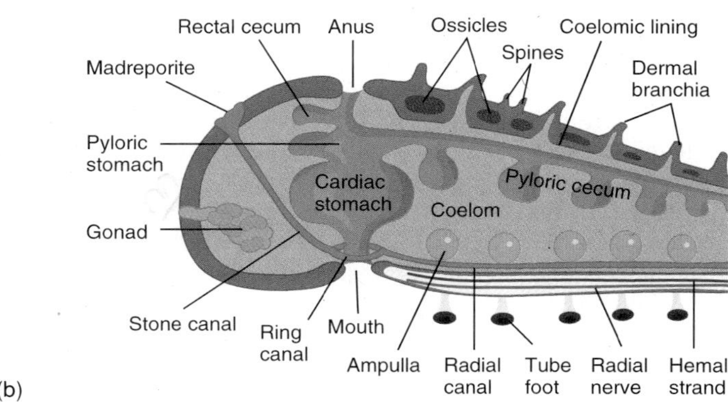

FIGURE 13.6
Digestive Structures in a Sea Star. A mouth leads to a large cardiac stomach and a pyloric stomach. Pyloric ceca extend into each arm. (*a*) Aboral view. (*b*) Lateral view through central disk and one arm.

the oral (cardiac) portion of the stomach is everted through the mouth and into the bivalve shell by increased coelomic pressure. Digestive enzymes are released, and partial digestion occurs in the bivalve shell. This digestion further weakens the bivalve's adductor muscles, and the shell eventually opens completely. Partially digested tissues are taken into the aboral (pyloric) portion of the stomach, and into the pyloric ceca for further digestion and absorption. After feeding and initial digestion, the stomach is retracted using stomach retractor muscles.

Transport of gases, nutrients, and metabolic wastes in the coelom occurs by diffusion and by the action of ciliated cells lining the body cavity. Gas exchange and excretion of metabolic wastes (principally ammonia) occur by diffusion across dermal branchiae, tube feet, and other membranous structures. A sea star's hemal system consists of strands of tissue that encircle the mouth near the ring canal, extend aborally near the stone canal, and run into the arms near radial canals (*see figure 13.4*).

The nervous system of sea stars consists of a nerve ring that encircles the mouth and radial nerves that extend into each arm. Radial nerves lie within the ambulacral groove, just oral to the radial canal of the water-vascular system and the radial strands of the hemal system (*see figure 13.5*). Radial nerves are essential for coordinating the functions of tube feet. Other nervous elements are in the form of a nerve net that is associated with the body wall.

Most sensory receptors are distributed over the surface of the body and tube feet. Sea stars respond to light, chemicals, and various mechanical stimuli. They often have specialized photoreceptors at the tips of their arms. These are actually tube feet that lack suction cups but have a pigment spot surrounding a group of ocelli.

REGENERATION, REPRODUCTION, AND DEVELOPMENT

Sea stars are well known for their powers of regeneration. Any part of a broken arm can be regenerated. In a few species, an entire starfish can be regenerated from a broken arm if the arm contains a portion of the central disk. Regeneration is a slow process, taking up to 1 year for complete regeneration. Asexual reproduction involves dividing the central disk, followed by regeneration of each half.

Sea stars are dioecious, but sexes are indistinguishable externally. Two gonads are present in each arm and increase in size to nearly fill an arm during the reproductive periods. Gonopores open between the bases of each arm.

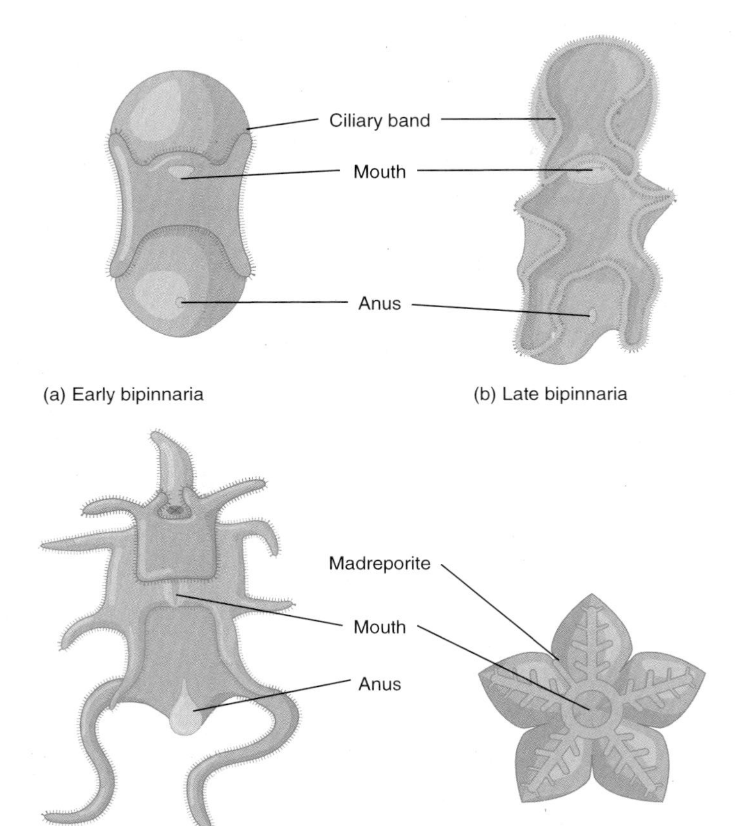

FIGURE 13.7

Development of a Sea Star. Later embryonic stages are ciliated and swim and feed in the plankton. In a few species, embryos develop from yolk stored in the egg during gamete formation. Following blastula and gastrula stages, larvae develop. (*a*) Early bipinnaria larva. (*b*) Late bipinnaria larva. (*c*) Brachiolaria larva. (*d*) Juvenile sea star.

The embryology of echinoderms has been studied extensively because of the relative ease of inducing spawning and maintaining embryos in the laboratory. External fertilization is the rule. Because gametes cannot survive long in the ocean, maturation of gametes and spawning must be coordinated if fertilization is to take place. The photoperiod (the relative length of light and dark in a 24-hour period) and temperature are environmental factors that are used to coordinate sexual activity. In addition, the release of gametes by one individual is accompanied by the release of spawning pheromones, which induce other sea stars in the area to spawn, increasing the likelihood of fertilization.

Embryos are planktonic, and cilia are used in swimming (figure 13.7). After gastrulation, bands of cilia differentiate, and a bilaterally symmetrical larva, called a bipinnaria larva, is formed. The larva usually feeds on planktonic protists. The development of larval arms results in a brachiolaria larva. Settling to the substrate is followed by attachment and metamorphosis to a juvenile sea star.

Stop and Ask Yourself

1. Why do many zoologists believe that ancestral echinoderms were bilaterally symmetrical?
2. What are the functions of the water-vascular system?
3. How does a sea star feed on a clam?
4. How is the nervous tissue arranged in a sea star?

CLASS OPHIUROIDEA

The class Ophiuroidea (o′fe-u-roi′de-ah) (Gr. *ophis*, snake + *oura*, tail + *oeides*, in the form of) includes the basket stars and the brittle stars. With over 2,000 species, this is the most diverse group of echinoderms. Ophiuroids, however, are often overlooked because of their small size and their tendency to occupy crevices in rocks and coral or to cling to algae.

The arms of ophiuroids are long and, unlike those of asteroids, are sharply set off from the central disk, giving the central disk a pentagonal shape. Brittle stars have unbranched arms and most range in size from 1 to 3 cm (figure 13.8*a*). Basket stars have arms that branch repeatedly (figure 13.8*b*). Neither dermal branchiae nor pedicellariae are present in ophiuroids. The tube feet of ophiuroids lack suction disks and ampullae, and contraction of muscles associated with the base of a tube foot causes the tube foot to be extended. Unlike the sea stars, the madreporite of ophiuroids is located on the oral surface.

The water-vascular system of ophiuroids is not used for locomotion. Instead, the skeleton is modified to permit a unique form of grasping and movement. Superficial ossicles, which originate on the aboral surface, cover the lateral and oral surfaces of each arm. The ambulacral groove—containing the radial nerve, hemal strand, and radial canal—is thus said to be "closed." Ambulacral ossicles are in the arm, forming a central supportive axis. (4) Successive ambulacral ossicles articulate with one another and are acted upon by relatively large muscles to produce snakelike movements, allowing the arms to be curled around a stalk of algae or hooked into a coral crevice. During locomotion, the central disk is held above the substrate and two arms are used to pull the animal along, while other arms extend forward and/or trail behind the animal.

MAINTENANCE FUNCTIONS

Ophiuroids are predators and scavengers. They use their arms and tube feet in sweeping motions to collect prey and particulate matter, which are then transferred to the mouth. Some ophiuroids are filter feeders that wave their arms and trap plankton on mucus-covered tube feet. Trapped plankton is passed from tube foot to tube foot along the length of an arm until it reaches the mouth.

(a)

(b)

Figure 13.8

Class Ophiuroidea. (*a*) This brittle star (*Ophiopholis aculeata*) uses its long, snakelike arms for crawling along its substrate and curling around objects in its environment. (*b*) Basket stars have five highly branched arms. The arms are waved in the water to capture planktonic organisms on mucous-covered tube feet.

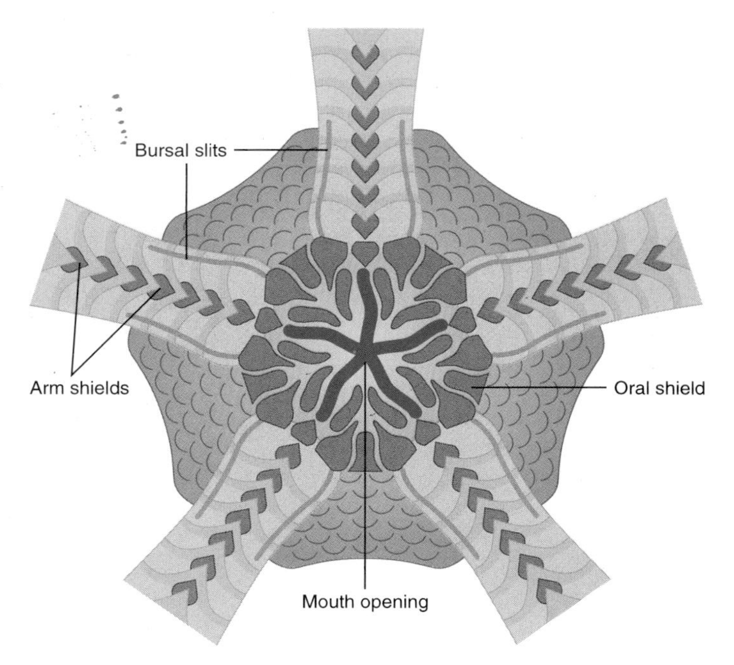

Figure 13.9

Class Ophiuroidea. Oral view of the disk of the brittle star, *Ophiomusium*. *Redrawn from L. Hyman,* The Invertebrates, *Volume IV. Copyright © 1959 McGraw-Hill, Inc. Used by permission.*

The mouth of ophiuroids is in the center of the central disk, and five triangular jaws form a chewing apparatus. The mouth leads to a saclike stomach. There is no intestine, and no part of the digestive tract extends into the arms.

The coelom of ophiuroids is reduced and is mainly confined to the central disk, but it still serves as the primary means for the distribution of nutrients, wastes, and gases. Coelomocytes aid in the distribution of nutrients and the expulsion of particulate wastes. Ammonia is the primary nitrogenous waste product and it is lost by diffusion. Diffusion occurs across tube feet and membranous sacs, called **bursae**, that invaginate from the oral surface of the central disk. Slits in the oral disk, near the base of each arm, allow cilia to move water into and out of the bursae (figure 13.9).

Regeneration, Reproduction, and Development

Ophiuroids, as sea stars, are able to regenerate lost arms. If a brittle star is grasped by an arm, the contraction of certain muscles may sever and cast off the arm—hence the common name brittle star. This process is called autotomy (Gr. *autos*, self + *tomos*, to cut) and is used in escape reactions. The arm is later regenerated. Some species also have a fission line across their central disk. When split into halves along this line, two ophiuroids will be regenerated.

Ophiuroids are dioecious. Males are usually smaller than females and often are carried about by females. The gonads are associated with each bursa, and gametes are released into the bursa. Eggs may be shed to the outside, or retained in the bursa, where they are fertilized and held through early development. Embryos are protected in the bursa and are sometimes nourished by the parent. A larval stage, called an ophiopluteus, is planktonic. Its long arms bear ciliary bands that are used to feed on plankton, and it undergoes metamorphosis before sinking to the substrate.

(a)

(b)

Figure 13.10

Class Echinoidea. (*a*) A sea urchin (*Strongylocentrotus*). (*b*) Sand dollars are specialized for living in soft substrates, where they often occur partially buried.

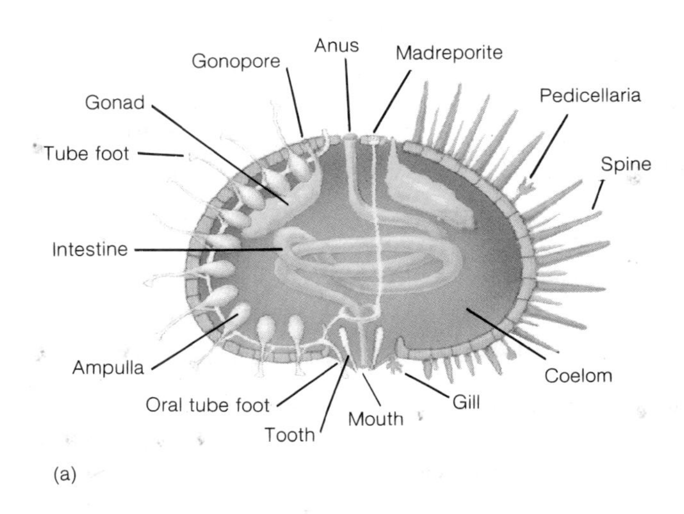

(a)

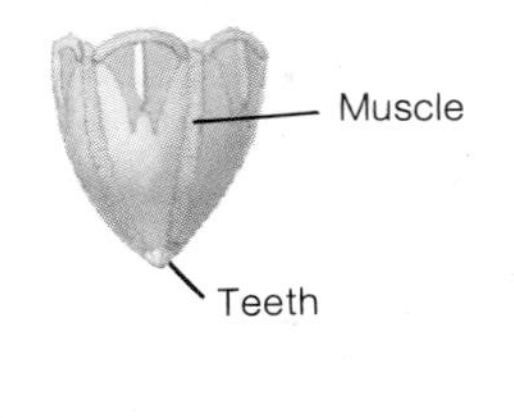

(b)

Figure 13.11

Internal Anatomy of a Sea Urchin. (*a*) Sectional view. (*b*) Aristotle's lantern is a chewing structure consisting of about 35 ossicles and associated muscles.

Class Echinoidea

The sea urchins, sand dollars, and heart urchins make up the class Echinoidea (ek′i-noi-de-ah) (Gr. *echinos*, spiny + *oeides*, in the form of). There are about 1,000 species widely distributed in nearly all marine environments. Sea urchins are specialized for living on hard substrates, often wedging themselves into crevices and holes in rock or coral (figure 13.10*a*). Sand dollars and heart urchins usually live in sand or mud and burrow just below the surface (figure 13.10*b*). They use tube feet to catch organic matter settling on them or passing over them. Sand dollars often occur in very dense beds, which favors efficient reproduction and feeding.

Sea urchins are rounded and have their oral end oriented toward the substrate. Their skeleton, called a test, consists of 10 closely fitting plates that arch between oral and aboral ends. Five ambulacral plates have openings for tube feet, and alternate with five interambulacral plates, which have tubercles for the articulation of spines. The base of each spine is a concave socket, and muscles at its base move the spine. Spines are often sharp and sometimes hollow and contain venom that is dangerous to swimmers. The pedicellariae of sea urchins have either two or three jaws and are connected to the body wall by a relatively long stalk (figure 13.11*a*). They are used for cleaning the body of debris and capturing planktonic larvae, which provide an extra source of food. Pedicellariae of some sea urchins contain venom sacs and are grooved or hollow to inject venom into a predator, such as a sea star.

The water-vascular system is similar to that of other echinoderms. Radial canals run along the inner body wall between the oral and the aboral poles. Tube feet possess ampullae and suction cups, and the water-vascular system opens to the outside through many pores in one aboral ossicle that serves as a madreporite.

Echinoids move by using spines for pushing against the substrate and tube feet for pulling. Sand dollars and heart urchins use spines to help burrow in soft substrates. (5) Some sea urchins burrow into rock and coral to escape the action of waves and strong currents. They form cup-shaped depressions and deeper burrows using the action of their chewing Aristotle's lantern, which is described next.

Maintenance Functions

Echinoids feed on algae, bryozoans, coral polyps, and dead animal remains. Food is manipulated by oral tube feet surrounding the mouth. A chewing apparatus, called **Aristotle's lantern,** projects from the mouth (figure 13.11*b*). It consists of about 35 ossicles and attached muscles and cuts food into small pieces for ingestion. The mouth cavity leads to a pharynx, an esophagus, and a long, coiled intestine that ends aborally at the anus.

Echinoids have a large coelom, and coelomic fluids are the primary circulatory medium. Small gills, found in a thin membrane surrounding the mouth, are outpockets of the body wall and are lined by ciliated epithelium. Gas exchange occurs by diffusion across this epithelium and across the tube feet. Ciliary currents, changes in coelomic pressure, and the contraction of muscles associated with Aristotle's lantern move coelomic fluids into and out of gills. Excretory and nervous functions are similar to those described for asteroids.

Reproduction and Development

Echinoids are dioecious. Gonads are located on the internal body wall of the interambulacral plates. During breeding season, they nearly fill the spacious coelom. One gonopore is located in each of five ossicles, called genital plates, at the aboral end of the echinoid, although the sand dollars usually have only four gonads and gonopores. Gametes are shed into the water, and fertilization is external. Development eventually results in a pluteus larva that spends several months in the plankton and eventually undergoes metamorphosis to the adult.

Stop and Ask Yourself

5. How is a sea star distinguished from a brittle star?
6. What is autotomy?
7. What is Aristotle's lantern?
8. What structures of echinoids are used for gas exchange?

Class Holothuroidea

There are approximately 1,500 species in the class Holothuroidea (hol′o-thu-roi′de-ah) (Gr. *holothourion,* sea cucumber + *oeides,* in the form of), and they are commonly called sea cucumbers. Sea cucumbers are found at all depths in all oceans, where they crawl over hard substrates or burrow through soft substrates (figure 13.12).

Figure 13.12
Class Holothuroidea. A sea cucumber (*Parastichopus californicus*).

Sea cucumbers have no arms, and they are elongate along the oral-aboral axis. They lie on one side, which is usually flattened as a permanent ventral side, giving them a secondary bilateral symmetry. Tube feet surrounding the mouth are elongate and referred to as tentacles. Most adults range in length between 10 and 30 cm. Their body wall is thick and muscular and lacks protruding spines or pedicellariae. Beneath the epidermis is the dermis, a thick layer of connective tissue in which ossicles are embedded. Ossicles of sea cucumbers are microscopic in size and do not function in determining body shape. A circle of larger ossicles forms a calcareous ring and encircles the oral end of the digestive tract, serving as a point of attachment for body wall muscles (figure 13.13). Beneath the dermis is a layer of circular muscles overlying longitudinal muscles. The body wall of sea cucumbers, when boiled and dried, is known as trepang in the Orient. It may be eaten as a main course item or be added to soups as flavoring and a source of protein.

The madreporite of sea cucumbers is internal, and the water-vascular system is filled with coelomic fluid. The ring canal encircles the oral end of the digestive tract and gives rise to one to ten Polian vesicles. Five radial canals and the canals to the tentacles branch from the ring canal. Radial canals and tube feet, with suction cups and ampullae, run between the oral and aboral poles. The side of a sea cucumber resting on the substrate contains three of the five rows of tube feet, which are primarily used for attachment. The two rows of tube feet on the upper surface may be reduced in size, or they may be absent.

Sea cucumbers are mostly sluggish burrowers and creepers, although some swim by undulating their body from side to side. Locomotion using tube feet is inefficient because the tube feet are not anchored by body wall ossicles. Locomotion more commonly results from contractions of body wall muscles, producing wormlike, locomotor waves that pass along the length of the body.

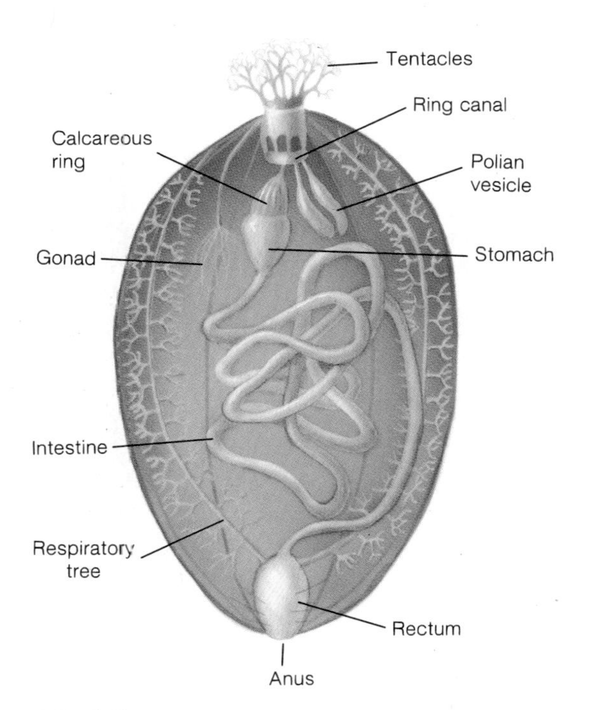

FIGURE 13.13

Internal Structure of a Sea Cucumber, *Thyone*. The mouth leads to a stomach that is supported by a calcareous ring. The calcareous ring is also the attachment site for longitudinal retractor muscles of the body. Contractions of these muscles pull the tentacles into the anterior end of the body. The stomach leads to a looped intestine. The intestine continues to the rectum and anus. (The anterior portion of the digestive tract is displaced aborally in this illustration.)

MAINTENANCE FUNCTIONS

Most sea cucumbers ingest particulate organic matter using their tentacles. Food is trapped in mucus covering the tentacles, either as the tentacles sweep across the substrate or while tentacles are held out in seawater. The digestive tract consists of a stomach; a long, looped intestine; a rectum; and an anus (figure 13.13). Tentacles are thrust into the mouth to wipe off trapped food. During digestion, coelomocytes move across the intestinal wall, secrete enzymes to aid in digestion, and engulf and distribute the products of digestion.

The coelom of sea cucumbers is large, and the cilia of the coelomic lining circulate fluids throughout the body cavity, distributing respiratory gases, wastes, and nutrients. The hemal system of sea cucumbers is well developed, with relatively large sinuses and a network of channels containing coelomic fluids. Its primary role is food distribution.

Respiratory trees are a pair of tubes that attach at the rectum and branch throughout the body cavity of sea cucumbers. Water circulates into these tubes by the pumping action of the rectum. When the rectum dilates, water moves through the anus into the rectum. Contraction of the rectum, along with contraction of an anal sphincter, forces water into the respiratory tree. Water exits the respiratory tree when tubules of the tree contract. Respiratory gases and nitrogenous wastes move between the coelom and seawater across these tubules.

The nervous system of sea cucumbers is similar to that of other echinoderms, but has additional nerves supplying the tentacles and pharynx. Some sea cucumbers have statocysts, and others have relatively complex photoreceptors.

Casual examination would lead one to believe that sea cucumbers are defenseless against predators. Many sea cucumbers, however, produce toxins in their body walls that act to discourage predators. In other sea cucumbers, tubules of the respiratory tree, called Cuverian tubules, can be everted through the anus. They contain sticky secretions and toxins capable of entangling and immobilizing predators. (6) In addition, contractions of the body wall may result in expulsion of one or both respiratory trees, the digestive tract, and the gonads through the anus. This process is called evisceration and is a defensive adaptation that may discourage predators. It is followed by regeneration of lost parts.

REPRODUCTION AND DEVELOPMENT

Sea cucumbers are dioecious. They possess a single gonad, located anteriorly in the coelom, and a single gonopore near the base of the tentacles. Fertilization is usually external, and embryos develop into planktonic larvae. Metamorphosis precedes settling to the substrate. In some species, eggs are trapped by a female's tentacles as they are released. After fertilization, eggs are transferred to the body surface, where they are brooded. Although it is rare, coelomic brooding also occurs. Eggs are released into the body cavity where fertilization (by an unknown mechanism) and early development occur. The young leave by a rupture in the body wall. Sea cucumbers can also reproduce by transverse fission, followed by regeneration of lost parts.

Stop and Ask Yourself

9. What structural features of sea cucumbers result in their having a secondary bilateral symmetry?
10. How is the water-vascular system of sea cucumbers modified for feeding?
11. What is a respiratory tree?
12. How does a sea cucumber defend itself against predators?

CLASS CRINOIDEA

Members of the class Crinoidea (krin-oi'de-ah) (Gr. *krinon*, lily + *oeides*, in the form of) include the sea lilies and the feather stars. They are the most primitive of all living echinoderms and are very different from any covered thus far. There are approximately

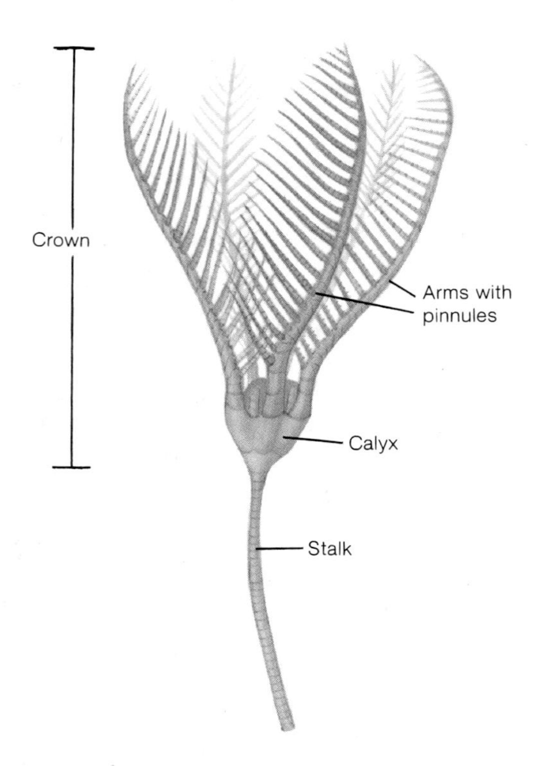

Figure 13.14
Class Crinoidea. A sea lily (*Ptilocrinus*).

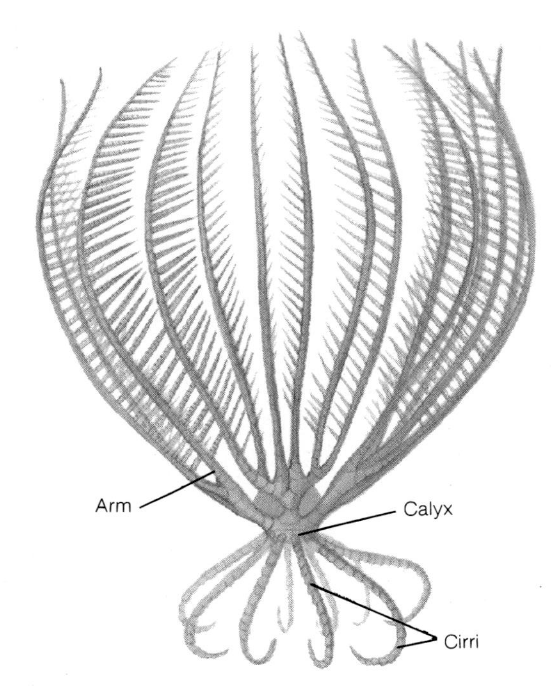

Figure 13.15
Class Crinoidea. A feather star (*Neometra*).

630 species living today; however, an extensive fossil record indicates that many more were present during the Paleozoic era, 200 to 600 million years ago.

Sea lilies are attached permanently to their substrate by a stalk (figure 13.14). The attached end of the stalk bears a flattened disk or rootlike extensions that are fixed to the substrate. Disklike ossicles of the stalk appear to be stacked on top of one another and are held together by connective tissues, giving a jointed appearance. The stalk usually bears projections, or cirri that are arranged in whorls around the stalk. The unattached end of a sea lily is called the crown. The aboral end of the crown is attached to the stalk and is supported by a set of ossicles, called the calyx. Five arms also attach at the calyx. They are branched, supported by ossicles, and bear smaller branches—giving them a featherlike appearance. Tube feet are located in a double row along each arm. Ambulacral grooves on the arms lead toward the mouth. The mouth and anus open onto the upper (oral) surface.

Feather stars are similar to sea lilies, except they lack a stalk and are swimming and crawling animals (figure 13.15). The aboral end of the crown bears a ring of rootlike cirri, which are used for clinging when the animal is resting on a substrate. Swimming is accomplished by raising and lowering the arms, and crawling results from using the tips of the arms to pull the animal over the substrate.

Maintenance Functions

Circulation, gas exchange, and excretion in crinoids are similar to these functions in other echinoderms. In feeding, however, crinoids differ from other living echinoderms. They use outstretched arms for suspension feeding. When a planktonic organism contacts a tube foot, it is trapped and carried to the mouth by cilia in ambulacral grooves. Although this method of feeding is different from the way other modern echinoderms feed, it probably reflects the original function of the water-vascular system.

Crinoids lack the nerve ring found in most echinoderms. Instead, a cup-shaped nerve mass below the calyx gives rise to radial nerves that extend through each arm and control the tube feet and arm musculature.

Reproduction and Development

Crinoids, as other echinoderms, are dioecious. Gametes form from germinal epithelium in the coelom and are released by rupturing the walls of the arms. Some species spawn into seawater, where fertilization and development occur. Other species brood embryos on the outer surface of the arms. Metamorphosis occurs after larvae have attached to the substrate. As with other echinoderms, crinoids can regenerate lost parts.

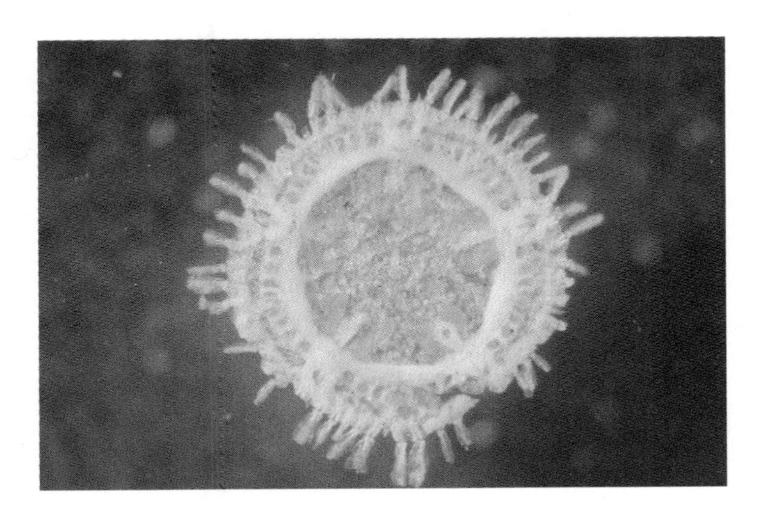

FIGURE 13.16
Class Concentricycloidea. Photograph of a preserved sea daisy (*Xyloplax medusiformis*). This specimen is 3 mm in diameter.

CLASS CONCENTRICYCLOIDEA

The class Concentricycloidea (kon-sen′tri-si-kloi″de-ah) (ME *consentrik*, having a common center + Gr. *kykloeides*, like a circle) contains a single described species, known as the sea daisy. 7 Sea daisies have been recently discovered on debris in deep oceans (figure 13.16). They lack arms and are less than 1 cm in diameter. The most distinctive feature of this species is two circular water-vascular rings that encircle the disklike body. The inner of the two rings probably corresponds to the ring canal of members of other classes because it has Polian vesicles attached. The outer ring contains tube feet and ampullae and probably corresponds to the radial canals of members of other classes. In addition, this animal lacks an internal digestive system. Instead, the surface of the animal that is applied to the substrate (e.g., decomposing wood) is covered by a thin membrane, called a velum, that digests and absorbs nutrients. Internally, there are five pairs of brood pouches where embryos are held during development. There are apparently no free-swimming larval stages. The mechanism for fertilization is unknown.

FURTHER PHYLOGENETIC CONSIDERATIONS

As described earlier, most zoologists believe that echinoderms evolved from bilaterally symmetrical ancestors. Radial symmetry probably evolved during the transition from active to more sedentary life-styles; however, the oldest echinoderm fossils, about 600 million years old, give little direct evidence of how this transition occurred.

Ancient fossils do give clues regarding the origin of the water-vascular system and the calcareous endoskeleton. Of all living echinoderms, the crinoids most closely resemble the oldest fossils. 8 Because crinoids use their water-vascular system for suspension feeding, it is believed that filter feeding, not locomotion, was probably the original function of the water-vascular system. As do crinoids, early echinoderms probably assumed a mouth-up position and were attached aborally. Arms and tube feet could have been used to capture food and move it to the mouth. The calcium carbonate endoskeleton may have evolved for support of extended filtering arms and for protection of these sessile animals.

Many modern echinoderms are more mobile. This free-living life-style is probably secondarily derived, as is the mouth-down orientation of most echinoderms. The mouth-down position would be advantageous for predatory and scavenging life-styles. Similarly, changes in the water-vascular system, such as the evolution of ampullae, suction disks, and feeding tentacles, can be interpreted as adaptations for locomotion and feeding in a more mobile life-style. The idea that the free-living life-style is secondary is reinforced by the observation that some echinoderms, such as the irregular echinoids and the holothuroids, have bilateral symmetry imposed upon a pentaradial body form.

The evolutionary relationships among the echinoderms are not clear. There are numerous fossils dating into the Cambrian period, but there is no definitive interpretation of the evolutionary relationships among living and extinct echinoderms. One interpretation of the evolutionary relationships among extant (with living members) echinoderm classes is shown in figure 13.17. Most taxonomists agree that the echinoids and holothuroids are closely related. Whether the ophiuroids are more closely related to the echinoid/holothuroid lineage or the asteroid lineage is debated. The position of the Concentricycloidea in echinoderm phylogeny is highly speculative and is not shown in the figure.

Stop and Ask Yourself

13. What is the function of the stalk, calyx, and cirri in crinoids?
14. How is the function of the crinoid water-vascular system different from that of other echinoderms?
15. How do the water-vascular systems and feeding mechanisms of sea daisies differ from those of other echinoderms?
16. What could account for the evolution of the calcium carbonate endoskeleton of echinoderms?

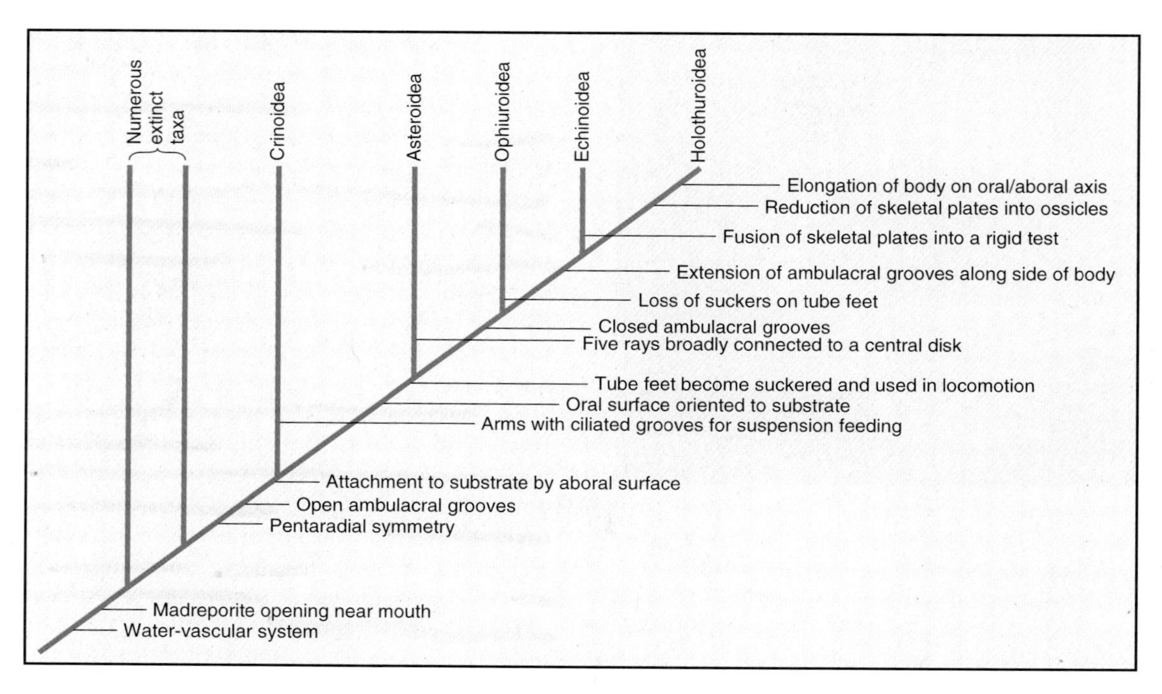

FIGURE 13.17

Echinoderm Phylogeny. The evolutionary relationships among echinoderms are not clear. The interpretation shown here depicts a relatively distant relationship between the Asteroidea and Ophiuroidea. Some taxonomists interpret the five-rayed body form as synapomorphy that links these two groups to a single ancestral lineage. The position of the Concentricycloidea is highly speculative and is not shown here.

SUMMARY

1. Echinoderms, chordates, and other deuterostomes share a common, but remote, ancestry. Modern echinoderms were probably derived from bilaterally symmetrical ancestors.
2. Echinoderms are pentaradially symmetrical, have an endoskeleton of interlocking calcium carbonate ossicles, and have a water-vascular system that is used for locomotion, food gathering, attachment, and exchanges with the environment.
3. Members of the class Asteroidea are the sea stars. They are predators and scavengers, and their arms are broadly joined to the central disk. Sea stars are dioecious, and external fertilization results in the formation of planktonic bipinaria and brachiolaria larvae. Sea stars also have remarkable powers of regeneration.
4. The brittle stars and basket stars make up the class Ophiuroidea. Arms are sharply set off from the central disk. Ophiuroids are dioecious. Externally fertilized eggs may either develop in the plankton, or they may be brooded.
5. The class Echinoidea includes the sea urchins, heart urchins, and sand dollars. They have a specialized chewing structure, called Aristotle's lantern. External fertilization results in a planktonic pluteus larva.
6. Members of the class Holothuroidea include the sea cucumbers. They rest on one side, are elongate along their oral-aboral axis, and their body wall contains microscopic ossicles. Many sea cucumbers eviscerate themselves when disturbed. Sea cucumbers are dioecious, and fertilization and development are external.
7. The class Crinoidea contains the sea lilies and feather stars. They are oriented oral side up and use arms and tube feet in suspension feeding. Crinoids are dioecious, and fertilization and development are external.
8. The class Concentricycloidea contains one recently discovered species that lives on wood and other debris in deep water.
9. Radial symmetry of echinoderms probably evolved during a transition to a sedentary, filter-feeding life-style. The water-vascular system and the calcareous endoskeleton are probably adaptations for that life-style. The evolution of a more mobile life-style has resulted in the use of the water-vascular system for locomotion and the assumption of a mouth-down position.

SELECTED KEY TERMS

ambulacral groove (*p. 224*)
Aristotle's lantern (*p. 230*)
dermal branchiae (*p. 224*)
pedicellariae (*p. 224*)
pentaradial symmetry (*p. 222*)
respiratory trees (*p. 231*)
tube feet (*p. 223*)
water-vascular system (*p. 223*)

CRITICAL THINKING QUESTIONS

1. What is pentaradial symmetry and why is it adaptive for echinoderms?
2. Why do zoologists think that pentaradial symmetry was not present in the ancestors of echinoderms?
3. Compare and contrast the structure and function of the water-vascular systems of asteroids, ophiuroids, echinoids, holothuroids, and crinoids.
4. In which of the above groups is the water-vascular system probably most similar in form and function to an ancestral condition? Explain your answer.
5. What physical process is responsible for gas exchange and excretion in all echinoderms? What structures facilitate these exchanges in each echinoderm class?

endpaper FOUR

Some Lesser Known Invertebrates: The Lophophorates

The Lophophorates: Brachiopoda, Ectoprocta (Bryozoa), and Phoronida

The three phyla (Brachiopoda, Ectoprocta [Bryozoa], and Phoronida) discussed in this endpaper share one major anatomical feature—the **lophophore** (Gr. *lophos*, crest or tuft + *phorein*, to bear). The lophophore is a circumoral (around the mouth) body region characterized by a circular or U-shaped ridge, with either one or two rows of ciliated, hollow tentacles (figure 1). The lophophore functions as a food-collecting organ and as a surface for gas exchange. It also has sensory cells receptive to chemicals and touch that are concentrated on its tentacles. The lophophore can usually be extended for feeding or withdrawn for protection.

All lophophorates are sessile or sedentary filter feeders that possess a U-shaped digestive tract and live in a secreted chitinous or calcareous tube. As in deuterostomes, these phyla have radial cleavage in embryonic stages and a coelom that is divided into compartments. As in protostomes, however, the embryonic mouth forms in the region of the blastopore. Although it has been debated, most zoologists have considered the lophophorates to be deuterostomes. Recent molecular biological studies involving the sequencing of ribosomal DNA, however, suggest that the lophophorates are protostomes.

The evolutionary relationships between the lophophorate phyla are very distant. The presence of a lophophore in all three phyla indicates common evolutionary ties. Other similarities—such as a reduced head, U-shaped digestive tract, and secreted protective covering—are correlated with adaptations for a sessile, filter-feeding existence. These common features may represent evolutionary convergence rather than close evolutionary relationships.

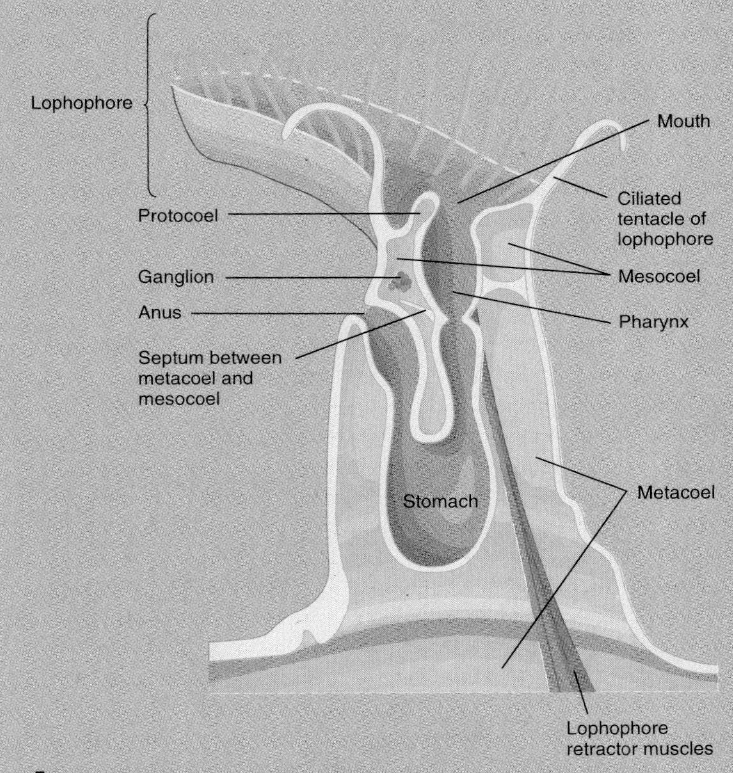

FIGURE 1 **Lophophorates.** Longitudinal section through the body of a lophophorate with the lophophore extended.

Phylum Brachiopoda: The Brachiopods, or Lampshells

The brachiopods (brak-i-op′ods) (Gr. *brachion*, arm + *podos*, foot) bear a superficial resemblance to the bivalve molluscs because they have a bivalved, calcareous and/or chitinous shell that is secreted by a mantle that encloses nearly all of the body. However, unlike

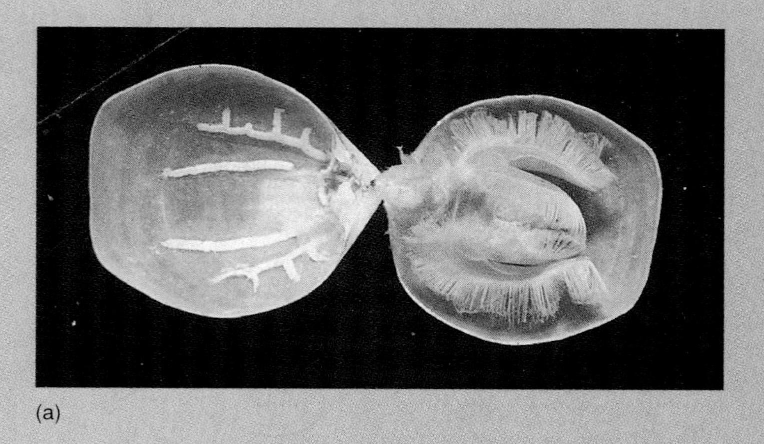

(a)

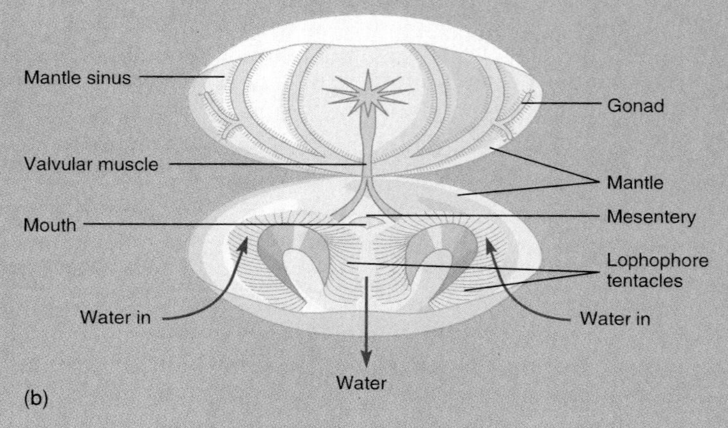

(b)

FIGURE 2 **Phylum Brachiopoda.** (*a*) Articulata brachiopod opened to show the two attached valves. (*b*) The internal anatomy of an articulate brachiopod.

the left and right valves in molluscs, the brachiopods have dorsal and ventral valves. In addition, molluscs filter with their gills, whereas brachiopods use a lophophore. Brachiopods are commonly called lampshells because they resemble ancient Roman oil lamps.

Brachiopods are exclusively marine; most species live from the tidal zone to the edge of the continental shelves (about 200 m deep). There are about 300 living species in the phylum. In the Articulata, the valves are composed primarily of calcium carbonate and have a hinge with interlocking teeth (figure 2*a*). The Inarticulata have unhinged valves that are composed primarily of calcium phosphate. The Inarticulata valves are held together only by muscles. Most members of both classes have a stalked pedicel that is usually attached to a hard surface. Some, such as *Lingula*, have a muscular pedicel used for burrowing and anchoring in mud or sand.

The large horseshoe-shaped lophophore in the anterior mantle cavity bears long, ciliated tentacles used in respiration and feeding (figure 2*b*). The cilia set up water currents that carry food particles (mainly organic detritus and algae) between the valves and over the lophophore into the mouth.

A solitary, dioecious brachiopod reproduces sexually by releasing gametes from multiple gonads into the metacoel and discharging them into the water by the nephridia. Fertilization is usually external, and the nonfeeding, ciliated, free-swimming larva is planktonic before settling and developing into an adult. Development is similar to deuterostomes, with radial, mostly equal, holoblastic cleavage, and enterocoelous coelom formation. In the Inarticulata, the juvenile resembles a small brachiopod with a coiled pedicel in the mantle cavity. There is no metamorphosis; when the juvenile settles to the bottom, the pedicel attaches to a solid object, and adult existence begins.

Phylum Ectoprocta (Bryozoa): Moss Animals

The ectoprocts (ek-to-proks) (Gr. *ektos*, outside + *proktos*, anus) or bryozoans superficially resemble hydroids or corals. Bryozoa (Gr. *bryon*, moss + *zoon*, animal) means moss animals and refers to the mosslike appearance of the colonies. The name ectoprocta is used to distinguish this group of coelomate animals, with the anus located outside the ring of tentacles, from the entoprocts in which the anus is within the ring of tentacles (*see figure 20.21*). There are about 4,000 living species of pseudocoelomate ectoprocts belonging to three classes. All species are aquatic (both fresh water and marine) and less than 1.5 mm in length.

Each body, or zooid, has a circular or horseshoe-shaped lophophore and is covered with a thin cuticle that encloses a calcified exoskeleton (*see figure 1 on page 235*). The feeding body (lophophore, digestive tract, muscles, nerves) is called the polypide, the exoskeleton plus body wall (epidermis) is the cystid, and the secreted, nonliving part (exoskeleton) is the zooecium (Gr. *zoo*, animal + *oceus*, house). Ectoprocts have an eversible lophophore that can be withdrawn into the body. Contraction of the retractor muscle rapidly withdraws the lophophore, whereas contraction of the muscles encircling the body wall exert pressure on the coelomic fluid, everting the lophophore.

Ectoprocts grow by budding. Thin portions of the body wall grow out as small vesicles or tubes and contain a complete zooid. The different budding patterns reflect the genetics of the individual animal and factors such as current flow and substrate. These factors determine the colony shape (e.g., thin sheets, convoluted folds, massive corallike heads, upright tangles) (figure 3*a*). Each colony can contain about 2 million zooids.

Most ectoprocts are monoecious. In some species, heterozooids produce either eggs or sperm in different colonies. In others, both sperm and eggs are produced in the same autozooid in simple gonads. Sperm are released into the coelomic cavity, exit through pores in the tips of the tentacles, and are caught by the tentacles of other colonies. Eggs are fertilized as they are released and brooded in the coelom; some species have a modified ovicell in which the embryo develops. Marine species have radial cleavage and a free-swimming, ciliated larva. This larva swims for a variable period of time, depending on the species, and then sinks and attaches to a rock or other substrate and grows into a zooid. A colony is formed by budding.

Some freshwater ectoprocts produce a dormant stage called a statoblast (figure 3*b*). A statoblast is a hard, resistant capsule containing a mass of germinative cells. Statoblasts are asexually produced and accumulate in the metacoel. They can survive long periods of freezing and drying, enabling a colony to survive many years in seasonally variable lakes and ponds. Some

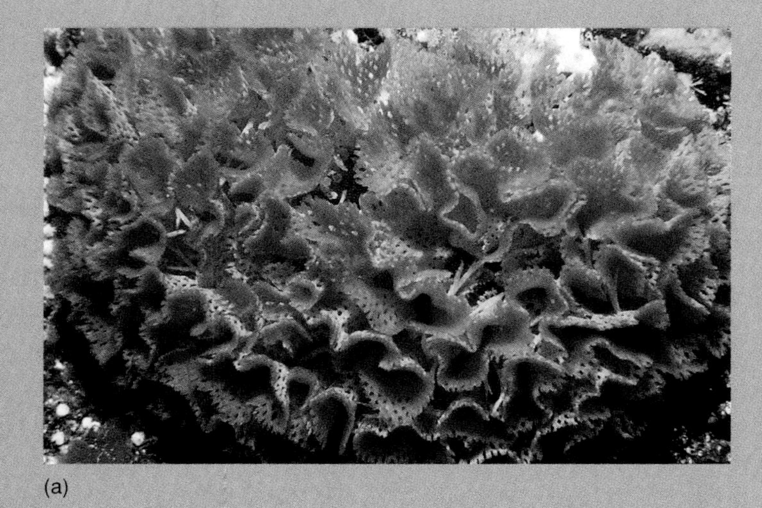

(a)

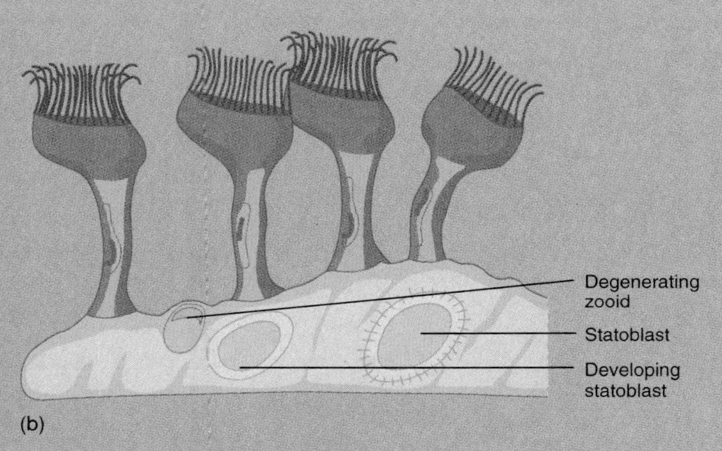

(b)

FIGURE 3 **Phylum Ectoprocta.** (*a*) Ectoprocts have a wide variety of shapes. *Philolopora* has lacy, delicate folds. (*b*) An ectoproct colony forming statoblasts.

float and are carried downstream, or are blown or carried from pond to pond, spreading ectoprocts over a large area. When environmental conditions become favorable, the statoblasts hatch and give rise to new polypides and eventually new colonies.

PHYLUM PHORONIDA: THE PHORONIDS

The phoronids (fo-ron-ids) consist of about one dozen marine species divided between two genera: *Phoronis* and *Phoronopsis*. These animals live in permanent, chitinous tubes either buried in muddy or sandy sediments, or attached to solid surfaces. A few species bore into mollusc shells or calcareous rock. Generally, only the tentacles extend into the overlying water. Most phoronids are small—less than 20 cm long.

The adult phoronid body consists of an anterior lophophore with two parallel rings of long tentacles (figure 4). The tentacles of the lophophore are filled with coelomic fluid that serves as a hydrostatic skeleton to hold them upright. The cilia on the tentacles drive water into the ring of tentacles from the top of the lophophore and out through the narrow spaces between the tentacles. Suspended food particles are directed toward the mouth. A flap of tissue called the epistome (Gr. *epi*, around + *stome*, mouth) covers the mouth.

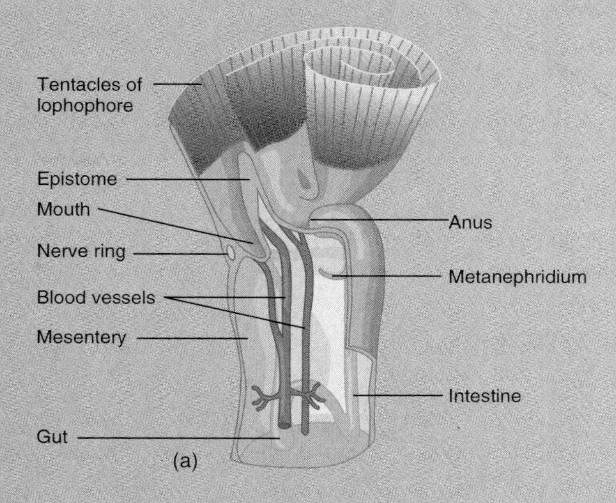

(a)

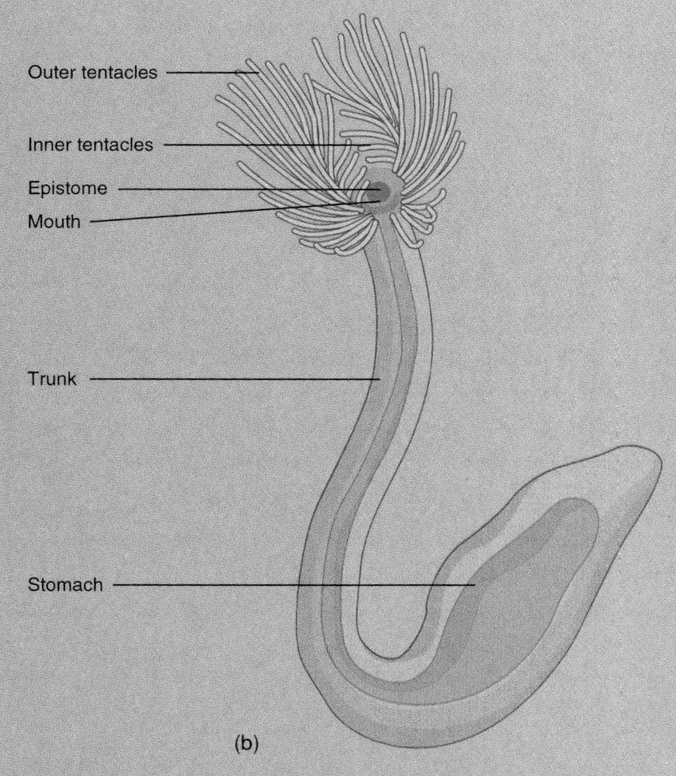

(b)

FIGURE 4 **Phylum Phoronida.** (*a*) A longitudinal view of the internal anatomy of the anterior portion of a phoronid. (*b*) A phoronid removed from its tube.

Some phoronids reproduce asexually by budding and transverse fission; however, the majority are hermaphroditic. The gonads are located in the coelom. Gametes pass from the coelom through the nephridiopore to the tentacles. Cross-fertilization is the rule, and the zygotes are either protected among the coils of the lophophore or released into the sea. Cleavage is radial, and a free-swimming larva called the actinotroch develops and feeds on plankton while drifting in the sea. It eventually settles to the bottom, metamorphoses, and begins to grow ventrally to form the body of the sedentary adult. As the animal grows, it burrows into the substrate. The body wall contains gland cells that eventually secrete the chitinous tube.

chapter 14

Hemichordata and Invertebrate Chordates

Outline

Concepts

1. Members of the phyla Echinodermata, Hemichordata, and Chordata are probably derived from a common diploblastic or triploblastic ancestor.
2. The phylum Hemichordata includes the acorn worms (class Enteropneusta) and the pterobranchs (class Pterobranchia). Hemichordates live in or on marine substrates and feed on sediment or suspended organic matter.
3. Animals in the phylum Chordata are characterized by a notochord, pharyngeal gill slits or pouches, a tubular nerve cord, and a postanal tail.
4. The urochordates are marine, and are called tunicates. They are attached or planktonic, and solitary or colonial as adults. All are filter feeders.
5. Members of the subphylum Cephalochordata are called lancelets. They are filter feeders that spend most of their time partly buried in marine substrates.
6. Motile, fishlike chordates may have evolved from sedentary, filter-feeding ancestors as a result of paedomorphosis in a motile larval stage.

Would You Like to Know:

1. what acorn worms are? *(p. 240)*
2. what characteristics are shared by all chordates at some time in their life history? *(p. 244)*
3. what animals deposit cellulose in their body walls? *(p. 245)*
4. why sea squirts are classified in the same phylum (Chordata) as humans? *(p. 247)*
5. why cephalochordates, such as amphioxus, are studied in introductory zoology laboratories? *(p. 247)*
6. how fishlike chordates could have evolved from filter-feeding ancestors? *(p. 248)*

These and other useful questions will be answered in this chapter.

This chapter contains evolutionary concepts, which are set off in this font.

Evolutionary Perspective

Some members of one of the phyla discussed in this chapter are more familiar to beginning students of zoology than members of any other group of animals. This familiarity is not without good reason, for zoologists themselves are members of one of these phyla—Chordata. Other members of these phyla, however, are much less familiar. Observations during a walk along a seashore at low tide may reveal coiled castings (sand, mud, and excrement) at the openings of U-shaped burrows, and excavating these burrows may reveal a wormlike animal that is one of the members of a small phylum—Hemichordata. Other members of this phylum include equally unfamiliar filter feeders called pterobranchs.

While at the seashore, one could also see animals clinging to rocks exposed by low tide. At first glance, they might be described as jellylike masses with two openings at their unattached end. Some are found as solitary individuals, others live in colonies. Handling these animals may be rewarded with a stream of water squirted from their openings. Casual observations provide little evidence that these small filter feeders, called sea squirts or tunicates, are chordates. However, detailed studies have made that conclusion certain. Tunicates and a small group of fishlike cephalochordates are often called the invertebrate chordates because they lack a vertebral column (figure 14.1).

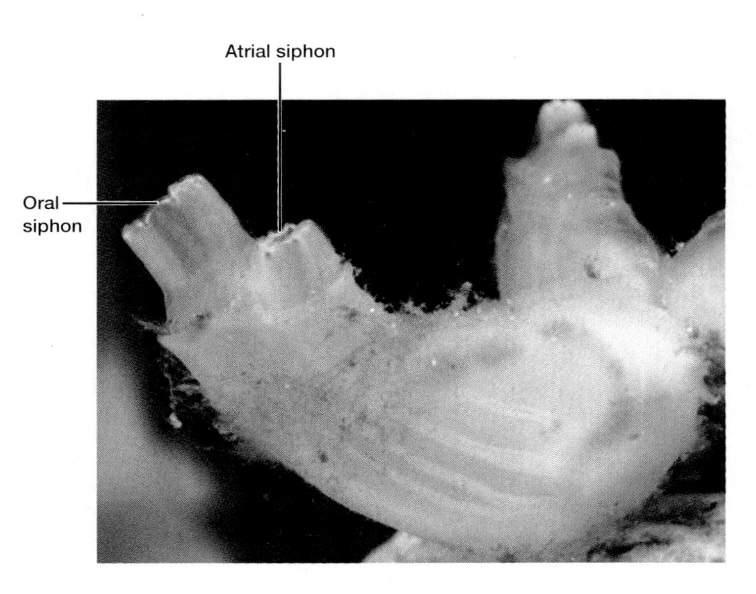

Figure 14.1
Phylum Chordata. This tunicate, or sea squirt (*Ciona intestinalis*), is an invertebrate chordate that is found attached to substrates in marine environments. Note the two siphons used in circulating water through a filter-feeding apparatus.

Phylogenetic Relationships

Animals in the phyla Hemichordata and Chordata share deuterostome characteristics with echinoderms (figure 14.2). Most zoologists, therefore, believe that ancestral representatives of these phyla were derived from a common, as yet undiscovered, triploblastic ancestor. The chordates are characterized by a dorsal, tubular nerve cord, a notochord, pharyngeal gill slits, and a postanal tail. The only characteristics that they share with the hemichordates are gill slits and, in some species, a dorsal, tubular nerve cord. Therefore, most zoologists agree that the evolutionary ties between the chordates and hemichordates are closer than those between echinoderms and either phylum. Chordates and hemichordates, however, probably diverged from widely separated points along the deuterostome lineage. This generalization is supported by the diverse body forms and life-styles present in these phyla.

Phylum Hemichordata

The phylum Hemichordata (hem′i-kor-da′tah) (Gr. *hemi*, half + L. *chorda*, cord) includes the acorn worms (class Enteropneusta) and the pterobranchs (class Pterobranchia) (table 14.1). Members of both classes live in or on marine sediments.

Characteristics of the phylum Hemichordata include the following:

1. Marine, deuterostomate animals with a body divided into three regions: proboscis, collar, and trunk; coelom divided into three cavities
2. Ciliated, pharyngeal gill slits
3. Open circulatory system
4. Complete digestive tract
5. Dorsal, sometimes hollow, nerve cord

Class Enteropneusta

Members of the class Enteropneusta (ent′er-op-nus″tah) (Gr. *entero*, intestine + *pneustikos*, for breathing) are marine worms that usually range in size between 10 and 40 cm, although some can be as long as 2 m. There are about 70 described species, and most occupy U-shaped burrows in sandy and muddy substrates between the limits of high and low tides. (1) The common name of the enteropneusts—acorn worms—is derived from the appearance of the proboscis, which is a short, conical projection at the anterior end of the worm. A ringlike collar is posterior to the proboscis, and an elongate trunk is the third division of the body (figure 14.3). Acorn worms are covered by a ciliated epidermis and gland cells. The mouth is located ventrally between the proboscis and the collar. A variable number of gill slits, from a few to several hundred, are positioned laterally on the trunk. Gill slits are openings between the anterior region of the digestive tract, called the pharynx, and the outside of the body.

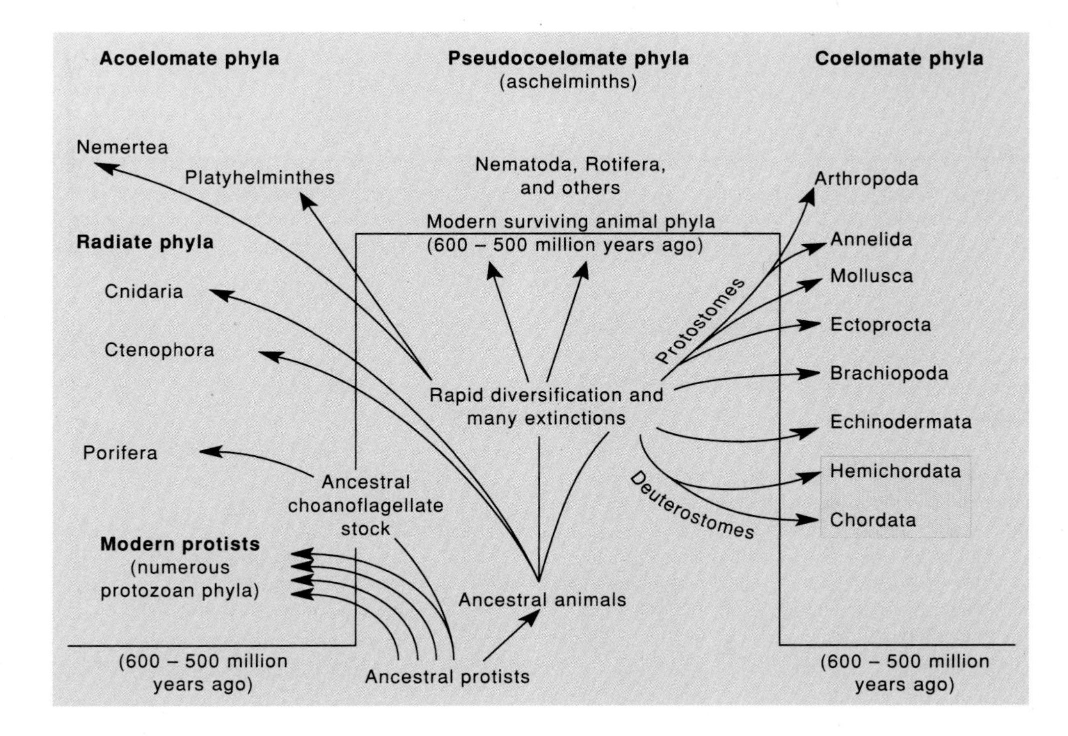

FIGURE 14.2
Phylogenetic Relationships among the Hemichordata and Chordata. Hemichordates and chordates (shaded in orange) are distantly related deuterostomes derived from a common, as yet undiscovered, diploblastic or triploblastic ancestor.

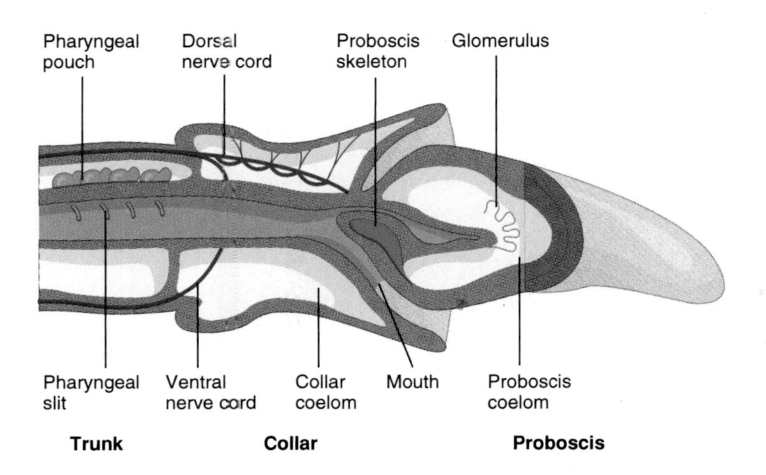

FIGURE 14.3
Class Enteropneusta. Longitudinal section showing the proboscis, collar, pharyngeal regions, and the internal structures. The arrow shows the path of water through a gill slit.

Maintenance Functions

Feeding of acorn worms is assisted by cilia and mucus. Detritus and other particles adhere to the mucus-covered proboscis. Tracts of cilia transport food and mucus posteriorly and ventrally. Ciliary tracts converge near the mouth and form a mucoid string that enters the mouth. Acorn worms may reject some substances trapped in the mucoid string by pulling the proboscis against the collar. Material to be rejected is transported along ciliary tracts of the collar and trunk and discarded posteriorly.

The digestive tract of enteropneusts is a simple tube. Digestion of food occurs as enzymes are released from diverticula of the gut, called hepatic sacs. The posterior end of the worm is extended out of the burrow during defecation. At low tide, one may see coils of fecal material, called castings, lying on the substrate at an opening of a burrow.

The nervous system of enteropneusts is ectodermal in origin and lies at the base of the ciliated epidermis. It consists of dorsal and ventral nerve tracts and a network of epidermal nerve cells, called a nerve plexus. In some species, the dorsal nerve is tubular and usually contains giant nerve fibers that rapidly transmit impulses. There are no major ganglia. Sensory receptors are unspecialized and widely distributed over the body.

Because acorn worms are small, exchanges of respiratory gases and metabolic waste products (principally ammonia) probably occur by diffusion across the body wall. In addition, respiratory gases are exchanged at the pharyngeal gill slits. Cilia associated with pharyngeal gill slits circulate water into the mouth and out of the body through gill slits. As water passes through gill slits, gases are exchanged by diffusion between water and blood sinuses surrounding the pharynx.

TABLE 14.1 CLASSIFICATION OF THE HEMICHORDATA AND CHORDATA

Phylum Hemichordata (hem'i-kor-da'tah)
Widely distributed in shallow, marine, tropical waters and deep, cold waters; soft bodied and wormlike; epidermal nervous system; most with pharyngeal gill slits.

Class Enteropneusta (ent'er-op-nus"tah)
Shallow water, wormlike animals; inhabit burrows on sandy shorelines; body divided into three regions: proboscis, collar, and trunk. Acorn worms (*Balanoglossus*, *Saccoglossus*).

Class Pterobranchia (ter'o-brang"ke-ah)
With or without gill slits; two or more arms; often colonial, living in an externally secreted encasement. *Rhabdopleura*.

Class Planctosphaeroidea (plank'to-sfer-roi'de-ah)
Body spherical with ciliary bands covering the surface; U-shaped digestive tract; coelom poorly developed; planktonic. Only one species is known to exist (*Planctosphaera pelagica*).

Phylum Chordata (kor-dat'ah) (L. *chorda*, cord)
Occupy a wide variety of marine, freshwater, and terrestrial habitats. A notochord, pharyngeal gill slits, a dorsal tubular nerve cord, and a postanal tail all present at some time in chordate life histories.

Subphylum Urochordata (u'ro-kor-dat'ah)
Notochord, nerve cord, and postanal tail present only in free-swimming larvae; adults sessile, or occasionally planktonic, and enclosed in a tunic that contains some cellulose; marine. Sea squirts or tunicates.

Class Ascidiacea (as-id'e-as"e-ah)
All sessile as adults; solitary or colonial; colony members interconnected by stolons.

Class Appendicularia (a-pen'di-ku-lar'e-ah) **(Larvacea)** (lar-vas'e-ah)
Planktonic; adults retain tail and notochord; lack a cellulose tunic; epithelium secretes a gelatinous covering of the body.

Class Sorberacea (sor'ber-as"e-ah)
Ascidianlike urochordates possessing dorsal nerve cords as adults; deep water, benthic; carnivorous. *Octacnemus*.

Class Thaliacea (tal'e-as"e-ah)
Planktonic; adults are tailless and barrel shaped; oral and atrial openings are at opposite ends of the tunicate; water currents produced by muscular contractions of the body wall.

Subphylum Cephalochordata (sef'a-lo-kor-dat'ah)
Body laterally compressed and transparent; fishlike; all four chordate characteristics persist throughout life. Amphioxus (*Branchiostoma*).

Subphylum Vertebrata (ver'te-bra'tah)
Notochord, nerve cord, postanal tail, and gill slits present at least in embryonic stages; vertebrae surround nerve cord and serve as primary axial support; skeleton modified anteriorly into a skull for protection of the brain.

Class Cephalaspidomorphi (sef-ah-las'pe-do-morf'e)
Fishlike; jawless; no paired appendages; cartilaginous skeleton; sucking mouth with teeth and rasping tongue. Lampreys.

Class Myxini (mik-sy-ny)
Fishlike; jawless; no paired appendages; mouth with four pairs of tentacles; olfactory sacs open to mouth cavity; 5 to 15 pairs of gill slits. Hagfishes.

Class Chondrichthyes (kon-drik'thi-es)
Fishlike; jawed; paired appendages and cartilaginous skeleton; no swim bladder. Skates, rays, sharks.

Class Osteichthyes (os'te-ik'thee-ez)
Bony skeleton; swim bladder and operculum present. Bony fishes.

Class Amphibia (am-fib'e-ah)
Skin with mucoid secretions; possess lungs and/or gills; moist skin serves as respiratory organ, aquatic developmental stages usually followed by metamorphosis to an adult. Frogs, toads, salamanders.

Class Reptilia (rep-til'e-ah)
Dry skin with epidermal scales; amniotic eggs; terrestrial embryonic development. Snakes, lizards, alligators.

Class Aves (a'vez)
Scales modified into feathers for flight; efficiently regulate body temperature (endothermic); amniotic eggs. Birds.

Class Mammalia (ma-may'le-ah)
Bodies at least partially covered by hair; endothermic; young nursed from mammary glands; amniotic eggs. Mammals.

The circulatory system of acorn worms consists of one dorsal and one ventral contractile vessel. Blood moves anteriorly in the dorsal vessel and posteriorly in the ventral vessel. Branches from these vessels lead to open sinuses. All blood flowing anteriorly passes into a series of blood sinuses, called the glomerulus, at the base of the proboscis. Excretory wastes may be filtered through the glomerulus, into the coelom of the proboscis, and released to the outside through one or two pores in the wall of the proboscis. The blood of acorn worms is colorless, lacks cellular elements, and distributes nutrients and wastes.

Reproduction and Development

Enteropneusts are dioecious. Two rows of gonads lie in the body wall in the anterior region of the trunk, and each gonad opens separately to the outside. Fertilization is external. Spawning by one worm induces others in the area to spawn—behavior that suggests the presence of spawning pheromones. Ciliated larvae, called **tornaria,** swim in the plankton for several days to a few weeks. The larvae settle to the substrate and are gradually transformed into the adult form (figure 14.4).

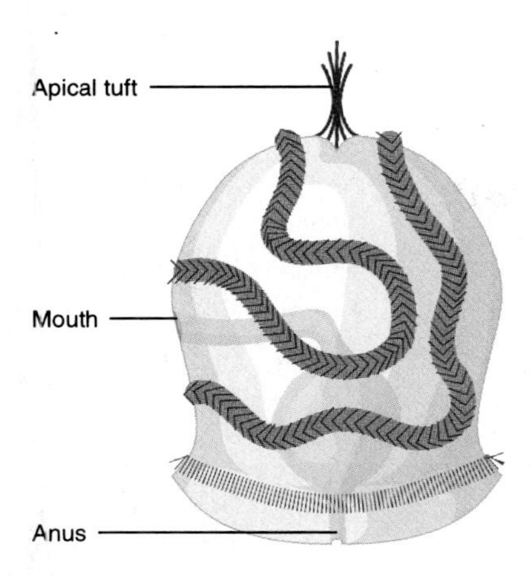

FIGURE 14.4
The Tornaria Larva (*Balanoglossus*). When larval development is complete, a tornaria locates a suitable substrate, settles, and begins to burrow and elongate.

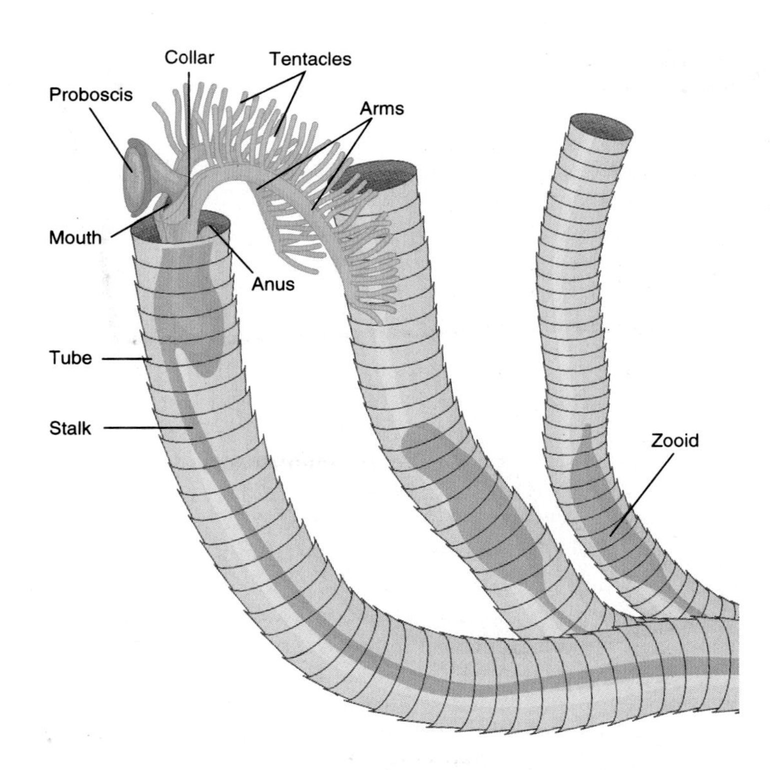

FIGURE 14.5
External Structure of *Rhabdopleura*. Ciliated tracts on tentacles and arms direct food particles toward the mouth.

CLASS PTEROBRANCHIA

Pterobranchia (ter′o-brang″ke-ah) (Gk. *pteron*, wing or feather + *branchia*, gills) is a small class of hemichordates found mostly in deep, oceanic waters of the Southern Hemisphere. A few are found in European coastal waters and in shallow waters near Bermuda. There are about 20 described species of pterobranchs.

Pterobranchs are small, ranging in size from 0.1 to 5 mm. Most live in secreted tubes in asexually produced colonies. As in enteropneusts, the pterobranch body is divided into three regions. The proboscis is expanded and shieldlike (figure 14.5). It secretes the tube and aids in movement in the tube. The collar possesses two to nine arms with numerous ciliated tentacles. The trunk is U-shaped.

Maintenance Functions

Pterobranchs use water currents generated by cilia on their arms and tentacles to filter feed. Food particles are trapped and transported by cilia toward the mouth. Although there is a single pair of pharyngeal gill slits in one genus, there is little need for either respiratory or excretory structures in animals as small as pterobranchs, because exchanges of gases and wastes occur by diffusion.

Reproduction and Development

Asexual budding is common in pterobranchs and is responsible for colony formation. Pterobranchs also possess one or two gonads in the anterior trunk. Most species are dioecious, and external fertilization results in the development of a planulalike larva that lives for a time in the tube of the female. This nonfeeding larva eventually leaves the female's tube, settles to the substrate, forms a cocoon, and metamorphoses into an adult.

Stop and Ask Yourself

1. What are the three body regions of a hemichordate? What is the function of these regions in acorn worms? In pterobranchs?
2. How are the feeding mechanisms of acorn worms and pterobranchs similar?
3. How do respiratory and excretory functions occur in hemichordates?

PHYLUM CHORDATA

Although the phylum Chordata (kor-dat′ah) (L. *chorda*, cord) does not have an inordinately large number of species (about 45,000), its members have been very successful at adapting to aquatic and terrestrial environments throughout the world. Sea squirts, members of the subphylum Urochordata, were briefly described in the introduction to this chapter. Other chordates include lancelets (subphylum Cephalochordata) and the vertebrates (subphylum Vertebrata) (*see table 14.1*). Characteristics of the phylum Chordata include the following:

1. Bilaterally symmetrical, deuterostomate animals
2. Four unique characteristics present at some stage in development: notochord, pharyngeal gill slits, dorsal tubular nerve cord, and postanal tail
3. Presence of an endostyle or thyroid gland
4. Complete digestive tract
5. Ventral, contractile blood vessel (heart)

2 Four of the characteristics listed above (no. 2) are unique to all chordates. They are discussed further in the following paragraphs.

The phylum is named after the **notochord** (Gr. *noton*, the back + L. *chorda*, cord), a supportive rod that extends most of the length of the animal dorsal to the body cavity and into the tail. It consists of a connective-tissue sheath that encloses cells, each of which contains a large, fluid-filled vacuole. This arrangement gives the notochord some turgidity, which prevents compression along the anteroposterior axis. At the same time, it is flexible enough to allow some freedom for lateral bending, as in the lateral undulations of a fish during swimming. In most adult vertebrates, the notochord is partly or entirely replaced by cartilage or bone.

Pharyngeal gill slits are a series of openings in the pharyngeal region between the digestive tract and the outside of the body. In some chordates, diverticula from the gut in the pharyngeal region never break through to form an open passageway to the outside. These diverticula are then called pharyngeal gill pouches. The earliest chordates used the gill slits for filter feeding; some living chordates still use them for feeding. Other chordates have developed gills in the pharyngeal pouches for gas exchange. The pharyngeal gill slits of terrestrial vertebrates are mainly embryonic features and may be incomplete.

The **tubular nerve cord** and its associated structures are largely responsible for the success of the chordates. The nerve cord runs along the longitudinal axis of the body, just dorsal to the notochord, and is usually expanded anteriorly as a brain. This central nervous system is associated with the development of complex systems for sensory perception, integration, and motor responses.

The fourth chordate characteristic is a **postanal tail.** (A postanal tail extends posteriorly beyond the anal opening.) The tail is either supported by the notochord or vertebral column.

Subphylum Urochordata

Members of the subphylum Urochordata (u′ro-kor-dat′ah) (Gr. *uro*, tail + L. *chorda*, cord) are the tunicates or sea squirts. The ascidians comprise the largest class of tunicates (table 14.1). They are sessile as adults and are either solitary or colonial. The appendicularians and thaliaceans are planktonic as adults (figure 14.6; box 14.1). In some localities, tunicates occur in large enough numbers to be considered a dominant life form.

Sessile urochordates attach their saclike bodies to rocks, pilings, hulls of ships, and other solid substrates. The unattached

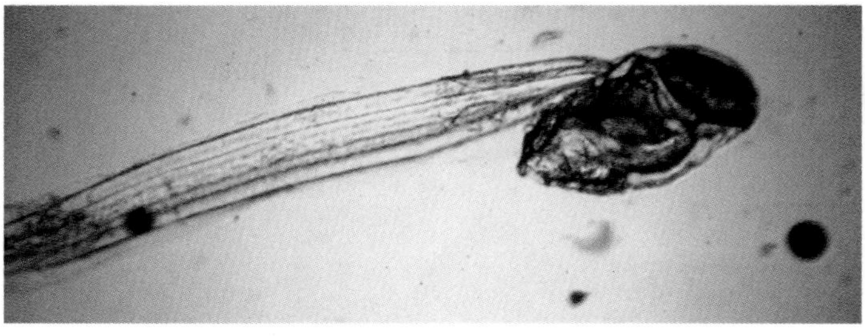

(a)

(b)

Figure 14.6
Subphylum Urochordata. (*a*) Members of the class Appendicularia are planktonic and have a tail and notochord that persist into the adult stage. (*b*) The thaliaceans are barrel-shaped, planktonic urochordates. Oral and atrial siphons are at opposite ends of the body, and muscles of the body wall contract to create a form of weak jet propulsion.

BOX 14.1 PLANKTONIC TUNICATES

Although the description of urochordates in this textbook is based primarily upon attached ascidian tunicates, planktonic species are numerous and important in marine food webs. Dense swarms of tunicates, hundreds of kilometers wide and many meters deep, are common in the open ocean. Swarms of larvaceans (*see figure 14.6a*) have been estimated to contain up to 25,000 animals per cubic meter! Larvaceans filter organisms as small as bacteria (0.1 μm in diameter), and in turn, are fed on by other plankton feeders, such as sardines and herring.

Thaliacean tunicates also occur in large, dense swarms in the open ocean (*see figure 14.6b*). Most are aggregations of solitary individuals; however, some form spectacular luminescent colonies. Pyrosome colonies, such as the one shown in figure 1, are found in many oceans. Colonies 10 m long and 1 m in diameter are common. Individuals are oriented with oral siphons pointed outward and atrial siphons directed toward the center of the colony. Ciliary currents and contractions of body-wall muscles direct water toward a central cavity of the colony and slowly move the entire colony through the water. When a part of the colony is stimulated by chemical or mechanical stimuli, it luminesces and ceases ciliary beating. The luminescence spreads over the entire colony, and the colony stops moving. This behavior may help the colony avoid unfavorable environments or confuse or frighten predators.

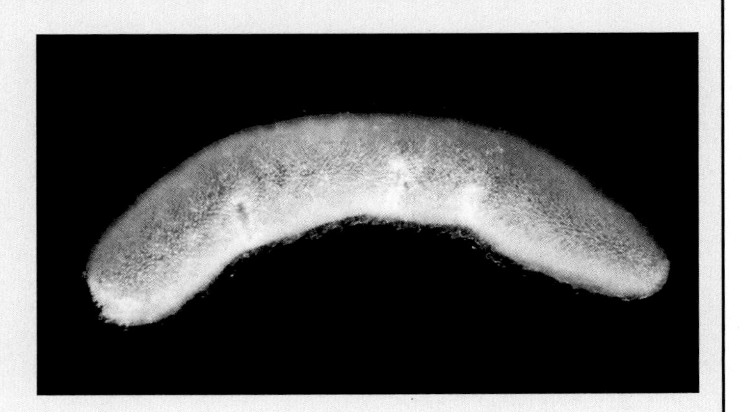

FIGURE 1 **Planktonic Tunicates.** A pyrosome colony (*Pyrosoma spinosum*).

end of urochordates contains two siphons that permit seawater to circulate through the body. One siphon is the oral siphon, which is the inlet for water circulating through the body and is usually directly opposite the attached end of the ascidian (figure 14.7). It also serves as the mouth opening. The second siphon, the atrial siphon, is the opening for excurrent water.

3 The body wall of most tunicates (L. *tunicatus*, to wear a tunic or gown) is a connective-tissuelike covering, called the tunic, that appears gellike, but is often quite tough. It is secreted by the epidermis and is composed of proteins, various salts, and cellulose. Some mesodermally derived tissues are incorporated into the tunic, including blood vessels and blood cells. Rootlike extensions of the tunic, called stolons, help anchor a tunicate to the substrate and may connect individuals of a colony.

Maintenance Functions

Longitudinal and circular muscles are present below the body wall epithelium and help to change the shape of the adult tunicate. They act against the elasticity of the tunic and the hydrostatic skeleton created by seawater confined to internal chambers.

The nervous system of tunicates is largely confined to the body wall. It forms a nerve plexus with a single ganglion located on the wall of the pharynx between the oral and atrial openings (figure 14.8*a*). This ganglion is not vital for coordinating bodily functions. Tunicates are sensitive to many kinds of mechanical and chemical stimuli, and receptors for these senses are distributed over the body wall, especially around the siphons. There are no complex sensory organs.

The most obvious internal structures of the urochordates are a very large pharynx and a cavity, called the atrium, that surrounds the pharynx laterally and dorsally (figure 14.8*b*). The pharynx of tunicates originates at the oral siphon and is continuous with the remainder of the digestive tract. The oral margin of the pharynx has tentacles that prevent large objects from entering the pharynx. The pharynx is perforated by numerous gill slits called stigmas. Cilia associated with the stigmas cause water to circulate into the pharynx, through the stigmas, and into the surrounding atrium. Water leaves the tunicate through the atrial siphon.

The digestive tract of adult tunicates continues from the pharynx and ends at the anus near the atrial siphon. During feeding, a mucous sheet is formed by cells of a ventral, ciliated

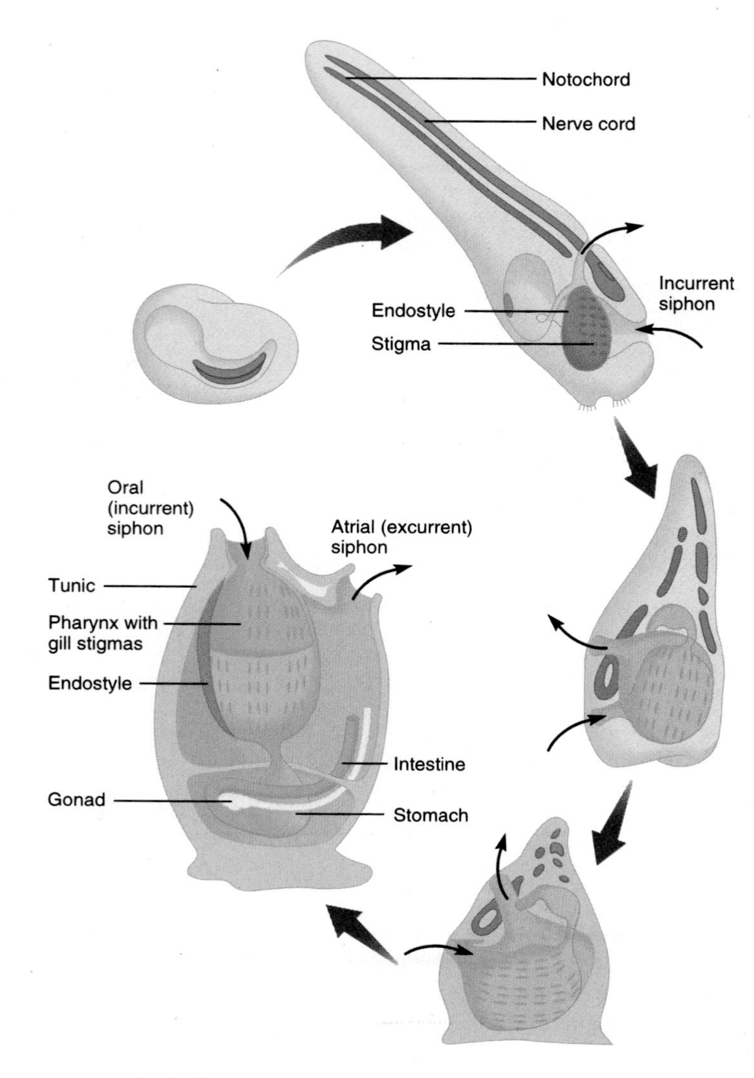

FIGURE 14.7

Tunicate Metamorphosis. Small arrows show the path of water through the body.

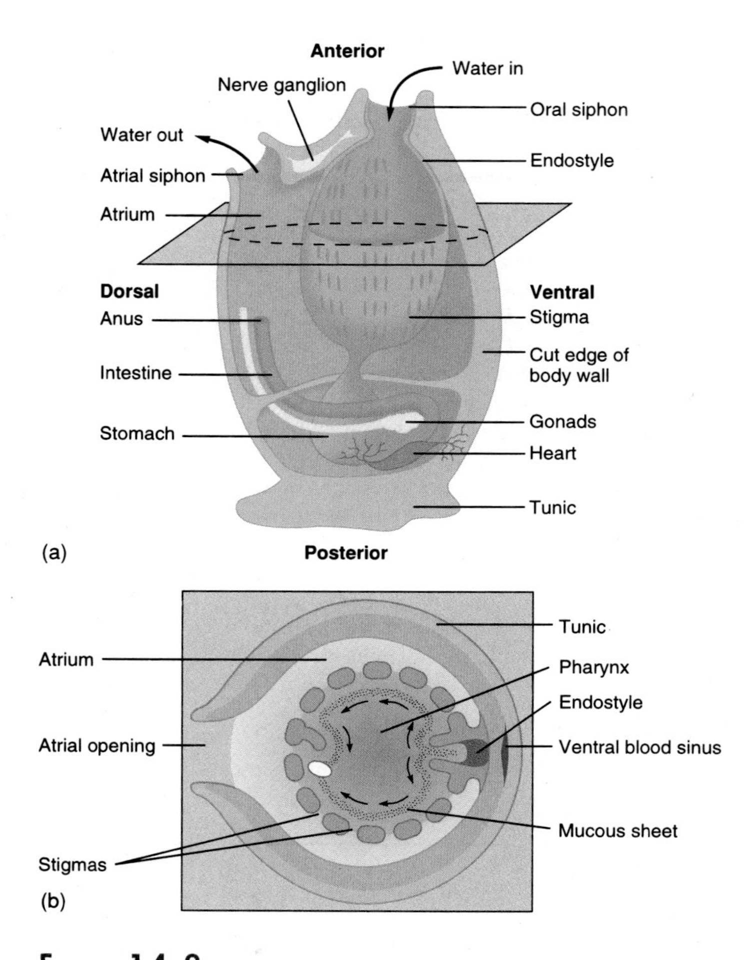

FIGURE 14.8

Internal Structure of a Tunicate. (*a*) Longitudinal section. Arrows show the path of water. (*b*) Cross section at the level of the atrial siphon. Arrows show movement of food trapped in mucus, which is produced by the endostyle.

groove, called the **endostyle** (figure 14.8*b*; box 14.2). Cilia move the mucous sheet dorsally across the pharynx. Food particles, brought into the oral siphon with incurrent water, are trapped in the mucous sheet and passed dorsally. Food is incorporated into a string of mucus that is moved by ciliary action into the next region of the gut tract. Digestive enzymes are secreted in the stomach, and most absorption occurs across the walls of the intestine. Digestive wastes are carried from the anus out of the atrial siphon with excurrent water.

In addition to its role in feeding, the pharynx also functions in gas exchange. Gases are exchanged between water circulating through the tunicate.

The tunicate heart lies at the base of the pharynx. One vessel from the heart runs anteriorly under the endostyle and another runs posteriorly to the digestive organs and gonads. Blood flow through the heart is not unidirectional. Peristaltic contractions of the heart may propel blood in one direction for a few beats, then the direction is reversed. The significance of this reversal is not understood. Tunicate blood plasma is colorless and contains various kinds of amoeboid cells.

Excretion is accomplished by the diffusion of ammonia into water that passes through the pharynx. In addition, amoeboid cells of the circulatory system accumulate uric acid and sequester it in the intestinal loop. Pyloric glands on the outside of the intestine are also thought to have excretory functions.

BOX 14.2 THE ENDOSTYLE AND THE VERTEBRATE THYROID GLAND

Neither the structure nor the function of the urochordate or cephalochordate endostyle give any clue to its fate in the vertebrates. Similarly, when examining adult vertebrates, one never sees a ciliated groove that once functioned to produce a sticky, mucous trap for filter-feeding ancestors. The study of the development of one group of vertebrates, the lampreys, has provided insight into the endostyle's evolutionary fate.

An endostyle is present in larval lampreys, where it produces a mucous filter, just as in the invertebrate chordates. In addition, it has the ability to bind iodine to the amino acid tyrosine. The significance of this second function is revealed in observing what happens to the endostyle when the larval lamprey metamorphoses to the adult and becomes a predator. Mucus-secreting functions of the endostyle become secondary, and the secretion of iodine-bound tyrosine derivatives becomes its primary function. During larval metamorphosis, the endostyle is transformed into an endocrine gland common to all vertebrates, the thyroid gland. The iodine-containing secretions of the thyroid gland of vertebrates regulate metamorphosis and metabolic rate.

The development of the thyroid gland of lampreys may reflect evolutionary events leading to the vertebrate thyroid gland. The endostyle of vertebrate ancestors may have had both mucus-secreting and endocrine functions. With the evolution of jaws and a more active, predatory life-style, endocrine functions were probably favored.

Reproduction and Development

Urochordates are monoecious. Gonads are located near the loop of the intestine, and genital ducts open near the atrial siphon. Gametes may be shed through the atrial siphon for external fertilization, or eggs may be retained in the atrium for fertilization and early development. Although self-fertilization occurs in some species, cross-fertilization is the rule. (4) Development results in the formation of a tadpolelike larva that possesses all four chordate characteristics. Metamorphosis begins after a brief free-swimming larval existence, during which the larva does not feed. The larva settles to a firm substrate and attaches by adhesive papillae located below the mouth. During metamorphosis, the tail is reduced by a shrinking of the outer epidermis that pulls the notochord and other tail structures internally for reorganization into adult tissues. The internal structures rotate 180°, resulting in the positioning of the oral siphon opposite the adhesive papillae and the bending of the digestive tract into a U shape (*see figure 14.7*).

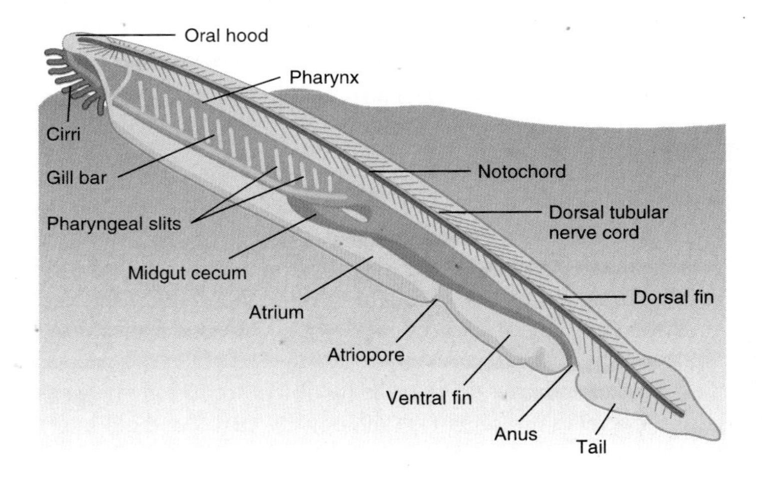

FIGURE 14.9
Subphylum Cephalochordata. Internal structure of *Branchiostoma* (amphioxus) shown in its partially buried feeding position.

SUBPHYLUM CEPHALOCHORDATA

Members of the subphylum Cephalochordata (sef′a-lo-kor-dat″ah) (Gr. *kephalo*, head + L. *chorda*, cord) are called lancelets. (5) They are almost universally studied in introductory zoology courses during introductions to chordate structure and function because they so clearly demonstrate the four chordate characteristics.

There are two genera, *Branchiostoma* (amphioxus) and *Asymmetron*, and about 45 species of cephalochordates. They are distributed throughout the world's oceans in shallow waters that have clean sand substrates.

Cephalochordates are small (up to 5 cm long), tadpolelike animals. They are elongate, laterally flattened, and nearly transparent. In spite of their streamlined shape, cephalochordates are relatively weak swimmers and spend most of their time in a filter-feeding position—partly to mostly buried with their anterior end sticking out of the sand (figure 14.9).

The notochord of cephalochordates extends from the tail to the head, giving them their name. Unlike the notochord of other chordates, most of the cells are muscle cells, making the notochord somewhat contractile. Both of these characteristics are probably adaptations to burrowing. Contraction of the muscle cells increases the rigidity of the notochord by compressing the fluids within, giving additional support when pushing into sandy substrates. Relaxation of these muscle cells allows increased flexibility for swimming.

Muscle cells are arranged on either side of the notochord and cause undulations that propel the cephalochordate through the water. Longitudinal, ventrolateral folds of the body wall help stabilize cephalochordates during swimming, and a median dorsal fin and a caudal fin also aid in swimming.

An oral hood projects from the anterior end of cephalochordates. Ciliated, fingerlike projections, called cirri, hang from the ventral aspect of the oral hood and are used in feeding. The posterior wall of the oral hood bears the mouth opening that leads to a large pharynx. Numerous pairs of gill slits perforate the pharynx and are supported by cartilaginous gill bars. Large folds of the body wall extend ventrally around the pharynx and fuse at the ventral midline of the body, creating the atrium, which is a chamber that surrounds the pharyngeal region of the body. It may protect the delicate, filtering surfaces of the pharynx from bottom sediments. The opening from the atrium to the outside is called the atriopore (figure 14.9).

Maintenance Functions

Cephalochordates are filter feeders. During feeding, they are partially or mostly buried in sandy substrates with their mouths pointed upwards. Water is brought into the mouth by the action of cilia on the lateral surfaces of gill bars. Water passes from the pharynx, through gill slits to the atrium, and out of the body through the atriopore. Initial sorting of food occurs at the cirri. Larger materials are caught on cilia of the cirri. As these larger particles accumulate, they are thrown off by contractions of the cirri. Smaller, edible particles are pulled into the mouth with water and are collected by cilia on the gill bars and in mucus secreted by the endostyle. As in tunicates, the endostyle is a ciliated groove that extends longitudinally along the midventral aspect of the pharynx. Cilia move food and mucus dorsally, forming a food cord that is moved by cilia to the gut. A ring of cilia rotates the food cord, and in the process, food is dislodged. Digestion is both extracellular and intracellular. A diverticulum off the gut, called the midgut cecum, extends anteriorly. It ends blindly along the right side of the pharynx and secretes digestive enzymes. An anus is on the left side of the ventral fin.

Cephalochordates do not possess a true heart. Blood is propelled by contractile waves in the walls of major vessels. Blood contains amoeboid cells and bathes tissues in open spaces.

Excretory tubules are modified coelomic cells that are closely associated with blood vessels. This arrangement suggests active transport of materials between the blood and excretory tubules.

The coelom of cephalochordates is reduced as compared to most other chordates. It is restricted to canals near the gill bars, the endostyle, and the gonads.

Reproduction and Development

Cephalochordates are dioecious. Gonads bulge into the atrium from the lateral body wall. Gametes are shed into the atrium and leave the body through the atriopore. External fertilization leads to a bilaterally symmetrical larva. Larvae are free swimming, but they eventually settle to the substrate before metamorphosing into adults.

Further Phylogenetic Considerations

The evolutionary relationships between the hemichordates and chordates are difficult to document with certainty. The dorsal, tubular nerve cord and pharyngeal gill slits of hemichordates are evidence of evolutionary ties between these phyla (figure 14.10). There are, however, questions regarding the homologies of these structures. Synapomorphies that distinguish chordates from hemichordates include tadpole larvae, notochord, postanal tail, and an endostyle.

Evolutionary relationships between members of the three chordate subphyla are also shown in figure 14.10. As discussed in chapter 13, the earliest echinoderms were probably sessile filter feeders.

The life-style of adult urochordates suggests a similar ancestry (perhaps from a common ancestor with echinoderms) for chordates. The evolution of motile chordates from attached ancestors may have involved the development of a tadpolelike larva. Increased larval mobility is often adaptive for species with sedentary adults because it promotes dispersal. (6) The evolution of motile adults could have resulted from paedomorphosis, which is the development of sexual maturity in the larval body form. (The occurrence of paedomorphosis is well documented in the animal kingdom, especially among amphibians.) Paedomorphosis could have led to a small, sexually reproducing, fishlike chordate that could have been the ancestor of higher chordates.

The largest and most successful chordates belong to the subphylum Vertebrata. They are characterized by bony or cartilaginous vertebrae that completely or partially replace the notochord. A high degree of cephalization is evidenced by the development of the anterior end of the nerve cord into a brain and the development of specialized sense organs

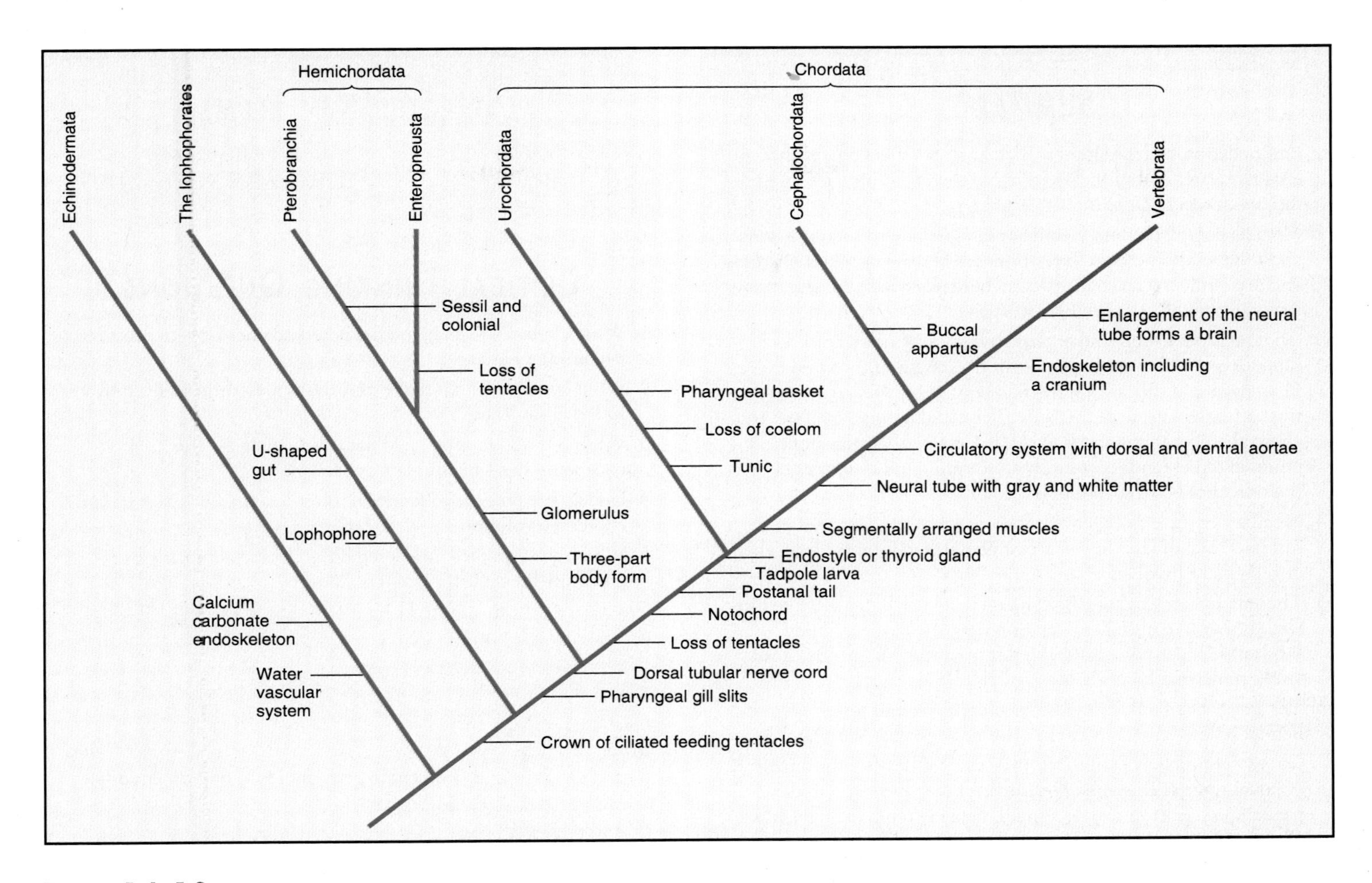

FIGURE 14.10

One Interpretation of Deuterostomate Phylogeny. The dorsal tubular nerve cord and pharyngeal gill slits are possible synapomorphies that link the Hemichordata and the Chordata. The notochord, postanal tail, and endostyle or thyroid gland are important characters that distinguish the Hemichordata and Chordata. Some of the synapomorphies that distinguish the chordate subphyla are shown.

on the head. The skeleton is modified anteriorly into a skull or cranium. There are eight classes of vertebrates (*see table 14.1*). Because of their cartilaginous and bony endoskeletons, vertebrates have left an abundant fossil record. Ancient jawless fishes were common in the Ordovician period, approximately 500 million years ago. Over a period of approximately 100 million years, fishes became the dominant vertebrates. Near the end of the Devonian period, approximately 400 million years ago, terrestrial vertebrates made their appearance. Since that time, vertebrates have radiated into most of the earth's habitats. Chapters 15 to 19 give an account of these events.

Stop and Ask Yourself

4. What are four characteristics shared by all chordates at some time in their life history?
5. What group of chordates deposit cellulose in their body wall?
6. What is an endostyle? What functions does it have in urochordates? In cephalochordates?
7. What is the function of the midgut cecum of cephalochordates?

Summary

1. Echinoderms, hemichordates, and chordates share deuterostome characteristics and are believed to have evolved from a common diploblastic or triploblastic ancestor.
2. Members of the phylum Hemichordata include the acorn worms and the pterobranchs. Acorn worms are burrowing, marine worms, and pterobranchs are marine hemichordates whose collar possesses arms with numerous ciliated tentacles.
3. Chordates have four unique characteristics. A notochord is a supportive rod that extends most of the length of the animal. Pharyngeal gill slits are a series of openings between the digestive tract and the outside of the body. The tubular nerve cord lies just above the notochord and is expanded anteriorly into a brain. A postanal tail extends posteriorly to the anus and is supported by the notochord or the vertebral column.
4. Members of the subphylum Urochordata are the tunicates or sea squirts. Urochordates are sessile or planktonic filter feeders. Their development involves a tadpolelike larva.
5. The subphylum Cephalochordata includes small, tadpolelike filter feeders that live in shallow, marine waters with clean sandy substrates. Their notochord extends from the tail into the head and is somewhat contractile.
6. The presence of gill slits and a tubular nerve cord link hemichordates and chordates to the same evolutionary lineage.
7. Chordates probably evolved from a sessile, filter-feeding ancestor. A larval stage of this sedentary ancestor may have undergone paedomorphosis to produce a small, sexually reproducing fishlike chordate.

Selected Key Terms

endostyle (*p. 246*)
notochord (*p. 244*)
pharyngeal gill slits (*p. 244*)
postanal tail (*p. 244*)
tubular nerve cord (*p. 244*)

Critical Thinking Questions

1. What evidence links hemichordates and chordates to the same evolutionary lineage?
2. What evidence of chordate affinities is present in adult tunicates? In larval tunicates?
3. What is paedomorphosis? What is a possible role for paedomorphosis in chordate evolution?
4. Discuss the possible influence of filter-feeding life-styles on early chordate evolution.
5. What selection pressures could have favored a foraging or predatory life-style for later chordates?

chapter 15

The Fishes: Vertebrate Success in Water

Outline

Evolutionary Perspective
- Phylogenetic Relationships

Survey of Fishes
- Superclass Agnatha
- Superclass Gnathostomata

Evolutionary Pressures
- Locomotion
- Nutrition and the Digestive System
- Circulation and Gas Exchange
- Nervous and Sensory Functions
- Excretion and Osmoregulation
- Reproduction and Development

Further Phylogenetic Considerations

Concepts

1. The earliest fossil vertebrates are 510-million-year-old ostracoderms.
2. Members of the superclass Agnatha include extinct ostracoderms, the lampreys, and the hagfishes. Agnathans lack jaws and paired appendages.
3. The superclass Gnathostomata includes the cartilaginous (class Chondrichthyes) and bony (class Osteichthyes) fishes.
4. Aquatic environments have selected for certain adaptations in fishes. These include the ability to move in a relatively dense medium, exchange gases with water or air, regulate buoyancy, detect environmental changes, regulate salt and water in their tissues, and successfully reproduce.
5. Adaptive radiation resulted in the large variety of fishes present today. Evolution of some fishes led to the terrestrial vertebrates.

Would You Like to Know:

1. what group of fishes probably gave rise to all other fishes? (*p. 252*)
2. how a shark's teeth are replaced? (*p. 258*)
3. what fishes have lungs and breathe air? (*p. 259*)
4. what fish living today is the closest relative of terrestrial vertebrates? (*p. 260*)
5. why some sharks never stop moving? (*p. 263*)
6. how a shark can find a flounder completely covered by sand? (*p. 265*)
7. what environmental conditions selected for adaptations in fishes that eventually led to terrestrial vertebrates? (*p. 269*)

These and other useful questions will be answered in this chapter.

This chapter contains evolutionary concepts, which are set off in this font.

Evolutionary Perspective

Over 70% of the earth's surface is covered by water, a medium that is buoyant and resistant to rapid fluctuations in temperature. Because life began in water, and living tissues are made mostly of water, it might seem that nowhere else would life be easier to sustain. This chapter describes why that is not entirely true.

You do not need to wear SCUBA gear to appreciate the fact that fishes are adapted to aquatic environments in a fashion unsurpassed by any other group of animals. If you spend recreational hours with hook and line, visit a marine theme park, or simply glance into a pet store when walking through a shopping mall, you can attest to the variety and beauty of fishes. This variety is evidence of adaptive radiation that began 500 million years ago and shows no sign of ceasing. Fishes not only dominate many watery environments, they are also the ancestors of all other members of the subphylum Vertebrata.

Phylogenetic Relationships

Fishes are members of the chordate subphylum Vertebrata; thus, they have vertebrae that surround their spinal cord and provide the primary axial support. They also have a skull that protects the brain (*see table 14.1*; figure 15.1).

Zoologists do not know what animals were the first vertebrates. Recent cladistic analysis of vertebrate evolution indicates that a group of fishes, called hagfishes, are the most primitive vertebrates known (living or extinct). Fossilized bony fragments indicate that bone was present at least 510 million years ago. These fossils are from bony armor that covered animals called ostracoderms. Ostracoderms were relatively inactive filter feeders that lived on the bottom of prehistoric lakes and seas. They possessed neither jaws nor paired appendages; however, the evolution of fishes resulted in both jaws and paired appendages as well as many other structures. The results of this adaptive radiation are described in this chapter.

Did ancestral fishes live in fresh water or in the sea? The answer to this question is not simple. The first vertebrates were probably marine because ancient stocks of other deuterostome phyla were all marine. Vertebrates, however, adapted to fresh water very early, and much of the evolution of fishes occurred there. Apparently, early vertebrate evolution involved the movement of fishes back and forth between marine and freshwater environments. The majority of the evolutionary history of some fishes took place in ancient seas, and most of the evolutionary history of others occurred in fresh water. The importance of fresh water in the evolution of fishes is evidenced by the fact that over 41% of all fish species are found in fresh water, even though freshwater habitats represent only a small percentage (0.0093% by volume) of the earth's water resources.

Figure 15.1
The Fishes. Five hundred million years of evolution have resulted in unsurpassed diversity in the fishes. The spines of this beautiful marine lionfish (*Pterois*) are extremely venomous.

Survey of Fishes

The taxonomy of fishes has been the subject of debate for many years. Modern cladistic analysis has resulted in complex revisions in the taxonomy of this group of vertebrates (figure 15.2). The system used in this textbook divides fishes into two superclasses based on whether they lack jaws and paired appendages (superclass Agnatha) or possess those structures (superclass Gnathostomata) (table 15.1).

Superclass Agnatha

Members of the superclass Agnatha (ag-nath′ah) (Gr. *a*, without + *gnathos*, jaw), in addition to lacking jaws and paired appendages, possess a cartilaginous skeleton and a notochord that persists into the adult stage. 1 Ancient agnathans are believed to be ancestral to all other fishes (*see figure 15.2*).

Ostracoderms are extinct agnathans that belonged to several classes. The fossils of predatory water scorpions (phylum Arthropoda [*see figure 11.7*]) are often found with fossil ostracoderms. As sluggish as ostracoderms apparently were, bony armor was probably their only defense. Ostracoderms were bottom dwellers, often about 15 cm long (figure 15.3). Most are believed to have been filter feeders, either filtering suspended organic matter from the water or extracting annelids and other animals from muddy sediments. Bony plates around the mouths of some ostracoderms may have been used in a jawlike fashion to crack gastropod shells or the exoskeletons of arthropods.

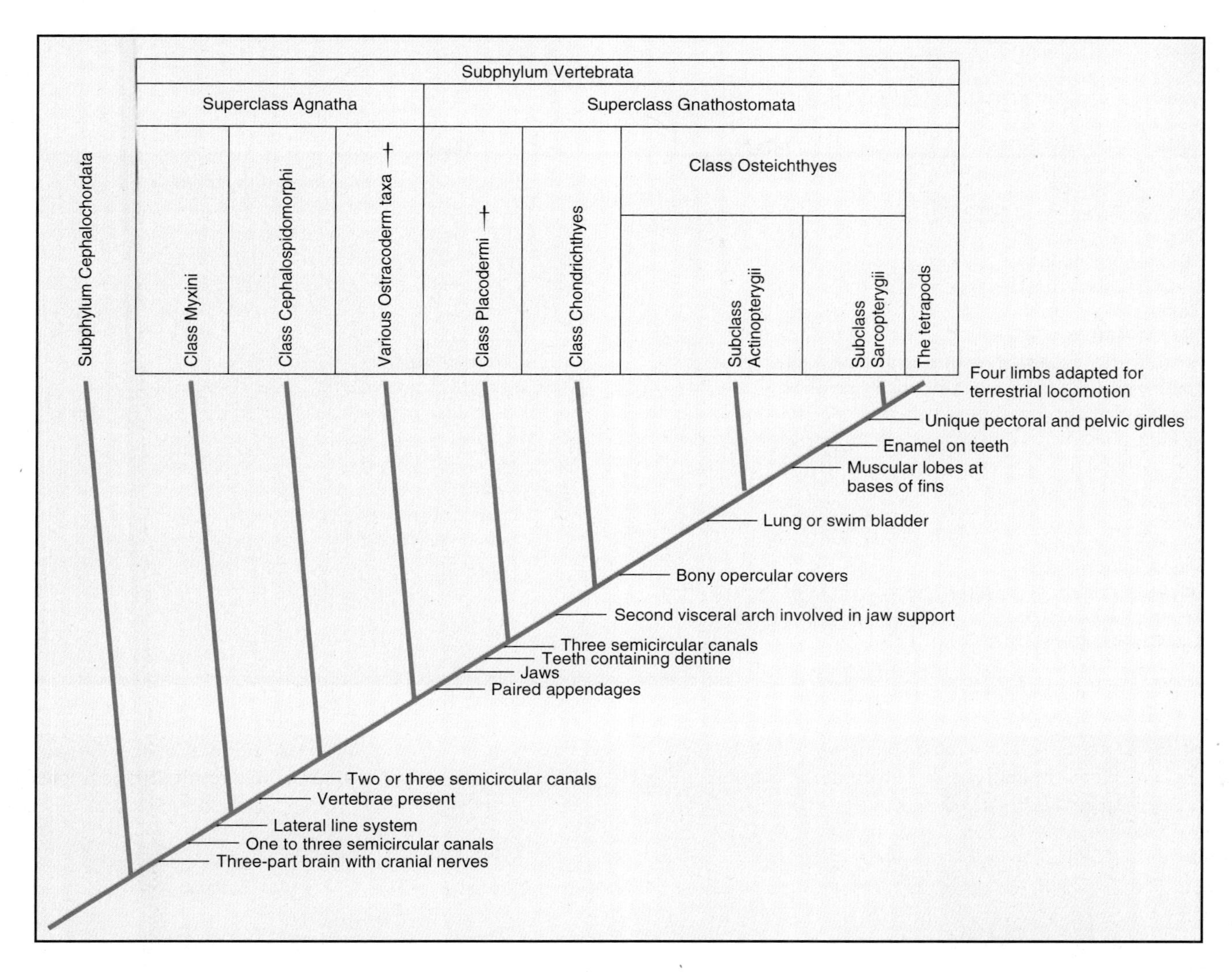

FIGURE 15.2

One Interpretation of the Phylogeny of the Fishes. The evolutionary relationships among the fishes are unsettled. A few selected ancestral and derived characters are shown in this cladogram. Each lower taxon has numerous synapomorphies that are not shown. The position of the lampreys in fish phylogeny is debated. Recent evidence indicates that both lampreys and ostracoderms are more closely related to jawed vertebrates than to hagfishes. The ostracoderms are considered a paraphyletic group (having multiple lineages) by most zoologists. Their representation as a monophyletic group is an attempt to simplify this presentation. Groups whose members are extinct are indicated with daggers (†).

Class Myxini

Hagfishes are members of the class Myxini (mik-sy′ny) (Gr. *myxa*, slime). Hagfishes live buried in the sand and mud of marine environments, where they feed on soft-bodied invertebrates and scavenge dead and dying fish (figure 15.4). When hagfishes find a suitable fish, they enter the fish through the mouth and eat the contents of the body, leaving only a sack of skin and bones. Anglers must contend with hagfishes because they will bite at a baited hook. Hagfishes have the annoying habit of swallowing a hook so deeply that the hook is frequently lodged near the anus. The excessively slimy bodies of hagfishes make all but the grittiest fishermen cut their lines and tie on a new hook. Most zoologists now consider the hagfishes to be the most primitive group of vertebrates.

Class Cephalaspidomorphi

Lampreys are agnathans in the class Cephalaspidomorphi (sef-ah-las′pe-do-morf′e) (Gr. *kephale*, head + *aspidos*, shield + *morphe*, form). They are common inhabitants of marine and freshwater environments in temperate regions. Most adult lampreys prey on other fishes, and the larvae are filter feeders. The mouth of an adult is suckerlike and surrounded by lips that

TABLE 15.1 CLASSIFICATION OF LIVING FISHES

Subphylum Vertebrata
Superclass Agnatha (ag-nath′ah)
Lack jaws and paired appendages; cartilaginous skeleton; persistent notochord; two semicircular canals. (Hagfishes have one semicircular canal that may represent a fusion of two canals.)
Class Myxini (mik-sy′ny)
Mouth with four pairs of tentacles; olfactory sacs open to mouth cavity; 5 to 15 pairs of gill slits. Hagfishes.
Class Cephalaspidomorphi (sef-ah-las′pe-do-morf′e)
Sucking mouth with teeth and rasping tongue; seven pairs of gill slits; blind olfactory sacs. Lampreys.
Superclass Gnathostomata (na′tho-sto′ma-tah)
Hinged jaws and paired appendages present; notochord may be replaced by vertebral column; three semicircular canals.
Class Chondrichthyes (kon-drik′thi-es)
Tail fin with large upper lobe (heterocercal tail); cartilaginous skeleton; lack opercula and a swim bladder or lungs. Sharks, skates, rays, ratfishes.
Subclass Elasmobranchii (e-laz-mo′bran′ke-i)
Cartilaginous skeleton may be partially ossified; placoid scales or no scales. Sharks, skates, rays.
Subclass Holocephali (hol′o-sef′a-li)
Operculum covers gill slits; lack scales; teeth modified into crushing plates; lateral-line receptors in an open groove. Ratfishes.
Class Osteichthyes (os′te-ik″the-es)
Most with bony skeleton; single gill opening covered by operculum; pneumatic sacs function as lungs or swim bladders. Bony fishes.
Subclass Sarcopterygii (sar-kop-te-rij′e-i)
Paired fins with muscular lobes; pneumatic sacs funtion as lungs. Lungfishes and coelacanths (lobe-finned fishes).
Subclass Actinopterygii (ak′tin-op″te-rig-e-i)
Paired fins supported by dermal rays; basal portions of paired fins not especially muscular; tail fin with approximately equal upper and lower lobes (homocercal tail); blind olfactory sacs. Ray-finned fishes.

FIGURE 15.3
Artist's Rendering of an Ancient Silurian Seafloor. Two ostracoderms, *Pteraspis* and *Anglaspis*, are shown in the background and a predatory water scorpion (phylum Arthropoda, class Merostomata) is shown in the foreground.

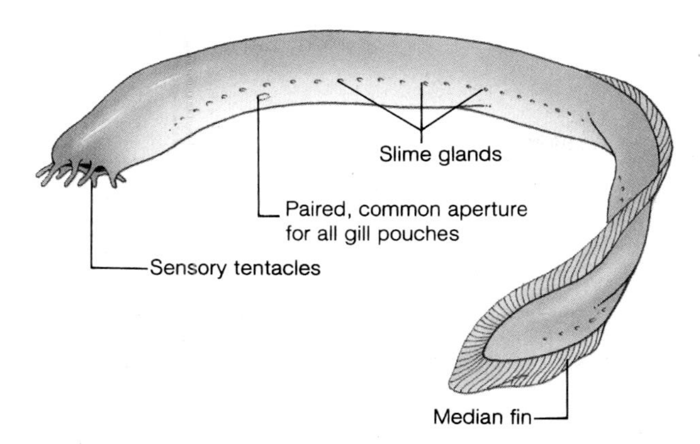

FIGURE 15.4
Class Myxini. Hagfish external structure.

FIGURE 15.5
Class Agnatha. A lamprey (*Petromyzon marinus*). Note the sucking mouth and teeth used to feed on other fish.

have sensory and attachment functions. Numerous epidermal teeth line the mouth and cover a movable tonguelike structure (figure 15.5). Adults attach to prey with their lips and teeth and use their tongue to rasp away scales. Lampreys have salivary glands with anticoagulant secretions and feed mainly on the blood of their prey (box 15.1). Some lampreys, however, are

BOX 15.1 LAMPREYS AND GREAT LAKES FISHERIES

Along with many other European immigrants to the United States in the nineteenth century came many Scandinavians. Many of these Scandinavian immigrants fished for a living, and they were understandably attracted to the Great Lakes Region. Through the early part of the twentieth century, fishing was very good. Catches included a variety of smaller fishes that inhabited the shoals and bays of the Great Lakes, but the prize catches were two deep water predators—lake trout (*Salvelinus namaycush*) and whitefish (*Corygonus clupeaformis*). The yearly catch of lake trout in each of the Great Lakes in the 1930s exceeded 2,000 metric tons.

These commercial fisheries, however, were doomed. Their fate was sealed years before many immigrants even left their homes in Scandinavia. In 1829, the Welland Canal was completed. It provided a shipping route around Niagara Falls between Lakes Ontario and Erie. Niagara Falls, however, had not only been a barrier to shipping, it had also been a barrier to the sea lamprey. After the Welland Canal was completed, sea lampreys slowly worked their way from Lake Ontario to the other Great Lakes. By 1937, spawning lampreys were recorded in Lake Michigan, and by the early 1940s Great Lakes fishermen had to work very hard to bring home a single lake trout with a day's catch of predominately smaller fishes. The lamprey, like humans, had a decided preference for the larger, cold-water fish species. In 1944, the annual catch of lake trout from Lake Michigan had been reduced to less than 100 metric tons. In 1953, the annual catch of lake trout in Lake Michigan was reduced to a few hundred kilograms!

Although the invasion of lampreys into the Great Lakes brought severe economic hardship to many fishermen, it also resulted in an important success story in fishery management. In the 1950s, an intensive lamprey control program was instituted by the states bordering the Great Lakes and by Canada. Control measures involved the use of mechanical weirs that prevented spawning migrations of lampreys into the tributaries of the Great Lakes. Electrical shocking devices were employed in an attempt to kill lampreys in spawning streams. Finally, chemical control measures were employed. Lamprey populations began to decline, and by the mid-1960s lamprey control measures were considered a success.

The void left by the decline of lake trout has been filled by a sport fishery. In the late 1960s, Coho (*Concorhynchus kisutch*) and Chinook (*Oncorhynchus tshawytscha*) salmon were stocked in the Great Lakes to create a sport fishery. Survival and growth of these salmon have been remarkable, and fewer than 5% of the salmon caught are marked by lamprey wounds. Whitefish and lake trout are again being caught. To preserve this fishery, lamprey control measures will be maintained in the future to prevent large-scale growth of lamprey populations.

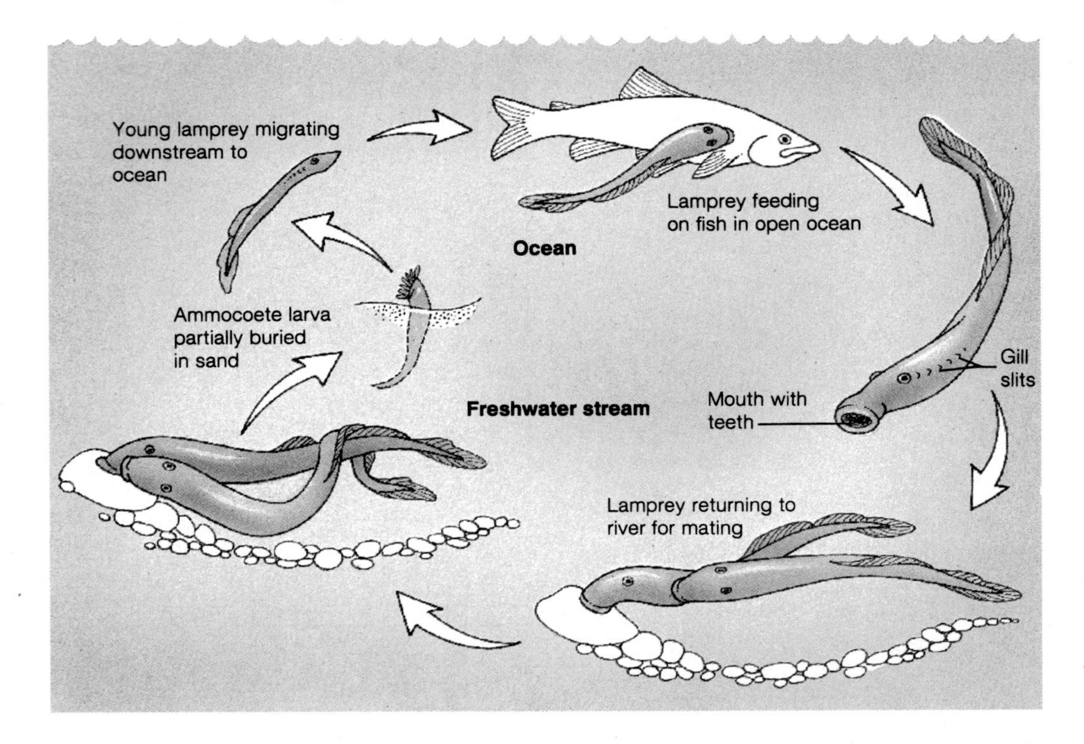

FIGURE 15.6

External Structure and Life History of a Sea Lamprey. Sea lampreys feed in the open ocean and near the end of their lives, migrate into freshwater streams where mating occurs. Eggs are deposited in nests on the stream bottom, and young ammocoete larvae hatch in about 3 weeks. Ammocoete larvae live as filter feeders until they attain sexual maturity.

not predatory. Members of the genus *Lampetra* are called brook lampreys. The larval stages of brook lampreys last for about 3 years, and the adults neither feed nor leave their stream. They reproduce soon after metamorphosis and then die.

Adult sea lampreys live in the ocean or the Great Lakes. Near the end of their lives, they undertake a migration that may take them hundreds of miles to a spawning bed in a freshwater stream. Once lampreys reach their spawning site, usually in relatively shallow water with swift currents, nest building begins. Lampreys make small depressions in the substrate. When the nest is prepared, a female usually attaches to a stone with her mouth. A male attaches to the female's head using his mouth, and wraps his body around the female (figure 15.6). Eggs are shed in small batches over a period of several hours, and fertilization is external. The relatively sticky eggs are then covered with sand.

Eggs hatch in approximately 3 weeks into ammocoete larvae. The larvae drift downstream to softer substrates, where they bury themselves in sand and mud and filter feed in a fashion similar to amphioxus (*see figure 14.9*).

Ammocoete larvae grow from 7 mm to about 17 cm over 3 years. During later developmental stages, the larvae metamorphose to the adult over a period of several months. The mouth becomes suckerlike, and the teeth, tongue, and feeding musculature develop. Lampreys eventually leave the mud permanently and begin a journey to the sea to begin life as predators. Adults will return only once to the headwaters of their stream to spawn and die.

Stop and Ask Yourself

1. What superclass of vertebrates contains the probable ancestors of all other vertebrates?
2. How are members of the superclass Agnatha characterized?
3. What animals make up the class Myxini?
4. How are the life-styles of adult and larval lampreys different?

SUPERCLASS GNATHOSTOMATA

Two major developments in vertebrate evolution were the appearance of jaws and paired appendages. These structures are first seen in members of the superclass Gnathostomata (na'tho-sto'ma-tah) (Gr. *gnathos*, jaw + *stoma*, mouth). Jaws are used in feeding and are partly responsible for a transition to more active, predatory life-styles. Pectoral fins of fishes are appendages that are usually located just behind the head, and

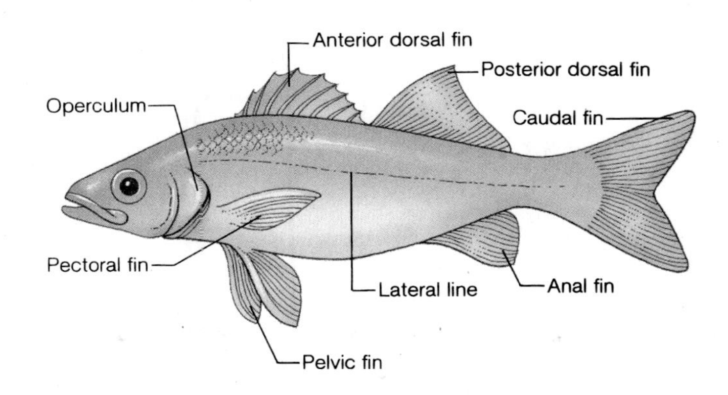

Figure 15.7

Paired Appendages. Appendages of a member of the superclass Gnathostomata. These appendages are secondarily reduced in some species.

pelvic fins are usually located ventrally and posteriorly (figure 15.7). Both sets of paired fins increase the agility of fishes by giving them a more precise steering mechanism.

Two classes of gnathostomes still have living members: the cartilaginous fishes (class Chondrichthyes) and the bony fishes (class Osteichthyes). Another class, the armored fishes, or placoderms, contained the earliest jawed fishes. They are now extinct and apparently left no descendants. A fourth group of ancient, extinct fishes, the acanthodians, may be more closely related to the bony fishes (*see figure 15.1*).

Class Chondrichthyes

Members of the class Chondrichthyes (kon-drik′thi-es) (Gr. *chondros*, cartilage + *ichthyos*, fish) include the sharks, skates, rays, and ratfishes (*see table 15.1*). Most chondrichthians are carnivores or scavengers, and most are marine species. In addition to their biting mouthparts and paired appendages, chondrichthians possess epidermal placoid scales and a cartilaginous endoskeleton.

There are about 700 species in the subclass Elasmobranchii (e-laz′mo-bran′ke-i) (Gr. *elasmos*, plate metal + *branchia*, gills), which includes the sharks, skates, and rays (figure 15.8*a–c*). Sharks arose from early jawed fishes midway through the Devonian period, about 375 million years ago. The absence of certain features characteristic of bony fishes (e.g., a swim bladder to regulate buoyancy, a gill cover, and a bony skeleton) is sometimes interpreted as evidence of the primitiveness of elasmobranchs. This interpretation is mistaken, as these characteristics simply resulted from different adaptations in the two groups to similar selection pressures. Some of these adaptations are described later in this chapter.

(a)

(b)

(c)

Figure 15.8

Class Chondrichthyes. (*a*) A gray reef shark (*Carcharhinus*). (*b*) A manta ray (*Manta hamiltoni*) with two remoras (*Remora remora*) attached to its ventral surface. (*c*) A bullseye stingray (*Urolophus concentricus*).

(a)

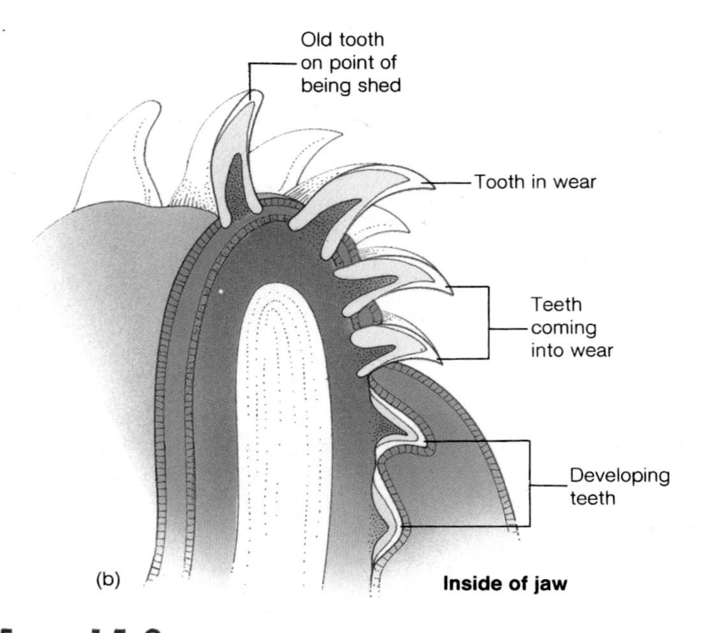

Figure 15.9

Scales and Teeth of Sharks. (*a*) A section of shark skin magnified to show posteriorly pointing placoid scales (scanning electron micrograph × 500). (*b*) The teeth of sharks develop as modified placoid scales. Older teeth are continuously replaced by newer teeth that move from the inside to the outside of the jaw.

Sharks are covered by tough skin with dermal, placoid scales (figure 15.9*a*). These scales project posteriorly and give the skin a tough, sandpaper texture. (In fact, dried sharkskin has been used for sandpaper.) Posteriorly pointed scales also reduce friction with the water when a shark is swimming.

The teeth of sharks are actually modified placoid scales. The row of teeth on the outer edge of the jaw is backed up by rows of teeth attached to a ligamentous band that covers the jaw cartilage inside the mouth. As the outer teeth wear and become useless, they are replaced by newer teeth moving into position from inside the jaw. In young sharks, this replacement is rapid, with a new row of teeth developing every 7 or 8 days (figure 15.9*b*). Crowns of teeth in different species may be adapted for shearing prey or for crushing the shells of molluscs.

Sharks range in size from less than 1 m (e.g., *Squalus*, the laboratory dissection specimen) to greater than 10 m (e.g., basking sharks and whale sharks). The largest sharks are not predatory but are filter feeders. They have gill-arch modifications that strain plankton. The fiercest and most feared sharks are the great white shark (*Carcharodon*) and the mako (*Isurus*). Extinct specimens may have reached lengths of 25 m or more (box 15.2)!

Skates and rays are specialized for life on the ocean floor. They usually inhabit shallow water, where they use their blunt teeth to feed on invertebrates. Their most obvious modification for life on the ocean floor is a lateral expansion of the pectoral fins into winglike appendages. Locomotion results from dorsoventral muscular waves that pass posteriorly along the fins. Frequently, elaborate color patterns on the dorsal surface of these animals provide effective camouflage (*see figure* 15.8c). The sting ray (*Raja*) has a tail modified into a defensive lash—the dorsal fin persists as a venomous spine. Also included in this group are the electric rays (*Narcine*) and manta rays (*Aetobatus*) (*see figure* 15.8b).

A second major group of chondrichthians, in the subclass Holocephali (hol′o-sef′a-li) (Gr. *holos*, whole + *kephalidos*, head), contains about 30 species. A frequently studied example, *Chimaera*, has a large head with a small mouth that is surrounded by large lips. A narrow tapering tail has resulted in the common name "ratfish." Holocephalans diverged from other chondrichthians nearly 300 million years ago. During this time, specializations not found in other elasmobranchs have evolved. These include a gill cover, called an **operculum,** and teeth modified into large plates that are used for crushing the shells of molluscs. Holocephalans lack scales.

Class Osteichthyes

Members of the class Osteichthyes (os′te-ik″the-es) (Gr. *osteon*, bone + *ichthyos*, fish) are characterized by having at least some bone in their skeleton and/or scales, an operculum covering the gill openings, and lungs or a swim bladder. Any group that has 20,000 species and is a major life-form in most of the earth's vast aquatic habitats must be judged very successful from an evolutionary perspective.

The first fossils of bony fishes are from late Silurian deposits (approximately 405 million years old). By the Devonian period (350 million years ago), the two subclasses were in the midst of their adaptive radiations (table 15.1; *see also figure* 15.2).

BOX 15.2 Jaws from the Past

Paleontological records can tell scientists much about life-forms from the past. We know most about ancient animals whose bodies contained hard parts that were resistant to decay and that were more likely to fossilize. The fossil record for fishes that contained substantial quantities of bone is fairly complete. Records of fishes that lack bone are usually harder to find.

In spite of the cartilaginous skeletons of sharks, our knowledge of ancient sharks, although not perfect, is more complete than might be expected. Much of what we know of ancient sharks comes from discoveries made in Ohio. Sharks in Ohio may seem surprising at first, but during the upper Devonian (about 350 million years ago), the sea extended southwest from the St. Lawrence River region, across the Great Lakes, and down to Arkansas. The floor of this ocean, in the region now occupied by Ohio, was made of soft, deep sediments—ideal for fossilization. Some elasmobranch specimens are so well preserved that not only are cartilaginous skeletons preserved, but details of gill structures, details of muscle organization, and even remains of a last meal are sometimes found intact.

Discoveries such as these provide a wealth of information about ancient sharks. The body form of ancient sharks allowed them to become efficient predators, and that basic form is retained in most modern sharks. The shape of fossilized teeth can sometimes be used to identify the kind of shark they came from. Mineral deposits build up on teeth as they rest on the ocean floor. Assuming mineral deposits accumulate at a constant rate, it is possible to determine how long teeth have been resting on the ocean floor. Assuming that a ratio between tooth size and body size is constant for a given species, estimates of the length of a shark can be obtained from a fossilized tooth.

Reconstruction of ancient sharks from all available evidence is the job of some museum scientists. The reconstruction shown in figure 1 is of an ancient shark that was found in a North Carolina quarry. (Because too few teeth were found to fill out the entire jaw, false teeth were constructed from hard rubber.) They are the teeth of a 30-million-year-old *Carcharodon megalodon*, that was the ancestor of the great white shark. The largest tooth of this specimen was about 15 cm (6 in.) long. Because 2.5 cm (1 in.) of tooth equals about 3 m (10 ft) of shark, it is estimated that this specimen was 18.5 m (60 ft) long! The model shown in figure 1 is on display in the Smithsonian's National Museum of Natural History.

FIGURE 1 **Ancient Jaws.** These reconstructed jaws of *Carcharodon megalodon* measure 1 m wide by 2 m high.

Members of the subclass Sarcopterygii (sar-kop-te-rij′e-i) (Gr. *sark*, flesh + *pteryx*, fin) have muscular lobes associated with their fins and usually use lungs in gas exchange. One group of sarcopterygians are the lungfishes. Only three genera survive today, and all are found in regions where seasonal droughts are common. When freshwater lakes and rivers begin to stagnate and dry, these fishes use lungs to breathe air (figure 15.10). Some (*Neoceratodus*) inhabit the fresh waters of Queensland, Australia. They survive stagnation by breathing air, but normally use gills and cannot withstand total drying. Others are found in freshwater rivers and lakes in tropical Africa (*Protopterus*) and tropical South America (*Lepidosiren*). They have completely lost the use of gills for gas exchange and can survive when rivers or lakes completely dry. When a lake or river has nearly dried, these lungfishes burrow into the mud. They keep a

FIGURE 15.10

Subclass Sarcopterygii. The lungfish, *Lepidosiren paradoxa*, has lungs that allow it to withstand stagnation and drying of its habitat.

FIGURE 15.11
A Sarcopterygian, the Coelacanth. *Latimeria* is the only known surviving coelacanth.

narrow air pathway open by bubbling air to the surface. After the substrate dries, the only evidence of a lungfish burrow is a small opening in the earth. Lungfishes may remain in aestivation for 6 months or more. (Aestivation is a dormant state that helps an animal withstand hot, dry periods.) When rain again fills the lake or riverbed, lungfishes emerge from their burrows to feed and reproduce.

A second group of sarcopterygians are the coelacanths. The most recent fossils of coelacanths are over 70 million years old. In 1938, however, people fishing in deep water off the coast of South Africa brought up fishes that were identified as coelacanths (figure 15.11). Since then, numerous other specimens have been caught in deep water around the Comoro Islands off Madagascar. 4 The discovery of this fish, *Latimeria chalumnae*, was a milestone event because *Latimeria* is probably the closest living fish relative of terrestrial vertebrates. It is large—up to 80 kg—and has heavy scales. Ancient coelacanths lived in freshwater lakes and rivers; thus, the ancestors of *Latimeria* must have moved from freshwater habitats to the deep sea.

A third group of sarcopterygians are entirely extinct. These fish, called rhipidistians, became extinct before the close of the Paleozoic period and are believed to have been the ancestors of ancient amphibians.

The subclass Actinopterygii (ak′tin-op″te-rig-e-i) (Gr. *aktis*, ray + *pteryx*, fin) contains fishes that are sometimes called the ray-finned fishes because their fins lack muscular lobes. They usually possess **swim bladders,** which are gas-filled sacs located along the dorsal wall of the body cavity and used to regulate buoyancy. Zoologists now realize that there have been many points of divergence in the evolution of the Actinopterygii. One modern classification system divides the Actinopterygii into two infraclasses.

(a)

(b)

FIGURE 15.12
Subclass Actinopterygii, the Chondrosteans. (*a*) A shovelnose sturgeon (*Scaphirhynchus platorynchus*). Sturgeons are covered anteriorly by heavy bony plates and posteriorly by scales. (*b*) The distinctive rostrum of a paddlefish (*Polydon spathula*) is densely innervated with sensory structures that are probably used to detect minute electric fields. Note the mouth in its open, filter-feeding position.

One group of actinopterygians, the chondrosteans, contains many species that lived during the Permian, Triassic, and Jurassic periods (215 to 120 million years ago), but only 25 species remain today. Ancestral chondrosteans had a bony skeleton, but living members, the sturgeons and paddlefishes, have cartilaginous skeletons. Chondrosteans also have a tail with a large upper lobe.

Most sturgeons live in the sea and migrate into rivers to breed (figure 15.12*a*). (Some sturgeons live in fresh water but maintain the migratory habits of their marine relatives.) They are very large (up to 1,000 kg) and have bony plates covering the anterior portion of the body. Heavy scales cover the tail. The mouth of a sturgeon is small, and its jaws are weak. Sturgeons feed on invertebrates that they stir up from the sea or riverbed using their snout. Because sturgeons are valued for their caviar (eggs), they have been severely overfished.

Paddlefishes are large, freshwater chondrosteans. They have a large, paddlelike rostrum that is innervated with sensory organs believed to detect weak electric fields (figure 15.12*b*). They swim through the water with their large mouths open, filtering crustaceans and small fishes. They are found mainly in lakes and large rivers of the Mississippi River basin and are also known from western North America.

The second group of actinopterygians flourished in the Jurassic period and succeeded most chondrosteans. Two very primative genera occur in temperate to warm fresh waters of North America. *Lepisosteus*, the garpike, has thick scales and long jaws that it uses to catch fishes. *Amia* is commonly referred to as the dogfish or bowfin. Most living fishes are members of this group and are referred to as teleosts or modern bony fishes. They have a symmetrical caudal fin and a swim bladder that has lost its connection to the digestive tract. After their divergence from ancient marine actinopterygians in the late Triassic period, a remarkable evolutionary diversification occurred. Teleosts adapted to nearly every available aquatic habitat (figure 15.13*a–c*). There are in excess of 20,000 species of teleosts.

Stop and Ask Yourself

5. What are characteristics of the superclass Gnathostomata? In what ways are members of this superclass considered more advanced than agnathans?
6. What class of fishes is characterized by jaws, paired appendages, a cartilaginous skeleton, and placoid scales?
7. What class of fishes is characterized by some bone in their skeleton, an operculum, and usually lungs or a swim bladder?
8. What is aestivation? What is the role of aestivation in the life of some lungfishes?

Evolutionary Pressures

Why is a fish fishlike? This apparently redundant question is unanswerable in some respects because some traits of animals are selectively neutral and, thus, neither improve nor detract from overall fitness. On the other hand, aquatic environments have physical characteristics that are important selective forces for aquatic animals. Although animals have adapted to aquatic environments in different ways, one can understand many aspects of the structure and function of a fish by studying the fish's habitat. The material presented in this section will help you appreciate the many ways that a fish is adapted for life in water.

(a)

(b)

(c)

Figure 15.13

Subclass Actinopterygii, the Teleosts. (*a*) Bottom fish, such as this winter flounder (*Pseudopleuronectes americanus*), have both eyes on one side of the head, and they often rest on their side fully or partially buried on the substrate. (*b*) Freshwater teleosts, such as this speckled darter (*Etheostoma stigmaeum*), are common in temperate streams. (*c*) Cichlid fish, including this harlequin cichlid (*Cichlasoma festae*), are common in tropical fresh waters.

Locomotion

Picture a young girl running full speed down the beach and into the ocean. She hits the water and begins to splash. At first, she lifts her feet high in the air between steps, but as she goes deeper, her legs encounter more and more resistance. The momentum of her upper body causes her to fall forward and she resorts to labored and awkward swimming strokes. The density of the water makes movement through it difficult and costly. For a fish, however, swimming is less energetically costly than running is for a terrestrial organism. Friction between a fish and the water is reduced by the streamlined shape of a fish and the mucoid secretions that lubricate the body surface. The buoyant properties of water also contribute to the efficiency of a fish's movement through the water. A fish needs to expend little energy in support against the pull of gravity.

Fishes move through the water using their fins and body wall to push against the incompressible surrounding water. Anyone who has eaten a fish filet probably realizes that muscle bundles of most fishes are arranged in a ≶ pattern. Because these muscles extend posteriorly and anteriorly in a zig-zag fashion, contraction of each muscle bundle can affect a relatively large portion of the body wall. Very efficient, fast-swimming fishes, such as tuna and mackerel, supplement body movements with a vertical caudal (tail) fin that is tall and forked. The forked shape of the caudal fin reduces surface area that could cause turbulence and interfere with forward movement.

Nutrition and the Digestive System

The earliest fishes were probably filter feeders and scavengers that sifted through the mud of ancient sea floors for decaying organic matter, annelids, molluscs, or other bottom-dwelling invertebrates. Dramatic changes in the nutrition of fishes came about when the evolution of jaws transformed early fishes into efficient predators.

Most modern fishes are predators and spend much of their life searching for food. The prey that different fishes eat vary tremendously. Some fishes feed on invertebrate animals floating or swimming in the plankton or living in or on the substrate. Many feed on other vertebrates. Similarly, the kinds of food that one fish eats at different times in its life varies. For example, a fish may feed on plankton as a larva but switch to larger prey, such as annelids or smaller fish, as an adult. Prey are usually swallowed whole. Teeth are often used to capture and hold prey, and some fishes have teeth that are modified for crushing the shells of molluscs or the exoskeletons of arthropods. Prey capture often utilizes the suction created by closing the opercula and rapidly opening the mouth, which develops a negative pressure that sweeps water and prey inside the mouth.

Other feeding strategies have also evolved in fishes. Herring, paddlefishes, and whale sharks are filter feeders. Long gill processes, called **gill rakers**, trap plankton while the fish is swimming through the water with its mouth open (*see figure 15.12b*). Other fishes, such as carp, feed on a variety of plants and small animals. A few, such as the lamprey, are external parasites for at least a portion of their life. A few are primarily herbivores, feeding on plants.

The digestive tract of a fish is similar to that of other vertebrates. An enlargement, called the stomach, is primarily used for storing large, often infrequent, meals. The small intestine, however, is the primary site for enzyme secretion and food digestion. Sharks and other elasmobranchs have a spiral valve in their intestine, and bony fishes possess outpockets of the intestine, called pyloric ceca, which increase absorptive and secretory surfaces.

Circulation and Gas Exchange

All vertebrates have a closed circulatory system in which blood, with red blood cells containing hemoglobin, is pumped by a heart through a series of arteries, capillaries, and veins. The evolution of lungs in fishes was paralleled by changes in vertebrate circulatory systems. These changes are associated with the loss of gills, delivery of blood to the lungs, and separation of oxygenated and unoxygenated blood in the heart.

The vertebrate heart develops from four embryological enlargements of a ventral aorta. In fishes, blood flows from the venous system through the sinus venosus, the atrium, the ventricle, the conus arteriosus, and into the ventral aorta (figure 15.14*a*). Five afferent vessels carry blood to the gills, where the vessels branch into capillaries. Blood is collected by efferent vessels, delivered to the dorsal aorta, and distributed to the body.

Even though lungfishes are not a transitional group, they provide a good example of how the presence of lungs alters the circulatory pattern. There is still circulation to gills, but a vessel to the lungs has developed as a branch off aortic arch VI (figure 15.14*b*). This vessel is now called the pulmonary artery. Blood returns to the heart through pulmonary veins and enters the left side of the heart. The atrium and ventricle of the lungfish heart are partially divided. These partial divisions help keep unoxygenated blood from the body separate from the oxygenated blood from the lungs. A spiral valve in the conus arteriosus helps direct blood from the right side of the heart to the pulmonary artery and blood from the left side of the heart to the remaining aortic arches. Thus, in the lungfishes, we see a distinction between a pulmonary circuit and a systemic circuit.

Gas Exchange

Fishes live in an environment that contains less than 2.5% of the oxygen present in air. To maintain adequate levels of oxygen in their bloodstream, fishes must pass large quantities of water across gill surfaces and extract the small amount of oxygen present in the water.

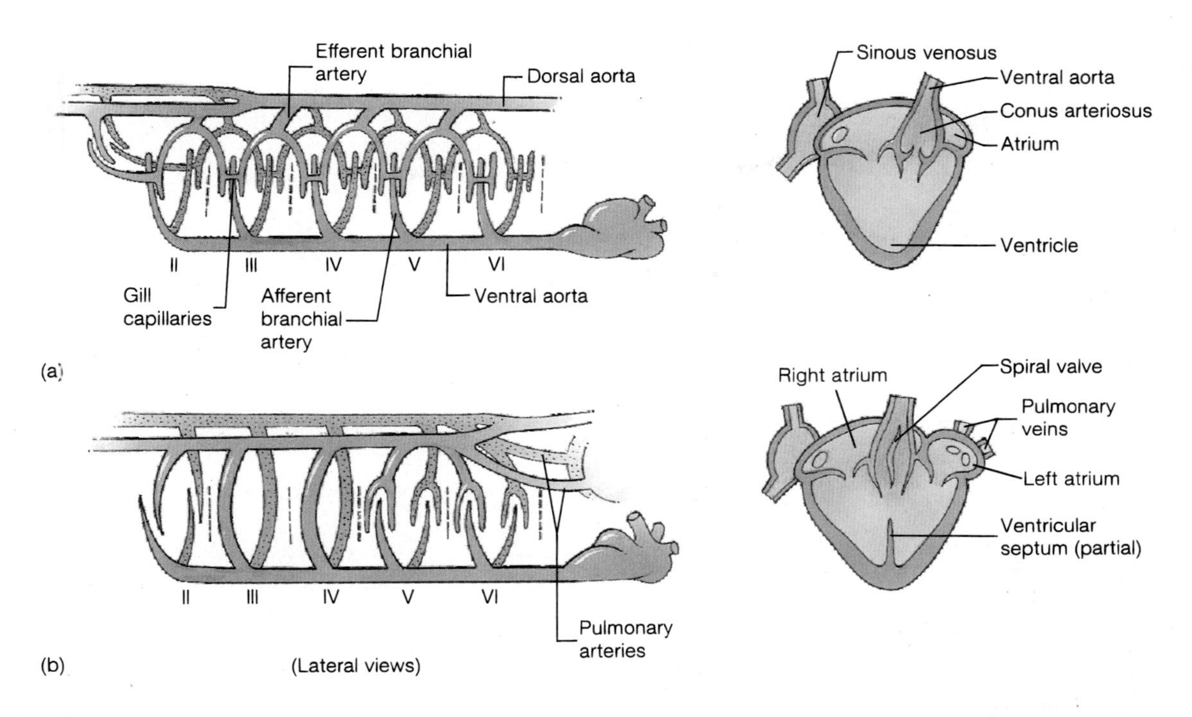

Figure 15.14

Circulatory System of Fishes. Diagrammatic representation of the circulatory systems of (*a*) bony fishes and (*b*) lungfishes. Hearts are drawn from a ventral view. Major branches of arteries carrying blood to and from the gills are called branchial arteries (or embryologically, aortic arches) and are numbered with Roman numerals. They begin with II because aortic arch I is lost during embryological development.

Most fishes use a muscular pumping mechanism to move the water into the mouth and pharynx, over the gills, and out of the fish through gill openings. This pump is powered by muscles surrounding the pharynx and the opercular cavity, which is between the gills and the operculum.

Some elasmobranchs and open-ocean bony fishes, such as the tuna, maintain water flow by holding their mouth open while swimming. This method is called **ram ventilation.** 5 Elasmobranchs do not have opercula to help pump water, and therefore, some sharks must keep moving to survive. Others can move water over their gills by a pumping mechanism similar to that described above. Rather than using an operculum in the pumping process, however, their gill bars have external flaps that close and form a cavity functionally similar to the opercular cavity of other fishes. Spiracles are modified gill slits that open just behind the eyes of elasmobranchs and are used as an alternate route for water entering the pharynx.

Gas exchange across gill surfaces is very efficient. Gills are supported by **gill (visceral) arches. Gill filaments** extend from each gill arch and include vascular folds of epithelium, called **gill lamellae** (figure 15.15*a*,*b*). Blood is carried to the gills and into gill filaments in branchial arteries. The arteries break into capillary beds in gill lamellae. Gas exchange occurs as blood and water move in opposite directions on either side of the lamellar epithelium. This **countercurrent exchange mechanism** provides very efficient gas exchange by maintaining a concentration gradient between the blood and the water over the entire length of the capillary bed (figure 15.15*c*,*d*).

Swim Bladders and Lungs

The Indian climbing perch spends its life almost entirely on land. These fishes, as most bony fishes, have gas chambers called **pneumatic sacs.** In nonteleost fishes, the pneumatic sacs connect to the esophagus or another part of the digestive tract by a **pneumatic duct**. Swallowed air enters these sacs, and gas exchange occurs across vascular surfaces. Thus, in the Indian climbing perch, lungfishes, and ancient rhipidistians, pneumatic sacs function(ed) as lungs. In other bony fishes, pneumatic sacs act as swim bladders.

Most zoologists believe that lungs are more primitive than swim bladders. Much of the early evolution of bony fishes occurred in warm, freshwater lakes and streams during the Devonian period. These bodies of water frequently became stagnant and periodically dried. Having lungs in these habitats could have meant the difference between life and death. On the other hand, later evolution of modern bony fishes occurred in marine and freshwater environments where stagnation was not a problem. In these environments, the use of pneumatic sacs in buoyancy regulation would have been adaptive (figure 15.16).

venous system → sinus venosus → atrium → ventricle → conus arteriosus → ventral aorta → → 5 afferent vessels to gills → capillaries → efferent vessels → dorsal aorta → body → pulmonary veins → Ⓛ side

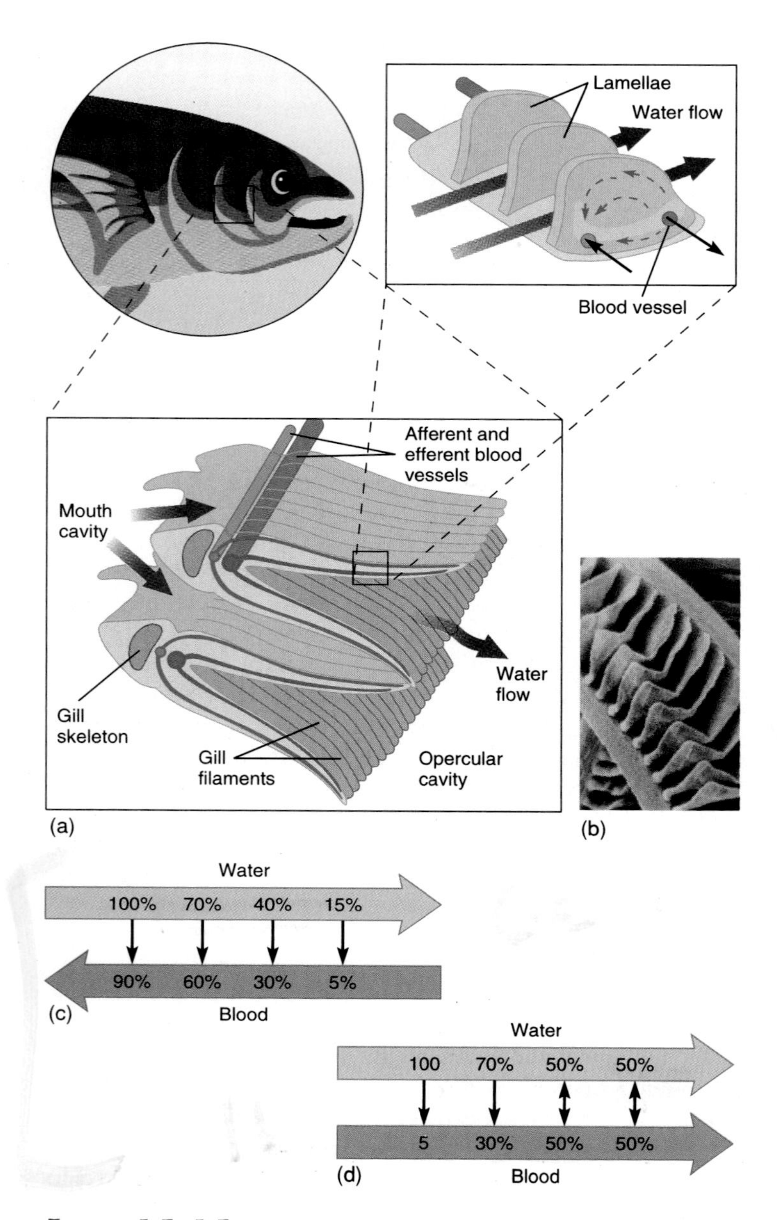

FIGURE 15.15

Gas Exchange at the Gill Lamellae. (*a*) The gill arches under the operculum support two rows of gill filaments. Blood flows into gill filaments through afferent branchial arteries, and these arteries break into capillary beds in the gill lamellae. Water and blood flow in opposite directions on either side of the lamellae. (*b*) Electron micrograph of the tip of a trout gill filament showing numerous lamellae. A comparison of countercurrent (*c*) and parallel (*d*) exchanges. Water entering the spaces between gill lamellae is saturated with oxygen in both cases. In countercurrent exchange, this water encounters blood that is almost completely oxygenated, but a diffusion gradient still favors the movement of more oxygen from the water to the blood. As water continues to move between lamellae, it loses oxygen to the blood, because it is continually encountering blood that has a lower oxygen concentration than is present in the water. Thus, a diffusion gradient is maintained along the length of the lamellae. If blood and water were to move in parallel fashion, diffusion of oxygen from water to blood would occur only until the concentration of oxygen in blood equalled the concentration of oxygen in water, and the exchange would be much less efficient.

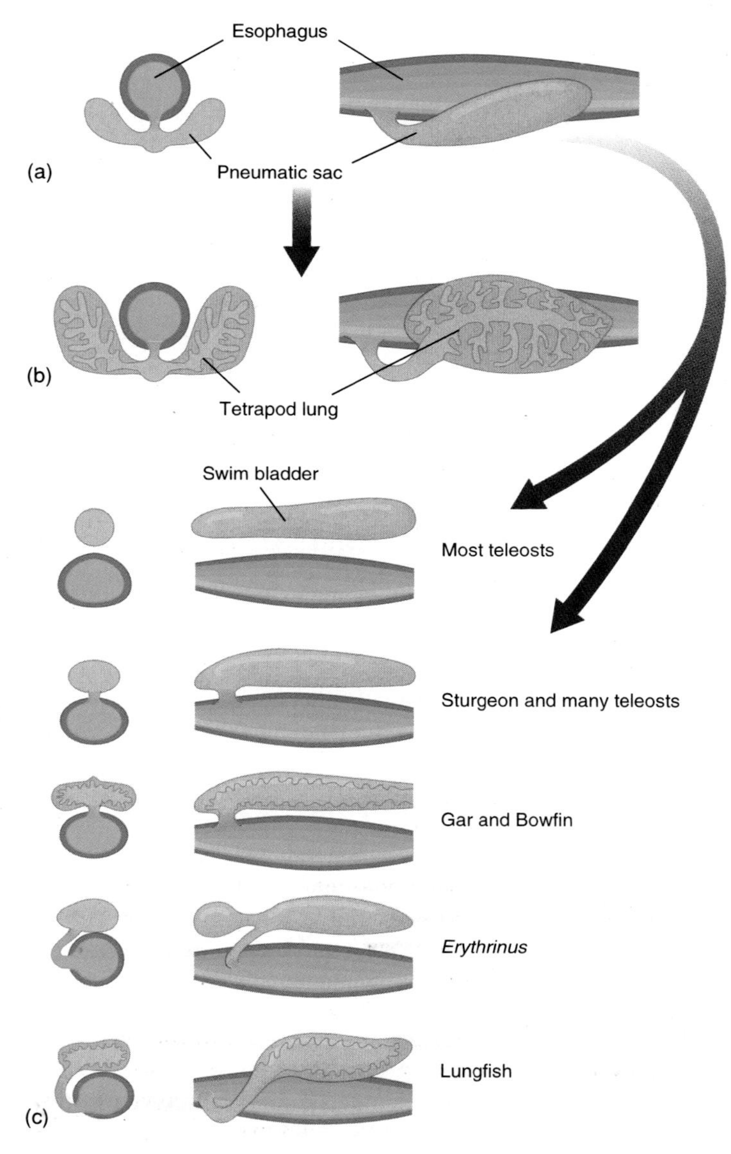

FIGURE 15.16

A Possible Sequence in the Evolution of Pneumatic Sacs. (*a*) Pneumatic sacs may have originally developed from ventral outgrowths of the esophagus. Many ancient fishes probably used pneumatic sacs as lungs. (*b*) Primitive lungs developed further during the evolution of land vertebrates. Internal compartmentalization increases surface area for gas exchange in land vertebrates. (*c*) In most bony fishes, pneumatic sacs are called swim bladders, and they are modified for buoyancy regulation. Swim bladders are in a dorsal position to prevent a tendency for the fish to "belly up" in the water. Pneumatic duct connections to the esophagus are frequently lost, and gases are transferred from the blood to the swim bladder.

Buoyancy Regulation

Did you ever consider why it is possible for you to float in water? Water is a supportive medium, but that is not sufficient to prevent you from sinking. Even though you are made mostly of water, other constituents of tissues are more dense than water. Bone, for example, has a specific gravity twice that of water. Why is it then that you can float? Largely because of two large, air-filled organs called lungs that allow you to float to the surface.

Fishes maintain their vertical position in a column of water in four ways. One way is to incorporate low-density compounds into their tissues. Fishes (especially their livers) are saturated with buoyant oils. A second way fishes maintain vertical position is to use fins to provide lift. The pectoral fins of a shark serve as planing devices that help to create lift as the shark moves through the water. Also, the large upper lobe of a shark's caudal fin provides upward thrust for the posterior end of the body (*see figure 15.8a*). A third adaptation is the reduction of heavy tissues in fishes. The bones of fishes are generally less dense than those of terrestrial vertebrates. One of the adaptive features of the elasmobranch cartilaginous skeleton probably results from cartilage being only slightly heavier than water. The fourth adaptation is the swim bladder. Using a swim bladder, buoyancy can be regulated to meet the day-to-day needs of a fish by precisely regulating the volume of gas in it. (You can mimic this adaptation while floating in water. How well do you float after forcefully exhaling as much air as possible?)

The swim bladders of garpike, sturgeons, and other primitive bony fishes connect to the esophagus or another part of the digestive tract by the pneumatic duct. These fishes gulp air at the surface to force air into their swim bladders.

Most teleosts have swim bladders that have lost a functional connection to the digestive tract. Gases (various mixtures of nitrogen and oxygen) are secreted into the swim bladder from the blood using a countercurrent exchange mechanism in a vascular network called the rete mirabile ("miraculous net"). Gases may be reabsorbed into the blood at the posterior end of the bladder.

Nervous and Sensory Functions

The central nervous system of fishes, as in other vertebrates, consists of a brain and a spinal cord. Sensory receptors are widely distributed over the body. In addition to generally distributed receptors for touch and temperature, fishes possess specialized receptors for olfaction, vision, hearing, equilibrium and balance, and for detecting water movements.

Openings in the snout of fishes, called external nares, lead to olfactory receptors. In most fishes, receptors are located in blind-ending olfactory sacs. In a few fishes, the external nares open to nasal passages that lead to the mouth cavity. Recent research has revealed that some fishes rely heavily on their sense of smell. For example, salmon and lampreys return to spawn in the streams in which they hatched years earlier. Their migrations to these streams often involve distances of hundreds of miles and are guided by the fishes' perception of the characteristic odors of their spawning stream.

The eyes of fishes are similar in most aspects of structure to those found in other vertebrates. They are lidless, however, and the lenses are round. Focusing is accomplished by moving the lens forward or backward in the eye. (Most other vertebrates focus by changing the shape of the lens.)

Receptors for equilibrium, balance, and hearing are located in the inner ears of fishes, and their functions are similar to those of other vertebrates. Semicircular canals detect rotational movements, and other sensory patches help with equilibrium and balance by detecting the direction of the gravitational pull. Fishes lack the outer and/or middle ear, which conducts sound waves to the inner ear in other vertebrates. As anyone who enjoys fishing knows, however, most fishes can hear. Vibrations may be passed from the water through the bones of the skull to the middle ear, and a few fishes have chains of bony ossicles (modifications of vertebrae) that connect the swim bladder to the back of the skull. Vibrations strike the fish, are amplified by the swim bladder, and sent through the ossicles to the skull.

Running along each side and branching over the head of most fishes is a **lateral-line system**. The **lateral-line system** consists of sensory pits in the epidermis of the skin that connect to canals that run just below the epidermis. In these pits are receptors that are stimulated by water moving against them. Lateral lines are used either to detect water currents or for detecting a predator or a prey that may be causing water movements in the vicinity of the fish. Low-frequency sound may also be detected.

Electric Fishes

A U.S. Navy pilot has just ejected from his troubled aircraft over shark-infested water! What measures can the pilot take to ensure survival under these hostile conditions? The Navy has considered this scenario. One of the solutions to the problem is a polyvinyl bag suspended from an inflatable collar. The polyvinyl bag helps conceal the downed flyer from a shark's vision and keen sense of smell. But is that all that is required to ensure protection?

All organisms produce weak electrical fields from the activities of nerves and muscles. **Electroreception** is the detection of electrical fields generated by the fish or another organism in the environment. Electroreception and/or electrogeneration has been demonstrated in over 500 species of fishes in seven families of Chondrichthyes and Osteichthyes. These fishes use their electroceptive sense for detecting prey and for orienting toward or away from objects in the environment.

Nowhere is prey detection with this sense better developed than in the rays and sharks. Spiny dogfish sharks, the common laboratory specimens, locate prey by electroreception. (6) A shark can find and eat a flounder that is buried in sand and it will try to find and eat electrodes that are creating electrical signals similar to those emitted by the flounder. On the other hand, a shark cannot find a dead flounder buried in the sand or a live flounder covered by an insulating polyvinyl sheet.

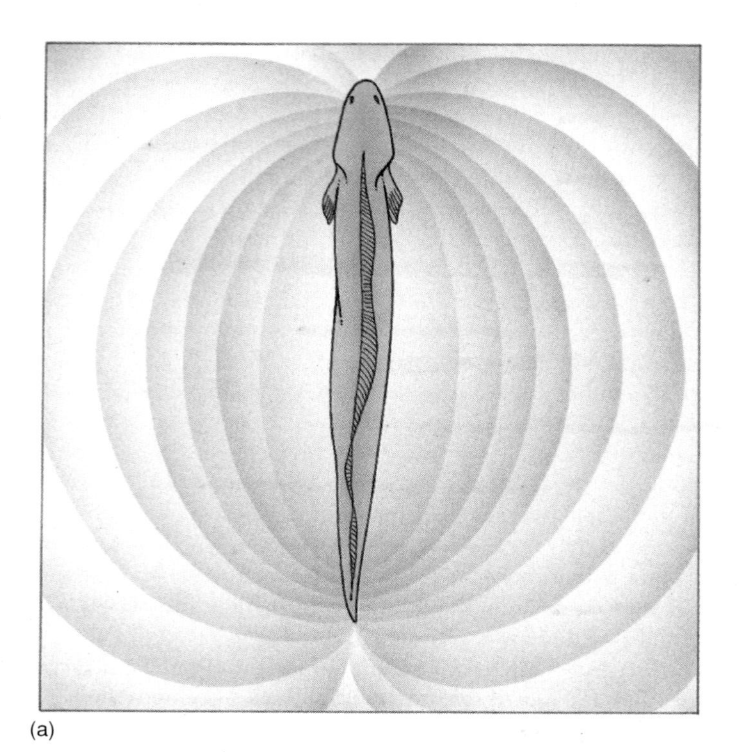

(a)

(b)

FIGURE 15.17

Electric Fishes. (*a*) The electric field of *Gymnarchus niloticus* is used to detect the presence of prey and other objects in the fish's murky environment. Currents circulate from electric organs in its tail to electroreceptors near its head. An object in this electrical field changes the pattern of stimulation of electroreceptors. (*b*) The electric fish (*Gymnarchus niloticus*).

Some fishes are not only capable of electroreception, but are also capable of generating electrical currents. An electric fish (*Gymnarchus niloticus*) lives in freshwater systems of Africa. Muscles near its caudal fin are modified into organs that produce a continuous electrical discharge. This current spreads between the tail and the head. Porelike perforations near the head contain electroreceptors. The electrical waves circulating between the tail and the head are distorted by objects in their field. This distortion is detected in changing patterns of stimulation of receptors (figure 15.17). The electrical sense of *Gymnarchus* is an adaptation to living in murky freshwater habitats where eyes are of limited value.

FIGURE 15.18

Electric Fishes. A lesser electric ray (*Narcine brasiliensis*).

The fishes best known for producing strong electric currents are the electric eel (a bony fish) and the electric ray (an elasmobranch). The electric eel (*Electrophorus*) occurs in rivers of the Amazon basin in South America. The organs used for producing electric currents are located in the trunk of the electric eel and can deliver shocks in excess of 500 volts. The electric ray (*Narcine*) (figure 15.18), has electric organs in its fins that are capable of producing pulses of 50 amperes at about 50 volts. Shocks produced by these fishes are sufficiently strong to stun or kill prey, discourage large predators, and teach unwary humans a lesson that will never need to be repeated.

Stop and Ask Yourself

9. What are two methods of gill ventilation used by fishes?
10. What is a countercurrent exchange mechanism, and why does it occur in fish gills?
11. Why do some fishes, such as lungfishes, need lungs?
12. What are four mechanisms used by fishes to improve their buoyancy?

Excretion and Osmoregulation

Fishes, as all animals, must maintain a proper balance of ions and water in their tissues. The regulation of these balances is called osmoregulation and is a major function of the kidneys and gills of fishes. Kidneys are located near the midline of the body, just dorsal to the peritoneal membrane that lines the body cavity. As with all vertebrates, the excretory structures in the kidneys are called **nephrons.** Nephrons filter bloodborne nitrogenous wastes, ions, water, and small organic compounds across a network of capillaries called a **glomerulus.** The filtrate then passes into a tubule system, where essential components may be reabsorbed into the blood. The filtrate remaining in the tubule system is then excreted.

Freshwater fishes live in an environment containing few dissolved substances. Osmotic uptake of water across gill, oral, and intestinal surfaces, and the loss of essential ions by excretion and defecation are constant. To control the excess buildup of water and loss of ions, freshwater fishes never drink, and only take in water when feeding. Also, the nephrons of freshwater fishes are numerous and frequently possess large glomeruli and relatively short tubule systems. Filtration is followed by reabsorption of some ions and organic compounds. Because the tubule system is relatively short, little water is reabsorbed. Thus, large quantities of very dilute urine are produced. Even though the urine of freshwater fishes is dilute, ions are still lost through the urine and by diffusion across gill and oral surfaces. Loss of ions is compensated for by active transport of ions into the blood at the gills. Freshwater fishes also get some salts in their food (figure 15.19*a*).

Marine fishes face the opposite problems. Their environment contains 3.5% ions, and their tissues contain approximately 0.65% ions. Marine fishes, therefore, must combat water loss and accumulation of excess ions. They drink water and eliminate excess ions by excretion, defecation, and active transport across gill surfaces. The nephrons of marine fishes frequently possess small glomeruli and long tubule systems. Much less blood is filtered than in freshwater fishes, and water is efficiently, although not entirely, reabsorbed from the nephron (figure 15.19*b*).

Elasmobranchs have a unique osmoregulatory mechanism. They convert some of their nitrogenous wastes into urea in the liver. This in itself is somewhat unusual, because most fishes excrete ammonia rather than urea. Even more unusual, however, is that urea is sequestered in tissues all over the body. Enough urea is stored to make body tissues isosmotic with seawater. (That is, the concentration of solutes in a shark's tissues is essentially the same as the concentration of ions in seawater.) Therefore, the problem most marine fishes have of losing water to their environment is much less severe for elasmobranchs. Energy that does not have to be devoted to water conservation can now be used in other ways. This adaptation required the development of tolerance to high levels of urea, because urea disrupts important enzyme systems in the tissues of most other animals.

In spite of this unique adaptation, elasmobranchs must still regulate the concentrations of ions in their tissues. In addition to having ion-absorbing and secreting tissues in their gills and kidneys, elasmobranchs possess a rectal gland that removes excess sodium chloride from the blood and excretes it into the cloaca. (A cloaca is a common opening for excretory, digestive, and reproductive products.)

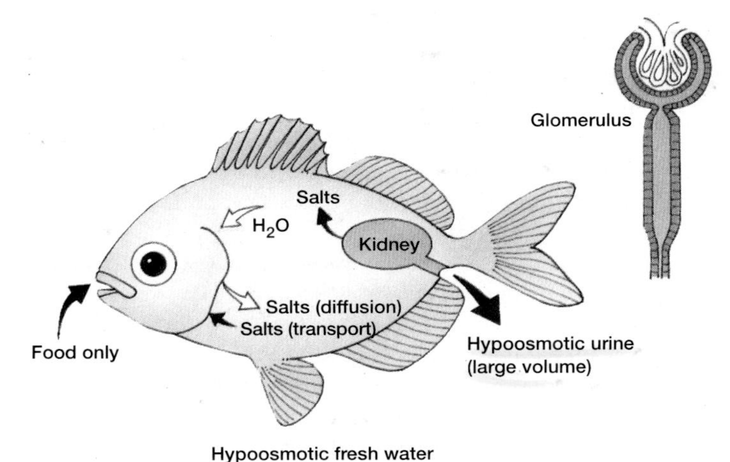

(a) Freshwater teleosts (hypertonic blood)

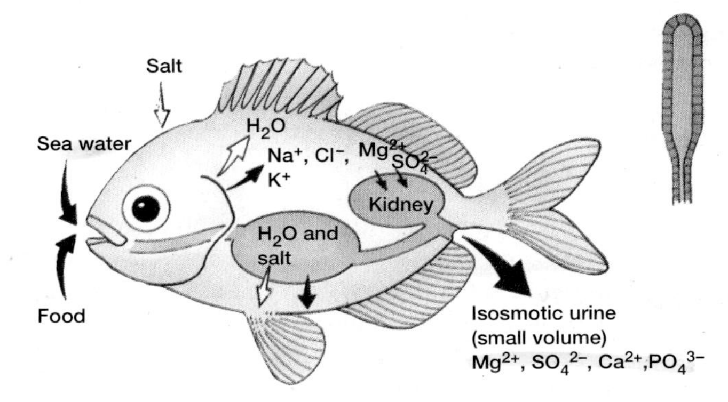

(b) Marine teleosts (hypotonic blood)

Figure 15.19

Osmoregulation by (*a*) Freshwater and (*b*) Marine Fishes. Large arrows indicate passive uptake or loss of water or ions through ingestion and excretion. Small solid arrows indicate active transport processes occurring at gill membranes and kidney tubules. Small open arrows indicate passive uptake or loss by diffusion through permeable surfaces. Insets of kidney nephrons depict adaptations within the kidney. Water, ions, and small organic molecules are filtered from the blood at the glomerulus of the nephron. Essential components of the filtrate can be reabsorbed within the tubule system of the nephron. Marine fishes conserve water by reducing the size of the glomerulus of the nephron, and thus reducing the quantity of water and ions filtered from the blood. Salts can be secreted from the blood into the kidney tubules. Marine fishes can produce urine that is isoosmotic with the blood. Freshwater fishes have enlarged glomeruli and short tubule systems. They filter large quantities of water from the blood, and tubules reabsorb some ions from the filtrate. Freshwater fishes produce a hypoosmotic urine.

Diadromous fishes migrate between freshwater and marine environments. Salmon (e.g., *Oncorhynchus*) and marine lampreys (*Petromyzon*) migrate from the sea to fresh water to spawn, and the freshwater eel (*Anguilla*) migrates from freshwater to marine environments to spawn. Diadromous migrations require that gills are capable of coping with both uptake and secretion of ions. Osmoregulatory powers needed for migration between marine and freshwater environments may not be developed in all life-history stages. Young salmon, for example, cannot enter the sea until certain cells on the gills develop ion-secreting powers.

Fishes have few problems getting rid of the nitrogenous by-products of protein metabolism. Up to 90% of nitrogenous wastes are eliminated as ammonia by diffusion across gill surfaces. Even though ammonia is very toxic, its use as an excretory product is possible in aquatic organisms because ammonia can diffuse into the surrounding water. The remaining 10% of nitrogenous wastes are excreted as urea, creatine, or creatinine. These wastes are produced in the liver and are excreted via the kidneys.

Figure 15.20

A Male Garibaldi (*Hypsypops rubicundus*) Guarding Eggs. The male cultivates a nest of filamentous red algae and then entices a female to lay eggs in the nest. This male is carrying off a bat star that came too close to the nest.

Reproduction and Development

Imagine, 45 kg of caviar from a single, 450 kg sturgeon! Admittedly, a 450 kg sturgeon is a very large fish (even for a sturgeon), but it is not unusual for a fish to produce millions of eggs in a single season. These numbers simply reflect the hazards of developing in aquatic habitats unattended by a parent. The vast majority of these millions of potential adults will never survive to reproduce. Many eggs will never be fertilized, many fertilized eggs may wash ashore and dry, many eggs and embryos may be smashed by currents and tides, and others will fall victim to predation. In spite of all of these hazards, if only four of the millions of embryos of each breeding pair survive and reproduce, the population will double.

Producing overwhelming numbers of eggs, however, is not the only way that fishes increase the chances that a few of their offspring will survive. Some fishes show mating behavior that helps ensure fertilization, or nesting behavior that protects eggs from predation, sedimentation, and fouling.

Mating may occur in large schools, and the release of eggs or sperm by one individual often releases spawning pheromones that induce many other adults to spawn. Huge masses of eggs and sperm released into the open ocean help to ensure fertilization of as many eggs as possible.

The vast majority of fishes are oviparous, meaning that eggs develop outside the female from stored yolk. Some elasmobranchs are ovoviviparous, and their embryos develop in a modified oviduct of the female. Nutrients are supplied from yolk stored in the egg. Other elasmobranchs, including gray reef sharks and hammerheads, are viviparous. A placentalike outgrowth of a modified oviduct diverts nutrients from the female to the yolksacs of developing embryos. Internal development of viviparous bony fishes usually occurs in ovarian follicles, rather than in the oviduct. In guppies (*Lebistes*), eggs are retained in the ovary, and fertilization and early development occur there. Embryos are then released into a cavity within the ovary and development continues, with nourishment coming partly from yolk and partly from secretions of the ovary.

Some fishes have specialized structures that aid in sperm transfer. Male elasmobranchs, for example, have modified pelvic fins called claspers. During copulation, a clasper is inserted into the cloaca of a female. Sperm travel along grooves of the clasper. Fertilization occurs in the reproductive tract of the female and usually results in a higher proportion of eggs being fertilized than in external fertilization. Thus, internal fertilization is usually accompanied by the production of fewer eggs.

In many fishes, care of the embryos is limited or nonexistent. Some fishes, however, construct and tend nests (figure 15.20), and some carry embryos during development. Clusters of embryos may be brooded by being attached to some part of the body in special pouches, or they may be brooded in the mouth. Some of the best-known brooders include the seahorses (*Hippocampus*) and pipefishes (e.g., *Syngnathus*). Males of these closely related fishes carry embryos throughout development in ventral pouches. The male Brazilian catfish (*Loricaria typhys*) broods embryos in an enlarged lower lip.

Most fishes do little, if any, caring for young after they have hatched. There are exceptions, however. Short-term care of posthatching young occurs in sunfishes and sticklebacks. Male sticklebacks collect fresh plant material and accumulate it into a mass. Young take refuge in this mass. If one wanders too far from the nest, the male will snap it up in its mouth and spit it back into the nest. Sunfish males do the same for young that wander from schools of recently hatched fish. Longer-term care occurs in the Chichlidae (*see figure 15.13c*). In some species, young are mouth brooded, and other species tend young in a nest. After hatching, the young venture from the parent's mouth or nest, but the young return quickly when the parent signals danger with a flicking of the pelvic fins.

Further Phylogenetic Considerations

Two important series of evolutionary events occurred during the evolution of the Osteichthyes. One of these was an evolutionary explosion that began about 150 million years ago and resulted in the vast diversity of teleosts that we see today. The last half of this chapter should have helped you appreciate some of these events.

To appreciate the second series of events, let us again consider the lungfishes. The lungfishes' life-style represents a survival strategy that must have been important for some early Devonian fishes. The current seasonal droughts of tropical South America and Africa must be similar to, but much less geographically extensive than, those in the Devonian period. As in the lungfishes, the ability to breathe air and aestivate must have been present in many freshwater Devonian fishes. Distinctive characteristics of the lungfish skeleton lead most zoologists to conclude that the lungfish evolutionary line gave rise to no other vertebrate taxa. Instead, the rhipidistians are thought to be a part of the evolutionary lineage that led to terrestrial vertebrates. Rhipidistians, like lungfishes, had lungs, could breathe air, and may have used their muscular fins to crawl across land looking for water when their own stream dried.

7 Adaptations that favored survival in Devonian streams and lakes may have preadapted some rhipidistians for life that would become increasingly terrestrial. While millions of years would elapse before any vertebrate could be considered terrestrial, brief excursions onto land allowed some vertebrates to exploit resources that for the previous 50 million years were available only to terrestrial arthropods.

Stop and Ask Yourself

13. Why is osmoregulation, not excretion, the major function of the kidneys of fishes?
14. What osmoregulatory problems are faced by a diadromous fish?
15. What is viviparity? How does viviparity of some elasmobranchs differ from that of some teleosts?
16. What adaptations to climatic conditions may have preadapted rhipidistian fishes for a partially terrestrial existence?

Summary

1. Ancient members of the vertebrate superclass Agnatha were probably the ancestors of all other vertebrates.
2. Agnathans lack jaws and paired appendages and include the extinct ostracoderms, lampreys, and hagfishes. Hagfishes are scavengers in marine environments. Lampreys have a life history involving migrations from the open ocean, or large body of fresh water, to freshwater spawning streams.
3. The superclass Gnathostomata includes fishes with jaws and paired appendages. The class Chondrichthyes includes the sharks, skates, rays, and ratfishes. The class Osteichthyes includes the bony fishes.
4. There are two subclasses of Osteichthyes. The subclass Sarcopterygii includes the lungfishes, the coelacanths, and the rhipidistians; and the subclass Actinopterygii includes the ray-finned fishes. In the Actinopterygii, the teleosts are the modern bony fishes. Members of this very large group have adapted to virtually every available aquatic habitat.
5. Fishes show numerous adaptations to living in aquatic environments. These adaptations include an arrangement of body-wall muscles that creates locomotor waves in the body wall; mechanisms that provide constant movement of water across gill surfaces; a countercurrent exchange mechanism to promote efficient gas exchange; buoyancy regulation; well-developed sensory receptors, including eyes, inner ears, and lateral line receptors; mechanisms of osmoregulation; and mechanisms that help ensure successful reproduction.
6. Two evolutionary lineages in the Actinopterygii are very important. One of these resulted in the adaptive radiation of modern bony fishes, the teleosts. The second evolutionary line probably diverged from the rhipidistians. Adaptations that favored the survival of rhipidistians in early Devonian streams preadapted some rhipidistians for terrestrial habitats.

Selected Key Terms

cloaca (*p. 267*)
countercurrent exchange mechanism (*p. 263*)
gill (visceral) arches (*p. 263*)
gill filaments (p. 263)
gill lamellae (*p. 263*)
lateral-line system (*p. 265*)
operculum (*p. 258*)
pneumatic sacs (*p. 263*)
ram ventilation (*p. 263*)
swim bladders (*p. 260*)

Critical Thinking Questions

1. What characteristic of water makes it difficult to move through, but also makes support against gravity a minor consideration? How is a fish adapted for moving through water?
2. Would it be possible for a fish to drown? Explain. Would it make a difference if the fish was an open-ocean fish, such as a tuna, or a fish such as a freshwater perch?
3. Why is it a mistake to consider the cartilaginous skeleton of chondrichthians a primitive characteristic?
4. Would swim bladders with functional pneumatic ducts work well for a fish that lives at great depths? Why or why not?
5. What would happen to a deep-sea fish that was rapidly brought to the surface? Explain your answer in light of the fact that gas pressure in the swim bladders of some deep-sea fishes is increased up to about 300 atmospheres.

chapter 16

Amphibians: The First Terrestrial Vertebrates

Outline

Concepts

1. Adaptations that favored survival of fishes during periodic droughts preadapted vertebrates to life on land. There were two lineages of ancient amphibians: one gave rise to modern amphibians, and the other lineage resulted in amniote vertebrates.
2. Modern amphibians belong to three orders. Caudata contains the salamanders, Gymnophiona contains the caecilians, and Anura contains the frogs and toads.
3. Although amphibians are restricted to moist habitats, most spend much of their adult life on land. Virtually all amphibian body systems show adaptations for living on land.
4. Evolution of eggs and developmental stages that were resistant to drying probably occurred in some ancient amphibians. This development was a major step in vertebrate evolution as it weakened vertebrate ties to moist environments.

Would You Like to Know:

1. how the skeleton of an amphibian is adapted for life on land? (*p. 277*)
2. how frogs, toads, and some salamanders catch prey with their tongues? (*p. 278*)
3. why an amphibian's skin is moist? (*p. 279*)
4. what terrestrial animals lack lungs? (*p. 279*)
5. how a frog's ear can filter out certain frequencies of sound? (*p. 281*)
6. what the functions of a frog's calls are? (*p. 283*)
7. what adaptations permitted life on land? (*p. 285*)

These and other useful questions will be answered in this chapter.

This chapter contains evolutionary concepts, which are set off in this font.

Evolutionary Perspective

Who, while walking along the edge of a pond or stream, has not been startled by the "plop" of an equally startled frog jumping to the safety of its watery retreat? Or who has not marveled at the sounds of a chorus of frogs breaking through an otherwise silent spring evening? These experiences and others like them have led some to spend their lives studying members of the class Amphibia (am-fib'e-ah) (L. *amphibia*, living a double life): frogs, toads, salamanders, and caecilians (figure 16.1). The class name implies that amphibians either move back and forth between water and land, or live one stage of their life in water, and another on land. One or both of these descriptions is accurate for most amphibians.

Amphibians are the first vertebrates we have encountered that are called **tetrapods** (Gr. *tetra*, four + *podos*, foot). It is a nontaxonomic designation that applies to all vertebrates other than fishes, and adaptations for life on land are found in most tetrapods.

Figure 16.1
Class Amphibia. Amphibians, like this tree frog (*Hyla andersoni*) are common vertebrates in most terrestrial and freshwater habitats. Their ancestors were the first terrestrial vertebrates.

Phylogenetic Relationships

During the first 250 million years of vertebrate history, adaptive radiation resulted in vertebrates filling most aquatic habitats. There were many active, powerful predators in the prehistoric waters. Land, however, was free of vertebrates and except for some arthropods, was free of predators. Animals that moved around the water's edge were not likely to be prey for other animals. With lungs for breathing air and muscular fins to scurry across mud, these animals probably found ample food in the arthropods that lived there. It is no surprise that the major component of the diet of most modern amphibians is arthropods.

The origin of amphibians from ancient sarcopterygians was described in chapter 15. Adaptive radiation of amphibians resulted in a much greater variety of forms than exists today. Later convergent and parallel evolution and widespread extinction have clouded our perceptions of evolutionary pathways. No one knows, therefore, what animal was the first amphibian, but the structure of limbs, skulls, and teeth suggests that *Ichthyostega* is probably similar to the earliest amphibians (figure 16.2). During the late Devonian and early Carboniferous periods, two lineages of early amphibians can be distinguished by details of the way the roof and the posterior portion of the skull are attached to each other. One lineage of amphibians became extinct late in the Carboniferous period. The development of an egg that was resistant to drying, an amniotic egg, occurred in this group. This lineage, called the **amniote lineage,** left as its descendants the reptiles, birds, and mammals (figure 16.3). A second lineage, flourished into the Jurassic period. Most of this lineage became extinct, but not before giving rise to the three orders of living amphibians. This lineage is called the **nonamniote lineage.**

Survey of Amphibians

Amphibians occur on all continents except Antarctica, but are absent from many oceanic islands. The 3,000 modern species are a mere remnant of this once-diverse group. Modern amphibians are divided into three orders: Caudata or Urodela, the salamanders; Anura, the frogs and toads; and Gymnophiona, the caecilians (table 16.1).

Order Caudata

Members of the order Caudata (kaw'dat-ah) (L. *cauda*, tail + Gr. *ata*, to bear) are the salamanders. They possess a tail throughout life, and both pairs of legs, when present, are relatively unspecialized (figure 16.4).

Approximately 115 of the 350 described species of salamanders occur in North America. Most terrestrial salamanders live in moist forest-floor litter and have aquatic larvae. A number of families are found in caves, where constant temperature and moisture conditions create a nearly ideal environment. Salamanders in the family Plethodontidae are the most fully terrestrial salamanders in that their eggs are laid on land, and the young hatch as miniatures of the adult. Members of the

FIGURE 16.2

***Ichthyostega*: An Early Amphibian.** Fossils of this early amphibian were discovered in eastern Greenland in late Devonian deposits. The total length of the restored specimen is about 65 cm. Terrestrial adaptations are heavy pectoral and pelvic girdles and sturdy limbs that probably aided in lifting the body off the ground. Strong jaws suggest that it was a predator in shallow water, perhaps venturing onto shore. Other features include a skull that is similar in structure to ancient sarcopterygian fishes and a finlike tail. Note that the tail fin is supported by bony rays dorsal to the spines of the vertebrae. This pattern is similar to the structure of the dorsal fins of fishes and is unknown in any other tetrapod.

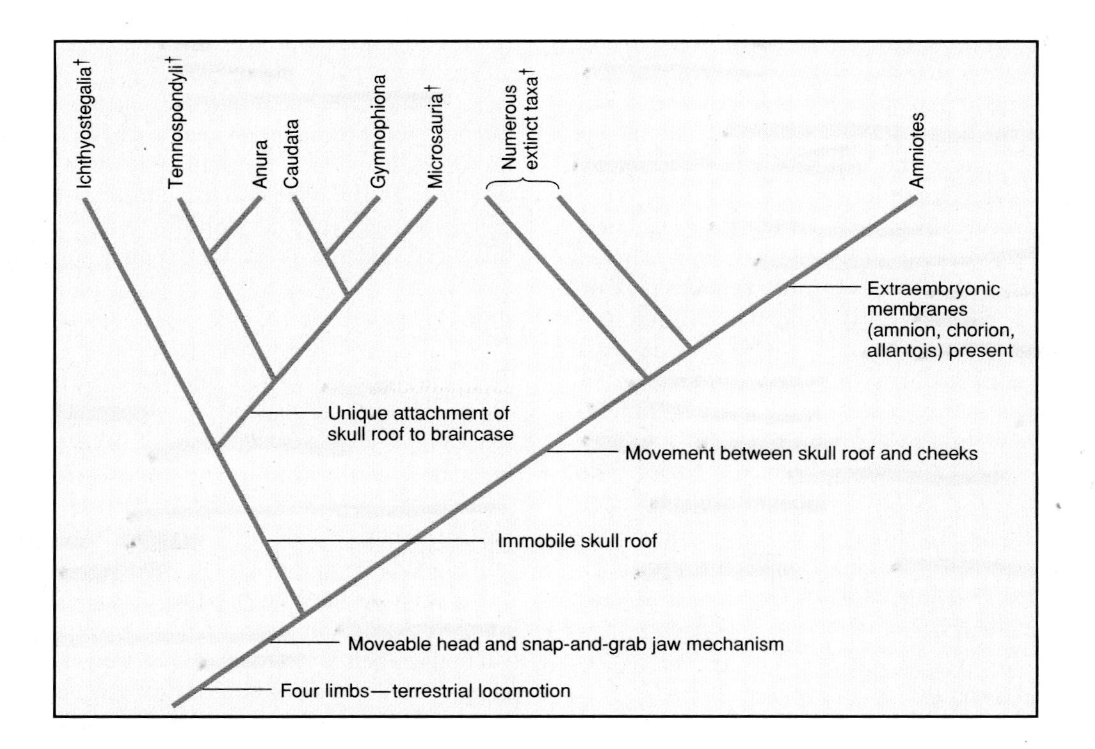

FIGURE 16.3

Evolutionary Relationships among the Amphibians. Earliest amphibians arose during the Devonian period. A nonamniotic lineage gave rise to three classes of modern amphibians and numerous extinct taxa. Some zoologists think that the class Amphibia is a paraphyletic group. If this is true, the three modern classes should be represented as monophyletic taxa. The amniotic lineage of early tetrapods gave rise to reptiles, birds, mammals, and other extinct taxa. Extinct taxa are indicated with a dagger (†). Synapomorphic characters for lower taxonomic groups are not indicated.

TABLE 16.1 CLASSIFICATION OF LIVING AMPHIBIANS

Class Amphibia (am-fib′e-ah)
Skin with mucoid secretions and lacks epidermal scales, feathers, or hair; larvae usually aquatic and undergo metamorphosis to the adult; two atrial chambers in the heart. One cervical and one sacral vertebra.

Order Caudata (kaw′dat-ah)
Long tail, two pairs of limbs; lack middle ear. Salamanders, newts.

Order Gymnophiona (jim′no-fy″o-nah)
Elongate, limbless; segmented by annular grooves; specialized for burrowing; tail short and pointed; rudimentary left lung. Caecilians.

Order Anura (ah-noor′ah)
Tailless; elongate hind limbs modified for jumping and swimming; five to nine presacral vertebrae with transverse processes (except the first); postsacral vertebrae fused into rodlike urostyle; tympanum and larynx well developed. Frogs, toads.

FIGURE 16.4
Order Caudata. Blue Ridge Spring salamander (*Gyrinophilus danielsi*).

family Salimandridae are commonly called newts. They spend most of their lives in water and frequently retain caudal fins. Salamanders range in length from only a few centimeters to 1.5 m (the Japanese giant salamander, *Andrias japonicus*). The largest North American salamander is the hellbender (*Cryptobranchus alleganiensis*), which reaches lengths of about 65 cm.

Most salamanders have internal fertilization. Males produce a pyramidal, gelatinous spermatophore that is capped with sperm and is deposited on the substrate. Females pick up the sperm cap with the cloaca, and the sperm are stored in a special pouch, the spermatheca. Eggs are fertilized as they pass through the cloaca and are usually deposited singly, in clumps, or in strings (figure 16.5*a*). Larvae are similar to adults but smaller. They often possess external gills, a tail fin, larval dentition, and a rudimentary tongue (figure 16.5*b*). The aquatic larval stage usually undergoes metamorphosis into a terrestrial adult (figure 16.5*c*). Many other salamanders undergo incomplete metamorphosis and are paedomorphic (e.g., *Necturus*); that is, they become sexually mature while still showing larval characteristics.

ORDER GYMNOPHIONA

Members of the order Gymnophiona (jim′no-fy″o-nah) (Gr. *gymnos*, naked + *ophineos*, like a snake) are the caecilians (figure 16.6). There are about 160 described species, which are confined to tropical regions. Caecilians are wormlike burrowers that feed on worms and other invertebrates in the soil. Caecilians appear segmented because of folds in the skin that overlie separations between muscle bundles. They have a retractile tentacle between their eyes and nostrils. The tentacle may transport chemicals from the environment to olfactory cells in the roof of the mouth. The eyes are covered by skin; thus, caecilians are probably nearly blind.

Fertilization is internal in caecilians. Larval stages are often passed within the oviducts, where they feed on the inner lining of the oviducts by scraping it with fetal teeth. The young emerge from the female as miniatures of the adults. Other caecilians lay eggs that develop into either aquatic larvae or embryos that undergo direct development on land.

ORDER ANURA

The order Anura (ah-noor″ah) (Gr. *a*, without + *oura*, tail) includes about 3,500 species of frogs and toads. Anurans are found in most moist environments, except in high latitudes and on some oceanic islands. A few even occur in very dry deserts. Adults lack tails, and caudal (tail) vertebrae are fused into a rodlike structure called the urostyle. Hind limbs are very long and muscular, and they end in webbed feet.

Anurans have diverse life histories. Fertilization is almost always external, and eggs and larvae are typically aquatic. Larval stages, called tadpoles, have well-developed tails. Their plump bodies lack limbs until near the end of their larval existence. Unlike adults, the larvae are herbivores and possess a proteinaceous, beaklike structure used in feeding. Anuran larvae undergo a drastic and rapid metamorphosis from the larval to the adult body form.

The distinction between "frog" and "toad" is more vernacular than scientific. "Toad" usually refers to anurans with relatively dry and warty skin, and they are more terrestrial than other members of the order. These characteristics are found in a number of distantly related taxa. True toads belong to the family Bufonidae (figure 16.7).

FIGURE 16.5
Order Caudata. (*a*) Eggs, (*b*) larva, and (*c*) adult of the spotted salamander, *Ambystoma maculatum*. Larvae are herbivores, and adults feed on worms and small arthropods.

FIGURE 16.6
Order Gymnophiona. A caecilian (*Ichthyophis glutinosus*).

FIGURE 16.7
Order Anura. American toad (*Bufo americanus*).

BOX 16.1 Poison Frogs of South America

A South American native stalks quietly through the jungle, peering into the tree branches overhead. A monkey's slight movements divulge its presence, and the hunter takes careful aim with what appears to be an almost toylike bow and arrow. The arrow sails true, and the monkey is hit. The arrow seems ineffectual at first, however, after a few moments, the monkey tumbles from the tree. Thousands of years of cultural evolution have taught these natives a deadly secret that makes effective hunting tools out of seemingly innocuous instruments.

All amphibians possess glandular secretions that are noxious or toxic to varying degrees. These glands are distributed throughout the skin and exude milky toxins that help ward off potential predators. Toxic secretions are frequently accompanied by warning (aposematic) coloration that signals to predators the presence of noxious secretions (figure 1).

Four genera of frogs (*Atopophryhnus*, *Colostethus*, *Dendrobates*, and *Phyllobates*) in the family Dendrobatidae live in tropical forests from Costa Rica to southern Brazil. South American natives use toxins from these frogs to tip their arrows. Frogs are killed with a stick and held over a fire. Granular glands in the skin release their venom, which is collected and allowed to ferment. Poisons collected in this manner are neurotoxins that prevent the transmission of nerve impulses between nerves and between nerves and muscles. Arrow tips dipped in this poison and allowed to dry contain sufficient toxin to paralyze a bird or small mammal.

Members of this family of frogs, in addition to being exploited by South American natives, have interesting reproductive habits. A female lays one to six large eggs in moist, terrestrial habitats. The female promptly abandons the eggs, but the male visits the clutch regularly and guards the eggs. The eggs hatch after approximately 2 weeks, and the tadpoles wiggle onto the male's back. The male then transports the tadpoles from the egg-laying site to water, where they are left to develop. The tadpoles metamorphose to the adult body form after approximately 6 weeks.

FIGURE 1 **A Poison Arrow Frog (*Dendrobates pumilo*).** The striking coloration shown here is an example of aposematic coloration.

Stop and Ask Yourself

1. What were two ancient lineages of amphibians? What groups of animals are the modern descendants of each lineage?
2. What animals are members of the order Caudata?
3. What order of amphibians is characterized by wormlike burrowing?
4. What order of amphibians is characterized by tail vertebrae fused into a urostyle?

Evolutionary Pressures

Most amphibians divide their lives between fresh water and land. This divided life is shown by adaptations to both environments that can be observed in virtually every body system. In the water, amphibians are supported by water's buoyant properties, they exchange gases with the water, and face the same osmoregulatory problems as freshwater fishes. On land, amphibians support themselves against gravity, exchange gases with the air, and tend to lose water to the air.

External Structure and Locomotion

Vertebrate skin protects against infective microorganisms, ultraviolet light, desiccation, and mechanical injury. As discussed later in this chapter, the skin of amphibians also functions in gas exchange, temperature regulation, and absorption and storage of water.

The skin of amphibians lacks a covering of scales, feathers, or hair. It is, however, highly glandular and its secretions aid in protection. These glands keep the skin moist to prevent drying. They also produce sticky secretions that help a male cling to a female during mating and produce toxic chemicals that discourage potential predators (box 16.1). The skin of many amphibians is smooth, although epidermal thickenings may produce warts, claws, or sandpapery textures, which are usually the result of keratin deposits or the formation of hard, bony areas.

Chromatophores are specialized cells in the epidermis and dermis of the skin and are responsible for skin color and color changes. Cryptic coloration, aposematic coloration (box 16.1), and mimicry are all common in amphibians.

Support and Movement

Water buoys and supports aquatic animals. The skeletons of fishes function primarily in protecting internal organs, providing

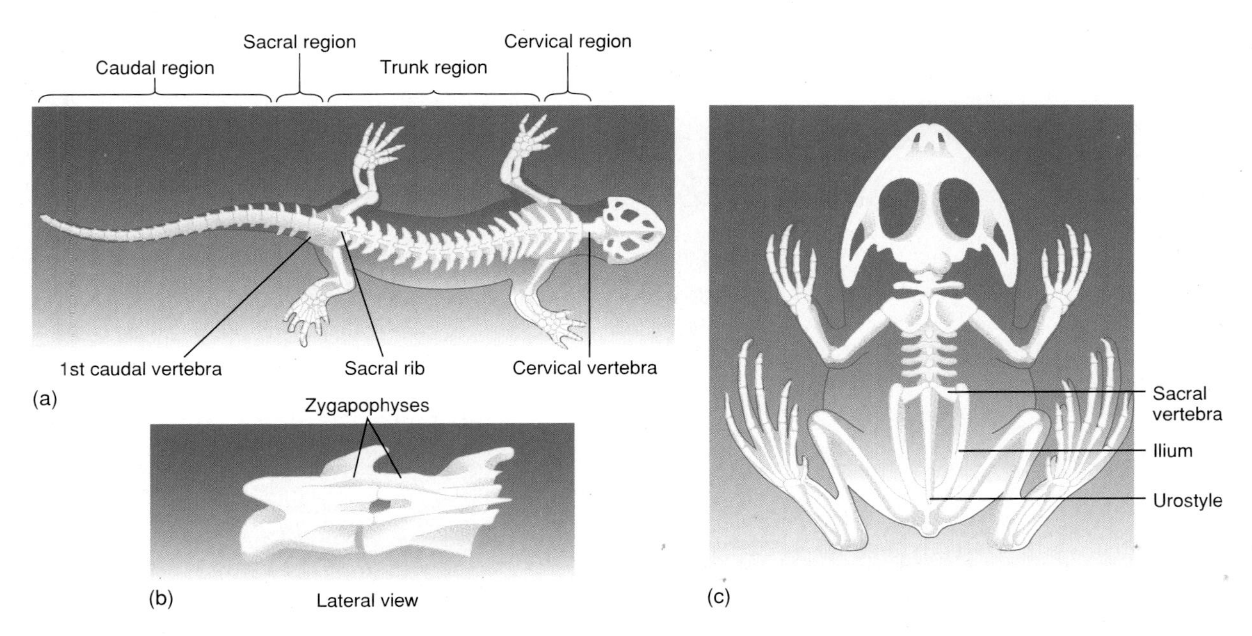

FIGURE 16.8

The Skeletons of Amphibians. (*a*) The salamander skeleton is divided into four regions: cervical, trunk, sacral, and caudal. (*b*) Interlocking processes, called zygapophyses prevent twisting between vertebrae. (*c*) The skeleton of a frog shows adaptations for jumping. Note the long back legs and the firm attachment of the back legs to the vertebral column through the ilium and urostyle.

points of attachment for muscles, and keeping the body from collapsing during movement. In terrestrial vertebrates, however, the skeleton is modified to provide support against gravity and must be strong enough to support the relatively powerful muscles that propel terrestrial vertebrates across land. 1 The skull of amphibians is flattened, is relatively smaller, and has fewer bony elements than the skull of fishes. These changes lighten the skull so it can be supported out of the water. Changes in jaw structure and musculature allow a crushing force to be applied to prey held in the mouth.

The vertebral column of amphibians is modified to provide support and flexibility on land (figure 16.8). It acts somewhat like the arch of a suspension bridge by supporting the weight of the body between anterior and posterior paired appendages. Supportive processes called zygapophyses on each vertebra prevent twisting. Unlike fishes, amphibians have a neck. The first vertebra is a cervical vertebra, which moves against the back of the skull and allows the head to nod vertically. The last trunk vertebra is a sacral vertebra. This vertebra is used to anchor the pelvic girdle to the vertebral column to provide increased support. A ventral plate of bone, called the sternum, is present in the anterior, ventral trunk region and provides support for the forelimbs and protection for internal organs. It is reduced or absent in the Anura.

The origin of the bones of vertebrate appendages is not precisely known; however, similarities in the structures of the bones of the amphibian appendages and the bones of the fins of ancient rhipidistian fishes suggest possible homologies (figure 16.9). The presence of joints at the shoulder, hip, elbow, knee, wrist, and ankle allows freedom of movement and better contact with the substrate. The pelvic girdle of amphibians consists of three

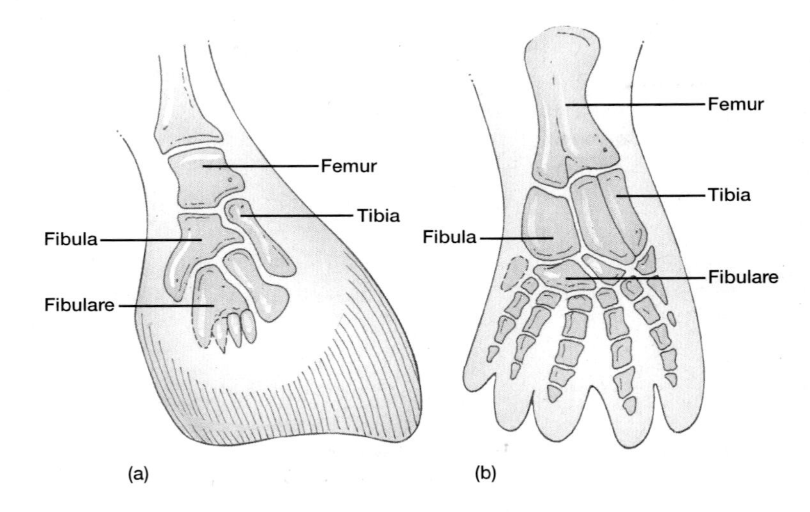

FIGURE 16.9

The Origin of Tetrapod Appendages. A comparison between the fin bones of a rhipidistian (*a*) and the limb bones of a tetrapod (*b*). This comparison suggests that the basic arrangements of bones seen in tetrapod limbs was already present in primitive fishes.

bones (the ilium, ischium, and pubis) that attach pelvic appendages firmly to the vertebral column. These bones, which are present in all tetrapods, but not fishes, are important for support on land.

Tetrapods depend more on appendages than the body wall for locomotion. Thus, body-wall musculature is reduced, and appendicular musculature predominates. (Contrast, for example, what one eats in a fish dinner as compared to a plate of frog legs.)

Figure 16.10
Salamander Locomotion. The pattern of leg movement in salamander locomotion. Arrows show leg movements.

Salamanders employ a relatively unspecialized form of locomotion that is reminiscent of the undulatory waves that pass along the body of a fish. Terrestrial salamanders also move by a pattern of limb and body movements in which the alternate movement of appendages results from muscle contractions that throw the body into a curve to advance the stride of a limb (figure 16.10). Caecilians move in an accordionlike movement in which adjacent parts of the body are pushed or pulled forward at the same time. The long hindlimbs and the pelvic girdle of anurans are modified for jumping. The dorsal bone of the pelvis (the ilium) extends anteriorly and is securely attached to the vertebral column, and the urostyle extends posteriorly and attaches to the pelvis (*see figure* 16.8). These skeletal modifications stiffen the posterior half of the anuran. Long hindlimbs and powerful muscles form an efficient lever system for jumping. Elastic connective tissues and muscles attach the pectoral girdle to the skull and vertebral column and function as shock absorbers for landing on the forelimbs.

Nutrition and the Digestive System

Most adult amphibians are carnivores that feed on a wide variety of invertebrates. The diets of some anurans, however, are more diverse. For example, a bullfrog will prey on small mammals, birds, and other anurans. The main factors that determine what amphibians will eat are prey size and availability. Larvae are herbivorous and feed on algae and other plant matter. Most amphibians locate their prey by sight and simply wait for prey to pass by. Olfaction plays an important role in prey detection by aquatic salamanders and caecilians.

Many salamanders are relatively unspecialized in their feeding methods, using only their jaws to capture prey. Anurans and plethodontid salamanders, however, use their tongue and jaws in a flip-and-grab feeding mechanism (figure 16.11). A true tongue is first seen in amphibians. (The "tongue" of fishes is simply a fleshy fold on the floor of the mouth. Fish food is swallowed whole and not manipulated by the "tongue.") 2 The tongue of amphibians is attached at the anterior margin of the jaw and lies folded back over the floor of the mouth. Mucous and buccal glands on the tip of the tongue exude sticky secretions. When a prey comes within range, an amphibian lunges forward and flicks out its tongue. The tongue turns over, and the lower jaw is depressed. The fact that the head can tilt on its single cervical vertebra aids in aiming the strike. The tip of the tongue entraps the prey, and the tongue and prey are flicked back inside the mouth. All of this may happen in 0.05 to

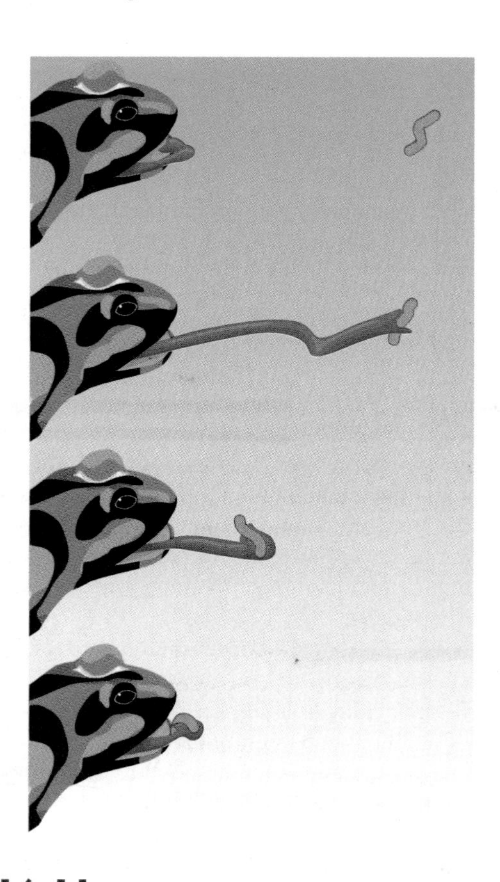

Figure 16.11
Flip-and-Grab Feeding in a Toad. The tongue is attached at the anterior margin of the toad's jaw and is flipped out to capture a prey on its sticky secretions.

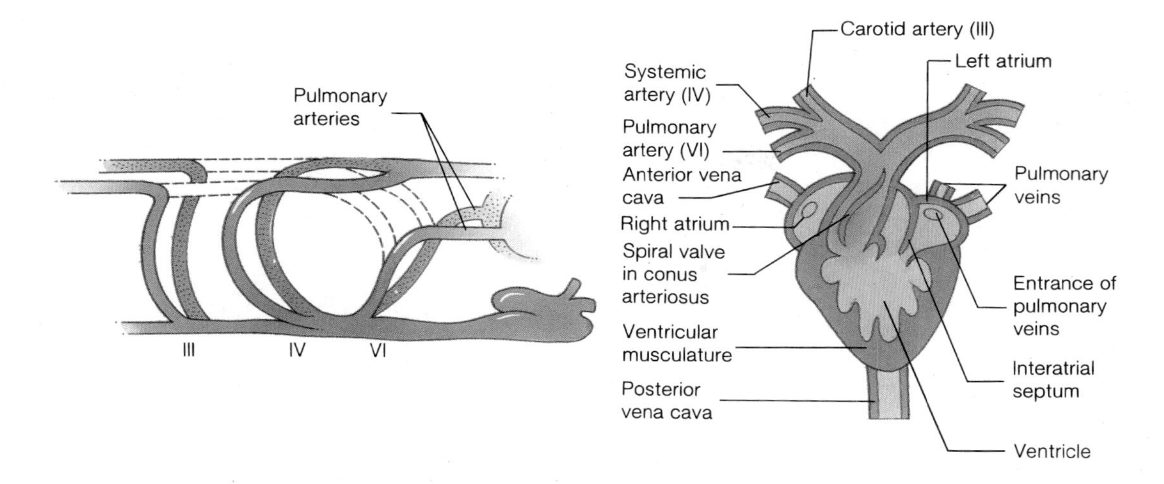

FIGURE 16.12
Diagrammatic Representation of an Anuran Circulatory System. (*a*) The Roman numerals indicate the various aortic arches. Vessels shown in dashed outline are lost during embryological development. (*b*) The heart is drawn in a ventral view.

0.15 second! The prey is held by pressing it against teeth on the roof of the mouth, and the tongue and other muscles of the mouth push food toward the esophagus. The eyes sink downward during swallowing, and help force food toward the esophagus.

CIRCULATION, GAS EXCHANGE, AND TEMPERATURE REGULATION

The circulatory system of amphibians shows remarkable adaptations for a life that is divided between aquatic and terrestrial habitats. The separation of pulmonary and systemic circuits is less efficient in amphibians than in lungfishes (figure 16.12; *see also figure 15.14b*). The atrium is partially divided in urodeles and completely divided in anurans. The ventricle has no septum. A spiral valve is present in the conus arteriosus or ventral aorta and helps direct blood into pulmonary and systemic circuits. As discussed later, gas exchange occurs across the skin of amphibians, as well as in the lungs. Therefore, blood entering the right side of the heart is nearly as well oxygenated as blood entering the heart from the lungs! When an amphibian is completely submerged, all gas exchange occurs across the skin and other moist surfaces; therefore, blood coming into the right atrium has a higher oxygen concentration than blood returning to the left atrium from the lungs. Under these circumstances, blood vessels leading to the lungs constrict, reducing blood flow to the lungs and conserving energy. This adaptation is especially valuable for those frogs and salamanders that overwinter in the mud at the bottom of a pond.

Fewer aortic arches are present in adult amphibians than in fishes. After leaving the conus arteriosus, blood may enter the carotid artery (aortic arch III), which takes blood to the head; the systemic artery (aortic arch IV), which takes blood to the body; or the pulmonary artery (aortic arch VI).

In addition to a vascular system that circulates blood, amphibians have a well-developed lymphatic system of blind-ending vessels that returns fluids, proteins, and ions filtered from capillary beds in tissue spaces to the circulatory system. Water absorbed across the skin is also transported by the lymphatic system. Unlike other vertebrates, amphibians have contractile vessels, called lymphatic hearts, that pump fluid through the lymphatic system. Lymphatic spaces are present between body-wall muscles and the skin. These spaces transport and store water absorbed across the skin.

Gas Exchange

Terrestrial animals need to expend much less energy moving air across gas-exchange surfaces than do aquatic organisms because air contains 20 times more oxygen per unit volume than does water. On the other hand, exchanges of oxygen and carbon dioxide require moist surfaces, and exposure of respiratory surfaces to air may result in rapid water loss.

Anyone who has searched pond and stream banks for frogs knows that the skin of amphibians is moist. Amphibian skin is also richly supplied with capillary beds. These two factors permit the skin to function as a respiratory organ. Gas exchange across the skin is called **cutaneous respiration** and can occur either in water or on land. This ability allows a frog to spend the winter in the mud at the bottom of a pond. In salamanders, 30 to 90% of gas exchange occurs across the skin. Gas exchange also occurs across the moist surfaces of the mouth and pharynx. This is called **buccopharyngeal respiration** and accounts for 1 to 7% of total gas exchange.

Most amphibians, except for plethodontid salamanders, possess lungs (figure 16.13*a*). The lungs of salamanders are relatively simple sacs. The lungs of anurans are subdivided, increasing the surface area for gas exchange. Pulmonary (lung) ventilation occurs by a **buccal pump** mechanism. Muscles of the mouth and pharynx create a positive pressure to force air into the lungs (figure 16.13*b–e*).

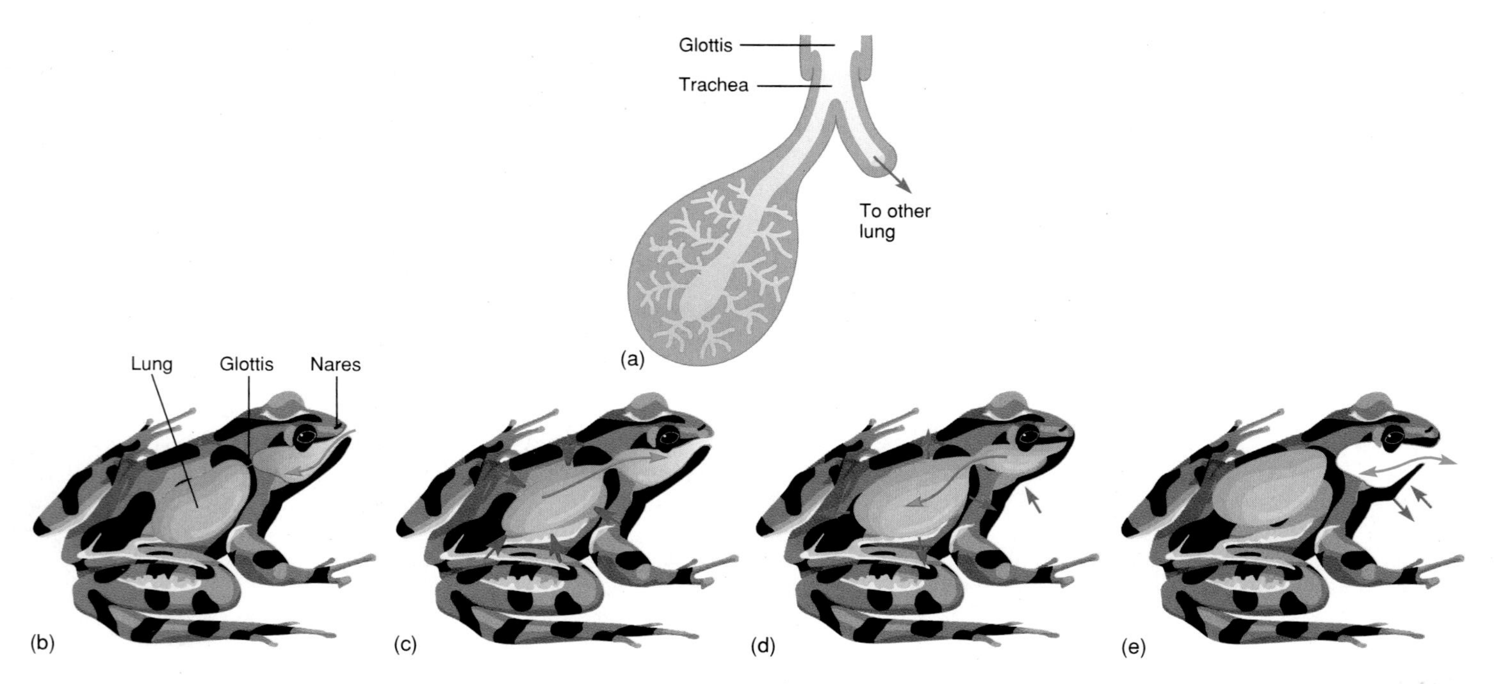

Figure 16.13

Amphibian Lung Structure, Buccal Pump, and Buccopharyngeal Ventilation. (*a*) Lung structure of a frog. (*b*) With the opening of the respiratory tract (the glottis) closed, the floor of the mouth is lowered, and air enters the mouth cavity. (*c*) The glottis opens, and the elasticity of the lungs and contraction of the body wall forces air out of the lungs, over the top of air just brought into the mouth. (*d*) The mouth and nares are closed, and the floor of the mouth is raised, forcing air into the lungs. (*e*) With the glottis closed, oscillations of the floor of the mouth exchange air in the mouth cavity to facilitate buccopharyngeal respiration. Blue arrows show air movements. Red arrows show movements of the lungs and body wall.

Cutaneous and buccopharyngeal respiration have a disadvantage in that the absolute contribution of these exchange mechanisms to total gas exchange is relatively constant. There is no way to increase the quantity of gas exchanged across these surfaces when the metabolic rate increases. Lungs, however, compensate for this shortcoming. As environmental temperature and activity increase, lungs contribute more to total gas exchange. At 5° C, approximately 70% of gas exchange occurs across the skin and mouth lining of a frog. At 25° C, the absolute quantity of oxygen exchanged across external body surfaces does not change significantly, but because pulmonary respiration is increased, exchange across skin and mouth surfaces accounts for only about 30% of total oxygen exchange.

Amphibian larvae and some adults respire using external gills. Three pairs of gills are supported by cartilaginous rods that are formed between embryonic gill slits. At metamorphosis, the gills are usually reabsorbed, gill slits close, and lungs become functional.

Temperature Regulation

Amphibians are ectothermic. (They depend upon external sources of heat to maintain body temperature.) Any poorly insulated aquatic animal, regardless of how much metabolic heat is produced, will lose heat as quickly as it is produced because of the powerful heat-absorbing properties of water. Therefore, when amphibians are in water, they take on the temperature of their environment. On land, however, their body temperatures can differ from that of the environment.

Temperature regulation is mainly behavioral. Some cooling results from evaporative heat loss. In addition, many amphibians are nocturnal and remain in cooler burrows or under moist leaf litter during the hottest part of the day. Amphibians may warm themselves by basking in the sun or on warm surfaces. Body temperatures may be raised 10° C above the air temperature. Basking after a meal is common, because increased body temperature increases the rate of all metabolic reactions—including digestive functions, growth, and the deposition of fats necessary to survive periods of dormancy.

Amphibians often experience wide daily and seasonal temperature fluctuations, and therefore have correspondingly wide temperature tolerances. Critical temperature extremes for some salamanders lie between −2 and 27° C, and for some anurans between 3 and 41° C.

Stop and Ask Yourself

5. What are the functions of amphibian skin?
6. What chambers are present in a frog's heart?
7. Why is the incomplete separation of atria and ventricles believed to be adaptive for an amphibian?
8. What are four forms of gas exchange in amphibians?

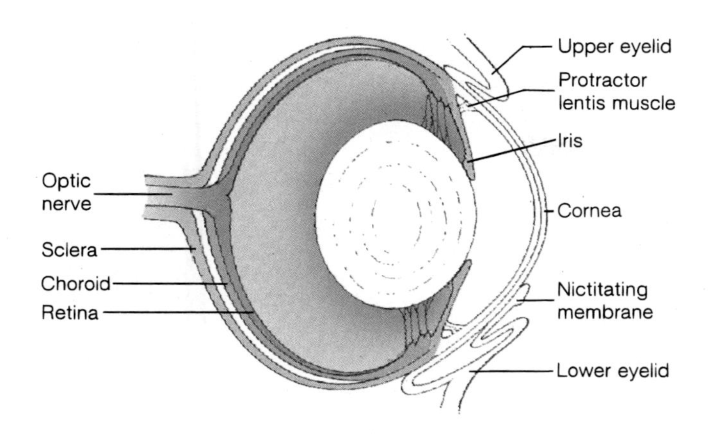

Figure 16.14

The Amphibian Eye. Longitudinal section of the eye of the leopard frog, *Rana pipiens*.

Nervous and Sensory Functions

The nervous system of amphibians is similar to that of other vertebrates. The brain of adult vertebrates develops from three embryological subdivisions. In amphibians, the forebrain contains olfactory centers and regions that regulate color change and visceral functions. The midbrain contains a region called the optic tectum, in which sensory information is assimilated and motor responses are initiated. Visual sensory information is also processed in the midbrain. The hindbrain functions in motor coordination and in regulating heart rate and mechanics of respiration.

Many amphibian sensory receptors are widely distributed over the skin. Some of these are simply bare nerve endings that respond to heat, cold, and pain. The lateral-line system is similar in structure to that found in fishes, and it is present in all aquatic larvae, aquatic adult salamanders, and some adult anurans. Lateral-line organs are distributed singly or in small groups along the lateral and dorsolateral surfaces of the body, especially the head. These receptors respond to low-frequency vibrations in the water and movements of the water relative to the animal. On land, however, lateral-line receptors are less important.

Chemoreception is an important sense for many amphibians. Chemoreceptors are located in the nasal epithelium, the lining of the mouth, on the tongue, and over the skin. Olfaction is used in mate recognition, as well as detecting noxious chemicals and in locating food.

Vision is one of the most important senses in amphibians because they are primarily sight feeders. (Caecilians are an obvious exception.) A number of adaptations allow the eyes of amphibians to function in terrestrial environments (figure 16.14). The fact that the eyes of some amphibians (i.e., anurans and some salamanders) are located on the front of the head provides the binocular vision and well-developed depth perception necessary for capturing prey. Other amphibians with smaller lateral eyes (some salamanders) lack binocular vision.

The lower eyelid is movable and functions to clean and protect the eye. Much of it is transparent and is called the **nictitating membrane.** When the eyeball is retracted into the orbit of the skull, the nictitating membrane is drawn up over the cornea. In addition, orbital glands lubricate and wash the eye. Together, eyelids and glands keep the eye free of dust and other debris. The lens is large and nearly round. It is set back from the cornea and is surrounded by a fold of epithelium called the iris. The iris can dilate or constrict to control the size of the pupil.

Focusing, or accommodation, involves bending (refracting) light rays to a focal point on the retina. Light waves moving from air across the cornea are refracted because of the change in density between the two media. Further refraction is accomplished by the lens. As are the eyes of most tetrapods, the amphibian eye is focused on distant objects when the eye is at rest. To focus on near objects, the lens must be moved forward by the protractor lentis muscle (figure 16.14). Receptors called rods and cones are found in the retina. Because cones are associated with color vision in some other vertebrates, their occurrence suggests that amphibians are capable of distinguishing between some wavelengths of light. The extent to which color vision is developed is unknown. The neuronal interconnections in the retina are very complex and allow an amphibian to distinguish between flying insect prey, shadows that may warn of an approaching predator, and background movements, such as blades of grass moving with the wind.

The auditory system of amphibians is clearly an evolutionary adaptation to life on land. It transmits both substrateborne vibrations and, in anurans, airborne vibrations. The ears of anurans consist of a tympanic membrane, a middle ear, and an inner ear. The tympanic membrane is a piece of integument stretched over a cartilaginous ring that receives airborne vibrations and transmits these vibrations to the middle ear, which is a chamber beneath the tympanic membrane. Abutting the tympanic membrane is a middle-ear ossicle (bone) called the stapes (columella), which transmits vibrations of the tympanic membrane into the inner ear. High-frequency (1,000 to 5,000 Hz) airborne vibrations are transmitted to the inner ear through the tympanic membrane. Low-frequency (100 to 1,000 Hz) substrateborne vibrations are transmitted through the front appendages and the pectoral girdle to the inner ear through a second ossicle called the operculum.

5 Muscles attached to the operculum and stapes can lock either or both of these ossicles, allowing an anuran to screen out either high- or low-frequency sounds. This mechanism is adaptive because low-and high-frequency sounds are used in different situations by anurans. Mating calls are high-frequency sounds that are of primary importance for only a part of the year (breeding season). At other times, low-frequency sounds may warn of approaching predators.

Salamanders lack a tympanic membrane and middle ear. They live in streams, ponds, caves, and beneath leaf litter. They have no mating calls, and the only sounds they hear are probably low-frequency vibrations transmitted through the substrate and skull to the stapes and inner ear.

(a)

(b)

FIGURE 16.15

Water Conservation by Anurans. (*a*) The daytime sleeping posture of the green tree frog, *Hyla cinerea*. Exposed surface area is reduced by the closely tucked appendages. (*b*) The Australian burrowing frog, *Cyclorana alboguttatus*, in its burrow and water-retaining skin.

The sense of equilibrium and balance is similar to that described for fishes in the previous chapter. The inner ear of amphibians has semicircular canals that help detect rotational movements and other sensory patches that respond to gravity. The latter detect linear acceleration and deceleration.

EXCRETION AND OSMOREGULATION

The kidneys of amphibians lie on either side of the dorsal aorta on the dorsal wall of the body cavity. A duct leads to the cloaca, and a storage structure, the urinary bladder, is a ventral outgrowth of the cloaca.

The nitrogenous waste product excreted by amphibians is either ammonia or urea. Amphibians that live in fresh water excrete ammonia. It is the immediate end product of protein metabolism; therefore, no energy is expended converting it into other products. The toxic effects of ammonia are avoided by its rapid diffusion into the surrounding water. Amphibians that spend more time on land excrete urea that is produced from ammonia in the liver. Although urea is less toxic than ammonia, it still requires relatively large quantities of water for its excretion. Unlike ammonia, urea can be stored in the urinary bladder. Some amphibians excrete ammonia when in water and urea when on land.

One of the biggest problems faced by amphibians is osmoregulation. In water, amphibians face the same osmoregulatory problems as freshwater fishes. They must rid the body of excess water and conserve essential ions. Amphibian kidneys produce large quantities of hypotonic urine, and the skin and walls of the urinary bladder transport Na^+, Cl^-, and other ions into the blood.

On land, amphibians must conserve water. Adult amphibians do not replace water by intentional drinking, nor do they have the impermeable skin characteristic of other tetrapods or kidneys capable of producing a hypertonic urine. Instead, amphibians limit water loss by behavior that reduces exposure to desiccating conditions. Many terrestrial amphibians are nocturnal. During daylight hours, they retreat to areas of high humidity, such as under stones, in logs, in leaf mulch, or in burrows. Water loss on nighttime foraging trips must be compensated for by water uptake across the skin while in the retreat. Diurnal amphibians usually live in areas of high humidity and rehydrate themselves by entering the water. Many amphibians reduce evaporative water loss by reducing the amount of body surface exposed to air. They may curl their bodies and tails into tight coils and tuck their limbs close to their bodies (figure 16.15*a*). Individuals may form closely packed aggregations to reduce overall surface area.

Some amphibians have protective coverings that reduce water loss. Hardened regions of skin are resistant to water loss and may be used to plug entrances to burrows or other retreat openings to maintain high humidity in the retreat. Other amphibians prevent water loss by forming cocoons that encase the body during long periods of dormancy. Cocoons are made from outer layers of the skin that detach and become parchmentlike. These cocoons open only at the nares or the mouth and have been found to reduce water loss 20 to 50% over noncocooned individuals (figure 16.15*b*).

Paradoxically, the skin—the most important source of water loss—is also the most important structure for rehydration. When an amphibian flattens its body on moist surfaces, the skin, especially in the ventral pelvic region, absorbs water. The permeability and vascularization of the skin and its epidermal sculpturing are all factors that promote water reabsorption. Minute channels increase surface area and spread water over surfaces not necessarily in direct contact with water.

Amphibians can also temporarily store water. Water accumulated in the urinary bladder and lymph sacs can be selectively reabsorbed to replace evaporative water loss. Amphibians living in very dry environments can store volumes of water equivalent to 35% of their total body weight.

Reproduction, Development, and Metamorphosis

Amphibians are dioecious, and ovaries and testes are located near the dorsal body wall. Fertilization is usually external, and because the developing eggs lack any resistant coverings, development is tied to moist habitats, usually water. A few anurans have terrestrial nests that are kept moist by being enveloped in foam or by being located near the water and subjected to flooding. In a few species, larval stages are passed in the egg membranes, and the immatures hatch into an adultlike body. The main exception to external fertilization in amphibians is the salamanders. Only about 10% of all salamanders have external fertilization. All others use spermatophores, and fertilization is internal. Eggs may be deposited in soil or water or retained in the oviduct during development. All caecilians have internal fertilization and 75% have internal development. Amphibian development has been studied extensively and usually includes the formation of larval stages called tadpoles. Amphibian tadpoles often differ from the adults in mode of respiration, form of locomotion, and diet. These differences reduce competition between adults and larvae.

The timing of reproductive activities is determined by interactions between internal (largely hormonal) controls and extrinsic factors. In temperate regions, breeding periods are seasonal and occur during spring and summer. In temperate areas, temperature seems to be the most important environmental factor that induces physiological changes associated with breeding. In tropical regions, breeding of amphibians is correlated with rainy seasons.

Courtship behavior helps individuals locate breeding sites, identify potential mates, prepare individuals for reproduction, and ensure that eggs are fertilized and deposited in locations that promote successful development.

Salamanders rely primarily on olfactory and visual cues in courtship and mating, whereas in anurans, vocalizations by the male and tactile cues are important. Many species congregate in one location during times of intense breeding activity. Calls by males are usually species specific, and they function in the initial attraction between mates. Once initial contact has been made, tactile cues become more important. The male grasps the female—his forelimbs around her waist—so that they are oriented in the same direction, and the male is dorsal to the female. This positioning is called **amplexus** and may last from 1 to 24 hours. During amplexus, the male releases sperm as the female releases eggs.

Little is known of caecilian breeding behavior. Males possess an intromittent organ that is a modification of the cloacal wall, and fertilization is internal.

Vocalization

6 Sound production is primarily a reproductive function of male anurans. Advertisement calls attract females to breeding areas, and announce to other males that a given territory is occupied. Advertisement calls are species specific, and any one species has a very limited repertoire of calls. They may also help induce psychological and physiological readiness to breed. Reciprocation calls are given by females in response to male calls to indicate receptiveness of a female. Release calls inform a partner that a frog is incapable of reproducing. They are given by unresponsive females during attempts at amplexus by a male, or by males that have been mistakenly identified as female by another male. Distress calls are not associated with reproduction, but are given by either sex in response to pain or being seized by a predator. These calls may be loud enough to cause a predator to release the frog. The distress call of the South American jungle frog, *Leptodactylus pentadactylus*, is a loud scream similar to the call of a cat in distress.

The sound-production apparatus of frogs consists of the larynx and its vocal cords. This laryngeal apparatus is well developed in males, who also possess a vocal sac. In the majority of frogs, vocal sacs develop as a diverticulum from the lining of the buccal cavity (figure 16.16). Air from the lungs is forced over the vocal cords and cartilages of the larynx, causing them to vibrate. Muscles control the tension of the vocal cords and are responsible for regulating the frequency of the sound. Vocal sacs act as resonating structures and increase the volume of the sound.

The use of sound to attract mates is especially useful in organisms that occupy widely dispersed habitats and must come together for breeding. Because many species of frogs often converge at the same pond for breeding, finding a mate of the proper species could be chaotic. Vocalizations help to reduce the chaos.

Parental Care

Parental care increases the chances of any one egg developing, but it requires large energy expenditures on the part of the parent. The most common form of parental care in amphibians is attendance of the egg clutch by either parent. Maternal care occurs in species with internal fertilization (predominantly salamanders and caecilians), and paternal care may occur in species with external fertilization (predominantly anurans). It may involve aeration of aquatic eggs, cleaning and/or moistening of terrestrial eggs, protection of eggs from predators, or removal of dead and infected eggs.

Transport of eggs may occur when development occurs on land. Females of the genus *Pipa* carry eggs on their back.

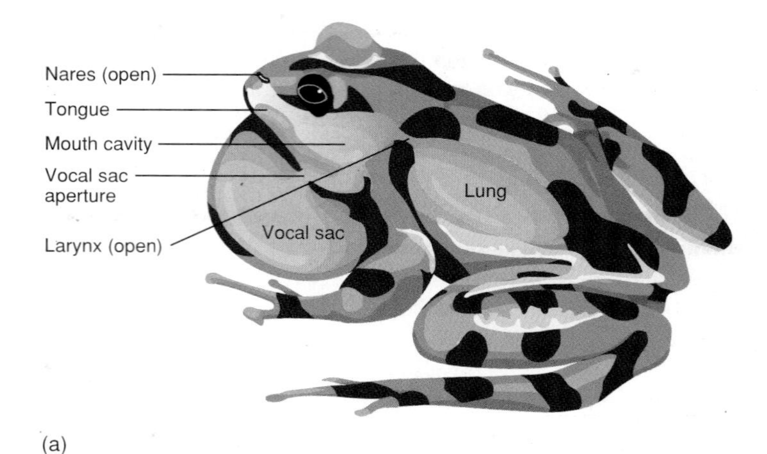

FIGURE 16.16
Anuran Vocalization. (*a*) Generalized vocal apparatus of anurans. (*b*) Inflated vocal sac of the Great Plains toad, *Bufo cognatus*.

FIGURE 16.17
Parental Care of Young. Female *Rheobatrachus* with young emerging from her mouth.

Rheobatrachus females carry developing eggs and tadpoles in their stomach, and the young emerge from the female's mouth (figure 16.17)! Viviparity and ovoviviparity occur primarily in salamanders and caecilians.

Metamorphosis

Metamorphosis is a series of abrupt structural, physiological, and behavioral changes that transform a larva into an adult. The time required for metamorphosis is influenced by a variety of environmental conditions, including crowding and food availability. Most directly, however, metamorphosis is under the control of neurosecretions of the hypothalamus, hormones of the anterior lobe of the pituitary gland (the adenohypophysis), and the thyroid gland.

Morphological changes associated with metamorphosis of caecilians and salamanders are relatively minor. Reproductive structures develop, gills are lost, and a caudal fin (when present) is lost. In the Anura, however, changes from the tadpole into the small frog are more dramatic (figure 16.18). Limbs and lungs develop, the tail is reabsorbed, the skin thickens, and marked changes in the head and digestive tract (associated with a new mode of nutrition) occur.

Paedomorphosis in amphibians can be explained based on the mechanisms of metamorphosis. Some salamanders are paedomorphic because of a failure of cells to respond to thyroid hormones, whereas others are paedomorphic because of a failure to produce the hormones associated with metamorphosis. Should environmental conditions change, they are able to metamorphose into the adult form.

FURTHER PHYLOGENETIC CONSIDERATIONS

One unresolved controversy concerning amphibian phylogeny is the relationship among the three orders of modern amphibians. Some zoologists place anurans, urodeles, and caecilians into a single subclass, Lissamphibia. This placement implies a common ancestry for modern amphibians and suggests that they are more closely related to each other than to any other group. Supporters of this classification point to common characteristics, such as the stapes/operculum complex, the importance of the skin in gas exchange, and aspects of the structure of the skull and teeth, as evidence of this close relationship. This interpretation is depicted in figure 16.3. Other zoologists think that modern amphibians were derived from at least two nonamniotic lineages. They note that fine details of other structures, such as the vertebral column, are different enough in the three orders to suggest separate origins. If this is true, the class Amphibia

Figure 16.18

Events of Metamorphosis in the Frog, *Rana temporaria*. (*a*) Before metamorphosis. Prolactin secretion, controlled by the hypothalamus and the adenohypophysis, promotes the growth of larval structures. (*b–d*) Metamorphosis. The median eminence of the hypothalamus develops and initiates the secretion of thyroid-stimulating hormone (TSH). TSH begins to inhibit prolactin release. TSH causes the release of large quantities of T_4 and T_3, which promote the growth of limbs, reabsorption of the tail, and other changes of metamorphosis, resulting eventually in a young, adult frog.

would be a paraphyletic group and should be dividied into multiple monophyletic taxa. This controversy is not likely to be settled soon.

In the next three chapters, our attention will turn to descendants of the amniote lineage (*see figure 16.3*). A group called anthracosaurs are often cited as amphibian ancestors of these animals, but support for this conclusion is weak. 7 Three sets of evolutionary changes occurred in amphibian lineages that allowed movement onto land. Two of these occurred early enough that they are found in all amphibians. One was the set of changes in the skeleton and muscles that allowed greater mobility on land. A second change involved a jaw mechanism and moveable head that permitted effective exploitation of insect resources on land. A jaw-muscle arrangement that permitted rhipidistian fishes to snap, grab, and hold prey was adaptive when early tetrapods began feeding on insects in terrestrial environments. The third set of changes occurred in the amniote lineage—the development of an egg that was resistant to drying. Although the **amniotic egg** is not completely independent of water, a series of **extraembryonic membranes** form during development that protect the embryo from desiccation, store wastes, and promote gas exchange. In addition, this egg has a leathery or calcified shell that is protective, yet porous enough to allow exchange of gases with the environment. These evolutionary events eventually resulted in the remaining three vertebrate groups: reptiles, birds, and mammals.

Stop and Ask Yourself

9. What region of the amphibian brain integrates sensory information and initiates motor responses?
10. What respiratory and excretory adaptations do amphibians possess that promote life in terrestrial environments?
11. What are four functions of anuran vocalizations?
12. In what ways are amniotic eggs adaptive for life on land?

Summary

1. Terrestrial vertebrates are called tetrapods and probably arose from sarcopterygians. Two lineages of ancient amphibians diverged. The nonamniote lineage gave rise to the three orders of modern amphibians. The amniote lineage gave rise to reptiles, birds, and mammals.
2. Members of the order Caudata are the salamanders. Salamanders are widely distributed, usually have internal fertilization, and may have aquatic larvae or direct development.
3. The order Gymnophiona contains the caecilians. Caecilians are tropical, wormlike burrowers. They have internal fertilization and many are viviparous.
4. Frogs and toads comprise the order Anura. Anurans lack tails and possess adaptations for jumping and swimming. External fertilization results in tadpole larvae, which metamorphose to adults.
5. The skin of amphibians is moist and functions in gas exchange, water regulation, and protection.
6. Skeletal and muscular systems of amphibians are adapted for movement on land.
7. Amphibians are carnivores that capture prey in their jaws or by using their tongue.
8. The circulatory system of amphibians is modified to accommodate the presence of lungs, gas exchange at the skin, and loss of gills in most adults.
9. Gas exchange is cutaneous, buccopharyngeal, and pulmonary. Pulmonary ventilation is accomplished by a buccal pump. A few amphibians retain gills as adults.
10. Sensory receptors of amphibians, especially the eye and ear, are adapted for functioning on land.
11. Amphibians excrete ammonia or urea. Ridding the body of excess water when in water and conserving water when on land are functions of the kidneys, the skin, and behavior.
12. Reproductive habits of amphibians are diverse. Many have external fertilization and development. Others have internal fertilization and development. Courtship, vocalizations, and parental care are common in some amphibians. Metamorphosis is under the control of the nervous and endocrine systems.
13. The evolution of an egg that is resistant to drying occurred in the amniote lineage, which is represented today by reptiles, birds, and mammals.

Selected Key Terms

amniote lineage (*p. 272*)
amplexus (*p. 283*)
buccal pump (*p. 279*)
buccopharyngeal respiration (*p. 279*)
cutaneous respiration (*p. 279*)
nictitating membrane (*p. 281*)
nonamniote lineage (*p. 272*)
tetrapods (*p. 272*)

Critical Thinking Questions

1. How are skeletal and muscular systems of amphibians adapted for life on land?
2. Would the buccal pump be more important for an active amphibian or for one that is becoming inactive for the winter? Explain your answer.
3. Why is the separation of oxygenated and nonoxygenated blood in the heart not very important for amphibians?
4. Explain how the skin of amphibians is used in temperature regulation, protection, gas exchange, and water regulation. Under what circumstances might cooling interfere with water regulation?
5. In what ways could anuran vocalizations have influenced the evolution of that order?

chapter 17

Reptiles: The First Amniotes

Outline

Concepts

1. Adaptive radiation of primitive amniotes resulted in the three or four lineages of reptiles. These lineages have given rise to four orders of modern reptiles, the birds, and the mammals.
2. The class Reptilia is divided into four orders: Testudines includes the turtles; Squamata includes the lizards, the snakes, and the worm lizards; Rhynchocephalia includes a single species, *Sphenodon punctatus;* and Crocodilia includes the alligators and crocodiles.
3. Reptiles possess adaptations that allow many members of the class to spend most of their lives apart from standing or flowing water. These include adaptations for support and movement, feeding, gas exchange, temperature regulation, excretion, osmoregulation, and reproduction.
4. Two reptilian evolutionary lineages gave rise to other vertebrate classes: Aves and Mammalia.

Would You Like to Know:

1. what living animals are most closely related to dinosaurs? (*p. 288*)
2. why turtles are vulnerable to extinction? (*p. 292*)
3. what animal "walks at both ends"? (*p. 294*)
4. why a lizard's tail breaks easily? (*p. 295*)
5. how a chameleon captures prey? (*p. 296*)
6. why reptiles divert blood away from their lungs? (*p. 297*)
7. what a median eye is? (*p. 299*)

These and other useful questions will be answered in this chapter.

This chapter contains evolutionary concepts, which are set off in this font.

Evolutionary Perspective

The earliest members of the class Reptilia (rep-til′e-ah) (L. *reptus*, to creep) were the first vertebrates to possess **amniotic eggs** (figure 17.1). Amniotic eggs have a series of extraembryonic membranes that protect the embryo from desiccation, cushion the embryo, promote gas transfer, and store waste materials. The amniotic eggs of reptiles and birds also have hard or leathery shells that protect the developing embryo, albumen that cushions and provides moisture and nutrients for the embryo, and yolk to supply food to the embryo. All of these features are adaptations for development on land. (The amniotic egg is not, however, the only kind of land egg—some arthropods, amphibians, and even a few fishes have eggs that develop on land.) The amniotic egg is the major synapomorphy that distinguishes the reptiles, birds, and mammals from vertebrates in the nonamniote lineage. Even though the amniotic egg has played an important role in the successful invasion of terrestrial habitats by vertebrates, it is one of many reptilian adaptations that have allowed members of this class to flourish on land. Living representatives of the class Reptilia include the turtles, lizards, snakes, worm lizards, crocodilians, and the tuatara (table 17.1).

Even though there are abundant fossil records of many reptiles, there is much to be learned of reptilian origins. As indicated by the circle at the base of the cladogram in figure 17.2, the ancestral amniote has not yet been discovered. The adaptive radiation of the early amniotes began in the late Carboniferous and early Permian periods. This time coincided with the adaptive radiation of terrestrial insects, the major prey of early amniotes. The adaptive radiation of the amniotes resulted in the lineages described below. One of the ways that these lineages are distinguished is by the structure of the skull, particularly the modifications in jaw muscle attachment (figure 17.3).

Reptiles in the subclass Anapsida (Gr. *an*, without + *hapsis*, arch) lack openings or fenestrae in the temporal (posterolateral) region of the skull. This lineage is represented today by the turtles. Recent evidence suggests that the anapsid lineage probably does not share close evolutionary ties to other reptiles. Changes have occurred in their long evolutionary history, but the fundamental form of their skull and shell is recognizable in 200-million-year-old fossils. Evidence of the anapsid lineage has been found in 245-million-year-old rocks from South Africa.

A second group of reptiles are diapsid (Gr. *di*, two). They have upper and lower openings in the temporal region of the skull. It is debated whether or not this condition reflects a single lineage (*see figure 17.2*). Some taxonomists prefer to divide this group into two subclasses. One subclass, the Lepidosauria, includes modern snakes, lizards, and tuataras. A second subclass, Archosauria, underwent extensive evolutionary radiation in the Mesozoic era and includes the dinosaurs. Most archosaurs are now extinct (box 17.1). 1 Living archosaurs include the crocodilians and the dinosaurs' closest living relatives—the birds.

Another group of reptiles are synapsid (Gr. *syn*, with). They possess a single dorsal opening in the temporal region of the skull. Although there are no living reptilian descendants of this group, they are important because a group of synapsids, called therapsids, gave rise to the mammals.

Figure 17.1
Class Reptilia. Members of the class Reptilia were the first vertebrates to possess amniotic eggs, which develop free from standing or flowing water. Numerous other adaptations have allowed members of this class to flourish on land. A Nile crocodile (*Crocodylus niloticus*) is shown here.

Table 17.1 Classification of Living Reptiles

Class Reptilia (rep-til′e-ah)
Skin dry, with epidermal scales; skull with one point of articulation with the vertebral column (occipital condyle); respiration via lungs; metanephric kidneys; internal fertilization; amniotic eggs.

Order Testudines (tes-tu′din-ez) or Chelonia (ki-lo′ne-ah)
Teeth absent in adults and replaced by a horny beak; body short and broad; shell consisting of a dorsal carapace and ventral plastron. Turtles.

Order Rhynchocephalia (rin′ko-se-fay′le-ah)
Contains very primitive, lizardlike reptiles; well-developed parietal eye. A single species, *Sphenodon punctatus*, survives in New Zealand. Tuataras.

Order Squamata (skwa-ma′tah)
Recognized by specific characteristics of the skull and jaws (temporal arch reduced or absent and quadrate movable or secondarily fixed); the most successful and diverse group of living reptiles. Snakes, lizards, worm lizards.

Order Crocodilia (krok′o-dil′e-ah)
Elongate, muscular, and laterally compressed; tongue not protrusible; complete ventricular septum. Crocodiles, alligators, caimans, gavials.

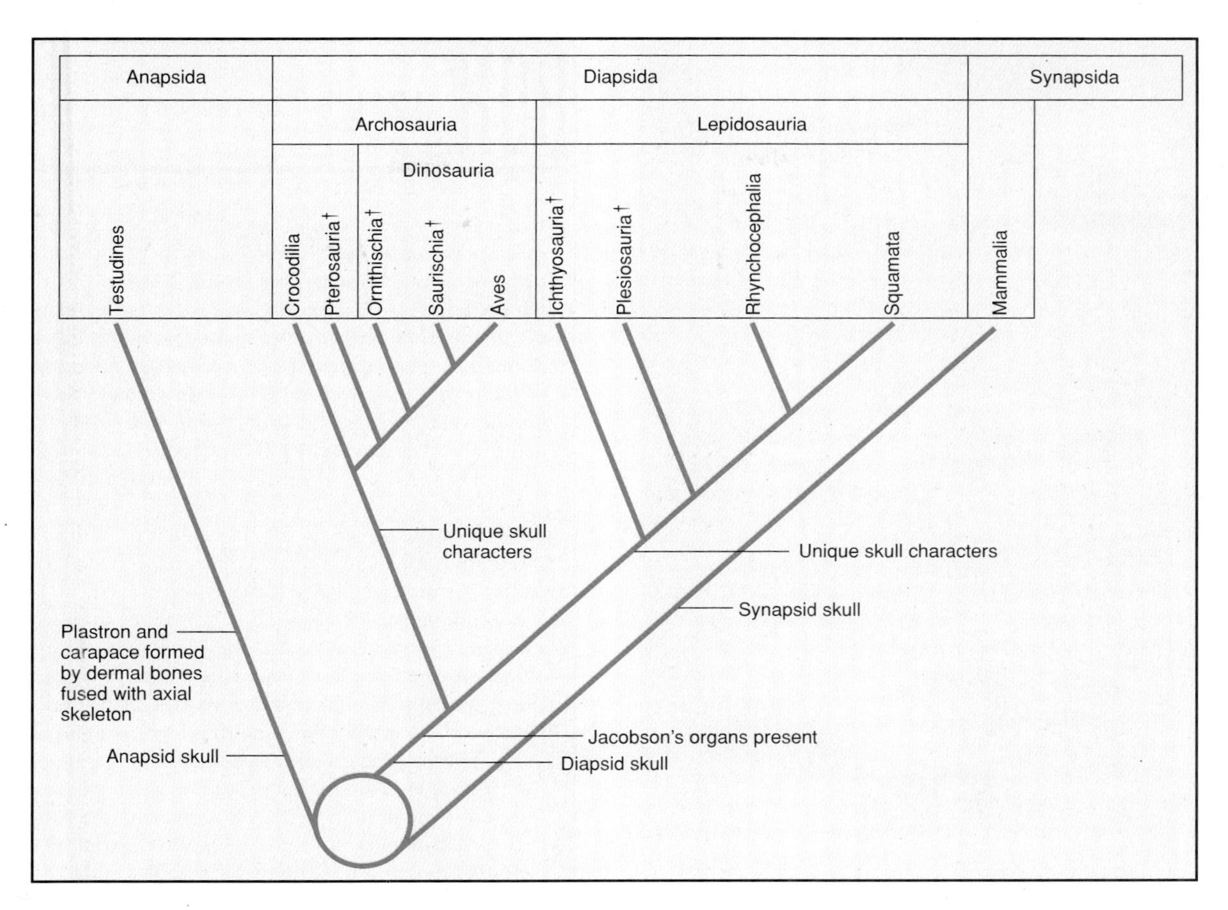

Figure 17.2

Amniote Phylogeny. This cladogram shows one interpretation of the phylogeny of the amniotes. The circle at the base of the cladogram indicates that the ancestral amniotes have not been described. Some researchers believe that the Diapsida is not a single lineage and that the Archosauria and Lepidosauria should be elevated to subclass status. Synapomorphies used to distinguish lower taxa are not shown. Some extinct taxa are indicated with a dagger (†). Numerous extinct other taxa are not shown.

Cladistic Interpretation of the Amniotic Lineage

Cladistic taxonomic methods have resulted in a reexamination and reinterpretation of the amniotic lineage. As shown in figure 16.3, the amniotic lineage is monophyletic. Figure 17.2 shows that the birds (traditionally the class Aves) and the mammals (traditionally the class Mammalia) share a common ancestor with the reptiles. The rules of cladistic analysis state that all descendants of a most recent common ancestor must be included in a particular taxon. Clearly that is not the case with the traditional class Reptilia—the birds and mammals are excluded even though they share a common ancestry with reptiles. According to cladistic interpretations, birds should be classified as "reptiles" with their closest relatives, the dinosaurs. Similarly, cladistic interpretations take into account the close relationships of the mammals and a group of ancient reptiles, the mammal-like reptiles.

Evolutionary systematists disagree with cladists' interpretations. They contend that both the birds and the mammals have important morphological, behavioral, and ecological characteristics (e.g., feathers and endothermy in the birds; hair, mammary glands, and endothermy in mammals) that make their assignment to separate classes warranted. In effect, evolutionary systematists weigh these characters and conclude that they are of overriding importance in the taxonomy of these groups.

The classification of the amniotes presented in this textbook follows the traditional interpretation. Students should realize, however, that the presentation of amniote systematics may change in future editions of this textbook since the disagreements between cladists and evolutionary systematists will probably continue.

Survey of the Reptiles

Reptiles are characterized by a skull with one surface (condyle) for articulation with the first neck vertebra, respiration by lungs, metanephric kidneys, internal fertilization, and amniotic eggs. Reptiles also have dry skin with keratinized

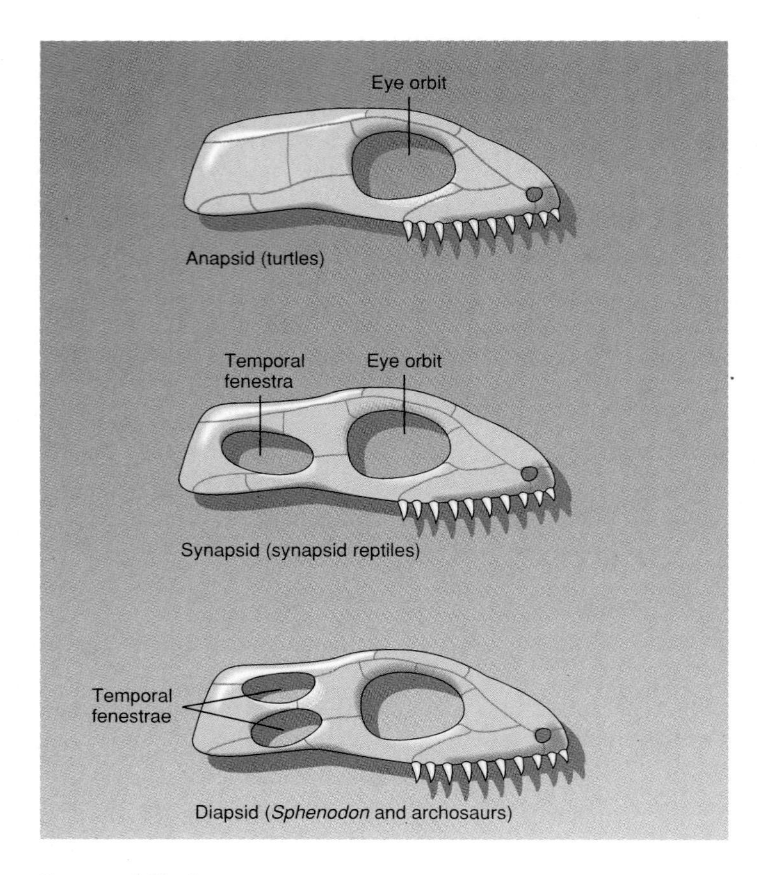

FIGURE 17.3

Amniote Skull Characteristics. Amniotes are classified based upon skull characteristics and the attachment of jaw muscles. (*a*) Anapsid skulls lack openings in the temporal region of the skull. This kind of skull is characteristic of turtles. (*b*) Synapsid skulls have a single temporal opening and are characteristic of the lineage of amniotes leading to mammals. (*c*) Diapsid skulls have two temporal openings. This kind of skull is characteristic of lizards, snakes, worm lizards, the tuatara, and birds.

epidermal scales. **Keratin** is a resistant protein found in epidermally derived structures of amniotes. It is protective, and when it is chemically bonded to phospholipids, it prevents water loss across body surfaces. Members of three of the four orders described on the following pages are found on all continents except Antarctica. However, reptiles are a dominant part of any major ecosystem only in tropical and subtropical environments. There are 17 orders of reptiles, but members of most orders are extinct. The four orders containing living representatives are described next (*see table 17.1*).

ORDER TESTUDINES OR CHELONIA

Members of the order Testudines (tes-tu′din-ez) (L. *tes-tudo*, tortise), or Chelonia (ki-lo′ne-ah) (Gr. *chelone*, tortise) are the turtles. There are about 225 species of turtles in the world, and they are characterized by a bony shell with limbs articulating

17.1 COLLISION OR COINCIDENCE?

Extinction is the eventual fate of all species. Indeed, 99% of all species ever on the earth are now extinct. Often, the rate of extinction is low; however, on several occasions in the history of the earth, rates of extinction have increased dramatically, resulting in the total extinction of many formerly successful taxa. For example, at the Permian/Triassic boundary, a mass extinction resulted in the loss of 80 to 90% of all animal species. About 65 million years ago, the dinosaurs, along with 50% of all other animal species became extinct. This Cretaceous/Tertiary extinction occurred over a period of tens of thousands to several million years. From a geological perspective, these extinctions are "sudden." How the age of the reptiles ended has been the subject of speculation for many years. Various hypotheses involving catastrophic or gradual change have been proposed.

The impact of asteroids or periods of intense volcanic activity are catastrophic events that have been suggested as causes of mass extinctions. Both events may have injected large quantities of dust into the air that shaded the earth's surface, reducing photosynthetic production, and thus food for animals.

The presence of the element iridium in rock strata from the Cretaceous/Tertiary boundary and other periods of mass extinction is the primary evidence supporting the catastrophic hypotheses. Iridium is a primarily extraterrestrial element that is deposited during asteroid impacts with the earth. Some deposits may also result from volcanic activity. Opponents of the catastrophic hypotheses do not deny that asteroid impacts or periods of volcanic activity resulted in iridium deposits on the earth. They question, however, whether the catastrophes are responsible for mass extinctions. They point out that the paleontological record indicates that extinctions are not as abrupt as implied in catastrophic hypotheses. Extinctions apparently occurred over tens of thousands of years, not tens of years. In addition, catastrophic events would be expected to affect all animal groups more or less equally, which was not the case with the Cretaceous/Tertiary extinction. For example, dinosaurs became extinct, but crocodiles, turtles, birds, and early mammals did not.

Numerous hypotheses propose gradual, selective changes as explanations for mass extinction. Some of these involve climatic changes that could have been induced by continental drift. In the Cretaceous period, 70% of the present land area was covered by warm, shallow seas. By the end of the Cretaceous period, these seas were reduced to 15% of the present land area, resulting in the reduction of habitat for shallow-water marine organisms, a decrease in atmospheric temperatures, and dissection of land areas by newly formed rivers. Climatic and habitat changes such as these could have resulted in extinctions over periods of tens of thousands of years.

Regardless of what hypothesis (or hypotheses) of mass extinction is correct, this question is a good example of how interest fueled by controversy stimulates scientific inquiry. The gradualism/catastrophism debate has led to new, innovative ideas on the origin, evolution, and extinction of taxa.

BOX 17.2 Bone and Scales

Bone is the primary skeletal tissue of vertebrates. In addition to making up the skeleton, it also is present in the scales of some vertebrates.

Developmentally, bone is derived from two sources. Dermal or membrane bone forms many of the superficial, flat bones of the top (dorsal) portion of the skull and some bones in the pectoral girdle. These bones were especially numerous in the roof of the skull of early vertebrates. Dermal bone is formed in the connective tissues of the dermis of the skin. During its formation, bone-forming cells called osteoblasts line up along connective-tissue fibers and begin depositing bone. Bony fibers coalesce into the latticework that makes up a flat bone.

Endochondral bone forms many of the long bones and the ventral and posterior bones of the skull. It develops by replacing the cartilage that formed early in development (figure 1*a*). The cartilage-based bone grows during development. As it does, osteoblasts lay down a collar of bone around the middle region, the diaphysis. Cartilage cells in the diaphysis begin to break down, beginning the formation of cavity called the marrow cavity. Eventually, this cavity will be filled with bone marrow, in which blood cells are formed. Bone formation proceeds toward the ends of a bone. The cartilage near the end of the bone, however, continues to grow and results in further bone elongation. Each end of the bone is an epiphysis. In mammals, a secondary ossification center occurs in each epiphysis and forms bony caps on the ends of the bone. However, a cartilaginous plate, the epiphyseal plate, remains between the epiphysis and the diaphysis and is the site of cartilage growth and bone elongation. Bone growth continues until maturity. Except for thin cartilages at the ends of bones that provide gliding surfaces for joints, all cartilage is replaced by bone at maturity, and growth stops.

The scales of fishes are composed, in part, of dermal bone. Osteoblasts in the dermis of the skin lay down a core of bone. Other dermal cells lay down a layer of dentine, which is similar to bone, around the bony core. Then, epidermal cells lay down a covering of enamel (figure 1*b*). Enamel is one of the hardest tissues in the vertebrate body and also occurs on teeth.

The scales of the skin of reptiles and the legs of birds are formed entirely in the epidermis of the skin and do not contain bone. These scales are composed of many layers of epidermal cells (figure 1*c*). Keratin and phospholipids are incorporated into the outer, horny layers of a scale to reduce water loss across the skin. As you will see in chapter 30, the feathers of birds are modified epidermal scales.

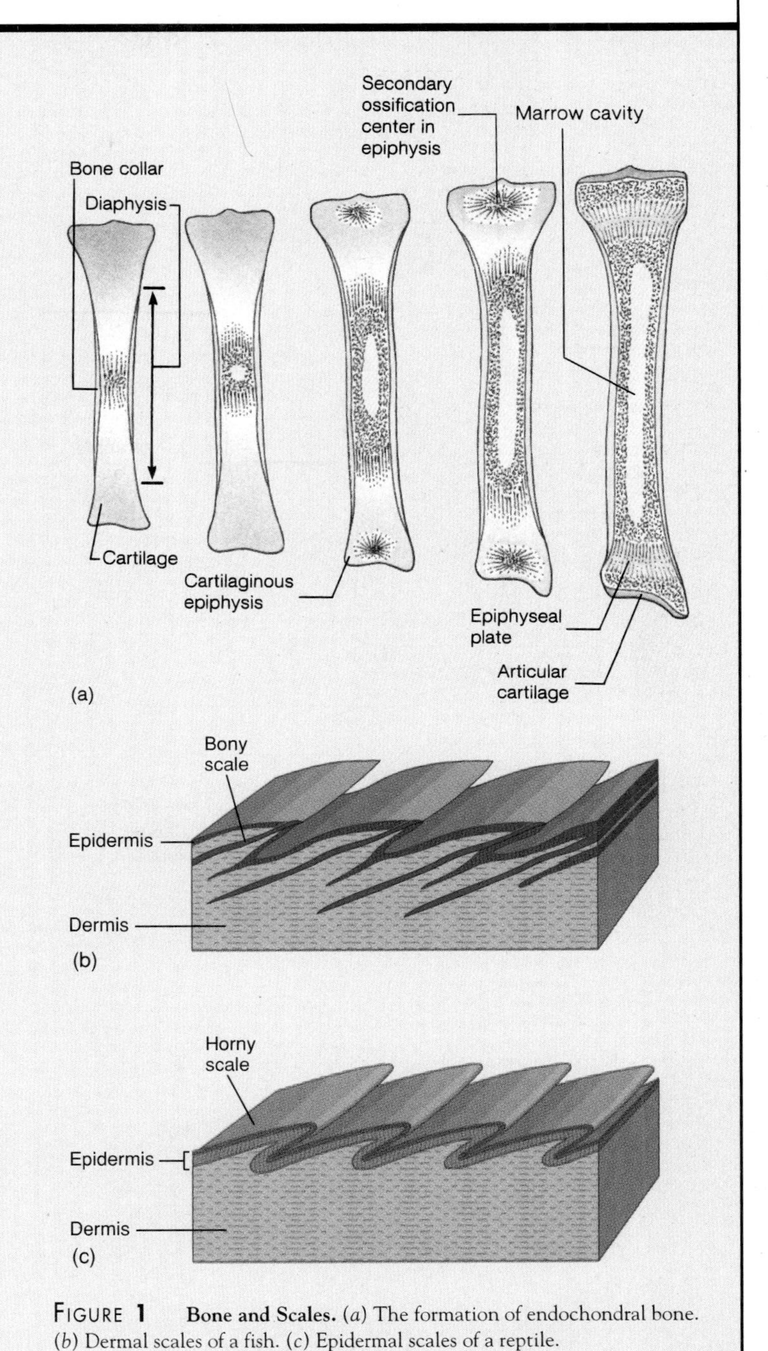

FIGURE 1 **Bone and Scales.** (*a*) The formation of endochondral bone. (*b*) Dermal scales of a fish. (*c*) Epidermal scales of a reptile.

internally to the ribs and a keratinized beak rather than teeth. The dorsal portion of the shell is the **carapace,** which is formed from a fusion of vertebrae, expanded ribs, and bones formed in the dermis of the skin (box 17.2). The bone of the carapace is covered by keratin. The ventral portion of the shell is the **plastron.** It is formed from bones of the pectoral girdle and dermal bone, and is also covered by keratin (figure 17.4). In some turtles, such as the North American box turtle (*Terrapene*), the shell has flexible areas, or hinges, that allow the anterior and posterior edges of the plastron to be raised. The hinge allows the shell openings to close when the turtle is withdrawn into the shell. Turtles have eight cervical vertebrae that can be articulated into an

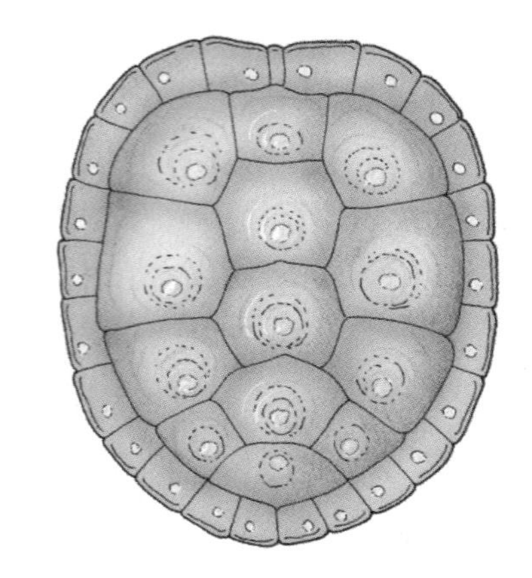

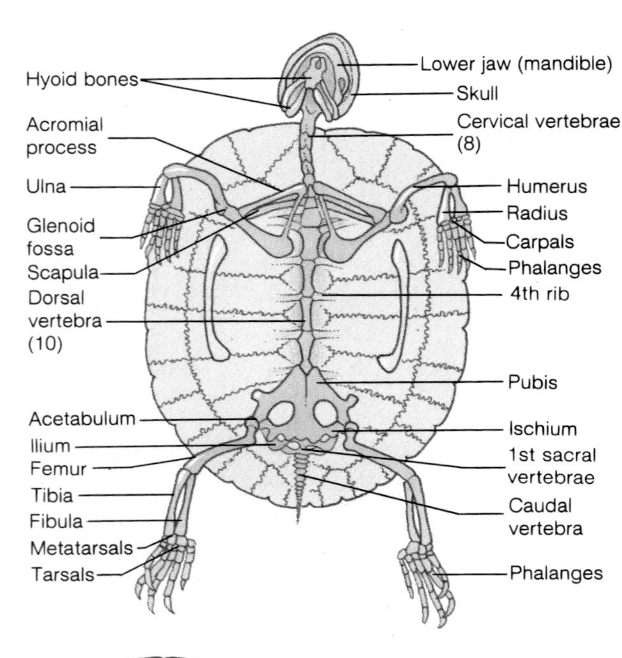

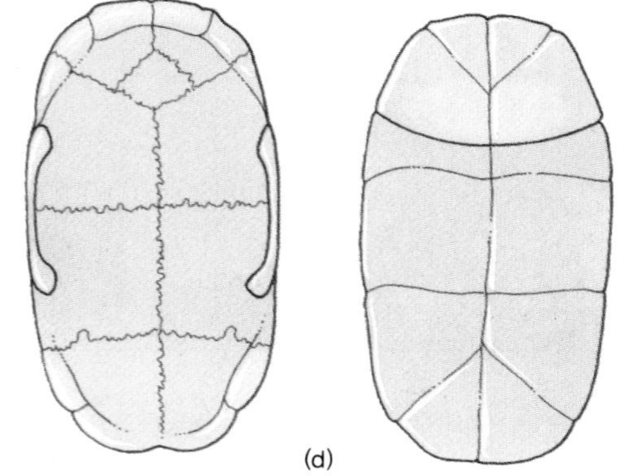

Figure 17.4
The Skeleton of a Turtle. (*a*) Dorsal view of the carapace. (*b*) Ventral view of the carapace and appendicular skeleton. The carapace is composed of fused vertebrae, expanded ribs, and dermal bone and is covered by keratin. (*c*) Dorsal view of the plastron. (*d*) Ventral view of the plastron. The plastron is formed from dermal bone and bone of the pectoral girdle. It is also covered by keratin.

"S-shaped" configuration, which allows the head to be drawn into the shell.

Turtles have long life spans. Most reach sexual maturity after 7 or 8 years and live 14 or more years. Large tortoises of the Galápagos Islands may live in excess of 100 years (*see figure 2.2*). All turtles are oviparous. Females use their hind limbs to excavate nests in the soil. Clutches of 5 to 100 eggs are laid and covered with soil. Development takes from 4 weeks to 1 year, and eggs are not attended by the parent during development. The young are independent of the parent at hatching.

In recent years, turtle conservation programs have been enacted. Slow rates of growth and long juvenile periods make turtles vulnerable to extinction in the face of high mortality rates. Turtle hunting and predation on young turtles and turtle nests by dogs and other animals has severely threatened some species. Predation on nests and young is made more serious by the fact that certain beaches are used year after year by nesting sea turtles. Conservation of sea turtles is complicated by the fact that they have ranges of thousands of square kilometers of ocean, so that protective areas must include waters under the jurisdiction of many different nations (figure 17.5).

Order Rhynchocephalia

The one surviving species of the order Rhynchocephalia (rin'ko-se-fay'le-ah) (Gr. *rhynchos*, snout + *kephale*, head) is the tuatara (*Sphenodon punctatus*) (figure 17.6). This superficially lizardlike reptile is virtually unchanged from extinct relatives that were present at the beginning of the Mesozoic era, nearly 200 million years ago. It is distinguished from other reptiles by tooth attachment and structure. Two rows of teeth on the upper jaw and a single row of teeth in the lower jaw produce a shearing bite that can decapitate a small bird. Formerly more widely distributed in New Zealand, the tuatara fell prey to human influences and domestic animals. It is now present only on remote offshore islands and is protected by New Zealand law. It is oviparous and shares underground burrows with ground-nesting seabirds. Tuataras venture out of their burrows at dusk and dawn to feed on insects or occasionally small vertebrates.

Order Squamata

The order Squamata (skwa-ma'tah) (L. *squama*, scale + *ata*, to bear) is divided into three suborders. Ancestral members of these suborders originated in the lepidosaur lineage about 150 million years ago and diverged into numerous modern forms.

Suborder Sauria—The Lizards

There are about 3,300 species of lizards in the suborder Sauria (sawr'e-ah) (Gr. *sauro*, lizard). In contrast to snakes, lizards usually have two pairs of legs. The few that are legless retain remnants of a pectoral girdle and sternum. Lizards vary in length

FIGURE 17.5
Order Testudines. Green sea turtles (*Chelonia mydas*) nest every 2 to 4 years and migrate many miles to nesting beaches in the Caribbean and South Atlantic Oceans.

FIGURE 17.7
Order Squamata. The gila monster (*Heloderma suspectum*) is a poisonous lizard of southwestern North America.

FIGURE 17.6
Order Rhynchocephalia. The tuatara (*Sphenodon punctatus*).

from only a few centimeters to as large as 3 m. Many lizards live on surface substrates and retreat under rocks or logs when necessary. Others are burrowers or tree dwellers. Most lizards are oviparous; some are ovoviviparous or viviparous. Eggs are usually deposited under rocks, debris, or in burrows.

Geckos, commonly found on the walls of human dwellings, are short and stout. They are nocturnal, and unlike most lizards, are capable of clicking vocalizations. Their large eyes, with pupils that contract to a narrow slit during the day and dilate widely at night, are adapted for night vision. Adhesive disks on their digits aid in clinging to trees and walls.

Iguanas have robust bodies, short necks, and distinct heads. This group includes the marine iguanas of the Galápagos Islands and the flying dragons (*Draco*) of Southeast Asia. The latter have lateral folds of skin that are supported by ribs. Like the ribs of an umbrella, the ribs of *Draco* can be expanded to form a gliding surface. When this lizard launches itself from a tree it can glide 30 m or more!

Another group of iguanas, the chameleons, is found mainly in Africa and India. They are adapted to arboreal lifestyles and use a long, sticky tongue to capture insects. *Anolis*, or the "pet-store chameleon," is also an iguanid, but is not a true chameleon. Chameleons and *Anolis* are well known for their ability to change color in response to illumination, temperature, or their behavioral state.

The only venomous lizards are the gila monster (*Heloderma suspectum*) (figure 17.7) and the Mexican beaded lizard (*Heloderma horridum*). These heavy-bodied lizards are native to southwestern North America. Venom is released into grooves on the surface of teeth and introduced into prey as the lizard chews. Lizard bites are seldom fatal to humans.

Suborder Serpentes—The Snakes

There are about 2,300 species in the suborder Serpentes (serpen'tez) (L. *serpere*, to crawl). Although the vast majority of snakes are not dangerous to humans, about 300 species are venomous. Worldwide, about 30,000 to 40,000 people die from snake bites each year. Most of these deaths are in Southeast Asia. In the United States, fewer than 100 people die each year from snake bites.

Snakes are elongate and lack limbs, although vestigial pelvic girdles and appendages are sometimes present. The skeleton may contain more than 200 vertebrae and pairs of ribs. Joints between vertebrae make the body very flexible. Snakes possess skull adaptations that facilitate swallowing large prey. Other differences between lizards and snakes include the mechanism for focusing the eyes and the morphology of the retina. Elongation and narrowing of the body has

Figure 17.8
Order Squamata. An amphisbaenian "worm lizard" (*Amphisbaenia alba*), sometimes called a two-headed snake.

resulted in the reduction or loss of the left lung and displacement of the gallbladder, the right kidney, and often the gonads. Most snakes are oviparous, although a few, such as the New World boas, give birth to live young.

The evolutionary origin of the snakes is debated. The earliest fossils are from 135-million-year-old Cretaceous deposits. Some zoologists believe that the earliest snakes were burrowers. Loss of appendages and changes in eye structure could be adaptations similar to those seen in caecilians (*see figure 16.6*). The loss of legs could also be adaptive if early snakes were aquatic or lived where densely tangled vegetation was common.

Suborder Amphisbaenia—Worm Lizards

There are about 135 species in the suborder Amphisbaenia (am′fis-be′ne-ah) (Gr. *amphi*, double + *baen*, to walk). They are specialized burrowers that live in soils of Africa, South America, the Caribbean, and the Mideast (figure 17.8). Most are legless, and their skulls are wedge or shovel shaped. They are distinguished from all other vertebrates by the presence of a single median tooth in the upper jaw. The skin of amphisbaenians has ringlike folds called annuli and is loosely attached to the body wall. Muscles of the skin cause it to telescope and bulge outward, forming an anchor against a burrow wall. 3 Amphisbaenians move easily forward or backward, thus the suborder name. They feed on worms and small insects and are oviparous.

Order Crocodilia

There are 21 species in the order Crocodilia (krok′o-dil′e-ah) (Gr. *krokodeilos*, lizard). Along with dinosaurs, crocodilians are derived from the archosaurs and distinguished from other reptiles by certain skull characteristics: openings in the skull in front of the eye, triangular rather than circular eye orbits, and laterally compressed teeth. Living crocodilians include the alligators, crocodiles, gavials, and caimans.

Crocodilians have not changed much over their 170-million-year history. The snout is elongate and often used to capture food by a sideways sweep of the head. The nostrils are at the tips of the snout, so the animal can breathe while mostly submerged. Air passageways of the head lead to the rear of the mouth and throat, and a flap of tissue near the back of the tongue forms a watertight seal that allows breathing without inhaling water in the mouth. A plate of bone, called the secondary palate, evolved in the archosaurs and separates the nasal and mouth passageways. The tail is muscular, elongate, and laterally compressed. It is used to swim, in offensive and defensive maneuvers, and to attack prey. Teeth are used only for seizing prey. Food is swallowed whole, but if a prey item is too large, crocodilians tear apart prey by holding onto a limb and rotating their bodies wildly until the prey is dismembered. The stomach is gizzardlike, and crocodilians swallow rocks and other objects as abrasives for breaking apart ingested food. Crocodilians are oviparous and display parental care of hatchlings that parallels that of birds. Nesting behavior and parental care may be traced back to the common ancestor of both groups.

Stop and Ask Yourself

1. What are the evolutionary lineages that diverged from the earliest amniotes?
2. What order of reptiles has members that lack teeth, have ribs and vertebrae incorporated into a bony shell, and are oviparous?
3. What adaptations for burrowing are shown by worm lizards?
4. What is a secondary palate? Why is it adaptive for crocodilians?

Evolutionary Pressures

The life-styles of most reptiles reveal striking adaptations for terrestrialism. To appreciate this, one might consider a lizard common to deserts of southwestern United States, the chuckwalla (*Sauromalus obesus*) (figure 17.9). Chuckwallas survive during late summer when temperatures exceed 40° C (104° F) and when arid conditions result in the withering of plants and blossoms upon which chuckwallas browse. To withstand these hot and dry conditions, chuckwallas disappear below ground and aestivate. Temperatures moderate during the winter, but little rain falls, so life on the desert surface is still not possible for the chuckwalla. The summer's sleep, therefore, merges into a winter's sleep. The chuckwalla will not emerge until March when rain falls, and the desert explodes with greenery and flowers. The chuckwalla browses and drinks, storing water in large

Figure 17.9
The Chuckwalla (*Sauromalus obesus*). Many reptiles, like this chuckwalla, possess adaptations that make life apart from standing or running water possible.

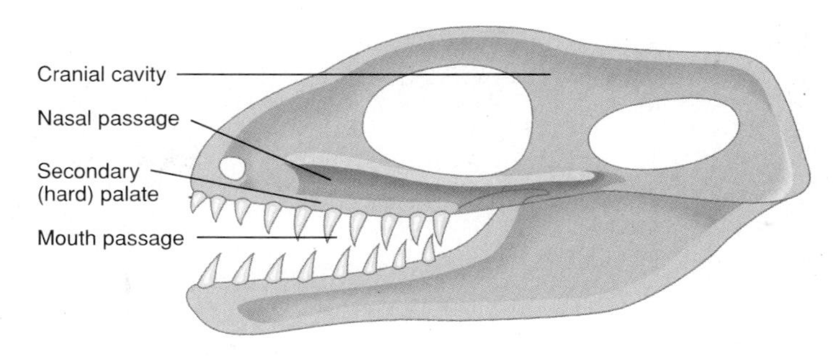

Figure 17.10
The Secondary Palate. A sagittal section of the skull of a synapsid reptile showing the secondary palate that separates the nasal and mouth cavities. Extension of the bones of the anterior skull forms the anterior portion of the secondary palate (the hard palate), and skin and soft connective tissues form the posterior portion of the secondary palate (the soft palate).

reservoirs under its skin. Chuckwallas are not easy prey. If threatened, the nearest rock crevice becomes a chuckwalla's refuge. The chuckwalla inflates its lungs with air, increasing its girth and wedging itself against the rock walls of its refuge. Friction of its body scales against the rocks makes it nearly impossible to dislodge.

Adaptations displayed by chuckwallas are not exceptional for reptiles. This section discusses some adaptations that make life apart from an abundant water supply possible.

External Structure and Locomotion

Unlike that of amphibians, the skin of reptiles has no respiratory functions. Their skin is thick, dry, and keratinized. Scales may be modified for various functions. The large belly scales of snakes provide contact with the substrate during locomotion. Although the skin of reptiles is much less glandular than that of amphibians, secretions include pheromones that function in sex recognition and defense.

All reptiles periodically shed the outer, epidermal layers of the skin in a process called **ecdysis**. (The term ecdysis is also used for a similar, though unrelated, process in arthropods [*see figure 11.5*].) Because the blood supply to the skin does not extend into the epidermis, the outer epidermal cells lose contact with the blood supply and die. Movement of lymph between the inner and outer epidermal layers loosens the outer epidermis. Ecdysis is generally initiated in the head region, and in snakes and many lizards, the epidermal layers come off in one piece. In other lizards, smaller pieces of skin flake off. The frequency of ecdysis varies from one species to another, and it is greater in juveniles than adults.

The chromatophores of reptiles are primarily dermal in origin and function much like those of amphibians. Cryptic coloration, mimicry, and **aposematic** coloration occur in reptiles. Color and color change also function in sex recognition and thermoregulation.

Support and Movement

There are modifications in the skeletons of snakes, amphisbaenians, and turtles; however, in its general form, the skeleton of reptiles is based on one inherited from ancient amphibians. The skeleton is highly ossified to provide greater support. The skull is longer than that of amphibians, and a plate of bone, the **secondary palate,** partially separates the nasal passages from the mouth cavity (figure 17.10). As described earlier, the secondary palate evolved in archosaurs, where it was an adaptation for breathing when the mouth was full of water or food. It is also present in other reptiles, although developed to a lesser extent. Longer snouts also permit greater development of olfactory epithelium and increased reliance on the sense of smell.

Reptiles have more cervical vertebrae than do amphibians. The first two cervical vertebrae (**atlas and axis**) provide greater freedom of movement for the head. An **atlas** articulates with a single **condyle** on the skull and facilitates nodding. An **axis** is modified for rotational movements. The atlas and axis are followed by a variable number of cervical vertebrae that provide additional flexibility for the neck.

The ribs of reptiles may be highly modified. Those of turtles and the flying dragon were described previously. The ribs of snakes have muscular connections to large belly scales to aid locomotion. The cervical vertebrae of cobras have ribs that may be flared in aggressive displays.

Two or more **sacral vertebrae** attach the pelvic girdle to the vertebral column. The caudal vertebrae of many lizards possess a vertical fracture plane. 4 When a lizard is grasped by the tail, caudal vertebrae can be broken, and a portion of the tail is lost. Tail loss, or **autotomy,** is an adaptation that allows a lizard to escape from a predator's grasp, or the disconnected, wiggling piece of tail may distract a predator from the lizard. The lost portion of the tail is later regenerated.

Locomotion in primitive reptiles is similar to that of salamanders. The body is slung low between paired, stocky

Figure 17.11
Order Squamata. A chameleon (*Chameleo chameleon*) using its tongue to capture prey. Note the prehensile tail.

appendages, which extend laterally and move in the horizontal plane. The limbs of other reptiles are more elongate and slender. Limbs are held closer to the body, the knee and elbow joints are rotated posteriorly, so the body is thus higher off the ground and weight is supported vertically. Many prehistoric reptiles were bipedal, meaning that they walked on hind limbs. They had a narrow pelvis and a heavy outstretched tail for balance. Bipedal locomotion freed the front appendages, which became adapted for prey capture or flight in some animals.

Nutrition and the Digestive System

Most reptiles are carnivores, although turtles will eat almost anything organic. The tongues of turtles and crocodilians are nonprotrusible and aid in swallowing. 5 The sticky tongues of some lizards and the tuatara are used to capture prey as do some anurans. The extension of the tongue of chameleons exceeds their body length (figure 17.11).

Probably the most remarkable adaptations of snakes involve modifications of the skull for feeding. The bones of the skull and jaws are loosely joined to each other and may spread apart to ingest prey much larger than a snake's normal head size (figure 17.12*a*). The bones of the upper jaw are movable on the skull, and the halves of both of the upper and lower jaws are loosely joined anteriorly by ligaments. Therefore, each half of the upper and lower jaws can be moved independently of one another. After a prey is captured, opposite sides of the upper and lower jaws are alternately thrust forward and retracted. Posteriorly pointing teeth prevent the escape of the prey and help force the food into the esophagus. The glottis, the respiratory opening, is far forward so that the snake can breathe while slowly swallowing its prey.

Vipers (family Viperidae) possess hollow fangs on the maxillary bone at the anterior margin of the upper jaw (figure 17.12*b*). These fangs connect to venom glands that inject venom when the viper bites. The maxillary bone (upper jaw bone) of vipers is hinged so that when the snake's mouth is closed, the fangs fold back and lie along the upper jaw. When the mouth opens, the maxillary bone rotates and causes the fangs to swing down (figure 17.12*c*). Because the fangs project outward from the mouth, vipers may strike at objects of any size. Rear-fanged snakes (family Colubridae) possess grooved rear teeth. In those that are venomous, venom is channeled along these grooves and worked into prey to quiet them during swallowing. These snakes usually do not strike and most are harmless; however, the African boomslang (*Dispholidus typus*) has caused human fatalities. Coral snakes, sea snakes, and cobras have fangs that are rigidly attached to the upper jaw in an erect position. When the mouth is closed, the fangs fit into a pocket in the outer gum of the lower jaw. Fangs are grooved or hollow, and venom is injected by contraction of muscles associated with venom glands. Some cobras can "spit" venom at their prey and, if not washed from the eyes, may cause blindness.

Venom glands are modified salivary glands. Most snake venoms are mixtures of neurotoxins and hemotoxins. The venoms of coral snakes, cobras, and sea snakes are primarily neurotoxins that attack nerve centers and cause respiratory paralysis. The venoms of vipers are primarily hemotoxins. They break up blood cells and attack blood vessel linings.

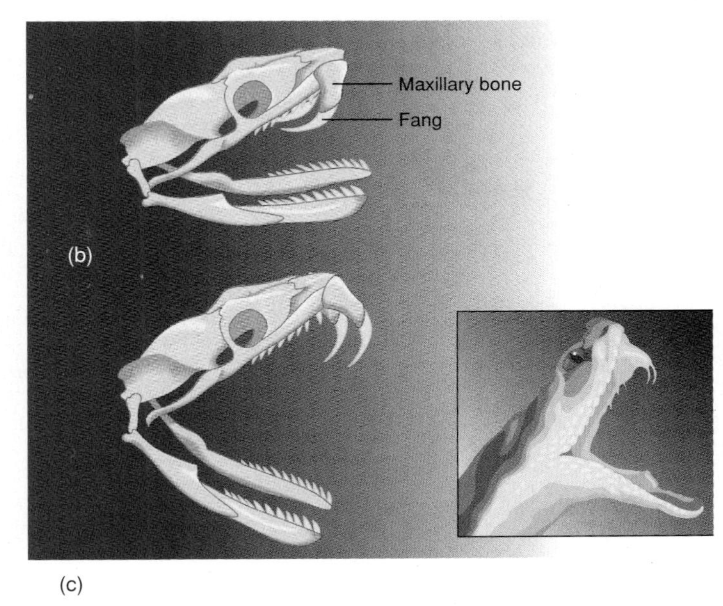

(a) (c)

Figure 17.12

Feeding Adaptations of Snakes. (*a*) A copperhead (*Ankistrodon*) ingesting a prey. The bones of the skull are joined by flexible joints that allow them to separate during feeding. Note the pit organ located just anterior to the eye. (*b*) The skull of a viper. The hinge mechanism of the jaw allows upper and lower bones on one side of the jaw to slide forward and backward alternately with bones of the other side. Posteriorly curved teeth hold prey as it is worked toward the esophagus. (c) Note that the maxillary bone, into which the fang is embedded, swings forward when the mouth is opened.

Circulation, Gas Exchange, and Temperature Regulation

The circulatory system of reptiles is based on that of amphibians. Because reptiles are, on average, larger than amphibians, their blood must travel under higher pressures to reach distant body parts. To take an extreme example, the blood of *Brachiosaurus* had to be pumped a distance of about 6 m from the heart to the head—mostly uphill! (The blood pressure of a giraffe is about double that of a human to move blood the 2 m from heart to head.)

Like amphibians, reptiles possess two atria that are completely separated in the adult and have veins from the body and lungs emptying into them. Except for turtles, the sinus venosus is no longer a chamber but has become a patch of cells that acts as a pacemaker. The ventricle of most reptiles is incompletely divided (figure 17.13). (Only in crocodilians is the ventricular septum complete.) The ventral aorta and the conus arteriosus divide during development and become three major arteries that leave the heart. A pulmonary artery leaves the ventral side of the ventricle and takes blood to the lungs. Two systemic arteries, one from the ventral side of the heart and the other from the dorsal side of the heart, take blood to the lower body and the head.

Blood low in oxygen enters the ventricle from the right atrium and leaves the heart through the pulmonary artery and moves to the lungs. Blood high in oxygen enters the ventricle from the lungs via pulmonary veins and the left atrium, and leaves the heart through left and right systemic arteries.

6 The incomplete separation of the ventricle permits shunting of some blood away from the pulmonary circuit to the systemic circuit by constriction of muscles associated with the pulmonary artery. This is advantageous because virtually all reptiles breathe intermittently. When turtles are withdrawn into their shell, their method of lung ventilation cannot function. They also stop breathing during diving. During periods of apnea ("no breathing"), the flow of blood to the lungs is limited, conserving energy and permitting more efficient use of the pulmonary oxygen supply.

Gas Exchange

Reptiles exchange respiratory gases across internal respiratory surfaces to prevent the loss of large quantities of water. A larynx is present; however, vocal cords are usually absent. The respiratory passages of reptiles are supported by cartilages, and lungs are partitioned into spongelike, interconnected chambers. Lung chambers provide a large surface area for gas exchange.

Lung ventilation occurs in most reptiles by a negative pressure mechanism. Expanding the body cavity by a posterior movement of the ribs and the body wall decreases pressure in the lungs and draws air into the lungs. Air is expelled by elastic recoil of the lungs and forward movements of the ribs and body wall, which compress the lungs. The ribs of turtles are a part of their shell; thus, movements of the body wall to which they are attached are impossible. Turtles exhale by contracting muscles that force the viscera upward, compressing the lungs.

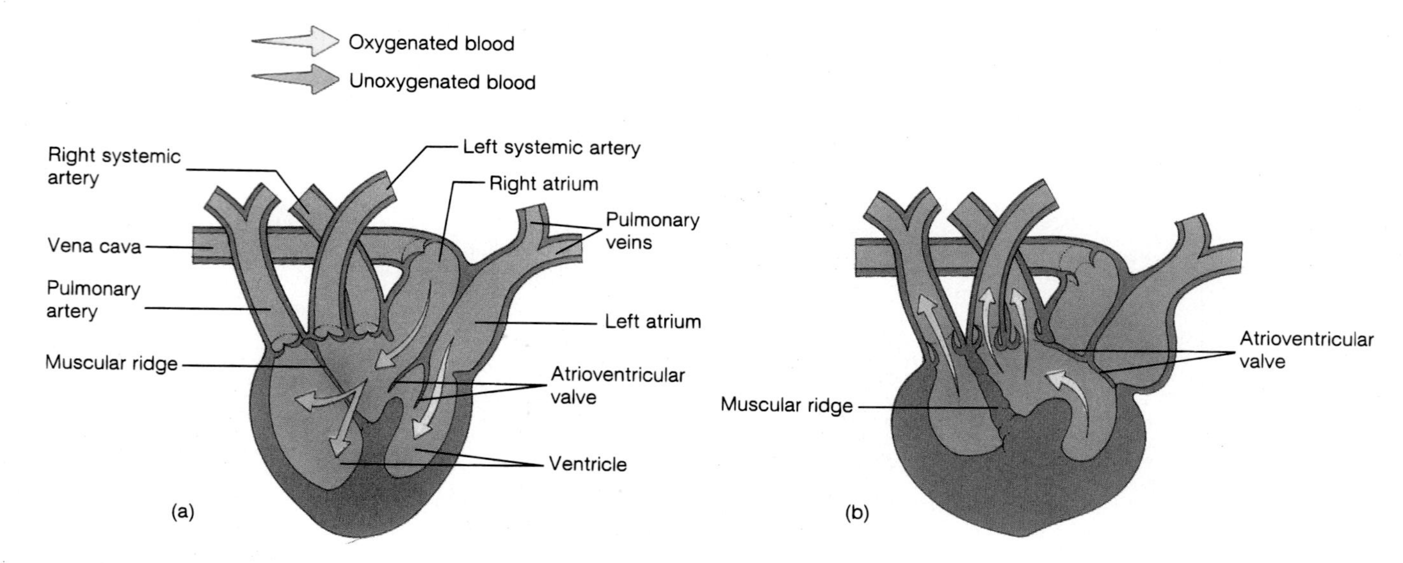

FIGURE 17.13

The Heart and Major Arteries of a Lizard. (*a*) When the atria contract, blood enters the ventricle. An atrioventricular valve prevents the mixing of oxygenated and unoxygenated blood across the incompletely separated ventricle. (*b*) When the ventricle contracts, a muscular ridge closes to direct oxygenated blood to the systemic arteries and unoxygenated blood to the pulmonary artery.

They inhale by contracting muscles that increase the volume of the visceral cavity, creating negative pressure to draw air into the lungs.

Temperature Regulation

Unlike aquatic animals, terrestrial animals may be faced with temperature extremes (–65 to 70° C) that are incompatible with life. Temperature regulation, therefore, is important for animals that spend their entire lives out of water. Most reptiles use external heat sources for thermoregulation and are, therefore, ectotherms. Exceptions include monitor lizards and brooding Indian pythons. Female pythons coil around their eggs and elevate their body temperature as much as 7.3° C above the air temperature using metabolic heat sources (box 17.3).

Some reptiles can survive wide temperature fluctuations (e.g., –2 to 41° C for some turtles). To sustain activity, however, body temperatures are regulated within a narrow range, between 25 and 37° C. If that is not possible, the reptile usually seeks a retreat where body temperatures are likely to remain within the range compatible with life.

Most thermoregulatory activities of reptiles are behavioral, and they are best known in the lizards. To warm itself, a lizard may orient itself at right angles to the sun's rays, often on a surface inclined toward the sun, and may press its body tightly to a warm surface to absorb heat by conduction. To cool itself, a lizard will orient its body parallel to the sun's rays, seek shade or burrows, or assume an erect posture (legs extended and tail arched) to reduce conduction from warm surfaces. In hot climates, many reptiles are nocturnal.

Various physiological mechanisms are also used to regulate body temperature. As temperatures rise, some reptiles begin panting, which releases heat through evaporative cooling. (Because their skin is dry, little evaporative cooling occurs across the skin of reptiles.) Marine iguanas divert blood to the skin while basking in the sun and warm up quickly. On diving into the cool ocean, however, heart rate and blood flow to the skin are reduced, which slow heat loss. Chromatophores also aid in temperature regulation. Dispersed chromatophores (thus a darker body) increase the rate of heat absorption.

In temperate regions, many reptiles withstand cold temperatures of winter by entering torpor when body temperatures and metabolic rates decrease. Individuals that are usually solitary may migrate to a common site to spend the winter. Heat loss from these groups, called hibernacula, is reduced because of a reduction in total surface area of many individuals clumped together compared to widely separated animals. Unlike true hibernators, body temperatures of reptiles in torpor are not regulated, and if the winter is too cold or the retreat is too exposed, the animals can freeze and die. Cold death is an important source of mortality for temperate reptiles.

NERVOUS AND SENSORY FUNCTIONS

The brain of reptiles is similar to the brains of other vertebrates. The cerebral hemispheres are somewhat enlarged compared to those of amphibians. This increased size is associated with an improved sense of smell. The optic lobes and the cerebellum are also enlarged, which reflects increased reliance on vision and more refined coordination of muscle functions.

BOX 17.3 Changing Perceptions of Ancient Life-Styles

In the years prior to 1960, paleontologists worked on assembling ancient lineages from fossil remains based on gross anatomical features. Now, it is possible to derive information about the life-styles of extinct animals from markings of blood vessels, muscles, and tendons present on fossils. In the 1970s, Robert T. Bakker began challenging previous perceptions of dinosaurs as lumbering giants and overgrown lizards. Bakker contended that that view of dinosaurs is difficult to reconcile with fossils, indicating that many dinosaurs were bipedal, or at least held their bodies well off the ground. Reptiles achieved many locomotor advancements long before similar advancements ever appeared in mammals. Unlike modern reptiles, dinosaurs had highly vascular bones, a condition typical of endothermic animals. Finally, many believe it unlikely that very large, strictly ectothermic animals could warm up rapidly enough to maintain high activity levels. Bakker proposed that many, if not all dinosaurs were endothermic.

Other scientists discount ideas that dinosaurs were endothermic. They maintain that dinosaurs may have been able to maintain active life-styles, not in spite of their large size, but because of it. As ectotherms increase in size, they resemble endotherms. Large body mass results in slower warming, but it also results in slower heat loss. Further, the climate during the Mesozoic era was considerably warmer than it is now. Even though dinosaurs may have been ectotherms, they could have had stable body temperatures and an active life-style. Opponents of the endothermy hypothesis point out that some ectotherms have highly vascular bones. They also note that large endotherms usually have obvious cooling devices (e.g., the ears of an elephant), and these were not found on most large dinosaurs.

There are no clear winners in this debate. There may not have been a single thermoregulatory strategy for the dinosaurs. The largest dinosaurs could have been ectothermic and, although not swift and agile, they would still have been able to find food and avoid predators. On the other hand, maintenance of speed and agility in smaller dinosaurs that lived in cool climates may have required endothermy.

The complexity of reptilian sensory systems is evidenced by a chameleon's method of feeding. Its protruding eyes that swivel independently each have a different field of view. Initially, the brain keeps both images separate, but when an insect is spotted, both eyes converge on the prey. Binocular vision then provides the depth perception used to determine whether or not the insect is within range of the chameleon's tongue (*see figure 17.11*).

Vision is the dominant sense in most reptiles, and their eyes are similar to those of amphibians (*see figure 16.14*). Snakes focus on nearby objects by moving the lens forward. Contraction of the iris places pressure on the gellike vitreous body in the posterior region of the eye, and displacement of this gel pushes the lens forward. In all other reptiles, focusing on nearby objects occurs when the normally elliptical lens is made more spherical, as a result of ciliary muscles pressing the ciliary body against the lens. Reptiles have a greater number of cones than do amphibians and probably have well-developed color vision.

Upper and lower eyelids, a nictitating membrane, and a blood sinus protect and cleanse the surface of the eye. In snakes and some lizards, the upper and lower eyelids become fused in the embryo to form a protective window of clear skin, called the **spectacle**. (During ecdysis, the outer layers of the spectacle become clouded and impair the vision of snakes.) The blood sinus, which is at the base of the nictitating membrane, swells with blood to help force debris to the corner of the eye, where it may be rubbed out. Horned lizards squirt blood from their eyes by rupturing this sinus in a defensive maneuver to startle predators.

7 Some reptiles possess a **median (parietal) eye** that develops from outgrowths of the roof of the forebrain. In the tuatara, it is an eye with a lens, a nerve, and a retina. In other reptiles, the parietal eye is less developed. Parietal eyes are covered by skin and probably cannot form images. They can, however, differentiate light and dark periods and are used in orientation to the sun.

There is variation in the structure of reptilian ears. The ears of snakes detect substrate vibrations. They lack a middle-ear cavity, a eustachian tube, and a tympanic membrane. A bone of the jaw articulates with the stapes and receives substrate vibrations. Snakes can also detect airborne vibrations. In other reptiles, a tympanic membrane may be on the surface or in a small depression in the head. The inner ear of reptiles is similar to that of amphibians.

Olfactory senses are better developed in reptiles than in amphibians. In addition to the partial secondary palate providing more surface for olfactory epithelium, many reptiles possess blind-ending pouches that open through the secondary palate into the mouth cavity. These pouches, called **Jacobson's (vomeronasal) organs,** are found in diapsid reptiles, however, they are best developed in the squamates. Jacobson's organs develop in embryonic crocodilians but are not present in adults of this group. Anapsids (turtles) lack these olfactory organs. The protrusible, forked tongues of snakes and lizards are accessory olfactory organs that are used to sample airborne chemicals. A snake's tongue is flicked out and then moved to the Jacobson's organs where odor molecules are perceived. In the tuatara, Jacobson's organs are used to taste objects held in the mouth.

Rattlesnakes and other pit vipers have heat-sensitive **pit organs** on each side of the face between the eye and nostril (*see figure 17.12*a). These depressions are lined with sensory epithelium and are used to detect objects with temperatures different from the snake's surroundings. Pit vipers are usually nocturnal, and their pits are used to locate small, warm-blooded prey.

Stop and Ask Yourself

5. How is the skin of a reptile adapted for preventing water loss?
6. What is a secondary palate? How is a secondary palate adaptive for reptiles?
7. How are the tongues of reptiles used in feeding? In sensory perception?
8. How is the heart of a reptile adapted for shunting blood away from the pulmonary circulation? Under what circumstances does this shunting occur?

Excretion and Osmoregulation

The kidneys of embryonic reptiles are similar to those of fishes and amphibians. Life on land, increased body size, and higher metabolic rates require kidneys capable of processing wastes with little water loss. The embryonic kidney is replaced during development by a kidney with many more blood-filtering units, called nephrons. The functional kidneys of adult reptiles are called metanephric kidneys. Their function depends on a circulatory system that delivers more blood at greater pressures to filter large quantities of blood.

Uric acid is the principal excretory product of most reptiles. It is nontoxic, and being relatively insoluble in water, it will precipitate in the excretory system. Water is reabsorbed by the urinary bladder or the cloacal walls, and the uric acid can be stored in a pastelike form. Utilization of uric acid as an excretory product also made possible the development of embryos in terrestrial environments because nontoxic uric acid can be concentrated in egg membranes.

In addition to water reabsorption by the excretory system, internal respiratory surfaces and relatively impermeable exposed surfaces reduce evaporative water loss. The behaviors that help regulate temperature also help conserve water. Nocturnal habits and avoiding hot surface temperatures during the day by burrowing reduce water loss. When water is available, many reptiles (e.g., chuckwallas) store large quantities of water in lymphatic spaces under the skin or in the urinary bladder. Many lizards possess salt glands below the eyes that are used to rid the body of excess salt.

Reproduction and Development

Vertebrates could never be truly terrestrial until their reproduction and embryonic development became separate from standing or running water. For vertebrates, internal fertilization and the amniotic egg made complete movement to land possible. The amniotic egg, however, is not completely independent of water. Pores in the eggshell that permit gas exchange also allow water to evaporate. Amniotic eggs require significant energy expenditures by parents. Parental care occurs in some reptiles and may involve maintaining relatively high humidity around the eggs. These eggs are often supplied with large quantities of yolk for long developmental periods, and parental energy and time is sometimes invested in posthatching care of dependent young.

Accompanying the development of amniotic eggs is the necessity for internal fertilization. Fertilization must occur in the reproductive tract of the female before protective egg membranes are laid down around an egg. All male reptiles, except tuataras, possess an intromittent organ for introducing sperm into the reproductive tract of a female. Lizards and snakes possess paired hemipenes that are located at the base of the tail and are erected by being turned inside out, like the finger of a glove.

Gonads lie in the abdominal cavity. In males, a pair of ducts delivers sperm to the cloaca. After copulation, sperm may be stored in a seminal receptacle in the reproductive tract of the female. Secretions of the seminal receptacle nourish and arrest the activity of the sperm. Sperm may be stored for up to 4 years in some turtles, and up to 6 years in some snakes! In temperate latitudes, sperm can be stored over winter. Copulation may take place in the fall when individuals congregate in hibernacula, and fertilization and development occur in the spring when temperatures favor successful development. Fertilization occurs in the upper regions of the oviduct, which leads from the ovary to the cloaca. Glandular regions of the oviduct are responsible for secreting albumen and the eggshell. The shell is usually tough yet flexible. In some crocodilians, the eggshell is calcareous and rigid, like the eggshells of birds.

Parthenogenesis has been described in six families of lizards and one species of snakes. In these species, no males have been found. Populations of parthenogenetic females have higher reproductive potential than bisexual populations. A population that suffers high mortality over a cold winter can repopulate its habitat rapidly because all surviving individuals can produce offspring. This apparently offsets disadvantages of genetic uniformity resulting from parthenogenesis.

Reptiles often have complex reproductive behaviors that may involve males actively seeking out females. As in other animals, courtship functions in sexual recognition and behavioral and physiological preparation for reproduction. Head-bobbing displays by some male lizards reveal bright patches of color on the throat and enlarged folds of skin. Courtship in snakes is based primarily on tactile stimulation. Tail-waving displays are followed by the male running his chin along the female, entwining his body around her, and creating wavelike contractions that pass posteriorly to anteriorly along his body. Recent research indicates that sex pheromones are also used by lizards and snakes. Vocalizations are important only in crocodilians. During the breeding season, males are hostile and may bark or cough as territorial warnings to other males. Roaring vocalizations also attract females, and mating occurs in the water.

After they are laid, reptilian eggs are usually abandoned (figure 17.14). Virtually all turtles bury their eggs in the ground or in plant debris. Other reptiles lay their eggs under rocks, in debris, or in burrows. About 100 species of reptiles have some degree of parental care of eggs. One example is the American alligator, *Alligator mississippiensis* (figure 17.15). The female builds

FIGURE 17.14
Reptile Eggs and Young. These fence lizards (*Sceloporus undulatus*) are hatching from their leathery eggs.

a mound of mud and vegetation about 1 m high and 2 m in diameter. The center of the mound is hollowed out and filled with mud and debris. Eggs are deposited in a cavity in the center of the mound and covered. The female remains in the vicinity of the nest throughout development to protect the eggs from predation. She frees hatchlings from the nest in response to their high-pitched calls and picks them up in her mouth to transport them to water. She may scoop shallow pools for the young and remain with them for up to 2 years. Young feed on scraps of food dropped by the female when she feeds and on small vertebrates and invertebrates that they catch on their own.

FURTHER PHYLOGENETIC CONSIDERATIONS

The archosaur and synapsid lineages of ancient reptiles diverged from ancient amniotes about 280 million years ago and are ancestral to animals described in the next two chapters (*see figure 17.2*). The archosaur lineage not only included the dinosaurs and gave rise to crocodilians, but also gave rise to two groups of fliers. The pterosaurs (Gr. *pteros,* wing + *sauros*, lizard) ranged from sparrow size to animals with wing spans of 13 m. Their membranous wings were supported by an elongation of the fourth finger, their sternum was adapted for the attachment of flight muscles, and their bones were hollow to lighten the skeleton for flight. As presented in chapter 18, these adaptations are paralleled by, though not identical to, adaptations in the birds—the descendants of the second lineage of flying archosaurs.

FIGURE 17.15
Parental Care in Reptiles. A female American alligator (*Alligator mississippiensis*) tending to her nest.

The synapsid lineage eventually gave rise to the mammals. The legs of synapsids were relatively long and held their body off the ground. Teeth and jaws were adapted for effective chewing and tearing. Additional bones were incorporated into the middle ear. These and other mammal-like characteristics developed between the Carboniferous and Triassic periods. The "Evolutionary Perspective" of chapter 19 describes more about the nature of this transition.

Stop and Ask Yourself

9. Why is uric acid an adaptive excretory product for reptiles?
10. What groups of reptiles contain parthenogenetic species?
11. In what group of reptiles are vocalizations an important part of reproductive activities?
12. What reptilian lineage gave rise to the birds? To the mammals?

SUMMARY

1. The earliest amniotes are classified as reptiles. The evolution of the amniotes resulted in lineages leading to the turtles; birds and dinosaurs; squamates (snakes, lizards, and worm lizards) and tuataras; and mammals.
2. The order Testudines contains the turtles. Turtles have a bony shell and lack teeth. All are oviparous.
3. The order Squamata contains the lizards, snakes, and worm lizards. Lizards usually have two pairs of legs and most are oviparous. Snakes lack developed limbs and have skull adaptations for swallowing large prey. Worm lizards are specialized burrowers. They have a single median tooth in the upper jaw and most are oviparous.
4. The order Rhynchocephalia contains one species, the tuatara. It is found only on remote islands of New Zealand.
5. The order Crocodilia contains alligators, crocodiles, caimans, and gavials. They have a well-developed secondary palate and display nesting behaviors and parental care.
6. The skin of reptiles is dry and keratinized, and provides a barrier to water loss. It also has epidermal scales and chromatophores.
7. The reptilian skeleton is modified for support and movement on land. Loss of appendages in snakes is accompanied by greater use of the body wall in locomotion.
8. Reptiles have a tongue that may be used in feeding. Bones of the skull of snakes are loosely joined and spread apart during feeding.
9. The circulatory system of reptiles is divided into pulmonary and systemic circuits and functions under relatively high blood pressures. Blood may be shunted away from the pulmonary circuit during periods of apnea.
10. Gas exchange occurs across convoluted lung surfaces. Ventilation of lungs occurs by a negative-pressure mechanism.
11. Reptiles are ectotherms and mainly use behavioral mechanisms to thermoregulate.
12. Vision is the dominant sense in most reptiles. Parietal eyes, ears, Jacobson's organs, and pit organs are important receptors in some reptiles.
13. Because uric acid is nontoxic and relatively insoluble in water, it can be stored and excreted as a semisolid. Internal respiratory surfaces and dry skin also promote conservation of water.
14. The amniotic egg and internal fertilization permit development on land. They are accompanied by significant energy expenditure on the part of the parent.
15. Some reptiles use visual, olfactory, and auditory cues for reproduction. Parental care is important in crocodilians.
16. Descendants of the diapsid evolutionary lineage include the birds. Descendants of the synapsid lineage are the mammals.

SELECTED KEY TERMS

amniotic eggs (*p.* 288)
Jacobson's organs (*p.* 299)
keratin (*p.* 290)
median (parietal) eye (*p.* 299)
pit organs (*p.* 299)

CRITICAL THINKING QUESTIONS

1. Explain the nature of the controversy between cladists and evolutionary systematists regarding the higher taxonomy of the amniotes. Do you think that the Reptilia should be retained as a formal class designation? If so, what groups of animals should it contain?
2. What characteristics of the life history of turtles make them vulnerable to extinction? What steps do you think should be taken to protect endangered turtle species?
3. What might explain the fact that parental care is common in crocodilians and birds?
4. Make a list of the adaptations that make life on land possible for a reptile. Explain why each is adaptive.
5. The incompletely divided ventricle of reptiles is sometimes portrayed as an evolutionary transition between the heart of primitive amphibians and the completely divided ventricles of birds and mammals. Do you agree with this portrayal? Why or why not?

chapter 18

Birds: Feathers, Flight, and Endothermy

Outline

Concepts

1. Fossils of the earliest birds clearly show reptilian features. Fossils of *Archaeopteryx* and *Sinornis* give clues to the origin of flight in birds.
2. Integumentary, skeletal, muscular, and gas exchange systems of birds are adapted for flight and endothermic temperature regulation.
3. Large regions of the brain of birds are devoted to integrating sensory information.
4. Complex mating systems and behavior patterns increase the chances of offspring survival.
5. Migration and navigation allow birds to live, feed, and reproduce in environments that are favorable to survival of adults and young.

Would You Like to Know:

1. how flight evolved in birds? (*p. 305*)
2. what ancient animal is the earliest known bird? (*p. 305*)
3. how a bird can sleep without falling from its perch? (*p. 309*)
4. how wings create enough lift to get a bird off the ground? (*p. 309*)
5. how an osprey diving for a fish keeps the fish in focus throughout the dive? (*p. 314*)
6. how birds navigate during migration? (*p. 318*)

These and other useful questions will be answered in this chapter.

This chapter contains evolutionary concepts, which are set off in this font.

Evolutionary Perspective

Drawings of birds on the walls of caves in southern France and Spain, bird images of ancient Egyptian and ancient American cultures, and the bird images in Biblical writings are evidence that humans have marveled at birds and bird flight for thousands of years. From the early drawings of flying machines by Leonardo da Vinci (1490) to the first successful powered flight by Orville Wright on December 17, 1903, humans have tried to take to the sky and experience what it would be like to soar like a bird.

The ability of birds to navigate long distances between breeding and wintering grounds is just as impressive as flight. For example, Arctic terns have a migratory route that takes them from the Arctic to the Antarctic and back again each year, a distance of approximately 35,000 km (22,000 mi) (figure 18.1). Their rather circuitous route takes them across the northern Atlantic Ocean, to the coast of Europe and Africa, and then across vast stretches of the southern Atlantic Ocean before reaching their wintering grounds.

Phylogenetic Relationships

Birds are traditionally classified as members of the class Aves (a′ves) (L. *avis*, bird). Cladistic interpretations of the relationship of birds to other amniotes were discussed in chapter 17 (*see figure 17.2*). The major characteristics of this class concern adaptations for flight, including appendages modified as wings, feathers, endothermy, a high metabolic rate, a vertebral column modified for flight, and bones that are lightened by numerous air spaces. In addition, modern birds possess a horny bill and lack teeth.

The similarities between birds and reptiles are so striking that birds are often referred to as "glorified reptiles." Like the crocodilians, birds have descended from ancient archosaurs (*see figure 17.2*). Other flying reptiles in this evolutionary lineage (e.g., pterosaurs and pterodactyls) are ruled out of bird ancestry because these reptiles lost an important avian feature, the clavicles, long before birds appeared. (The clavicles, or "wishbone" serves as one of the attachment points for flight muscles. Thus, these reptiles could not have been as strong fliers as modern birds.) Rather than having feathered wings, the flight surfaces of the wings of these primitive reptiles were membranous folds of skin.

Ancient Birds and the Evolution of Flight

In 1861, one of the most important vertebrate fossils was found in a slate quarry in Bavaria, Germany (figure 18.2). It was a fossil of a pigeon-sized animal that lived during the Jurassic period, about 150 million years ago. It had a long, reptilian tail and clawed fingers. The complete head of this specimen was not

(a)

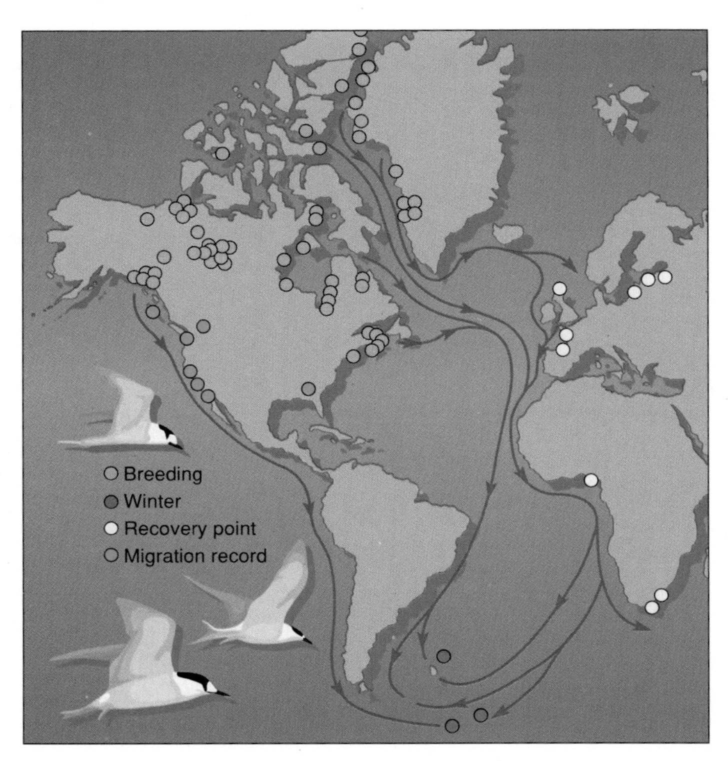

(b)

Figure 18.1

Class Aves. (*a*) The birds were derived from the archosaur lineage of ancient reptiles. Adaptations for flight include appendages modified as wings, feathers, endothermy, a high metabolic rate, a vertebral column modified for flight, and bones that are lightened by numerous airspaces. Flight has given birds, like this Arctic tern (*Sterna arctica*), the ability to exploit resources that are unavailable to other vertebrates. (*b*) The migration route of the Arctic tern is shown here. Arctic terns breed in northern North America, Greenland, and the Arctic. Migrating birds cross the Atlantic Ocean on their trip to wintering grounds in Antarctica. In the process, they fly about 22,000 miles each year.

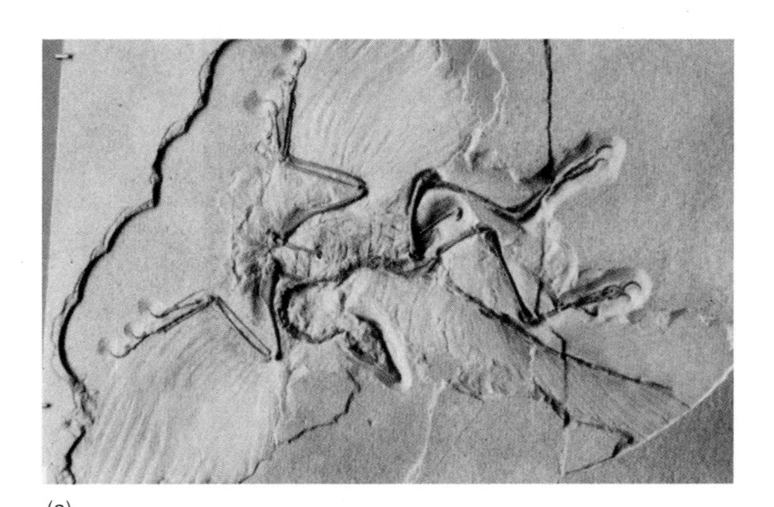

(a)

(b)

FIGURE 18.2

Archaeopteryx, An Ancient Bird. (*a*) *Archaeopteryx* fossil. (*b*) Artist's representation. Some zoologists think that *Archaeopteryx* was a ground dweller rather than the tree dweller depicted here.

preserved, but imprints of feathers on the tail and on short rounded wings were the main evidence that this was the fossil of an ancient bird. It was named *Archaeopteryx* (Gr. *archaios*, ancient + *pteron*, wing). Sixteen years later, a more complete fossil was discovered, revealing teeth in beaklike jaws. Four later discoveries of *Archaeopteryx* fossils have reinforced the ideas of reptilian ancestry for birds.

Interpretations of the life-style of *Archaeopteryx* have been important in the development of hypotheses on the origin of flight. The clavicles (wishbone) of *Archaeopteryx* are well developed and probably provided points of attachment for wing muscles. The sternum, another site for attachment of flight muscles in modern birds, is less developed. These observations indicate that *Archaeopteryx* may have primarily been a glider that was capable of flapping flight over short distances.

1 Some zoologists think that the clawed digits of the wings may have been used to climb trees and cling to branches. A sequence in the evolution of flight may have involved jumping from branch to branch or branch to ground. At some later point, gliding evolved. Still later, weak flapping, supplemented gliding and finally wing-powered flight evolved.

Other zoologists note that the structure of the hindlimbs of the earliest birds suggests that they may have been bipedal, running and hopping along the ground. Their wings may have functioned in batting flying insects out of the air or in trapping insects and other prey against the ground. The teeth and claws, which resemble talons of modern predatory birds, may have been used to grasp prey. Wings would have been useful in providing stability during horizontal jumps when pursuing prey, and they would also have allowed flight over short distances. The benefits of such flight may have led eventually to wing-powered flight.

Other ancient fossil birds have been found. One fossil, named *Protoavis*, has been discovered in Texas and dates back 225 million years. This fossil, therefore, predates *Archaeopteryx* by 75 million years. This description has caused considerable controversy. Many paleontologists doubt that *Protoavis* is really a fossil of a bird. They consider it to be a fossilized ancient reptile. Other paleontologists insist that *Protoavis* was a bird, and that its antiquity means that the origin of birds was much earlier than previously assumed. If this is true, *Archaeopteryx* would simply represent an ancient offshoot of the bird lineage. 2 At least for now, *Archaeopteryx* is considered by most zoologists to be the oldest bird yet discovered and very close to the main line of evolution between reptiles and birds.

A third ancient bird (*Sinornis*), which was recently discovered in China, fits well with the view that *Archaeopteryx* was closely related to ancestral bird stocks. *Sinornis* fossils are 135 million years old—only 15 million years younger than *Archaeopteryx*. In addition to having has some very primitive dinosaurlike characteristics, *Sinornis* had characteristics similar to modern birds. These characteristics included a shortened body and tail and a sternum with a large surface area for flight muscles. The claws were reduced, and the forelimbs were modified to permit the folding of wings at rest. These characteristics all indicate that powered flight was well developed in birds 135 million years ago.

DIVERSITY OF MODERN BIRDS

Archaeopteryx and *Sinornis* provide the only evidence of the transition between reptiles and birds. We do not know, however, whether or not either bird is the direct ancestor of modern birds. There are a variety of fossil birds found for the period between 100 million and 70 million years ago. Some of these birds were large, flightless birds; others were adapted for swimming and diving, and some were fliers. Most, like *Archaeopteryx*, had reptilelike teeth. Most of the lineages represented by these fossils became extinct, along with the dinosaurs, at the end of the Mesozoic era.

Some of the few birds that survived into the Tertiary period were the ancestors of modern, toothless birds. The phylogeny of

TABLE 18.1 CLASSIFICATION OF THE BIRDS

Class Aves (a′ves) (L. *avis*, bird)*
Adaptations for flight include: foreappendages modified as feathered wings, endothermic, high metabolic rate, neck flexible, posterior vertebrae fused, and bones lightened by numerous air spaces. The skull is lightened by a reduction in bone and the presence of a horny bill that lacks teeth. The birds.

Order Sphenisciformes (sfe-nis′i-for′mez)
Heavy bodied; flightless, flipperlike wings for swimming; well insulated with fat. Penguins.

Order Struthioniformes (stroo′the-oni-for′mez)
Large, flightless birds; wings with numerous fluffy plumes. Ostriches.

Order Rheiformes (re′i-for′mez)
Large, flightless birds; degenerate wings with soft, loose plumes. Rheas.

Order Casuariiformes (kaz′u-ar′e-i-for′mez)
Wings reduced; plumage coarse and hairlike. Cassowaries, emus.

Order Gaviiformes (ga′ve-i-for′mez)
Strong, straight bill; diving adaptations include legs far back on body, bladelike tarsus, webbed feet, and heavy bones. Loons.

Order Podicipediformes (pod′i-si-ped′i-for′mez)
Wings short; plumage soft and dense; feet webbed with flattened nails. Grebes.

Order Procellariiformes (pro-sel-lar-e-i-for′mez)
Tubular nostrils, large nasal glands; wings long and narrow. Albatrosses, shearwaters, petrels.

Order Pelecaniformes (pel′e-can-i-for′mez)
Four toes joined in common web; nostrils rudimentary or absent; large gular sac. Pelicans, boobies, cormorants, anhingas, frigatebirds.

Order Ciconiiformes (si-ko′ne-i-for′mez)
Neck long, often folded in flight; long-legged waders. Herons, egrets, storks, wood ibises, flamingos.

Order Anseriformes (an′ser-i-for′mez)
South American screamers, ducks, geese and swans. The latter possess a wide flat bill and an undercoat of dense down. Webbed feet. Worldwide.

Order Falconiformes (fal′ko-ni-for′mez)
Strong, hooked beak; wings large; raptorial feet. Vultures, secretarybirds, hawks, eagles, ospreys, falcons.

Order Galliformes (gal′li-for′mez)
Short beak; short, concave wings; feet and claws strong. Curassows, grouse, quail, pheasants, turkeys.

Order Gruiformes (gru′i-for′mez)
Order characteristics variable and not diagnostic. Marsh birds including cranes, limpkins, rails, coots.

Order Charadriiformes (ka-rad′re-i-for′mez)
Order characteristics variable. Shorebirds, gulls, terns, auks.

Order Columbiformes (co-lum′bi-for′mez)
Dense feathers loosely set in skin; well-developed crop. Pigeons, doves, sandgrouse.

Order Psittaciformes (sit′ta-si-for′mez)
Maxilla hinged to skull; tongue thick; fourth toe reversible; usually brightly colored. Parrots, lories, macaws.

Order Cuculiformes (ku-koo′li-for′mez)
Fourth toe reversible; skin soft and tender; Plantaineaters, roadrunners, cuckoos.

Order Strigiformes (strij′i-for′mez)
Large head with fixed eyes directed forward; raptorial foot. Owls.

Order Caprimulgiformes (kap′ri-mul′ji-for′mez)
Owllike head and plumage, but weak bill and feet; beak with wide gape; insectivorous. Whippoorwills, other goatsuckers.

Order Apodiformes (a-pod′i-for′mez)
Long wings; weak feet. Swifts, hummingbirds.

Order Coraciiformes (kor′ah-si′ah-for′mez)
Large head; large beak; metallic plumage. Kingfishers, todies, bee eaters, rollers.

Order Piciformes (pis′i-for′mez)
Beak usually long and strong; legs and feet strong with fourth toe permanently reversed in woodpeckers. Woodpeckers, toucans, honeyguides, barbets.

Order Passeriformes (pas′er-i-for′mez)
Largest avian order; 69 families of perching birds; perching foot; variable external features. Swallows, larks, crows, titmice, nuthatches, and many others.

*Selected bird orders are described.

modern birds is very controversial. It is sufficient to say that adaptive radiation has resulted in about 9,100 species of living birds, which are divided into about 27 orders (table 18.1). (The number of orders varies depending on the classification system used.) The orders are distinguished from one another by characteristic behaviors, songs, anatomical differences, and ecological niches.

Stop and Ask Yourself

1. What are characteristics of the class Aves?
2. What characteristics of *Archaeopteryx* are reptilelike?
3. What are two hypotheses for the origin of flight in birds? What aspects of the structure of *Archaeopteryx* are used to support each hypothesis?

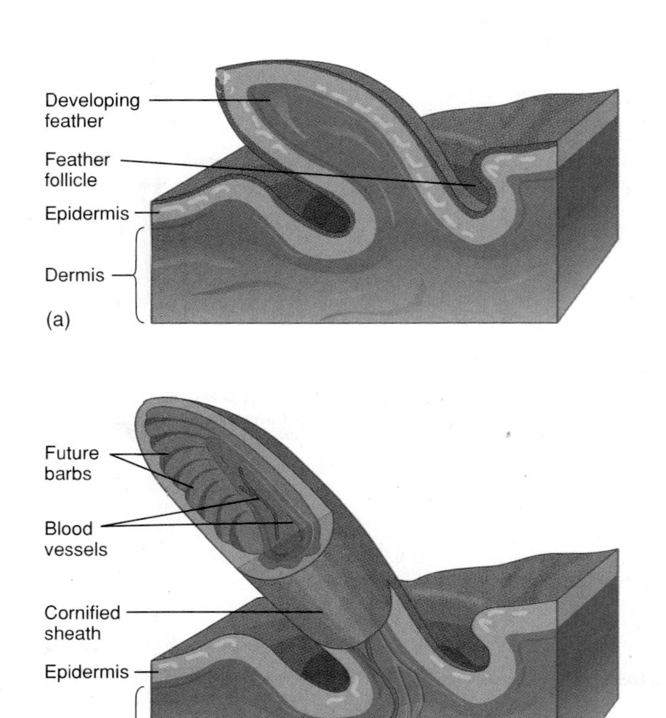

FIGURE 18.3

The Formation of Bird Feathers during Embryonic Development. (*a*) Feathers form from epidermal evaginations. (*b*) Later in development, the blood supply to the feather is cut off and the feather becomes a dead, keratinized epidermal structure seated in a feather follicle.

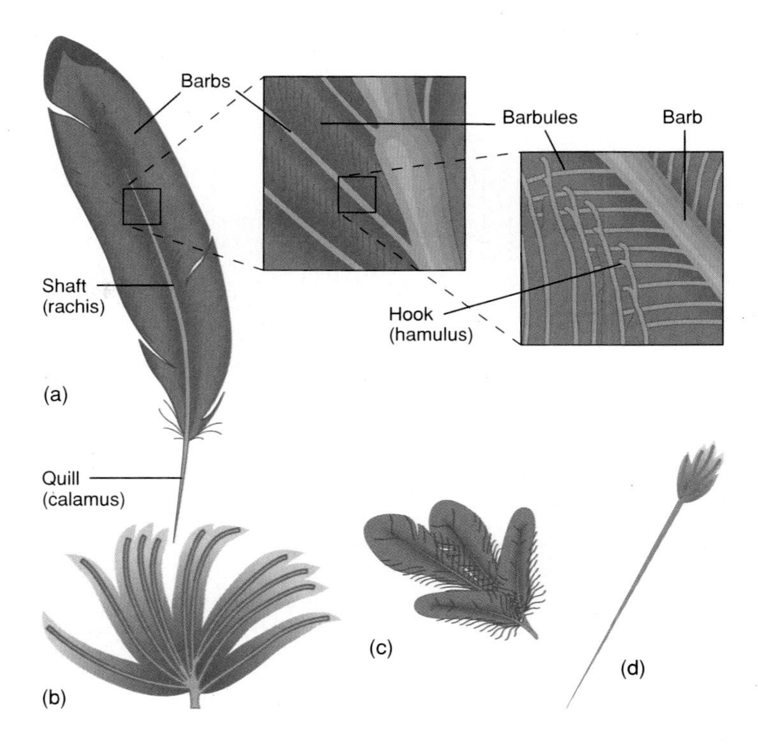

FIGURE 18.4

Anatomy of Selected Feather Types. (*a*) Anatomy of a contour feather showing enlargements of barbs and barbules. (*b*) A down feather. Various types of down feathers provide insulation for adult and immature birds. (*c*) A contour feather with an aftershaft. (*d*) A filoplume. Filoplume feathers are usually covered by contour feathers and are associated with nerve endings in the skin, thus serving as sensory structures.

EVOLUTIONARY PRESSURES

Virtually every body system of a bird shows some adaptation for flight. Endothermy, feathers, acute senses, long flexible necks, and lightweight bones are a few of the many adaptations described in this section.

EXTERNAL STRUCTURE AND LOCOMOTION

The covering of feathers on a bird is called the plumage. Feathers have two primary functions essential for flight. They form the flight surfaces that provide lift and aid steering, and they prevent excessive heat loss, permitting the endothermic maintenance of high metabolic rates. Feathers also have roles in courtship, incubation, and waterproofing.

Feathers develop in a fashion similar to epidermal scales of reptiles, and this similarity is one source of evidence that demonstrates the evolutionary ties between reptiles and birds (figure 18.3; *see box 17.2*). Only the inner pulp of feathers contains dermal elements, such as blood vessels, which supply nutrients and pigments for the growing feather. As feathers mature, their blood supply is cut off, and the feathers become dead, keratinized, epidermal structures seated in epidermal invaginations of the skin called feather follicles.

The most obvious feathers are **contour feathers,** which cover the body, wings, and tail (figure 18.4*a, c*). Contour feathers consist of a vane with its inner and outer webs, and a supportive shaft. Feather barbs branch off the shaft, and barbules branch off the barbs. Barbules of adjacent barbs overlap one another. The ends of barbules are locked together with hooklike hamuli (s., hamulus). Interlocking barbs keep contour feathers firm and smooth. Other types of feathers include **down feathers,** which function as insulating feathers, and **filoplume feathers** (pinfeather), which have sensory functions (figure 18.4*b, d*).

Birds maintain a clean plumage to rid the feathers and skin of parasites. Preening, which is done by rubbing the bill over the feathers, keeps the feathers smooth, clean, and in place. Hamuli that become dislodged can be rehooked by running a feather through the bill. Secretions from an oil gland at the base of the tail of many birds are spread over the feathers during preening to keep the plumage water repellant and supple. The secretions also lubricate the bill and legs to prevent chafing. Anting is a maintenance behavior common to many songbirds and involves picking up ants in the bill and rubbing them over the feathers. The formic acid secreted by ants is apparently toxic to feather mites.

Most colors in a bird's plumage are produced by feather pigments deposited during feather formation. Other colors,

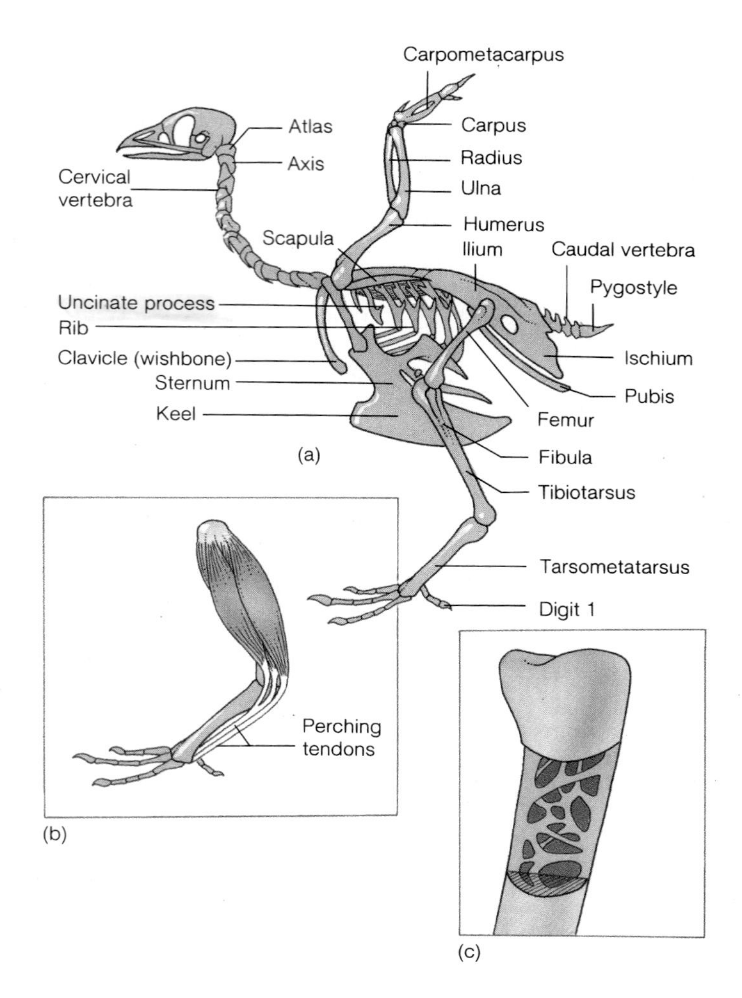

FIGURE 18.5
The Bird Skeleton. (*a*) The skeleton of a pigeon. (*b*) Perching tendons run from the toes across the back of the ankle joint, which cause the foot to grip a perch. (*c*) The internal structure of the humerus.

termed structural colors, arise from irregularities on the surface of the feather that diffract white light. For example, blue feathers are never blue because of the presence of blue pigment. A porous, nonpigmented outer layer on a barb reflects blue wavelengths of light. The other wavelengths pass into the barb and are absorbed by the dark pigment melanin. Iridescence results from the interference of light waves caused by a flattening and twisting of barbules. An example of iridescence is the perception of interchanging colors on the neck and back of hummingbirds and grackles. Color patterns are involved in cryptic coloration, species and sex recognition, and sexual attraction.

Mature feathers receive constant wear; thus, all birds undergo a periodic renewal of their feathers by shedding and replacing them in a process called **molting.** The timing of molt periods varies in different taxa. The following is a typical molting pattern for songbirds. After hatching, a chick is covered with down. Down is replaced with juvenile feathers at the postnatal molt. A postjuvenile molt usually occurs in the fall and results in plumage similar to that of the adult. Once sexual maturity is attained, a prenuptial molt occurs in later winter or early spring, prior to the breeding season. A postnuptial molt usually occurs between July and October. Flight feathers are frequently lost in a particular sequence so that birds are not wholly deprived of flight during molt periods. However, many ducks, coots, and rails cannot fly during molt periods and hide in thick marsh grasses.

The Skeleton

The bones of most birds are lightweight yet strong. Some bones, such as the humerus (forearm bone), have large air spaces and internal strutting (reinforcing bony bars), which increase strength (figure 30.5*c*). (Engineers take advantage of this same principle. They have discovered that a strutted girder is stronger than a solid girder of the same weight.) Birds also have a reduced number of skull bones, and teeth are replaced with a lighter, keratinized sheath called a bill. The demand for lightweight bones for flight is countered in some birds with other requirements. For example, some aquatic birds (e.g., loons) have dense bones, which help reduce buoyancy during diving.

The appendages involved in flight cannot manipulate nesting materials or feed young. These activities are possible because of the bill and very flexible neck. The cervical vertebrae have saddle-shaped articular surfaces that permit great freedom of movement. In addition, the first cervical vertebra (the atlas) has a single point of articulation with the skull (the occipital condyle), which permits a high degree of rotational movement between the skull and the neck. (The single occipital condyle is another characteristic shared with reptiles.) This flexibility allows the bill and neck to function as a fifth appendage.

The pelvic girdle, vertebral column, and ribs are strengthened for flight. The thoracic region of the vertebral column contains ribs, which attach to thoracic vertebrae. The ribs have posteriorly directed uncinate processes that overlap the next rib to strengthen the rib cage (figure 18.5*a*). (Uncinate processes are also present on the ribs of most reptiles and are additional evidence of their common ancestry.) Posterior to the thoracic region is the lumbar region. The **synsacrum** is formed by the fusion of the posterior thoracic vertebrae, all the lumbar and sacral vertebrae, and the anterior caudal vertebrae. Fusion of these bones helps maintain the proper flight posture and supports the hind appendages during landing, hopping, and walking. The posterior caudal vertebrae are fused into a **pygostyle,** which helps support the tail feathers that are important in steering.

The sternum of most birds bears a large, median keel for the attachment of flight muscles. (Exceptions to this include some flightless birds, such as ostriches.) It attaches firmly to the rest of the axial skeleton by the ribs. Paired clavicles are fused medially and ventrally into a furcula (wishbone).

The appendages of birds have also been modified. Some bones of the front appendages have been lost or fused and serve as points of attachment of flight feathers. The rear appendages are used for hopping, walking, running, and perching. Perching tendons run from the toes across the back of the ankle joint to muscles of the lower leg. When the ankle joint is flexed, as in landing on a perch, tension on the perching tendons is increased, and the foot grips the perch (figure 18.5*b*). This automatic grasp helps a bird perch even while sleeping. The muscles of the lower leg can increase the tension on these tendons, for example, when an eagle grasps a fish in its talons.

Muscles

The largest, strongest muscles of most birds are the flight muscles. They attach to the sternum and clavicles and run to the humerus. The muscles of most birds are adapted physiologically for flight. Flight muscles must contract quickly and fatigue very slowly. These muscles have many mitochondria and produce large quantities of ATP to provide the energy required for flight, especially long-distance migrations. Domestic fowl have been selectively bred for massive amounts of muscle that is well liked by humans as food, but is poorly adapted for flight because it contains rapidly contracting fibers with few mitochondria and poor vascularization.

Flight

The wings of birds are adapted for different kinds of flight. However, regardless of whether a bird soars, glides, or has a rapid flapping flight, the mechanics of staying aloft are similar. Bird wings form an **airfoil.** The anterior margin of the wing is thicker than the posterior margin. The upper surface of the wing is slightly convex, and the lower surface is flat or slightly concave. Air passing over the wing travels farther and faster than air passing under the wing, decreasing air pressure on the upper surface of the wing and creating lift (figure 18.6*a*). The lift created by the wings must overcome the bird's weight, and the forces that propel the bird forward must overcome the drag created by the friction of the bird moving through the air. Lift can be increased by increasing the angle the leading edge of the wing makes with the oncoming air (the angle of attack). As the angle of attack increases, however, the flow of air over the upper surface becomes turbulent, reducing lift (figure 18.6*b*). Turbulence can be reduced by forming slots at the leading edge of the wing through which air can flow rapidly, thus smoothing air flow once again. Slotting the feathers at the wing tips and the presence of an alula on the anterior margin of the wing reduce turbulence. The **alula** is a group of small feathers supported by bones of the medial digit. During takeoff, landing, and hovering flight, the angle of attack is increased, and the alula is elevated (figure 18.6*c*,*e*). During soaring and fast flight, the angle of attack is decreased, and slotting is reduced.

Most of the propulsive force of flight is generated by the distal part of the wing. Because it is farther from the shoulder joint, the distal part of the wing moves farther and faster than the proximal part of the wing. During the downstroke (the powerstroke), the leading edge of the distal part of the wing is oriented slightly downward and creates a thrust somewhat analogous to the thrust created by a propeller on an airplane (figure 18.6*d*). During the upstroke (the recovery stroke), the distal part of the wing is oriented upward to decrease resistance. Feathers on a wing overlap so that on the downstroke, air presses the feathers at the wing margins together, allowing little air to pass

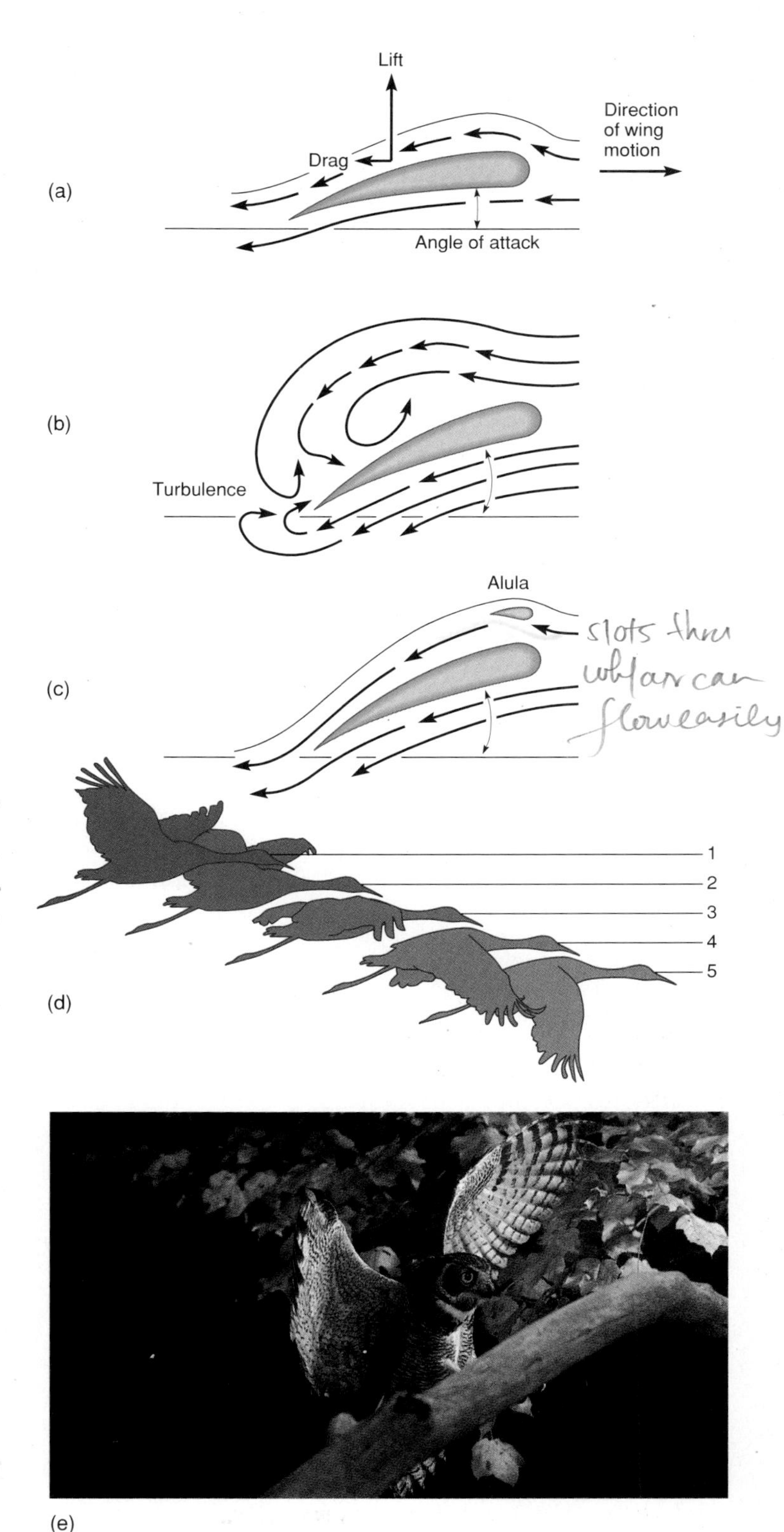

Figure 18.6

Mechanics of Bird Flight. (*a*) A bird's wing acts as an airfoil. Air passing over the top of the wing travels farther and faster than air passing under the wing, creating lift. (*b*) Increasing the angle of attack increases lift but also increases turbulence. (*c*) Turbulence is reduced by the alula. (*d*) The orientation of the wing during a downstroke. (*e*) Note the alula on the wings of the great horned owl (*Bubo virginianus*).

BOX 18.1 SAILORS' CURSE—GLIDERS' ENVY

They are known by many, not very complimentary names. Dutch sailors called them mollymawks ("stupid gull"), the English call them goonies (another reference to stupidity), and the Japanese call them bakadori ("foul birds"). The names of albatrosses are probably the result of inexpressive facial features and awkward movements on land. The albatross's reputation for stupidity is accompanied by another reputation. Their appearance alongside a ship was believed to be a sure sign of changing winds. As the old sailor with "a long grey beard and glittering eye" discovered, to kill an albatross brought extremely bad luck. (*The Rime of the Ancient Mariner* by Samuel Coleridge, 1798.) His deadly aim with a crossbow caused the winds to die, and all sailors on the becalmed ship, except the mariner, died of thirst. The mariner was forced to sail on alone with the albatross hung around his neck.

And I had done an hellish thing
And it would work 'em woe:
For all averr'd, I had kill'd the Bird
That made the Breeze to blow.
Ah wretch! said they, the bird to slay,
That made the breeze to blow!

The origin of the superstition that associated the albatross with breezes is not difficult to understand. The albatross has relatively poorly developed flight muscles and relies primarily on soaring flight and wind to keep aloft (figure 1). Most species of albatross are found around the Antarctic where breezes are almost constant, and they can launch themselves from cliffs into the air with minimal flapping flight. They soar swiftly downwind, picking up speed and losing altitude. Just above the water's surface, they turn sharply into the wind and use the oncoming wind to soar higher. When air speed drops, they turn again to move downwind. Under favorable winds, an albatross can follow a ship for many miles, zig-zagging upwind and downwind, without flapping their long, narrow wings. The sight of a soaring albatross usually does mean a favorable sailing breeze!

FIGURE 1 **An Albatross in Flight.** An albatross (*Diomedea irrorata*) uses wind currents to soar around the Southern Hemisphere.

When grounded in calm winds, an albatross experiences great difficulty becoming airborne again. They must run along the ground, flapping their wings, until air speed adequate for takeoff has been achieved.

Albatrosses feed on fishes and invertebrates near the ocean's surface and on refuse tossed from ships. After a courtship dance and mating, a single egg is laid in a mud nest. Incubation may last as long as 85 days, and parental care another 3 to 9 months. After leaving the nest, young albatrosses depart the nesting grounds and circle their newly discovered world many times. (For most albatrosses "their world" is the entire Southern Hemisphere!) Sexual maturity is reached after about 7 years, and some may live for up to 30 years.

between them, enhancing both lift and propulsive forces. Feathers part slightly on the upstroke, allowing air to pass between them, which reduces resistance during the recovery stroke.

The tail of a bird serves a variety of balancing, steering, and braking functions during flight. During horizontal flight, spreading the tail feathers increases lift at the rear of the bird and causes the head to dip for descent. Closing the tail feathers has the opposite effect. Tilting the tail sideways causes the bird to turn. When a bird lands, its tail is deflected downward, serving as an air brake.

Different kinds of flight are used by different birds or by the same bird at different times. During gliding flight, the wing is stationary, and a bird loses altitude. Waterfowl coming in for a landing use gliding flight. Flapping flight generates the power for flight and is the most common type of flying. Many variations in wing shape and flapping patterns result in species-specific speed and maneuverability. Soaring flight allows some birds to remain airborne with little energy expenditure. During soaring, wings are essentially stationary, and the bird utilizes updrafts and air currents to gain altitude. Hawks, vultures, and other soaring birds are frequently observed circling along mountain valleys, soaring downwind to pick up speed and then turning upwind to gain altitude. As the bird slows and begins to lose altitude, it turns downwind again. The wings of many soarers are wide and slotted to provide maximum maneuverability at relatively low speeds. Oceanic soarers, such as albatrosses and frigate birds, have long, narrow wings that provide maximum lift at high speeds, but they compromise maneuverability and ease of takeoff and landing (box 30.1). Hummingbirds perform hovering flight. They hover in still air by fanning their wings back and forth (50 to 80 beats per second) to remain suspended in front of a flower or feeding station.

Stop and Ask Yourself

4. Describe feather maintenance behavior.
5. What adaptations of the bird skeleton promote flight?
6. How do bird wings provide lift and forward propulsion during flight?
7. What function is served by the alula?

Figure 18.7
Bird Flight and Feeding Adaptations. This ruby-throated hummingbird (*Archilochus colubris*) hovers while feeding on flower nectar. The beak of hummingbirds often matches the length and curvature of the flower from which it extracts nectar.

Nutrition and the Digestive System

Most birds have ravenous appetites! This appetite supports a high metabolic rate that makes endothermy and flight possible. For example, hummingbirds feed almost constantly during the day. In spite of high rates of food consumption, their rapid metabolism often cannot be sustained overnight and they may become torpid, with reduced body temperature and respiratory rate, until they can feed again in the morning.

Bird bills and tongues are modified for a variety of feeding habits and food sources (figures 18.7 and 18.8). For example, the tongue of a woodpecker is barbed for extracting grubs from the bark of trees. Sapsuckers excavate holes in trees and use a brushlike tongue for licking the sap that accumulates in these holes. The tongues of hummingbirds and other nectar feeders are rolled into a tube and used for extracting nectar from flowers.

In many birds, a diverticulum of the esophagus, called the crop, is a storage structure that allows birds to quickly ingest large quantities of locally abundant food and then seek safety while digesting their meal. The crop of pigeons produces "pigeon's milk," a cheesy secretion formed by the proliferation and sloughing of cells lining the crop. Young pigeons (squabs) are fed pigeon's milk until they are able to eat grain. Cedar waxwings, vultures, and birds of prey use their esophagus for similar storage functions. Crops are less well developed in insect-eating birds because insectivorous birds feed throughout the day on sparsely distributed food.

The stomach of birds is modified into two regions. The proventriculus secretes gastric juices that initiate digestion (figure 18.9). The ventriculus (gizzard) has muscular walls to abrade

(a)

(b)

(c)

Figure 18.8
Some Specializations of Bird Bills. (*a*) The bill of a bald eagle (*Haliacetus leucocephalus*) is specialized for tearing prey. (*b*) The thick, powerful bill of this cardinal (*Carcinalis cardinalis*) is used to crack through tough seeds. (*c*) The bill of a flamingo (*Phoenicopterus ruber*) is used for straining food from the water in a head-down feeding posture. The upper and lower mandibles are fringed with large bristles. As water is sucked into the bill, larger particles are filtered and left outside. Inside the bill, smaller algae and animals are filtered on tiny inner bristles. The tongue is used to remove food from the bristles.

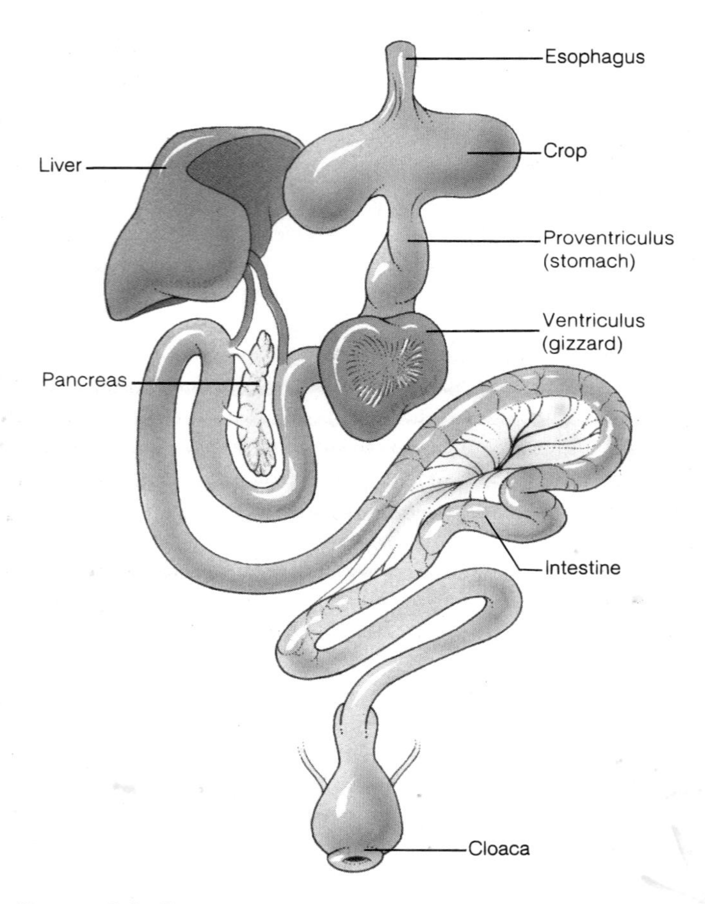

Figure 18.9
The Digestive System of a Pigeon. Birds have very high metabolic rates that require a nearly constant supply of nutrients.

and crush seeds or other hard materials. Sand and other abrasives may be swallowed to aid digestion. The bulk of enzymatic digestion and absorption occurs in the small intestine, aided by secretions from the pancreas and liver. Paired ceca may be located at the union of the large and small intestine. These blind-ending sacs contain bacteria that aid in the digestion of cellulose. Undigested food is usually eliminated through the cloaca; however, owls form pellets of bone, fur, and feathers that are ejected from the ventriculus through the mouth. Owl pellets accumulate in and around owl nests and are useful in studying their food habits.

It is common practice to group birds by their feeding habits. It is somewhat artificial, however, because birds may eat different kinds of food at different stages in their life history, or they may change diets simply because of changes in food availability. Robins, for example, feed largely on worms and other invertebrates when these foods are available. In the winter, however, robins may feed on berries.

In some of their feeding habits, birds have come into direct conflict with human interests. Bird damage to orchard and grain crops is tallied in the millions of dollars each year. Flocking and roosting habits of some birds, such as European starlings and redwing blackbirds, concentrate millions of birds in local habitats, and fields of grain can be devastated. Recent monocultural practices tend to aggravate problems with grain-feeding birds by encouraging the formation of very large flocks.

In spite of commonly held beliefs, the impact of birds of prey on poultry, game birds, and commercial fisheries is minimal. Unfortunately, birds of prey have been killed with guns and poisons because of the mistaken impression that they are responsible for significant losses.

Circulation, Gas Exchange, and Temperature Regulation

The circulatory system of birds is similar to that of reptiles, except that the heart has completely separated atria and ventricles, resulting in separate pulmonary and systemic circuits. **In vertebrate evolution, the sinus venosus has undergone a gradual reduction in size.** It is a separate chamber in fishes, amphibians, and turtles and receives blood from the venous system. In other reptiles, it is a group of cells in the right atrium that serves as the pacemaker for the heart. In birds, the sinus venosus also persists only as a patch of pacemaker tissue in the right atrium. The bird heart is relatively large (up to 2.4% of total body weight), and its rate of beating is rapid. Rates in excess of 1,000 beats per minute have been recorded for hummingbirds under stress. Larger birds have relatively smaller hearts and slower heart rates. The heart rate of an ostrich, for example, varies between 38 and 176 beats per minute. **A large heart, rapid heart rate, and complete separation of oxygenated from unoxygenated blood are important adaptations for delivering the large quantities of blood required for endothermy and flight.**

Gas Exchange

Because of high metabolic rates associated with flight, birds have a greater rate of oxygen consumption than any other vertebrate. When other vertebrates inspire and expire, air passes into and out of respiratory passageways in a simple back and forth cycle. Ventilation is interrupted during expiration and there is a considerable quantity of "dead air" in the lungs because not all air is forced out during expiration. Because of their unique structure, bird lungs provide a nearly continuous movement of fresh air over respiratory surfaces during inspiratory and expiratory cycles. The quantity of "dead air" in the lungs is sharply reduced compared with other vertebrates.

Most air is taken in through external nares, which lead to nasal passageways and the pharynx. The trachea is supported by bone and cartilage. A larynx is undifferentiated, but a special voicebox, called the syrinx, is located where the trachea divides into bronchi. The muscles of the syrinx and bronchi, as well as characteristics of the trachea are responsible for bird vocalizations.

The lungs of birds are made of small air tubes called **parabronchi** (figure 18.10). Air capillaries about 10 μm in diameter branch from the parabronchi and provide gas-exchange

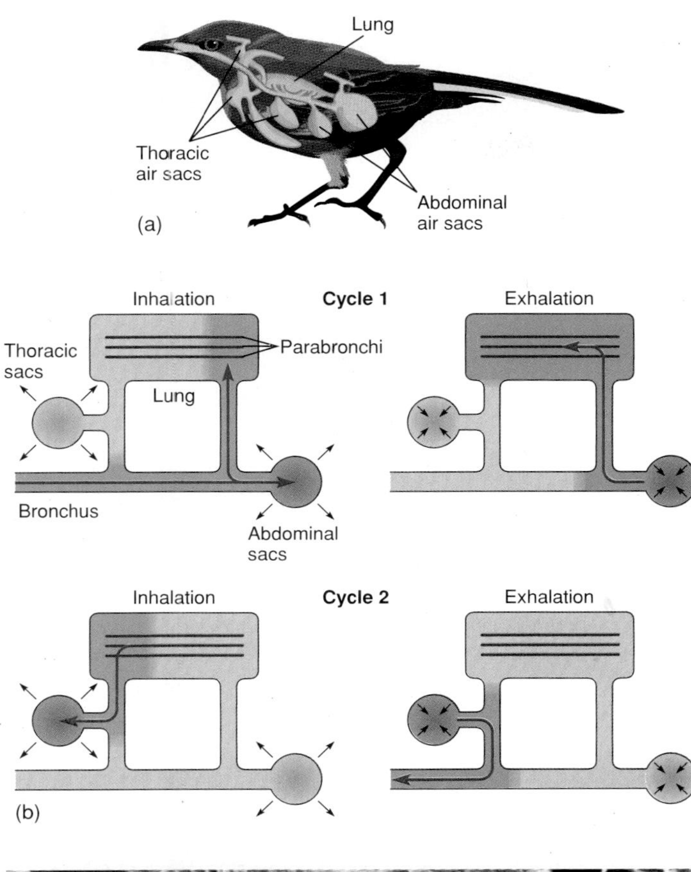

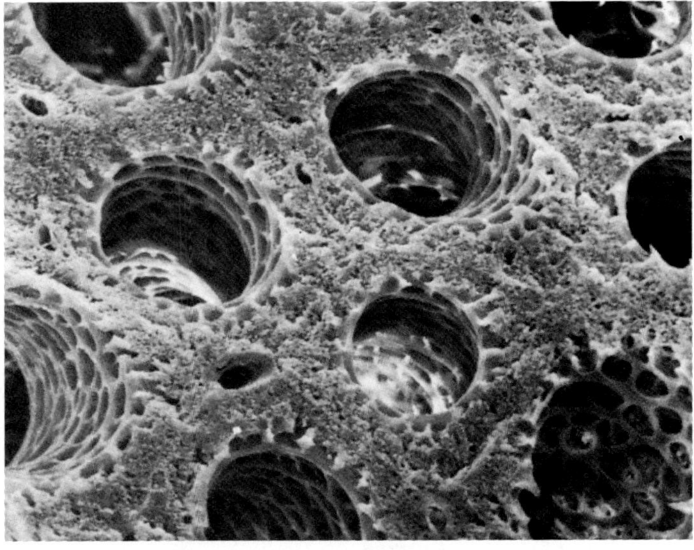

Figure 18.10

The Respiratory System of a Bird. (*a*) Air sacs branch from the respiratory tree. (*b*) Air flow during inspiration and expiration. Air flows through the parabronchi during both inspiration and expiration. The shading represents the movement through the lungs of one inspiration. (*c*) Scanning electron micrograph showing parabronchi.

surfaces. Inspiration and expiration are accomplished by the expansion and compression of thin-walled air sacs that ramify throughout the body cavity and even penetrate some bones, such as the humerus of the wing.

Ventilation occurs by alternate compression and expansion of the air sacs during flight and other activities. When breathing, the movement of the sternum and the posterior ribs compresses the thoracic air sacs. X-ray movies of European starlings in a wind tunnel show that the wishbone is distorted when flight muscles contract. Alternate distortion and recoiling helps compress and expand the air sacs located between the bone's two shafts. During inspiration, air moves into the abdominal air sacs. At the same time, air already in the lungs moves through parabronchi to the thoracic air sacs. During expiration, the air in the thoracic air sacs moves out of the respiratory system, and the air in the abdominal air sacs moves into parabronchi. At the next inspiration, the air moves into the thoracic air sacs, and is expelled during the next expiration. It takes two ventilatory cycles to move a particular volume of air through the respiratory system of a bird.

Thermoregulation

Birds regulate their body temperatures between 38 and 45° C. Lethal extremes are lower than 32 and higher than 47° C. On a cold day, resting birds fluff their feathers to increase their insulating properties, as well as the dead air space within them. They also tuck their bills into their feathers to reduce heat loss from the respiratory tract. The most exposed parts of a bird are the feet and tarsi which have neither fleshy muscles nor a rich blood supply. Temperatures in these extremities are allowed to drop near freezing to prevent heat loss. Countercurrent heat exchange between the warm blood flowing to the legs and feet, and the cooler blood flowing to the body core from the legs and feet, prevent excessive heat loss at the feet by returning heat to the body core before it goes to the extremities and is lost to the environment. Shivering is also used to generate heat in extreme cold. Increases in metabolism during winter months require additional food.

Some birds become torpid and allow their body temperatures to drop on cool nights. For example, whippoorwills allow their body temperatures to drop from about 40° C to near 16° C, and respiratory rates become very slow.

Muscular activity during flight produces large quantities of heat. Excess heat can be dissipated by panting. Evaporative heat loss from the floor of the mouth, is enhanced by fluttering vascular membranes of this region.

Nervous and Sensory Systems

A mouse skitters across the floor of a barn enveloped in the darkness of night. An owl in the loft overhead turns in the direction of the faint sounds made by tiny feet. As the mouse reaches a

Figure 18.11
A Barn Owl (*Tyto alba*). A keen sense of hearing allows barn owls to find prey in spite of the darkness of night.

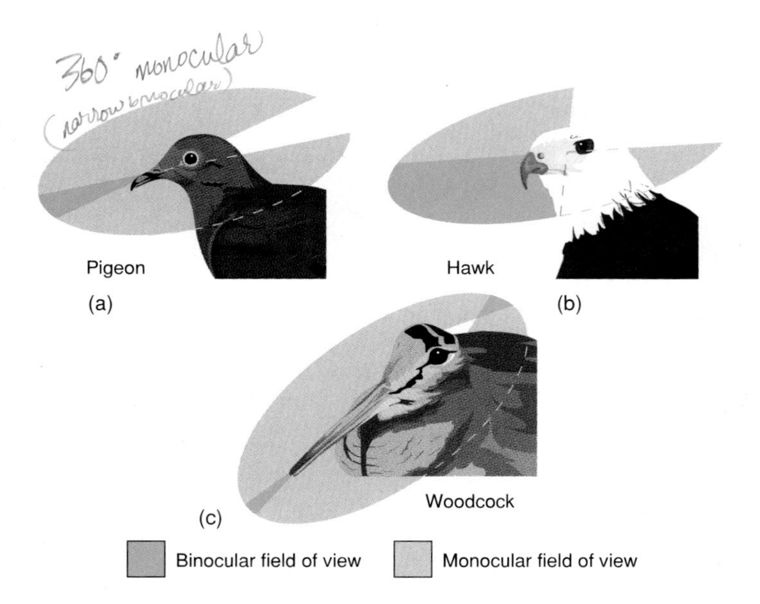

Figure 18.12
Avian Vision. The fields of view of a pigeon (*a*), a hawk (*b*), and a woodcock (*c*). Woodcocks have eyes located far posteriorly and have a narrow field of binocular vision in front and behind. They can focus on predators that might be circling above them while probing mud with their long beaks.

sack of grain, and the sounds made by hurrying feet change to a scratchy gnawing of teeth on the sack of feed, the barn owl dives for its prey (figure 18.11). Fluted tips of flight feathers make the owl's approach imperceptible to the mouse, and the owl's ears, not its eyes, provide information to guide the owl to its prey. Under similar circumstances, barn owls will successfully locate and capture prey in over 75% of attempts! This ability is just one example of the many sensory adaptations found in birds.

The forebrain of birds is much larger than that of reptiles due to the enlargement of the cerebral hemispheres, including a region of gray matter, the corpus striatum. The corpus striatum functions in visual learning, feeding, courtship, and nesting. A pineal body is located on the roof of the forebrain. It appears to play a role in stimulating ovarian development and in regulating other functions influenced by light and dark periods. The optic tectum (the roof of the midbrain), along with the corpus striatum, plays an important role in integrating sensory functions. The midbrain also receives sensory input from the eyes. As in reptiles, the hindbrain includes the cerebellum and the medulla oblongata, which coordinate motor activities and regulate heart and respiratory rates, respectively.

Vision is an important sense for most birds. The structure of bird eyes is similar to that of other vertebrates, but they are much larger in proportion to body size than in other vertebrates (*see figure 16.13*). The eyes are usually somewhat flattened in an anteroposterior direction; however, the eyes of birds of prey protrude anteriorly because of a bulging cornea. Birds have a unique, double-focusing mechanism. Padlike structures (similar to those of reptiles) control the curvature of the lens, and ciliary muscles change the curvature of the cornea. 5 Double, nearly instantaneous focusing allows an osprey, or other bird of prey, to remain focused on a fish throughout a brief, but breathtakingly fast, descent.

The retina of a bird's eye is thick and contains both rods and cones. Rods are active under low light intensities and cones under high light intensities. Cones are especially concentrated ($1{,}000{,}000/mm^2$) at a focal point called the fovea. Unlike other vertebrates, some birds have two foveae per eye. The one at the center of the retina is sometimes called the "search fovea" because it gives the bird a wide angle of monocular vision. The other fovea is at the posterior margin of the retina. It functions with the posterior fovea of the other eye to allow binocular vision. The posterior fovea is called the "pursuit fovea," because binocular vision is necessary for depth perception, which is necessary to capture prey. "Search" and "pursuit" are not meant to imply that the two foveae are found only in predatory birds. Other birds use the "search fovea" to observe the landscape below them during flight and the "pursuit" fovea when depth perception is needed, as in landing on a branch of a tree.

The position of the eyes on the head also influences the degree of binocular vision (figure 18.12). Pigeons have eyes located well back on the sides of their head, giving them a nearly 360° monocular field, but a narrow binocular field. They do not have to pursue their food (grain), and a wide monocular field of view helps them stay alert to predators while feeding on the ground. Hawks and owls have eyes farther forward on the head. Their binocular field of view is increased, and their monocular field of view is correspondingly decreased.

Like reptiles, birds have a nictitating membrane that is drawn over the surface of the eye to cleanse and protect the eye.

Olfaction apparently plays a minor role in the lives of most birds. External nares open near the base of the beak, but the olfactory epithelium is poorly developed. Exceptions include turkey vultures, which locate their dead and dying prey largely by smell.

In contrast, hearing is well developed in most birds. The external ear opening is covered by loose, delicate feathers called auriculars. Middle- and inner-ear structures are similar to those of reptiles. The sensitivity of the avian ear (100 to 15,000 Hz) is similar to that of the human ear (16 to 20,000 Hz).

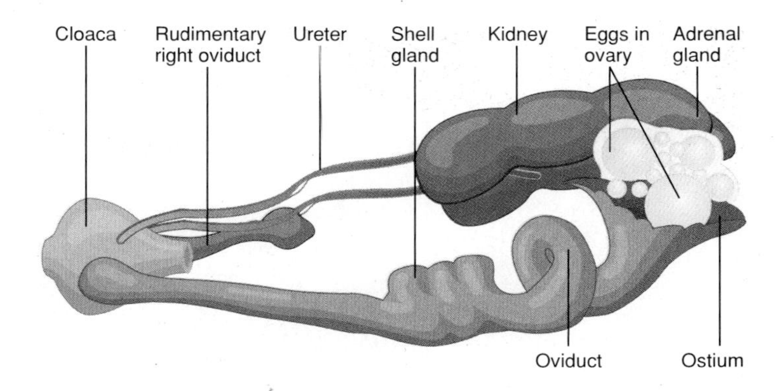

Figure 18.13

The Urogenital System of a Female Pigeon. The right ovary and oviduct are rudimentary in most female birds.

Excretion and Osmoregulation

Birds and reptiles face essentially identical excretory and osmoregulatory demands. Like reptiles, birds excrete uric acid, which is temporarily stored in the cloaca. Water reabsorption also occurs in the cloaca. As with reptiles, the excretion of uric acid conserves water and promotes development of embryos in terrestrial environments. In addition, some birds have supraorbital salt glands that drain excess sodium chloride through the nasal openings to the outside of the body. These are especially important in marine birds that drink seawater and feed on invertebrates containing large quantities of salt in their tissues. Salt glands can secrete salt in a solution that is about two to three times more concentrated than other body fluids. Salt glands, therefore, compensate for the kidney's inability to concentrate salts in the urine.

Stop and Ask Yourself

8. What evolutionary remnant of the vertebrate sinus venosus is found in the heart of a bird? What is the function of this remnant?
9. How are the lungs of birds adapted to provide continuous, one-way movement of air across gas exchange surfaces?
10. How does a bird cool itself?
11. What is the value of separate "search" and "pursuit" foveae for a bird of prey?

Reproduction and Development

Sexual activities of birds have been observed more closely than those of any other group of animals. These activities include establishing territories, finding mates, constructing nests, incubating eggs, and feeding young.

All birds are oviparous. Gonads are located in the dorsal abdominal region, next to the kidneys. Testes are paired, and coiled tubules (vasa deferentia) conduct sperm to the cloaca. An enlargement of the vasa deferentia, the seminal vesicle, is a site for temporary storage and maturation of sperm prior to mating. Testes enlarge during the breeding season. Except for certain waterfowl and ostriches, birds have no intromittent organ, and sperm transfer occurs by cloacal contact during brief mounts by the male on the female.

In females, two ovaries form during development, but usually only the left ovary fully develops (figure 18.13). A large, funnel-shaped opening of the oviduct envelopes the ovary and receives eggs after ovulation. Fertilization of the egg occurs in the upper portions of the oviduct, and the zygote is gradually surrounded by albumen secreted from glandular regions of the oviduct wall as the egg completes its passage. A shell is added by a shell gland in the lower region of the oviduct. The oviduct opens into the cloaca.

Territories are established by many birds prior to mating. Although size and function vary greatly among species, territories generally allow birds to mate without interference. They provide nest locations and sometimes food resources for adults and offspring. Breeding birds defend their territories and expel intruders of the same sex and species. Threats are common, but actual fighting is minimized.

Mating may follow the attraction of a mate to a territory. For example, female woodpeckers are attracted to the drumming by males on trees. Male ruffed grouse fan their wings on logs and create sounds that can be heard for many miles. Cranes have a courtship dance that includes stepping, bowing, stretching, and jumping displays. Mating occurs when a mate's call or posture signals readiness. It is accomplished quickly, but occurs repeatedly to assure fertilization of all the eggs that will be laid.

Over 90% of birds are **monogamous:** a single male pairs with a single female during the breeding season. Some birds (swans, geese, eagles) pair for life. Frequent mating apparently strengthens the pair bonds that develop. Monogamy is common when resources are widely and evenly distributed, and one bird cannot control access to resources. Monogamy is also

Figure 18.14
Courtship Displays. A male, greater prairie chicken (*Tympanuchus cupido*) displaying in a lek.

advantageous because both parents usually participate in nest building and care of the young. One parent can incubate and protect the eggs or chicks while the other searches for food.

Some birds are **polygynous.** Males mate with more than one female, and the females care for the eggs and chicks. Polygyny tends to occur in species whose young are less dependent at hatching and in situations where patchy resource distribution may attract many females to a relatively small breeding area. Prairie chickens are polygynous, and males display in groups called leks. In prairie chicken leks, the males in the center positions are preferred and attract the majority of females (figure 18.14).

A few bird species are **polyandrous,** and the females mate with more than one male. Polyandry occurs in spotted sandpipers. Females are larger than males and establish and defend their territories from other females. They lay a clutch of eggs for each male that is attracted to and builds a nest in her territory. If a males loses his eggs to a predator, the female replaces those eggs. Polyandry results in the production of more eggs than monogamous matings. It is thought to be advantageous when food is plentiful but, because of predation or other threats, the chances of successfully rearing young are low.

Nest construction usually begins after pair formation. It is an instinctive behavior and is usually initiated by the female. A few birds do not make nests. Emperor penguins, for example, breed on the snow and ice of Antarctica where no nest materials are available. Their single egg is incubated on the web of the foot (mostly the male's foot) tucked within a fold of abdominal skin.

Nesting Activities

The nesting behavior of birds is often species specific. Some birds choose nest sites away from other members of their species, and other birds nest in very large flocks. Unfortunately, predictable nesting behavior has led to the extinction of some species of birds (box 18.2).

The number of eggs laid by a female is usually variable. Most birds incubate their eggs, and some birds have a featherless, vascularized incubation or brood patch that helps keep the eggs at temperatures between 33 and 37° C. Eggs are turned to prevent the adherence of egg membranes in the egg and deformation of the embryo. Adults of some species sprinkle the eggs with water to cool and humidify them. The Egyptian plover carries water from distant sites in the breast feathers. The incubation period lasts between 10 and 80 days and is correlated with egg size and degree of development at hatching. One or two days before hatching, the young bird penetrates an air sac at the blunt end of its egg, inflates it lungs, and begins breathing. Hatching occurs as the young bird pecks the shell with a keratinized egg tooth on the tip of the upper jaw and struggles to free itself.

Some birds are helpless at hatching, others are more independent. Those that are entirely dependent on their parents are said to be **altricial** (L. *altricialis,* to nourish), and they are often naked at hatching (figure 18.15*a*). Altricial young must be brooded constantly at first because endothermy is not developed. They grow rapidly, and when they leave the nest they are nearly as large as their parents. (For example, American robins weigh 4 to 6 g at hatching and leave the nest 13 days later weighing 57 g.) **Precocial** (L. *praecoci,* early ripe) young are alert and lively at hatching (figure 18.15*b*). They are usually covered by down and can walk, run, swim, and feed themselves—although one parent is usually present to lead the young to food and shelter.

Young altricial birds have huge appetites and keep one or both parents continually searching for food. They may consume a mass of food that equals their own weight each day. Adults bring food to the nest or regurgitate food stored in the crop or esophagus. Vocal signals or color patterns on the bills or throats of adults initiate feeding responses in the young. Parents instinctively feed gaping mouths, and many hatchlings have brightly colored mouth linings that attract a parent's attention. The first-hatched young is fed first and most often because it is usually the largest and can stretch its neck higher than can its nestmates.

Life is usually brief for birds. About 50% of eggs laid yield birds that leave the nest. Most birds, if kept in captivity, have a potential life span of 10 to 20 years. Natural longevity is much shorter. The average American robin lives 1.3 years, and the average black-capped chickadee lives less than 1 year. Mortality is high in the first year from predators and inclement weather.

Migration and Navigation

Over twenty centuries ago, Aristotle described birds migrating to escape the winter cold and summer heat. He had the mistaken impression that some birds disappear during winter because they hibernate, and that others transmutate to another species. It is now known that some birds migrate long distances. Modern zoologists study the timing of migration, the stimuli for

BOX 18.2 Bright Skies and Silent Thunder

There is no sadder tale in the history of human interactions with wildlife than that of *Ectopistes migratorius*, the passenger pigeon. Although the full story of the extinction of this bird may never be known, humans will never be able to escape the responsibility for the excessive slaughter that led to the decline of a vulnerable species.

In the nineteenth century, flocks of passenger pigeons were so vast in eastern North America that their numbers were estimated at up to 2 billion birds. In 1813, the famed naturalist John Audubon watched a flock of passenger pigeons pass overhead for 3 days. The flock was said to completely darken the sky by day, and their wings sounded like thunder.

How could a species so abundant become extinct in just a few decades? What made these birds so vulnerable was their social behavior. Passenger pigeons nested, roosted in the evening, and foraged for food in huge aggregations that covered thousands of acres. It is said that trees became so laden with birds that their branches often broke under the weight. One nest site in Wisconsin was said to be 100 miles long! On feeding forays, flocks would travel hundreds of miles and strip trees of acorns, beechnuts, and other nuts. These colonies were apparently not just convenient associations, but were essential for survival. Highly organized flocks were also an important social stimulus for breeding.

Unfortunately, this behavior made the birds vulnerable. Nesting sites were predictable and hunters came to them year after year. Professional hunters are said to have traveled from nest site to nest site during the nesting season. Birds were blinded at night with lights and knocked out of trees. They were captured in huge nets or choked by burning sulfur. Their trees were felled and birds harvested. They were shot by the thousands with guns. In 1861, 14,850,000 passenger pigeons were shipped to big city markets from a single nesting site near Petoskey, Michigan.

How could any bird be expected to successfully produce and rear their young under this onslaught? The single egg tended by a pair of passenger pigeons had little chance for survival during these unfortunate years. Flocks gradually dwindled. By the late 1890s people began to realize that the passenger pigeons were in trouble. In addition to rampant killing and the resultant breakup of the social organization needed for successful mating, deforestation was contributing to the passenger pigeon's decline. Laws that prohibited further killing were too late. After 1900, no passenger pigeons were seen in the wild. "Martha," the last passenger pigeon in captivity (Cincinnati Zoological Gardens), died on September 1, 1914, after living for 29 years.

(a)

(b)

Figure 18.15

Altricial and Precocial Chicks. (*a*) An American robin (*Turdus migratorius*) feeding nestlings. Robins have altricial chicks that are helpless at hatching. (*b*) Killdeer (*Charadrius vociferus*) have precocial chicks that are down covered and can move about.

migration, and the physiological changes that occur during migration, as well as migration routes and how birds navigate over huge expanses of land or water.

Migration (as used here) refers to periodic round trips between breeding and nonbreeding areas. Most migrations are annual, with nesting areas in northern regions and wintering grounds in the south. (Migration is more pronounced for species found in the Northern Hemisphere because about 70% of the earth's land is in the Northern Hemisphere.) Migrations occasionally involve east/west movements or altitude changes. Migration allows birds to avoid climatic extremes, and to secure adequate food, shelter, and space throughout the year.

Birds migrate in response to species-specific physiological conditions. Innate (genetic) clocks and environmental factors influence preparation for migration. The photoperiod is often cited as an important migratory cue for many birds, particularly for birds in temperate zones. The changing photoperiod initiates seasonal changes in gonadal development that often serve as migratory stimuli. Increasing day length in the spring promotes gonadal development, and decreasing day length in the fall promotes regression of gonads. In many birds, the changing photoperiod also appears to promote fat deposition, which acts as an energy reserve for migration. The anterior lobe of the pituitary gland and the pineal body have been implicated in mediating photoperiod responses.

The mechanics of migration are species specific. Some long-distance migrants may store fat equal to 50% of their body weight and make nonstop journeys. Other species that take a more leisurely approach to migration begin their journeys early and stop frequently to feed and rest. In clear weather, many birds fly at altitudes greater than 1,000 m, which reduces the likelihood of hitting tall obstacles. Many birds have very specific migration routes (*see figure 18.1*).

Navigation

Homing pigeons have served for many years as a pigeon postal service. As long ago as the ancient Egyptian times, and as recently as World War II, pigeons were used to return messages from the battlefield.

6 Two forms of navigation are used by birds. Route-based navigation involves keeping track of landmarks (visual or auditory) on an outward journey so that those landmarks can be used in a reverse sequence on the return trip. Location-based navigation involves establishing the direction of the destination from information available at the journey's site of origin. It involves the use of sun compasses, other celestial cues, and/or the earth's magnetic field.

Birds' lenses are transparent to ultraviolet light, and their photoreceptors respond to it, allowing them to orient using the sun, or even on cloudy days. This orientation cue is referred to as a sun compass. Because the sun moves through the sky between sunrise and sunset, birds use internal clocks to perceive that the sun rises in the east, is approximately overhead at noon, and sets in the west. The biological clocks of migratory birds can be altered. For example, birds ready for northward migration can be held in a laboratory in which the "laboratory sunrise" occurs later than the natural sunrise. When released to natural light conditions, they fly in a direction they perceive to be north, but which is really northwest. Night migrators can also orient using the sun by flying in the proper direction from the sunset.

Celestial cues other than the sun can be used to navigate. Humans recognize that in the Northern Hemisphere, the north star lines up with the axis of rotation of the earth. The angle between the north star and the horizon decreases as one moves toward the equator. Birds may use a similar information to determine latitude. Experimental rotations of the night sky in a planetarium have altered the orientation of birds in test cages.

There has long been speculation that birds employ magnetic compasses to detect the earth's magnetic field, and thus determine direction. Typically, these ideas have been met with skepticism, but direct evidence of their existence has been uncovered. Magnets strapped to the heads of pigeons severely disorient them. European robins and a night migrator, the garden warbler, orient using the earth's magnetic field. However, no discrete magnetic receptors have been found in either birds or other animals. Early reports of finding a magnetic iron, magnetite, in the head and necks of pigeons did not lead to a greater understanding of magnetic compasses. Further experiments failed to demonstrate magnetic properties in these regions. Magnetic iron has been found in bacteria and a variety of animal tissues. None is clearly associated with a magnetic sense, although the pineal body of pigeons has been implicated in the use of a sun compass and in responses to magnetic fields.

There is redundancy in bird navigational mechanisms, which suggests that under different circumstances, different sources of information are probably used.

Stop and Ask Yourself

12. Why is monogamy advantageous for most birds?
13. Why are birds with precocial young more likely to be polygynous than birds with altricial young?
14. What environmental cue is important in preparing birds for migration?
15. What is route-based navigation?

Summary

1. Birds are members of the archosaur lineage. Fossils of ancient birds, *Archaeopteryx* and *Sinornis*, show reptilian affinities and give clues into the origin of flight.
2. Feathers evolved from reptilian scales and function in flight, insulation, sex recognition, and waterproofing. Feathers are maintained and periodically molted.
3. The bird skeleton is light and made more rigid by fusion of bones. The neck and bill are used as a fifth appendage.
4. Bird wings form airfoils that provide lift. Propulsive force is generated by tilting the wing during flapping. Gliding, flapping, soaring, and hovering flight are used by different birds or by the same bird at different times.
5. Birds feed on a variety of foods as reflected in the structure of the bill and other parts of the digestive tract.
6. The heart of birds consists of two atria and two ventricles. A very rapid heart rate, and rapid blood flow, support the high metabolic rate of birds.
7. The respiratory system of birds provides one-way, nearly constant, air movement across respiratory surfaces.
8. Birds are able to maintain high body temperatures endothermically because of insulating fat deposits and feathers.
9. Cerebral hemispheres of a bird are enlarged by the development of the corpus striatum. Vision is the most important avian sense.
10. Birds are oviparous. Reproductive activities include the establishment and defense of territories, courtship, and nest building.
11. Eggs are usually incubated by either or both parents, and one or both parents feed the young. Altricial chicks are helpless at hatching, and precocial chicks are alert and lively shortly after hatching.
12. Migration allows some birds to avoid climatic extremes, and to secure adequate food, shelter, and space throughout the year. The photoperiod is the most important migratory cue for birds.
13. Birds use both route-based navigation and location-based navigation.

Selected Key Terms

airfoil (*p. 309*)
altricial (*p. 316*)
molting (*p. 308*)
monogamous (*p. 315*)
polyandrous (*p. 316*)
polygynous (*p. 316*)
precocial (*p. 316*)

Critical Thinking Questions

1. Birds are sometimes called "glorified reptiles." Discuss why this description is appropriate.
2. What adaptations of birds promote endothermy and flight? Why is endothermy important for birds?
3. Birds are, without exception, oviparous. Why do you think that is true?
4. What are the advantages that offset the great energy expenditure required by migration?
5. Compare and contrast the advantages of monogamy, polygyny, and polyandry for birds. In what ways are the advantages and disadvantages of each related to the abundance and utilization of food and other resources?

chapter 19

MAMMALS: SPECIALIZED TEETH, ENDOTHERMY, HAIR, AND VIVIPARITY

Outline

Concepts

1. Mammalian characteristics evolved gradually over a 200-million-year period in the synapsid lineage.
2. Two subclasses of mammals evolved during the Mesozoic era—Prototheria and Theria. Modern mammals include monotremes, marsupial mammals, and placental mammals.
3. The skin of mammals is thick and protective and has an insulating covering of hair.
4. Adaptations of teeth and the digestive tract allow mammals to exploit a wide variety of food resources.
5. Efficient systems for circulation and gas exchange support the high metabolic rate associated with endothermy.
6. The brain of mammals has an expanded cerebral cortex that processes information from various sensory structures.
7. Metanephric kidneys permit mammals to excrete urea without excessive water loss.
8. Complex behavior patterns enhance survival.
9. Most mammals are viviparous and have reproductive cycles that help ensure internal fertilization and successful development.

Would You Like to Know:

1. what group of ancient reptiles is most closely related to mammals? (*p. 322*)
2. what group of mammals is oviparous? (*p. 323*)
3. why mammal hair stands on end when a mammal is frightened? (*p. 326*)
4. how teeth of mammals are specialized for different feeding habits? (*p. 329*)
5. why the four-chambered hearts of mammals and birds are an example of convergent evolution? (*p. 331*)
6. why salt glands are not found in mammals? (*p. 334*)
7. why a domestic cat rubs its face on furniture around the house? (*p. 336*)

These and other useful questions will be answered in this chapter.

This chapter contains evolutionary concepts, which are set off in this font.

Evolutionary Perspective

The beginning of the Tertiary period, about 70 million years ago, was the start of the "age of mammals." It coincided with the extinction of many reptilian lineages, which led to the adaptive radiation of the mammals. To trace the roots of the mammals, however, we must go back to the Carboniferous period, when the synapsid lineage diverged from other amniote lineages (*see figure 17.2*).

Mammalian characteristics evolved gradually over a period of 200 million years (figure 19.1). The early synapsids were the pelycosaurs. Some were herbivores; others showed skeletal adaptations that reflect increased effectiveness as predators (figure 19.2*a*). The anterior teeth of their upper jaw were large and were separated from the posterior teeth by a gap that accommodated the enlarged anterior teeth of the lower jaw when the jaw was closed. The palate was arched, which gave additional strength to the upper jaw and allowed air to pass over prey held in the mouth. Their legs were longer and slimmer than those of earlier amniotes.

By the middle of the Permian period, other successful mammallike reptiles had arisen from the pelycosaurs. They were a diverse group known as the therapsids. Some were predators and others were herbivores. In the predatory therapsids, teeth were concentrated at the front of the mouth and enlarged for holding and tearing prey. The posterior teeth were reduced in size and number. The jaws of some therapsids were elongate and generated a large biting force when they snapped closed. The teeth of the herbivorous therapsids were also mammallike. Some had a large space, called the diastema, separating the anterior and posterior teeth. The posterior teeth had ridges (cusps) and cutting edges that were probably used to shred plant material. Unlike other reptiles, the hind limbs of therapsids were held directly beneath the body and moved parallel to the long axis of the body. Changes in the size and shape of the ribs suggest the separation of the trunk into thoracic and abdominal regions and a breathing mechanism similar to that of mammals. The last therapsids were a group called the cynodonts (figure 19.2*b*). Some of these were as large as a big dog, but most were small and little different from the earliest mammals.

Figure 19.1
Class Mammalia. The decline of the ruling reptiles about 70 million years ago permitted mammals to radiate into diurnal habitats previously occupied by dinosaurs and other reptiles. Mammals are characterized by hair, endothermy, and mammary glands. The lowland gorilla (*Gorilla gorilla graueri*, order Primates) is shown here.

Figure 19.2
Members of the Subclass Synapsidia. (*a*) *Dimetrodon* was a 3 m long pelycosaur. It probably fed on other reptiles and amphibians. The large sail may have served as a recognition signal and a thermoregulatory device. (*b*) *Cynognathus* was a mammallike reptile that probably foraged for small animals, much like a badger does today. The badger-sized animal was a cyncodont with the order of Therapsida, the stock from which mammals arose during the mid-Triassic period.

The first mammals were small (less than 10 cm long) with delicate skeletons. Most of our knowledge of early mammalian phylogeny comes from the study of their fossilized teeth and skull fragments. These studies suggest that the mammals of the Jurassic and Cretaceous periods were mostly predators that fed on other vertebrates and arthropods. A few were herbivores, and others combined predatory and herbivorous feeding habits. Changes in the structure of the middle ear and the regions of the brain devoted to hearing and olfaction indicate that these senses were important during the early evolution of mammals.

Although it is somewhat speculative, some zoologists think that the small size, well-developed olfactory and auditory abilities, and the lack of color vision in most mammals suggest that early mammals were nocturnal. This habit may have allowed them to avoid competition with the much larger dinosaurs and the smaller diurnal (day-active [L. *diurnalis,* daily]) reptiles living at the same time. Again it is speculative, but nocturnal habits could have led to endothermy. Endothermy would have allowed small mammals to maintain body temperatures above that of their surroundings after the sun had set and the air temperature began to fall.

Diversity of Mammals

Modern members of the class Mammalia (ma-ma'le-ah) (L. *mamma,* breast) are characterized by hair, mammary glands, specialized teeth, three middle-ear ossicles, and other characteristics listed in table 19.1. It is currently not possible to determine the extent to which all of these characteristics were developed in the earliest mammals. Although there is some disagreement among zoologists regarding subclass-level classification, most zoologists consider mid-Cretaceous (about 130 million years ago) mammals to have diverged into two subclasses (figure 19.3). Until recently, monotremes (the duckbilled platypus and the echidna) were classified in the subclass Prototheria (Gr. *protos,* first + *therion,* wild beast). Recent fossil evidence showing monotreme dentition that is characteristic of the subclass Theria has resulted in this group being reassigned to the latter subclass. The Prototheria, therefore, contains only extinct forms.

The subclass Theria diverged into three infraclasses by the late Cretaceous period. The infraclass Ornithodelphia (Gr. *ornis,* bird + *delphia,* birthplace) contains the monotremes (Gr. *monos,* one + *trema,* opening). This name refers to the fact that, unlike other mammals, monotremes possess a cloaca. 2 Monotremes are distinguished from all other mammals by the fact that they are oviparous (figure 19.4*a,b*). There are six species of monotremes found in Australia and New Guinea.

The infraclass Metatheria (Gr. *meta,* after) contains the marsupial mammals. They are viviparous, but have very short gestation periods. A protective pouch, called the marsupium, covers the mammary glands of the female. The young crawl into the marsupium after birth, where they feed and complete development. There are about 250 species of marsupials that live in the Australian region and the Americas (figure 19.4*c*; *see also figure 19.17*).

The other therian infraclass, Eutheria (Gr. *eu,* true), contains the placental mammals. They are usually born at an advanced stage of development, having been nourished within the uterus. Exchanges between maternal and fetal circulatory systems occur by diffusion across an organ called the **placenta,** which is composed of both maternal and fetal tissue. There are about 3,800 species of eutherians that are classified into 17 orders (figures 19.5 and 19.6; *see also figures 19.11 and 19.15–19.17*).

Stop and Ask Yourself

1. What characteristics of synapsid reptiles are mammallike?
2. What source has provided the most information about early mammals?
3. Why do some zoologists think that early mammals were nocturnal?
4. What are two subclasses of mammals?

Evolutionary Pressures

Mammals are naturally distributed on all continents except Antarctica and in all oceans. The many adaptations that have accompanied their adaptive radiation are discussed in this section.

External Structure and Locomotion

The skin of a mammal, like that of other vertebrates, consists of epidermal and dermal layers. It protects from mechanical injury, invasion by microorganisms, and the sun's ultraviolet light. Skin is also important in temperature regulation, sensory perception, excretion, and water regulation.

Hair is a keratinized derivative of the epidermis of the skin and is uniquely mammalian. It is seated in an invagination of the epidermis, called a hair follicle. A coat of hair, called pelage, usually consists of two kinds of hair. Long guard hairs protect a dense coat of smaller, insulating underhairs.

Because hair is composed largely of dead cells, it must be periodically molted. In some mammals (e.g., humans), molting occurs gradually and may not be noticed. In others, hair loss occurs rapidly and may result in altered pelage characteristics. In the fall, many mammals acquire a thick coat of insulating underhair, and the pelage color may change. For example, the Arctic fox takes on a white or cream color with its autumn molt, which helps conceal the fox in a snowy environment. With its spring molt, the Arctic fox acquires a gray and yellow pelage (*see figure 19.6*).

Hair is also important for the sense of touch. Mechanical displacement of a hair stimulates nerve cells associated with the hair root. Guard hairs may sometimes be modified into thick-

TABLE 19.1 Classification of Mammals

Class Mammalia (ma-ma′le-ah)
Mammary glands; hair; diaphragm; three middle-ear ossicles; heterodont dentition; sweat, sebaceous, and scent glands; four-chambered heart; large cerebral cortex.

Subclass Prototheria (pro′to-ther′e-ah)
This subclass formerly contained the monotremes. Monotremes have recently been reclassified and now this subclass contains only extinct species.

Subclass Theria (ther′e-ah)
Members of this subclass are distinguished by very technical characteristics of the skull.

Infraclass Ornithodelphia (or′ne-tho-del′fe-ah)
Members of this infraclass are distinguished by very technical characteristics of the skull. Monotremes.

Infraclass Metatheria (met′ah-ther′e-ah)
Viviparous, primitive placenta; young are born very early and often are carried in a marsupial pouch on the belly of the female. Marsupials.

Infraclass Eutheria (u-ther′-e-ah)*
Complex placenta. Young develop to advanced stage prior to birth. Placentals.

Order Insectivora (in-sec-tiv′or-ah)
Diverse group of small, primitive mammals; third largest mammalian order. Hedgehogs, tenrecs, moles, shrews.

Order Chiroptera (ki-rop′ter-ah)
Cosmopolitan, but especially abundant in the tropics. Bones of the arm and hand are elongate and slender. Flight membranes extend from the body, between digits of forelimbs, to the hindlimbs. Most are insectivorous, but some are fruit eaters, fish eaters, and blood feeders. Bats.

Order Primates (pri-ma′tez)
Adaptations of primates reflect adaptations for increased agility in arboreal (tree-dwelling) habitats. Omnivorous diets, unspecialized teeth, grasping digits, freely movable limbs, nails on digits, reduced nasal cavity, enlarged eyes and cerebral hemispheres. Lemurs (Madagascar and the Comoro Islands), tasiers (jungles of Sumatra and the East Indies), monkeys, gibbons, great apes (apes and humans).

Order Edentata (e′den-ta′tah)
Incisors and canines absent; cheek teeth, when present, lack enamel. Braincase is long and cylindrical. Hind foot is four toed; forefoot with two or three prominent toes with large claws. Limbs specialized for climbing or digging. Anteaters, tree sloths, armadillos.

Order Lagamorpha (lag′o-mor′fah)
Two pairs of upper incisors, one pair of lower incisors. Incisors are ever-growing and slowly worn down by feeding on vegetation. Rabbits, pikas.

Order Rodentia (ro-den′che-ah)
Largest mammalian order. Upper and lower jaws bear a single pair of ever-growing incisors. Squirrels, chipmunks, rats, mice, beavers, porcupines, woodchucks, lemmings.

Order Cetacea (se-ta′she-ah)
Streamlined, nearly hairless, and insulated by thick layers of fat (blubber). Sebaceous glands are absent. Forelimbs modified into paddlelike flippers for swimming; hindlimbs reduced and not visible externally. Tail fins (flukes) are flattened horizontally. External nares (blowhole) are located on the top of the skull. Toothed whales (beaked whales, narwhals, sperm whales, dolphins, porpoises, killer whales); toothless, filter-feeding whales (right whales, gray whales, blue whales, and humpback whales).

Order Carnivora (kar-niv′o-rah)
Predatory mammals; usually have a highly developed sense of smell and a large braincase. Premolars and molars modified into carnassial apparatus; three pairs of upper and lower incisors usually present, and canines are well developed. Dogs, cats, bears, raccoons, minks, sea lions, seals, walruses, otters.

Order Proboscidea (pro′bah-sid′e-ah)
Long, muscular proboscis (trunk) with one or two fingerlike processes at the tip. Skull is short, with the second incisor on each side of the upper jaw modified into tusks. Six cheek teeth are present in each half of each jaw. Teeth erupt (grow into place) in sequence from front to rear, so that one tooth in each jaw is functional. African and Indian elephants.

Order Sirenia (si-re′ne-ah)
Large, aquatic herbivores that weigh in excess of 600 kg. Nearly hairless, with thick, wrinkled skin. Heavy skeleton. Forelimb is flipperlike, and hindlimb is vestigial. Horizontal tail fluke is present. Teeth lack enamel. Manatees (coastal rivers of the Americas and Africa), dugongs (western Pacific and Indian Oceans).

Order Perissodactyla (pe-ris′so-dak′ti-lah)
Skull usually elongate. Large molars and premolars. (The Artiodactyla also have hoofs. Artiodactyls and perissodactyls are, therefore, called *ungulates*. "Ungula" is from the Latin, for hoof.) Primarily grazers. Horses, rhinoceroses, zebras, tapirs.

Order Artiodactyla (ar′te-o-dak′ti-lah)
Hoofed. Axis of support passes between third and fourth digits. Digits one, two, and five reduced or lost. Primarily grazing and browsing animals. (Pigs are an obvious exception.) Pigs, hippopotamuses, camels, antelope, deer, sheep, giraffes, cattle.

*Selected eutherian orders are described.

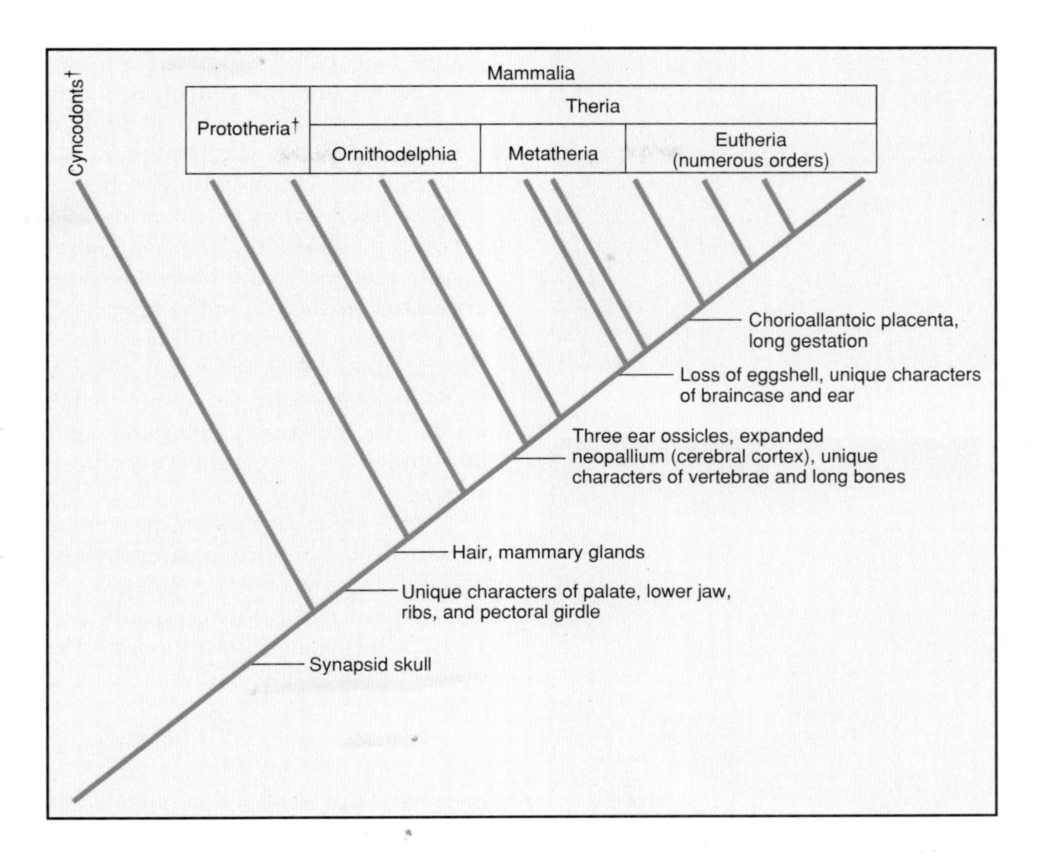

FIGURE 19.3

Mammalian Phylogeny. A cladogram showing the evolutionary relationships among the mammals. Selected characters are shown. Some extinct taxa are indicated with a dagger (†), however, numerous extinct groups have been omitted from the cladogram. The names of the 17 orders of eutherians have also been omitted (*see table 31.1*).

FIGURE 19.4

Representatives of the Mammalian Infraclasses Ornithodelphia and Metatheria. The infraclass Ornithodelphia. (*a*) A duckbilled platypus (*Ornithorhychus anatinus*). (*b*) An echidna or spiny anteater (*Tachyglossus aculeatus*). The infraclass Metatheria. (c) The koala (*Phascolarctos cinereus*) feeds on *Eucalyptus* leaves in Australia.

(a)

(b)

Figure 19.5
Order Edentata. (*a*) A giant anteater (*Myrmecophaga tridactyla*). Anteaters lack teeth. They use powerful forelimbs to tear into an insect nest and a long tongue covered with sticky saliva to capture prey. (*b*) An armadillo (*Dasypus novemcinctus*).

Figure 19.6
Order Carnivora. An Arctic fox (*Alopex lagopus*) with its spring and summer pelage. With its autumn molt, the Arctic fox acquires a white or cream color coat.

shafted hairs called vibrissae. Vibrissae occur around the legs, nose, mouth, and eyes of many mammals. Their roots are richly innervated and very sensitive to displacement.

Air spaces in the hair shaft and air trapped between hair and the skin provide an effective insulating layer. A band of smooth muscle, called the arrector pili muscle, runs between the hair follicle and the lower epidermis. When the muscle contracts, the hairs stand upright, increasing the amount of air trapped in the pelage and improving its insulating properties. Arrector pili muscles are under the control of the autonomic nervous system, which also controls a mammal's "fight-or-flight" response. 3 In threatening situations, the hair (especially on the neck and tail) stands on end and may give the perception of increased size and strength.

Hair color depends on the amount of pigment (melanin) deposited in it and the quantity of air in the hair shaft. The pelage of most mammals is dark above and lighter underneath. This pattern makes them less conspicuous under most conditions. Some mammals advertise their defenses using aposematic (warning) coloration. The contrasting markings of a skunk are a familiar example.

Pelage is reduced in large mammals from hot climates (e.g., elephants and hippopotamuses) and in some aquatic mammals (e.g., whales) that often have fatty insulation.

Claws are present in all amniote classes. They are used for locomotion and offensive and defensive behavior. Claws are formed from accumulations of keratin that cover the terminal phalanx (bone) of the digits. In some mammals, they are specialized to form nails or hooves (figure 19.7).

Glands develop from the epidermis of the skin. **Sebaceous (oil) glands** are associated with hair follicles, and their oily secretion lubricates and waterproofs the skin and hair. Most mammals also possess **sudoriferous (sweat) glands.** Small sudoriferous glands (eccrine glands) release watery secretions used in evaporative cooling. Larger sudoriferous glands (apocrine glands) secrete a mixture of salt, urea, and water, which are converted to odorous products by microorganisms on the skin.

Scent or musk glands are located around the face, feet, or anus of many mammals. These glands secrete pheromones, which may be involved with defense, species and sex recognition, and territorial behavior.

Mammary glands are functional in female mammals and are present, but nonfunctional, in males. The milk that they secrete contains water, carbohydrates (especially the sugar lactose), fat, protein, minerals, and antibodies. Mammary glands are probably derived evolutionarily from apocrine glands and usually contain substantial fatty deposits.

Monotremes have mammary glands that lack nipples. The glands discharge milk into depressions on the belly, where it is lapped up by the young. In other mammals, mammary glands open via nipples or teats, and the young suckle for their nourishment (figure 19.8).

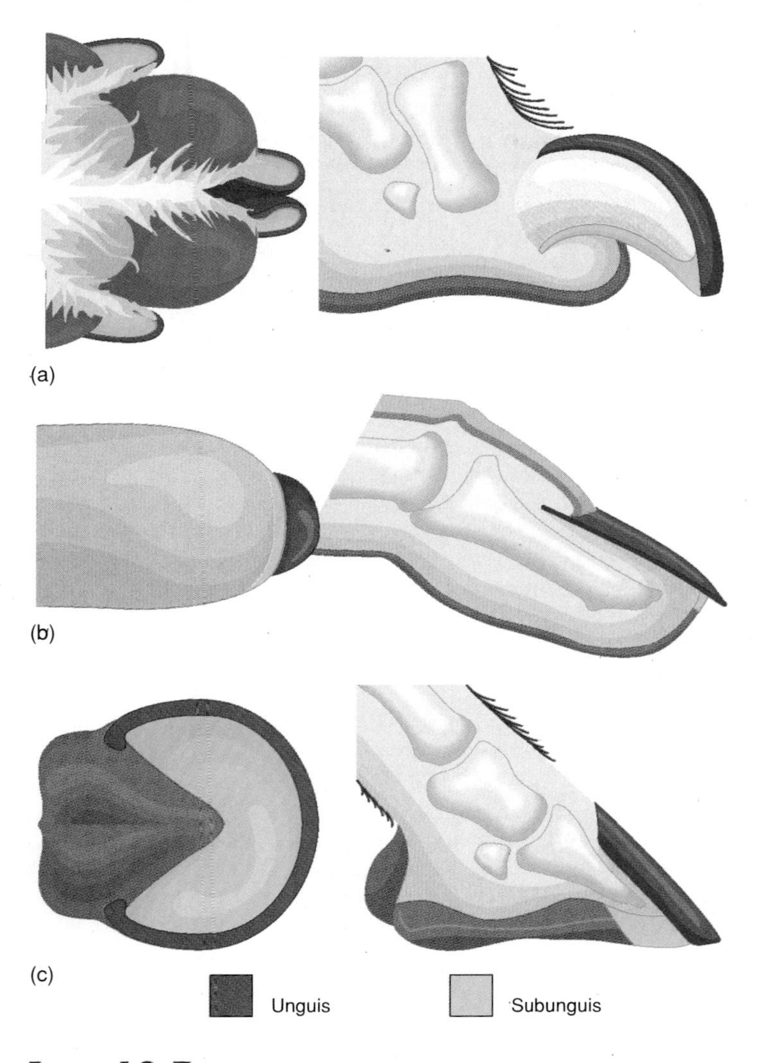

FIGURE 19.7

The Structure of Claws, Nails, and Hooves. (*a*) Claws. (*b*) Nails are flat, broad claws found on the hands and feet of primates and are an adaptation for arboreal habits, where grasping is essential. (*c*) Hooves are characteristic of ungulate mammals. The number of toes is reduced, and the animals walk or run on the tips of the remaining digits. The unguis is a hard, keratinized dorsal plate, and the subunguis is a softer ventral plate.

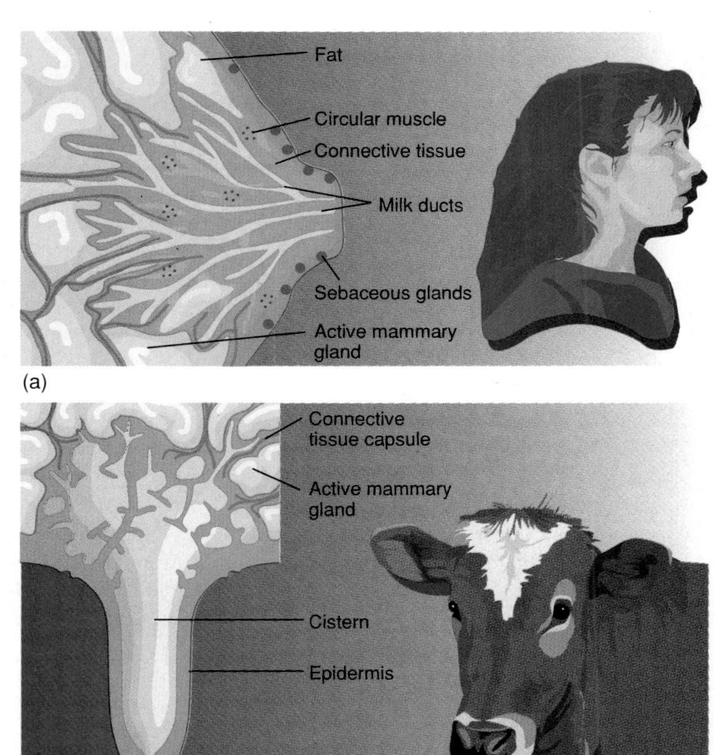

FIGURE 19.8

Mammary Glands. Mammary glands are specialized to secrete milk following the birth of young. (*a*) Many ducts lead from the glands to a nipple. Parts of the duct system are enlarged to store milk. Suckling by an infant initiates a hormonal response that causes the mammary glands to release milk. (*b*) Some mammals (e.g., cattle) have teats that are formed by the extension of a collar of skin around the opening of mammary ducts. Milk collects in a large cistern prior to its release. The number of nipples or teats varies with the number of young produced.

The Skull and Teeth

The skulls of mammals show important modifications of the reptilian pattern (box 19.1). One feature used by zoologists to distinguish reptilian from mammalian skulls is the method of jaw articulation. In reptiles, the jaw articulates at two small bones at the rear of the jaw. In mammals, these bones have moved into the middle ear, and along with the stapes, form the middle-ear ossicles. Jaw articulation in mammals is by a single bone of the lower jaw.

A secondary palate evolved twice in vertebrates—in the archosaur lineage (*see figure 17.2*) and in the synapsid lineage.

In some therapsids, small, shelflike extensions of bone (the hard palate) partially separated the nasal and oral passageways (*see figure 17.10*). In mammals, the secondary palate is extended posteriorly by a fold of skin, called the soft palate, which almost completely separates the nasal passages from the mouth cavity. Unlike other vertebrates that swallow food whole or in small pieces, some mammals chew their food. The more extensive secondary palate allows mammals to breathe while chewing. Breathing needs to stop only briefly during swallowing (figure 19.9).

The structure and arrangement of teeth are important indicators of mammalian life-styles. In reptiles, the teeth are uniformly conical. This condition is referred to as **homodont.** In mammals, the teeth are often specialized for different functions, a condition called **heterodont.** Reptilian teeth are attached along the top or inside of the jaw, whereas in mammals, the teeth are set into sockets of the jaw. Most mammals have two sets of teeth during their life. The first teeth emerge before or shortly after birth and are called deciduous or milk teeth. These teeth are lost and replaced by permanent teeth.

BOX 19.1 Horns and Antlers

Horns were surely a familiar sight in prehistoric landscapes, 100 million years before they became common in mammals. *Triceratops* had three horns, one nasal horn and two above the eyes. It also had a horny shield along the posterior margin of the head. *Styracosaurus* had a 0.7 m nasal spike. These early reptilian horns probably provided a very effective defense against fierce prehistoric carnivores.

Mammalian horns are a carryover from their reptilian heritage. They are most common in a group of hoofed mammals, the artiodactyls (e.g., cattle, sheep, and goats). A horn is a spike of bone that arises from the frontal bone of the skull and is covered with the protein keratin (figure 1*a*,*b*). This bony spike, the "os cornu," grows slowly from youth to adulthood, and its marrow core is highly vascularized. Filaments of keratin arise from folliclelike structures in the skin. Keratin filaments are cemented together and completely cover the os cornu. There is no blood supply to the outer horn layers.

Horns are defensive structures. They exist in symmetrical pairs and are present in both sexes. As any farmer or rancher knows, horns are not regenerated if they are cut off. Horns are usually not shed. One exception is the prong-horn antelope (*Antilocapra americana*). Every year, a new horn grows on the os cornu beneath the old horn, and the latter is eventually pushed off.

Another kind of head ornamentation, the antler, is common in deer, elk, moose, and caribou. Antlers are highly branched structures made of bone, but are not covered by keratin. Unlike horns, they are usually present only in males and are shed and reformed every year (figure 1c). Caribou are an exception because antlers are present in both sexes. Antlers are more recent than horns; the earliest records of antlered animals are from the Miocene epoch (mid-Tertiary period), and by the Pleistocene epoch, they had become common.

Antler development is regulated by seasonal changes in the level of the male hormone testosterone. Antlers of male elk begin to form in April as skin-covered buds from the frontal bone. The primordial cells that initiate antler growth are left behind from the previous year when the antlers were lost. Antlers begin to branch after only 2 weeks. By May, they are well formed, and by August, they are mature. Each year, antlers become more complexly branched. Throughout the spring and summer, they are covered with delicate, vascular tissue called velvet. In August, the bone at the base of the antler becomes progressively more dense and cuts off blood flow to the center of the antler. Later, blood flow to the velvet is cut off, and the velvet begins to dry. It is shed in strips as the antlers are rubbed against the ground or tree branches. Breeding activities commence after the velvet is shed, and the antlers are used in jousting matches as rival males compete for groups of females. (Rarely do these jousting matches lead to severe injury.) Selection by females of males with large antlers may explain why they can get so large in some species. (Although now extinct, the Giant Stag, *Cervis megaceros*, had antlers with a 3 m spread and a mass of 70 kg.) Later in the fall, or in early winter, the base of the antler is weakened as bone is reabsorbed at the pedicels of the frontal bone. Antlers are painlessly cast off when an antler strikes a tree branch or other object.

Other hornlike structures are present in some mammals. Rhinos are the only perissodactyls (e.g., horses, rhinos, and tapirs) to have hornlike structures (*see figure* 19.11). Their "horns" consist of filamentous secretions of keratin cemented together and mounted to the skin of the head. There is no bony core, and thus they are not true horns. Rhino "horns" are prized in the Orient for their presumed aphrodisiac and medicinal properties and as dagger handles in certain mideastern cultures. These demands have led to very serious overhunting of rhinos; in many regions they are almost extinct (*see figure* 19.11).

The horns of giraffes are skin-covered bony knobs. Zoologists do not understand their function.

There are up to four kinds of teeth in adult mammals. **Incisors** are the most anterior teeth in the jaw. They are usually chisellike and used for gnawing or nipping. **Canines** are often long, stout, and conical, and are usually used for catching, killing, and tearing prey. Canines and incisors have single roots. **Premolars** are positioned next to canines, have one or two roots, and truncated surfaces for chewing. **Molars** have broad chewing surfaces and two (upper molars) or three (lower molars) roots.

Mammalian species have characteristic numbers of each kind of adult tooth. A **dental formula** is an important tool used by zoologists to characterize taxa. It is an expression of the number of teeth of each kind in one-half of the upper and lower jaws. The teeth of the upper jaw are listed above those of the lower jaw and they are indicated in following order: incisors, canine, premolars, and molars. For example:

$$\text{Human: } \frac{2\cdot1\cdot2\cdot3}{2\cdot1\cdot2\cdot3} \qquad \text{Beaver: } \frac{1\cdot0\cdot1\cdot3}{1\cdot0\cdot1\cdot3}$$

Mammalian teeth (**dentition**) may be specialized for particular diets. In some mammals, the dentition is reduced, sometimes to the point of having no teeth. For example, **armadillos** and the **giant anteater** (**order Edentata**) feed on termites and ants, and their teeth are reduced.

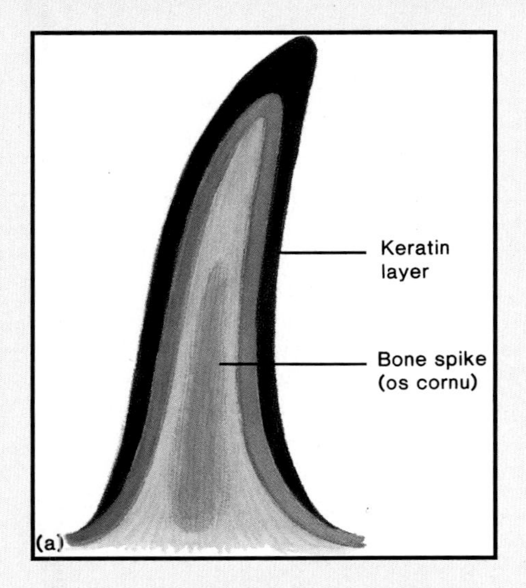

FIGURE 1 **Horns and Antlers.** (*a*) The structure of a horn. (*b*) The horn of a male bighorn sheep (*Orvis canadensis*) (order Artiodactyla). (*c*) Development of deer antlers.

4 Some mammals (e.g., humans, order Primates; and pigs, order Artiodactyla) are omnivorous; they feed on a variety of plant and animal materials. They have anterior teeth with sharp ripping and piercing surfaces, and posterior teeth with flattened grinding surfaces for rupturing plant cell walls (figure 19.10*a*).

Mammals that eat plant material often have flat, grinding posterior teeth and incisors, and sometimes canines, that are modified for nipping plant matter (e.g., horses, order Perissodactyla; deer, order Artiodactyla) or gnawing (e.g., rabbits, order Lagamorpha; beavers, order Rodentia) (figure 19.10*b*,*c*). In rodents, the incisors grow throughout life. Although most mammals have enamel covering the entire tooth, rodents have enamel only on the front surfaces of their incisors. They are kept sharp by slower wear in front than in back. The anterior food-procuring teeth are separated from the posterior grinding teeth by a gap, called the diastema. The diastema results from elongation of the snout that allows the anterior teeth to reach close to the ground or into narrow openings to procure food. The posterior teeth have a high, exposed surface (crown) and continuous growth, which allows these teeth to withstand years of grinding tough vegetation.

Canines and incisors of predatory mammals are used for catching, killing, and tearing prey. In members of the order Carnivora (e.g., coyotes, dogs, and cats), the fourth upper premolars and first lower molars form a scissorlike shearing surface, called the carnassial apparatus, that is used for cutting flesh from prey (figure 19.10*d*).

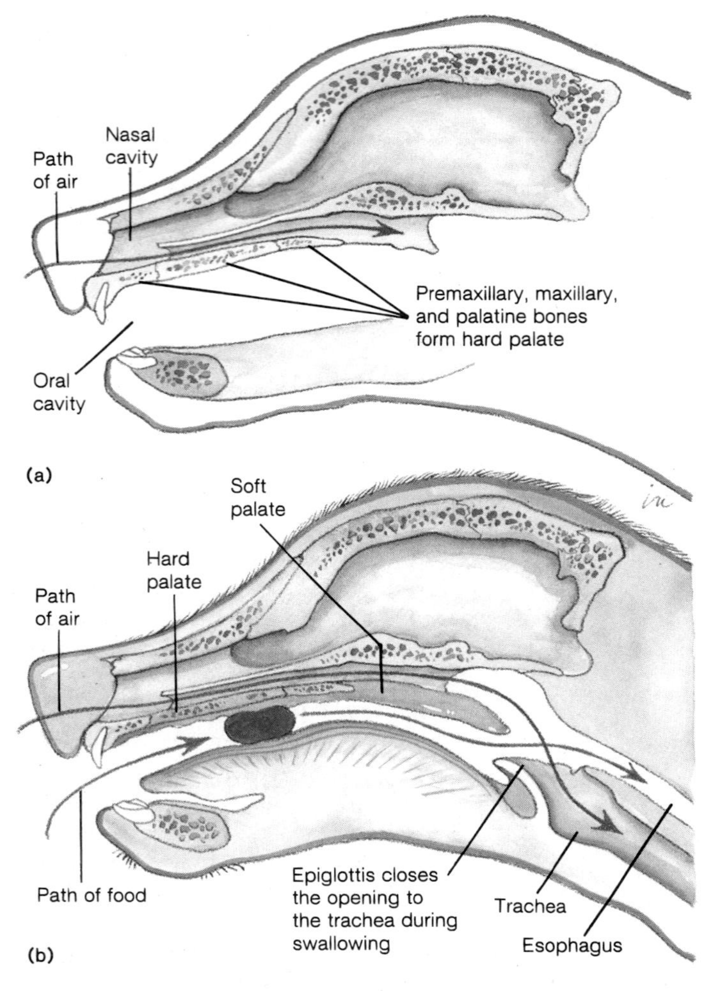

FIGURE 19.9

The Secondary Palate. (*a*) The secondary palate of a mammal provides a nearly complete separation between nasal and oral cavities. (*b*) Breathing stops only momentarily during swallowing.

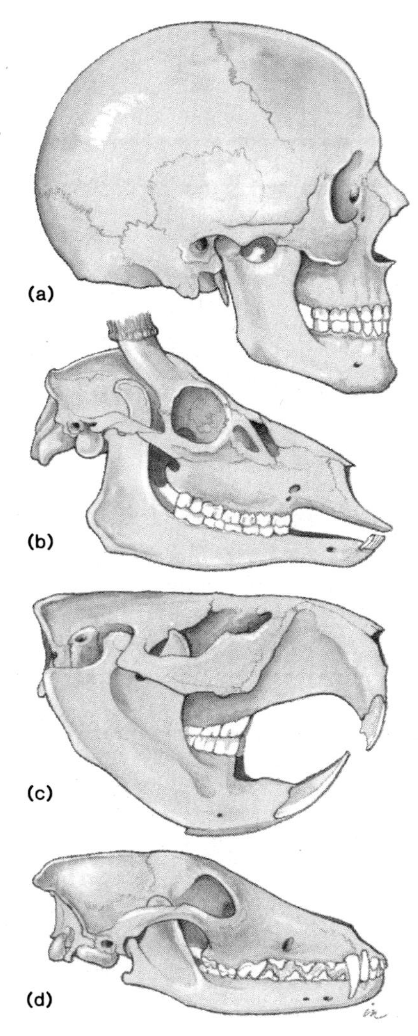

FIGURE 19.10

Specializations of Teeth. (*a*) An omnivore (*Homo sapiens*). (*b*) An herbivore, the male fallow deer (*Dama dama*). (*c*) A rodent, the beaver (*Castor canadensis*). (*d*) A carnivore, the coyote (*Canis latrans*).

The Vertebral Column and Appendicular Skeleton

The vertebral column of mammals is divided into five regions. As with reptiles and birds, the first two cervical vertebrae are the atlas and axis. These are usually followed by five other cervical vertebrae. Even the giraffe and the whale have seven neck vertebrae, which are greatly elongate or compressed, respectively.

The trunk is divided into thoracic and lumbar regions, as is the case for birds. In mammals, the division is correlated with their method of breathing. The thoracic region contains the ribs, which connect to the sternum via costal cartilage and protect the heart and lungs. The articulation between the thoracic vertebrae provides the flexibility needed in turning, climbing, and lying on the side to suckle young. Lumbar vertebrae have interlocking processes that give support, but little freedom of movement.

The appendicular skeleton of mammals is rotated under the body so that the appendages are usually directly beneath the body. Joints usually limit the movement of appendages to a single anteroposterior plane, causing the tips of the appendages to move in long arcs. The bones of the pelvic girdle are highly fused in the adult, a condition that is advantageous for locomotion, but presents problems during the birth of offspring. In a pregnant female, the ventral joint between the halves of the pelvis—the pubic symphysis—loosens before birth, allowing the pelvis to spread apart during birth.

Muscles

Because the appendages are directly beneath the body of most mammals, the weight of the body is borne by the skeleton. Muscle mass is concentrated in the upper appendages

Figure 19.11
Order Perissodactyla. The black rhino (*Diceros bicornis*) is an endangered herbivore that roams the plains of Africa. There are fewer than 4,000 living in Africa today.

and girdles. Many running mammals (e.g., deer, order Artiodactyla) have little muscle in their lower leg that would slow leg movement. Instead, tendons run from muscles high in the leg to cause movement at the lower joints.

Nutrition and the Digestive System

The digestive tract of mammals is similar to that of other vertebrates. There are, however, many specializations for different feeding habits. Some specializations of teeth have already been described.

It is difficult to make accurate generalizations regarding the feeding habits of mammals. Feeding habits are reflections of the ecological specializations that have evolved. For example, most members of the order Carnivora feed on animal flesh and are, therefore, carnivores. Other members of the order, such as bears, feed on a variety of plant and animal products and are omnivores. Some carnivorous mammals are specialized for feeding on arthropods or soft-bodied invertebrates and are often referred to (rather loosely) as insectivores. These include animals in the orders Insectivora (e.g., shrews), Chiroptera (bats), and Edentata (anteaters) (*see figure* 19.5a). Herbivores such as deer (order Artiodactyla) and rhinos (order Perissodactyla) (figure 19.11) feed mostly on vegetation, but their diet also includes invertebrates inadvertently ingested while feeding.

Specializations in the digestive tract of most herbivores reflect the difficulty of digesting food rich in cellulose. Horses, rabbits, and many rodents have an enlarged cecum at the junction of large and small intestines. A cecum serves as a fermentation pouch where microorganisms aid in the digestion of cellulose. Sheep, cattle, and deer are called ruminants (L. *ruminare*, to chew the cud). Their stomachs are modified into four chambers. The first three chambers are storage and fermentation chambers and contain microorganisms that synthesize a cellulose-digesting enzyme (cellulase). Gases produced by fermentation are periodically belched, and some plant matter (cud) is regurgitated and rechewed. Other microorganisms convert nitrogenous compounds in the food into new proteins.

Stop and Ask Yourself

5. What are the functions of hair in mammals?
6. What are the four kinds of glands found in mammals?
7. What is the usefulness of a dental formula?
8. What specializations of teeth and the digestive system are present in some herbivorous mammals?

Circulation, Gas Exchange, and Temperature Regulation

The hearts of birds and mammals are superficially similar. Both are four-chambered pumps that keep blood in the systemic and pulmonary circuits separate and both evolved from the hearts of ancient reptiles. Their similarities, however, are a result of adaptations to active life-styles. (5) The evolution of similar structures in different lineages is called convergent evolution. The evolution of the mammalian heart occurred in the synapsid reptilian lineage, whereas the avian heart evolved in the archosaur lineage (figure 19.12).

One of the most important adaptations in the circulatory system of eutherian mammals concerns the distribution of respiratory gases and nutrients in the fetus (figure 19.13*a*). Exchanges between maternal and fetal blood occur across the placenta. Although there is intimate association between maternal and fetal blood vessels, no actual mixing of blood occurs. Nutrients, gases, and wastes simply diffuse between fetal and maternal blood supplies.

Blood entering the right atrium of the fetus is returning from the placenta and is highly oxygenated. Because fetal lungs are not inflated, resistance to blood flow through the pulmonary arteries is high. Therefore, most of the blood entering the right atrium bypasses the right ventricle and passes instead into the left atrium through a valved opening between the atria (the foramen ovale). Some blood from the right atrium, however, does enter the right ventricle and the pulmonary artery. Because of the resistance at the uninflated lungs, most of this blood is shunted to the aorta through a vessel connecting the aorta and pulmonary artery (the ductus arteriosus). At birth, the placenta is lost, and the lungs are inflated. Resistance to blood flow through the lungs is reduced, and blood flow to them increases. Flow through the ductus arteriosus decreases, and the vessel is gradually reduced to a ligament. Blood flow back to the left atrium from the lungs correspondingly increases, and the valve of foramen ovale closes and gradually fuses with the tissue separating the right and left atria.

Gas Exchange

High metabolic rates are accompanied by adaptations for efficient gas exchange. The separate nasal and oral cavities and lengthening of the snout of most mammals provide an increased surface area for warming and moistening inspired air.

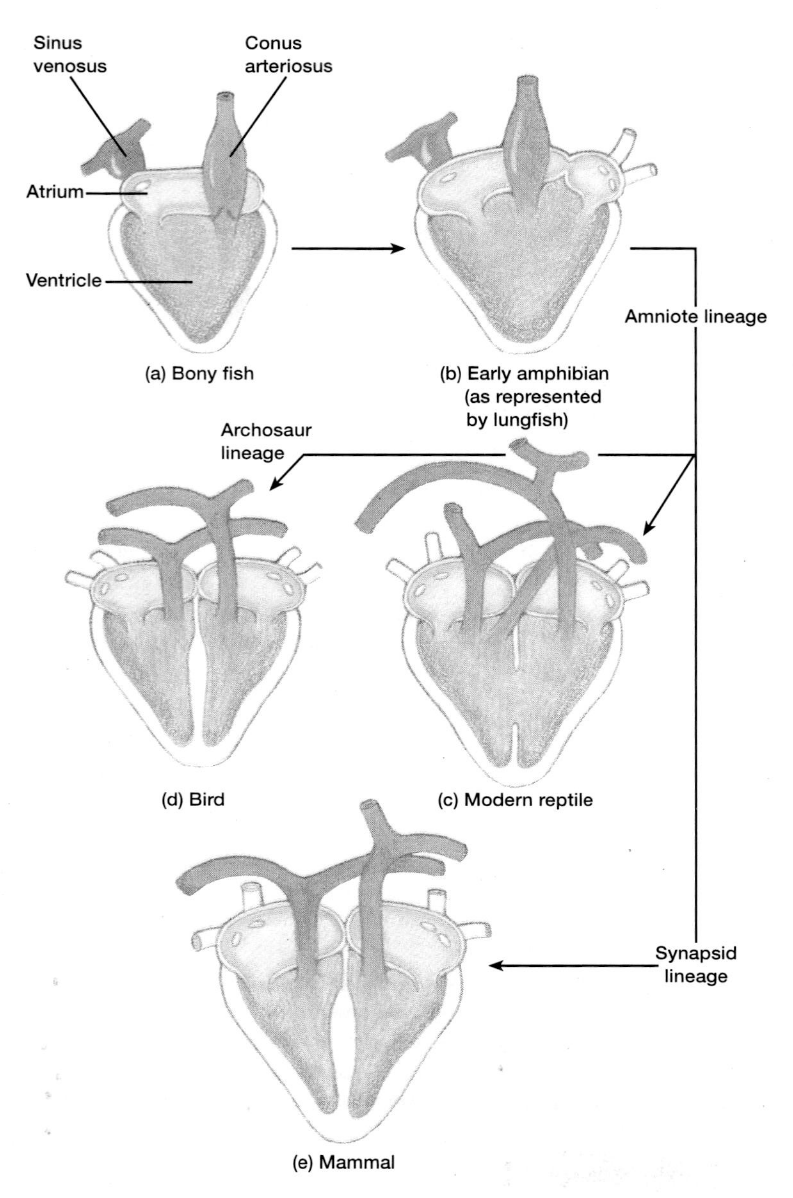

Figure 19.12

A Possible Sequence in the Evolution of the Vertebrate Heart. (*a*) Diagrammatic representation of a bony fish heart. (*b*) In lungfish, partially divided atria and ventricles separate pulmonary and systemic circuits. This heart was probably similar to that in primitive amphibians and early amniotes. (*c*) The hearts of modern reptiles were derived from the pattern shown in (*b*). (*d*) The archosaur and (*e*) synapsid lineages resulted in completely separated, four-chambered hearts.

Respiratory passageways are highly branched, and large surface areas exist for gas exchange. Mammalian lungs resemble a highly vascular sponge, rather than the saclike structures of amphibians and a few reptiles.

The lungs, like those of reptiles, are inflated using a negative-pressure mechanism. Unlike reptiles and birds, however, mammals possess a muscular **diaphragm** that separates thoracic and abdominal cavities. Inspiration results from contraction of the diaphragm and expansion of the rib cage, both of which decrease the intrathoracic pressure and allow air to enter the lungs. Expiration is by elastic recoil of the lungs. Forceful exhalation can be accomplished by contraction of thoracic and abdominal muscles.

Inspira^n contrac^n Expan^n

Temperature Regulation

Mammals are widely distributed over the earth, and some face harsh environmental temperatures. Nearly all face temperatures that require them to dissipate excess heat at some times and conserve and generate heat at other times.

Heat-producing mechanisms of mammals are divided into two categories. Shivering thermogenesis is muscular activity that results in the generation of large amounts of heat, but little movement. Nonshivering thermogenesis involves heat production by general cellular metabolism, and the metabolism of special fat deposits called brown fat.

Heat production is effective in thermoregulation because mammals are insulated by their pelage and/or fat deposits. Fat deposits are also sources of energy to sustain high metabolic rates.

Mammals without a pelage can conserve heat by allowing the temperature of surface tissues to drop. A walrus in cold, arctic waters has a surface temperature near 0° C; however, a few centimeters below the skin surface, body temperatures are about 35° C. Upon emerging from the icy water, the skin warms quickly by increasing peripheral blood flow. Most tissues cannot tolerate such rapid and extreme temperature fluctuations. Further investigations are likely to reveal some very unique biochemical characteristics of these skin tissues.

Even though most of the body of arctic mammals is unusually well insulated, appendages often have thin coverings of fur as an adaptation to changing thermoregulatory needs. Even in winter, an active mammal sometimes produces more heat than is required to maintain body temperature. Patches of poorly insulated skin allow excess heat to be dissipated. During periods of inactivity or extreme cold, however, heat loss from these exposed areas must be reduced, often by assuming heat-conserving postures. Mammals sleeping in cold environments conserve heat by tucking poorly insulated appendages and their faces under well-insulated body parts.

Countercurrent heat-exchange systems may help regulate heat loss from exposed areas (figure 19.14). Arteries passing peripherally through the core of an appendage are surrounded by veins that carry blood back toward the body. When blood returns to the body through these veins, heat is transferred from arterial blood to venous blood and returned to the body rather than lost to the environment. When excess heat is produced, blood is shunted away from the countercurrent veins toward peripheral vessels, and excess heat is radiated to the environment.

Mammals have few problems getting rid of excess heat in cool, moist environments. Heat can be radiated into the air from vessels near the surface of the skin or lost by evaporative cooling from either sweat glands or respiratory surfaces during panting.

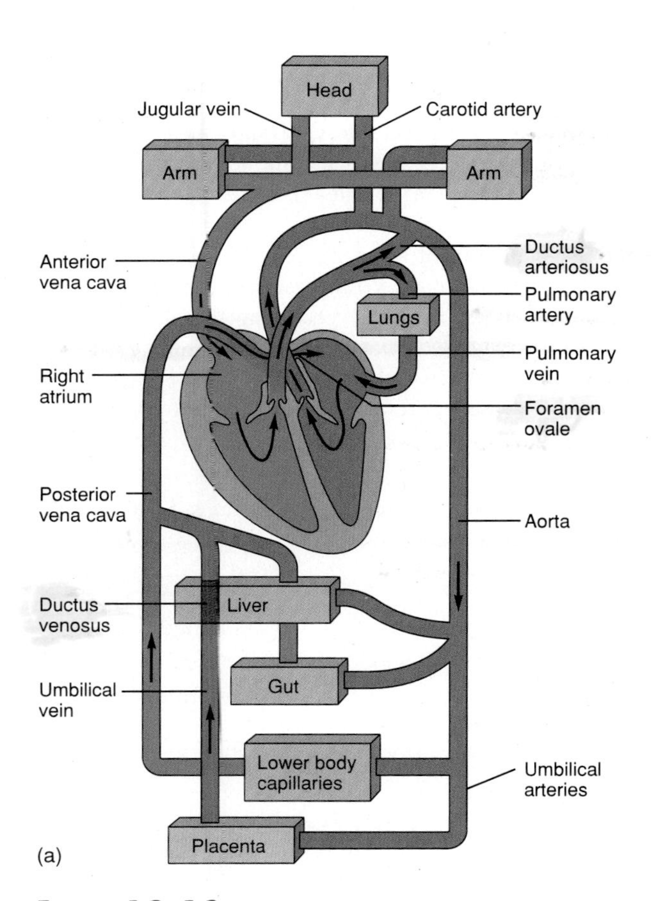

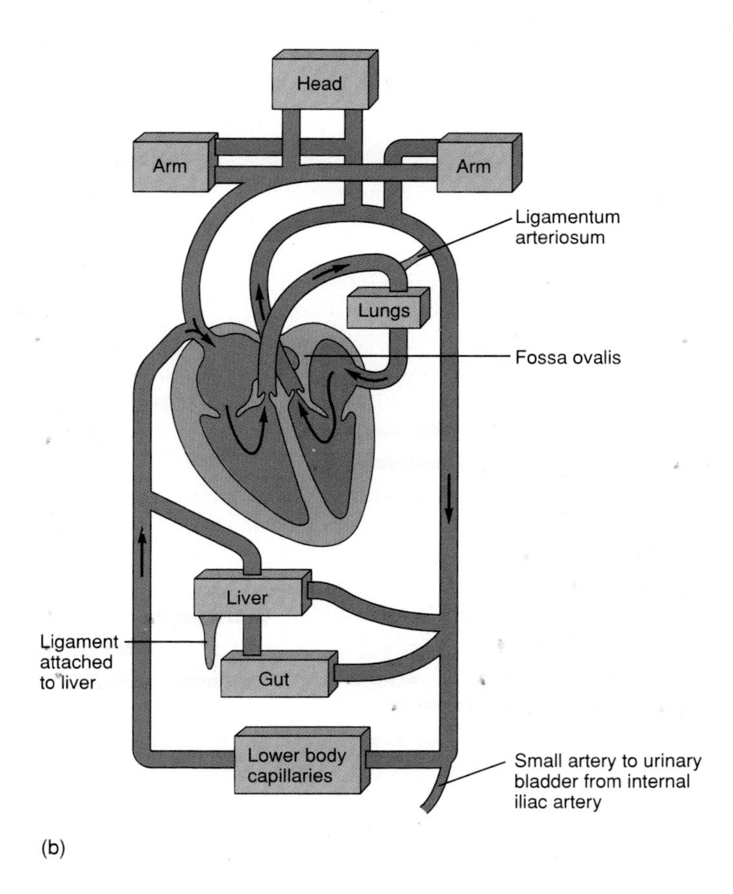

FIGURE 19.13
Mammalian Circulatory Systems. The circulatory patterns of (*a*) fetal and (*b*) adult mammals. Highly oxygenated blood is shown in red, and poorly oxygenated blood shown in blue.

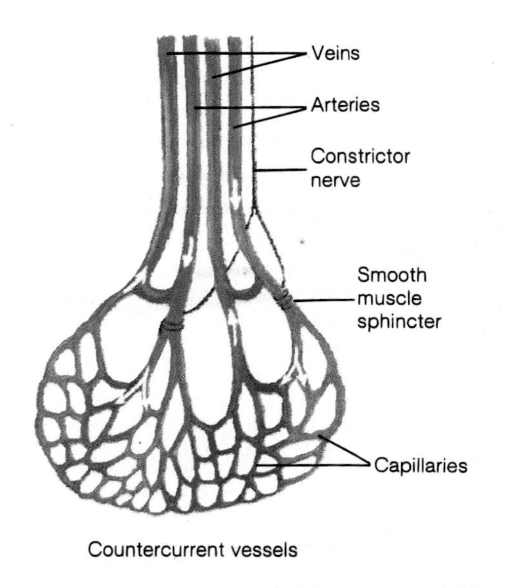

FIGURE 19.14
Countercurrent Heat Exchange. Countercurrent heat exchangers conserve body heat in animals adapted to cold environments. Systems similar to the one depicted here are found in the legs of reindeer (*Rangifer tarandus*) and in the flippers of dolphins. Venous blood returning from an extremity is warmed by heat transferred from blood moving peripherally in arteries. During winter, the lower part of a reindeer's leg may be at 10° C, while body temperature is about 40° C. Arrows indicate direction of blood flow.

Hot, dry environments present far greater problems because evaporative cooling may upset water balances. The large ears of jackrabbits and elephants are used to radiate heat. Small mammals often avoid the heat by remaining in burrows during the day and foraging for food at night. Other mammals seek shade or watering holes for cooling.

Winter Sleep and Hibernation

Mammals react in various ways to environmental extremes. Caribou migrate to avoid extremes of temperature, and wildebeest migrate to avoid seasonal droughts. Other mammals retreat to burrows under the snow where they become less active, but are still relatively alert and easily aroused—a condition called **winter sleep.** For example, bears and raccoons retreat to dens in winter. Their body temperatures and metabolic rates decrease somewhat, but they do not necessarily remain inactive all winter.

Hibernation is a period of winter inactivity in which the hypothalamus of the brain slows the metabolic, heart, and respiratory rates. True hibernators include the monotremes (echidna and duckbill platypus) and many members of the Insectivora (e.g., moles and shrews), Rodentia (e.g., chipmunks and woodchucks), and Chiroptera (bats). In preparation for hibernation, mammals usually accumulate large quantities of body fat. After retreating to a burrow or a nest, the hypothalamus sets the body's thermostat to about 2° C. The respiratory rate of a hibernating ground squirrel falls from 100 to 200 breaths per minute to about 4 breaths per minute. The heart rate falls from 200 to 300 beats per minute to about 20 beats per minute. During hibernation, a mammal may lose ⅓ to ½ of its body weight. Arousal from hibernation occurs by metabolic heating, frequently using brown fat deposits, and it takes several hours to raise body temperature to near 37° C.

Nervous and Sensory Functions

The basic structure of the vertebrate nervous system is retained in mammals. The development of complex nervous and sensory functions goes hand-in-hand with active life-styles and is most evident in the enlargement of the cerebral hemispheres and the cerebellum of mammals. The enlargement of the cerebral cortex (neocortex) is accompanied by most integrative functions being shifted to this region.

In mammals, the sense of touch is well developed. Receptors are associated with the bases of hair follicles and are stimulated when a hair is displaced.

Olfaction was apparently an important sense in early mammals, because fossil skull fragments show elongate snouts, which would have contained olfactory epithelium. Cranial casts of fossil skulls show enlarged olfactory regions. Olfaction is still an important sense for many mammals. Olfactory stimuli can be perceived over long distances during either the day or night and are used to locate food, recognize members of the same species, and avoid predators.

Auditory senses were similarly important to early mammals. More recent adaptations include an ear flap (the pinna) and the auditory tube leading to the tympanum that directs sound to the middle ear. The sensory patch of the inner ear that contains receptors for sound is long and coiled and is called the cochlea. This structure provides more surface area for receptor cells and allows mammals greater sensitivity to pitch and volume than is present in reptiles (box 19.2). Cranial casts of early mammals show well-developed auditory regions.

Vision is an important sense in many mammals, and the structure of the eye is similar to that described for other vertebrates. Accommodation occurs by changing the shape of the lens. Color vision is less well developed in mammals than in reptiles and birds. The fact that the retinas of most mammals are dominated by rods supports the hypothesis that early mammals were nocturnal. Primates, squirrels, and a few other mammals have well-developed color vision.

Stop and Ask Yourself

9. In what way are the hearts of birds and mammals an example of convergent evolution?
10. What are two ways that mammals can generate metabolic heat?
11. How is excessive heat loss from the poorly insulated legs of a reindeer prevented?
12. What is the difference between hibernation of a ground squirrel and the winter sleep of a bear?

Excretion and Osmoregulation

Mammals, like all amniotes, have a metanephric kidney. Unlike reptiles and birds, which excrete mainly uric acid, mammals excrete urea. Urea is less toxic than ammonia and does not require large quantities of water in its excretion. Unlike uric acid, however, urea is highly water soluble and cannot be excreted in a semisolid form; thus, some water is lost. Excretion in mammals is always a major route for water loss.

In the nephron of the kidney, fluids and small solutes are filtered from the blood through the walls of a group of capillary-like vessels, called the glomerulus. The remainder of the nephron consists of tubules that reabsorb water and essential solutes and secrete particular ions into the filtrate.

The primary adaptation of the mammalian nephron is a portion of the tubule system called the loop of the nephron. The transport processes in this loop and the remainder of the tubule system allow mammals to produce urine that is 2 to 22 times more concentrated than blood (e.g., beavers and Australian hopping mice, respectively). 6 This accomplishes the same function that nasal and orbital salt glands do in reptiles and birds.

Water loss varies greatly depending on activity, physiological state, and environmental temperature. Water is lost in urine, feces, evaporation from sweat glands and respiratory surfaces, and during nursing. Mammals in very dry environments have numerous behavioral and physiological mechanisms to

BOX 19.2 Mammalian Echolocation

Imagine a pool's water made so murky that a human is only able to see a few centimeters below the surface. Also imagine a clear Plexiglas sheet with a dolphin-sized opening in the middle, dividing the pool in half. At one end of the pool is an eager dolphin; at the other end is a trainer. The trainer throws a dead fish into the water, and on signal, the dolphin unhesitatingly finds its way through the murky water to the opening in the Plexiglas and then finds the fish at the other end of the pool.

Although the dolphin in the above account was trained to find the fish, it relied on a sense that it shares with a few other mammals. Toothed whales, bats, and some shrews use the return echoes of high-frequency sound pulses to locate objects in their environment. This mechanism is called **echolocation.**

Echolocation has been studied in bats more completely than in any other group of mammals. The Italian scientist Lassaro Spallanzani discovered in the late 1700s that blinded bats could navigate successfully at night, whereas bats with plugged ears could not. Spallanzani believed that echoes of the sounds made by beating wings were used in echolocation. In 1938, however, ultrasonic bat cries (inaudible to humans) were electronically recorded, and their function in echolocation was described.

Insect-eating bats navigate through their caves and the night sky and locate food by echolocation. During normal cruising flight, ultrasonic (100 to 20 kHz) "clicks" are emitted approximately every 50 milliseconds. As insect prey is detected, the number of clicks per second increases, the duration between clicks decreases, and the wavelength of the sound decreases, increasing directional precision and making small flying insects more easily detected. On final approach, the sound becomes buzzlike, and the bat scoops up the insect with its wings or in the webbing of its hind legs (figure 1).

FIGURE 1 **Bat Echolocation.** A greater horseshoe bat (*Rhinophus*) capturing a moth.

Modifications of the bat ear and brain allow bats to perceive faint echoes of their vocalizations and to precisely determine direction and distance. Enlarged ear flaps funnel sounds toward particularly thin eardrums and very sensitive ear ossicles. The auditory regions of the bat brain are very large, and special neural pathways enhance a bat's ability to determine the direction of echoes.

Bats must distinguish echoes of their own cries from the cries themselves and from other noises. Leaflike folds of the nostrils direct sound emitted from the nostrils forward, rather than in all directions from the head, much like the megaphone of a cheerleader. The ears, therefore, receive little stimulation from direct vocalizations. Fat and blood sinuses surrounding the middle and inner ears reduce transmission of sound from the mouth and pharynx. Some bats temporarily turn off their hearing during sound emission by making the ear ossicles insensitive to sound waves and then turn on their hearing an instant later when the reflected sound is returning to the bat.

reduce water loss. The kangaroo rat, named for its habit of hopping on large hind legs, is capable of extreme water conservation (figure 19.15). It is native to the southwestern deserts of the United States and survives without drinking water. Its feces are almost dry, and evaporative water loss is reduced by its nocturnal habits. Respiratory water loss is minimized by condensation as warm air in the respiratory passages encounters the cooler nasal passages. Excretory water loss is minimized by a diet low in protein, which reduces the production of urea. The nearly dry seeds that the kangaroo rat eats are rich sources of carbohydrates and fats. Metabolic oxidation of carbohydrates produces water as a by-product.

Behavior

Mammals have complex behaviors that enhance survival. Visual cues are often used in communication. The bristled fur, arched back, and open mouth of a cat communicates a clear message to curious dogs or other potential threats. A tail-wagging display of a dog has a similarly clear message. A wolf defeated in a fight with other wolves lies on its back and exposes its vulnerable throat. Similar displays may allow a male already recognized as being subordinate to another male to avoid conflict within a social group.

Pheromones are used to recognize members of the same species, members of the opposite sex, and the reproductive state

(a)

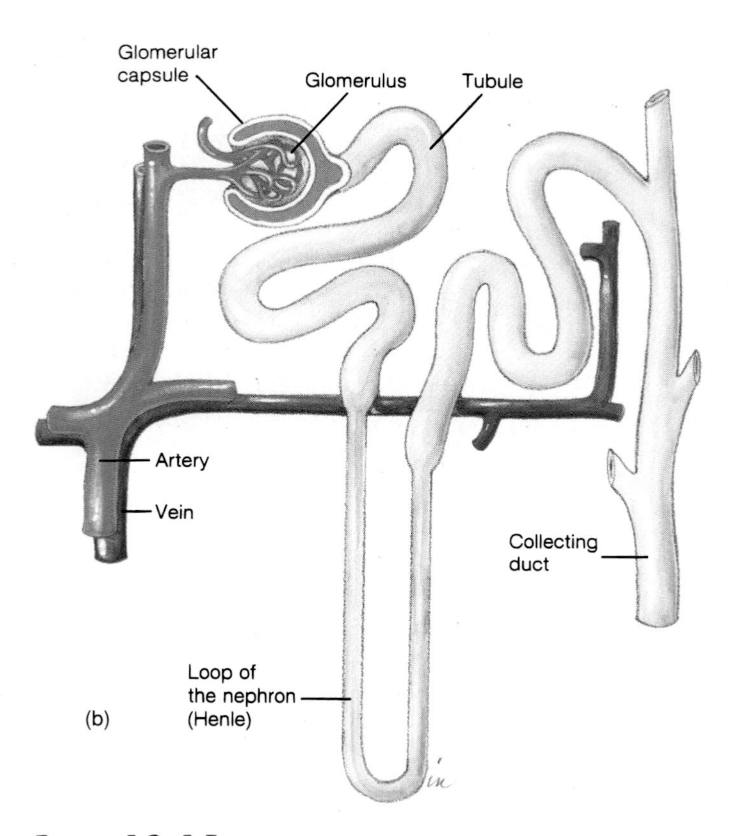

Figure 19.15
Order Rodentia. (*a*) The kangaroo rat (*Dipodomys ordii*). (*b*) The long loop of the nephron of this desert animal acts to conserve water, preventing dehydration.

of a member of the opposite sex. Pheromones may also induce sexual behavior, help establish and recognize territories, and ward off predators. The young of many mammalian species recognize their parents, and parents recognize their young, by smell. Bull elk smell the rumps of females during the breeding season to recognize those in their brief receptive period. They also urinate on their own belly and underhair to advertise their reproductive status to females and other males. Male mammals urinate on objects in the environment to establish territories and to allow females to become accustomed to their odors. Rabbits and rodents spray urine on a member of the opposite sex to inform the second individual of the first's readiness to mate. Skunks use chemicals to ward off predators.

Auditory and tactile communication are also important in the lives of mammals. Herd animals are kept together and remain calm as long as the array of familiar sounds (e.g., bellowing, hooves walking over dry grasses and twigs, and rumblings from ruminating stomachs) are uninterrupted. Unfamiliar sounds may cause alarm and flight.

Vocalizations and tactile communication are important in primate social interactions. Tactile communication ranges from precopulatory "nosing" that occurs in many mammals to grooming. Grooming does much more than help maintain a healthy skin and pelage. It reinforces important social relationships within primate groups.

Territoriality

Many mammals mark and defend certain areas from intrusion by other members of the same species. 7 (When cats rub their face and neck on us or on furniture in our homes, we like to think the cat is being affectionate. Cats, however, are really staking claim to their territory, using odors from facial scent glands.) Some territorial behavior attracts females to, and excludes other males from, favorable sites for mating and rearing young.

Male California sea lions (*Zalophus californianus*) establish territories on shorelines where females come to give birth to young. For about 2 weeks, males engage in vocalizations, displays, and sometimes serious fighting to stake claim to favorable territories (figure 19.16). Older, dominant bulls are usually most successful in establishing territories, and young bulls generally swim and feed just offshore. When they arrive at the beaches, females select a site for giving birth. Selection of the birth site also selects the bull that will father next year's offspring. Mating occurs approximately 2 weeks after the birth of the previous year's offspring. Development is arrested for the 3 months during which the recently born young do most of their nursing. This mechanism is called embryonic diapause. Thus, even though actual development takes about 9 months, the female carries the embryo and fetus for a period of 1 year.

Reproduction and Development

In no other group of animals has viviparity developed to the extent it has in mammals. It requires a large expenditure of energy on the part of the female during development and on the part of one or both parents caring for young after they are born. Viviparity is advantageous because females are not necessarily tied to a single nest site, but can roam or migrate to find food or a proper climate. Viviparity is accompanied by the evolution of a portion of

FIGURE 19.16
Order Carnivora. California sea lions (*Zalophus californianus*) on a beach during the breeding season. The adult males in the foreground are vocalizing and posturing.

the reproductive tract where the young are nourished and develop. In viviparous mammals, the oviducts are modified into one or two uteri (s., uterus).

Reproductive Cycles

Most mammals have a definite time or times during the year in which ova (eggs) mature and are capable of being fertilized. Reproduction usually occurs when climatic conditions and resource characteristics favor successful development. Mammals living in environments with few seasonal changes and those that exert considerable control of immediate environmental conditions (e.g., humans) may reproduce at any time of the year. However, they are still tied to physiological cycles of the female that determine when ova can be fertilized.

Most female mammals undergo an **estrus** (Gr. *oistros*, a vehement desire) **cycle**, which includes a time during which the female is behaviorally and physiologically receptive to the male. During the estrus cycle, hormonal changes stimulate the maturation of ova in the ovary and induce ovulation (release of one or more mature ova from an ovarian follicle). A few mammals (e.g., rabbits, ferrets, and mink) are induced ovulators; ovulation is induced by **coitus.**

Hormones also mediate changes in the uterus and vagina. As the ova are maturing, the inner lining of the uterus proliferates and becomes more vascular in preparation for receiving developing embryos. Proliferation of vaginal mucosa is accompanied by external swelling in the vaginal area and increased glandular discharge. During this time, males show heightened interest in females, and females are receptive to males. If fertilization does not occur, the above changes in the uterus and vagina are reversed until the next cycle begins. No bleeding or sloughing of uterine lining usually occurs.

Many mammals are **monestrus** and go through only a single yearly estrus cycle that is sharply seasonal. Wild dogs, bears, and sea lions are monestrus; domestic dogs are diestrus. Other mammals are **polyestrus**. Rats and mice have estrus cycles that are repeated every 4 to 6 days.

The menstrual cycle of female humans, apes, and monkeys is similar to the estrus cycle in that it results in a periodic proliferation of the inner lining of the uterus and is correlated with the maturation of an ovum. If fertilization does not occur before the end of the cycle, **mensus**—the sloughing of the uterine lining—occurs.

Fertilization usually occurs in the upper ⅓ of the oviduct within hours of copulation. In a few mammals, fertilization may be delayed. In some bats, for example, coitus occurs in autumn, but fertilization is delayed until spring. Females store sperm in the uterus for periods in excess of 2 months. This example of **delayed fertilization** is apparently an adaptation to winter dormancy. Fertilization can occur immediately after females emerge from dormancy rather than having to wait until males attain their breeding state.

In many other mammals, fertilization occurs right after coitus, but development is arrested after the first week or two. This **embryonic diapause** was described previously for sea lions, and also occurs in some bats, bears, martens, and marsupials. The adaptive significance of embryonic diapause varies with species. In the sea lion, embryonic diapause allows the mother to give birth and mate within a short interval, but not have her resources drained by both nursing and pregnancy. It also allows young to be born at a time when resources favor their survival. In some bats, it allows fertilization to occur in the fall before hibernation, but birth is delayed until resources become abundant in the spring.

Modes of Development

Monotremes are **oviparous**. Ova are released from the ovaries with large quantities of yolk. After fertilization, shell glands in the oviduct deposit a shell around the ovum, forming an egg. Female echidnas incubate eggs in a ventral pouch. Platypus eggs are laid in their burrows.

All other mammals nourish young by a placenta through at least a portion of their development. Nutrients are supplied from the maternal bloodstream, not yolk.

In **marsupials**, most nourishment for the fetus comes from "uterine milk" secreted by uterine cells. Some nutrients diffuse from maternal blood into a highly vascular yolk sac that makes contact with the uterus. This connection in marsupials is a primitive placenta. The **gestation period** (the length of time young develop within the female reproductive tract) varies between 8 and 40 days in different species. The short gestation period is a result of the inability to sustain the production of hormones that maintain the uterine lining. After birth, tiny young crawl into the marsupium, and attach to a nipple, where they suckle for an additional 60 to 270 days (figure 19.17).

In **eutherian mammals**, the embryo implants deeply into the uterine wall. Embryonic and uterine tissues grow rapidly and become highly folded and vascular, forming the placenta. Although maternal and fetal blood do not mix, nutrients, gases, and wastes diffuse between the two bloodstreams. Gestation periods of eutherian mammals vary widely between 20 days (some

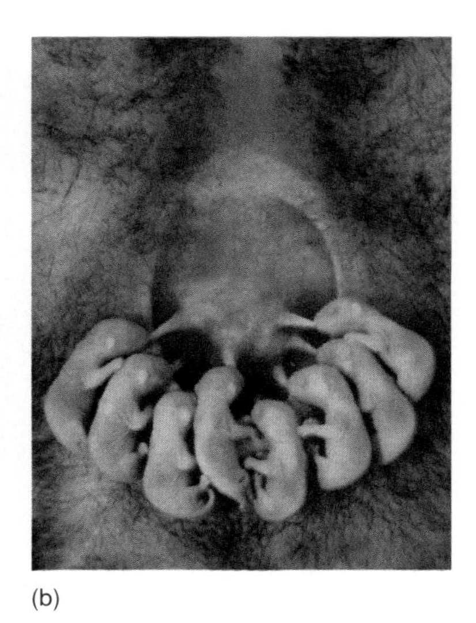

FIGURE 19.17
Order Marsupialia. (*a*) An opossum (*Didelphis marsupialis*) with young. (*b*) Opossum young nursing in a marsupial pouch.

rodents) and 19 months (the African elephant). Following birth, the placenta and other tissues that surrounded the fetus in the uterus are expelled as "afterbirth." The newborns of many species are helpless at birth (e.g., humans); others can walk and run shortly after birth (e.g., deer and horses).

Stop and Ask Yourself

13. What adaptation of the kidney allows mammals to excrete urine that is hypertonic to the blood?
14. What are the advantages of viviparity for mammals? In what way is viviparity costly?
15. What is embryonic diapause?
16. Why is the gestation period of a marsupial so short?

SUMMARY

1. Mammalian characteristics evolved in the synapsid lineage over a period of about 200 million years. Mammals evolved from a group of synapsids called therapsids.
2. Modern mammals include the monotremes, marsupial mammals, and placental mammals.
3. Hair is uniquely mammalian. It functions in sensory perception, temperature regulation, and communication.
4. Sebaceous, sudoriferous, scent, and mammary glands are present in mammals.
5. The teeth and digestive tracts of mammals are adapted for different feeding habits. Herbivores are characterized by flat, grinding teeth and fermentation structures for digesting cellulose. Predatory mammals have sharp teeth for killing and tearing prey.
6. The heart of mammals has four chambers, and circulatory patterns are adapted for viviparous development.
7. Mammals possess a diaphragm that alters intrathoracic pressure, which helps ventilate the lungs.
8. Metabolic heat production, insulating pelage, and behavior are used in mammalian thermoregulation.
9. Mammals react to unfavorable environments by migration, winter sleep, and hibernation.
10. The nervous system of mammals is similar to that of other vertebrates. Olfaction and hearing were important for early mammals. Vision, hearing, and smell are the dominant senses in many modern mammals.
11. The nitrogenous waste of mammals is urea, and the kidney is adapted for excreting a concentrated urine.
12. Mammals have complex behavior to enhance survival. Visual cues, pheromones, and auditory and tactile cues are important in mammalian communication.
13. Most mammals have specific times during the year when reproduction occurs. Estrus or menstrual cycles are present in female mammals. Monotremes are oviparous. All other mammals nourish young by a placenta.

Selected Key Terms

cecum (*p. 331*)
delayed fertilization (*p. 337*)
dental formula (*p. 328*)
diaphragm (*p. 332*)
embryonic diapause (*p. 337*)
estrus cycle (*p. 337*)
gestation period (*p. 337*)
heterodont (*p. 327*)
homodont (*p. 327*)

Critical Thinking Questions

1. Why is tooth structure important in the study of mammals?
2. What does the evolution of secondary palates have in common with the evolution of completely separated, four-chambered hearts?
3. Why is the classification of mammals by feeding habits not particularly useful to phylogenetic studies?
4. Under what circumstances is endothermy disadvantageous for a mammal?
5. Discuss the possible advantages of embryonic diapause for marsupials that live in climatically unpredictable regions of Australia.

glossary

A

abdomen (ab′do-men) 1. The portion of a tetrapod's body between the thorax and pelvic girdle. 2. The region of an arthropod's body behind the thorax. It contains the visceral organs. 206

aboral (ab-or′al) The end of a radially symmetrical animal opposite the mouth. 58

acanthella (a-kan′thel-a) Developing acanthocephalan larva between an acanthor and a cystacanth, in which the definitive organ systems are developed; develops in the intermediate host. 136

Acanthocephala (a-kan′tho-sef-a-la) The phylum of aschelminths commonly called the spiny-headed worms. 128

acanthor (a-kan′thor) Acanthocephalan larva (first larval stage) that hatches from the egg. The larva has a rostellum with hooks that are used in penetrating the host's tissues. 136

acetabulum (as″e-tab′u-lum) Sucker; the ventral sucker of a fluke; a sucker on the scolex of a tapeworm. 114

Acetospora (ah-seat-o-spor′ah) The protozoan phylum characterized by members having multicellular spores; all parasitic in invertebrates. Examples: acetosporans (*Paramyxa, Halosporidium*). 72

acid rain The combination of sulfur dioxide and nitrogen oxides with water in the atmosphere. This combination produces acidic precipitation called acid rain. The burning of fossil fuels is a major contributor to acid rain. 19

acoelomate (a-se′lah-māt) Without a body cavity. 59

adaptation Structures or processes that increase an organism's potential to successfully reproduce in a specified environment. 30

adaptive radiation Evolutionary change that results in the formation of a number of new characteristics from an ancestral form. 43

Adenophorea (a-den″o-for′e-a) The class of nematodes formerly called Aphasmidia. Examples: *Trichinella, Trichuris*. 123

adhesive gland Attachment glands in Turbellaria that produce a chemical that attaches part of the turbellarian to a substrate. 109

aestivation (es′te-va-shun) The condition of dormancy or torpidity during the hot summer months. 9

age structure The proportion of a population that is in prereproductive, reproductive, and postreproductive classes. 18

Agnatha (ag-nath′ah) A superclass of vertebrates whose members lack jaws and paired appendages and possess a cartilaginous skeleton and a persistent notochord. Lampreys and hagfishes. 237

airfoil A surface, such as a wing, that provides lift by using currents of air it moves through. 309

allopatric speciation (al′o-pat′rik spe′se-a′shun) Speciation that occurs in populations separated by geographical barriers. 43

altricial (al-trish′al) An animal that is helpless at hatching or birth. 316

alula (al′u-lah) A group of feathers on the wing of a bird that is supported by the bones of the medial digit. The alula reduces turbulent airflow over the upper surface of the wing. 309

ambulacral groove (am′byul-ac″ral groov) The groove along the length of the oral surface of a sea star arm. Ambulacral grooves contain tube feet. 224

ametabolous metamorphosis (a′me-tab′a-lus met′ah-mor′fe-sis) Development in which the number of molts is variable; immature stages resemble adults, and molting continues into adulthood. 212

amictic (ē-mik′tic) Pertaining to female rotifers that produce only diploid eggs that cannot be fertilized. The eggs develop directly into amictic females. 128

amniote lineage (am′ne-ōt lin′e-ij) The evolutionary lineage of vertebrates leading to modern reptiles, birds, and mammals. 272

amniotic egg (am′ne-ot-ik) The egg of reptiles, birds, and mammals. It possesses a series of extraembryonic membranes that help prevent desiccation, store wastes, and promote gas exchange. These adaptations allowed vertebrates to invade terrestrial habitats. 285, 288

Amphibia (am-fib′e-ah) The class of vertebrates whose members are characterized by skin with mucoid secretions, which serves as a respiratory organ. Developmental stages are aquatic and are usually followed by metamorphosis to an amphibious adult. Frogs, toads, and salamanders. 253

amphid (am″fed) One of a pair of chemosensory organs found on the anterior end of certain nematodes. 130

amplexus (am-plek′sus) The positioning of a male amphibian dorsal to female amphibian, his forelimbs around her waist. During amplexus, the male releases sperm as the female releases eggs. 283

analogous (a-nal′a-ges) Structures that have similar functions in two organisms but have not evolved from a common ancestral form. 7

Animalia (an′i-mal′eah) The kingdom of organisms whose members are multicellular, eukaryotic, and heterotrophic. The animals. 51

Annelida (ah-nel′i-dah) The phylum of triploblastic, coelomate animals whose members are metameric (segmented) and wormlike. Annelids have a complete digestive tract and a ventral nerve cord. 151

annuli (an′u-li) Secondary divisions of each body segment of a leech (phylum Annelida, class Hirudinea). 173

anterior The head end; usually the end of a bilateral animal that meets its environment. 58

Anthozoa (an′tho-zo″ah) The class of cnidarians whose members are solitary or colonial polyps. Medusae absent; gametes originate in the gastrodermis; mesenteries divide the gastrovascular cavity. Sea anemones and corals. 90

Apicomplexa (a′pi-kom-plex′ah) The protozoan phylum characterized by members having an apical complex used for penetrating host cells; cilia and flagella lacking, except in certain reproductive stages. Examples include the gregarines (*Monocystis*), coccidians (*Eimeria, Isospora, Sarcocystis, Toxoplasma*), *Pneumocystis, and Plasmodium*. 70

Aplacophora (a′pla-kof″o-rah) The class of molluscs whose members lack a shell, mantle, and foot. Wormlike, burrowing animals with head poorly developed. Some authors divide this group into two classes: Caudofoveata and Solengasters. 149

aposematic coloration (ah′pos-mat′ik) Sharply contrasting colors of an animal that warn other animals of unpleasant or dangerous effects. 12

Arachnida (ah-rak′ni-dah) The class of chelicerate arthropods whose members are mostly terrestrial, possess book lungs, or tracheae, and usually have four pairs of walking legs as adults. Spiders, scorpions, ticks, mites, and harvestmen. 174

Aristotle's lantern The series of ossicles making up the jawlike structure of echinoid echinoderms. 230

Arthropoda (ar-thra-po′dah) The phylum of animals whose members possess metamerism with tagmatization, a jointed exoskeleton, and a ventral nervous system. Insects, crustaceans, spiders, and related animals. 169

aschelminth (ask′hel-minth) Any animal in the phyla Gastrotricha, Rotifera, Kinorhyncha, Nematoda, Nematomorpha, Acanthocephala, Loricitera, Priapulida, or Entoprocta. 124

Ascidiacea (as-id′e as″e-ah) A class of urochordates whose members are sessile as adults, and solitary or colonial. 225

ascon (as′kon) The simplest of the three sponge body forms. Asconoid sponges are vaselike, with choanocytes directly lining the spongocoel. 87

Asteroidea (as′te-roi″de-ah) The class of echinoderms whose members have rays that are not sharply set off from the central disk; ambulacral grooves with tube feet; and suction disks on tube feet. Sea stars. 210

asymmetry (a-sim′i-tre) Without a balanced arrangement of similar parts on either side of a point or axis. 57

asynchronous flight *See* **indirect flight.**

auricle (aw′re-kl) The portion of the external ear not connected within the head. Also used to designate an atrium of a heart. In the class Turbellaria, the sensory lobes that project from the side of the head. 111

autotomy (au-tot′o-me) The self amputation of an appendage. For example, the casting off of a section of a lizard's tail caught in the grasp of a predator. The autotomized appendage is usually regenerated. 295

autotrophic (au′to-trōf″ic) Having the ability to synthesize food from inorganic compounds. 9

Aves (a′vez) A class of vertebrates whose members are characterized by scales modified into feathers for flight, endothermy, and amniotic eggs. The birds. 283

axopodium (ak′se-pōd-eum) Fine, needlelike pseudopodium that contains a central bundle of microtubules. Also called axopod. Found in certain sarcodine protozoa. 71

B

balanced polymorphism (pol′e-morf-ism) Occurs when different phenotypic expressions are maintained at a relatively stable frequency in a population. 42

Bdelloidea (del-oid′e-a) A class of rotifers containing members where there are no males; anterior end retractile and bearing two disks; mastax adapted for grinding; paired ovaries; cylindrical body. Example: *Rotaria*. 120

bilateral symmetry (bi-lat′er-al sim′i-tre) A form of symmetry in which only the midsagittal plane will divide an organism into mirror images; bilateral symmetry is characteristic of actively moving organisms that have definite anterior (head) and posterior (tail) ends. 58

binary fission (bi′ne-re fish′en) Asexual reproduction in protists in which mitosis is followed by cytoplasmic division, producing two new organisms. 65

biodiversity The variety of organisms in an ecosystem. 19

biogeochemical cycles The cycling of elements between reservoirs of inorganic compounds and living matter in an ecosystem. 14

biogeography The study of the distribution of life on earth. 4

biological magnification The concentration of substances in animal tissues as the substances are passed through ecosystem food webs. 19

biramous appendages (bi-ra′mus ah-pen′dij-ez) Appendages having two distal processes connected to the body by a single proximal process. 185

Bivalvia (bi″val′ve-ah) The class of molluscs whose members are enclosed in a shell consisting of two dorsally hinged valves, lack a radula, and possess a wedge-shaped foot. Clams, mussels, oysters. 149

bladder worm The unilocular hydatid cyst of a tapeworm. *See* **cysticercus.** 119

book gill Modifications of a horseshoe crab's exoskeleton into a series of leaflike plates that serve as a surface for gas exchange between the arthropod and the water (phylum Arthropoda, class Merostomata). 187

book lung Modification of the arthropod exoskeleton into a series of internal plates that provide surfaces for exchange of gases between the blood and air. 188

bothria (both-re-ah) Dorsal or ventral grooves, which may be variously modified, on the scolex of a cestode. 119

bottleneck effect Changes in gene frequency that result when numbers in a population are drastically reduced, and genetic variability is reduced as a result of the population being built up again from relatively few surviving individuals. 40

Brachiopoda (bra-ke-op′o-dah) A phylum of marine animals whose members possess a bivalved calcareous and/or chitinous shell that is secreted by a mantle and encloses nearly all of the body. Unlike the molluscs, the valves are dorsal and ventral. Possess a lophophore. Lampshells. 235

buccal pump (buk′el) The mechanism by which lung ventilation occurs in amphibians; muscles of the mouth and pharynx create positive pressure to force air into the lungs. 279

buccopharyngeal respiration (buk′o-fah-rin′je-al res′pah-ra′shun) The diffusion of gases across moist linings of the mouth and pharynx of amphibians. 279

budding The process of forming new individuals asexually in many different invertebrates. 65

bursa (bur′sah) A membranous sac that invaginates from the oral surface of ophiuroid echinoderms. Functions in diffusion of gases and waste material. 228

C

Calcarea (kal-kar′ea) The class of sponges whose members are small and possess monaxon, triaxon, or tetraaxon calcium carbonate spicules. 87

calyx (ka′liks) 1. A boat-shaped or cuplike central body of an entoproct or crinoid. The body and tentacles of an entoproct. 2. A cuplike set of ossicles that support the crown of a sea lily or feather star (class Crinoidea, phylum Echinodermata). 138, 232

camouflage (kam′ah-flazh) The presence of body color patterns and/or shapes that hide or disguise an animal. *See* **countershading** and **cryptic coloration.** 12

carapace (kar′ah-pās) The dorsal portion of the shell of a turtle. Formed from a fusion of vertebrae, ribs, and dermal bone. 291

carrying capacity The maximum population size that an environment can support. 10

caste (kast) One of the distinct kinds of individuals in a colony of social insects (e.g., queens, drones, and workers in a honeybee colony). 214

caudal (kaw′dal) Having to do with, or toward, the tail of an animal. 58

Caudofoveata (kaw′do-fo′ve-at-ah) The class of molluscs characterized by a wormlike shell-less body and scalelike calcareous spicules; lack eyes, tentacles, statocysts, crystalline style, foot, and nephridia. Deep-water marine burrowers. *Chaetoderma.* 158

cecum (se′kum) 1. Each arm of the blind-ending Y-shaped digestive tract of trematodes (phylum Platyhelminthes). 2. A region of the vertebrate digestive tract where fermentation can occur. It is located at the proximal end of the large intestine. 331

Cephalaspidomorphi (sef′a-las′pe-do-morf′e) The class of vertebrates characterized by the absence of paired appendages and the presence of sucking mouthparts with teeth and a rasping tongue. Lampreys. 242

cephalic (se-fal′ik) Having to do with, or toward, the head of an animal. 58

cephalization (sef′al-iz-a″shun) The development of a head with an accumulation of nervous tissue into a brain. 58

Cephalochordata (sef′a-lo-kor-dat′ah) The subphylum of chordates whose members possess a laterally compressed, transparent body. They are fishlike and possess all four chordate characteristics throughout life. Amphioxus. 247

cephalothorax (sef′al-o-thor″aks) The fused head and thoracic regions of crustaceans and some arachnids. 186

cercaria (ser-kar-e-a) Juvenile digenetic trematode, produced by asexual reproduction within a sporocyst or redia. Cercaria are freeswimming and have a digestive tract, suckers, and a tail. They develop into a metacercaria. 115

Cestoidea (ses-toid′e-ah) The class of platyhelminthes that has members that are all parasitic with no digestive tract; have great reproductive potentials. Tapeworms. 118

chelicerae (ke-lis′er-ae) One of the two pairs of anterior appendages of arachnids, may be pincerlike or modified for piercing and sucking or other functions. 186

Chelicerata (ke-lis″e-ra′tah) The subphylum of arthropods whose members have a body that is divided into prosoma and opisthoma. The first pair of appendages are feeding appendages called chelicerae. Spiders, scorpions, mites, and ticks. 186

Chilopoda (ki′le-pod′ah) The class of uniramous arthropods whose members have one pair of legs per segment and whose body is oval in cross section. Centipedes. 203

chitin (ki′tin) The polysaccharide found in the exoskeleton of arthropods. 183

chloragogen tissue (klor′ah-gog′en tish′u) Cells covering the dorsal blood vessel and digestive tract of annelids; function in glycogen and fat synthesis and urea formation. 172

choanocytes (ko-an′o-sitz) Cells of sponges that create water currents and filter food. 85

Chondrichthyes (kon-drik′thi-es) The class of vertebrates whose members are fishlike, possess paired appendages and a cartilaginous skeleton, and lack a swim bladder. Skates, rays, and sharks. 242

Chordata (kor-dat′ah) A phylum of animals whose members are characterized by a notochord, pharyngeal gill slits, a dorsal tubular nerve cord, and a postanal tail. 243

chromatophores (kro-mah-tah-forz) Cells containing pigment that, through contraction and expansion, produce temporary color changes. 156

chrysalis (kris′ah-lis) The pupal case of a butterfly that forms from the exoskeleton of the last larval instar. 213

Ciliophora (sil-o-of′or-ah) The protozoan phylum characterized by members with simple or compound cilia at some stage in their life history; heterotrophs with a well-developed cytostome and feeding organelles; at least one macronucleus and micronucleus present. Examples: *Paramecium*, *Stentor*, *Vorticella*, *Balantidium*. 76

cirri (ser′i) Any of various slender or filamentous, usually flexible appendages, such as one of the compound organelles composed of groups of fused cilia seen in certain peritrichious ciliate protozoa that are used for locomotion; an eversible penis in flatworms; a fingerlike projection of a polychete parapodium. 76

Cirripedia (sir′i-ped′eah) The class of crustaceans whose members are sessile and highly modified as adults. Enclosed by calcium carbonate valves. Barnacles. 198

cladistics (klad-is-tiks) *See* **phylogenetic systematics.** 53

cladograms (klad′o-gramz) Diagrams depicting the evolutionary history of taxa, which are derived from phylogenetic systematics (cladistics). 54

class A level of classification between phylum and order. 50

climax community A final, relatively stable stage in an ecological succession. 14

clitellum (klit′el-um) The region of an annelid responsible for secreting mucus around two worms in copula and for secreting a cocoon to protect developmental stages. 170

cloaca (klo-a-kah) A common opening for excretory, digestive, and reproductive systems. 267

closed circulatory system A circulatory system in an animal in which blood is confined to vessels throughout its circuit. 155

Cnidaria (ni-dar′e-ah) The phylum of animals whose members are characterized by radial or biradial symmetry, diploblastic organization, a gastrovascular cavity, and nematocysts. Jellyfish, sea anemones, and their relatives. 90

cnidocytes (ni-do-sītz) The cell that produces and discharges the stinging organelles (nematocysts) in members of the phylum Cnidaria. 90

cocoon The protective covering of a resting or developmental stage; sometimes refers to both the covering and the contents. 112, 213

coelom (se′lom) A fluid-filled body cavity lined by mesoderm. 59

coevolution (ko-ev′ah-loo″shun) The evolution of ecologically related species such that each species exerts a strong selective influence on the other. 11

colloblasts (kol′ah-blasts) Adhesive cells on the tentacles of ctenophorans used to capture prey. 101

colonial hypothesis A hypothesis formulated to explain the origin of multicellularity from protist ancestors; animals may have been derived when protists associated together and cells became specialized and interdependent. 82

comb rows Rows of cilia that serve as the locomotor organs of ctenophorans. 101

commensalism (kah-men'sal-izm) Living within or on an individual of another species without harm. 12

community The different kinds of organisms living in an area. 12

community diversity The number of different kinds of organisms living in an area. 13

comparative anatomy The study of animal structure in an attempt to deduce evolutionary pathways in particular animal groups. 5

comparative embryology (em'bre-ol"o-je) The study of animal development in an attempt to deduce evolutionary pathways in particular animal groups. 142

compound eye An eye consisting of many individual lens systems (ommatidia). Present in many members of the phylum Arthropoda. 209

Concentricycloidea (kon-sen'tri-si-kloi"de-ah) The class of echinoderms whose members are characterized by two concentric water-vascular rings encircling a disklike body; no digestive system; and internal brood pouches. Sea daisies. 233

conjugation (kon'ju-ga"shun) A form of sexual union used by ciliates involving a mutual exchange of haploid micronuclei. 78

continental drift The breakup and movement of land masses of the earth. The earth had a single landmass about 250 million years ago. This mass broke apart into continents, which have moved slowly to their present positions. 32

contour feathers Feathers that cover the body, wings, and tail of a bird. Contour feathers provide flight surfaces and are responsible for plumage colors. 307

contractile vacuole (kon-trak'til vak"u-ōl') An organelle that collects and discharges water in protists and a few lower metazoa. It takes up and releases water in a cyclical manner in order to accomplish osmoregulation and some excretion. 65

convergent evolution Evolutionary changes that result in members of one species resembling members of a second unrelated (or distantly related) species. 7

coracidium (kor"ah-sid'e-um) Larva with a ciliated epithelium hatching from the egg of certain cestodes; a ciliated free-swimming oncosphere. 120

coralline algae (kor'ah-lin al'je) Any red alga that is impregnated with calcium carbonate. Coralline algae often contribute to coral reefs. 99

corona (ko-ro'nah) A crown; an encircling structure. The ciliated organ at the anterior end of rotifers used for swimming or feeding. 126

countercurrent exchange mechanism The passive exchange of something between fluids moving in opposite directions past each other. 263

countershading Contrasting coloration that helps conceal the animal (e.g., the darkly pigmented top and lightly pigmented bottom of frog embryos). 12

coxal glands (koks'el) An organ of excretion found in some arthropods. 188

Crinoidea (kri-noi'de-ah) The class of echinoderms whose members are attached by a stalk of ossicles or are free living. Possess a reduced central disk. Sea lilies and feather stars. 231

Crustacea (krus-tas'eah) The subphylum of mandibulate arthropods whose members are characterized by having two pairs of antennae, one pair of mandibles, two pairs of maxillae, and biramous appendages. Crabs, crayfish, lobsters. 193

cryptic coloration (kript'ik) Occurs when an animal takes on color patterns of its environment. 12

crystalline style A proteinaceous, rodlike structure in the digestive tract of a bivalve (Mollusca) that rotates against a gastric shield and releases digestive enzymes. 151

Ctenophora (te-nof'er-ah) The phylum of animals whose members are characterized by biradial symmetry, diploblastic organization, colloblasts, and meridionally arranged comb rows. Comb jellies. 100

Cubozoa (ku'bo-zo"ah) The class of cnidarians whose members have prominent cuboidal medusae with tentacles that hang from the corner of the medusa. Small polyp, gametes gastrodermal in origin. *Chironex*. 96

cutaneous respiration (kyoo-ta'ne-us res'pah-ra'shun) Exchange of gases across thin, moist surfaces of the skin. Also cutaneous exchange or integumentary exchange. 279

cuticle (ku-tikel) A noncellular, protective, organic layer secreted by the external epithelium (hypodermis) of many invertebrates; refers to the epidermis or skin in higher animals. 125, 165, 183

cystacanth (sis"ta-kanth) Juvenile acanthocephalan that is infective to its definitive host. 136

cysticercosis (sis"ti-ser-ko'sis) Infection with the larval forms (*Cysticercus cellulosae*) of *Taenia solium*. 119

cysticercus (sis"ti-ser'kus) Metacestode developing from the oncosphere in most Cyclophyllidea; usually has a tail and a well-formed scolex and characterized by a fluid-filled oval body with an invaginated scolex; cysticercoid. 119

cytopharynx (si'to-far'inks) A region of the plasma membrane and cytoplasm of some ciliated and flagellated protists specialized for endocytosis. A permanent oral canal. 65

cytopyge (si'to-pij) A region of the plasma membrane and cytoplasm of some ciliated protists specialized for exocytosis of undigested wastes. 65

D

daughter sporocysts In digenetic trematodes the embryonic cells that develop from sporocysts and give rise to rediae. 115

definitive host The host in the life cycle of a parasite that harbors the adult stage or sexual stage of the parasite. 67

delayed fertilization Occurs when fertilization of an egg does not occur immediately following coitus, but may be delayed for weeks or months. 337

deme (deem) A small, local subpopulation. Isolated subpopulations sometimes display genetic changes that may contribute to evolutionary change in the subpopulation. 44

Demospongiae (de-mo-spun'je-e) The class of poriferans whose members have monaxon or tetraaxon siliceous spicules or spongin. Leuconoid body forms are present and vary in size from a few centimeters to 1 m in height. 87

dental formula A notation that indicates the number of incisors, canines, premolars, and molars in the upper and lower jaw of a mammal. 328

dermal branchiae (der'mal branch'e-ae) Thin folds of the body wall of a sea star that extend between ossicles and function in gas exchange and other exchange processes. 224

deuterostomes (du'te-ro-stōms") Animals in which the anus forms from, or in the region of, the blastopore; often characterized by enterocoelous coelom formation, radial cleavage, and the presence of a dipleurulalike larval stage. 142

diaphragm (di"ah-fram') The domed respiratory muscle between thoracic and abdominal compartments of mammals. 332

dioecious (di-es'eus) Having separate (male and female) sexes. 69

diploblastic (dip″lo-blas′tik) Animals whose body parts are organized into layers that are derived embryologically from two tissue layers: ectoderm and endoderm. Animals in the phyla Cnidaria and Ctenophora are diploblastic. 59

Diplopoda (dip′le-pod′ah) The class of arthropods whose members are characterized by having two pairs of legs per apparent segment and a body that is round in cross section. Millipedes. 203

direct flight Insect flight that is accomplished by flight muscles acting on wing bases and in which a single nerve impulse results in a single wing cycle; also called synchronous flight. *See* **indirect (asynchronous) flight.** 206

directional selection Natural selection that occurs when individuals at one phenotypic extreme have an advantage over individuals with more common phenotypes. 40

disruptive selection Natural selection that occurs when individuals of the most common phenotypes are at a disadvantage; produces contrasting subpopulations. 41

distal Away from the point of attachment of a structure on the body (e.g., the toes are distal to the knee). 58

dominant species (dom′ah-nent spe′shēz) A species that exerts an overriding influence in determining the characteristics of a community. 13

dorsal (dor′sal) The back of an animal; usually the upper surface; synonymous with posterior for animals that walk upright. 58

down feathers Feathers that provide insulation for adult and immature birds. 307

E

ecdysis (ek-dis′is) (1) The shedding of the arthropod exoskeleton to accommodate increased body size or a change in morphology (as may occur in molting from immature to adult); to molt; may also refer to the shedding of the outer epidermis of the skin of reptiles. (2) The shedding of the cuticle in aschelminths in order to grow. Also called **molting.** 125, 184

Echinodermata (i-ki′na-dur″ma-tah) The phylum of coelomate animals whose members are pentaradially symmetrical as adults and possess an endoskeleton covered by epithelium and a water-vascular system. Sea stars, sea urchins, sea cucumbers, sea lilies. 222

Echinoidea (ek′i-noi″de-ah) The class of echinoderms whose members are globular or disk shaped, possess moveable spines, and a skeleton of closely fitting plates. Sea urchins and sand dollars. 229

Echiura (ek-ee-yur′iah) A phylum of protostomate, marine animals whose members burrow in mud or sand or live in rock crevices. They possess a spatula-shaped proboscis and are 15 to 50 cm in length. Spoon worms. 179

echolocation (ek′o-lo-ka′shun) A method of locating objects by determining the time it takes for an echo to return and the direction from which it returns. As in bat echolocation. 210, 335

ecological niche (ek′o-loj′i-kal nich) The role of an organism in a community. 13

ecology (ek′ol′ah-je) The study of the relationships between organisms and their environment. 8

ecosystems (ek-o-sis′temz) All of the populations of organisms living in a certain area plus their physical environment. 14

ectoplasm (ek′to-plaz-em) The outer, viscous cytoplasm of a protist; contrasts with **endoplasm.** 65

Ectoprocta (ek-to-prok′tah) A phylum of animals whose members are colonial and fresh water or marine. Anus ends outside a ring of tentacles. Lophophore used in feeding. Moss animals or bryozoans. 235

egestion vacuole (e-jes′chen vak′u-ol) A membrane-bound vesicle within the cytoplasm of a protist that functions in expelling wastes. 65

electroreception (i-lek′tro-re-sep′shun) The ability to detect weak electrical fields in the environment. 265

elephantiasis (el″e-fan-ti′ah-sis) A chronic filarial disease most commonly occurring in the tropics due to infection of the lymphatic vessels with the nematode *Wuchereria* spp. 135

embryology (em′bre-ol′a-je) The study of development from the egg to the point that all major organ systems have formed. 3

embryonic diapause (em′bre-on′ik di′ah-pauz′) The arresting of early development to allow young to hatch, or be born, when environmental conditions favor survival. 337

endoplasm (en′do-plaz-em) The inner, fluid cytoplasm of a protist; contrasts with **ectoplasm.** 65

endopodite (end-op′o-dīt) The medial ramus of the biramous appendage of crustaceans and trilobites (phylum Arthropoda). 193

endostyle (en″do-stīl′) A ciliated tract within the pharynx of some chordates that is used in forming mucus for filter feeding. 246

energy budget An accounting of the way in which energy coming into an ecosystem from the sun is lost or processed by organisms of the ecosystem. 9

Enteropneusta (ent′er-op-nus″tah) A class of hemichordates whose members live in burrows in shallow marine water. Their bodies are divided into three regions: proboscis, collar, and trunk. Acorn worms. 240

Entoprocta (en′to-procta) A phylum of aschelminths commonly called entoprocts. 138

ephyra (e-fi′rah) Miniature medusae produced by asexual budding of a scyphistoma (class Scyphoza, phylum Cnidaria). Ephyrae mature into sexually mature medusae. 96

epidermis (ep′i-durm′is) A sheet of cells covering the surface of an animal's body. In invertebrates, a single layer of ectodermal epithelium. 90

epitoky (ep′i-to′ke) The formation of a reproductive individual (epitoke) that differs from the nonreproductive (atoke) form of that species. 169

estrus cycle (es′trus si′kel) A recurrent series of changes in the reproductive physiology of female mammals other than primates; females are receptive, physiologically and behaviorally, to the male only at certain times in this cycle. 337

eutely (u′te-le) Condition where the body is composed of a constant number of cells or nuclei in all adult members of a species (e.g., rotifers, some nematodes, and acanthocephalans). 125

evolution Change over time. Organic or biological evolution is a series of changes in the genetic composition of a population over time. *See also* **natural selection** and **punctuated equilibrium model.** 24

evolutionary systematics The study of the classification of, and evolutionary relationships among, animals; evolutionary systematists attempt to reconstruct evolutionary pathways based on resemblances between animals that result from common ancestry. 51

exopodite (eks-op′o-dīt) The lateral ramus of the biramous appendages of a crustacean or trilobite (phylum Arthropoda). 193

exoskeleton (eks′o-skel″e-ton) A skeleton that forms on the outside of the body (e.g., the exoskeleton of an arthropod). 183

exponential growth (ek′spo-nen″shal) Population growth in which the number of individuals doubles in each generation. 10

F

family The level of classification between order and genus. 50

fibrillar flight muscle (fi′bra-lar) Insect flight muscle responsible for indirect flight. A single nerve impulse results in many cycles of flight muscle contraction and relaxation. 207

filoplume feather (fil′o-ploom) A small thin feather that probably has sensory functions in birds (pinfeather). 307

filopodium (fi′li-po-de-um) Pseudopodeum that is slender, clear, and sometimes branched. 71

flame cell Specialized, hollow excretory or osmoregulatory structure consisting of one to several cells containing a tuft of cilia (the "flame") and located at the end of a minute tubule; flame bulb. 111

fluke (flook) Any trematode worm; a member of the class Trematoda or class Monogenea. 113

food chain A linear sequence of organisms through which energy is transferred in an ecosystem from producers through several levels of consumers. 14

food vacuole (food vak′yoo-ol) An organelle in the cell that functions in intracellular digestion. 65

food web A sequence of organisms through which energy is transferred in an ecosystem; rather than being a linear series, a food web has highly branched energy pathways. 14

fossil Any remains, impressions, or traces of organisms of a former geological age. 5

founder effect Changes in gene frequency that occur when a few individuals from a parental population colonize new habitats; the change is a result of founding individuals not having a representative sample of the parental population's genes. 38

Fungi (fun′ji) The kingdom of life whose members are characterized by being eukaryotic, multicellular, and saprophytic (mushrooms, molds). 51

G

Galápagos Islands (gah-lah″pe-gos′) An archipelago on the equator in the Pacific Ocean about 1,000 km west of Ecuador. Charles Darwin's observations of the plant and animal life of these islands were important in the formulation of the theory of evolution by natural selection. 26

gametogony (ga′mēt-o-gony) Multiple fission that forms gametes that fuse to form a zygote. Also called gamogony. Occurs in the class Sporozoea. 74

gastric shield A chitinized plate in the stomach of a bivalve (phylum Mollusca) on which the crystalline style is rotated. 151

gastrodermis (gas-tro-derm′is) The endodermally derived lining of the gastrovascular cavity of Cnidaria. 90

Gastropoda (gas-trop′o-dah) The class of molluscs characterized by torsion. A shell, when present, is usually coiled. Snails. 146

Gastrotricha (gas-tro-tri′ka) A small phylum of marine and freshwater species of gastrotrichs that inhabit the spaces between bottom sediments. 126

gastrovascular cavity (gas′tro-vas′ku-lar kav′i-te) The large central cavity of cnidarians and flatworms that serves as a chamber for receiving and digesting food. 91

gastrozooid (gas′tro-zo′oid) A feeding polyp in a colonial hydrozoan (phylum Cnidaria). 93

gemmule (jem′yool) Resistant, overwintering capsule formed by freshwater, and some marine, sponges that contains masses of mesenchyme cells; amoeboid mesenchyme cells are released and organize themselves into a sponge. 88

gene flow Changes in gene frequency in a population that result from emigration or immigration. 40

gene pool The sum of all genes in a population. 36

genetic drift (je-net′ik) Occurs when chance events influence evolution; also called **neutral selection.** 38

genetics (je-net′iks) The study of the mechanisms of transmission of genes from parents to offspring. 3

genus (je′nus) The level of classification between species and family. 50

gerontology (jer″on-tol′o-je) The scientific problems of aging in all their aspects, including clinical, biological, and sociological. 125

giardiasis (je″ar-di′ah-sis) A common infection of the lumen of the small intestine with the flagellated protozoan *Giardia lamblia*, and spread via contaminated food and water and by direct person-to-person contact. 70

gill arches Bony or cartilaginous gill supports of some vertebrates; also called **visceral arches.** 263

gill filaments A thin-walled, fleshy extension of a gill arch that contains vessels carrying blood to and from gas exchange surfaces. 263

gill lamellae (la-mel′a) Thin plates of tissue on gill filaments that contain the capillary beds across which gases are exchanged. 263

gill slit One of several openings in the pharyngeal region of chordates. Gill slits allow water to pass from the pharynx to the outside of the body. In the process, water passes over gills or suspended food is removed in a filter-feeding mechanism. 244

glochidium (glo-kid′e-um) A larval stage of freshwater bivalves in the family Unionidae; it lives as a parasite on the gills or fins of fishes. 152

glomerulus (glo-mer′u-lus) A capillary tuft located within the capsule (Bowman's) of a nephron. 267

Gnathostomata (na′tho-sto′ma-tah) A superclass of vertebrates whose members possess hinged jaws and paired appendages. Notochord may be replaced by the vertebral column. 256

gonozooid (gon′o-zo″id) A polyp of a hydrozoan cnidarian that produces medusae. 93

Gordian worm *See* **horsehair worms.**

greenhouse effect The warming of a global climate due to the accumulation of carbon dioxide in the atmosphere. Carbon dioxide accumulates as a result of burning fossil fuels. 19

H

habitat The native environment of an organism. 8

Hardy-Weinberg equilibrium (har′de wīn′berg e′kwe-lib′re-em) The condition in which the frequency of genes in a population does not change from one generation to another; the conditions defined by Hardy-Weinberg equilibrium define the conditions under which evolution does not occur. 37

head-foot The body region of a mollusc that contains the head and is responsible for locomotion as well as retracting the visceral mass into the shell. 145

heartworm disease A parasitic infection in dogs caused by the nematode *Dirofilaria immitis*. 135

hectocotylus (hek′to-kot′i-lus) A modified arm of some male cephalopods that is used in sperm transfer. 156

hemal system (he′mal sis′tem) Strands of tissue found in echinoderms. The hemal system is of uncertain function. It may aid in the transport of large molecules or coelomocytes, which engulf and transport waste particles within the body. 224

Hemichordata (hem'i-kor-da'tah) The phylum of marine, wormlike animals whose members have an epidermal nervous system and pharyngeal gill slits. Acorn worms and pterobranchs. 240

hemimetabolous metamorphosis (hem'i-met-ab''ol-us met-ah-morf'a-sis) A type of insect metamorphosis in which immature insects are different in form and habitats from the adult. It is different from holometabolous metamorphosis in that there is a gradual series of changes in form during the transition from immature to adult. 213

hemocoel (hem'o-sēl) Large tissue spaces within arthropods that contain blood; derived from the blastocoel of the embryo. 188

heterotrophic (het''er-o-trofic) The type of nutrition in which organisms derive energy from the oxidation of organic compounds either by consumption of or absorption of other organisms. 9

Hexactinellida (hex-act'in-el'id-ah) The class of sponges whose members are characterized by triaxon siliceous spicules, which are sometimes formed into an intricate lattice. Cup or vase shaped. Scyconoid body form. Glass sponges. 87

Hexapoda (hex'sah-pod'ah) The class of mandibulate arthropods whose members are characterized by having three pairs of legs. Commonly called insects. Hexapods often have wings and a body divided into head, thorax, and abdomen. Insecta has been used as an alternate class name. 206

hibernation Condition of mammals that involves passing the winter in a torpid state in which the body temperature drops to nearly freezing and the metabolism drops close to zero. 9, 334

Hirudinea (hi'roo-din''eah) The class of annelids whose members are characterized by bodies with 34 segments, each of which is subdivided into annuli. Anterior and posterior suckers present. Leeches. 173

holometabolous metamorphosis (hol'o-met-ab''ol-us met-ah-morf'a-sis) A type of insect metamorphosis in which immatures, called larvae, are different in form and habitats from the adult; the last larval molt results in the formation of a pupa; radical cellular changes in the pupal stage end in adult emergence. 213

Holothuroidea (hol'o-thu-roi''de-ah) The class of echinoderms whose members are elongate along the oral-aboral axis, have microscopic ossicles embedded in a muscular body wall, and have circumoral tentacles. Lack rays. Sea cucumbers. 230

homodont (ho'mo-dont) Having a series of similar, unspecialized teeth. 327

homologous (ho-mol'o-ges) Structures that have a common evolutionary origin; the wing of a bat and the arm of a human are homologous; each can be traced back to a common ancestral appendage. 7

horsehair worms Pseudocoelomate animals that belong to the phylum Nematomorpha. Also known as Gordian worms or hairworms (*Gordius* is the name for an ancient king who tied an intricate knot). 135

host An animal or protist that harbors or nourishes another organism (parasite). 66

hydraulic skeleton (hi-dro'lik) The use of body fluids in open circulatory systems to give support and facilitate movement; muscles contracting in one part of the body force body fluids into some distant tissue space, thus causing a part of the body to extend or become turgid. *See* **hydrostatic skeleton.** 148

hydrostatic skeleton (hi'dro-stat'ik) The use of body cavity fluids, confined by the body wall, to give support (e.g., the hydrostatic skeleton of nematodes and annelids). Also called hydroskeleton. 92

hydrothermal vents (hi'dro-thur-mal) Deep, oceanic regions where the tectonic plates of the earth's crust are moving apart. They are characterized by occasional lava flows and hot water springs. These vents support a rich community by chemolithotrophy. 15

Hydrozoa (hi'dro-zo-ah) The class of cnidarians whose members have epidermally derived gametes, mesoglea without wandering amoeboid cells, and gastrodermis without nematocysts. Medusae, when present, with a velum. *Hydra, Obelia, Physalia.* 93

I

indirect flight Insect flight accomplished by flight muscles acting on the body wall. Changes in shape of the thorax cause wing movements. A single nerve impulse results in many cycles of the wings; also called asynchronous flight. *See* **direct (synchronous) flight.** 206

inferior Below a point of reference (e.g., the mouth is inferior to the nose in humans). 58

Insecta (in-sekt'ah) *See* **Hexapoda.**

intermediate host (in'ter-me''de-it host) The organism in the life cycle of a parasite that harbors an immature stage of the parasite and where asexual reproduction usually occurs. 67

introvert (in'tro-vert) The anterior narrow portion that can be withdrawn (introverted) into the trunk of a sipunculid worm. A loriciferan, or a bryozoan. 137

J

Jacobson's (vomeronasal) organ Olfactory receptor present in most reptiles; blind-ending sacs that open through the secondary palate into the mouth cavity; they are used to sample airborne chemicals. 299

Johnston's organ Mechanoreceptor (auditory receptor) found at the base of the antennae of male mosquitoes and midges. 209

K

keratin (ker'a-tin) A tough, water-resistant protein found in the epidermal layers of the skin. Found in hair, feathers, hoofs, nails, claws, bills, etc. 290

kingdom The highest level of classification of life; the most widely accepted classification system includes five kingdoms: Monera, Protista, Fungi, Plantae, and Animalia. 50

Kinorhyncha (kin'o-rink-ah) The phylum of aschelminths that contains members called kinorhynchs; small elongate worms found exclusively in marine environments where they live in mud and sand. 128

L

labial palp (la'be-al palp) 1. Chemosensory appendage found on the labium of insects (Arthropoda). 2. Flaplike lobe surrounding the mouth of bivalve molluscs that directs food toward the mouth. 150

labium (la'be-um) The posterior mouthpart of insects. It is often referred to as the "lower lip," is chemosensory, and was derived evolutionarily from paired head appendages (Hexapoda, Arthropoda). 207

larva (lar'vah) 1. The immature, feeding stage of an insect that undergoes holometabolous metamorphosis. 2. The immature stage of any animal species in which adults and immatures are different in body form and habitat. 88

Larvacea (lar-vas'e-ah) The class of urochordates whose members are planktonic and whose adults retain a tail and notochord. With a gelatinous covering of the body. 242

larval instars (lar' val' in'starz) Any of the different immature feeding stages of an insect that undergoes holometabolous metamorphosis. 212

lateral (lat′er-al) Away from the plane that divides a bilateral animal into mirror images. 58

lateral-line system 1. A line of sensory receptors along the side of some fishes and amphibians used to detect water movement (phylum Chordata). 2. The external manifestation of a lateral excretory canal of nematodes (phylum Nematoda). 265

leucon (lu′kon) The sponge body form that has an extensively branched canal system; the canals lead to chambers lined by choanocytes. 87

lobopodium (lo′bo-po-de-um) A blunt, lobelike pseudopodium that is commonly tubular, and is composed of both ectoplasm and endoplasm. 71

logistic population growth (lo-jis′tik pop′yu-la″shun groth) Population growth that is characterized by an exponential growth phase followed by lowered growth rates due to resources limiting population size. The population size that an environment can support is called the environment's carrying capacity (symbolized by K). Logistic growth curves usually are usually sigmoid (flattened S) in shape. 10

lophophore (lof′a-for) Tentacle-bearing ridge or arm within which is an extension of the coelomic cavity in lophophorate animals (e.g., brachiopods, ectoprocts, phoronids). 235

lorica (lo′re-ka) The protective external case found in rotifers and some protozoa. It is formed by a thickened cuticle. 127

Loricifera (lor′a-sif-er-a) A phylum of aschelminths. The most recent animal phylum to be described; members are commonly called loriciferans. 137

M

macronucleus (mak′ro-nuk″le-us) A large nucleus found within the Ciliata (Protista) that regulates cellular metabolism. Directly responsible for the phenotype of the cell. 77

Malacostraca (mal-ah-kos′trah-kah) The class of crustaceans whose members are characterized by having appendages modified for crawling along the substrate, as in lobsters, crayfish, and crabs. Alternatively, the abdomen and body appendages may be used in swimming, as in shrimp. 193

malpighian tubules (mal-pig′e-an tu′bulz) The blind-ending excretory tubules that join the midgut of insects and some other arthropods. 188

Mammalia (ma-may′le-ah) The class of vertebrates whose members are at least partially covered by hair, have specialized teeth, and are endothermic. Young are nursed from mammary glands. The mammals. 242

mammary gland (mam′ar-e) The breast. In female mammals, the mammary glands produce and secrete milk to nourish developing young. 326

mandible (man′dib-el) 1. The lower jaw of vertebrates. 2. The paired, grinding and tearing mouthparts of arthropods, which were derived from anterior head appendages. 194

mantle (man′tel) The outer fleshy tissue of molluscs that secretes the shell. The mantle of cephalopods may be modified for locomotion. 145

mantle cavity (man′tel kav′i-te) The space between the mantle and the visceral mass of molluscs. 145

manubrium (me-nub′re-um) A structure that hangs from the oral surface of a cnidarian medusa and surrounds the mouth. 93

mastax (mas′tax) The pharyngeal apparatus of rotifers used for grinding ingested food. 128

Mastigophora (mas-ti-gof′o-rah) The protozoan subphylum where members possess one or more flagella that are used for locomotion; autotrophic, heterotrophic, or saprozoic. 67

maxilla (maks′il-ah) One member of a pair of mouthparts located just posterior to the mandibles of many arthropods. 194

medial (me′de-al) On or near the plane that divides a bilateral animal into mirror images. Also median. 58

median (parietal) eye (me′de-an) A photoreceptor located middorsally on the head of some vertebrates; it is associated with the vertebrate epithalamus. 299

medusa (me-du′sah) Usually, the sexual stage in the life cycle of cnidarians; the jellyfish body form. 91

Merostomata (mer′o-sto′mah-tah) The class of arthropods whose members are aquatic and possess book gills on the opisthosoma. Eurypterids (extinct) and horseshoe crabs. 186

mesoglea (mez-o-gle′ah) A gel-like matrix found between the epidermis and gastrodermis of cnidarians. 90

mesohyl (mez-o-hīl′) A jellylike layer between the outer (pinacocyte) and inner (choanocyte) layers of a sponge. Contains wandering amoeboid cells. 85

mesothorax (mes′o-thor″aks) The middle of the three thoracic segments of an insect; usually contains the second pair of legs and the first pair of wings. 206

Mesozoa (mes′o-zo″ah) A phylum of animals whose members are parasites of marine invertebrates. With a two-layered body organization. Dioecious, complex life histories. Orthonectids and dicyemids. 105

metacercaria (me′ta-ser-ka′re-ah) Stage between the cercaria and adult in the life cycle of most digenetic trematodes; usually encysted and quiescent. 115

metamerism (met-tam″a-riz′em) A segmental organization of body parts. Metamerism occurs in the Annelida, Arthropoda, and other smaller phyla. 163

metamorphosis (met″ah-mor′fo-sis) Change of shape or structure, particularly a transition from one developmental stage to another as from larva to adult form. 185

metanephridium (met′ah-ne-frid′e-um) An excretory organ found in many invertebrates; it consists of a tubule that has one end opening at the body wall and the opposite end in the form of a funnel-like structure that opens to the body cavity. 168

metathorax (met′ah-thor″aks) The posterior of the three segments of an insect thorax; it usually contains the third pair of walking legs and the second pair of wings (Arthropoda). 206

microfilaria (mi″kro-fi-lar′e-ah) The prelarval stage of filarial worms. Found in the blood of humans and the tissues of the vector. 135

micronucleus (mi′kro-nuk″le-us) A small body of DNA that contains the hereditary information of ciliates (Protista); exchanged between protists during conjugation. It undergoes meiosis before functioning in sexual reproduction. 77

Microspora (mi-cro-spor′ah) The protozoan phylum characterized by members having unicellular spores; intracellular parasites in nearly all major animal groups. Examples: microsporeans (*Nosema*). 75

mictic eggs (mik′tik) Pertaining to the haploid eggs of rotifers. If it isn't fertilized, the egg develops parthenogenetically into a male; if fertilized, mictic eggs secrete a heavy shell and become dormant, hatching in the spring into amictic females. 128

migration Periodic round trips of animals between breeding and nonbreeding areas or to and from feeding areas. 318

mimicry (mim′ik-re) When one species resembles one or more other species; often protection is afforded the mimic species. 12

miracidium (mi-rah-sid'e-um) The ciliated, free-swimming first stage larva of a digenean trematode that undergoes further development in the body of a snail. 114

modern synthesis The combination of principles of population genetics and Darwinian evolutionary theory. 31

molecular biology The study of the biochemical structure and function of organisms. 8

Mollusca (mol-lus'kah) The phylum of coelomate animals whose members possess a head-foot, visceral mass, mantle, and mantle cavity. Most molluscs also possess a radula and a shell. The molluscs. Bivalves, snails, octopuses, and related animals. 142

molting *See* **ecdysis.**

Monera (mon'er-ah) The kingdom of life whose members are characterized by having cells that lack a membrane-bound nucleus, as well as other internal, membrane-bound organelles (they are prokaryotic); bacteria. 51

monoecious (mon-es'e-es) An organism in which both male and female sex organs occur in the same individual. 69

monogamous (mah-nog'ah-mus) Having one mate at a time. 315

Monogenea (mon'oh-gen'ee-uh) The class of Platyhelminthes that has members that are called monogenetic flukes; most ectoparasites on vertebrates (usually on fishes, occasionally on turtles, frogs, copepods, squids); one life-cycle form in only one host; bear an opisthaptor. Examples: *Disocotyle, Gyrodactylus, Polystoma.* 113

Monogononta (mon'o-go-non'ta) A class of rotifers containing members that possess one ovary; mastax not designed for grinding; produce mictic and amictic eggs. Example: *Notommata.* 127

monophyletic group (mon'o-fi-let'ik) A group of organisms descended from a single ancestor. 51

Monoplacophora (mon'o-pla-kof''o-rah) The class of molluscs whose members have a single, arched shell; a broad, flat foot; and certain serially repeated structures. *Neopilina.* 157

mosaic evolution (mo-za-ik ev'ah-loo''shun) A change in a portion of an organism (e.g., a bird wing) while the basic form of the organism is retained. 46

Muller's larva A free-swimming ciliated larva that resembles a modified ctenophore, characteristic of many marine polyclad turbellarians. 113

multiple fission (mul'te-pel fish'on) Asexual reproduction by the splitting of a cell or organism into many cells or organisms. *See* **schizogony.** 65

musk gland *See* **scent gland.**

mutation pressure (myoo-ta'shun presh'er) A measure of the tendency for gene frequencies to change through mutation. 40

mutualism (myoo'choo-ah-liz-em) A relationship between two species in which both members of the relationship benefit. 12

myriapods (mir'e-a-podz) Members of the four noninsect classes of the subphylum Uniramia. Includes centipedes, millipedes, pauropods, and symphylans. 203

Myxini (mik-sy-ny) The class of vertebrates whose members are fishlike, jawless, without paired appendages, and possess four pairs of tentacles around the mouth. Hagfishes. 242

Myxozoa (myx-o-zo-a) The protozoan phylum characterized by members having spores of multicellular origin; the myxozoans. 75

N

naiad (na'ad) The aquatic immature stage of any hemimetabolous insect. 213

natural selection A theory, conceived by Charles Darwin and Alfred Wallace of how some evolutionary changes occur. 29

nematocyst (ni-mat'ah-sist) An organelle characteristic of the Cnidaria that is used in defense, food gathering, and attachment. 91

Nematoda (nem-a-to-dah) The phylum of aschelminths that contains members commonly called either roundworms or nematodes. Tripoblastic, bilateral, vermiform, unsegmented, and pseudocoelomate. 130

Nematomorpha (nem'a-to-mor-pha) The phylum of aschelminths commonly called horsehair worms. 135

Nemertea (nem-er'te-a) The phylum that has members commonly called the proboscis worms; elongate, flattened worms found in marine mud and sand; triploblastic; complete digestive tract with anus; closed circulatory system. 120

neo-Darwinism (ne'o-dar'wi-niz'um) *See* **modern synthesis.**

nephridiopore (ne-frid-i-o'por) The opening to the outside of a nephridium. 111

nephron (nef'ron) The functional unit of a kidney, consisting of a renal corpuscle and a renal tubule. 267

neutral selection *See* **genetic drift.**

nictitating membrane (nik'ti-tat-ing mem-brān) The thin, transparent lower eyelid of amphibians and reptiles. 281

nomenclature (no'men-kla-cher) The study of the naming of organisms in the fashion that reflects their evolutionary relationships. 50

nonamniote lineage (non-am'ne-ōt lin'e-ij) The vertebrate lineage leading to modern amphibians. 272

notochord (no''ta-kord') A rodlike, supportive structure that runs along the dorsal midline of all larval chordates and many adult chordates. 244

Nuda (nuda) The class of ctenophorans whose members lack tentacles and have a flattened body with a highly branched gastrovascular cavity. 101

numerical taxonomy (noo'mer'i-kal tak-son'ah-me) A system of classification in which there is no attempt to distinguish true and false similarities. 53

nutrient cycling (nu'tre-ent si'kling) The movement of any element essential for life through an ecosystem. Gaseous nutrient cycles involve elements such as carbon that have a nonliving reservoir in the atmosphere. Sedimentary nutrient cycles involve elements such as sulfur that have a nonliving reservoir in the earth's crust. 14

nymph (nimf) The immature stage of a paurometabolous insect; resembles the adult but is sexually immature and lacks wings (Arthropoda). 213

O

ocellus (o-sel-as) A simple eye or eyespot in many invertebrates; a small cluster of photoreceptors. 111

odontophore (o-dont''o-for') The cartilaginous structure that supports the radula of molluscs. 146

Oligochaeta (ol''i-go-ket'ah) The class of annelids whose members are characterized by having few setae and no parapodia. Monoecious with direct development. The earthworm (*Lumbricus*) and *Tubifex.* 170

ommatidia (om'ah-tid''e-ah) The sensory units of the arthropod compound eye. 209

onchosphere (ong'ko-sfer) The larva of the tapeworm contained within the external embryonic envelope and armed with six hooks and cilia. Typically referred to as a coracidium when released into the water. 119

oncomiracidium (on′ko-mir-a-sid′e-um) Ciliated larva of a monogenetic trematode. 113

Onychophora (on-y-kof′o-rah) A phylum of terrestrial animals with 14 to 43 pairs of unjointed legs, oral papillae, and two large antennae. Onycophorans live in humid tropical areas of the world. Their ancestors may have been an evolutionary transition between annelids and arthropods. Velvet worms or walking worms. 219

Opalinata (op′ah-li-not′ah) The protozoan subphylum where members are cylindrical; covered with cilia. Examples: *Opalina, Zelleriella*. 67

open circulatory system A circulatory system in which blood is not confined to vessels in a part of its circuit within an animal; blood bathes tissues in blood sinuses. 148

operculum (o-per′ku-lum) A cover. 1. The cover of a gill chamber of a bony fish (Chordata). 2. The cover of the genital pores of a horseshoe crab (Meristomata, Arthropoda). 3. The cover of the aperature of a snail shell (Gastropoda, Mollusca). 114, 147, 258

Ophiuroidea (o-fe-u-roi″de-ah) The class of echinoderms whose members have arms sharply set off from the central disk. Tube feet without suction disks. Brittle stars. 227

opisthaptor (a′pis-thap′ter) Posterior attachment organ of a mongenetic trematode. 114

opisthosoma (a′pis-tho-so″mah) The portion of the body of a chelicerate arthropod that contains digestive, reproductive, excretory, and respiratory organs. 186

oral Having to do with the mouth. The end of an animal containing the mouth. 58

oral sucker The sucker on the anterior end of a tapeworm, fluke, or leech. 114

order The level of classification between class and family. 50

organic evolution The change in an organism over time; a change in the sum of all genes in a population. 24

Osteichthyes (os′te-ik′thee-ez) The class of fishes whose members are characterized by the presence of a bony skeleton, a swim bladder, and an operculum. Bony fishes. 242

oviparous (o-vip′er-us) Organisms that lay eggs that develop outside the body of the female. 190

ovipositor (ov-i-poz′it-or) A modification of the abdominal appendages of some female insects that is used for depositing eggs in or on some substrate (Arthropoda, Hexapoda). 212

ovoviviparous (o′vo-vi-vip′er-us) Organisms with eggs that develop within the reproductive tract of the female and are nourished by food stored in the egg. 190

P

paedomorphosis (pe′dah-mor′fo-sis) The development of sexual maturity in the larval body form. 108

paleontology (pa′le-on-tol′o-je) The study of early life-forms on earth. 5

parabronchi (par′ah-brong″ke) The tiny air tubes within the lung of a bird across which gas exchange occurs. 312

parapatric speciation (par′ah-pat′rik spe′she-a′shun) Speciation that occurs in small, local populations, called **demes.** 44

parapodia (par′ah-pod″e-ah) Paired lateral extensions on each segment of polychaetes (Annelida); may be used in swimming, crawling, and burrowing. 165

parasitism (par′ah-si′tiz-em) A relationship between two species in which one member lives at the expense of the second. 12

parenchyma (pa″ren′ka-ma) A spongy mass of mesenchyme cells filling spaces around viscera, muscles, or epithelia in acoelomate animals. Depending on the species, parenchyma may function in providing skeletal support, nutrient storage, motility, reserves of regenerative cells, transport of materials, structural interactions with other tissues, modifiable tissue for morphogenesis, oxygen storage, and perhaps other functions that have yet to be determined. 109

parietal eye (pah-ri′e-tal) *See* **median eye.**

paurometabolous metamorphosis (por′o-me-tab′a-lus met′ah-morf′a-sis) A form of insect development in which immatures resemble parents, and molting is restricted to the immature stages. 213

Pauropoda (por′e-pod′ah) A class of arthropods whose bodies are small, soft, 11 segmented, and have 9 pairs of legs. 205

pebrine (pa-brēn) An infectious disease of silkworms caused by the protozoan *Nosema bombicis*. 75

pedicellariae (ped′e-sel-ar″i-ae) Pincerlike structures found on the body wall of many echinoderms. They are used in cleaning and defense. 224

pedipalps (ped′e-palps) The second pair of appendages of chelicerate arthropods. These appendages are sensory in function. 186

pellicle (pel-ik-el) A thin, frequently noncellular covering of an animal (e.g., the protective and supportive pellicle of protists occurs just below the plasma membrane); may be composed of a cell membrane, cytoskeleton, and other organelles. 65

pentaradial symmetry (pen′tah-ra′de-al sim′i-tre) A form of radial symmetry found in the echinoderms in which body parts are arranged in fives around an oral-aboral axis. 222

Pentastomida (pent-ta-stom′id-ah) A phylum of worms that are all endoparasites in the lungs or nasal passageways of carnivorous vertebrates. Tongue worms. 220

peristomium (per″i-stom′e-um) The segment of the body of an annelid that surrounds the mouth. 165

pharyngeal gill slits (far-in′je-al) *See* **gill slit.**

phasmid (faz-mid) Sensory pit on each side near the end of the tail of nematodes of the class Phasmidea. 130

Phoronida (fo-ron-i-dah) A phylum of marine animals whose members live in permanent chitinous tubes in muddy, sandy, or solid substrates. Feed via an anterior lophophore with two parallel rings of long tentacles. 237

phyletic gradualism (fi-let′ik graj′oo-el-izm) The idea that evolutionary change occurs at a slow, constant pace over millions of years. 44

phylogenetic systematics (fi-lo-je-net′ik sis-tem′at-iks) The study of the phylogenetic relationships among organisms in which true and false similarities are differentiated; cladistics. 53

phylum (fi′lum) The level of classification between kingdom and class; members are considered a monophyletic assemblage derived from a single ancestor. 50

physiology The branch of science that deals with the function of living organisms. 3

Phytomastigophorea (fi′to-mas-ti-go-for′ah) The protozoan class where members usually have chloroplasts; mainly autotrophic, some heterotrophic. Examples: *Euglena, Volvox, Chlamydomonas*. 68

pilidium larva (pi-lid-e-um lar′va) Free-swimming, hat-shaped larva of nemertean worms characterized by an apical tuft of cilia. 121

pinacocyte (pin′ah-ko′sīt) Thin, flat cell covering the outer surface, and some of the inner surface, of poriferans. 85

pinfeather *See* **filoplume feather.**

pioneer community The first community to become established in an area. 14

pit organ Receptor of infrared radiation (heat) on the head of some snakes (pit vipers). 299

placenta (plah-sen′tah) Structure by which an unborn child or animal is attached to its mother's uterine wall and through which it is nourished. 323

placid (plac-id) Plates on Kinorhyncha. 129

Placozoa (plak′o-zo″ah) A phylum of small, flattened, marine animals that feed by forming a temporary digestive cavity. *Tricoplax adherans*. 105

Plantae (plant′a) One of the five kingdoms of life; characterized by being eukaryotic and multicellular, and having rigid cell walls and chloroplasts. 51

planula (plan′u-lah) A ciliated, free-swimming larva of most cnidarians. The planula develops following sexual reproduction and metamorphoses into a polyp. 93

plastron (plas′tron) The ventral portion of the shell of a turtle. Formed from bones of the pectoral girdle and dermal bone. 291

plate tectonics (tek-ton′iks) The study of the movement of the earth's crustal plates. These movements are called continental drift. 32

Platyhelminthes (plat″e-hel-min′thez) The phylum of flatworms; bilateral acoelomates. 109

plerocercoid larva (ple″ro-ser′koid) Metacestode that develops from a procercoid larva; it usually shows little differentiation. 120

pneumatic sacs (noo-mat′ik saks) Gas-filled sacs that arise from the esophagus, or another part of the digestive tract, of fishes. Pneumatic sacs are used in buoyancy regulation (swim bladders) or gas exchange (lungs). 263

Pogonophora (po′go-nof′e-rah) A phylum of protostomate, marine animals that are distributed throughout the world's oceans. Live in secreted, chitinous tubes in cold water at depths exceeding 100 m. Lack a mouth and digestive tract. Nutrients absorbed across the body wall and from endosymbiotic bacteria that they harbor. Beard worms. 180

polyandrous (pol″e-an′drous) Having more than one male mate. Polyandry is advantageous when food is plentiful but, because of predation or other factors, the chances of successfully rearing young are low. 316

Polychaeta (pol″e-kēt′ah) The class of annelids whose members are mostly marine and are characterized by a head with eyes and tentacles and a body with parapodia. Parapodia bear numerous setae. Examples: *Nereis, Arenicola*. 165

polygynous (pa-lij′a-nus) Having more than one female mate. Polygyny tends to occur in species whose young are relatively independent at birth or hatching. 316

polyp (pol′ip) The attached, usually asexual, stage of a cnidarian. 91

polyphyletic group (pol′e-fi-let′ik) An assemblage of organisms that includes multiple evolutionary lineages. Polyphyletic assemblages usually reflect insufficient knowledge regarding the phylogeny of a group of organisms. 51

Polyplacophora (pol′e-pla-kof′o-rah) The class of molluscs whose members are elongate, dorsoventrally flattened, and have a shell consisting of eight dorsal plates. 156

population A group of individuals of the same species that occupy a given area at the same time and share a unique set of genes. 9

population genetics The study of events occurring in gene pools. 36

Porifera (po-rif′er-ah) The animal phylum whose members are sessile and either asymmetrical or radially symmetrical. Body organized around a system of water canals and chambers. Cells are not organized into tissues or organs. Sponges. 84

porocytes (por′o-sītz) Tubular cells found in a sponge body wall that create a water channel to an interior chamber. 85

postanal tail (post-an′al) A tail that extends posterior to the anus; one of the four unique characteristics of chordates. 244

postmating isolation (post-mat′ing i′sah-la′shun) Isolation that occurs when fertilization is prevented even though mating has occurred. 43

preadaptation (pre-a-dap-ta′shun) Occurs when a structure or a process present in members of a species proves useful in promoting reproductive success when an individual encounters new environmental situations. 188

precocial (pre-ko′shel) Having developed to a high degree of independence at the time of hatching or birth. 316

premating isolation (pre-mat′ing i′sah-la′shun) When behaviors or other factors prevent animals from mating. 43

Priapulida (pri′a-pyu-lida) A phylum of aschelminths commonly called priapulids. 138

Primates The order of mammals whose members include humans, monkeys, apes, lemurs, and tarsiers. 324

procercoid larva (pro-ser′koid lar′va) Cestode developing from a coracidium in some orders; it usually has a posterior cercomer. Developmental stage between oncosphere and plerocercoid. 120

proglottid (pro-glot′id) One set of reproductive organs in a tapeworm strobila; usually corresponds to a segment. One of the linearly arranged segmentlike sections that make up the strobila of a tapeworm. 118

prosoma (pro′soma) A sensory, feeding, and locomotor tagma of chelicerate arthropods. 186

prostomium (pro-stōm′e-um) A lobe lying in front of the mouth, as found in the Annelida. 165

protandrous (pro-tan′drus) *See* **protandry.**

protandry (pro-tan′dre) The condition in a monoecious organism in which male gonads mature before female gametes; prevents self-fertilization. 99

prothorax (pro′thor′aks) The first of the three thoracic segments of an insect; usually contains the first pair of walking appendages. 206

Protista (pro-tist′ah) The kingdom whose members are characterized by being eukaryotic and unicellular or colonial. 51

protonephridium (pro′to-ne-frid′e-um) Primitive osmoregulatory or excretory organ composed of a tubule terminating internally with a flame bulb or solenocyte; the unit of a flame bulb system. Protonephridia are specialized for ultrafiltration. 111

protopodite (pro′to-po′dīt) The basal segment of a biramous appendage of a crustacean. 193

protostome (pro′to-stōm″) Animal in which the embryonic blastopore becomes the mouth; often possesses a trochophore larva, schizocoelous coelom formation, and spiral embryonic cleavage. 142

protostyle (pro′to-stīl″) A rotating mucoid mass into which food is incorporated in the gut of a gastropod (phylum Mollusca). 148

protozoa (pro″to-zo′ah) A subkingdom (formerly a phylum) comprising the simplest organisms called protista; divided into seven phyla. 65

protozoologist (pro′to-zo′ol-o-jist) A person who studies protozoa. 67

proximal (proks′em-al) Toward the point of attachment of a structure on an animal (e.g., the hip is proximal to the knee). 58

pseudocoelom (soo″do-se′lom) A body cavity between the mesoderm and endoderm; a persistent blastocoele that is not lined with peritoneum. Also pseudocoel. 59

pseudocoelomate (soo″do-sēl′o-māt) Animals having a pseudocoelom, as the aschelminths. 125

pseudopodia (soo′dah-po′de-ah) Temporary cytoplasmic extensions of amoebas that are used in feeding and locomotion. 71

Pterobranchia (ter′o-brang″ke-ah) The class of hemichordates whose members lack gill slits and have two or more arms. Colonial, living in externally secreted encasements. 243

punctuated equilibrium model (pungk′choo-at′ed e′kwe-lib′riam mod′el) The idea that evolutionary change can occur rapidly over periods of thousands of years and that these periods of rapid change are interrupted by periods of constancy (stasis). 44

pupa (pu′pa) A nonfeeding immature stage in the life cycle of holometabolous insects. It is a time of radical cellular changes that result in a change from the larval to the adult body form. 213

puparium (pu-par′e-um) A pupal case formed from the last larval exoskeleton. *See* **pupa.** 213

Pycnogonida (pik′no-gon″i-dah) The class of chelicerate arthropods whose members have a reduced abdomen and four to six pairs of walking legs. Without special respiratory or excretory structures. Sea spiders. 192

pygostyle (pig′o-stīl) The fused posterior caudal vertebrae of a bird; helps support tail feathers that are important in steering. 308

pyrenoid (pi′re-noid) Part of the chloroplast that synthesizes and stores polysaccharides. 68

R

radial symmetry A form of symmetry in which any plane passing through the oral-aboral axis divides an organism into mirror images. 57

radula (raj′oo-lah) The rasping, tonguelike structure of most molluscs that is used for scraping food; composed of minute chitinous teeth that move over a cartilaginous odontophore. 145

ram ventilation The movement of water across gills as a fish swims through the water with its mouth open. 263

range of optimum The range of values for a condition in the environment that is best able to support survival and reproduction of an organism. 8

redia (re′de-ah) A larval, digenetic trematode produced by asexual reproduction within a miracidium, sporocyst, or mother redia. 115

releaser gland A gland in turbellarians that secretes a chemical that dissolves the attachment of the organism from a substrate. 109

Remipedia (re-mi-pe′de-ah) A class of crustaceans whose members possess about 30 body segments and uniform, biramous appendages. This class contains a single species of cave-dwelling crustaceans from the Bahamas. 184

renette (re′net) An excretory structure found in some worms. 132

reproductive isolation Occurs when individuals are prevented from mating, even though they may occupy overlapping ranges. *See* **premating** and **postmating isolation.** 43

Reptilia (rep-til′e-ah) The class of vertebrates whose members have dry skin with epidermal scales and amniotic eggs that develop in terrestrial environments. Snakes, lizards, and alligators. 243

respiratory tree A pair of tubules attached to the rectum of a sea cucumber that branch through the body cavity and function in gas exchange. 231

reticulopodium (re-tik′u-lo-po′de-um) A pseudopodium that forms a threadlike branched mesh and contains axial microtubules. 71

rhabdite (rab′dīt) A rodlike structure in the cells of the epidermis or underlying parenchyma in certain tubellarians that are discharged in mucous secretions. 109

rhopalium (ro-pal′e-um) A sensory structure at the margin of the scyphozoan medusa. It consists of a statocyst and a photoreceptor (phylum Cnidaria). 96

rhynchocoel (ring′ko-sēl) In nemerteans, the fluid-filled coelomic cavity that contains the inverted proboscis. 121

Rotifera (ro-tif′era) The phylum of aschelminths that has members with a ciliated corona surrounding a mouth; muscular pharynx (mastax) present with jawlike features; nonchitinous cuticle; parthenogenesis common; both freshwater and marine species. 126

S

Sarcodina (sar′ko-din′ah) The protozoan subphylum where members have pseudopodia for movement and food gathering; naked or with shell or test; mostly free living. 71

Sarcomastigophora (sar′ko-mas-ti-gof′o-rah) The protozoan phylum where members possess flagella, pseudopodia, or both for locomotion and feeding; single type of nucleus. 67

scalid (sca-lid) A set of complex spines found on the kinorhynchs, loriciferans, priapulans, and larval nematomorphs, with sensory, locomotor, food capture, or penetrant function. 129

Scaphopoda (ska-fop′o-dah) A class of molluscs whose members have a tubular shell that is open at both ends. Possess tentacles but no head. *Dentalium.* 157

scent gland A gland located around the feet, face, or anus of many mammals; secretes pheromones, which may be involved with defense, species and sex recognition, and territorial behavior. Musk gland. 326

schizogony (skiz-og′on-e) A form of fission involving multiple nuclear divisions and the formation of many individuals from the parental organism. *See* **multiple fission.** 65

scolex (sko′leks) The attachment or holdfast organ of a tapeworm, generally considered the anterior end; it is used to adhere to the host. 118

scyphistoma (si-fis′to-mah) The polyp stage of a scyphozoan (phylum Cnidaria); develops from a planula and produces ephyrae by budding. 96

Scyphozoa (si′fo-zo″ah) A class of cnidarians whose members have prominent medusae. Gametes are gastrodermal in origin and are released to the gastrovascular cavity. Nematocysts are present in the gastrodermis. Polyps are small. *Aurelia.* 95

sebaceous (oil) gland (se-ba′shus) Gland of the skin that secretes sebum; oil gland. 326

Secernentea (ses-er-nen′te-a) The class of nematodes formerly called Phasmidea. Examples: *Ascaris, Enterobius, Necator, Wuchereria.* 130

secondary palate A plate of bone that separates the nasal and oral cavities of mammals and some reptiles. 295

Seisonidea (sy′son-id′ea) A class of rotifers containing members that are commensals of crustaceans; large and elongate body with rounded corona. Example: *Seison.* 127

selection pressure The tendency for natural selection to occur; natural selection occurs whenever some genotypes are more fit than other genotypes. 40

seminal receptacle (sem′i-nal ri-sep′tah-kel) A structure in the female reproductive system that stores sperm received during copulation (e.g., many insects and annelids). 172

seminal vesicle (sem′i-nal ves′i-kel) 1. One of the paired accessory glands of the reproductive tract of male mammals. It secretes the fluid medium for sperm ejaculation (phylum Chordata). 2. A structure associated with the male reproductive tract that stores sperm prior to its release (e.g., earthworms—phylum Annelida). 172

sensilla (sen-cil′ah) Modifications of the exoskeleton of an arthropod that, along with nerve cells, form sensory receptors. 188

seral stage (ser′al) A successional stage in an ecosystem. 14

sere (ser) An entire successional sequence in an ecosystem (e.g., the sequence of stages in the succession of a lake to a climax forest). 14

serially homologous (ser′e-al-e ho-mol′o-ges) Metameric structures that have evolved from a common form; the biramous appendages of crustaceans are serially homologous. 194

seta (se′tah) Hairlike modifications of an arthropod's exoskeleton that may be set into a membranous socket. Displacement of a seta initiates a nerve impulse in an associated cell. 165

siphon (si′fon) A tubular structure through which fluid flows; siphons of some molluscs allow water to enter and leave the mantle cavity. 148

Sipuncula (sigh-pun′kyu-lah) A phylum of protostomate worms whose members burrow in soft marine substrates throughout the world's oceans. Range in length from 2 mm to 75 cm. Peanut worms. 180

speciation (spe′she-a′shun) The process by which two or more species are formed from a single ancestral stock. 43

species A group of populations in which genes are actually, or potentially, exchanged through multiple generations; numerous problems with this definition make it difficult to apply in all circumstances. 42

species diversity *See* **community diversity.**

spermatophores (sper-mat″ah-fors) Encapsulated sperm that can be deposited on a substrate by a male and picked up by a female, or transferred directly to a female by a male. 156

spicules (spik′ulz) Skeletal elements secreted by some mesenchyme cells of a sponge body wall; may be made of calcium carbonate or silica. 87

spiracle (spi′rah-kel) An opening for ventilation. The opening(s) of the tracheal system of an arthropod or an opening posterior to the eye of a shark, skate, or ray. 188

spongin (spun′jin) A fibrous protein that makes up the supportive framework of some sponges. 87

sporocyst (spor′oo-sist) (1) Stage of development of a sporozoan protozoan, usually with an enclosing membrane, the oocyst. (2) An asexual stage of development in some digenean trematodes that arises from a miracidium and gives rise to rediae. 115

sporogony (spor-og-a-ne) Multiple fission that produces sporozoites after zygote formation. Occurs in the class Sporozoea. 74

stabilizing selection Natural selection that results in the decline of both extremes in a phenotypic range; results in a narrowing of the phenotypic range. 42

statocyst (stat″o-sist) An organ of equilibrium and balance found in many invertebrates. Statocysts usually consist of a fluid-filled cavity containing sensory hairs and a mineral mass called a statolith. The statolith stimulates the sensory hairs, which helps orient the animal with regard to the pull of gravity. 93

statolith *See* **statocyst.**

stigma (stig′ma) The mass of bright red photoreceptor granules found in certain flagellated protozoa (Euglena) that serves as a shield for the photoreceptor. Also the spiracle of certain terrestrial arthropods. 68

strobila (stro-bi′lah) The chain of proglottids constituting the bulk of the body of adult tapeworms. 118

succession A sequence of community types that occurs during the maturation of an ecosystem. 14

sudoriferous gland (su″do-rif′er-us) A sweat gland. 326

superior Above a point of reference (e.g., the neck is superior to the chest of humans). 58

swim bladder A gas-filled sac, which is usually located along the dorsal body wall of bony fishes; it is an outgrowth of the digestive tract and regulates buoyancy of a fish. 246

sycon (si′kon) A sponge body form characterized by choanocytes lining radial canals. 87

symbiosis (sim″bi-o′sis) The biological association of two individuals or populations of different species, classified as mutualism, commensalism, or parasitism, depending on the advantage or disadvantage derived from the relationship. 12

symmetry (sim′i-tre) A balanced arrangement of similar parts on either side of a common point or axis. 57

sympatric speciation (sim′pat′rik spe′she-a′shun) Speciation that occurs in populations that have overlapping ranges. 44

Symphyla (sim-fi′lah) A class of arthropods whose members are characterized by having long antennae, 10 to 12 pairs of legs, and centipedelike bodies. Occupy soil and leaf mold. 204

symplesiomorphies (sim-ples′e-o-mor′fēz) Taxonomic characters that are common to all members of a group of organisms. These characters indicate common ancestry but cannot be used to describe relationships within the group. 54

synapomorphies (sin-ap′o-mor′fēz) Characters that have arisen within a group since it diverged from a common ancestor. Synapomorphies are used to indicate degrees of relatedness within a group. Also called shared, derived characters. 54

synchronous flight *See* **direct flight.**

syncytial hypothesis (sin-sit′e-al hi-poth′e-sis) The idea that multicellular organisms could have arisen by the formation of cell boundaries within a large multinucleate protist. 82

synsacrum (sin-sak′rum) The fused posterior thoracic vertebrae, all lumbar and sacral vertebrae, and anterior caudal vertebrae of a bird; helps maintain proper flight posture. 308

systematics The study of the classification and phylogeny of organisms. *See* **taxonomy.** 3

T

tagmatization (tag′mah-ti-za″shun) The specialization of body regions of a metameric animal for specific functions. The head of an arthropod is specialized for feeding and sensory functions, the thorax is specialized for locomotion, and the abdomen is specialized for visceral functions. 164

Tardigrada (tar-di-gray′dah) A phylum of animals whose members live in marine and freshwater sediments and in water films on terrestrial lichens and mosses. Possess four pairs of unsegmented legs and a proteinaceous cuticle. Water bears. 220

taxon (tak′son) A group of organisms that are genetically (evolutionarily) related. 50

taxonomy (tak′son′ah-me) The description of species and the classification of organisms into groups that reflect evolutionary relationships. *See* **phylogenetic systematics, evolutionary systematics,** and **numerical taxonomy.** Also **systematics.** 50

tegument (teg′u-ment) The external epithelial covering in cestodes, trematodes, and acanthocephalans; once called a cuticle. 114

Tentaculata (ten-tak′u-lata) The class of ctenophorans with tentacles that may or may not be associated with sheaths into which tentacles can be retracted. *Pleurobranchia.* 101

test A shell or hardened outer covering, typically covered externally by cytoplasm or living tissue. 71

tetrapods (te′trah-podz) A nontaxonomic designation used to refer to amphibians, reptiles, birds, and mammals. 272

Thaliacea (tal′e-as″e-ah) A class of urochordates whose members are planktonic. Adults are tailless and barrel shaped. Oral and atrial openings are at opposite ends of the tunicate. Water currents are produced by muscular contractions of the body wall and result in a weak form of jet propulsion. 242

theory of evolution by natural selection A theory conceived by Charles Darwin and Alfred Russell Wallace on how some evolutionary changes occur. 24

theory of inheritance of acquired characteristics The mistaken idea that organisms develop new organs, or modify existing organs as environmental problems present themselves, and that these traits are passed on to offspring. 24

tolerance range The range of variation in an environmental parameter that is compatible with the life of an organism. 8

tornaria (tor-nar′iah) The ciliated larval stage of an acorn worm (class Enteropneusta, phylum Hemichordata). 242

torpor A time of decreased metabolism and lowered body temperature that occurs in daily activity cycles. 9

torsion (tor′shun) A developmental twisting of the visceral mass of a gastropod mollusc that results in an anterior opening of the mantle cavity and a twisting of nerve cords and the digestive tract. 146

tracheae (tra′che-e) The small tubes that carry air from spiracles through the body cavity of an arthropod; arthropod tracheae are modifications of the exoskeleton. 188

tracheal system *See* **tracheae.**

Trematoda (trem′a-to′da) The class of platyhelminthes that has members that are all parasitic; several holdfast devices present; have complicated life cycles involving both sexual and asexual reproduction. 113

trichinosis (trik″i-no′sis) A disease resulting from infection by *Trichinella spiralis* (Nematoda) larvae by eating undercooked meat; characterized by muscular pain, fever, edema, and other symptoms. 135

trichocysts (trik′o-sists) An anchoring structure present in the ectoplasm of some ciliates. A bottle-shaped extrusible organelle of the ciliate pellicle. 76

Trilobitamorpha (tri″lo-bit′a-mor′fah) The subphylum of arthropods whose members had bodies divided into three longitudinal lobes. Head, thorax, and abdomen present. One pair of antennae and biramous appendages. Entirely extinct. 185

triploblastic (trip′lo-blas″tik) Animals whose body parts are organized into layers that are derived embryologically from three tissue layers: ectoderm, mesoderm, and endoderm. Platyhelminthes and all coelomate animals are triploblastic. 59

trochophore larva (trok″o-for lar′va) A larval stage characteristic of many molluscs, annelids, and some other protostomate animals. 142

trophic levels (trōf′ik lev′elz) The feeding level of an organism in an ecosystem; green plants and other autotrophs function at producer trophic levels; animals function at the consumer trophic levels. 14

tube feet Muscular projections from the water-vascular system of echinoderms that are used in locomotion, gas exchange, feeding, and attachment. 223

tubular nerve cord A hollow nerve cord that runs middorsally along the back of chordates; one of four unique chordate characteristics; also called the neural tube and, in vertebrates, the spinal cord. 244

Turbellaria (tur′bel-lar′e-a) The class of Platyhelminthes that has members that are mostly free living and aquatic; external surface usually ciliated; predaceous; possess rhabdites; protrusable proboscis; mostly hermaphroditic. Examples: *Convoluta, Notoplana, Dugesia.* 109

tympanal (tympanic) organs (tim-pan′al) Auditory receptors present on the abdomen or legs of some insects. 209

U

umbo (um′bo) The rounded prominence at the anterior margin of the hinge of a bivalve (Mollusca) shell; it is the oldest part of the shell. 149

unicellular (cytoplasmic) organization The life-form in which all functions are carried out within the confines of a single plasma membrane; members of the kingdom Protista display unicellular organization; also called **cytoplasmic organization.** 58

uniformitarianism (yoo′nah-for′mi-tar′e-an-ism) The idea that today the earth is shaped by forces of wind, rain, rivers, volcanoes, and geological uplift, just as it was formed in the past. 27

Uniramia (yoo′ne-ram′eah) The subphylum of arthropods whose members are characterized by a head with one pair of antennae and one pair of mandibles. All appendages are uniramous. 202

Urochordata (u′ro-kor-dat′ah) The subphylum of chordates whose members have all four chordate characteristics as larvae. Adults are sessile or planktonic and enclosed in tunic that usually contains cellulose. Sea squirts or tunicates. 244

V

valves 1. Devices that permit a one-way flow of fluids through a vessel or chamber. 2. The halves of a bivalve (Mollusca) shell. 149

veliger larva (vel′i-jer lar′va) The second free-swimming larval stage of many molluscs; develops from the trochophore and forms rudiments of the shell, visceral mass, and head-foot before settling to the substrate and undergoing metamorphosis. 149

ventral The belly of an animal; usually the lower surface; synonymous with anterior for animals that walk upright. 58

Vertebrata (ver′te-bra′tah) The subphylum of chordates whose members are characterized by cartilaginous or bony vertebrae surrounding a nerve cord. The skeleton is modified anteriorly into a skull for protection of the brain. 242

vestigial structures (ve-stij′e-al) Visible evidence of a structure that was present in an earlier stage in the evolution of an organism. One of the sources of evidence for evolution. 7

visceral arches *See* **gill arches.**

visceral mass (vis'er-al mas) The region of a mollusc's body that contains visceral organs. 145

viviparous (vi-vip'er-us) Having eggs that develop within the female reproductive tract and are nourished by the female. 190

vomeronasal organ *See* **Jacobson's organ.**

W

water-vascular system (wah'ter vas'ku-lar) A series of water-filled canals and muscular tube feet present in echinoderms; provides the basis for locomotion, food gathering, and attachment. 223

winter sleep A period of inactivity in which a mammal's body temperature remains near normal and the mammal is easily aroused. 9, 334

Z

zonite (zo-nīt) The individual body unit of a member of the phylum Kinorhyncha. 129

zooid (zo-oid) An individual member of a colony of animals, such as colonial cnidarians and ectoprocts, produced by incomplete budding. 112

zoology (zo-ol'-o-je) The study of animals. 2

Zoomastigophorea (zo'o-mas-ti-go-for'ah) The protozoan class where members lack chloroplasts; heterotrophic or saprozoic. Examples: *Trypanosoma*, *Trichonympha*, *Trichomonas*, *Giardia*. 69

zooxanthellae (zo'o-zan-thel"e) A group of dinoflagellates that live in mutualistic relationships with some cnidarians. They promote high rates of calcium carbonate deposition in coral reefs. 99

suggested READINGS

BOOKS

Aidley, D. J. (ed.). 1980. *Animal Migration*. Cambridge: Cambridge University Press.

Ayala, F. J. 1982. *Population and Evolutionary Genetics: A Primer*. Menlo Park: Benjamin Cummings Publishing Company, Inc.

Baker, R. 1984. *Bird Navigation: The Solution of a Mystery?* New York: Holmes & Meier Publishers, Inc.

———. (ed.). 1981. *The Mystery of Migration*. Minneapolis: The Viking Press.

Bakker, R. T. 1986. *The Dinosaur Heresies*. New York: William Morrow & Co., Inc.

Barnes, R., Calow, P., and Olive, P. 1993. *The Invertebrates: A New Synthesis*. 2nd ed. Boston: Blackwell Scientific.

Barrington, E. J. W. 1965. *The Biology of Hemichordata and Protochordata*. San Francisco: W. H. Freeman.

Binyon, J. 1972. *Physiology of Echinoderms*. New York: Pergamon Press, Inc.

Bliss, D. E. (ed.). 1982. *The Biology of Crustacea*, Vols. 1–5. San Diego: Academic Press, Inc.

Bogitsh, B., and Cheng, T. 1990. *Human Parasitology*. Philadelphia: W. B. Saunders Co.

Boitani, L., and Bartoli, S. 1983. *Simon and Schuster's Guide to Mammals*. New York: Simon and Schuster.

Borror, D. J., Tripplehorn, C. A., and Johnson, N. F. 1989. *An Introduction to the Study of Insects*. 6th ed. Philadelphia: Saunders College Publishing.

Boss, K. J. 1982. Mollusca. In S. P. Parker (ed.), *Synopsis and Classification of Living Organisms*, Vol. 1. New York: McGraw-Hill Book Co., Inc.

Bowler, P. 1984. *Evolution: The History of an Idea*. Berkeley: University of California Press.

Brinkhurst, R. O., and Jamieson, B. B. 1972. *Aquatic Oligochaeta of the World*. Toronto: Toronto University Press.

Bristow, W. S. 1971. *The World of Spiders*. Minneapolis: William Collins Sons & Co., Ltd.

Brusca, R. C., and Brusca, G. J. 1990. *Invertebrates*. Sunderland, Mass.: Sinauer Associates, Inc.

Chapman, J. A., and Feldhamer, G. A. (eds.). 1982. *Wild Mammals of North America: Biology, Management, and Economics*. Baltimore: The Johns Hopkins University Press.

Corliss, J. D. 1979. *The Ciliated Protozoa: Characterization, Classification, and Guide to the Literature*. 2d ed. Elmford, New York: Pergamon Press, Inc.

Cousteau, J., and Diol'e, P. 1973. *Octopus and Squid, the Soft Intelligence*. Garden City, Mich.: Doubleday & Co.

Darwin, C. 1894. *On the Origin of Species*. Reprint. 1975. Cambridge: Cambridge University Press.

Dethier, V. G. 1976. *The Hungry Fly*. Cambridge: Harvard University Press.

Dodson, E. O., and Dodson, P. 1985. *Evolution: Process and Product*. Belmont: Wadsworth Publishing Co.

Duellman, W. E., and Trueb, L. 1986. *Biology of Amphibians*. New York: McGraw-Hill Book Co.

Duncan, T., and Stuessy, T. F. 1984. *Cladistics: Perspectives on the Reconstruction of Evolutionary History*. New York: Columbia University Press.

Dunn, D. F. 1982. Cnidaria. In S. P. Parker (ed.), *Synopsis and Classification of Living Organisms*, Vol. 1. New York: McGraw-Hill Book Co.

Eisenberg, J. F. 1981. *The Mammalian Radiations: An Analysis of Trends in Evolution, Adaptation, and Behavior*. Chicago: University of Chicago Press.

Eldredge, N., and Cracraft, J. 1980. *Phylogenetic Patterns and the Evolutionary Process, Method and Theory in Comparative Biology*. New York: Columbia University Press.

Endler, J. A. 1986. *Natural Selection in the Wild*. Princeton: Princeton University Press.

Faaborg, J. 1988. *Ornithology: An Ecological Approach*. Englewood Cliffs, N.J.: Prentice-Hall.

Farmer, J. N. 1980. *The Protozoa: Introduction to Protozoology*. St. Louis: C. V. Mosby Co.

Feduccia, A. 1980. *The Age of Birds*. Cambridge: Harvard University Press.

Fell, H. B. 1982. Echinodermata. In S. P. Parker (ed.), *Synopsis and Classification of Living Animals*, Vol. 2. New York: McGraw-Hill Book Co.

Fenchel, T. 1987. *Ecology of Protozoa*. New York: Springer-Verlag.

Foelix, R. F. 1982. *Biology of Spiders*. Cambridge: Harvard University Press.

Foreman, R. E., Gorbman, A., Dodd, J. M., and Olsson, R. (eds.). 1985. *Evolutionary Biology of Fishes*. NATO ASI Series, Vol. 103. New York: Plenum Press.

Frangsmyr, T. 1983. *Linnaeus: The Man and His Work*. Berkeley: University of California Press.

Futuyma, D. J. 1986. *Evolutionary Biology*, 2d ed. Sunderland: Sinauer Associates, Inc.

Gibson, R. 1972. *Nemerteans*. London: Hutchinson University Library.

———. 1982. *British Nemerteans: Keys and Notes for Identification of the Species*. New York: Cambridge University Press.

Gill, F. B. 1990. *Ornithology*. New York: W. H. Freeman.

Gillot, Cedric. 1980. *Entomology*. New York: Plenum Press.

Godfrey, L. R. 1985. *What Darwin Began*. Old Tappan: Allyn and Bacon, Inc.

Goin, C. J., and Goin, O. B. 1978. *Introduction to Herpetology*. 3d ed. San Francisco: W. H. Freeman.

Gould, S. J. 1989. *Wonderful Life: The Burgess Shale and the Nature of the History*. New York: W. W. Norton.

Grant, P. R. 1986. *Ecology and Evolution of Darwin's Finches*. Princeton: Princeton University Press.

Grell, K. G. 1973. *Protozoology*. Berlin: Springer-Verlag.

Griffiths, M. 1978. *The Biology of Monotremes*. San Diego: Academic Press, Inc.

Halstead, L. B. 1968. *The Pattern of Vertebrate Evolution*. San Francisco: W. H. Freeman.

Harbison, G. R., and Madin, L. P. 1982. Ctenophora. In S. P. Parker (ed.), *Synopsis and Classification of Living Organisms*, Vol. 1. New York: McGraw-Hill Book Co.

Hardisty, M. W. 1979. *Biology of the Cyclostomes*. New York: Chapman and Hall, Ltd.

Hasler, A. D., and Scholz, A. T. 1983. *Olfactory Imprinting and Homing in Salmon*. New York: Springer-Verlag, Zoophysiology Series.

Hecht, M. K., Wallace, B., and Prance, G. T. (eds.). 1967–1988. *Evolutionary Biology*, Vols. 1–22. New York: Plenum Press.

Holldobler, B., and Wilson, E. O. 1990. *The Ants*. Cambridge: Harvard University Press.

Horridge, G. A. (ed.) 1975. *The Compound Eye and Vision of Insects*. Atlanta: Clarendon Group, Inc.

Hyman, L. H. 1940. *The Invertebrates*. New York: McGraw-Hill Book Co., Inc.

———. 1951. *The Invertebrates*. Vol. II. *Platyhelminthes and Rhynchocela*. New York: McGraw-Hill.

Jacques, H. E. 1978. *How to Know the Insects*. 3d ed. Dubuque: Wm. C. Brown Publishers.

Jahn, T., Bovee, R., and Jahn, F. 1979. *How to Know the Protozoa*. Dubuque: Wm. C. Brown Publishers.

Kaston, B. J. 1972. *How to Know the Spiders*. 2d ed. Dubuque: Wm. C. Brown Publishers.

King, F. W., and Behler, J. 1979. *The Audubon Society Field Guide to North American Reptiles and Amphibians*. New York: Alfred A. Knopf, Inc.

King, P. E. 1973. *Pycnogonids*. New York: St. Martins Press, Inc.

Kozloff, E. 1990. *Invertebrates*. Philadelphia: W. B. Saunders Co.

Laybourn-Parry, J. 1984. *A Functional Biology of Free-Living Protozoa*. Berkeley: University of California Press.

Lee, J., Hunter, S., and Bovee, E. 1985. *An Illustrated Guide to the Protozoa*. Lawrence, Kan.: Allen Press, Society of Protozoologists.

Levandowsky, M., and Hunter, S. H. 1979. *Biochemistry and Physiology of Protozoa*, Vols. I–III. New York: Academic Press.

Levi, H. W. 1982. Crustacea. In S. P. Parker (ed.), *Synopsis and Classification of Living Organisms*, Vol. 2. New York: McGraw-Hill Book Co.

Lovtrup, S. 1977. *The Phylogeny of Vertebrata*. New York: John Wiley & Sons, Inc.

Margulis, L., and Schwartz, K. V. 1987. *Five Kingdoms: An Illustrated Guide to the Phyla of Life on Earth*. 2d ed. San Francisco: W. H. Freeman.

Matthews, T. W., and Matthews, J. R. 1978. *Insect Behavior*. New York: John Wiley & Sons, Inc.

Mayr, E. 1982. *The Growth of Biological Thought: Diversity, Evolution, and Inheritance*. Cambridge: Harvard University Press.

McCafferty, W. P. 1981. *Aquatic Entomology: A Fisherman's and Ecologist's Illustrated Guide to Insects and Their Relatives*. Providence: Science Books International.

Meglitsch, P. A., and Schram, F. R. 1991. *Invertebrate Zoology*. 3d ed. New York: Oxford University Press.

Menard, H. W. 1986. *The Ocean of Truth: A Personal History of Global Tectonics*. Princeton, N.J.: Princeton University Press.

Moyle, P. B., and Cech, J. J. 1982. *Fishes: An Introduction to Ichthyology*. Englewood Cliffs: Prentice-Hall, Inc.

Moynihan, M. H., and Rodaniche, A. F. 1977. Communication, crypsis, and mimicry among cephalopods. In T. A. Sebeok (ed.), *How Animals Communicate*. Indiana University Press.

Nelson, J. S. 1984. *Fishes of the World*. 2d ed. New York: John Wiley & Sons, Inc.

Nisbet, B. 1984. *Nutrition and Feeding Strategies in Protozoa*. London: Croom Helm Ltd.

Otte, D., and Endler, J. A. (eds.). 1989. *Speciation and its Consequences*. Sunderland: Sinauer Associates, Inc.

Pearse, V., Pearse, J., Buchsbaum, M., and Buchsbaum, R. 1987. *Living Invertebrates*. New York: Blackwell Scientific Publications.

Pechenik, J. A. 1991. *Biology of the Invertebrates*. 2d ed. Dubuque: Wm. C. Brown Publishers.

Pennak, R. W. 1989. *Freshwater Invertebrates of the United States*. 3d ed. New York: John Wiley & Sons, Inc.

Peterson, R. T. 1980. *Field Guide to the Birds*. 2d ed. Boston: Houghton Mifflin Co.

Pettibone, M. H. 1982. Annelida. In S. P. Parker (ed.), *Synopsis and Classification of Living Organisms*. Vol. 2. New York: McGraw-Hill Book Co.

Poinar, G. O. 1983. *The Natural History of Nematodes*. Englewood Cliffs, N.J.: Prentice-Hall.

Pough, F. H., Heiser, J. B., and McFarland, W. N. 1989. *Vertebrate Life*. 3d ed. New York: Macmillan Publishing Co.

Rice, M. E., and Todorovic, M. (eds.). 1975. *Proceedings of the International Symposium on the Biology of the Sipuncula and Echiura*. Washington, D.C.: American Museum of Natural History.

Ridley, M. 1986. *Evolution and Classification: The Reformation of Cladism*. New York: Longman.

Riser, N. W., and Morse, M. P. (eds.). 1974. *Biology of the Turbellaria*. New York: McGraw-Hill.

Romer, A. S., and Parsons, T. S. 1986. *The Vertebrate Body*. 6th ed. Philadelphia: Saunders College Publishing.

Ruppell, G. 1975. *Bird Flight*. New York: Van Nostrand Reinhold Co.

Ruppert, E. E., and Barnes, R. D. 1994. *Invertebrate Zoology*. 6th ed. Philadelphia: Saunders College Publishing.

Russell-Hunter, W. D. 1979. *A Life of Invertebrates*. New York: Macmillan Publishing Co., Inc.

Savage, R. J. G., and Long, M. T. 1986. *Mammal Evolution*. New York: Facts On File Publications.

Savory, T. H. 1977. *Arachnida*. 2d ed. San Diego: Academic Press, Inc.

Schmidt, G. D., and Roberts, L. S. 1996. *Foundations of Parasitology*. Dubuque: Wm. C. Brown Publishers.

Schmidt, K. P., and Inger, R. F. 1975. *Living Reptiles of the World*. New York: Doubleday & Co., Inc.

Sleigh, M. 1989. *Protozoa and Other Protists*. New York: Routledge, Chapman, and Hall.

Smith, J. M. 1989. *Evolutionary Genetics*. New York: Oxford University Press.

Smith, J. M. (ed.). 1982. *Evolution Now: A Century After Darwin*. New York: W. H. Freeman.

Smith, R. J. F. 1985. *The Control of Fish Migration*. New York: Springer-Verlag, Zoophysiology Series.

Solem, A. 1974. *The Shell Makers: Introducing Mollusks*. New York: John Wiley & Sons, Inc.

Stahl, B. J. 1974. *Vertebrate History: Problems in Evolution*. New York: McGraw-Hill Book Co.

Stebbins, G. L. 1982. *Darwin to DNA, Molecules to Humanity*. New York: W. H. Freeman.

Steele, R. S. 1985. *Sharks of the World*. New York: Facts on File Publications.

Stonehouse, B., and Gilmore, D. (eds.). 1977. *The Biology of Marsupials*. New York: Macmillan Publishing Co.

Terres, J. K. 1980. *The Audubon Society Encyclopedia of North American Birds*. New York: Alfred A. Knopf, Inc.

Thomas, R. D. K., and Olson, E. C. (eds.). 1980. *A Cold Look at the Warm-Blooded Dinosaurs*. AAAS Selected Symposium Series. Boulder: Westview Press, Inc.

Travola, W. N., Popper, A. N., and Fay, R. R. 1981. *Hearing and Sound Communication in Fishes*. New York: Springer-Verlag.

Tytler, P., and Calow, P. 1985. *Fish Energetics: New Perspectives*. Baltimore: Johns Hopkins University Press.

Vandenbergh, J. G. 1983. *Pheromones and Reproduction in Mammals*. San Diego: Academic Press.

Vaughan, T. A. 1978. *Mammalogy*. Philadelphia: W. B. Saunders Company.

Volpe, E. P. 1985. *Understanding Evolution*. 5th ed. Dubuque: Wm. C. Brown Publishers.

Walker, W. F. 1987. *Functional Anatomy of the Vertebrates*. Philadelphia: Saunders College Publishing.

Welty, J. C. 1988. *The Life of Birds*. 4th ed. Philadelphia: W. B. Saunders Co.

Wigglesworth, V. B. 1982. *Principles of Insect Physiology*. 7th ed. New York: John Wiley & Sons, Inc.

Wiley, E. O. 1981. *Phylogenetics: The Theory and Practice of Phylogenetic Systematics*. New York: John Wiley and Sons.

Wilford, J. N. 1985. *The Riddle of the Dinosaur*. New York: Alfred A. Knopf.

Wilson, E. O. 1971. *The Insect Societies*. Cambridge: Harvard University Press.

Young, J. Z., and Hobbs, M. J. 1975. *The Life of Mammals: Their Anatomy and Physiology*. 2d ed. New York: Oxford University Press.

Zuckerman, B. M. (ed.). 1980. *Nematodes as Biological Models*. Vol. II: *Aging and Other Model Systems*. New York: Academic Press.

ARTICLES

Aldhous, P. 1993. Malaria: Focus on mosquito genes. *Science* 261(5121):546–548.

Alexander, R. M. How dinosaurs ran. *Scientific American* April, 1991.

Alldredge, A. L., and Madin, L. P. 1982. Pelagic tunicates: Unique herbivores in the marine plankton. *BioScience* 32:655–663.

Allegre, C. J., and Schneider, S. H. The evolution of the earth. *Scientific American* October, 1994.

Altman, S. A. 1989. The monkey and the fig. *American Scientist* 77(3):256–263.

Alverez, W., and Asaro, F. What caused mass extinction—an extra-terrestrial impact? *Scientific American* October, 1990.

Amato, I. 1987. Tics in the tocks of molecular clocks: Comparing the DNA, RNA and proteins of different species may reveal the entire tree of life, but obstacles are emerging. *Science News* 131:74–75.

Bakker, R. T. Dinosaur renaissance. *Scientific American* April, 1975.

Beard, J. 1992. Warding off bullets by a spider's thread. *New Scientist* 136(1847):18.

Beebee, T. J. C. 1992. Amphibian decline? *Nature* 355(6356):120.

Birkeland, C. 1989. The faustian traits of the crown-of-thorns starfish. *American Scientist* 72(2):154–163.

Blaustein, A. R. and Wake, D. B. The puzzle of declining amphibian populations. *Scientific American* April, 1995.

Bogan, A. E. 1993. Freshwater bivalve extinction (Mollusca: Unionoida): A search for causes. *American Zoologist* 33(6):599–610.

Boycott, B. B. Learning in the octopus. *Scientific American* March, 1965.

Briggs, D. E. G. 1991. Extraordinary fossils. *American Scientist* 79(2):130–141.

Brown, S. C. 1975. Biomechanics of water pumping by *Chaetopterus variopedatus* Renier: Skeletomusculature and mechanics. *Biological Review* 149:136–156.

Brownlee, S. 1987. Jellyfish aren't out to get us. *Discover* 8:42–52.

Buck, J., and Buck, E. Synchronous fireflies. *Scientific American* May, 1976.

Burnett, A. L. 1960. The mechanism employed by the starfish *Asterias forbesi* to gain access to the interior of the bivalve *Venus mercenaria*. *Ecology* 4:583–584.

Calder, W. A., III. The Kiwi. *Scientific American* July, 1978.

Callagan, C. A. 1987. Instances of observed speciation. *The American Biology Teacher* 49(1):34–36.

Camhi, J. M. The escape system of the cockroach. *Scientific American* December, 1980.

Carson, H. L. November 1987. The process whereby species originate. *BioScience* 37:715–720.

Carter, C. S., and Getz, L. L. Monogamy and the prairie vole. *Scientific American* June, 1993.

Cherfas, J. 1991. Ancient DNA: Still busy after death. *Science* 253:1354–1356.

Cloney, R. A. 1982. Ascidian larvae and the events of metamorphosis. *American Zoologist* 22:817–826.

Cole, J. Unisexual lizards. *Scientific American* January, 1984.

Conniff, R. 1987. The little suckers have made a comeback. *Discover* 8:84–93.

Corliss, J. 1984. The kingdom Protista and its 45 phyla. *Biosystems* 17:87–126.

Cortillot, V. What caused mass extinction?—a volcanic eruption. *Scientific American* October, 1990.

Coyne, J. A., and Barton, N. H. 1988. What do we know about speciation? *Nature* 331:485–486.

Cox, F. E. G. 1988. Which way for malaria? *Nature* 331:486–487.

Cracraft, J. 1988. Early evolution of birds. *Nature* 335:630–632.

Dalziel, I. W. D. Earth before Pangea. *Scientific American* January, 1995.

Dando, P. R., Southward, A. J., Southward, E. C., Dixon, D. R., Crawford, A., and Crawford, A. 1992 Shipwrecked tube worms. *Nature* 356(6371):667.

Daves, N. B., and Brooke, M. Coevolution of the cuckoo and its host. *Scientific American* January, 1991.

Day, S. 1992. A moving experience for sponges. *New Scientist* 136(1847):15–16.

Deaming, C., and Ferguson, F. 1989. In the heat of the nest. *New Scientist* 121:33–38.

D'Hondt, J. L. 1971. Gastroticha. *Annual Review Oceanography and Marine Biology* 9:141–150.

Duellman, W. E. 1985. Systematic Zoology: Slicing the Gordian knot with Ockham's razor. *American Zoologist* 25:751–762.

Dunelson, J., and Turner, M. How the trypanosome changes its coat. *Scientific American* February, 1985.

Evans, H. E., and O'Neill, K. M. Beewolves. *Scientific American* August, 1991.

Ewing, T. 1988. Thorny problem as Australia's coastline faces invasion. *Nature* 333:387.

Feder, H. M. Escape responses in marine invertebrates. *Scientific American* July, 1972.

Finchel, T., and Finlay, B. J. 1994. The evolution of life without oxygen. *American Scientist* 82 (1):22–29.

Fingerman, M. (ed.). 1992. The compleat crab (symposium). *American Zoologist* 32(2):359–542.

Fischer, E. A., and Peterson, C. W. 1987. The evolution of sexual patterns in the seabasses. *BioScience* 37:482–489.

FitzGerald, G. J. The reproductive behavior of the stickleback. (fish) *Scientific American* April 1993.

Fleming, T. H. 1993. Plant-visiting bats. *American Scientist* 81(5):460–467.

Forey, P. L. 1988. Golden Jubilee for the coelacanth *Latimeria chalumnae*. *Nature* 336:727–732.

Franks, N. 1989. Army ants: A collective intelligence. *American Scientist* 77(2):138–145.

Freedman, W. L. The expansion rate and size of the universe. *Scientific American* November, 1992.

Fricke, Hans. 1988. Coelacanth: The fish that time forgot. *National Geographic* 173(6):824–838.

Funk, D. H. The mating of tree crickets. *Scientific American* August, 1989.

Gaffney, E. S., Hutchison, J. H., Jenkings, F. A., Jr., and Meeker, L. J. 1987. Modern turtle origins: The oldest known cryptodire. *Science* 237:289–291.

Gee, H. 1988. Taxonomy bloodied by cladistic wars. *Nature* 335:585.

Gerhard, S. The mammals of island Europe. *Scientific American* February, 1992.

Gibbons, J. W. 1987. Why do turtles live so long? *BioScience* 37(4):262–268.

Gilbert, J. J. 1984. To build a worm. *Science* 84(5):62–70.

Gore, R. Extinctions. *National Geographic* June, 1989.

Goreau, T. F., Goreau, N. I., and Goreau, T. J. Corals and coral reefs. *Scientific American* August, 1979.

Gorniak, G. C., and Gans, C. 1982. How does the toad flip its tongue? Test of two hypotheses. *Science* 216:1335–1337.

Gorr, T., and Kleinschmidt, T. 1993. Evolutionary relationships of the coelacanth. *American Scientist* 81(1):72–82.

Gould, J. L. 1980. The case for magnetic sensitivity in birds and bees (such as it is). *American Scientist* 68:256–267.

Gould, J. L., and Marler, P. Learning by instinct. *Scientific American* January, 1987.

Gould, S. J. 1976. The five kingdoms. *Natural History* 85(6):30.

Gould, S. J. The evolution of life on the earth. *Scientific American* October, 1994.

Grant, P. R. Natural selection and Darwin's finches. *Scientific American* October, 1991.

Greenwood, J. J. D. 1992. In the pink. (snails) *Nature* 357(6375):192.

Griffiths, M. The Platypus. *Scientific American* May, 1988.

Grober, M. S. 1988. Brittle-star bioluminescence functions as an aposematic signal to deter crustacean predators. *Animal Behaviour* 36:493–501.

Gutzke, W. H. N., and Crews, D. 1988. Embryonic temperature determines adult sexuality in a reptile. *Nature* 332:832–834.

Handel, S. N., and Beattie, A. J. Seed dispersal by ants. *Scientific American* August, 1990.

Hanken, J. 1989. Development and evolution in amphibians. *American Scientist* 77(4):336–343.

Harvey, P. H., and Partridge, L. 1988. Of cuckoo clocks and cowbirds. *Nature* 335:630–632.

Hawking, F. The clock of the malaria parasite. *Scientific American* June, 1970.

Heinrich, B., and Esch, H. 1994. Thermoregulation in bees. *American Scientist* 82(2):164–170.

Herbert, S. Darwin as a geologist. *Scientific American* May, 1986.

Heslinga, G. A., and Fitt, W. K. 1987. The domestication of reef-dwelling clams. *BioScience* 37:332–339.

Horgan, J. In the beginning . . . *Scientific American* February, 1991.

Horn, M. H., and Gibson, R. N. Intertidal fishes. *Scientific American* January, 1988.

Jaeger, R. G. 1988. A comparison of territorial and non-territorial behaviour in two species of salamanders. *Animal Behavior* 36:307–400.

Jermeij, G. J. 1991. When biotas meet: Understanding biotic interchange. *Science* 253:1099–1104.

Joyce, G. F. Directed molecular evolution. *Scientific American* December, 1992.

Kabnick, K. S., and Peattie, D. A. 1991. *Giardia:* A missing link between prokaryotes and eukaryotes. *American Scientist* 79(1):34–43.

Kantor, F. S. Disarming Lyme disease. *Scientific American* September, 1994.

Kanwisher, J. W., and Ridgeway, S. H. The physiological ecology of whales and porpoises. *Scientific American* June, 1983.

Keeton, W. T. The mystery of pigeon homing. *Scientific American* December, 1974.

Kelly-Borges, M. M. 1995. Zoology: Sponges out of their depth. *Nature* 273(6512):284–285.

Kerstitch, A. 1992. Primates of the sea. (octupuses) *Discover* 13(2):34–37.

Kirchner, W. H., and Towne, W. F. The sensory basis of the honeybee's dance language. *Scientific American* June, 1994.

Kirshner, R. P. The earth's elements. *Scientific American* October, 1994.

Klimley, A. P. 1994. The predatory behavior of the white shark. *American Scientist* 82(2):122–133.

Knoll, A. H. End of the Proterozoic eon. *Scientific American* October, 1991.

Konishi, M. Listening with two ears. *Scientific American* April, 1993.

Kristensen, R. M. 1983. Loricifera, a new phylum with Aschelminthes characters from the meiobenthos. *Zeitschrift Zoologie Systumatiks Evolution-Fforschung* 21:163–180.

Langston, W., Jr. Pterosaurs. *Scientific American* February, 1981.

Lemche, H. 1957. A new living deep-sea mollusk of the Cambro-Devonian class Monoplacophora. *Nature* 179:413.

Lenhoff, H. M., and Lenhoff, S. G. Trembly's polyps. *Scientific American* April, 1988.

Lent, C. M., and Dickinson, M. H. The neurobiology of feeding in leeches. *Scientific American* June, 1988.

Levine, N. D. 1980. A newly revised classification of the protozoa. *Journal of Protozoology* 27:37–58.

Lewontin, R. C. Adaptation. *Scientific American* September, 1978.

Linde, A. The self-reproducing inflationary universe. *Scientific American* November, 1994.

Linsle, R. M. 1978. Shell form and evolution of gastropods. *American Scientist* 66:432–441.

Lizotte, R. S., and Rovner, J. S. 1988. Nocturnal capture of fireflies by lycosid spiders: Visual versus vibratory stimuli. *Animal Behaviour* 36:1809–1815.

Lohman, K. J. How sea turtles navigate. *Scientific American* January, 1992.

Mackenzie, D. 1991. Where earthworms fear to tread. *New Scientist* 131(1781):31–34.

Mangum, C. 1970. Respiratory physiology in annelids. *American Scientist* 58(6):641–647.

Mann, J. 1992. Sponges to wipe away pain. *Nature* 358(6387):540.

Manzel, R., and Erber, J. Learning and memory in bees. *Scientific American* July, 1978.

Marshall, L. G. 1988. Land mammals and the great American interchange. *American Scientist* 76(4):380–388.

Marshall, L. G. The terror birds of South America. *Scientific American* February, 1994.

May, M. 1991. Aerial defense tactics of flying insects. *American Scientist* 79(4):316–328.

May, R. R. The evolution of ecological systems. *Scientific American* September, 1987.

Mayr, E. 1981. Biological classification: Toward a synthesis of opposing methodologies. *Science* 241:510–516.

McClanahan, L. L., Ruibal, R., and Shoemaker, V. H. Frogs and toads in deserts. *Scientific American* March, 1994.

McMasters, J. H. 1989. The flight of the bumblebee and related myths of entomological engineering. *American Scientist* 72(2):164–169.

McMenamin, M. A. S. The emergence of animals. *Scientific American* April, 1987.

Messing, C. G. 1988. Sea lilies and feather stars. *Sea Frontiers* 34:236–241.

Millar, R. H. 1971. The biology of ascidians. *Advances Marine Biology* 9:1–100.

Miller, J. A. 1984. Spider silk, stretch, and strength. *Science News* 125:391.

Miller, L. H., Howard, R. J., Carter, R., Good, M. F., Nussenzwieg, V., and Nussenzwieg, R. S. 1986. Research toward malaria vaccines. *Science* 234:1349–1355.

Milne, L. J., and Milne, M. The social behavior of burying beetles. *Scientific American* August, 1978.

———. Insects of the water surface. *Scientific American* April, 1978.

Milner, A. 1989. Late extinctions of amphibians. *Nature* 338:117.

Mitchell, T. 1988. Coral-killing starfish meet their mesh. *New Scientist* 120:28.

Mock, D. W., Drummond, H., and Stinson, H. 1990. Avian siblicide. *American Scientist* 78(5):438–449.

Moffett, M. W. 1991. All eyes on jumping spiders. *National Geographic* 180(3):43–63.

Moore, J. Parasites that change the behaviour of their host. *Scientific American* January, 1984.

Morse, A. N. C. 1991. How do planktonic larvae know where to settle? *American Scientist* 79(2):154–167.

Myers, N. 1985. The ends of the lines. *Natural History* 94:2–6.

Newman, E. A., and Hartline, P. H. 1982. The infrared "vision" of snakes. *Scientific American* March 246:116–127.

Nordell, D. 1988. Milking leeches for drug research. *New Scientist* 117:43–44.

Ostrom, J. H. 1979. Bird flight: How did it happen? *American Scientist* 67:46–56.

Partridge, B. L. The structure and function of fish schools. *Scientific American* June, 1982.

Peebles, J. E., Schramm, D. N., Turner, E. L., and Kron, R. G. The evolution of the universe. *Scientific American* October, 1994.

Quicke, D. 1988. Spiders bite their way towards safer insecticides. *New Scientist* 120:38–41.

Raeburn, P. 1988. Ancient survivor. *National Wildlife* 26(5):36–38.

Ramos, V. A. 1989. The birth of southern South America. *American Scientist* 77(5): 444–459.

Rebek, J. Synthetic self-replicating molecules. *Scientific American* July, 1994.

Rennie, J. Insects are forever. *Scientific American* November, 1993.

Reynolds, C. V. 1994. Warm blood for cold water. *Discover* 15(1): 42–43.

Ricciuti, E. R. 1986. A genuine monster. *Audubon* 88:22–24.

Rinderer, T. E., Oldroyd, B. P., and Sheppard, W. S. Africanized bees in the U.S. *Scientific American* December, 1993.

Rinderer, T. E., Stelzer, J. A., Oldroyd, B. P., Buco, S. M., and Rubink, W. L. 1991. Hybridization between European and Africanized bees in neotropical Yucatan Peninsula. *Science* 253(5017):309–311.

Rismiller, P. D., and Seymour, R. S. The echidna. *Scientific American* February, 1991.

Roberts, L. 1988. Corals remain baffling. *Science* 239:256.

———. 1990. The worm project. *Science* 248:1310–1313.

Robinson, M. H. 1987. In a world of silken lines, touch must be exquisitely fine. *Smithsonian* 18:94–102.

Roe, P., and Norenburg, J. L. (eds.). 1985. Symposium on the comparative biology of nemertines. *American Zoologist* 25(1):1–151.

Rome, L. C., Swank, D., and Corda, D. 1993. How fish power swimming. *Science* 261(5119):340–343.

Roper, C. R. E., and Boss, K. J. The giant squid. *Scientific American* April, 1982.

Ross, P. E. Eloquent remains. *Scientific American* May, 1992.

Russell, D. A. 1982. The mass extinctions of the late Mesozoic. *Scientific American* January, 1982.

Ruthen, R. Adapting to complexity. *Scientific American* January, 1993.

Ryan, M. J. 1990. Signals, species, and sexual selection. *American Scientist* 78(1):46–52.

Seeley, T. D. How honeybees find a home. *Scientific American* October, 1982.

———. 1989. The honeybee colony as a superorganism. *American Scientist* 77(6):546–553.

Shapiro, D. Y. 1987. Differentiation and evolution of sex change in fishes; a coral reef fish's social environment can control its sex. *BioScience* 490–497.

Shear, W. A. 1994. Untangling the evolution of the web. *American Scientist* 82(3):256–266.

Sheldon, P. 1988. Making the most of evolution diaries. *New Scientist* 117:52–54.

Shimek, R. B. 1987. Sex among the sessile: With the onset of spring in cool northern Pacific waters, even sea cucumbers bestir themselves. *Natural History* 96:60–63.

Simmons, L. W. 1988. The calling song of the field cricket, *Gryllus bimaculatus* (De Geer): Constraints on transmission and its role in intermale competition and female choice. *Animal Behaviour* 36:380–394.

Sitwell, N. 1993. The grub and the Galapagos. (sea cucumbers) *New Scientist* 40(1903):32–35.

Stowe, M. K., Tumilinson, J. H., and Heath, R. R. 1987. Chemical mimicry: Bolas spiders emit components of moth prey species sex pheromones. *Science* 236:964–968.

Stuller, J. 1988. With the gales in their sails. *Audubon* 90:84–85.

Suga, N. Biosonar and neural computations in bats. *Scientific American* June, 1990.

Sutherland, W. J. 1988. The heritability of migration. *Nature* 334:471–472.

Tangley, L. 1987. Malaria: Fighting the African scourge. *BioScience* 37(2):94–98.

Tattersall, I. Madagascar's lemurs. *Scientific American* January, 1993.

Taylor, M. 1994. Amphibians that came to stay. *New Scientist* 141(1912):21–24.

Toner, M. 1992. Spin doctor. *Discover* 13(5):32–36.

Topoff, H. 1990. Slave-making ants. *American Scientist* 78(6):520–528.

Tumlinson, J. H., Lewis, W. J., and Vet, L. E. M. How parasitic wasps find their hosts. *Scientific American* March, 1993.

Tuttle, R. H. 1990. Apes of the world. *American Scientist* 78(2):115–125.

Verell, P. 1988. The chemistry of sexual persuasion. *New Scientist* 118:40–43.

Vickers-Rich, P., and Rich, T. H. Australia's polar dinosaurs. *Scientific American* July, 1993.

Vogel, S. 1988. How organisms use flow-induced pressures. *American Scientist* 76:92–94.

Vollrath, F. 1992. Spider webs and silks. *Scientific American* March, 1992.

Von Frisch, K. 1974. Decoding the language of the bee. *Science* 185:663–668.

Walsh, J. 1979. Rotifers, nature's water purifiers. *National Geographic* 155:286–292.

Ward, P., Greenwald, L., and Greenwald, O. E. The buoyancy of the chambered nautilus. *Scientific American* October, 1980.

Weinberg, S. Life in the universe. *Scientific American* October, 1994.

Wellnhofer, P. *Archaeopteryx. Scientific American* May, 1990.

West, M. J., and King, A. P. 1990. Mozart's starling. *American Scientist* 78(2):106–114.

Wheatley, D. 1988. Whale size quandary. *Nature* 336:626.

Whittaker, R. H. 1969. New concept of kingdoms of organisms. *Science* 163:150–160.

Wicksten, M. D. Decorator crabs. *Scientific American* February, 1980.

Wilczynski, W., and Brenowitz, E. A. 1988. Acoustic cues mediate inter-male spacing in a neotropical frog. *Animal Behaviour* 36:1054–1063.

Wilkinson, C. R. 1987. Interocean differences in size and nutrition of coral reef sponge populations. *Science* 236:1654–1657.

Wood, R. 1990. Reef-building sponges. *American Scientist* 78(3):224–235.

Wootton, R. J. The mechanical design of insect wings. *Scientific American* November, 1990.

Yager, J. 1981. Remipedia, a new class of Crustacea from a marine cave in the Bahamas. *Journal of Crustacean Biology* 1:328–333.

Yonge, C. M. Giant clams. *Scientific American* April, 1975.

York, D. The earliest history of the earth. *Scientific American* January, 1993.

Zapol, W. M. Diving adaptations of the Weddell Seal. *Scientific American* June, 1987.

credits

Photos

Chapter 1

Opener: Photo Courtesy of Digital Stock Images, Animals CD; **1.1:** © Robert Hitchman/Unicorn Stock Photos; **1.3a:** © Renne Lynn/Photo Researchers, Inc.; **1.3b:** © Gerald and Buff Corsi/Tom Stack & Associates; **1.5:** © Sinclair Stammers/SPL/Photo Researchers, Inc.; **1.8a,b:** © Stephen Dalton/Animals Animals/Earth Scenes; **1.13a:** © Robert C. Simpson/Tom Stack & Associates; **1.13b:** © S. Maslowski/Visuals Unlimited; **1.13c:** © Ron Austing/Photo Researchers, Inc.; **1.13d:** © Johann Schumacher/Peter Arnold, Inc.; **1.13e:** © S. Maslowski/Visuals Unlimited; **1.14:** © E. R. Degginer/Color-Pic, Inc.; **1.15:** © Paul Oppler/Visuals Unlimited; **1.17:** © Richard Thom/Visuals Unlimited; **Box 1.1b:** © ScienceVU/WHOI, J. Edward/Visuals Unlimited: **1.22a:** © Doug Wechsler/Animals Animals/Earth Scenes; **1.22b:** © Dr. Nigel Smith/Animals Animals/Earth Scenes

Chapter 2

Opener: Photo Courtesy of Digital Stock Images, Undersea Life CD; **2.2a:** © Walt Anderson/Visuals Unlimited; **2.2b:** © Joe McDonald/Visuals Unlimited; **2.4a:** © Alan L. Deitrick/Photo Researchers, Inc.; **2.4b:** © Walt Anderson/Visuals Unlimited

Chapter 3

Opener: Photo Courtesy of Digital Stock Images, Undersea Life CD; **3.2a:** © Don W. Fawcett/Visuals Unlimited; **3.3:** © Kevin Schafer/Tom Stack & Associates; **3.5 a,b:** © Michael Tweedie/Photo Researchers, Inc.

Chapter 4

Opener: Photo Courtesy of Digital Stock Images, Animals CD; **4.1:** © Kevin Schafer/Martha Hill/Tom Stack & Associates; **4.6:** © Daniel W. Gotshall/Visuals Unlimited; **4.7:** © Dave B. Fleetham/Visuals Unlimited

Chapter 5

Opener: Photo Courtesy of Digital Stock Images, Undersea Life CD; **5.1:** © David John/Visuals Unlimited; **5.5 a-d:** © Dennis Diener/Visuals Unlimited; **5.6:** © Terry Hazen/Visuals Unlimited; **5.8a:** © James W. Richards/Visuals Unlimited; **Box 5.1:** Courtesy of Dr. Stanley Erlandsen; **5.9b:** © John D. Cunningham/Visuals Unlimited; **5.11a:** © M. Abbey/Visuals Unlimited; **5.13:** © A.M. Siegelman/Visuals Unlimited; **5.14a:** © M. Schliwa/Visuals Unlimited; **5.14b:** © G. Shih - R. Kessel/Visuals Unlimited; **5.17a:** © Karl Auffderheide/Visuals Unlimited; **5.18:** © M. Abbey/Visuals Unlimited; **5.19:** © Biophoto Associates/Photo Researchers, Inc.; **5.20:** © D.J. Patterson/OSF/Animals Animals/Earth Scenes

Chapter 6

Opener: Photo Courtesy of Digital Stock Images, Undersea Life CD; **6.1:** © John D. Cunningham/Visuals Unlimited; **6.4a:** © Nancy Sefton/Photo Researchers, Inc.; **6.4b:** © Daniel W. Gotshall/Visuals Unlimited; **Box 6.1:** Field Museum of Natural History Neg. #80872, Chicago; **6.6, 6.13a:** © Carolina Biological/Visuals Unlimited; **6.14a:** © Edward Hodgson/Visuals Unlimited; **6.14b:** © N. G. Daniel/Tom Stack & Associates; **6.17:** © Neville Coleman/Visuals Unlimited; **6.18a:** © Daniel W. Gotshall/Visuals Unlimited; **18.18b:** © Milton H. Tierney, Jr./Visuals Unlimited; **6.21a:** © Daniel W. Gotshall/Visuals Unlimited; **6.21b:** © Marty Snyderman/Visuals Unlimited; **Box 6.2:** © Edward Hodgson/Visuals Unlimited; **6.22a:** © R. De Goursey/Visuals Unlimited

Endpaper 1

1: © Tom J. Ulrich/Visuals Unlimited

Chapter 7

Opener: Photo Courtesy of Digital Stock Images, Undersea Life CD; **7.1:** © Gary R. Robinson/Visuals Unlimited; **7.17b:** © Drs. Kessel & Shih/Peter Arnold, Inc.

Chapter 8

Opener: Photo Courtesy of Digital Stock Images, Undersea Life CD; **8.1:** © Lauritz Jensen/Visuals Unlimited; **8.4a:** © Peter Parks/OSF/Animals Animals/Earth Scenes; **8.5a:** © Cabisco/Visuals Unlimited; **8.14b:** Steve Miller (author); **8.15:** © Science VU-AFIP/Visuals Unlimited; **8.17:** © R. Calentine/Visuals Unlimited; **8.18b:** © Lauritz Jensen/Visuals Unlimited

Chapter 9

Opener: Photo Courtesy of Digital Stock Images, Animals CD; **9.1:** © Gary Miburn/Tom Stack & Associates; **9.6b:** © Science VU-Polaroid/Visuals Unlimited; **9.8a:** © William J. Weber/Visuals Unlimited; **9.9a:** © OSF/Animals Animals/Earth Scenes; **9.9b:** © Daniel W. Gotshall/Visuals Unlimited; **Box 9.1:** © Scott Camazine/Photo Researchers, Inc.; **9.15c:** © Ed Reschke; **9.17a:** © William E. Ferguson; **9.17b:** © Michael Di Spezio; **9.21a:** © Robert A. Ross; **9.22:** © Kjell B. Sandved/Photo Researchers, Inc.; **9.24:** © James Culter/Visuals Unlimited

Chapter 10

Opener: Photo Courtesy of Digital Stock Images, Undersea Life CD; **10.1:** © Marty Snyderman/Visuals Unlimited; **10.5:** © R. DeGoursey/Visuals Unlimited; **Box 10.1:** © C. P. Hickman/Visuals Unlimited

Endpaper 2

2: © WHOI/D. Foster/Visuals Unlimited

Chapter 11

Opener: Photo Courtesy of Digital Stock Images, Animals CD; **11.1:** © Roger Klocek/Visuals Unlimited; **11.6:** © John Cancalosi/Tom Stack & Associates; **11.8a:** © Francois Gohier/Photo Researchers, Inc.; **11.11a:** © Tom McHugh/Photo Researchers, Inc.; **11.13:** © Ken Highfill/Photo Researchers, Inc.; **11.14:** © David Scharf/Peter Arnold, Inc.; **Box 11.1a:** © S. Masiowski/Visuals Unlimited; **Box 11.1b:** © Richard Walters/Visuals Unlimited; **11.15a:** © R. Calentine/Visuals Unlimited; **11.15b:** © Scott Camazine/Photo Researchers, Inc.; **11.16:** © Frank T.Awbrey/Visuals Unlimited; **11.18:** © Andrew J. Martinez/Photo Researchers, Inc.; **11.21a:** © John D. Cunningham/Visuals Unlimited; **11.21b:** © Biophoto Assoc./Photo Researchers, Inc.; **11.22a:** © David M. Dennis/Tom Stack & Associates; **11.22b:** © Alan Desbonnet/Visuals Unlimited; **11.23:** © James Bell/Photo Researchers, Inc.; **11.24b:** © Tom Stack/Tom Stack & Associates

Chapter 12

Opener: Photo Courtesy of Digital Stock Images, Animals CD; **12.1:** © Patti Murray/Animals Animals/Earth Scenes; **12.3a:** © Bill Beatty/Visuals Unlimited; **12.3b:** © Glenn M. Oliver/Visuals Unlimited; **12.10a:** © S.L. Flegler/Visuals Unlimited; **12.14a:** © Treat Davidson/Photo Researchers, Inc.

Endpaper 3

1: © Tom J. Ulrich/Visuals Unlimited; **2:** Photomicrograph by John Ubelaker; **3:** Courtesy Diane R. Nelson, Ph.D.

Chapter 13

Opener: Photo Courtesy of Digital Stock Images, Undersea Life CD; **13.1:** © Carl Roessler/Tom Stack & Associates; **13.3a:** © Michael DiSpezio; **Box 13.1:** © Daniel W. Gotshall/Visuals Unlimited; **13.8a:** © Robert L. Dunne/Photo Researchers, Inc.; **13.8b:** © Harold W. Pratt/Biological Photo Service; **13.10a:** © C. McDaniel/Visuals Unlimited; **13.10b:** © Bruce Iverson/Visuals Unlimited; **13.12:** © Daniel W. Gotshall/Visuals Unlimited; **13.16:** Courtesy of the Museum of New Zealand Te Papa Tongarewa, Alan N. Baker, photographer

Endpaper 4

2: © John D. Cunningham/Visuals Unlimited; **3:** © Daniel W. Gotshall/Visuals Unlimited

Chapter 14

Opener: Photo Courtesy of Digital Stock Images, Undersea Life CD; **14.1:** © Michael DiSpezio; **Box 14.1:** © Gary R.Robinson/Visuals Unlimited; **14.6a:** © William C. Jorgenson/Visuals Unlimited; **14.6b:** © Daniel W. Gotshall/Visuals Unlimited

Chapter 15

Opener: Photo Courtesy of Digital Stock Images, Undersea Life CD; **15.1:** © Albert Copley/Visuals Unlimited; **15.5:** © Russ Kinne/Photo Researchers, Inc.; **15.8a:** © Ed Robinson/Tom Stack & Associated; **15.8 b,c:** © Daniel W. Gotshall/Visuals Unlimited; **15.9a:** © Science VU/NOAA/Visuals Unlimited; **Box 15.1:** © Roger Klocek/Visuals Unlimited; **15.10:** © John D. Cunningham/Visuals Unlimited; **15.11:** © Tom Stack/Tom Stack & Associates; **15.12 a,b:** © Patrice Ceisel/Visuals Unlimited; **15.13a:** © R. DeGoursey/Visuals Unlimited; **15.13b:** © Fred Rohde/Visuals Unlimited; **15.13c:** © Patrice Ceisel/Visuals Unlimited; **15.15b:** © Fred Hossler/Visuals Unlimited; **15.17b:** © Tom McHugh/Photo Researchers, Inc.; **15.18, 15.20:** © Daniel W. Gotshall/Visuals Unlimited

Chapter 16

Opener: Photo Courtesy of Digital Stock Images, Undersea Life CD; **16.1:** © Joe McDonald/Visuals Unlimited; **16.4:** © John D. Cunningham/Visuals Unlimited; **16.5 a,b:** © Dwight Kuhn; **16.5c:** © Joel Arrington/Visuals Unlimited; **16.6:** © Tom McHugh/Photo Researchers, Inc.; **16.7:** © John Serrao/Visuals Unlimited; **Box 16.1:** © Dennis Paulson/Visuals Unlimited; **16.15a:** © Nada Pecnik/Visuals Unlimited; **16.15b:** © Edward S. Ross; **16.16b:** © Dan Kline/Visuals Unlimited; **16.17:** © M.J. Tyler, The University of Adelaid, South Australia; **16.18 a-d:** © Jane Burton/Bruce Coleman, Inc.

Chapter 17

Opener: Photo Courtesy of Digital Stock Images, Animals CD; **17.1:** © E.R. Degginger/Photo Researchers, Inc.; **17.5:** © Valorie Hodgson/Visuals Unlimited; **17.6:** © Nathan W. Cohen/Visuals Unlimited; **17.7:** © Joe McDonald/Visuals Unlimited; **17.8:** © Thomas Gula/Visuals Unlimited; **17.9:** © John D. Cunningham/Visuals Unlimited; **17.11:** © Stephen Dalton/Animals Animals/Earth Scenes; **17.12a:** © Joe McDonald/Visuals Unlimited; **17.14:** © Carolina Biological/Visuals Unlimited; **17.15:** © Robert Hermes/Photo Researchers, Inc.

Chapter 18

Opener: Photo Courtesy of Digital Stock Images, Animals CD; **18.1a:** © Tim Hauf/Visuals Unlimited; **18.2a:** © John D. Cunningham/Visuals Unlimited; **18.2b:** Courtesy Department of Library Services, American Museum of Natural History, Neg. #K-Set3 #29.; **18.6e:** © Steve Maslowski/Visuals Unlimited; **Box 18.1:** © Tim Hauf/Visuals Unlimited; **18.7:** © Robert A. Lubeck/Animals Animals/Earth Scenes; **18.8a:** © Thomas Kitchin/Tom Stack & Associates; **18.8b:** © Kirtley-Perkins/Visuals Unlimited; **18.8c:** © Brian Parker/Tom Stack & Associates; **18.10c:** Courtesy of Dr. Hans-Rainer Duncker; **18.11:** © Karl Maslowski/Visuals Unlimited; **18.14:** © John D. Cunningham/Visuals Unlimited; **18.15a:** © Karl Maslowski/Visuals Unlimited; **18.15b:** © John D. Cunningham/Visuals Unlimited

Chapter 19

Opener: Photo Courtesy of Digital Stock Images, Animals CD; **19.1:** © M. Long/Visuals Unlimited; **19.4a:** © Tom McHugh/Photo Researchers, Inc.; **19.4b:** © J. Alcock/Visuals Unlimited; **19.4c:** © Bruce Berg/Visuals Unlimited; **19.5a:** © Walt Anderson/Visuals Unlimited; **19.5b:** © William J. Weber/Visuals Unlimited; **19.6:** © Bruce Cushing/Visuals Unlimited; **Box 19.1b:** © Tom J. Ulrich/Visuals Unlimited; **19.11:** © Glenn M. Oliver/Visuals Unlimited; **Box 19.2:** © Stephen Dalton/Animals Animals/Earth Scenes; **19.15a:** © Dennis Schmidt/Valan Photos; **19.16:** © Walt Anderson/Visuals Unlimited; **19.17 a,b:** © John D. Cunningham/Visuals Unlimited

Line Art

Chapter 1

1.4, 1.16: From Leland G. Johnson, *Biology*, 2d ed. Copyright © 1987 Wm. C. Brown Communications, Inc. Reprinted by permission of Times Mirror Higher Education Group, Inc., Dubuque, Iowa. All Rights Reserved. **1.18, 1.20:** From Ricki Lewis, *Life*, 2d ed. Copyright © 1995 Wm. C. Brown Communications, Inc. Reprinted by permission of Times Mirror Higher Education Group, Inc., Dubuque, Iowa. All Rights Reserved.

Chapter 2

2.1: From Ricki Lewis, *Life*, 2d ed. Copyright © 1995 Wm. C. Brown Communications, Inc. Reprinted by permission of Times Mirror Higher Education Group, Inc., Dubuque, Iowa. All Rights Reserved.

Chapter 3

3.2b: From Ricki Lewis, *Life*, 2d ed. Copyright © 1995 Wm. C. Brown Communications, Inc. Reprinted by permission of Times Mirror Higher Education Group, Inc., Dubuque, Iowa. All Rights Reserved; **3.7:** From Leland G. Johnson, *Biology*, 2d ed. Copyright © 1987 Wm. C. Brown Communications, Inc. Reprinted by permission of Times Mirror Higher Education Group, Inc., Dubuque, Iowa. All Rights Reserved.

Chapter 4

4.10: From Leland G. Johnson, *Biology*, 2d ed. Copyright © 1987 Wm. C. Brown Communications, Inc. Reprinted by permission of Times Mirror Higher Education Group, Inc., Dubuque, Iowa. All Rights Reserved.

Chapter 5

5.16: From Theodore Jahn, et al., *How to Know the Protozoa*, 3d ed. Copyright © 1979 Wm. C. Brown Communications, Inc. Reprinted by permission of Times Mirror Higher Education Group, Dubuque, Iowa. All Rights Reserved.

Chapter 6

6.15a: Adapted from Charles Lytle and J.E. Wodsedalek, *General Zoology Laboratory Guide*, 11th ed., Complete Version. Copyright © 1991 Wm. C. Brown Communications, Inc. Reprinted by permission of Times Mirror Higher Education Group, Inc., Dubuque, Iowa. All Rights Reserved; **6.19:** From Charles Lytle and J.E. Wodsedalek, *General Zoology Laboratory Guide*, 11th ed., Complete Version. Copyright © 1991 Wm. C. Brown Communications, Inc. Reprinted by permission of Times Mirror Higher Education Group, Inc., Dubuque, Iowa. All Rights Reserved.

Endpaper 1

1.2: From K.G. Grell, *Zeitung fur Morphologie der Tiere*, 73:297-314, 1972. Copyright Springer-Verlag, Heidelberg. Used by permission.

Chapter 7

7.17a: From Jan A. Pechenik, *Biology of the Invertebrates*, 2d ed. Copyright © 1991 Wm. C. Brown Communications, Inc. Reprinted by permission of Times Mirror Higher Education Group, Inc., Dubuque, Iowa. All Rights Reserved.

Chapter 10

10.4: Adapted from R.E. Snodgrass: *Principles of Insect Morphology*. Copyright © by Ellen Burden and Ruth Roach. Used by permission of the publisher, Cornell University Press; **10.9:** From: *A Life of Invertebrates* © 1979 W.D. Russell-Hunter.

Chapter 12

12.5: Based on C. Gillott, *Entomology*. Copyright © 1980 Plenum Publishing Corporation. Used by permission; **12.12, 12.13:** © Carolina Biological Supply Company. Used with permission of Phototake, New York.

Chapter 15

15.9b: From R. Steale, *Sharks of the World*. Copyright © 1985 Cassell PLC, London. Used by permission; **15.15:** From: *Animal Physiology 3/E* by Eckert, Randall, and Augustine. Copyright © 1988 by W.H. Freeman and Company. Used with permission.

Chapter 17

17.13: From N. Heisler, *Journal of Experimental Biology*, 1983. Copyright 1983 The Company of Biologists Limited, Cambridge, UK. Used by permission.

Illustrators

DIPHRENT STROKES: 1.2, 1.3C, 1.9, 1.10, 1.11, 1.16, 1.21, BOX 2.21, 3.1, 3.4, 4.4, 4.5, 4.8, 5.4, 5.8B, 5.9A, 5.12, 5.22, 6.8, 6.9, 6.13B, 6.15A-B, 6.20, 6.22B-C, 6.23, 7.4, 7.5, 7.6, 7.7, 7.8, 7.9, 7.17A, 7.21, 8.3, 8.4B, 8.19, 8.21, 9.5, 9.11, 9.12, 9.13, 9.14, 9.15A-B, 9.18, 9.19, 9.20, 9.25, 10.8, 10.14, 10.16, 10.18, 11.3, 11.5, 11.7, 11.8B, 11.9, 11.10, 11.11B, 11.12, 11.17, 11.19, 11.20, 11.24A, 12.4, 12.5, 12.6, 12.7, 12.8, 12.9, 12.10B-C, 12.11, 12.14D, 12.15A-B, 13.5, 13.6, 13.7, 13.9, 13.17, 14.3, 14.4, 14.5, 14.7, 14.8, 14.9, 14.10, 15.2, 15.15(art), 15.16, 16.3, 16.8, 16.10, 16.11, 16.13, 16.16A, BOX 17.1B-C, 17.2, 17.3, 17.10, 17.12(art), 18.1B, 18.3, 18.4, 18.10A-B, 18.12, 18.13, 19.3, 19.7, 19.8, 19.13
MARSHA HARTSOCK: 15.3, 15.4, 15.6, 15.7, 15.9B, 15.14A-B, 15.17A, 15.19A-B, 16.2, 16.9, 16.12, 16.14, BOX 17.1A, 17.4, 17.13, 18.5, 18.9
HILL-WERNER: 13.4, 13.11, 13.13, 13.14, 13.15
CARLYN IVERSON: 5.17B
RUTH KRABACH: 7.3
MARJORIE LEGGITT: 2.3, 2.5
IRIS NICHOLS: 19.9, 19.10, 19.12, 19.14, 19.15B, BOX 19.1(art)
LAURIE O'KEEFE: 1.18, 1.20, 4.2
ROLIN GRAPHICS: BOX 1.1A, 1.4, 1.6, 1.7, 1.12, 1.19, 2.1, BOX 3.1, 3.2B, 3.6, 3.7, 4.3, 5.2, 5.11B, 5.16, 6.2, 7.2, 7.10, 7.11, 7.13, 7.14, 8.2, 8.5B, 8.6, 8.7, 8.8, 8.9, 8.10, 9.2, 10.2, 10.9, 11.2, 11.4, 12.2, 12.12, 12.13, 13.2, 13.3B, 14.2, 18.6A-D, 19.2
NADINE SOKOL: 4.9, 4.10, 5.3, 5.7, 5.10, 5.15, 5.21, 6.3, 6.5, 6.7, 6.10, 6.11, 6.12, 6.16, 6.19, 7.12, 7.15, 7.16, 7.18, 7.19, 8.11, 8.12, 8.13, 8.14A, 8.16, 8.18A, 8.20
KEVIN SOMERVILLE: 9.3, 9.4, 9.6A, 9.7, 9.8B, 9.10, 9.16, 9.21B-C, 9.23, 10.3, 10.4, 10.6, 10.7, 10.10, 10.11, 10.12, 10.13, 10.15, 10.17

index

A

D

E

F

G

H

I

J

K

Q

R

S

T

U

V

W

X

Y

Z